Viruses in food and water

Related titles:

Advances in microbial food safety Volume 1 (ISBN 978-0-85709-438-4)

Determining mycotoxins and mycotoxigenic fungi in food and feed (ISBN 978-1-84569-674-0)

Tracing pathogens in the food chain (ISBN 978-1-84569-496-8)

Details of these books and a complete list of titles from Woodhead Publishing can be obtained by:

- visiting our web site at www.woodheadpublishing.com
- contacting Customer Services (e-mail: sales@woodheadpublishing.com; fax: +44 (0) 1223 832819; tel.: +44 (0) 1223 499140 ext. 130; address: Woodhead Publishing Limited, 80, High Street, Sawston, Cambridge CB22 3HJ, UK)
- in North America, contacting our US office (e-mail: usmarketing@woodheadpublishing.com; tel.: (215) 928 9112; address: Woodhead Publishing, 1518 Walnut Street, Suite 1100, Philadelphia, PA 19102–3406, USA

If you would like e-versions of our content, please visit our online platform: www.woodheadpublishingonline.com. Please recommend it to your librarian so that everyone in your institution can benefit from the wealth of content on the site.

We are always happy to receive suggestions for new books from potential editors. To enquire about contributing to our Food science, technology and nutrition series, please send your name, contact address and details of the topic/s you are interested in to nell.holden@woodheadpublishing.com. We look forward to hearing from you.

The team responsible for publishing this book:

Commissioning Editor: Nell Holden
Publications Coordinator: Adam Davies
Project Editor: Sarah Lynch
Editorial and Production Manager: Mary Campbell
Production Editor: Richard Fairclough
Project Manager: Newgen Knowledge Works Pvt Ltd
Copyeditor: Newgen Knowledge Works Pvt Ltd
Proofreader: Newgen Knowledge Works Pvt Ltd
Cover Designer: Terry Callanan

Woodhead Publishing Series in Food Science, Technology and Nutrition:
Number 249

Viruses in food and water

Risks, surveillance and control

Edited by
Nigel Cook

WOODHEAD PUBLISHING

Oxford Cambridge Philadelphia New Delhi

Published by Woodhead Publishing Limited,
80 High Street, Sawston, Cambridge CB22 3HJ, UK
www.woodheadpublishing.com
www.woodheadpublishingonline.com

Woodhead Publishing, 1518 Walnut Street, Suite 1100, Philadelphia,
PA 19102–3406, USA

Woodhead Publishing India Private Limited, G-2, Vardaan House, 7/28 Ansari Road,
Daryaganj, New Delhi – 110002, India
www.woodheadpublishingindia.com

First published 2013, Woodhead Publishing Limited

British Library Cataloguing in Publication Data
A catalogue record for this book is available from the British Library.

Library of Congress Control Number: 2013933543

ISBN 978-0-85709-430-8 (print)
ISBN 978-0-85709-887-0 (online)
ISSN 2042-8049 Woodhead Publishing Series in Food Science, Technology and Nutrition (print)
ISSN 2042-8057 Woodhead Publishing Series in Food Science, Technology and Nutrition (online)

The publisher's policy is to use permanent paper from mills that operate a sustainable forestry policy, and which has been manufactured from pulp which is processed using acid-free and elemental chlorine-free practices. Furthermore, the publisher ensures that the text paper and cover board used have met acceptable environmental accreditation standards.

Cover images courtesy of Professor Rosina Girones and Professor Silvia Bofil-Mas, University of Barcelona, Spain

Typeset by Newgen Knowledge Works Pvt Ltd, Chennai, India

Contents

Contributor contact details .... *xiii*
Woodhead Publishing Series in Food Science, Technology and Nutrition .... *xix*

Part I An introduction to food and environmental virology **1**

1 An introduction to food- and waterborne viral disease **3**
N. Cook, Food and Environment Research Agency, UK and G. P. Richards, Delaware State University, USA
1.1 Introduction to enteric viruses 3
1.2 Food and water as vehicles of virus transmission 5
1.3 Outbreaks of food- and waterborne viral illness 6
1.4 Virus detection 9
1.5 Control of virus contamination of food and water 10
1.6 References 11

2 Prevalence of viruses in food and the environment **19**
T. Petrović, Scientific Veterinary Institute 'Novi Sad', Serbia
2.1 Introduction 19
2.2 The prevalence of virus contamination in food and water 21
2.3 Gaps in current knowledge 35
2.4 Conclusion and future trends 37
2.5 Acknowledgements 40
2.6 References 40

Part II Detection, surveillance and risk assessment of viruses in food and water **47**

3 Molecular detection of viruses in foods and food-processing environments **49**
D. Rodríguez-Lázaro, University of Burgos, Spain, and K. Kovac and M. Hernández, Instituto Tecnológico Agrario de Castilla y León (ITACyL), Spain
3.1 Introduction 49
3.2 Molecular detection of viruses in foods: the process 50
3.3 Current issues in molecular detection of viruses in foods 60
3.4 Conclusion 68
3.5 References 69

4 Sampling strategies for virus detection in foods, food-processing environments, water and air **79**
A. Rzeżutka, National Veterinary Research Institute, Poland and A. Carducci, University of Pisa, Italy
4.1 Introduction 79
4.2 Virus monitoring at different levels of the food supply chain 80
4.3 The significance of water, air and surface sampling during food chain monitoring 84
4.4 Sampling strategy in relation to food- and waterborne outbreaks 87
4.5 Conclusion 90
4.6 Sources of further information and advice 91
4.7 References 91
4.8 Appendix: sampling from food and air 96

5 Molecular detection of viruses in water and sewage **97**
G. La Rosa and M. Muscillo, Istituto Superiore di Sanità, Italy
5.1 Introduction 97
5.2 Sample treatment: adsorption-elution methods 98
5.3 Sample treatment: ultrafiltration and ultracentrifugation 104
5.4 Key assays for virus detection 106
5.5 Advantages and disadvantages of polymerase chain reaction (PCR) and related methods 110
5.6 Current applications and results 113
5.7 References 116

6 Quality control in the analytical laboratory: analysing food- and waterborne viruses **126**
M. S. D'Agostino, Food and Environment Research Agency (Fera), UK
6.1 Introduction 126
6.2 Controls for the sample treatment step 129

6.3 Controls for the nucleic acid extraction step 131
6.4 Controls for the amplification step 131
6.5 Additional recommended controls 135
6.6 Reference materials 136
6.7 Conclusion 136
6.8 References 137

7 Tracing the sources of outbreaks of food- and waterborne viral disease and outbreak investigation using molecular methods 139
M.B. Taylor, University of Pretoria & National Health Laboratory Service, South Africa
7.1 Introduction 139
7.2 Challenges in food- and waterborne outbreak tracing and investigation 141
7.3 Microbial source tracking 141
7.4 Molecular-based source tracking 144
7.5 Molecular tracing in outbreaks 146
7.6 Conclusion 150
7.7 References 151

8 Quantitative risk assessment for food- and waterborne viruses 159
A. M. de Roda Husman and M. Bouwknegt, National Institute for Public Health and the Environment (RIVM), The Netherlands
8.1 Introduction 159
8.2 Quantitative microbiological risk assessments (QMRAs) and their outcomes 161
8.3 Data gaps and needs 164
8.4 Future trends 170
8.5 Conclusion 171
8.6 References 172

Part III Virus transmission routes and control of food and water contamination 177

9 Natural persistence of food- and waterborne viruses 179
P. Vasickova and K. Kovarcik, Veterinary Research Institute, Czech Republic
9.1 Introduction 179
9.2 Methods for studying persistence 181
9.3 General factors affecting the natural persistence of viruses .. 186
9.4 Persistence in aquatic environments 189
9.5 Persistence in soils 191
9.6 Persistence on food-related surfaces 193
9.7 Persistence in food 196
9.8 Acknowledgement 198
9.9 References 198

10 Occurrence and transmission of food- and waterborne viruses by fomites **205**
C. P. Gerba, University of Arizona, USA
10.1 Introduction: the role of fomites in virus transmission 205
10.2 Occurrence and survival of viruses on fomites 206
10.3 Virus transfer and modeling transmission 210
10.4 Disinfection and other interventions to prevent fomite transmission 213
10.5 Future trends 214
10.6 References 214

11 Viral contamination by food handlers and recommended procedural controls **217**
I. L. A. Boxman, Food and Consumer Product Safety Authority (NVWA), The Netherlands
11.1 Introduction 217
11.2 Role of food handlers in virus transmission 218
11.3 Current knowledge and hygiene practices among food handlers 220
11.4 Guidance documents on food hygiene 222
11.5 Guidelines on the application of general principles of food hygiene to the control of viruses in food 224
11.6 Designing Hazard Analysis and Critical Control Points (HACCP) with the viruses NoV and HAV in mind 229
11.7 Conclusion and future trends 231
11.8 Acknowledgement 232
11.9 References 232

12 Foodborne virus inactivation by thermal and non-thermal processes. **237**
L. Baert, Ghent University, Belgium
12.1 Introduction 237
12.2 Thermal processes 238
12.3 Non-thermal processes 242
12.4 Appropriateness of surrogates 249
12.5 Future trends 253
12.6 Sources of further information and advice 254
12.7 References 254

13 Preventing and controlling viral contamination of fresh produce **261**
S. Bidawid, Health Canada, Canada
13.1 Introduction: why food contamination occurs 261
13.2 Contamination of produce 263
13.3 Attachment, adsorption and internalization 267
13.4 Prevention 268
13.5 Recommendations 269

13.6 Additional intervention strategies 271
13.7 Future trends 272
13.8 Sources of further information and advice 272
13.9 References 273

14 Preventing and controlling viral contamination of shellfish 281
J. W. Woods and W. Burkhardt III, US Food and Drug Administration, USA
14.1 Introduction 281
14.2 Human enteric viruses in the environment 282
14.3 Enteric viruses in sewage and shellfish 283
14.4 Survival of enteric viruses in the environment 284
14.5 Mitigation strategies and depuration 286
14.6 Current regulations 287
14.7 Conclusion 289
14.8 References 289

15 Viral presence in waste water and sewage and control methods 293
C. P. Gerba, M. Kitajima and B. Iker, University of Arizona, USA
15.1 Introduction: virus occurrence in wastewater 293
15.2 Natural treatment systems 300
15.3 Disinfection of wastewaters 306
15.4 Future trends 310
15.5 References 311

Part IV Particular pathogens and future directions 317

16 Advances in understanding of norovirus as a food- and waterborne pathogen and progress with vaccine development 319
D. J. Allen, Health Protection Agency, UK, M. Iturriza-Gómara, University of Liverpool, UK and D. W. G. Brown, Health Protection Agency, UK
16.1 Introduction 320
16.2 Norovirus virology and clinical manifestations 321
16.3 Susceptibility, immunity and diagnosis 324
16.4 Epidemiology of norovirus gastroenteritis associated with food, water and the environment 328
16.5 Prevention and control 335
16.6 Conclusion 337
16.7 References 338

17 Advances in understanding of hepatitis A virus as a food- and waterborne pathogen and progress with vaccine development 349
R.M. Pintó and A. Bosch, University of Barcelona, Spain
17.1 Introduction: hepatitis A infection 349
17.2 Susceptibility in different sectors of the population 350

Contributor contact details

(* = main contact)

Editor
Nigel Cook
Food and Environment Research Agency
Sand Hutton
York, YO41 1LZ, UK

E-mail: nigel.cook@fera.gsi.gov.uk

Chapter 1
Nigel Cook*
Food and Environment Research Agency
Sand Hutton
York, YO41 1LZ, UK

E-mail: nigel.cook@fera.gsi.gov.uk

Gary P. Richards
United States Department of Agriculture
Agricultural Research Service
Delaware State University
1200 North DuPont Hwy
James W.W. Baker Center
Dover
Delaware 19901, USA

E-mail: gary.richards@ars.usda.gov

Chapter 2
Tamaš Petrović
Scientific Veterinary Institute 'Novi Sad'
Rumenački put 20
21000, Novi Sad
Serbia

E-mail: tomy@niv.ns.ac.rs

Chapter 3
David Rodríguez-Lázaro*
Microbiology Section
Faculty of Sciences
University of Burgos
Plaza Misael Bañuelos s/n
09001 Burgos, Spain

E-mail: drlazaro@ubu.es

Katarina Kovac and Marta Hernández
Instituto Tecnológico Agrario de Castilla y León (ITACyL)
Junta de Castilla y León
Carretera de Burgos km 119 s/n
47071 Valladolid, Spain

E-mail: ita-herperma@itacyl.es

Chapter 4

Artur Rzeżutka*
National Veterinary Research Institute
Department of Food and Environmental Virology
Al. Partyzantów 57
24–100 Puławy, Poland

E-mail: arzez@piwet.pulawy.pl

Annalaura Carducci
University of Pisa
Department of Biology
Via S. Zeno 35–39
56127 Pisa, Italy

E-mail: acarducci@biologia.unipi.it

Chapter 5

Giuseppina La Rosa*
and
Michele Muscillo
Istituto Superiore di Sanità
Department of Environment and Primary Prevention
Viale Regina Elena 299
00161 Rome, Italy

E-mail: giuseppina.larosa@iss.it

Chapter 6

M. S. D'Agostino
Food and Environment Research Agency
Food and Environmental Microbiology
Sand Hutton
York, YO41 1LZ, UK

E-mail: martin.dagostino@fera.gsi.gov.uk

Chapter 7

Maureen Beatrice Taylor
Department of Medical Virology
Faculty of Health Sciences
University of Pretoria
National Health Laboratory Service Tshwane Academic Division
Private Bag X323
Arcadia 0007, South Africa

E-mail: maureen.taylor@up.ac.za

Chapter 8

Ana Maria de Roda Husman*
and
Martijn Bouwknegt
Centre for Zoonoses and Environmental Microbiology
National Institute for Public Health and the Environment (RIVM)
PO Box 1
3720 BA Bilthoven
The Netherlands

E-mail: ana.maria.de.roda.husman@rivm.nl

Chapter 9
Petra Vasickova*
and
Kamil Kovarcik
Veterinary Research Institute
Hudcova 70
621 00 Brno, Czech Republic

E-mail: vasickova@vri.cz; kovarcik@vri.cz

Chapter 10
Charles P. Gerba
Department of Soil, Water and Environmental Science
University of Arizona
Tucson
Arizona 85721, USA

E-mail: gerba@ag.arizona.edu

Chapter 11
Ingeborg L. A. Boxman
Food and Consumer Product Safety Authority (NVWA)
Laboratory for Feed and Food Safety, Department of Microbiology
Akkermaalsbos 4
6708 WB Wageningen
The Netherlands

E-mail: Ingeborg.boxman@vwa.nl

Chapter 12
Leen Baert
Ghent University
Faculty of Bioscience Engineering
Department of Food Safety and Food Quality
Laboratory of Food Microbiology and Food Preservation
Coupure links 653
9000 Ghent, Belgium

Currently at:
Nestlé Research Center
Vers-Chez-Les-Blanc
1000 Lausanne 26
Switzerland

E-mail: Leen.Baert@rdls.nestle.com

Chapter 13
Sabah Bidawid
Health Canada
Food Directorate
Bureau of Microbial Hazards
Microbiology Research Division
251 Sir F.G. Banting Driveway
Postal Locator 2204E, Rm E401
Ottawa
Ontario, K1A 0K9, Canada

E-mail: sabah.bidawid@hc-sc.gc.ca

Chapter 14
Jacquelina W. Woods* and William Burkhardt III
US Food and Drug Administration
Center for Food Safety and Applied Nutrition
Gulf Coast Seafood Lab
1 Iberville Drive
Dauphin Island
Alabama 36528, USA

E-mail: Jacquelina.woods@fda.hhs.gov

Chapter 15
Charles P. Gerba*, Masaaki Kitajima and Brandon Iker
Department of Soil, Water and Environmental Science
University of Arizona
Tucson
Arizona 85721, USA

E-mail: gerba@ag.arizona.edu

Chapter 16
David James Allen and David W. G. Brown*
Virus Reference Department
Microbiology Services Division – Colindale
Health Protection Agency
61 Colindale Avenue
London, NW9 5EQ, UK

E-mail: David.Brown@HPA.org.uk

Miren Iturriza-Gómara
Institute of Infection of Global Health
University of Liverpool, UK

Chapter 17
Rosa M. Pintó* and Albert Bosch
Enteric Virus Laboratory
Department of Microbiology and Institute of Nutrition and Food Safety
University of Barcelona
Diagonal 643
08028 Barcelona, Spain

E-mail: rpinto@ub.edu

Chapter 18
Franco Maria Ruggeri*
Dept. Veterinary Public Health & Food Safety
Istituto Superiore di Sanità,
V. le Regina Elena, 299
00161 Rome, Italy

E-mail: franco.ruggeri@iss.it

Lucia Fiore
National Center for Immunobiological Research and Evaluation
Istituto Superiore di Sanità
V. le Regina Elena, 299
00161 Rome, Italy

Chapter 19
Wim H. M. van der Poel*
Central Veterinary Institute
Wageningen University and Research Centre Department of Virology
Edelhertweg 15
8219 PH Lelystad
The Netherlands

E-mail: wim.vanderpoel@wur.nl

Alessandra Berto
Health Protection Agency
Blood Borne Viruses Unit
Virus Reference Department
Microbiology Services Colindale
61 Colindale Avenue
London NW9 5EQ
DX 6530006
UK

Formerly at
Central Veterinary Institute
Wageningen University and Research
Centre Department of Virology
Edelhertweg 15
8219 PH Lelystad
The Netherlands

Chapter 20
Ricardo Santos*
Laboratório de Análises
Instituto Superior Técnico
Avenida Rovisco Pais
1049–001 Lisbon, Portugal

E-mail: ricardosantos@ist.utl.pt; silvia.monteiro@ist.utl.pt

Chapter 21
Carl-Henrik von Bonsdorff* and Leena Maunula
Department of Food Hygiene and Environmental Health
Faculty of Veterinary Medicine
P.O. Box 66
00014 University of Helsinki
Finland

E-mail: Carl-Henrik.vonBonsdorff@helsinki.fi

Chapter 22
Rosina Girones* and Sílvia Bofill-Mas
Laboratory of Virus Contaminants of Water and Food
Department of Microbiology
Faculty of Biology
University of Barcelona
Av. Diagonal 643
08028 Barcelona, Spain

E-mail: rgirones@ub.edu

Woodhead Publishing Series in Food Science, Technology and Nutrition

1 **Chilled foods: A comprehensive guide** *Edited by C. Dennis and M. Stringer*
2 **Yoghurt: Science and technology** *A. Y. Tamime and R. K. Robinson*
3 **Food processing technology: Principles and practice** *P. J. Fellows*
4 **Bender's dictionary of nutrition and food technology Sixth edition** *D. A. Bender*
5 **Determination of veterinary residues in food** *Edited by N. T. Crosby*
6 **Food contaminants: Sources and surveillance** *Edited by C. Creaser and R. Purchase*
7 **Nitrates and nitrites in food and water** *Edited by M. J. Hill*
8 **Pesticide chemistry and bioscience: The food-environment challenge** *Edited by G. T. Brooks and T. Roberts*
9 **Pesticides: Developments, impacts and controls** *Edited by G. A. Best and A. D. Ruthven*
10 **Dietary fibre: Chemical and biological aspects** *Edited by D. A. T. Southgate, K. W. Waldron, I. T. Johnson and G. R. Fenwick*
11 **Vitamins and minerals in health and nutrition** *M. Tolonen*
12 **Technology of biscuits, crackers and cookies Second edition** *D. Manley*
13 **Instrumentation and sensors for the food industry** *Edited by E. Kress-Rogers*
14 **Food and cancer prevention: Chemical and biological aspects** *Edited by K. W. Waldron, I. T. Johnson and G. R. Fenwick*
15 **Food colloids: Proteins, lipids and polysaccharides** *Edited by E. Dickinson and B. Bergenstahl*
16 **Food emulsions and foams** *Edited by E. Dickinson*
17 **Maillard reactions in chemistry, food and health** *Edited by T. P. Labuza, V. Monnier, J. Baynes and J. O'Brien*
18 **The Maillard reaction in foods and medicine** *Edited by J. O'Brien, H. E. Nursten, M. J. Crabbe and J. M. Ames*
19 **Encapsulation and controlled release** *Edited by D. R. Karsa and R. A. Stephenson*
20 **Flavours and fragrances** *Edited by A. D. Swift*
21 **Feta and related cheeses** *Edited by A. Y. Tamime and R. K. Robinson*
22 **Biochemistry of milk products** *Edited by A. T. Andrews and J. R. Varley*
23 **Physical properties of foods and food processing systems** *M. J. Lewis*
24 **Food irradiation: A reference guide** *V. M. Wilkinson and G. Gould*
25 **Kent's technology of cereals: An introduction for students of food science and agriculture Fourth edition** *N. L. Kent and A. D. Evers*
26 **Biosensors for food analysis** *Edited by A. O. Scott*

27 **Separation processes in the food and biotechnology industries: Principles and applications** *Edited by A. S. Grandison and M. J. Lewis*
28 **Handbook of indices of food quality and authenticity** *R. S. Singhal, P. K. Kulkarni and D. V. Rege*
29 **Principles and practices for the safe processing of foods** *D. A. Shapton and N. F. Shapton*
30 **Biscuit, cookie and cracker manufacturing manuals Volume 1: Ingredients** *D. Manley*
31 **Biscuit, cookie and cracker manufacturing manuals Volume 2: Biscuit doughs** *D. Manley*
32 **Biscuit, cookie and cracker manufacturing manuals Volume 3: Biscuit dough piece forming** *D. Manley*
33 **Biscuit, cookie and cracker manufacturing manuals Volume 4: Baking and cooling of biscuits** *D. Manley*
34 **Biscuit, cookie and cracker manufacturing manuals Volume 5: Secondary processing in biscuit manufacturing** *D. Manley*
35 **Biscuit, cookie and cracker manufacturing manuals Volume 6: Biscuit packaging and storage** *D. Manley*
36 **Practical dehydration Second edition** *M. Greensmith*
37 **Lawrie's meat science Sixth edition** *R. A. Lawrie*
38 **Yoghurt: Science and technology Second edition** *A. Y. Tamime and R. K. Robinson*
39 **New ingredients in food processing: Biochemistry and agriculture** *G. Linden and D. Lorient*
40 **Benders' dictionary of nutrition and food technology Seventh edition** *D. A. Bender and A. E. Bender*
41 **Technology of biscuits, crackers and cookies Third edition** *D. Manley*
42 **Food processing technology: Principles and practice Second edition** *P. J. Fellows*
43 **Managing frozen foods** *Edited by C. J. Kennedy*
44 **Handbook of hydrocolloids** *Edited by G. O. Phillips and P. A. Williams*
45 **Food labelling** *Edited by J. R. Blanchfield*
46 **Cereal biotechnology** *Edited by P. C. Morris and J. H. Bryce*
47 **Food intolerance and the food industry** *Edited by T. Dean*
48 **The stability and shelf-life of food** *Edited by D. Kilcast and P. Subramaniam*
49 **Functional foods: Concept to product** *Edited by G. R. Gibson and C. M. Williams*
50 **Chilled foods: A comprehensive guide Second edition** *Edited by M. Stringer and C. Dennis*
51 **HACCP in the meat industry** *Edited by M. Brown*
52 **Biscuit, cracker and cookie recipes for the food industry** *D. Manley*
53 **Cereals processing technology** *Edited by G. Owens*
54 **Baking problems solved** *S. P. Cauvain and L. S. Young*
55 **Thermal technologies in food processing** *Edited by P. Richardson*
56 **Frying: Improving quality** *Edited by J. B. Rossell*
57 **Food chemical safety Volume 1: Contaminants** *Edited by D. Watson*
58 **Making the most of HACCP: Learning from others' experience** *Edited by T. Mayes and S. Mortimore*
59 **Food process modelling** *Edited by L. M. M. Tijskens, M. L. A. T. M. Hertog and B. M. Nicolaï*
60 **EU food law: A practical guide** *Edited by K. Goodburn*
61 **Extrusion cooking: Technologies and applications** *Edited by R. Guy*
62 **Auditing in the food industry: From safety and quality to environmental and other audits** *Edited by M. Dillon and C. Griffith*
63 **Handbook of herbs and spices Volume 1** *Edited by K. V. Peter*
64 **Food product development: Maximising success** *M. Earle, R. Earle and A. Anderson*

65 **Instrumentation and sensors for the food industry Second edition** *Edited by E. Kress-Rogers and C. J. B. Brimelow*
66 **Food chemical safety Volume 2: Additives** *Edited by D. Watson*
67 **Fruit and vegetable biotechnology** *Edited by V. Valpuesta*
68 **Foodborne pathogens: Hazards, risk analysis and control** *Edited by C. de W. Blackburn and P. J. McClure*
69 **Meat refrigeration** *S. J. James and C. James*
70 **Lockhart and Wiseman's crop husbandry Eighth edition** *H. J. S. Finch, A. M. Samuel and G. P. F. Lane*
71 **Safety and quality issues in fish processing** *Edited by H. A. Bremner*
72 **Minimal processing technologies in the food industries** *Edited by T. Ohlsson and N. Bengtsson*
73 **Fruit and vegetable processing: Improving quality** *Edited by W. Jongen*
74 **The nutrition handbook for food processors** *Edited by C. J. K. Henry and C. Chapman*
75 **Colour in food: Improving quality** *Edited by D. MacDougall*
76 **Meat processing: Improving quality** *Edited by J. P. Kerry, J. F. Kerry and D. A. Ledward*
77 **Microbiological risk assessment in food processing** *Edited by M. Brown and M. Stringer*
78 **Performance functional foods** *Edited by D. Watson*
79 **Functional dairy products Volume 1** *Edited by T. Mattila-Sandholm and M. Saarela*
80 **Taints and off-flavours in foods** *Edited by B. Baigrie*
81 **Yeasts in food** *Edited by T. Boekhout and V. Robert*
82 **Phytochemical functional foods** *Edited by I. T. Johnson and G. Williamson*
83 **Novel food packaging techniques** *Edited by R. Ahvenainen*
84 **Detecting pathogens in food** *Edited by T. A. McMeekin*
85 **Natural antimicrobials for the minimal processing of foods** *Edited by S. Roller*
86 **Texture in food Volume 1: Semi-solid foods** *Edited by B. M. McKenna*
87 **Dairy processing: Improving quality** *Edited by G. Smit*
88 **Hygiene in food processing: Principles and practice** *Edited by H. L. M. Lelieveld, M. A. Mostert, B. White and J. Holah*
89 **Rapid and on-line instrumentation for food quality assurance** *Edited by I. Tothill*
90 **Sausage manufacture: Principles and practice** *E. Essien*
91 **Environmentally-friendly food processing** *Edited by B. Mattsson and U. Sonesson*
92 **Bread making: Improving quality** *Edited by S. P. Cauvain*
93 **Food preservation techniques** *Edited by P. Zeuthen and L. Bøgh-Sørensen*
94 **Food authenticity and traceability** *Edited by M. Lees*
95 **Analytical methods for food additives** *R. Wood, L. Foster, A. Damant and P. Key*
96 **Handbook of herbs and spices Volume 2** *Edited by K. V. Peter*
97 **Texture in food Volume 2: Solid foods** *Edited by D. Kilcast*
98 **Proteins in food processing** *Edited by R. Yada*
99 **Detecting foreign bodies in food** *Edited by M. Edwards*
100 **Understanding and measuring the shelf-life of food** *Edited by R. Steele*
101 **Poultry meat processing and quality** *Edited by G. Mead*
102 **Functional foods, ageing and degenerative disease** *Edited by C. Remacle and B. Reusens*
103 **Mycotoxins in food: Detection and control** *Edited by N. Magan and M. Olsen*
104 **Improving the thermal processing of foods** *Edited by P. Richardson*
105 **Pesticide, veterinary and other residues in food** *Edited by D. Watson*
106 **Starch in food: Structure, functions and applications** *Edited by A.-C. Eliasson*
107 **Functional foods, cardiovascular disease and diabetes** *Edited by A. Arnoldi*

108 **Brewing: Science and practice** *D. E. Briggs, P. A. Brookes, R. Stevens and C. A. Boulton*
109 **Using cereal science and technology for the benefit of consumers: Proceedings of the 12th International ICC Cereal and Bread Congress, 24–26th May, 2004, Harrogate, UK** *Edited by S. P. Cauvain, L. S. Young and S. Salmon*
110 **Improving the safety of fresh meat** *Edited by J. Sofos*
111 **Understanding pathogen behaviour: Virulence, stress response and resistance** *Edited by M. Griffiths*
112 **The microwave processing of foods** *Edited by H. Schubert and M. Regier*
113 **Food safety control in the poultry industry** *Edited by G. Mead*
114 **Improving the safety of fresh fruit and vegetables** *Edited by W. Jongen*
115 **Food, diet and obesity** *Edited by D. Mela*
116 **Handbook of hygiene control in the food industry** *Edited by H. L. M. Lelieveld, M. A. Mostert and J. Holah*
117 **Detecting allergens in food** *Edited by S. Koppelman and S. Hefle*
118 **Improving the fat content of foods** *Edited by C. Williams and J. Buttriss*
119 **Improving traceability in food processing and distribution** *Edited by I. Smith and A. Furness*
120 **Flavour in food** *Edited by A. Voilley and P. Etievant*
121 **The Chorleywood bread process** *S. P. Cauvain and L. S. Young*
122 **Food spoilage microorganisms** *Edited by C. de W. Blackburn*
123 **Emerging foodborne pathogens** *Edited by Y. Motarjemi and M. Adams*
124 **Benders' dictionary of nutrition and food technology Eighth edition** *D. A. Bender*
125 **Optimising sweet taste in foods** *Edited by W. J. Spillane*
126 **Brewing: New technologies** *Edited by C. Bamforth*
127 **Handbook of herbs and spices Volume 3** *Edited by K. V. Peter*
128 **Lawrie's meat science Seventh edition** *R. A. Lawrie in collaboration with D. A. Ledward*
129 **Modifying lipids for use in food** *Edited by F. Gunstone*
130 **Meat products handbook: Practical science and technology** *G. Feiner*
131 **Food consumption and disease risk: Consumer-pathogen interactions** *Edited by M. Potter*
132 **Acrylamide and other hazardous compounds in heat-treated foods** *Edited by K. Skog and J. Alexander*
133 **Managing allergens in food** *Edited by C. Mills, H. Wichers and K. Hoffman-Sommergruber*
134 **Microbiological analysis of red meat, poultry and eggs** *Edited by G. Mead*
135 **Maximising the value of marine by-products** *Edited by F. Shahidi*
136 **Chemical migration and food contact materials** *Edited by K. Barnes, R. Sinclair and D. Watson*
137 **Understanding consumers of food products** *Edited by L. Frewer and H. van Trijp*
138 **Reducing salt in foods: Practical strategies** *Edited by D. Kilcast and F. Angus*
139 **Modelling microorganisms in food** *Edited by S. Brul, S. Van Gerwen and M. Zwietering*
140 **Tamime and Robinson's Yoghurt: Science and technology Third edition** *A. Y. Tamime and R. K. Robinson*
141 **Handbook of waste management and co-product recovery in food processing Volume 1** *Edited by K. W. Waldron*
142 **Improving the flavour of cheese** *Edited by B. Weimer*
143 **Novel food ingredients for weight control** *Edited by C. J. K. Henry*
144 **Consumer-led food product development** *Edited by H. MacFie*
145 **Functional dairy products Volume 2** *Edited by M. Saarela*
146 **Modifying flavour in food** *Edited by A. J. Taylor and J. Hort*

147 **Cheese problems solved** *Edited by P. L. H. McSweeney*
148 **Handbook of organic food safety and quality** *Edited by J. Cooper, C. Leifert and U. Niggli*
149 **Understanding and controlling the microstructure of complex foods** *Edited by D. J. McClements*
150 **Novel enzyme technology for food applications** *Edited by R. Rastall*
151 **Food preservation by pulsed electric fields: From research to application** *Edited by H. L. M. Lelieveld and S. W. H. de Haan*
152 **Technology of functional cereal products** *Edited by B. R. Hamaker*
153 **Case studies in food product development** *Edited by M. Earle and R. Earle*
154 **Delivery and controlled release of bioactives in foods and nutraceuticals** *Edited by N. Garti*
155 **Fruit and vegetable flavour: Recent advances and future prospects** *Edited by B. Brückner and S. G. Wyllie*
156 **Food fortification and supplementation: Technological, safety and regulatory aspects** *Edited by P. Berry Ottaway*
157 **Improving the health-promoting properties of fruit and vegetable products** *Edited by F. A. Tomás-Barberán and M. I. Gil*
158 **Improving seafood products for the consumer** *Edited by T. Børresen*
159 **In-pack processed foods: Improving quality** *Edited by P. Richardson*
160 **Handbook of water and energy management in food processing** *Edited by J. Klemeš, R. Smith and J.-K. Kim*
161 **Environmentally compatible food packaging** *Edited by E. Chiellini*
162 **Improving farmed fish quality and safety** *Edited by Ø. Lie*
163 **Carbohydrate-active enzymes** *Edited by K.-H. Park*
164 **Chilled foods: A comprehensive guide Third edition** *Edited by M. Brown*
165 **Food for the ageing population** *Edited by M. M. Raats, C. P. G. M. de Groot and W. A Van Staveren*
166 **Improving the sensory and nutritional quality of fresh meat** *Edited by J. P. Kerry and D. A. Ledward*
167 **Shellfish safety and quality** *Edited by S. E. Shumway and G. E. Rodrick*
168 **Functional and speciality beverage technology** *Edited by P. Paquin*
169 **Functional foods: Principles and technology** *M. Guo*
170 **Endocrine-disrupting chemicals in food** *Edited by I. Shaw*
171 **Meals in science and practice: Interdisciplinary research and business applications** *Edited by H. L. Meiselman*
172 **Food constituents and oral health: Current status and future prospects** *Edited by M. Wilson*
173 **Handbook of hydrocolloids Second edition** *Edited by G. O. Phillips and P. A. Williams*
174 **Food processing technology: Principles and practice Third edition** *P. J. Fellows*
175 **Science and technology of enrobed and filled chocolate, confectionery and bakery products** *Edited by G. Talbot*
176 **Foodborne pathogens: Hazards, risk analysis and control Second edition** *Edited by C. de W. Blackburn and P. J. McClure*
177 **Designing functional foods: Measuring and controlling food structure breakdown and absorption** *Edited by D. J. McClements and E. A. Decker*
178 **New technologies in aquaculture: Improving production efficiency, quality and environmental management** *Edited by G. Burnell and G. Allan*
179 **More baking problems solved** *S. P. Cauvain and L. S. Young*
180 **Soft drink and fruit juice problems solved** *P. Ashurst and R. Hargitt*
181 **Biofilms in the food and beverage industries** *Edited by P. M. Fratamico, B. A. Annous and N. W. Gunther*

210 **Food and beverage stability and shelf life** *Edited by D. Kilcast and P. Subramaniam*
211 **Processed Meats: Improving safety, nutrition and quality** *Edited by J. P. Kerry and J. F. Kerry*
212 **Food chain integrity: A holistic approach to food traceability, safety, quality and authenticity** *Edited by J. Hoorfar, K. Jordan, F. Butler and R. Prugger*
213 **Improving the safety and quality of eggs and egg products Volume 1** *Edited by Y. Nys, M. Bain and F. Van Immerseel*
214 **Improving the safety and quality of eggs and egg products Volume 2** *Edited by F. Van Immerseel, Y. Nys and M. Bain*
215 **Animal feed contamination: Effects on livestock and food safety** *Edited by J. Fink-Gremmels*
216 **Hygienic design of food factories** *Edited by J. Holah and H. L. M. Lelieveld*
217 **Manley's technology of biscuits, crackers and cookies Fourth edition** *Edited by D. Manley*
218 **Nanotechnology in the food, beverage and nutraceutical industries** *Edited by Q. Huang*
219 **Rice quality: A guide to rice properties and analysis** *K. R. Bhattacharya*
220 **Advances in meat, poultry and seafood packaging** *Edited by J. P. Kerry*
221 **Reducing saturated fats in foods** *Edited by G. Talbot*
222 **Handbook of food proteins** *Edited by G. O. Phillips and P. A. Williams*
223 **Lifetime nutritional influences on cognition, behaviour and psychiatric illness** *Edited by D. Benton*
224 **Food machinery for the production of cereal foods, snack foods and confectionery** *L.-M. Cheng*
225 **Alcoholic beverages: Sensory evaluation and consumer research** *Edited by J. Piggott*
226 **Extrusion problems solved: Food, pet food and feed** *M. N. Riaz and G. J. Rokey*
227 **Handbook of herbs and spices Second edition Volume 1** *Edited by K. V. Peter*
228 **Handbook of herbs and spices Second edition Volume 2** *Edited by K. V. Peter*
229 **Breadmaking: Improving quality Second edition** *Edited by S. P. Cauvain*
230 **Emerging food packaging technologies: Principles and practice** *Edited by K. L. Yam and D. S. Lee*
231 **Infectious disease in aquaculture: Prevention and control** *Edited by B. Austin*
232 **Diet, immunity and inflammation** *Edited by P. C. Calder and P. Yaqoob*
233 **Natural food additives, ingredients and flavourings** *Edited by D. Baines and R. Seal*
234 **Microbial decontamination in the food industry: Novel methods and applications** *Edited by A. Demirci and M.O. Ngadi*
235 **Chemical contaminants and residues in foods** *Edited by D. Schrenk*
236 **Robotics and automation in the food industry: Current and future technologies** *Edited by D. G. Caldwell*
237 **Fibre-rich and wholegrain foods: Improving quality** *Edited by J. A. Delcour and K. Poutanen*
238 **Computer vision technology in the food and beverage industries** *Edited by D.-W. Sun*
239 **Encapsulation technologies and delivery systems for food ingredients and nutraceuticals** *Edited by N. Garti and D. J. McClements*
240 **Case studies in food safety and authenticity** *Edited by J. Hoorfar*
241 **Heat treatment for insect control: Developments and applications** *D. Hammond*
242 **Advances in aquaculture hatchery technology** *Edited by G. Allan and G. Burnell*
243 **Open innovation in the food and beverage industry** *Edited by M. Garcia Martinez*

244 **Trends in packaging of food, beverages and other fast-moving consumer goods (FMCG)** *Edited by Neil Farmer*
245 **New analytical approaches for verifying the origin of food** *Edited by P. Brereton*
246 **Microbial production of food ingredients, enzymes and nutraceuticals** *Edited by B. McNeil, D. Archer, I. Giavasis and L. Harvey*
247 **Persistent organic pollutants and toxic metals in foods** *Edited by M. Rose and A. Fernandes*
248 **Cereal grains for the food and beverage industries** *E. Arendt and E. Zannini*
249 **Viruses in food and water: Risks, surveillance and control** *Edited by N. Cook*
250 **Improving the safety and quality of nuts** *Edited by L. J. Harris*
251 **Metabolomics in food and nutrition** *Edited by B. Weimer and C. Slupsky*
252 **Food enrichment with omega-3 fatty acids** *Edited by C. Jacobsen, N. Skall Nielsen, A. Frisenfeldt Horn and A.-D. Moltke Sørensen*
253 **Instrumental assessment of food sensory quality: A practical guide** *Edited by D. Kilcast*
254 **Food microstructures: Microscopy, measurement and modelling** *Edited by V. J. Morris and K. Groves*
255 **Handbook of food powders: Processes and properties** *Edited by B. R. Bhandari, N. Bansal, M. Zhang and P. Schuck*
256 **Functional ingredients from algae for foods and nutraceuticals** *Edited by H. Domínguez*
257 **Satiation, satiety and the control of food intake: Theory and practice** *Edited by J. E. Blundell and F. Bellisle*
258 **Hygiene in food processing: Principles and practice Second edition** *Edited by H. L. M. Lelieveld, J. Holah and D. Napper*

Part I

An introduction to food and environmental virology

1

An introduction to food- and waterborne viral disease

N. Cook, Food and Environment Research Agency, UK and G. P. Richards, Delaware State University, USA

DOI: 10.1533/9780857098870.1.3

Abstract: Enteric viruses are the principal cause of food- and waterborne illnesses throughout the world. Among the enteric viruses are the noroviruses, sapovirus, hepatitis A and E viruses, Aichi virus, enteric adenoviruses, rotaviruses, and astroviruses. This chapter introduces the reader to food- and waterborne viruses including the diseases they cause, modes of transmission including potential zoonotic spread, documented outbreaks, implicated foods, virus detection methods, and control strategies.

Key words: virus, food and water, disease transmission, outbreaks, detection and control.

1.1 Introduction to enteric viruses

Viral diseases have plagued mankind since before the dawn of civilization. Only in the past century have technological advances led to a characterization of the etiological agents and their epidemiology. Today, the significance of food- and waterborne viral illnesses throughout the world remains underestimated in large part, because of the difficulties in amassing accurate incidence data for outbreaks. In some cases, enteric viruses cause short-term illness or asymptomatic infection, while in other cases these viruses produce high mortalities.

The most common enteric viruses from the standpoint of number of cases are human calicivirus species, particularly human norovirus. Currently, human caliciviruses are divided into two groups, the noroviruses and the sapoviruses. The incubation period for norovirus is 10–51 h (Dolin *et al.*, 1982; Green *et al.*,

2001; Wyatt *et al.*, 1974), and symptoms range from mild to severe diarrhea, vomiting, and dehydration. Symptoms generally last for 24–48 h. Viruses may persist in stool for 1–2 weeks and can be detected in asymptomatic carriers (Graham *et al.*, 1994; Richards *et al.*, 2004; White *et al.*, 1996). The existence of norovirus genogroups and strains having animal hosts raises the possibility of zoonotic transmission (Mattison *et al.*, 2007), and surveillance of circulating noroviruses in the human population has revealed the presence of several uncommon genotypes, which may represent zoonotic strains (Verhoef *et al.*, 2010); however zoonotic transmission has not been proven in any case. Sapovirus causes diarrhea and vomiting similar to norovirus, generally more often in young children than in adults; however, less is known about the incubation period for sapovirus or the duration of fecal shedding. Sapovirus can be present in pigs, but the strains are not genetically similar to those which infect humans (Reuter *et al.*, 2010), and zoonotic transmission is unknown.

Hepatitis A and hepatitis E viruses are both transmitted by food and water. Hepatitis A virus, a member of the Picornaviridae family, is a formidable pathogen capable of eliciting liver disease and death, although in most cases, the illness may go unnoticed and without apparent sequellae. The incubation period is 15–45 days and viral shedding may occur for months. Symptoms may include nausea, vomiting, anorexia (loss of appetite), fatigue, and fever. Hepatitis A virus infects the liver, leading to possible pain in the upper right quadrant, jaundice of the eyes or skin, and dark urine. The illness is potentially lethal, particularly in individuals with pre-existing liver disease. Hepatitis A virus is found only in humans and some simians, and no zoonotic transmission is known to occur. Hepatitis E virus, on the other hand, has four genotypes. Genotypes 1 and 2 are associated with human illness, while genotypes 3 and 4 are animal strains which are occasionally transferred to humans. For instance, genotypes 3 and 4 have been shown to spread zoonotically from pigs and deer to humans (Aggarwal and Naik, 2009). Hepatitis E virus is the sole member of the Hepeviridae family. Like hepatitis A virus it infects the liver and its symptoms are similar; however, it causes a higher incidence of death, particularly among pregnant woman where the death rate approaches 25% (Mast and Krawczynski, 1996), possibly from hormonal differences or other factors (Navaneethan *et al.*, 2008). The incubation period of hepatitis E virus ranges between 2 and 8 weeks, and may be dependent on the virus dose (Anderson and Shrestha, 2002; Li *et al.*, 1994; Tsarev *et al.*, 1994).

Rotavirus is a major pathogen causing infantile diarrhea. It is responsible for an estimated 800 000 deaths annually (Parashar *et al.*, 1998), mostly in developing countries where rehydration therapy is not available. The incubation period for rotavirus is variable, from 11 h to 6 days and symptoms include diarrhea, anorexia, dehydration, depression, and occasional vomiting (Bishop, 1994; Kapikian and Chanock, 1990; Saif *et al.*, 1994). Virus shedding in feces was shown to last for up to 57 days (Richardson *et al.*, 1998) and longer in children with immunodeficiencies (Saulsbury *et al.*, 1980). Rotaviruses are generally species-specific, but cross-species transmission is possible, and low level input of rotavirus strains or sequences into the human population from the animal population may be common. Food- or waterborne zoonotic transmission is a possibility (Cook *et al.*, 2004).

Among the many serotypes of adenovirus, types 40 and 41 are the typical enteric forms capable of eliciting food- and waterborne illness. Most enteric adenovirus cases are mild and transient except in immunocompromised individuals. Acute adenovirus gastroenteritis is characterized by watery diarrhea possibly containing mucus, fever, vomiting, abdominal pain and dehydration and may be accompanied by respiratory illness (Ruuskanen *et al.*, 2002). Adenovirus strains with genetic similarity to human adenovirus have been found in non-human primates, raising the possibility of zoonotic transmission through the consumption of primate meat (Wevers *et al.*, 2011). Other adenovirus serotypes are believed to be predominantly respiratory organisms. Adenovirus infection at an early age imparts long-term immunity against the particular serotype.

Astroviruses cause mild and usually self-limiting illness, often in children. Symptoms include vomiting and diarrhea, and occasional fever, abdominal pain, and anorexia. Astrovirus has an incubation period generally between 24 and 96-h lasting for up to four days in most patients with fecal shedding for a week; however, immunocompromised and malnourished individuals are at greater risk of more severe symptoms and death from dehydration, and can shed viruses in their stools for up to a month. No zoonotic transmission of astrovirus has been observed, although the level of similarity between some human and animal strains suggests that it may be possible (Kapoor *et al.*, 2009).

Two other viral agents responsible for foodborne illness are Aichi virus (a picornavirus) and tick-borne encephalitis virus (a flavivirus). Aichi virus causes acute gastroenteritis in humans with symptoms including diarrhea, abdominal pain, vomiting, nausea, and fever (Yamashita *et al.*, 1991, 2001). Viruses with similarity to Aichi virus can be found in some livestock species (Reuter *et al.*, 2011), but zoonotic transmission has not been recorded. Tick-borne encephalitis virus is most commonly transmitted directly via bites; however, the disease can also be transmitted indirectly via the gastrointestinal route by consumption of unpasteurized milk products and infected dairy animals (Dumpis *et al.*, 1999; Kriz *et al.*, 2009).

1.2 Food and water as vehicles of virus transmission

Water represents an important vehicle for the transmission of enteric viruses. Rivers, lakes, streams, and coastal waters are regularly contaminated by septic tanks, storm water runoff, and effluents from inefficiently operated sewage treatment plants or from overflows from treatment plants impacted by flooding events. Swimmers at bathing beaches also contribute to some level of viral contamination of the water. When contaminated waters are directly used for drinking purposes, they represent a significant hazard to the consumer. Likewise, when such waters are used for the irrigation of crops, the processing of foods, or the production of ice, virus transfer to foods and beverages, and ultimately to humans, can occur. Food may also become contaminated directly by the unsanitized hands of harvesters in the farm fields, truckers, processors, and those who prepare and serve foods in restaurants and at home.

Fecally polluted marine waters may also lead to the contamination of oysters, clams, mussels, and cockles, which represent a significant cause of illness to those who consume raw or undercooked shellfish. This is in large part due to the ability of bivalve mollusks to concentrate and retain viruses from the water column within their edible tissues as a normal part of their filter feeding abilities. Virus levels within the shellfish can increase at least 100-fold over the concentrations in the water. Berry fruits and leafy green vegetables are also common vehicles for virus transmission, particularly when they are consumed raw. Thorough washing of the surfaces of produce is a useful intervention to reduce virus levels, but some foods, such as raspberries, strawberries, and crinkly lettuce, are difficult to thoroughly wash. Foods that are extensively handled are more likely to become contaminated. A case in point is doughnuts which are often handled while frostings are applied or fillings are injected. Alternatively even foods that are cooked can become vehicles for transmission, if they subsequently come into contact with contaminated foods or surfaces: for example, a hamburger may become contaminated by the bun (if it was handled by a person with contaminated hands), or the lettuce, tomato, or onion toppings if they were previously contaminated during production or preparation. Likewise, ill food workers may contaminate food processing equipment, contact surfaces, or foods directly by feces or vomit. Aerosols from vomiting can be transmitted long distances.

1.3 Outbreaks of food- and waterborne viral illness

Enteric viral illnesses often go unrecognized or unreported for several reasons. In some cases, the symptoms are mild enough that no medical intervention is required. In other instances, as in the case of norovirus gastroenteritis, severe diarrhea and projectile vomiting deters patients from traveling to the doctor's office until symptoms resolve, but by then, remission of symptoms is rapid and a doctor may no longer be needed. Some situations preclude a visit to the doctor because of a lack of health insurance, inability to pay, or the inaccessibility of a medical center. In cases where the sick do seek medical treatment, the cause of the illness may not be readily discerned and treatment options are often based on their symptoms. In many locations, viral diseases are not reportable and there is little or no tracking of illnesses or outbreaks. Testing for specific etiological agents of viral illness is difficult, time-consuming, costly and seldom performed, except when larger outbreaks occur. In such cases, epidemiological traceback to food or water is possible.

Sporadic publications are the only way to document many of these outbreaks. Below are some such outbreaks reported for various food commodities.

1.3.1 Produce

A wide variety of fruits and vegetables have been associated with outbreaks of hepatitis A and norovirus illness. Such produce is generally consumed raw or after minimal processing, and any contaminating viruses can remain in an infectious state up to the point of consumption. Outbreaks of hepatitis A have been linked with tomatoes (Petrignani *et al.*, 2010), green onions (Centers for Disease Control and Prevention, 2003; Dentinger *et al.*, 2001; Wheeler *et al.*, 2005), raspberries (Ramsay and Upton, 1989; Reid and Robinson, 1987), strawberries (Niu *et al.*, 1992), blueberries (Calder *et al.*, 2003), fruit juices (Frank *et al.*, 2007), and other produce. Norovirus was responsible for major outbreaks from celery (Warner, 1992; Warner *et al.*, 1991), fruits and berries including fresh and frozen raspberries (Cotterelle *et al.*, 2005; Hjertqvist *et al.*, 2006), lettuce (Alexander *et al.*, 1986; Ethelberg *et al.*, 2010), coleslaw and green leaf salad (Zomer *et al.*, 2009); radishes (Yu *et al.*, 2010), cantaloupe (Bowen *et al.*, 2006), pumpkin salad (Götz *et al.*, 2002), and tropical fruits and juices (Straun *et al.*, 2011; Visser *et al.*, 2010). Other viruses, for example rotavirus, undoubtedly contribute to produce-associated illness, but are not fully recognized because the product is not routinely tested or because cases of illness occur sporadically rather than in large outbreaks. Contamination of fruits and vegetables may occur through fertilization of crops with sewage sludge, irrigation with wastewater, harvesting and handling with unsanitized hands, rinsing with contaminated water, and cross-contamination during preparation (Richards, 2001).

1.3.2 Shellfish

Molluscan shellfish can become readily contaminated with enteric viruses and have been associated with some of the largest outbreaks on record. Contaminated clams were reportedly responsible for 293 000 cases of hepatitis A in China (Halliday *et al.*, 1991). Other outbreaks of shellfish-associated hepatitis A have been noted worldwide (Conaty *et al.*, 2000; Guillois-Bécel *et al.*, 2009; Pintó *et al.*, 2009; Richards, 1985) and are currently problematic in some regions of the world. Norovirus remains a common cause of shellfish-associated illness and is credited with widespread outbreaks (Berg *et al.*, 2000; Simmons *et al.*, 2001; Webby *et al.*, 2007; Westrell *et al.*, 2010). In 1983, the United States had multiple shellfish-associated norovirus outbreaks affecting over 2000 shellfish consumers in New York and New Jersey, most of which occurred over a 3-month period (Richards, 1985). Shellfish-associated norovirus outbreaks continue to occur around the world. Sapovirus has also been responsible for oyster-related outbreaks (Nakagawa-Okamoto *et al.*, 2009). Hepatitis E virus-contaminated shellfish have been associated with several outbreaks (Cacopardo *et al.*, 1997; Said *et al.*, 2009; Tomar, 1998), perhaps because of the ability of shellfish to concentrate viruses within their edible tissues. Adenoviruses are commonly detected in shellfish, but have not led to recognized outbreaks, probably because only some of the serotypes of adenovirus are capable of eliciting gastrointestinal illness, and children, who are most susceptible, are not frequent consumers of raw shellfish (Lees, 2000). Astrovirus, Aichi virus, and rotavirus have also led to outbreaks of shellfish-borne illness (Ambert-Balay *et al.*, 2008; Le Guyader *et al.*, 2008; Yamashita *et al.*, 2000). Although thorough cooking inactivates enteric viruses, the incidence of shellfish-borne viral illness is high, in large part because many consumers prefer their shellfish raw or only lightly cooked.

1.3.3 Bakery products

Large outbreaks of norovirus have been associated with bakery products. An outbreak affecting 2700 guests at 46 weddings was attributed to the contamination of wedding cakes handled by two ill workers at one bakery (Friedman *et al.*, 2005). In a similar event, an estimated 3000 individuals developed norovirus after eating frosted cake prepared by an ill worker (Kuritsky *et al.*, 1984). Rolls prepared by an ill baker led to 231 illnesses at a lunch buffet (de Wit *et al.*, 2007). Outbreaks of hepatitis A have also been attributed to bakery products when frostings and glazes were contaminated by ill workers (Schoenbaum *et al.*, 1976; Weltman *et al.*, 1996). Filled doughnuts and other pastries were associated with outbreaks where two employees were diagnosed with hepatitis A (Schenkel *et al.*, 2006). Although pastries may be cooked and cooking inactivates enteric viruses, contamination after cooking poses a threat of enteric virus illness.

1.3.4 Meats and dairy products

Chicken, pork, and beef products have been associated with norovirus illness, usually as a result of contamination after cooking (Vivancos *et al.*, 2009; Zomer *et al.*, 2010). Delicatessen meats were linked to an outbreak of norovirus (Malek *et al.*, 2009) and hepatitis A (Gustafson *et al.*, 1983; Schmid *et al.*, 2009). Cheese was also related to a norovirus outbreak (Vivancos *et al.*, 2009) and to hepatitis A (Gustafson *et al.*, 1983). Transmission has been associated with product handling by an ill worker. Hepatitis A has been associated with milk consumption (Murphy *et al.*, 1946; Raska *et al.*, 1966). Milk and cheeses are also vectors for tick-borne encephalitis, often from sheep and goat milk from Europe and Asia where tick-borne encephalitis is endemic (Bogovic *et al.*, 2010; Gresikova *et al.*, 1975; Holzmann *et al.*, 2009; Kerbo *et al.*, 2005). Although the transmission of rotavirus illness was linked to tuna and chicken salad (Centers for Disease Control and Prevention, 2000), relatively few cases of foodborne rotavirus diarrhea have been reported.

1.3.5 Water

Contaminated drinking water is likely responsible for the majority of cases of enteric virus illness. One report indicates that an estimated 60% of the norovirus illnesses in the United States are from contaminated water (Mead *et al.*, 1999). Noroviruses have been listed as the primary cause of waterborne illness worldwide (Leclerc *et al.*, 2002). Waterborne rotavirus infections are responsible for significant morbidity and mortality, particularly in children (Divizia *et al.* 2004; Glass *et al.*, 2001; Villena *et al.*, 2003). Both rotavirus and caliciviruses were detected in 40% of the sick individuals during a waterborne outbreak involving sewage-contaminated drinking water (Räsänen *et al.*, 2010). Occasional waterborne outbreaks of hepatitis E have been documented (Corwin *et al.*, 1996; Naik *et al.*, 1992; Rab *et al.*, 1997). Astrovirus has been associated with outbreaks of gastroenteritis from contaminated drinking water (Gofti-Laroche *et al.*, 2003; Kukkula *et al.*, 1997); however, in 89 outbreaks of waterborne illness affecting 4321 people in England and Wales, astrovirus was associated with only 1% of the outbreaks (Smith *et al.*, 2006). The transmission of most enteric viruses via water and foods is not fully recognized due to poor detection efforts and inadequate reporting practices.

In addition to the foodborne route of transmission, viral disease can also be acquired through environmental exposure. Viruses such as norovirus and adenovirus may be highly prevalent in sewage-polluted recreational waters (Wyn-Jones *et al.*, 2011), and enteric viruses have frequently been implicated in disease outbreaks or cases linked to swimming, canoeing, etc. (Sinclair *et al.*, 2009).

1.4 Virus detection

Virus screening of various food products and water is seldom performed because of the cost and difficulty in conducting the assays. Detection of foodborne viruses is challenging and requires the use of complex methods. These methods are composed of sample treatment and assay stages. Sample treatment is a multi-step process aimed at removing viruses from a food sample and concentrating them for delivery to the assay, principally nucleic acid amplification by PCR. Methods are improving and some protocols are available for extracting, concentrating, and analyzing viruses from shellfish, meat, and some ready-to-eat foods (Richards *et al.*, in press). Currently, a group (CEN TC 275/WG6/TAG4) set up by the Committee for European Standardization is developing methods to detect norovirus and hepatitis A virus in salad vegetables, shellfish, and soft fruit (Lees and CEN WG6 TAG4, 2010). Publication of these international standards is scheduled for 2013. Methods are also becoming available to test for viruses in the rinse from fruits and vegetables (Richards *et al.*, in press). Extracts and rinses often contain substances inhibitory to the molecular methods used to identify the presence of viruses, namely real-time reverse transcription polymerase chain reaction (RT-PCR); therefore, numerous controls must be included to ensure the effectiveness of the extraction process and the validity of the assay (D'Agostino *et al.*, 2011). Such methods are most often employed in response to a suspected food- or waterborne outbreak to facilitate an epidemiological investigation of the incident. Virus extraction and analysis is generally labor-intensive, requires the use of expensive equipment and reagents, and demands a qualified technician to perform such procedures; therefore, viral assays are not performed routinely on any food product.

A potential breakthrough was recently published showing that many enteric viruses may become sequestered within the shellfish hemocytes and that extracting and testing the hemocytes for viruses may offer a simpler approach for virus detection from live shellfish (Provost *et al.*, 2011). The use of hemocytes alone may simplify the separation and concentration of viruses from shellfish, since hemocytes may be a source of already concentrated viruses. If the hemocytes contain the majority of viruses, then testing just the digestive tissues would likely miss the majority of the viral contaminants.

The evaluation of infectivity of viruses is problematic, particularly using molecular diagnostic techniques (Richards, 1999). The salient issue therefore is whether viruses detected by molecular assay are actually infectious (Cliver, 2009). A recent study has demonstrated a potential breakthrough in that human noroviruses that are likely to be infectious bind to porcine mucin, whereas noroviruses that have been inactivated by heat, UV irradiation, and high pressure processing appear not to bind to mucin (Dancho *et al.*, 2012). Thus, binding of NoV to mucin followed by RT-PCR testing may lead the way to better identifying and quantifying infectious NoV.

1.5 Control of virus contamination of food and water

Both physical (e.g., disinfection) and procedural (e.g., compliance with guidelines and regulations) measures can assist in reducing the potential for transmission of viruses via food and water. Currently, most treatment of drinking water or water for use in food production is performed by chlorination, and the effectiveness of this treatment is borne out by the evidence that outbreaks of disease due to contaminated drinking water generally only occur when the water treatment or distribution system has failed, for example, a breach in a water pipe allowing the ingress of sewage. In most countries, the effectiveness of potable water treatment is generally evaluated by determining the presence of coliform bacteria in compliance with national regulations. This is also the case for recreational waters or waters used in shellfish production. Since the presence of coliform bacteria has not been proven to consistently or reliably correlate with the presence of enteric viruses, the consideration of adoption of virus standards has been recommended (Wyn-Jones *et al.*, 2011).

As discussed above, the foodstuffs most at risk of being vehicles for the transmission of viral disease are those which are eaten raw or after only minimal processing. Consequently, they are seldom if ever subject to thorough disinfection. The effect of processes which are currently used in the food industry may not be sufficient to eliminate all infectious viruses which may contaminate the treated foodstuff (Koopmans and Duizer, 2004). The most reliable control of foodborne viral contamination will therefore be to prevent the contamination from occurring in the first place, and this should be achieved by effective guidelines made available to the food industry. Over the past decade, the role of viruses as the most prevalent agents of foodborne disease has become widely recognized, and consequently, there are current international efforts aimed at tackling the problem of contamination of foods by pathogenic viruses by provision of formal guidance. For instance, the Codex Alimentarius Commission Committee on Food Hygiene has developed guidelines on the control of viruses in food (Codex Alimentarius Commission, 2011). The European Framework 7 research project 'Integrated monitoring and control of foodborne viruses in European food supply chains' has produced basic guidance sheets on prevention of virus contamination of berry fruits, leafy greens, and pork products (available at http://www.eurovital.org/GuidanceSheets1.htm). These are intended to complement the Codex guidelines and be used in concordance with them.

In conclusion, enteric viruses transmitted by food and water pose significant challenges to public health globally. These challenges will be met by ongoing research into the mechanisms and conditions underlying how viruses contaminate food and environmental matrices, which should ultimately allow the determination of means to break the route of transmission from source to target. Research on enteric virus inactivation processes will also lead to the development of enhanced processing strategies to reduce viral contaminants in the food industry. In the subsequent sections and chapters of this book, extensive details will be provided on key aspects of virus contamination of food and waters, such as the prevalence, persistence, detection, and control of these pathogens, and issues which may become significant in the future will be discussed. This information will provide a comprehensive and current overview of the challenges of food- and waterborne viruses.

1.6 References

AGGARWAL R and NAIK S (2009), Epidemiology of hepatitis E: current status. *J Gastroenterol Hepatol*, **24**, 1484–1493.

ALEXANDER W J, HOLMES J R, SHAW J F, RILEY W E and ROPER W L (1986), Norwalk virus outbreak at a college campus. *South Med J*, **79**, 33–36, 40.

AMBERT-BALAY K, LORROT M, BON F, GIRAUDON H, KAPLON J, WOLFER M, LEBON P, GENDREL D and POTHIER P (2008), Prevalence and genetic diversity of Aichi virus strains in stool samples from community and hospitalized patients. *J Clin Microbiol*, **46**, 1252–1258.

ANDERSON D A and SHRESTHA I I (2002), Hepatitis E virus. In: *Clinical Virology*, 2nd ed., D D RICHMAN, R J WHITNEY and F G HAYDEN eds, ASM Press, Washington, DC, pp. 1061–1074.

BERG D E, KOHN M A, FARLEY T A and MCFARLAND L M (2000), Multi-state outbreak of acute gastroenteritis traced to fecal-contaminated oysters harvested in Louisiana. *J Infect Dis*, **181**(Suppl 2), S381–S386.

BISHOP R F (1994), Natural history of rotavirus infections. In: *Viral Infections of the Gastrointestinal Tract*. 2nd ed., A Z KAPIKIAN ed., MARCEL DEKKER, New York. pp. 131–167.

BOGOVIC P, LOTRIC-FURLAN S and STRLE F (2010), What tick-borne encephalitis may look like: clinical signs and symptoms. *Travel Med Infect Dis*, **8**, 246–250.

BOWEN A, FRY A, RICHARDS G and BEUCHAT L (2006), Infections associated with cantaloupe consumption: a public health concern. *Epidemiol Infect*, **134**, 675–685.

CACOPARDO B, RUSSO R, PREISER W, BENANTI F, BRANCATI G and NUNNARI A (1997), Acute hepatitis E in Catania (eastern Sicily) 1980–1994. The role of hepatitis E virus. *Infection*, **25**, 313–316.

CALDER L, SIMMONS G, THORNLEY C, TAYLOR P, PRITCHARD K, GREENING G and BISHOP J (2003), An outbreak of hepatitis A associated with consumption of raw blueberries. *Epidemiol Infect*, **131**, 745–751.

CENTERS FOR DISEASE CONTROL and PREVENTION (2000), Foodborne outbreak of Group A rotavirus gastroenteritis among college students – District of Columbia, March-April 2000. *MMWR Morb Mort Wkly Rep*, **49**, 1131–1133.

CENTERS FOR DISEASE CONTROL and PREVENTION (2003), Hepatitis A outbreak associated with green onions at a restaurant – Monaca, Pennsylvania, 2003. *MMWR Morb Mortal Wkly Rep*, **52**, 1155–1157.

CLIVER DO (2009), Capsid and infectivity in virus detection. *Food Environ Virol*, **1**, 123–128.

Codex Alimentarius Commission (2011). *Report of the Forty-Third Session of The Codex Committee on Food Hygiene, Miami, USA, 5–9 December 2011*. FAO/WHO, Geneva.

CONATY S, BIRD P, BELL G, KRAA E, GROHMANN G and MCANULTY J M (2000), Hepatitis A in New South Wales, Australia from consumption of oysters: the first reported outbreak. *Epidemiol Infect*, **124**, 121–130.

COOK N, BRIDGER J, KENDALL K, ITURRIZA-GÓMARA M, EL-ATTAR L and GRAY J (2004), The zoonotic potential of rotavirus. *J Inf*, **48**, 289–302.

CORWIN A L, KHIEM H B, CLAYSON E T, PHAM K S, VO T T, VU T Y, CAO T T, VAUGHN D, MERVEN J, RICHIE T L, PUTRI M P, HE J, GRAHAM R, WIGNALL F S and HYAMS K C (1996), A waterborne outbreak of hepatitis E virus transmission in Southwestern Vietnam. *Am J Trop Med Hyg*, **54**, 559–562.

COTTERELLE B, DROUGARD C, ROLLAND J, BECAMEL M, BOUDON M, PINEDE S, TRAORÉ O, BALAY K, POTHIER P and ESPIÉ E (2005), Outbreak of norovirus infection associated with the consumption of frozen raspberries, France, March 2005. *Euro Surveill*, **10**, E050428 050421.

D'AGOSTINO M, COOK N, RODRIGUEZ-LAZARO D and RUTJES S (2011), Nucleic acid amplification-based methods for detection of enteric viruses: definition of controls and interpretation of results. *Food Environ Virol*, **2**, 55–60.

DANCHO B A, CHEN H and KINGSLEY D H (2012), Discrimination between infectious and non-infectious human norovirus using porcine gastric mucin. *Int J Food Microbiol*, **155**, 222–226.

DE WIT M A, WIDDOWSON M A, VENNEMA H, de BRUIN E, FERNANDEZ T and KOOPMANS M (2007), Large outbreak of nrovirus: the baker who should have known better. *J Infect*, **55**, 188–193.

DENTINGER C M, BOWER W A, NAINAN O V, COTTER S M, MYERS G, DUBUSKY L M, FLOWER S, SALEHI E D and BELL B P (2001), An outbreak of hepatitis A associated with green onions. *J Infect Dis*, **183**, 1273–1276.

DIVIZIA M, GABRIELI R, DONIA D, MACALUSO A, BOSCH A, GUIX S, SÁNCHEZ G, VILLENA C, PINTÓ R M, PALOMBI L, BUONUOMO E, CENKO F, LENO L, BEBECI D and BINO S (2004), Waterborne gastroenteritis outbreak in Albania. *Water Sci Technol*, **50**, 57–61.

DOLIN R, REICHMAN R C, ROESSNER K D, TRALKA T S, SCHOOLEY R T, GARY W and MORENS D (1982), Detection by immune electron microscopy of the Snow Mountain agent of acute viral gastroenteritis. *J Infect Dis*, **146**, 184–189.

DUMPIS U, CROOK D and OKSI J (1999), Tick-borne encephalitis. *Clin Infect Dis*, **28**, 882–890.

ETHELBERG S, LISBY M, BOTTIGER B, SCHULTZ A C, VILLIF A, JENSEN T, OLSEN K E, SCHEUTZ F, KJELSO C and MULLER L (2010), Outbreaks of gastroenteritis linked to lettuce, Denmark, January 2010. *Euro Surveill* **15**(6), 19484.

FRANK C, WALTER J, MUEHLEN M, JANSEN A, VAN TREECK U, HAURI A M, ZOELLNER, I, RAKHA M, HOEHNE M, HAMOUDA O, SCHREIER E and STARK K (2007), Major outbreak of hepatitis A associated with orange juice among tourists, Egypt, 2004. *Emerg Infect Dis*, **1**, 156–158.

FRIEDMAN D S, HEISEY-GROVE D, ARGYROS F, BERL E, NSUBUGA J, STILES T, FONTANA J, BEARD R S, MONROE S, MCGRATH M E, SUTHERBY H, DICKER R C, DEMARIA A and MATYAS B T (2005), An outbreak of norovirus gastroenteritis associated with wedding cakes. *Epidemiol Infect*, **133**, 1057–1063.

GLASS R I, BRESEE J, JIANG B, GENTSCH J, ANDO T, FANKHAUSER R, NOEL J, PARASHAR U, ROSEN B and MONROE S S (2001), Gastroenteritis viruses: an overview. *Novartis Found Symp*, **238**, 5–25.

GOFTI-LAROCHE L, GRATACAP-CAVALLIER B, DEMANSE D, GENOULAZ O, SEIGNEURIN J M and ZMIROU D (2003), Are waterborne astrovirus implicated in acute digestive morbidity (E. MI. R.A. study)? *J Clin Virol*, **27**, 74–82.

GÖTZ H, DE JONG B, LINDBÄCK J, PARMENT P A, HEDLUND K O, TORVÉN M and EKDAHL K (2002), Epidemiological investigation of a foodborne gastroenteritis outbreak caused by Norwalk-like virus in 30 day-care centres. *Scand J Infect Dis*, **34**, 115–121.

GRAHAM, D Y, JIANG X, TANAKA T, OPEKUN A R, MADORE H P and ESTES M K (1994), Norwalk virus infection of volunteers: new insights based on improved assays. *J Infect Dis*, **180**, 34–43.

GREEN K Y, CHANOCK R M and KAPIKIAN A Z (2001), Human caliciviruses. In: *Fields Virology*, Vol. 1, D M KNIPE and P M HOWLEY, eds, Lippinscott Williams & Wilkins, Philadelphia, PA. pp. 841–74.

GRESIKOVA M, SEKEYOVA M, STÚPALOVA S and NECAS S (1975), Sheep milk-borne epidemic of tick-borne encephalitis in Slovakia. *Intervirology*, **5**, 57–61.

GUILLOIS-BÉCEL Y, COUTURIER E, LE SAUX J C, ROQUE-AFONSO A M, LE GUYADER, F S, LE GOAS A, PERNÈS J, LE BECHEC S, BRIAND A, ROBERT C, DUSSAIX E, POMMEPUY M and VAILLANT V (2009), An oyster-associated hepatitis A outbreak in France in 2007. *Euro Surveill*, **14**, 19144.

GUSTAFSON T L, HUTCHESON R L JR, FRICKER R S and SCHAFFNER W (1983), An outbreak of foodborne hepatitis A: the value of serological testing and matched case-control analysis. *Am J Public Health*, **73**, 1199–1201.

HALLIDAY L M, KANG L Y, ZHOU T K, HU M D, PAN Q C, FU T Y, HUANG Y S and HU S L (1991), An epidemic of hepatitis A attributable to the ingestion of raw clams in Shanghai, China. *J Infect Dis*, **164**, 852–859.

HJERTQVIST M, JOHANSSON A, SVENSSON N, ABOM P E, MAGNUSSON C, OLSSON M, HEDLUND K O and ANDERSSON Y (2006), Four outbreaks of norovirus gastroenteritis after consuming raspberries, Sweden, June–August 2006. *Euro Surveill*, **11**, E060907 060901.

HOLZMANN H, ABERLE S W, STIASNY K, WERNER P, MISCHAK A, ZAINER B, NETZER M, KOPPI S, BECHTER E and HEINZ F X (2009), Tick-borne encephalitis from eating goat cheese in a mountain region of Austria. *Emerg Infect Dis*, **15**, 1671–1673.

KAPIKIAN A Z and CHANOCK R M (1990), Rotaviruses. In: *Virology*, 2nd ed., B N Fields and D M Knipe, eds, Raven Press, New York. pp. 1353–404.

KAPOOR A, LI L, VICTORIA J, ODERINDE B, MASON C, PANDEY P, ZAIDI S Z and DELWART E (2009), Multiple novel astrovirus species in human stool. *J Gen Virol*, **90**, 2965–72.

KERBO N, DONCHENKO I, KUTSAR K and VASILENKO V (2005), Tick-borne encephalitis in Estonia linked to raw goat milk, May–June 2005. *Euro Surveill*, **10**, E050623 2.

KOOPMANS M and DUIZER E (2004), Foodborne viruses: an emerging problem. *Int J Food Microbiol*, **90**, 23–41.

KRIZ B, BENES C and DANIEL M (2009), Alimentary transmission of tick-borne encephalitis in the Czech Republic (1997–2008). *Epidemiol Mikrobiol Imunol*, **58**, 98–103.

KUKKULA M, ARSTILA P, KLOSSNER M L, MAUNULA L, BONSDORFF C H V and JAATINEN P (1997), Waterborne outbreak of viral gastroenteritis. *Scand J Infect Dis*, **29**, 415–18.

KURITSKY J N, OSTERHOLM M T, GREENBERG, H G, KORLATH J A, GODES J R, HEDBERG C W, FORFANG J C, KAPIKIAN A Z, MCCULLOUGH J C and WHITE K E (1984), Norwalk gastroenteritis: a community outbreak associated with bakery product consumption. *Ann Intern Med*, **100**, 519–21.

LECLERC H, SCHWARTZBROD L and DEI-CAS E (2002), Microbial agents associated with waterborne diseases. *Crit Rev Microbiol*, **28**, 371–409.

LEES D (2000), Viruses and bivalve shellfish. *Intl J Food Microbiol*, **59**, 81–116.

LEES D and CEN WG6 TAG4 (2010), International Standardisation of a method for detection of human pathogenic viruses in molluscan shellfish. *Food Environ Virol*, **2**, 146–55.

LE GUYADER F S, LE SAUX J C, AMBERT-BALAY K, KROL J, SERAIS O, PARNAUDEAU S, GIRAUDON H, DELMAS G, POMMEPUY M, POTHIER P and ATMAR R L (2008), Aichi virus, norovirus, astrovirus, enterovirus, and rotavirus involved in clinical cases from a French oyster-related gastroenteritis outbreak. *J. Clin. Microbiol*, **46**, 4011–17.

LI F, ZHUANG H, KOLIVAS S, LOCARNINI S and ANDERSON D (1994), Persistence and transient antibody responses to hepatitis E virus detected by Western immunoblot using open reading frame 2 and 3 and glutathione S-transferase fusion proteins. *J Clin Microbiol*, **32**, 2060–6.

MALEK M, BARZILAY E, KRAMER A, CAMP B, JAYKUS L A, ESCUDERO-ABARCA B, DERRICK G, WHITE P, GERBA C, HIGGINS C, VINJE J, GLASS R, LYNCH M and WIDDOWSON M A (2009), Outbreak of norovirus infection among river rafters associated with packaged delicatessen meat, Grand Canyon, 2005. *Clin Infect Dis*, **48**, 31–7.

MAST E E and KRAWCZYNSKI K (1996), Hepatitis E: an overview. *Annu Rev Med*, **47**, 257–266.

MATTISON K, SHUKLA A, COOK A, POLLARI F, FRIENDSHIP R, KELTON D, BIDAWID S and FARBER J M (2007), Human noroviruses in swine and cattle. *Emerg Inf Dis*, **13**, 1184–8.

MEAD P S, SLUTSKER L, DIETZ V, MCCAIG L F, BRESEE J S, SHAPIRO C, GRIFFIN P M and TAUXE R V (1999), Food-related illness and death in the United States. *Emerg Infect Dis*, **5**, 607–25.

MURPHY W J, PETRIE L M and WORK S D (1946), Outbreak of infectious hepatitis, apparently milk-borne. *Am J Public Health Nations Health*, **36**, 169–73.

NAIK S R, AGGARWAL R, SALUNKE P N and MEHROTRA N N (1992), A large waterborne hepatitis E epidemic in Kanpur, India. *Bull World Health Organ*, **70**, 597–604.

NAKAGAWA-OKAMOTO R, ARITA-NISHIDA T, TODA S, KATO H, IWATA H, AKIYAMA M, NISHIO O, KIMURA H, NODA M, TAKEDA N and OKA T (2009), Detection of multiple sapovirus genotypes and genogroups in oyster-associated outbreaks. *Jpn J Infect Dis*, **62**, 63–6.

NAVANEETHAN U, Al MOHAJER M, and SHATA M T (2008), Hepatitis E and pregnancy: understanding the pathogenesis. *Liver Int*, **28**, 1190–9.

NIU M T, POLISH L B, ROBERTSON B H, KHANNA B K, WOODRUFF B A, SHAPIRO C N, MILLER M A, SMITH J D, GEDROSE J K, ALTER M J and MARGLIS H S (1992), Multistate outbreak of hepatitis A associated with frozen strawberries. *J Infect Dis*, **166**, 518–24.

PARASHAR U D, BRESSE J S, GENTSCH J R and GLASS R I (1998). Rotavirus. *Emerg Infect Dis*, **4**, 561–70.

PETRIGNANI M, HARMS M, VERHOEF L, VAN HUNEN R, SWAAN C, VAN STEENBERGEN J, BOXMAN I, PERAN I SALA R, OBER H, VENNEMA H, KOOPMANS M and VAN PELT W (2010), Update: a foodborne outbreak of hepatitis A in the Netherlands related to semi-dried tomatoes in oil, January-February 2010. *Euro Surveill*, **15**(20), 19572.

PINTÓ R M, COSTAFREDA M I and BOSCH A (2009), Risk assessment in shellfish-borne outbreaks of hepatitis A. *Appl Environ Microbiol*, **75**, 7350–5.

PROVOST K, DANCHO B A, OZBAY G, ANDERSON R S, RICHARDS G P and KINGSLEY D K (2011), Hemocytes are sites of enteric virus persistence within oysters. *Appl Environ Microbiol*, **77**, 8360–9.

RAB M A, BILE M K, MUBARIK M M, ASGHAR H, SAMI Z, SIDDIQI S, DIL A S, BARZGAR M A, CHAUDHRY M A and BURNEY M I (1997), Waterborne hepatitis E virus epidemic in Islamabad, Pakistan: a common source outbreak traced to malfunction of a modern treatment plant. *Am J Trop Med Hyg*, **57**, 151–7.

RAMSAY C N and UPTON P A (1989), Hepatitis A and frozen raspberries. *Lancet*, **1**, 43–4.

RÄSÄNEN S, LAPPALAINEN S, KAIKKONEN S, HÄMÄLÄINEN M, SALMINEN M and VESIKARI T (2010). Mixed viral infections causing acute gastroenteritis in children in a waterborne outbreak. *Epidemiol Infect*, **138**, 1227–34.

RASKA K, HELCL J, JEZEK J, KUBELKA Z, LITOV M, NOVÁK K, RADKOVSKÝ J, SERÝ V, ZEJDL J and ZIKMUND V (1966), A milk-borne infectious hepatitis epidemic. *J Hyg Epidemiol Microbiol Immunol*, **10**, 413–28.

REID T M and ROBINSON H G (1987), Frozen raspberries and hepatitis A. *Epidemiol Infect*, **98**, 109–12.

REUTER G, BOROS A and PANKOVICS P (2011), Kobuviruses – a comprehensive review. *Rev Med Virol*, **21**, 32–41.

REUTER G, ZIMSEK-MIJOVSKI J, POLJSAK-PRIJATELJ M, DI BARTOLO I, RUGGERI F M, KANTALA T, MAUNULA L, KISS I, KECSKEMETI S, HALAIHEL N, BUESA J, JOHNSEN C, HJULSAGER C K, LARSEN L E, KOOPMANS M and BOTTIGER B (2010), Incidence, diversity, and molecular epidemiology of sapoviruses in swine across Europe. *J Clin Microbiol*, **48**, 363–8.

RICHARDS G P (1985), Outbreaks of shellfish-associated enteric virus illness in the United States: requisite for development of viral guidelines. *J Food Prot*, **48**, 815–23.

RICHARDS G P (1999), Limitations of molecular biological techniques for assessing the virological safety of foods. *J Food Prot*, **62**, 691–7.

RICHARDS G P (2001), Enteric virus contamination of foods through industrial practices: a primer on intervention strategies. *J Indust Microbiol Biotechnol*, **27**, 117–25.

RICHARDS G P, CLIVER D O and GREENING G E (in press), Food-borne viruses, In: *Compendium of Methods for the Microbiological Examination of Foods*, American Public Health Association, Washington, DC.

RICHARDS G P, WATSON M A and KINGSLEY D H (2004), A SYBR green real-time RT-PCR method to detect and quantitate Norwalk virus in stools. *J Virol Method*, **116**, 63–70.

RUUSKANEN O, MEURMAN O and AKUSJÄRVI G (2002), Adenoviruses. In: *Clinical Virology*, 2nd ed., D D RICHMAN, R J WHITLEY and F G HAYDEN, eds, ASM Press, Washington, DC, pp. 515–35.

SAID B, IJAZ S, KAFATOS G, BOOTH L, THOMAS H L, WALSH A, RAMSAY M, MORGAN D and HEPATITIS E INCIDENT INVESTIGATION TEAM (2009), Hepatitis E virus outbreak on cruise ship. *Emerg Infect Dis*, **15**, 1738–44.

SAIF L J, ROSEN B I and PARWANI A V (1994), Animal rotaviruses. In: *Viral Infections of the Gastrointestinal Tract*, 2nd ed., A Z KAPIKIAN, ed., Marcel Dekker, New York, pp. 279–367.

SAULSBURY F T, WINKELSTEIN J A and YOLKEN R H (1980), Chronic rotavirus infections in immunodeficiency. *J Pediatr*, **97**, 61–5.

SCHENKEL K, BREMER V, GRABE C, VAN TREECK U, SCHREIER E, HÖHNE M, AMMON A and ALPERS K (2006), Outbreak of hepatitis A in two federal states of Germany: bakery products as vehicle of infection. *Epidemiol Infect*, **134**, 1292–8.

SCHMID D, FRETZ R, BUCHNER G, KÖNIG C, PERNER H, SOLLAK R, TRATTER A, HELL M, MAASS M, STRASSER M and ALLERBERGER F (2009), Foodborne outbreak of hepatitis A, November 2007–January 2008, Austria. *Eur J Clin Microbiol Infect Dis*, **28**, 385–91.

SCHOENBAUM S C, BAKER O and JEZEK Z (1976), Common-source epidemic of hepatitis due to glazed and iced pastries. *Am J Epidemiol*, **104**, 74–80.

SIMMONS G, GREENING G, GAO W and CAMPBELL D (2001), Raw oyster consumption and outbreaks of viral gastroenteritis in New Zealand: evidence for risk to the public's health. *Aust N Z J Public Health*, **25**, 234–40.

SINCLAIR R G, JONES E L and GERBA C P (2009), Viruses in recreational waterborne disease outbreaks: a review. *J Appl Microbiol*, **107**, 1769–80.

SMITH A, REACHER M, SMERDON W, ADAK G K, NICHOLS G and CHALMERS R M (2006), Outbreaks of waterborne infectious intestinal disease in England and Wales, 1992–2003. *Epidemiol Infect*, **134**, 1141–9.

STRAUN L K, SCHNEIDER K R and DANYLUK M D (2011), Microbial safety of tropical fruits. *Crit Rev Food Sci Nutr*, **51**, 132–45.

TOMAR B S (1998), Hepatitis E in India. *Zhonghua Min Guo Xiao Er Ke Yi Xue Hui Za Zhi*, **39**, 150–6.

TSAREV S A, TSAREV T S, EMERSON S U, YARBOUGH P O, LEGTERS L J, MOSKAL T and PURCELL R H (1994), Infectivity titration of a prototype strain of hepatitis E virus in cynomolgus monkeys. *J Med Virol*, **43**, 135–42.

VERHOEF L, VENNEMA H, VAN PELT W, LEES D, BOSHUIZEN H, HENSHILWOOD K, KOOPMANS M and The Foodborne Viruses in Europe Network (2010), Use of norovirus genotypes to differentiate origins of foodborne outbreaks. *Emerg Inf Dis*, **16**, 617–24.

VILLENA C, GABRIELI R, PINTÓ R M, GUIX S, DONIA D, BUONOMO E, PALOMBI L, CENKO F, BINO S, BOSCH A and DIVIZIA M (2003), A large infantile gastroenteritis outbreak in Albania caused by multiple emerging rotavirus genotypes. *Epidemiol Infect*, **131**, 1105–10.

VISSER H, VERHOEF L, SCHOP W and GÖTZ H M (2010), Outbreak investigation in two groups of coach passengers with gastroenteritis returning from Germany to the Netherlands in February 2009. *Euro Surveill*, **15**, 19615.

VIVANCOS R, SHROUFI A, SILLIS M, AIRD H, GALLIMORE C I, MYERS L, MAHGOUB H and NAIR P (2009), Food-related norovirus outbreak among people attending two barbeques: epidemiological, virological, and environmental investigation. *Int J Infect Dis*, **13**, 629–35.

WARNER R D (1992), A large nontypical outbreak of Norwalk virus. *Int Food Safety News*, **1**, 55.

WARNER R D, CARR R W, MCCLESKEY F K, JOHNSON P C, ELMER L M and DAVISON V E (1991), A large nontypical outbreak of Norwalk virus: gastroenteritis associated with exposing celery to nonpotable water and with Citrobacter freundii. *Arch Intern Med*, **151**, 2419–24.

WEBBY R J, CARVILLE K S, KIRK M D, GREENING G, RATCLIFF R M, CRERAR S K, DEMPSEY K, SARNA M, STAFFORD R, PATEL M and HALL G (2007), Internationally distributed frozen oyster meat causing multiple outbreaks of norovirus infection in Australia. *Clin Infect Dis*, **44**, 1026–31.

WELTMAN A C, BENNETT N M, ACKMAN D A, MISAGE J H, CAMPANA J J, FINE L S, DONIGER A S, BALZANO G J and BIRKHEAD G S (1996), An outbreak of hepatitis A associated with a bakery, New York, 1994: the 1968 "West Branch, Michigan" outbreak repeated. *Epidemiol Infect*, **117**, 333–41.

WESTRELL T, DUSCH, V, ETHELBERG S, HARRIS J, HJERTQVIST M, JOURDAN-DA SILVA N, KOLLER A, LENGLET A, LISBY M and VOLD L (2010), Norovirus outbreaks linked to oyster consumption in the United Kingdom, Norway, France, Sweden and Denmark. *Euro Surveill*, **15**, 19524.

WEVERS D, METZGER S, BABWETEERA F, BIEBERBACH M, BOESCH C, CAMERON K, COUACY-HYMANN E, CRANFIELD M, GRAY M, HARRIS L A, HEAD J, JEFFERY K, KNAUF S, LANKESTER F, LEENDERTZ S A J, LONSDORF E, MUGISHA L, NITSCHE A, REED P, ROBBINS M, TRAVIS D A, ZOMMERS Z, LEENDERTZ F H and EHLERS B (2011), Novel adenoviruses in wild primates: a high level of genetic diversity and evidence of zoonotic transmissions. *J Virol*, **85**, 10774–84.

WHEELER C, VOGT T M, ARMSTRONG G L, VAUGHAN G, WELTMAN A, NAINAN O V, DATO V, XIA G, WALLER K, AMON J, LEE T M, HIGHBAUGH-BATTLE A, HEMBREE C, EVENSON S, RUTA M A, WILLIAMS I T, FIORE A E and BELL B P (2005), An outbreak of hepatitis A associated with green onions. *N Engl J Med*, **353**, 890–7.

WHITE K E, OSTERHOLM M T, MARIOTTI J A, KORLATH J A, LAWRENCE D H, RISTINEN T L and GREENBERG H L (1996), A foodborne outbreak of Norwalk virus gastroenteritis: evidence for post-recovery transmission. *Am J Epidemiol*, **124**, 120–6.

WYATT R G, DOLIN R, BLACKLOW N R, DUPONT H L, BUSCHO R F, THORNHILL T S and KAPIKIAN A Z (1974), Comparison of three agents of acute infectious nonbacterial gastroenteritis by cross-challenge of volunteers. *J Infect Dis*, **129**, 709–14.

WYN-JONES A, CARDUCCI A, COOK N, D'AGOSTINO M, DIVIZIA M, FLEISCHER J, GANTZER C, GAWLER A, GIRONES R, HÖLLER C, DE RODA HUSMAN A M, KAY D, KOZYRA I, LÓPEZ-PILA J, MUSCILLO M, SÃO JOSÉ NASCIMENTO M, PAPAGEORGIOU G, RUTJES S, SELLWOOD J, SZEWZYK R and WYER M (2011), Surveillance of adenoviruses and noroviruses in European recreational waters. *Water Res*, **45**, 1025–38.

YAMASHITA T, ITO M, TSUZUKI H and SAKAE K (2001), Identification of Aichi virus infection by measurement of immunoglobulin responses in an enzyme-linked immunosorbent assay. *J Clin Microbiol*, **39**, 4178–80.

YAMASHITA T, KOBAYASHI S, SAKAE K, NAKATA S, CHIBA S, ISHIHARA Y and ISOMURA S. (1991), Isolation of cytopathic small round viruses with BS-C-1 cells from patients with gastroenteritis. *J Infect Dis*, **164**, 954–7.

YAMASHITA T, SUGIYAMA M, TSUZUKI Y, SAKAE K, SUZUKI Y and MIYAZAKI Y (2000), Application of a reverse transcription-PCR for identification and differentiation of

Aichi virus, a new member of the picornavirus family associated with gastroenteritis in humans. *J Clin Microbiol*, **38**, 2955–61.
YU J H, KIM N Y, KOH Y J and LEE H J (2010), Epidemiology of foodborne Norovirus outbreak in Incheon, Korea. *J Korean Med Sci*, **25**, 1128–33.
ZOMER T P, DE JONG B, KÜHLMANN-BERENZON S, NYRÉN O, SVENUNGSSON B, HEDLUND K O, ANCKER C, WAHL T and ANDERSSON Y (2010), A foodborne norovirus outbreak at a manufacturing company. *Epidemiol Infect*, **138**, 501–6.

2

Prevalence of viruses in food and the environment

T. Petrović, Scientific Veterinary Institute 'Novi Sad', Serbia

DOI: 10.1533/9780857098870.1.19

Abstract: The chapter presents the results of surveys on the presence of viruses in different kinds of food commodities (fresh fruit and vegetables, pork meat and shellfish) and water environments (drinking water, sewage, surface and groundwater, and recreational waters). Where known, connections to outbreaks of viruses in the population are also discussed. Finally, the chapter looks at the principal gaps in current knowledge and major future expectations regarding detection and surveillance.

Key words: virus prevalence in foods, virus prevalence in water environments, food- and waterborne virus outbreaks, food- and waterborne zoonotic viruses.

2.1 Introduction

Enteric viruses may be naturally present in aquatic environments, but are more commonly introduced as a result of human activities through routes such as leakage from sewers and septic systems, urban runoff, agricultural runoff and runoff of animal manure used in agriculture. In sea water, sewage outfall and vessel wastewater discharge are also contributory factors (Fong and Lipp, 2005; Bosch *et al.*, 2006; Wyn-Jones *et al.*, 2011). However, more direct faecal contamination of the environment by humans and animals also occurs, for example by bathers or due to defecation by both farm and wild animals onto soil or surface waters (Rodrıguez-Lazaro *et al.*, 2011). Over 100 types of pathogenic viruses are excreted in human and animal wastes (Melnick, 1984), and can be transferred to the environment via a number of routes including groundwater, rivers, recreational waters, and irrigated vegetables and fruit, with subsequent risks of their reintroduction into human and animal populations (Rodrıguez-Lazaro *et al.*, 2011). Virus concentrations of 5000–100 000 pfu/L are commonly reported in raw sewage (Rao and Melnick, 1986) but may be greatly reduced during treatment. However, current treatment practices are unable to provide virus-free wastewater effluents (Vantarakis and Papapetropoulou, 1999; Bosch *et al.*, 2006) with an average of 50–100 pfu/L normally found in treated effluent (Rao and Melnick, 1986; Bosch *et al.*, 2006). Human exposure to even low levels of these pathogenic viruses in the environment can cause infection and diseases, particularly with viruses such as norovirus (NoV) where fewer than 10 virus particles are required to cause infection (Teunis *et al.*, 2008).

Human enteric viruses can contaminate fresh fruit and vegetables at different stages of the food production chain. During pre-harvest and harvesting, the main source of contamination is contact with polluted water and inadequately treated or untreated sewage sludge used for irrigation and fertilization; handling by virus-infected individuals is a further potential source of contamination at this stage. Contamination can also occur during post-harvest handling, storage, processing, distribution and shipping and at the point of sale, as a result of contact with infected people or a contaminated environment. Finally, produce may also be contaminated in the consumer's own home.

The proportion of all foodborne outbreaks associated with raw produce has increased over the last two decades. Multiple factors have contributed to these overall increases, including an improvement in detection. However, it is also possible that the reported increase in the number of illnesses associated with consumption of fresh produce reflects a true increase in contamination (Berger *et al.*, 2010). Advances in agronomic, processing, preservation, packaging, shipping and marketing technologies on a global scale have enabled the fresh fruit and vegetable industry to supply consumers with a wide range of high-quality produce year round, but with these advances have come new areas of risk. Changes in processing mean that more cutting and coring may be performed in the field at the time of harvest; moreover, as agriculture becomes more intensive, produce fields may be close to animal production zones, and the ecological connections between wild animals, farm animals and produce may be closer (Lynch *et al.*, 2009). The use of manure rather than chemical fertilizer, as well as the use of untreated sewage or irrigation water containing viruses, undoubtedly contributes to this increased risk. Another factor with a direct impact is the increase in (largely exotic) fruit and vegetable imports from abroad in order to meet consumers' expectations of a choice of products year round. Since the hygiene standards of the exporting countries can be subject to wide variation both at harvest and during storage, as well as in irrigation water, the potential for contamination of produce may be increased and consumers may be exposed to higher numbers and different strains of pathogens (Heaton and Jones, 2008).

2.2 The prevalence of virus contamination in food and water

2.2.1 Prevalence of viruses in food: survey results

The European Food Safety Authority (EFSA) BIOHAZ panel has identified three viruses of significance for foodborne transmission: norovirus (NoV), hepatitis A virus (HAV), and hepatitis E virus (HEV) (EFSA, 2011). However, estimating the prevalence of these and other viruses in different types of food items is not currently possible, principally due to a lack of data. No systematic surveillance has to date been carried out on national or international levels: the existing data were collected only partly from research project-based studies and mainly from studies carried out after specific outbreaks.

Viruses in fresh fruit and vegetables

In the online database of the European Rapid Alert System for Food and Feed (RASFF) (http://ec.europa.eu/food/food/rapidalert/rasff_portal_database_en.htm) a total of 24 cases are listed (up to 23 March 2012) in which enteric viruses were present in fruit and vegetables. The main virus detected was NoV (22/24; 91.67%), occurring principally in fruit (20/22), most often frozen raspberries (18/20) and in only two cases in lettuce (from Germany and France). The two instances of HAV-positive fruit and vegetables from 2010 and 2011 originated in dates from Algeria. The NoV-positive raspberries originated from Serbia (11/20), Poland (6/20), China (2/20) and Chile (1/20). The first report on the presence of NoV in fruit and vegetables was listed in the database on 3 June 2005, and related to NoV in raspberries from Poland. In the following 5 years (i.e. up to 2010) only eight cases of NoV-positive fruit and vegetables were reported, but in the period between January 2010 and March 2012 a total of 12 cases were listed. This increase in the number of reported cases of NoV-positive fruit and vegetables is likely to be due to improvements in diagnostic methods and the application of these methods in many more laboratories and countries, as well as to the now widespread knowledge of the possibility of virus infection through this type of food.

NoV outbreaks have been linked to fresh soft red fruits and leafy greens. In 76 foodborne NoV outbreaks from 1998 to 2000 in the USA, the main foods implicated in NoV transmission were salads (26%), raw produce/fruit (17%) and sandwiches (13%). The other implicated foods were meat, fish, oysters, bakery products and others (Widdowson *et al.*, 2005). Between 1992 and 2000, 1518 outbreaks of foodborne infectious intestinal disease were reported in England and Wales. Of those, 83 (5.5%) were associated with the consumption of salad vegetables or fruit. The pathogens most frequently reported were salmonellas (41.0%) and NoV (15.7%). In total 3438 people were affected; 69 were admitted to hospital and one person died. Most outbreaks were linked to commercial catering premises (67.5%) (Long *et al.*, 2002).

In South Korea, Cheong *et al.* (2009) examined 29 groundwater and 30 vegetable samples and found that five (17%) groundwater and three (10%) vegetable samples tested positive for enteric viruses. Adenoviruses (AdVs) were the most frequently detected viruses in four groundwater and three vegetable samples, while enteroviruses (EVs) and NoVs were detected in only one groundwater sample and one vegetable sample (spinach), respectively. Human rotavirus (HRoV) was not detected. A survey of salad vegetables conducted in Slovakia in Spring 2008 found that 5 out of 60 samples were contaminated with NoV; these samples were of lettuce, leeks, spring onions and mixed vegetables, and all were collected in large retail stores (EFSA, 2011).

Recently, Sarvikivi *et al.* (2012) reviewed the data regarding all notified food-borne outbreaks in 2009 in Finland and found that 13 NoV outbreaks affecting about 900 people could be linked to imported frozen raspberries. Two raspberry samples from two different batches of raspberries were found to be NoV-positive. These two batches were shown to have been the likely source of 6 of the 13 outbreaks. Analytical studies were not conducted for all outbreaks, and virological test results were inconclusive in two. During April and May 2009, Stals *et al.* (2011) examined 75 fruit products in Belgium, and despite the good bacteriological quality of the samples, found NoV genotype 1 (GI) and/or genotype 2 (GII) in 4 of 10 raspberry, 7 of 30 cherry tomato, 6 of 20 strawberry and 1 of 15 fruit salad samples. The level of detected NoV genomic copies ranged between 2.5 and 5.0 log per 10 g. No associated illness or virus outbreaks were reported.

According to Baert *et al.* (2011), Belgium, Canada and France were the first countries to provide data concerning the prevalence of NoV in fresh fruit and vegetables. In total, 867 samples of leafy greens, 180 samples of fresh soft red fruits and 57 samples of other types of fresh produce (tomatoes, cucumber and fruit salads) were analyzed in those three countries. NoV was detected by real-time reverse transcription polymerase chain reaction (RT-PCR) in 28.2% (n = 641), 33.3% (n = 6) and 50% (n = 6) of leafy greens tested in Canada, Belgium and France, respectively. Soft red fruits were found to be NoV-positive by real-time RT-PCR in 34.5% (n = 29) of the samples tested in Belgium and in 6.7% (n = 150) of the samples tested in France. Moreover, 55.5% (n = 18) of the other fresh produce types analyzed in Belgium were found to be NoV-positive. Sequencing was carried out to confirm the positive results, but was successful in only 34.6% (18/52) of the positive PCR products. With the increase in sensitivity of the detection methodology, and the absence of outbreaks linked to consumption of the identified virus-positive foods, there is increasing concern regarding the interpretation of positive NoV results using real-time amplification.

A total of 56 people were affected by NoV gastroenteritis after attending a one-day meeting in a hotel in Oslo, Norway, at the end of January 2011. The results of the epidemiological investigation suggested that food items served during the meeting may have been the vehicle of infection (Guzman-Herrador *et al.*, 2011). Boxman *et al.* (2011) conducted a year-long study of the prevalence of NoV in catering companies with no recently reported outbreaks of gastroenteritis, and compared the results to the observed prevalence in catering companies with recently reported outbreaks. Swab samples were collected from surfaces in kitchens and staff bathrooms for in 832 randomly chosen companies, and were analyzed for the presence of NoV RNA. In total, 42 out of 2496 (1.7%) environmental swabs from 35 (4.2%) of the catering companies with no recently reported outbreaks tested positive. In contrast, NoV was detected in 147 of the 370 samples (39.7%) from 44 of the 72 establishments (61.1%) associated with outbreaks of gastroenteritis. NoV-positive swabs have been more frequently found in winter, in specific types of establishments such as care homes and lunchrooms, and in establishments with separate bathrooms for staff. Holtby *et al.* (2001) described a NoV outbreak that affected 49 people who had eaten in the same restaurant. The subsequent investigation concluded that eating salad was strongly associated with infection. The food handler dealing with salad in the food preparation area, who had returned to work after being absent with symptoms of gastrointestinal illness, was found to be the source of salad contamination by NoV.

The need for a better surveillance system – not only for fruit and vegetables but for all food products – has been recognized by the Foodborne Viruses in Europe (FBVE) network, which has conducted virus-specific surveillance of gastroenteritis outbreaks since 1999 (Koopmans *et al.*, 2003). In the FBVE network database a total of 1639 NoV outbreaks occurring during a five-year period from January 2002 to December 2006 were reported by the countries involved in the network (Verhoef *et al.*, 2009). The final dataset comprised 77% (1254/1639) of outbreaks, as the remaining 23% were excluded due to a lack of laboratory confirmation of NoV. A sufficient level of evidence for food-relatedness was confirmed for 24 of 224 outbreaks (11%) and was found to be probable for 200 (89%) more outbreaks. Thirty food categories were associated with outbreaks, including shellfish, fruit, fancy cakes, buffets, sandwiches, and salads. In one foodborne outbreak, poor personal hygiene was mentioned as a contributory factor; an infected food handler was reported in 16 outbreaks, with one cook being involved in two outbreaks; and hygiene rather than preparation or consumption of food was mentioned in two outbreaks. Foodborne outbreaks have occurred more often in households or restaurants and less often in healthcare settings, with genogroup (G) II genotype 4 (GII.4) strains being involved relatively more frequently (Verhoef *et al.*, 2009).

Rotavirus (RoV) causes an estimated 2.7–3.9 million illnesses, 49 000–50 000 hospitalizations, and about 30 deaths per year in the United States (Mead *et al.*, 1999), and an estimated 800 000 deaths per year worldwide (Parashar *et al.*, 1998). However, it has been estimated that only 1% of RoV cases are foodborne (Mead *et al.*, 1999). Market lettuce was found to be contaminated with RoV and HAV at a time when there was a high incidence of rotaviral diarrhea in Costa Rica (Hernandez *et al.*, 1997). During 2009, Mattison *et al.* (2010) tested 328 samples of packaged leafy greens (representing 12–14 different lots from 3 to 6 companies per week) for NoV or RoV RNA in Canada. Of 275 samples, 148 (54%) were found to be NoV-positive, and 1 (0.4%) was found to be positive for RoV group A. During the confirmatory RT-PCR test, just 40 samples (15% of the total) ware confirmed NoV-positive. The one RoV-positive sample was confirmed by sequencing. Brassard *et al.* (2012) described the presence of pathogenic human and zoonotic viruses on irrigated, field-grown strawberries. NoV genogroup I, RoV and swine hepatitis E virus genogroup 3 were detected in the strawberries, and irrigation water was suspected to be the origin of contamination.

Viruses in pork products

Frequent zoonotic transmission of hepatitis E virus (HEV) has been suspected, particularly involving pork products, but there is still a relative lack of data in support of an animal origin of autochthonous cases. Norder *et al.* (2009) sequenced within ORF2 63 HEV strains originating from human blood sera collected between 1993 and 2007, and found that genotype 3 strains were responsible for infecting patients in Europe. In order to identify the connection between human and swine HEV, Norder *et al.* (2009) additionally sequenced the HEV strains from 18 piglets in 17 herds in Sweden and Denmark. Phylogenetic analyses of the genotype 3 strains showed geographical clades and high similarity between strains from patients and pigs from the same area, so the authors concluded that autochthonous HEV cases of animal origin are present in Scandinavia.

Bouquet *et al.* (2011) also assessed the genetic identity of HEV strains found in humans and swine during an 18-month period in France. HEV sequences identified in patients with autochthonous HEV infection (n = 106) were compared with sequences amplified from swine livers collected in slaughterhouses (n = 43). Phylogenetic analysis showed the same proportions of subtypes 3f (73.8%), 3c (13.4%) and 3e (4.7%) in human and swine populations. Furthermore, a similarity of >99% was found between HEV sequences of human origin and those of swine origin. These results indicate that the consumption of some pork products, such as raw liver, is a major source of exposure to autochthonous HEV infection (Bouquet *et al.*, 2011). HEV is found not only in domestic pigs, but also frequently in wild boars. The prevalence of HEV RNA in the livers of hunted wild boars ranges from 3% to 25% in several studies carried out in Europe and Japan (Pavio *et al.*, 2010).

Numerous survey studies have estimated the prevalence of HEV RNA in marketed livers. HEV RNA was detected in 1.9% of 363 livers from supermarkets in Japan (Yazaki *et al.*, 2003), and in 6% of 62 samples of packaged liver in the Netherlands (Bouwknegt *et al.*, 2007). In the United Kingdom, however, Banks *et al.* (2007) reported that all 80 packs of liver from supermarkets were free from HEV. Feagins *et al.* (2007) examined a total of 127 packages of commercial pig liver sold in local grocery stores in the United States for the presence of HEV RNA, and found that 14 (11%) tested positive for HEV RNA. Sequencing and phylogenetic analyses revealed that the 14 isolates all belonged to genotype 3. Subsequent experimental infection of pigs inoculated with positive pig liver homogenates demonstrated that HEV in pig livers was infective. Leblanc *et al.*, (2010) examined the presence of HEV in the tissues of 43 adult pigs, randomly selected from an experimental herd at slaughter in Canada. HEV RNA was detected in 14 of the 43 animals tested. Although no HEV RNA was detected in any of the muscle samples tested, 20.9% of the liver samples obtained at the slaughterhouse did test positive for HEV RNA, with viral loads of 10^3–10^7 copies/g in positive liver and bile samples.

Di Bartolo *et al.* (2012) evaluated the prevalence of HEV in the pork production chain in the Czech Republic, Italy, and Spain during 2010. Samples of faeces, liver and meat were taken, and HEV RNA was detected in at least one of the samples in 36 of the 113 pigs (32%) examined in slaughterhouses. HEV RNA was detected more frequently in slaughterhouse samples in Italy (18/34 – 53%) and Spain (15/39 – 38%) than in the Czech Republic, where the prevalence of HEV RNA in slaughterhouses was lower (3/40 – 8%). Pig faeces showed the highest HEV RNA presence (27%), followed by liver (4%) and meat (3%). Out of 313 sausages sampled at processing stage and at point of sale in supermarkets, HEV was detected only in Spain (6% – 6/93). HEV sequencing confirmed only genotype 3 HEV strains.

In the UK, Berto *et al.* (2012) detected HEV in 6 out of 63 (10%) sausages and in 1 of 40 (2.5%) livers. All 40 examined pig muscle samples were negative for HEV. The same authors also identified the presence of HEV at all three stages in the pork food supply chain: production, processing and point of sale. 1 out of 4 swabs (25%) of workers' hands at the slaughterhouse, a surface swab from a metal point used to hook the carcasses at the processing unit, and 2 of 8 (25%) surface swabs (from the knife and slicer) at point of sale were HEV-positive. The available data suggests therefore that the consumption of raw or undercooked sausage is a potential route of HEV transmission.

Li *et al.* (2009) tested liver samples in a Chinese abbatoir, and found that 3.5% of the samples tested were positive for HEV RNA. During 2009, the Centre for Food Safety in Hong Kong obtained a total of 100 fresh pig liver samples from pigs slaughtered at a local slaughterhouse. Around half of these were collected from roaster pigs (around 4 months old) and the other half were collected from porker pigs (around 6 months old). Of the collected samples, 16 out of 51 (31%) roaster liver samples but none of the 49 porker liver samples tested positive for HEV. The positive rates were 22% (6/27) and 42% (10/24) for roaster pigs sourced from farms in two different regions in mainland China. Partial ORF2 sequences of some HEV isolates from roaster pigs were found to be the same as those from seven local human cases from 2009 and from other local cases recorded in the past. This study suggests that, in addition to contaminated water or food such as raw or undercooked shellfish, pigs and the consumption of pork products could be a source of human HEV in endemic regions (Anon, 2010).

Viruses in shellfish and other bivalve molluscs

In commercially distributed shellfish the reported prevalence of NoV varies from 0% to 79%, while that of HAV ranges from 0% to 43%. A similar situation was observed in non-commercial shellfish: NoVs were detected with a prevalence of 0–60% and HAV was detected with a prevalence of 0–49% (EFSA, 2011). In recent studies NoV has been detected in 5–55% of oyster samples collected by random sampling from markets and oyster farms in Europe and the USA (Boxman *et al.*, 2006; Costantini *et al.*, 2006; Gentry *et al.*, 2009). RNA of enteric viruses has also been detected in shellfish from commercial and non-commercial harvesting areas, as well as in products available on the market for direct consumption and in shellfish associated with disease outbreaks (Boxman, 2010). Boxman (2010) published a detailed review of the presence of enteric viruses in humans and their prevalence in bivalve molluscs collected from European waters or markets. The data regarding the presence of enteric viruses in shellfish in this review were classified according to the geographic location (countries of origin or/and countries in which the study was carried out). The review included studies and reports on the presence of human enteric viruses in shellfish from 1990 until 2006, summarising data from 26 peer-reviewed articles covering 4260 shellfish samples from European waters or markets. The data suggest a high prevalence of different human enteric viruses, principally NoV, HAV, EV, human adenovirus (HAdV), and HRoV. The viruses were present in shellfish from polluted areas, in depurated shellfish and even in shellfish classified as class A, as well as those for human consumption.

In a recent UK study, NoV was detected in 76.2% of oyster samples (643/844), with a similar prevalence in the two different species of oysters tested: 76.1% (468/615) for *Crassostrea gigas* and 76.4% (175/229) for *Ostrea edulis*. A clear seasonality was observed, with a positivity rate of 90.0% (379/421) for samples taken between October and March compared to 62.4% (264/423) for samples taken between April and September (Anon, 2011a). The majority (52.1% – 335/643) of samples were shown to be below the limit of quantification of 100 detectable genome copies/g digestive tissues for both genogroups. However, a few samples (1.4% of all positives – 9/463) contained levels of over 10 000 copies/g. In all 39 examined sites at least one NoV-positive result was found, although the prevalence varied from 21% (5/24 samples) to 100% (20/20 samples). NoV levels varied markedly between sites with some sites scoring consistently over 1000 copies/g during the winter while others rarely or never exceeded 100 copies/g. More positive results were found for NoV GII than for NoV GI (Anon, 2011a). Diez-Valcarce *et al.* (2012) examined the prevalence of different enteric viruses in commercial mussels at the retail level in Finland, Greece and Spain. A total of 153 mussel samples of different origins were analyzed for the presence of NoV genogroups I and II, HAV and HEV, as well as for HAdV as an indicator of human faecal contamination. HAdV was the most prevalent virus detected (36%), followed by NoV genogroup II (16%), HEV (3%) and NoV genogroup I (0.7%). HAV was not detected. The estimated number of PCR-detectable units varied between 24 and $1.4 \times 10^3\ g^{-1}$ of the digestive tract of the mussel. No significant correlation was found between the presence of HAdV and human NoV, HAV, and HEV.

In New Zealand, the prevalence of enteric viruses (HAdV, NoV and EV) and bacteriophages in shellfish was studied over a two-year period from 2004 to 2006. (Greening and Lewis, 2007). Oysters, pipi (*Paphies australis*), cockles and mussels were collected from 28 sites around New Zealand, including harvesting sites and several sites downstream of a sewage outfall. Of the 360 pooled shellfish samples analyzed, 174 (48.3%) tested positive for one or more human enteric viruses. All samples from main sites tested positive for at least one virus during the study, and those collected close to a sewage outfall were 100% positive for viruses over the whole study period.

Mesquita *et al.* (2011) tested different shellfish collected in Portugal between March 2008 and February 2009 for the presence of NoV, HAV and EV. NoV was detected in 37% of the shellfish batches, EV in 35%, and HAV in 33%. Overall, 69% of all the batches analyzed were found to be contaminated by at least one of the viruses under investigation, while simultaneous contamination with two viruses was also found: 6% of batches had NoV/HAV contamination; 8% had NoV/EV contamination; and a further 8% had EV/HAV contamination.

Umesha *et al.* (2008) examined the prevalence of human enteric viruses in bivalve molluscan shellfish and shrimp collected off the south-west coast of India. Out of 194 samples analyzed, 37% of oyster samples, 46% of clam samples and 15% of shrimp samples tested positive for EV. Adenoviruses were found in 17% of oyster and 27% of clam samples. Seasonality was again a factor: a particularly high prevalence of EV and HAdV was observed in the period from May to December. In Korea, a total of 156 raw oyster samples were collected from 23 supermarkets and open-air fish markets in six towns and cities between February and April. NoV was detected in 14.1% (22/156) of the samples, of which 12.2% (19/156) were genogroup I (GI) and 1.9% (3/156) genogroup II (GII) (Moon *et al.*, 2011). HEV has also been found in shellfish, although there are currently few systematic studies available of the prevalence of this virus. The presence of HEV genotype 3 has been reported for 2 out of 32 Japanese samples of *Yamato-Shijimi* (a type of clam) studied by Li *et al.* (2007).

2.2.2 Prevalence of viruses in water: survey results

Water is essential for life of all living organisms. Throughout the world, millions of people do not have access to microbiologically safe water for drinking and other essential purposes (Gibson *et al.*, 2011). It is estimated that 884 million people, or 1/6 of the whole world's population, do not have access to improved sources of drinking water (Mara, 2003; WHO and UNICEF, 2010), and that poor water quality with limited sanitation and inadequate hygiene results in 3.5 billion diarrheal episodes per year causing 1.87 million childhood deaths per year worldwide (Arnold and Colford, 2007; Boschi-Pinto *et al.*, 2008). Water related diseases are associated with exposure to water environments in many ways. These include waters used for drinking and recreation purposes but also waters used for agricultural purposes such as crop irrigation and food processing (Bosch, 1998; Bosch *et al.*, 2008). Enteric viruses have been isolated from and linked to outbreaks originating from contaminated drinking water sources, recreational waters, urban rivers, and shellfish harvested from contaminated waters (Dewailly *et al.*, 1986; Cecuk *et al.*, 1993; Lipp and Rose, 1997; Jiang *et al.*, 2001; Lee and Kim, 2002; Fong and Lipp, 2005). Even from the late 1970s in the United States (Cliver, 1984) water is still considered to be the main vehicle in outbreaks of vehicle-associated viral disease.

Viruses in sewage, surface water, groundwater and drinking water

Data relating to the presence and prevalence of viruses in different water environments are routinely collected and reported as part of national or regional surveillance systems, as part of project-based studies, and in studies carried out after outbreaks of human infections. In one study in South Africa, HAdVs were detected in about 22% of river water samples and about 6% of treated water samples (van Heerden *et al.*, 2005). In another study undertaken in South Africa, about 29% of river water samples and 19% of treated drinking water samples had detectable levels of EVs (Ehlers *et al.*, 2005).

An accident in the city of Nokia in Finland caused about 450 000 l of treated sewage water to be allowed to run into the drinking water supplies of the city's 30 000 inhabitants over the course of two days. Over the next five weeks around 1000 people sought medical help because of gastroenteritis. A sample of the drinking water from the city tested positive for NoV, astrovirus (AstV), HRoV, EVs and HAdV. NoVs were also found in 29.8% of stool samples from affected patients, while AstV, HAdV, HRoV and EVs were detected in 19.7%, 18.2%, 7.5% and 3.7% of the specimens, respectively. HAV and HEV were not found: this was expected because HAV and HEV infection has rarely been linked to waterborne outbreaks in non-endemic regions such as Finland (Maunula *et al.*, 2009).

Virus concentrations in raw water that receives faecal waste such as sewage are often high (Okoh *et al.*, 2010). Patients suffering from viral gastroenteritis or viral hepatitis may excrete about 10^5–10^{11} virus particles per gram of stool (Bosch, 1998), comprising various genera such as HAdV, AstV, NoV, HEV, parvoviruses, EV (Coxsackie viruses, echoviruses and polioviruses), HAV and HRoV (Carter, 2005), so these viruses may often be present in urban sewage. HAdV and polyomaviruses are highly prevalent in all geographical areas (Pina *et al.*, 1998). Enteroviruses, NoV, HRoV and AstV show differing levels of prevalence, depending on the time of year and the occurrence of outbreaks in the population (Girones *et al.*, 2010). The presence of HAV varies between different geographical areas, but in endemic areas it is frequently detected in urban sewage throughout the year. HEV is more common in developing countries where sanitation is poor; however, autochthonous strains of HEV have been reported in urban sewage in several highly developed countries, as well as related cases of sporadic acute hepatitis caused by these non-imported strains (Pina *et al.*, 2000).

A number of recent studies have been carried out with the aim of estimating the risks from viral contamination related to the release of wastewater into surface waters. In a study performed by Carducci *et al.* (2008), the efficiency of virus removal by an urban sewage plant was evaluated by screening the inlet and outlet for the presence of HAV and HAdV. The results showed the presence of adenovirus DNA in 100% of collected samples; Torque Teno virus (TTV) DNA was present in 95% (19/20) of raw sewage samples and in 85% (17/20) of the samples taken at the outlet, while HAV was detected only in 2 of 40 examined samples (5%). The study reported a consistent presence of viruses, a reduction rate of about 2 $\log_{10}$ and the inability of bacterial indicators to assess the presence or removal of viruses (Carducci *et al.* 2008). In a study performed shortly afterwards, the presence of HAdV, TTV, HAV, RoV, EV and NoV genogroups I and II was confirmed, and HAdV was found to be the best indicator for virus inactivation in recycled waters (Carducci *et al.* 2009). In addition, in another study, the presence of different types of viruses including HAV, HAdV, HRoV, AstV, NoV and EV was detected in 67% of sewage samples (Anastasi *et al.*, 2008). In Norway, Myrmel *et al.* (2006) detected HAdVs in 96% of inlet and 94% of outlet samples, supporting the potential of these viruses as indicators of viral contamination from sewage. NoVs were detected in 43%, 53% and 24% of the inlet samples and 26%, 40% and 21% of the outlet samples of the three sewage treatment plants examined, while HAV and circoviruses were found only rarely.

In study supported by the UK Food Standards Agency (FSA) (Anon, 2011b), regular samples of influent and effluent wastewater from a typical large municipal secondary sewage treatment station (serving a population of > 100 000) were tested over a 2-year period from 2009 to 2011. Both GI and GII NoV were detected in all 41 pairs of samples tested. Levels of GII were found to be significantly higher than GI, with up to > 1 million detectable genome copies/mL in crude influent and up to 64 000 copies/mL in treated final effluent. In crude samples GI levels ranged between 48 and 94 541 copies/mL (geometric mean 3691 copies/mL), while GII levels ranged between 1399 and 1 141 478 copies/mL (geometric mean 25 504 copies/mL). For final effluent samples GII levels ranged between 2 and 18 747 copies/mL (geometric mean 202 copies/mL), while GII levels ranged between 18 and 64 406 copies/mL (geometric mean 585 copies/mL). Significant winter-spring seasonality was observed in GII levels, with higher levels recorded between December and May than between June and November. This seasonality was similar to that observed in laboratory reports of NoV illness identified in national surveillance. In contrast, GI NoV levels did not display the same seasonality, although the lowest levels were again observed during the summer months. Average log reductions of NoV GI and GII from crude to final effluent samples were 1.26 and 1.64, respectively. This study highlights the potential of municipal secondary sewage treatment stations to contaminate the water supply, and if situated upstream of shellfish production areas, for example, to lead in turn to contamination of shellfish (Anon, 2011b).

An 8-month survey was recently conducted in order to evaluate the presence of EV, HAdV and HAV in untreated and treated sewage samples collected from a primary treatment municipal wastewater plant located in north-eastern Greece (Kokkinos *et al.*, 2010). EVs, HAdVs and HAV were detected in 40% (10/25), 40% (10/25) and 4% (1/25) respectively of the samples collected from the plant's inlet, and in 12% (3/25), 44% (11/25) and 0% (0/25) of the samples collected from the plant's outlet (Kokkinos *et al.*, 2010). Adenovirus types 3 (Ad3), 10 (Ad10) and 41 (Ad41), and HAV type H2 were identified, along with EVs of the Coxsackie type A2 and Echovirus types 27 and 30. The results suggest that treated sewage may still contain human viruses, thereby representing a potential health hazard. Moreover, it was concluded that the possible reuse of treated sewage in agriculture or other applications should be cause for concern (Kokkinos *et al.*, 2010). Petrinca *et al.* (2009) examined raw wastewater and effluent samples that were collected from three wastewater treatment plants in Italy, in three different sampling periods. An analysis of the results showed the circulation of cultivable EVs and differences in the seasonal-geographical distribution. HAV was found with only two genotypes: IA and IB. HRoV was present in 11.11%, 24.14% and 2.78% of the samples in the 1st, 2nd and 3rd sampling periods, respectively; AstV in 33.33%, 6.9% and 25%; HAdV in 7.41%, 3.45% and 2.78%; and NoV in 7.41%, 10.34% and 5.56%.

Pinto *et al.* (2007) studied the presence of viruses in sewage from Egypt and Spain. Raw sewage samples (n = 35) from three sewage treatment plants in Cairo were collected on a monthly basis between November 1998 and October 1999. Raw sewage from a sewage treatment plant in Barcelona was sampled twice a month over a period of 4 years, from June 1998 to December 2002 (n = 249). HAV was detected in 25 out of 35 sewage samples from Cairo (71%), with no seasonality observed. In contrast, only 28 out of 249 sewage samples from Barcelona (11%) were HAV-positive. Per year, the percentage of positive HAV samples was 14.7 in 1998, 11.7 in 1999, 15.4 in 2000, 5.9 in 2001 and 6.7 in 2002. The difference between the percentage of HAV-positive sewage samples from Cairo (71%) and from Barcelona (11%) reflects the different infection patterns found in those regions. In Cairo, large numbers of asymptomatic excretors lead to the discharge of large amounts of HAV in sewage. In Barcelona, on the other hand, there are fewer instances of asymptomatic infection and a small number of acute cases, leading to a low level of virus spread (Pinto *et al.*, 2007). Pina *et al.* (1998) carried out an earlier study of 15 raw domestic sewage samples collected monthly in Spain during 1994 and 1995. The viruses most frequently detected were HAdV, which were present in 14 (93%) of the raw sewage samples and in two of the three effluent samples. Enteroviruses were present in 6 (40%) of the influent sewage samples, and in none of the effluent samples. Four raw sewage samples (27%), but no effluent samples, were positive for HAV.

In another study carried out in Spain, Rodrigez-Manzano *et al.* (2010) analyzed the presence of HAV and HEV in raw urban sewage in two areas over a period of several years. It was concluded that the percentage of HAV-positive samples in urban sewage fell from 57.4% to 3.1% over a 5–10 year period in the two areas under investigation. Around 30% of the urban sewage samples were positive for HEV in the absence of agricultural sources of contamination. This reduction in HAV in raw urban sewage observed in eastern Spain was connected to general improvements in sanitation. However, these improvements have not had the same effect on the presence of HEV: this can be explained by the influence of animal reservoirs for HEV (Rodrigez-Manzano *et al.*, 2010).

In India, Vaidya *et al.* (2002) examined the presence and prevalence of viruses in influent and effluent samples that were collected twice a week for one year from a sewage treatment plant in Pune. The overall virus prevalence was 24.42% for HAV (21/86), 10.98% for HEV (9/82) and 12.7% for TTV (8/63). The prevalence of HAV was significantly higher than that of HEV. Significantly higher levels of HAV RNA were noted during the summer months. In treated sewage samples a substantial reduction in HAV RNA positivity (15/48 *vs* 2/48) was recorded, but HEV RNA and TTV DNA positivity did not reduce significantly.

For surface waters, the main source of contamination is sewage discharge from urban areas (Maalouf *et al.*, 2010). In addition, rural run-off and run-off caused by heavy rains or melting snow can also contribute to the contamination of surface water by viruses. HAdVs are the most common type of virus found in surface water (Wyn-Jones *et al.*, 2011). Bosch *et al.* (2006) offer a comprehensive discussion of studies regarding the presence of human enteric viruses in freshwater environments in the 1970s and 1980s. In six studies conducted in five countries, EVs were reported in five cases, and HAdVs and HEV in one case each. More recently, to assess the presence of HRoV, AstV and NoV in surface water, a year-long study was carried out in Beijing, China. A total of 108 urban surface water samples were collected from nine collection sites in rivers or lakes from September 2006 to August 2007. A total of 63 virus strains were detected, with HRoV (48.1%) identified as the most prevalent of these, followed by AstV (5.6%) and NoV (4.6%). The concentration of HRoV ranged from 0 to 18.27 genome copies per L^{-1}, and seasonal variation was observed: it was found to be highly prevalent in cold weather (from September to February) and less prevalent in warm weather (from March to August) or even absent (from April to July). The results from this study confirmed the existing clinical data (He *et al.*, 2011). During the winter season in Beijing, the reported frequency of infection in the population by HRoV was 58%, while for AstV and NoV it was 8% and 6%, respectively (Liu *et al.*, 2006).

In Poland, 60 water samples were collected in 2007 from four sampling sites situated along the river Wieprz. Human pathogenic viruses were detected in 35% of samples. HAdVs were detected in 28.3% of samples, and were present throughout the whole year; 11.6% of the samples tested positive for NoVs, but, somewhat unusually, NoV was detected only during summer. Molecular identification revealed that the NoV strains belonged to genogroups I and II (Kozyra *et al.*, 2011). In Spain, 23 river water samples from two sites with different levels of faecal pollution were tested, and the human virus most often detected (Pina *et al.*, 1998) was HAdV, being present in 15 of the 23 samples (65%). Enterovirus was detected in five samples (22%) and HAV in 10 (43%). A study was also carried out in Germany between October 2002 and September 2003, examining a mining lake and the supplying surface waters, which were located downstream of a sewage plant (Pusch *et al.*, 2005). One hundred and twenty-three water specimens collected at different sites downstream of the waste water treatment plant, as well as nine samples from the sewage plant influent, were tested for the presence of enteric viruses. The virus prevalence in the sewage plant effluents and surface water samples varied according to the specific sampling site and were: 29–76% for EV, 24–42% for AstV, 15–53% for NoV, 3–24% for HRoV, 5–20% for HAV and 20% for HAdV. AstV was present at between 3.7×10^3 and 1.2×10^8 detectable units per litre, depending on the sampling location, and the average number of NoV genome particles ranged from 1.8×10^4 to 9.7×10^5 units/L. Three out of 18 EV PCR-positive samples were found to be infective on cell culture. Although microbiological parameters such as *Escherichia coli*, enterococci and coliphages indicated acceptable microbiological water quality, the data collected on the presence of viruses suggested that the surface waters investigated could constitute a source of enteric viral infections and thus a public health risk (Pusch *et al.*, 2005).

Enteric viruses can easily contaminate groundwater. Their extremely small size allows them to infiltrate soils from contamination sources such as broken sewage pipes and septic tanks, and consequently contaminate groundwater aquifers. Viruses can move considerable distances in the subsurface environment with penetration up to 67 m and horizontal migration up to 408 m (Borchardt *et al.*, 2003). In one study in the USA, 72% of examined groundwater sites tested positive for human enteric viruses (Fout *et al.*, 2003). In the USA, groundwater has been frequently implicated in the transmission of waterborne infectious disease, with about 80% of waterborne outbreaks, mostly of NoV and HAV, attributed to contaminated drinking well water (Beller *et al.*, 1997). Borchardt *et al.* (2003) examined 50 wells in seven hydrogeologic districts in Wisconsin. The wells were sampled four times over a year, once each season. Four of the wells (8%) tested positive for viruses. Three wells were positive for HAV, and the fourth well was positive for both HRoV and NoV in one sample and for EV in another sample.

Gibson *et al.* (2011) examined large volume (100 L) groundwater (n = 4), surface water (n = 9), and finished drinking water (n = 6) samples in order to evaluate human enteric viruses and bacterial indicators in Ghana. Human enteric viruses including NoV GI and GII, HAdV, and human polyomavirus (HPyV) were detected in one groundwater sample, three surface waters, and one drinking water sample. NoV was found in one ground water sample, one surface water sample and one drinking water sample; HAdV was found in two surface water samples, and HPyV in one drinking water sample. The total number of coliforms and *E. coli* assessed for each sample revealed a lack of correlation between bacterial indicators and the presence of human enteric viruses (Gibson *et al.*, 2011).

HEV has been recognized as a major cause of acute hepatitis in young adults in Asia, Africa and Latin America. The highest rates of infection occur in regions with the poorest sanitation (Khuroo *et al.*, 2004). Data on the presence of HEV in sewage in India are limited; however, Ippagunta *et al.* (2007) carried out a study of 192 sewage specimens collected from 2004 to 2006, and found that 79 (41%) tested positive for human HEV RNA. The positivity rate was higher during the second year (52/103, or 51%) than during the first year of study (27/89 or 30%). Seasonal variation in HEV positivity was also observed, with 28 out of 61 (46%) in the winter months, 36 of 66 (55%) in the summer months, and 15 of 65 (23%) during the monsoon months (July to October). Swine HEV RNA was not detected in any of the samples. The study suggested that HEV infection and faecal viral excretion may be common in HEV-endemic regions throughout the year even during non-epidemic periods (Ippagunta *et al.*, 2007). Ahmed *et al.* (2010) recently examined 86 raw sewerage water samples taken from drainage outlets at different locations in Islamabad and Rawalpindi, Pakistan. In total, 35 samples were found to be HEV RT-PCR-positive. From that number 19 out of 47 samples (44.7%) from Rawalpindi and 16 out of 39 samples (41.02%) from Islamabad were HEV-positive. All the positive samples were found in areas with high population density. This was the first study carried out on the presence of HEV in sewerage waste water from Pakistan, and the results showed a relatively high prevalence of HEV (Ahmad *et al.*, 2010).

Aichi viruses (AiVs) have emerged as viral agents associated with food and waterborne gastroenteritis in humans, with seroprevalence studies carried out in Japan and in Europe suggesting that infections are quite frequent (Alcalá *et al.*, 2010). Kitajima *et al.* (2011) examined the prevalence and genetic diversity of AiVs in wastewater and river water in Japan over one year (2005–2006). Influent (n = 12) and effluent (n = 12) samples were collected monthly from a wastewater treatment plant, and 60 river water samples were collected monthly from five sites along the Tamagawa River. AiV genomes were found in all 12 (100%) influent and in 11 (92%) effluent wastewater samples, as well as in 36 (60%) river water samples. Of the 260 AiV strains identified, 255 were genotype A and 5 were genotype B. This was the first report of the molecular detection and characterization of AiVs in aquatic environments in Japan. In Tunisia, Sdiri-Loulizi *et al.* (2010) identified AiVs in 15 of 250 (6%) sewage samples. The low prevalence of AiV found in this study could be the result of the detection method used, which was less sensitive. Alcalá *et al.* (2010) examined a major river polluted with sewage discharges in an urban area in Caracas, Venezuela. Five out of the 11 water samples studied tested positive for AiV genotype B.

Hundesa *et al.* (2006) examined environmental samples including urban sewage, slaughterhouse wastewater and river water samples for the presence of HAdVs, porcine adenoviruses (PAdVs), bovine adenoviruses (BAdVs) and bovine polyomaviruses (BPyVs). PAdVs and BPyVs were detected in a very high percentage of samples that could potentially have been affected by either porcine or bovine faecal contamination respectively. However, BAdVs were detected in only one sample, showing a lower prevalence than BPyV in the samples analyzed. Twenty-two slaughterhouse samples with faecal contamination of animal origin tested negative for HAdV. In contrast, the river water samples tested positive for both human and animal adenoviruses and polyomaviruses, indicating that the contamination must have originated from a variety of sources (Hundesa *et al.*, 2006).

Viruses in recreational water

The viruses responsible for waterborne infections in humans following activity on or in recreational water are not usually identified at the time of the outbreak. Strong connections have only occasionally been demonstrated between the presence of a virus or viruses in the water and of the same virus in the faeces of affected individuals (Hoebe *et al.*, 2004; Wyn-Jones *et al.*, 2011). In regions with temperate climates, peaks of human enteric virus infections and consecutive excretions can usually be found in summer and in early fall. This coincides with an increase in water-based recreational activities and human/water contact (Nairn and Clements, 1999; Kocwa-Haluch, 2001; Sedmak *et al.*, 2003). In tropical climates, the presence of human enteric viruses, especially EVs, can be evenly spread throughout the year and is sometimes more prevalent during rainy seasons (Fong and Lipp, 2005). It has been observed that the predominant clinical serotype in one season is also the predominant sewage serotype for that season. For example, in 1998, echovirus 30 accounted for 50% of sewage isolates and 46.1% of clinical cases, while in 1990, 79.7% of sewage isolates and 60.3% of clinical cases were due to echovirus 11 (Sedmak *et al.*, 2003). However, a direct connection cannot always be established. Viruses identified in sewage during the spring were the predominant clinical strains in the summer (Sedmak *et al.*, 2003).

Recently, a large European surveillance study was carried out to determine the frequency of occurrence of two human enteric viruses in recreational water. Out of 1410 analyzed samples, 553 (39.2%) tested positive for one or more of the target viruses. Adenoviruses, detected in 36.4% of samples, were more prevalent than NoV (9.4%), with 3.5% GI and 6.2% GII positive samples. Some samples tested positive for both GI and GII NoV genogroups. Out of 513 HAdV-positive samples, 63 (12.3%) were also NoV-positive, whereas 69 (7.7%) NoV-positive samples were HAdV-negative. The viruses were found more frequently in freshwater samples than in marine water samples (Wyn-Jones *et al.*, 2011). Approximately 25% of a small number of HAdV-positive samples were found to contain infective virus particles. These studies supported the idea that HAdV should be considered as an indicator of bathing water quality (Wyn-Jones *et al.*, 2011).

In Korea in January 2008, an outbreak of acute gastroenteritis at a water park was reported to the Bundang-gu Public Health Center in Seongnam. A total of 67 (31.0%) students and teachers developed acute gastroenteritis. NoV GI strains were detected in the stool samples of patients who had severe diarrhoea, and in groundwater samples taken from the water park. All the GI NoVs from the patients and the groundwater samples were identified as strain I.4 with a 97% homology. It was therefore concluded that the outbreak was closely connected with NoV-contaminated groundwater (Koh *et al.*, 2011). Finally, Xagoraraki *et al.* (2007) reported HAdV at concentrations of up to 10^3 virus particles per litre at recreational beaches in Lake Michigan.

Seawater is also significant in the transmission of enteric viruses to humans. Human enteric viruses have frequently been detected in coastal waters with treated wastewater effluents. Human pathogenic viruses enter the marine environment through several routes, including direct discharge of treated or untreated sewage effluents, unintentional discharges by urban and rural run-off, through rivers when wastewater discharges take place in fresh water, or by dumping from barges (Bosch *et al.*, 2006; Bosch and Le Guyader, 2010). Enteric viruses can survive for long periods in the marine environment, which increases the probability of human exposure through recreational contact and accumulation in shellfish (Lipp and Rose, 1997). Recently, Mocé-Llivina *et al.* (2005) detected EV in 55% of samples from beaches in Spain. These results, along with the data from many other studies, indicate that the occurrence of viruses in coastal waters leads to an increased risk of infection for swimmers and divers.

2.3 Gaps in current knowledge

Environmental virology, particularly food virology, is a very new scientific discipline; extensive data on the presence and prevalence of viruses in different matrices are consequently lacking, with only a few studies having been carried out, and the majority of existing data originating from studies carried out after outbreaks. This type of study only very occasionally contains data on the presence of viruses in foods.

Despite the fact that viruses are the most common pathogens transmitted via food, causing 66.6% of food-related illnesses in the United States, compared with 9.7% and 14.2% for salmonella and campylobacter, respectively (Mead *et al.*, 1999), no systematic inspection system exists, and no legislation has been drawn up that would set up virological criteria for food safety (Koopmans and Duizer, 2004). Data from systematic virus surveillance in foods are missing mainly because this type of systematic surveillance simply does not exist at national levels or beyond. Routine monitoring is not yet feasible because there are no harmonized and standardized methods, because end-product testing is not sufficiently reliable to assure food safety on statistical grounds, and because diagnostic methods for food or water are not routinely available in food microbiology laboratories (Koopmans and Duizer, 2004).

The relative contributions to the prevalence of foodborne illness of different sources, such as shellfish, fresh produce, food handlers (including asymptomatic shedders) and the food handling environment, have not been determined (EFSA, 2011). With regard to the contamination of foods, the majority of available data relates to HAV and NoV in bivalve molluscs. Nonetheless, little quantitative data exists relative to virus load. In the case of fresh produce and prepared foods, there is significantly less information available (FAO and WHO, 2008). While there is some information on foodborne viral infections caused by the consumption of contaminated fresh produce, and also on the detection of viruses in fresh produce implicated in outbreaks, there is little information on the general prevalence of viruses in different fruits and vegetables, mainly because there is no routine or regular monitoring of fresh produce for the presence of viral contaminants. Moreover, quantitative data on the viral load is scarce, making the establishment of microbiological criteria for these food categories a difficult task (EFSA, 2011).

Despite the fact that fresh produce is increasingly implicated as a vehicle for human pathogens, there is currently limited knowledge with regard to the point in the supply chain at which virus contamination occurs and the mechanism by which viruses survive on fruits and vegetables (Berger *et al.*, 2010). Prevalence data on emerging viruses such as HEV is particularly limited. This is clearly a major barrier to conducting risk assessment, particularly when the assessment is designed to evaluate mitigation strategies (FAO and WHO, 2008).

The current methods used to monitor foods and the use of *E. coli* as a microbiological indicator do not consistently provide results that correlate with the presence or absence of viruses in foods. As a consequence, food industry and food safety authorities lack the tools to monitor virological quality control, even though bacteriological contamination is well-monitored. For shellfish, standardized and validated protocols for virus detection are in the final stages of development, but similar processes for other foodstuffs remain unavailable (EFSA, 2011). There are many reported studies related to viral presence in foods and water environments but the type of consistency in the sample size and sampled material that is seen in shellfish studies is lacking; this means that there is a risk of the true aetiological agent of an outbreak not being identified, or of the infectious dose being under- or overestimated (Rodrıguez-Lazaro *et al.*, 2011).

Viruses in water environments have clearly emerged as primary health hazards, but a systematic procedure for the monitoring and control of these agents is still missing. Surveillance of waterborne pathogens is a complex and difficult task and is partly dependent on legislative measures concerning water quality that define the exact nature of the surveillance. Viruses have so far been considered too difficult and complex to handle and are therefore excluded from most legislation (von Bonsdorff and Maunula, 2008). As regular monitoring of viruses in water is not mandatory, viral analyses are only performed in outbreak situations or where an outbreak is suspected (Maunula *et al.*, 2009). Surveillance of levels of infectious disease does take place in several countries, with monitoring and mandatory reporting common in many parts of the world. However, viral diarrheal diseases are not usually among the regularly monitored diseases (von Bonsdorff and Maunula, 2008). Even in highly developed countries such as the USA, reports of waterborne disease outbreaks to the Center for Disease Control and Prevention (CDC) are voluntary on the part of the states. For this reason, many outbreaks go unreported and are not accounted for by the surveillance system (Sinclair *et al.*, 2009). Furthermore, current safety standards for determining food and water quality typically do not specify what level of virus contamination should be considered acceptable (Okoh *et al.*, 2010). While a great deal of research has been carried out into the public health impacts of inadequately treated wastewater effluents in developed countries, this is not the case in developing countries, where levels of infectious diseases are high and are usually caused by pollution of water with wastewater effluent discharges, most of which remains undocumented, unreported and not properly investigated (Okoh *et al.*, 2010).

The performance of wastewater treatment systems is currently monitored through the use of bacterial indicator organisms (Okoh *et al.*, 2010). However, infectious viruses have been isolated from aquatic environments that meet bacterial indicator standards, and in some cases these viruses have been connected with outbreaks (Karmakar *et al.*, 2008; He *et al.*, 2011). The use of bacterial indicators, therefore, can be said to be an unreliable indicators for the presence of viruses.

Although there are many studies of the occurrence of human enteric viruses in the environment and their role in waterborne transmission, little information is available on the environmental transmission of enteric viruses in animals, and additional research is needed (Fong and Lipp, 2005). Sewage, especially from slaughterhouses, may contain animal adenoviruses, sapoviruses, and HEV, which may also be zoonotic (Wyn-Jones *et al.*, 2011). The prevalence and distribution of animal-specific viruses in environmental waters must be determined in order to validate the use of these viruses for source tracking purposes (Fong and Lipp, 2005).

2.4 Conclusion and future trends

It is clear that food- and waterborne viral infections will become a greater challenge to public health in the future. At the same time, food microbiologists, virologists and epidemiologists will face the similarly challenging task of sharing and expanding their knowledge in this area and actively contributing to the prevention of virus infections transmitted through water and food (EFSA, 2011).

The key element in reducing the foodborne spread of viruses is continued surveillance and awareness. Enhancing risk-based laboratory surveillance will improve strategies for the prevention and control of virus contamination in foods and will consequently lessen the associated risks. No routine monitoring of viruses in foodstuffs is currently performed; however, this type of surveillance would be highly beneficial, along with a system in which data from food and environment monitoring could be epidemiologically compared with data relating to outbreaks in the population. The recent increase in RASFF notifications for suspected viral contamination is remarkable, and may be a result of increasing awareness. However, RASFF notifications are not representative and are not based on common notification criteria. These data must therefore be interpreted with care, but do still show at least a tendency towards greater awareness of viral threats in foodstuffs (EFSA, 2011).

Specific targeted studies with correct sampling strategies are clearly required in order to collect data that will be useful for quantitative microbial risk assessment (QMRA). These data include consumer habits, virus contamination levels in food and other reservoirs, virus transfer rates, natural persistence on/in foods (at the pre-harvest and post-harvest levels), and human dose–response relations (EFSA, 2011). Studies regarding the prevalence and levels of virus contamination in foods commonly implicated in outbreaks must be carried out, and are essential for QMRA (FAO and WHO, 2008). Satisfactory exposure assessment requires a reliable quantification of the virus present in the environmental material. This in turn requires the detection efficiency of the assay used to be determined, and appropriate controls to be employed to determine the true virus concentration in the environment (Rodrıguez-Lazaro *et al.*, 2011).

The expert group of the EFSA Panel on Biological Hazards (BIOHAZ) recommended the introduction of virus microbiological criteria for the classification of areas used for the production of high-risk bivalve molluscs (to be consumed raw); the findings of an EFSA sanitary survey determined that a virus monitoring programme for compliance with these criteria should be risk-based (EFSA, 2011). Regulation (EC) 2073/2005 indicates that criteria for pathogenic viruses in live bivalve molluscs should be established when adequate analytical methods are developed. Moreover, regulation (EC) No 853/2004 provides for the possible implementation of additional health standards for live bivalve molluscs including virus testing procedures and standards. In light of this, the BIOHAZ expert group has suggested that with the available quantitative data on viral load, it would be possible to establish criteria for NoV in bivalve molluscs (EFSA, 2011). Methods for detecting viruses in fresh produce are available, but prevalence studies are limited, and quantitative data on viral load is scarce, making the establishment of microbiological criteria for these food categories currently a difficult proposition. Although there are documented cases of illness caused by viruses in fresh produce, the relative contribution of fresh produce to the overall public health risk of foodborne viruses has not been established. Future research is required in order to gather more quantitative data on the presence and prevalence of viruses in different kinds of fresh produce to allow the establishment of standard microbiological criteria for viruses. Another task for the future is the assessment of the distribution of infectious HEV in meat and meat products in Europe in order to investigate the routes of HEV transmission to humans and to identify the risk factors for human hepatitis E (EFSA, 2011).

As food production and distribution increasingly take place on a global scale, a network approach is required, with partners in developed and developing countries, to undertake studies intended to fill in gaps in the current data (FAO and WHO, 2008). Since current surveillance systems are incomplete, a global system for the routine harmonized surveillance of water- and foodborne virus outbreaks is essential, with rapid detection and reporting networks implemented as part of standard surveillance systems. These networks should combine laboratory and epidemiological information, and a reporting strategy for international outbreaks should also be established (Koopmans and Duizer, 2004).

The need for a better surveillance system has also been recognized by the Foodborne Viruses in Europe (FBVE) network (http://www.noronet.nl/fbve/), which has conducted virus-specific surveillance of gastroenteritis outbreaks since 1999 (Koopmans *et al.*, 2003). The network investigates outbreaks from all modes of transmission to obtain an overview of viral activity in the community, focusing mainly on NoV. A total of 13 countries are participating in the FBVE surveillance network (Verhoef *et al.*, 2009), which is currently being merged into the NoroNet network (http://www.noronet.nl/noronet/index.jsp), an informal network of scientists working in public health institutes or universities sharing virological, epidemiological and molecular data on NoV. The network maintains a shared database accessible via the internet for data entry and sharing and analysis (Verhoef *et al.*, 2009). The aim of NoroNet is to increase the knowledge base with respect to geographical and temporal trends in the emergence and spread of NoV variants, and to design standardized nomenclature for existing and emerging NoV genotypes and variants.

In the USA in 2009, the Center for Disease Control and Prevention developed CaliciNet (http://www.cdc.gov/norovirus/php/reporting.html), a network of public health and food regulatory laboratories that submit norovirus sequences identified from outbreaks into a national database. The information is used to link NoV outbreaks that may be caused by common sources (such as food), monitor trends and identify emerging norovirus strains. As of February 2012, public health laboratories from 25 American states have been certified by the CDC to participate in the CaliciNet network. This network – and perhaps others – may be expected to increase in size in the future, establishing harmonized and well-connected national and international networks for the systematic surveillance of food- and waterborne viruses, and combining the epidemiologic assessment of the outbreaks with molecular typing of virus isolates to discover and track potential links between outbreaks.

2.5 Acknowledgements

The author would like to acknowledge funding from the EU Framework 7 Programme Project: 'Integrated Monitoring and Control of Foodborne Viruses in European Food Supply Chains'- VITAL (Grant agreement No. 213178), and from project No. TR31084, funded by the Ministry of Education Science and Technological Development of the Republic of Serbia. Special thanks go to Dr Nigel Cook, FERA, UK, who introduced the interest in food and environmental virology research to the Serbian virologist community.

2.6 References

AHMAD, T., WAHEED, Y., TAHIR, S., SAFI, S.Z., FATIMA, K., AFZAL, M.S., FAROOQI, Z.U. and QADRI, I. (2010). Frequency of HEV contamination in sewerage waters in Pakistan. *Infect. Dev. Ctries*, **4**(12), 842–5.

ALCALÁ, A., VIZZI, E., RODRÍGUEZ-DÍAZ, J., ZAMBRANO, J.L., BETANCOURT, W. and LIPRANDI, F. (2010). Molecular detection and characterization of Aichi viruses in sewage-polluted waters of Venezuela. *Appl. Environ. Microbiol.*, **76**, 4113–15.

ANASTASI, P., BONANNI, E., CECCHINI, G., DIVIZIA, M., DONIA, D., DI GIANFILIPPO, F., GABRIELI R., PENTRICA, A.R. and ZANOBINI, A. (2008). Virus removal in conventional wastewater treatment process. *Igiene e Sanita Pubblica*, **64**(3), 313–30.

ANON (2010). Hepatitis E virus in fresh pig livers. *Risk Assessment Studies Report No. 44*. Centre for Food Safety of the Food and Environmental Hygiene Department of the Government of the Hong Kong Special Administrative Region, 39.

ANON (2011a). Investigation into the prevalence, distribution and levels of norovirus titre in oyster harvesting areas in the UK. *Food Standard Agency project (FSA Project Code FS235003 (P01009)*; Cefas ref: C3027, Project report.

ANON (2011b). Investigation into the levels of norovirus in influent and treated wastewater samples from a sewage treatment works. *Food Standard Agency project (FSA Project Code FS235003 (P01009);* Cefas ref: C3027, Project report.

ARNOLD, B.F. and COLFORD J.M. (2007). Treating water with chlorine at point-of-use to improve water quality and reduce child diarrhea in developing countries: a systematic review and metaanalysis. *Am. J. Trop. Med. Hyg.*, **76**, 354–364.

BAERT, L., MATTISON, K., LOISY-HAMON, F., HARLOW, J., MARTYRES, A., LEBEAU, B., STALS, A., VAN COILLIE, E., HERMAN, L. and UYTTENDAELE, M. (2011). Review: norovirus prevalence in Belgian, Canadian and French fresh produce: a threat to human health? *Int. J. Food Microbiol.*, **151**(3), 261–9.

BANKS, M., GRIERSON, S., FELLOWS, H.J., STABLEFORTH, W., BENDALL, R. and DALTON, H.R. (2007). Transmission of hepatitis E virus. *Vet. Rec.*, **160**(6), 202.

BELLER, M., ELLIS, A., LEE, S.H., DREBOT, M.A., JENKERSON, S.A., FUNK, E., SOBSEY, M.D., SIMMONS, O.D., MONROE, S.S., ANDO, T., NOEL, J., PETRIC, M., HOCKIN, J., MIDDAUGH, J.P. and SPIKA, J.S. (1997). Outbreak of viral gastroenteritis due to a contaminated well – international consequences. *J. Am. Med. Assoc.*, **278**, 563–8.

BERGER, C.N, SODHA, S.V., SHAW, R.K., GRIFFIN, P.M., PINK, D., HAND, P. and FRANKEL, G. (2010). Fresh fruit and vegetables as vehicles for the transmission of human pathogens. *Environ. Microbiol.*, **12**(9), 2385–97.

BERTO, A., MARTELLI, F., GRIERSON, S., BANKS, M. (2012). Hepatitis E virus in pork food chain, United Kingdom, 2009–10. *Emerg. Infect. Dis.*, **18**(8), 1358–60.

BORCHARDT, M.A., BERTZ, P.D., SPENCER, S.K. and BATTIGELLI, D.A. (2003). Incidence of enteric viruses in groundwater from household wells in Wisconsin. *Appl. Environ. Microbiol.*, **69**, 1172–80.

BOSCH, A., (1998). Human enteric viruses in the water environment. A minireview. *Int. Microbiol.*, **1**, 191–6.

BOSCH, A., PINTÓ, R.M. and ABAD, F.X. (2006). Survival and transport of enteric viruses in the environment. In *Viruses in Food*, S. M. GOYAL (ed.), *Food Microbiology and Food Safety Series*, Springer, New York, 151–87.

BOSCH, A., GUIX, S., SANO, D. and PINTO, R.M. (2008). New tools for the study and direct surveillance of viral pathogens in water. *Curr. Opin. Biotech.*, **19**, 295–301.

BOSCH, A. and LE GUYADER, S.F. (2010). Introduction: viruses in shellfish. *Food Environ. Virol.*, **2**(3), 115–16.

BOSCHI-PINTO, C., VELEBIT, L. and SHIBUYA, K. (2008). Estimating child mortality due to diarrhoea in developing countries. *Bull. World Health Organ.*, **86**, 710–717.

BOUQUET, J., TESSE, S., LUNAZZI, A., ELOIT, M., ROSE, N., NICAND, E. and PAVIO, N. (2011). Close similarity between sequences of hepatitis E virus recovered from humans and swine, France, 2008–2009. *Emerg. Infect. Dis.*, **17**(11), 2018–25.

BOUWKNEGT, M., LODDER-VERSCHOOR, F., VAN DER POEL, W.H.M., RUTJES, S.A. and DE RODA HUSMAN, A.M. (2007). Hepatitis E Virus RNA in commercial porcine livers in the Netherlands. *J. Food Protect.*, **70**(12), 2889–95.

BOXMAN, I.L.A., TILBURG, J.J., TE LOEKE, N.A., VENNEMA, H., JONKER, K., DE BOER, E. and KOOPMANS, M. (2006). Detection of noroviruses in shellfish in the Netherlands. *Int. J. Food Microbiol.*, **108**(3), 391–6.

BOXMAN, I.L.A. (2010). Human enteric viruses occurrence in shellfish from European Markets. *Food Environ. Virol.*, **2**, 156–66.

BOXMAN, I.L.A., VERHOEF, L., DIJKMAN, R., HÄGELE, G., TE LOEKE, N.A.J.M and KOOPMANS, M. (2011). Year-round prevalence of norovirus in the environment of catering companies without recently reported outbreak of gastroenteritis. *Appl. Environ. Microbiol.*, **77**(9), 2968–2974. doi: 10.1128/AEM.02354-10.

BRASSARD, J., GAGNE, M.J., GENEREUX, M. and COTE, C. (2012). Detection of human foodborne and zoonotic viruses on irrigated, field-grown strawberries. *Appl. Environ. Microbiol.*, doi: 10.1128/AEM.00251–12.

CARDUCCI, A., MORICI, P., PIZZI, F., BATTISTINI, R., ROVINI, E. and VERANI, M., (2008). Study of the viral removal efficiency in a urban wastewater treatment plant. *Water Sci. Technol.*, **58**(4), 893–7.

CARDUCCI, A., BATTISTINI, R., ROVINI, E. and VERANI, M. (2009). Viral removal by wastewater treatment: monitoring of indicators and pathogens. *Food Environ. Virol.*, **1**, 85–91.

CARTER, M.J. (2005). Enterically infecting viruses: pathogenicity, transmission and significance for food and waterborne infection. *J. Appl. Microbiol.*, **98**, 1354–80.

CECUK, D., KRUZIC, V., TURKOVIC, B. and GREE, M. (1993). Human viruses in the coastal environment of a Croatian harbor. *Rev. Epidemiol. Sante*, **41**, 487–493.

CHEONG, S., LEE, C., SONG, S.W., CHOI, W.C., LEE, C.H. and KIM, S.J. (2009). Enteric viruses in raw vegetables and groundwater used for irrigation in South Korea. *Appl. Environ. Microbiol.*, **75**(24), 7745–51.

CLIVER, D.O. (1984). Significance of water and the environment in the transmission of viral disease. In: J. L. MELNICK (ed.), *Enteric Viruses in Water*, vol. 15, pp. 30–42. Karger, Basel, Switzerland.

COSTANTINI, V., LOISY, F., JOENS, L., LE GUYADER, F.S. and SAIF, L.J. (2006). Human and animal enteric caliciviruses in oysters from different coastal regions of the United States. *Appl. Environ. Microbiol.*, **72**(3), 1800–09.

DEWAILLY, E., POIRIER, C. and MEYER, F.M. (1986). Health hazards associated with windsurfing on polluted waters. *Am. J. Public Health*, **76**, 690–691.

DI BARTOLO, I., DIEZ-VALCARCE, M., VASICKOVA, P., KRALIK, P., HERNANDEZ, M., ANGELONI, G., OSTANELLO, F., BOUWKNEGT, M., RODRÍGUEZ-LÁZARO, D., PAVLIK, I. and RUGGERI, F.M. (2012). Hepatitis E Virus in Pork Production Chain in Czech Republic, Italy, and Spain, 2010. *Emerg. Infect. Dis.*, **18**(8), 1282–9.

DIEZ-VALCARCE, M., KOKKINOS, P., SÖDERBERG, K., BOUWKNEGT, M., WILLEMS, K., DE RODA-HUSMAN, A.M., VON BONSDORFF, C-H., BELLOU, M., HERNÁNDEZ, M., MAUNULA, L., VANTARAKIS, A., RODRÍGUEZ-LÁZARO, D. (2012). Occurrence of human enteric viruses in commercial mussels at retail level in Three European Countries. *Food Environ. Virol.*, **4**, 73–80. DOI 10.1007/s12560–012–9078–9.

EFSA PANEL ON BIOLOGICAL HAZARDS (BIOHAZ) (2011). Scientific opinion on an update on the present knowledge on the occurrence and control of foodborne viruses. *EFSA Journal*, **9** (7), 2190. (96 pp.). Available online: www.efsa.europa.eu/efsajournal (Accessed December 2011).

EHLERS, M.M., GRABOW, W.O.K. and PAVLOV, D.N. (2005). Detection of enteroviruses in untreated and treated drinking water supplies in South Africa. *Wat. Res.*, **39**, 2253–8.

FAO/WHO (2008). Viruses in Food: Scientific Advice to Support Risk Management Activities. Meeting Report. *Microbiological Risk Assessment Series*, No. 13.

FEAGINS, A.R., OPRIESSNIG, T., GUENETTE, D.K., HALBUR, P.G. and MENG, X.-J. (2007). Detection and characterization of infectious Hepatitis E virus from commercial pig livers sold in local grocery stores in the USA. *J. Gen. Virol.*, **88**, 912–17.

FONG, T.Z. and LIPP, E.K. (2005). Enteric viruses of humans and animals in aquatic environments: health risks, detection, and potential water quality assessment tools. *Microbiol. Mol. Biol. R.*, **69**(2), 357–71.

FOUT, G.S., MARTINSON, B.C., MOYER, M.W.N. and DAHLING, D.R. (2003). A multiplex reverse transcription-PCR method for detection of human enteric viruses in groundwater. *Appl. Environ. Microbiol.*, **69**, 3158–64.

GENTRY, J., VINJE, J., GUADAGNOLI, D. and LIPP, E.K. (2009). Norovirus distribution within an estuarine environment. *Appl. Environ. Microbiol.*,**75** (17), 5474–80.

GIBSON, K.E., OPRYSZKO, M.C., SCHISSLER, J.T., GUO, Y. and SCHWAB, K.J. (2011). Evaluation of human enteric viruses in surface water and drinking water resources in Southern Ghana. *Am. J. Trop. Med. Hyg.*, **84**(1), 20–29.

GIRONES, R., FERRU, M.A., ALONSO, J.L., RODRIGEZ-MANZANO, J., CALGUA, B., CORRĚA ADE, A., HUNDESA, A., CARRATELA, A. and BOFILL-MAS, S. (2010). Molecular detection of pathogens in water – The pros and cons of molecular techniques. *Wat. Res.*, **44**, 4325–39.

GREENING, G.E. and LEWIS, G.D. (2007). Virus prevalence in NZ shellfish, *Report – FRST Programme C03X0301 Safeguarding Environmental Health and Market Access for NZ Foods*, July 2007.

GUZMAN-HERRADOR, B., HEIER, B.T., OSBORG, E.J., NGUYEN, V.H. and VOLD, L. (2011). Outbreak of norovirus infection in a hotel in Oslo, Norway, January 2011. *Euro. Surveill.*, **16** (30), pii=19928. Available online: http://www.eurosurveillance.org/ViewArticle.aspx?ArticleId=19928.

HE, X., WEI, Y., CHENG, L., ZHANG, D. and WANG, Z. (2011). Molecular detection of three gastroenteritis viruses in urban surface waters in Beijing and correlation with levels of fecal indicator bacteria. *Environ. Monit. Assess.*, DOI 10.1007/s10661–011–2362–6.

HEATON, J.C. and JONES, K. (2008). Microbial contamination of fruit and vegetables and the behaviour of enteropathogens in the phyllosphere: a review. *J. Appl. Microbiol.*, **104**, 613–26.

HERNANDEZ, F., MONGE, R., JIMENEZ, C. and TAYLOR, L. (1997). Rotavirus and hepatitis A virus in market lettuce *(Latuca sativa)* in Costa Rica. *Int. J. Food Microbiol.*, **37**, 221–3.

HOEBE, C.J.P.A., VENNEMA, H., DE RODA HUSMAN, A.M. and VAN DUYNHOVEN, Y.T.H. P. (2004). Norovirus outbreak among primary schoolchildren who had played in a recreational water fountain. *J. Infect. Dis.*, **189**, 699–705.

HOLTBY, I., TEBBUTT, G.M., GREEN, J., HEDGELEY, J., WEEKS, G. and ASHTON V. (2001). Outbreak of Norwalk-like virus infection associated with salad provided in restaurant. *Commun. Dis. Public Health*, **4**, 305–10.

HUNDESA, A., MALUQUER DE MOTES, C., BOFILL-MAS, S., ALBINANA-GIMENEZ, N. and GIRONES, R. (2006). Identification of human and animal adenoviruses and polyomaviruses for determination of sources of fecal contamination in the environment. *Appl. Environ. Microbiol.*, **72**, 7886–93.

IPPAGUNTA, S.K., NAIK, S., SHARMA, B., and AGGARWAL, R. (2007). Presence of hepatitis E virus in sewage in Northern India: frequency and seasonal pattern. *J. Med. Virol.*, **79**(12), 1827–31.

JIANG, S., NOBLE, R. and CHUI, W.P. (2001). Human adenoviruses and coliphages in urban runoff-impacted coastal waters of Southern California. *Appl. Environ. Microbiol.*, **67**, 179–184.

KARMAKAR, S., RATHORE, A.S., KADRI, S.M., DUTT, S., KHARE, S. and LAL, S. (2008). Post-earthquake outbreak of rotavirus gastroenteritis in Kashmir (India): an epidemiological analysis. *Public Health*, **122**, 981–9.

KHUROO, M.S., KAMILI, S. and YATTOO, G.N. (2004). Hepatitis E virus infection may be transmitted through blood transfusions in an endemic area. *J. Gastroenterol. Hepatol.*, **19**(7), 778–84.

KITAJIMA, M., HARAMOTO E., PHANUWAN C. and KATAYAMA H. (2011). Prevalence and genetic diversity of aichi viruses in wastewater and river water in Japan. *Appl. Environ. Microbiol.*, **77**(6), 2184–7.

KOCWA-HALUCH, R. (2001). Waterborne enteroviruses as a hazard for human health. *Polish J. Environ. Studies*, **10**, 485–7.

KOH, S.J., CHO, H.G., KIM, B.H. and CHOI, B.Y. (2011). An outbreak of gastroenteritis caused by norovirus-contaminated groundwater at a waterpark in Korea. *J. Korean Med. Sci.*, **26**, 28–32.

KOKKINOS, P., FILIPPIDOU, S., KARLOU, K and VANTARAKIS, A. (2010). Molecular typing of enteroviruses, adenoviruses, and hepatitis A viruses in untreated and treated sewage of a biological treatment plant in Greece. *Food. Environ. Virol.*, **2**, **89**–96.

KOOPMANS, M., VENNEMA, H., HEERSMA, H., VAN STRIEN, E., VAN DUYNHOVEN, Y., BROWN, D., REACHER, M. and LOPMAN, B. (2003). For the European Consortium on Foodborne Viruses: early identification of common-source foodborne virus outbreaks in Europe. *Emerg. Infect. Dis.*, **9**(9), 1136–42.

KOOPMANS, M. and DUIZER, E. (2004). Foodborne viruses: an emerging problem. *Int. J. Food Microbiol.*, **90**, 23–41.

KOZYRA, I., KAUPKE, A. and RZEZUTKA, A. (2011). Seasonal occurrence of human enteric viruses in river water samples collected from rural areas of South-East Poland. *Food Environ. Virol.*, **3**, 115–20.

LEBLANC, D., POITRAS, E., GAGNÉ, M.-J., WARD, P. and HOUDE, A. (2010). Hepatitis E virus load in swine organs and tissues at slaughterhouse determined by real-time RT-PCR. *Int. J. Food Microbiol.*, **139**, 206–9.

LEE, S.H. and KIM, S.J. (2002). Detection of infectious enteroviruses and adenoviruses in tap water in urban areas in Korea. *Water Res.*, **36**, 248–256.

LI, T.C., MIYAMURA, T. and TAKEDA, N. (2007). Detection of hepatitis E virus RNA from the bivalve Yamato-Shijimi (Corbicula japonica) in Japan. *Am. J. Trop. Med. Hyg.*, **76**, 170–2.

LI, W., SHE, R., WEI, H., ZHAO, J., WANG, Y., SUN, Q., ZHANG, Y., WANG, D. and LI, R. (2009). Prevalence of hepatitis E virus in swine under different breeding environment and abattoir in Beijing, China. *Vet. Microbiol.*, **133**, 75–83.

LIPP, E.K. and ROSE, J.B. (1997). The role of seafood in foodborne diseases in the United States of America. *Rev. Sci. Tech. Office Int. Epizooties*, **16**, 620–40.

LIU, C., GRILLNER, L., JONSSON, K., LINDE, A., SHEN, K., LINDELL, A.T., WIRGART, B.Z. and JOHANSEN, K. (2006). Identification of viral agents associated with diarrhea in young children during a winter season in Beijing. *China J. Clin. Virol.*, **35**, 69–72.

LONG, S.M., ADAK, G.K., BRIEN, S.J. and GILLESPIE, I.A. (2002). General outbreaks of infectious intestinal disease linked with salad vegetables and fruit, England and Wales, 1992–2000. *Commun. Dis. Public Health*, **5**, 101–5.

LYNCH, M.F., TAUXE, R.V. and HEDBERG, C.W. (2009). The growing burden of foodborne outbreaks due to contaminated fresh produce: risks and opportunities. *Epidemiol. Infect.*, **137**, 307–15.

MAALOUF, H., POMMEPUY, M. and LE GUYADER, F.S. (2010). Environmental conditions leading to shellfish contamination and related outbreaks. *Food Environ. Virol.*, **2**(3), 136–45.

MARA, D.D. (2003). Water, sanitation and hygiene for the health of developing nations. *Public Health*, **117**, 452–456.

MATTISON, K., HARLOW, J., MORTON, V., COOK, A., POLLARI, F., BIDAWID, S. and FARBER, J.M. (2010). Enteric viruses in ready-to-eat packaged leafy greens (letter). *Emerg. Infect. Dis.*, **16**(11), 1815–1817 (serial on the internet). 2010 Nov. http://dx.doi.org/10.3201/eid1611.100877.

MAUNULA, L., KLEMOLA, P., KAUPPINEN, A., SÖDERBERG, K., NGUYEN, T., PITKÄNEN, T., KAIJALAINEN, S., SIMONEN, M.L., MIETTINEN, I.T., LAPPALAINEN, M., LAINE, J., VUENTO, R. and KUUSI, M., ROIVAINEN, M. (2009). Enteric viruses in a large waterborne outbreak of acute gastroenteritis in Finland. *Food Environ. Virol.*, **1**, 31–6.

MEAD, P.S., SLUTSKER, L., DIETZ, V., MCCAIG, L.F., BRESEE, J.S., SHAPIRO, C., GRIFFIN, P.M. and TAUXE, R.V. (1999). Food- related illness and death in the United States. *Emerg. Infect. Dis.*, **5**, 607–25.

MELNICK, J.L. (1984). Etiologic agents and their potential for causing waterborne virus diseases, 1–16. In J. L. MELNICK (ed.), *Enteric Viruses in Water, 15*. Karger, Basel, Switzerland.

MESQUITA, J.R, VAZA, L., CERQUEIRA, S., CASTILHO, F., SANTOS, R., MONTEIRO, S., MANSO, C.F., ROMALDE, J.L. and NASCIMENTO, M.S.J. (2011). Norovirus, hepatitis A virus and enterovirus presence in shellfish from high quality harvesting areas in Portugal, *Food Microbiol.*, **28**, 936–41.

MOCÉ-LLIVINA, L., LUCENA, F. and JOFRE, J. (2005). Enteroviruses and bacteriophages in bathing Waters. *Appl. Environ. Microbiol.*, **71**, 6838–44.

MOON, A., HWANG, I.G. and CHOI, W.S. (2011). Prevalence of Noroviruses in Oysters in Korea. *Food Sci. Biotechnol.*, **20**(4), 1151–4.

NAIRN, C. and CLEMENTS, G.B. (1999). A study of enterovirus isolations in Glasgow from 1977 to 1997. *J. Med. Virol.*, **58**, 304–12.

NORDER, H., SUNDQVIST, L., MAGNUSSON, L., ØSTERGAARD BREUM, S., LÖFDAHL, M., LARSEN, L.E., HJULSAGER, C.K., MAGNIUS, L., BÖTTIGER, B.E. and WIDÉN, F. (2009). Endemic hepatitis E in two Nordic countries. *Eurosurveillance*, **14**(19), 1–9.

OKOH, A.I., SIBANDA, T. and GUSHA, S.S. (2010). Inadequately treated wastewater as a source of human enteric viruses in the environment. *Int. J. Environ. Res. Public Health*, **7**, 2620–37.

PARASHAR, U.D., DOW, L., FANKHAUSER, R.L., HUMPHREY, C.D., MILLER, J., ANDO, T., WILLIAMS, K.S., EDDY, C.R., NOEL, J.S., INGRAM, T., BRESEE, J.S., MONROE, S.S. and GLASS, R.I. (1998). An outbreak of viral gastroenteritis associated with consumption of sandwiches: implications for the control of transmission by food handlers. *Epidemiol. Infect.*, **121**, 615–21.

PAVIO, N., MENG, X.J. and RENOU, C. (2010). Zoonotic hepatitis E: animal reservoirs and emerging risks. *Vet. Res.*, **41**, 46.

PETRINCA, A.R., DONIA, D., PIERANGELI, A., GABRIELI, R., DEGENER, A.M., BONANNI, E., DIACO, L., CECCHINI, G., ANASTASI, P. and DIVIZIA, M. (2009). Presence and environmental circulation of enteric viruses in three different wastewater treatment plants. *J. Appl. Microbiol.*, **106**(5), 1608–17.

PINA, S., PUIG, M., LUCENA, F., JOFRE, J. and GIRONES, R. (1998). Viral pollution in the environment and in shellfish: human adenovirus detection by PCR as an index of human viruses. *Appl. Environ. Microbiol.*, **64** (9), 3376–82.

PINA, S., BUTI, M., COTRINA, M., PIELLA, J. and GIRONES, R. (2000). HEV identified in serum from humans with acute hepatitis and in sewage of animal origin in Spain. *J. Hepatol.*, **33**(5), 826–33.

PINTO, R., ALEGRE, D., DOMINGUEZ, A., EL-SENOUSY, W., SANCHEZ, G., VILLENA, C., COSTAFREDA M. I., ARAGONES L. and BOSCH A. (2007). Hepatitis A virus in urban sewage from two Mediterranean countries. *Epidemiol. Infect.*, **135**(2), 270–3.

PUSCH, D., OH, D.Y., WOLF, S., DUMKE, R., SCHROTER-BOBSIN, U., HOHNE, M., ROSKE, I. and SCHREIR, E. (2005). Detection of enteric viruses and bacterial indicators in German environmental waters. *Arch. Virol.*, **150**(5), 929–47.

RAO, V.C. and MELNICK, J.L. (1986). Environmental virology. In J.A. COLE, C. J. KNOWLES and D. CHLESSINGER (eds), *Aspects of Microbiology 13*, American Society for Microbiology, Washington, DC.

RODRIGUEZ-LAZARO, D., COOK, N., RUGGERI, F.M., SELLWOOD, J., NASSER, A., SAO JOSE NASCIMENTO, M., D'AGOSTINO, M., SANTOS, R., SAIZ, J.C., RZEZUTKA, A., BOSCH, A., GIRONES, R., CARDUCCI, A., MUSCILLO, M., KOVAC, K., DIEZ-VALCARCE, M., VANTARAKIS, A., BONSDORFF, C.-H., DE RODA HUSMAN, A.M., HERNANDEZ, M. and VAN DER POEL, W.H.M. (2011). Virus hazards from food, water and other contaminated environments. *FEMS Microbiol. Rev.*, 1–29. DOI: 10.1111/j.1574–6976.2011.00306.x.

RODRIGEZ MANZANO, J., MIAGOSTOVICH, M., HUNDESA, A., CLEMENTE-CASARES, P., CARRATELA, A., BUTI, M., JARDI, R. and GIRONES, R. (2010). Analysis of the evolution in the circulation of HAV and HEV in eastern Spain by testing urban sewage samples. *J. Water Health*, **8**(2), 346–54.

SARVIKIVI, E., ROIVAINEN, M., MAUNULA, L., NISKANEN, T., KORGONEN, T., LAPPALAINEN, M. and KUUSI, M. (2012). Multiple norovirus outbreaks linked to imported frozen raspberries. *Epidemiol. Infect.*, **140**(2), 260–7.

SEDMAK, G., BINA, D. and MACDONALD, J. (2003). Assessment of an enterovirus sewage surveillance system by comparison of clinical isolates with sewage isolates from Milwaukee, Wisconsin, collected August 1994 to December 2002. *Appl. Environ. Microbiol.*, **69**, 7181–7.

SINCLAIR, R.G., JONES, E.L. and GERBA, C.P. (2009). Viruses in recreational water-borne disease outbreaks: A review. *J. Appl. Microbiol.*, **107**, 1769–80.

STALS, A., BAERT, L., JASSON, V., VAN COILLIE, E. and UYTTENDAELE M. (2011). Screening of fruit products for norovirus and the difficulty of interpreting positive PCR results. *J. Food Prot.*, **74**(3), 425–31.

TEUNIS, P.F., MOE, C.L., LIU, P., MILLER, S.E., LINDESMITH, L., BARIC, R.S., LE PENDU, J. and CALDERON, R.L. (2008). Norwalk virus: how infectious is it? *J. Med. Virol.*, **80**, 1468–76.

UMESHA, K.R., BHAVANI, N.C., VENUGOPAL, M.N., KARUNASAGAR, I., KROHNE, G. and KARUNASAGAR, I. (2008). Prevalence of human pathogenic enteric viruses in bivalve molluscan shellfish and cultured shrimp in south west coast of India. *Int. J. Food Microbiol.*, **122**, 279–86.

VAIDYA, S.R., CHITAMBAR, S.D. and ARANKALLE, V.A. (2002). Polymerase chain reaction-based prevalence of hepatitis A, hepatitis E and TT viruses in sewage from an endemic area. *J. Hepatol.*, **37**(1), 131–6.

VAN HEERDEN, J., EHLERS, M.M. and GRABOW, W.O.K. (2005). Detection and risk assessment of adenoviruses in swimming pool water. *J. Appl. Microbiol.*, **99**, 1256–64.

VANTARAKIS, A. and PAPAPETROPOULOU, M. (1999). Detection of enteroviruses, adenoviruses and hepatitis A viruses in raw sewage and treated effluents by nested-PCR. *Water Air Soil Poll.*, **114**, 85–93.

VERHOEF, L., KRONEMAN, A., VAN DUYNHOVEN, Y., BOSHUIZEN, H., van PELT, W. and KOOPMANS, M. (2009). On behalf of the Foodborne Viruses in Europe Network: selection tool for foodborne norovirus outbreaks. *Emerg. Infect. Dis.*, **15**(1), 31–8.

VON BONSDORFF, C-H. and MAUNULA, L. (2008). Waterborne viral infections. In E.A. PALOMBO and C.D. KIRKWOOD (eds), *Viruses in the Environment*, Kirkwood, Research Signpost. ISBN: 978-81-308-0235-0.

WIDDOWSON, M.A., SULKA, A., BULENS, S.N., BEARD, R.S., CHAVES, S.S., HAMMOND, R., SALEHI, E.D.P., SWANSON, E., TOTARO, J., WORON, R., MEAD, P.S., BRESEE, J.S., MONROE, S.S. and GLASS, R.I. (2005). Norovirus and foodborne disease, United States, 1991–2000. *Emerg. Infect. Dis.*, **11**(1), 95–102.

WORLD HEALTH ORGANIZATION and UNICEF (2010). Progress on sanitation and drinking water: 2010 update. Available at: http://www.who.int/water_sanitation_health/publications/9789241563956/en/index.html. Accessed February 18, 2012.

WYN-JONES, A., CARDUCCI, A., COOK, N., D'AGOSTINO, M., DIVIZIA, M., FLEISCHER, J., GANTZER, C., GAWLER, A., GIRONES, R., HÖLLER, C., DE RODA HUSMAN, A.M., KAY, D., KOZYRA, I., LÓPEZ-PILA, J., MUSCILLO, M., SÃO JOSÉ NASCIMENTO, M., PAPAGEORGIOU, G., RUTJES, S., SELLWOOD, J., SZEWZYK, R. and WYER, M. (2011). Surveillance of adenoviruses and noroviruses in European recreational waters. *Wat. Res.*, **45**(3), 1025–38.

YAZAKI, Y., MIZUO, H., TAKAHASHI, M., NISHIZAWA, K., SASAKI, N. and GOTANDA, Y. (2003). Sporadic acute or fulminant hepatitis E in Hokkaido, Japan, maybe foodborne, as suggested by the presence of hepatitis E virus in pig liver as food. *J. Gen. Virol.*, **84**, 2351–7.

XAGORARAKI, I., KUO, D.H.W., WONG, K., WONG, M. and ROSE, J.B. (2007). Occurrence of human adenoviruses at two recreational beaches of the great lakes. *Appl. Environ. Microbiol.*, **73**, 7874–81.

Part II

Detection, surveillance and risk assessment of viruses in food and water

3

Molecular detection of viruses in foods and food-processing environments

D. Rodríguez-Lázaro, University of Burgos, Spain, and K. Kovac and M. Hernández, Instituto Tecnológico Agrario de Castilla y León (ITACyL), Spain

DOI: 10.1533/9780857098870.2.49

Abstract: Monitoring the presence of enteric viruses in food is a challenging task, and molecular-based methods have become the reference detection methodology. This chapter describes in detail the main steps of the analytical process of detection of foodborne viruses by molecular methods, paying special attention to key aspects such as the interpretation of test results, the use of controls, and the implication for public health of the results obtained by molecular methods.

Key words: real-time PCR, molecular methods, detection, controls, standardization.

3.1 Introduction

Detection of the presence of enteric viruses in food, particularly norovirus (NoV), hepatitis A and E virus (HAV and HEV) and adenovirus, is an important issue in food safety and requires rapid and robust diagnostic methodology (Bosch *et al.*, 2011; Croci *et al.*, 2008). Several different approaches can be used to detect human enteric viruses in food samples, including direct observation by electron microscopy, detection of cytopathic effects in specific cell lines or detection using immunological or molecular methods. Direct observation by electron microscopy is time-consuming, subjective and of limited sensitivity (Atmar and Estes, 2001). The observation of cytopathic effects produced in specific cell lines is not always possible as some enteric viruses, notably NoV and HEV, cannot be propagated in mammalian cell lines. Even when it is possible, the detection of viruses using a cell culture is not a simple or cost-effective technique. This technique may also require the adaptation of the virus in order for it to grow effectively (Pintó and Bosch, 2008). Immunological tests do exist, and many are commercially available for the main enteric viruses. However, their analytical sensitivity is still too poor for effective testing of food samples.

To overcome these limitations, quantitative real-time PCR (q-PCR) has become the method of choice for the detection of enteric viruses. This approach has been reinforced by the recommendation of the international ISO/CEN committee CEN/TC275/WG6/TAG 4 that real-time PCR should serve as the basis for the forthcoming international standards for detection of NoV and HAV (Lees and CEN WG6 TAG4, 2010). A large number of scientific studies using molecular methods for the detection of main foodborne viruses have already been published (Table 3.1).

Other detection options include the combination of cell culture or immunological methods and a molecular technique. The combination of a cell culture step and subsequent detection by a molecular technique such as RT-PCR or nucleic acid sequence-based amplification (NASBA) reduces the incubation period and also allows the detection of viruses that grow without causing cytopathic effects (Dubois *et al.*, 2002; Duizer *et al.*, 2004).

3.2 Molecular detection of viruses in foods: the process

Detection of viruses in food samples is a challenging task due to the large variety and complexity of samples, the possible heterogeneous distribution of a small number of viruses and the presence of components that may inhibit or interfere with virus detection (Rodríguez-Lázaro *et al.*, 2012). A general flow chart for the detection process (from sampling to final identification) for the detection of human enteric viruses is given in Fig. 3.1. The features of the ideal procedure used for extraction and detection of foodborne viruses are presented in Table 3.2.

3.2.1 Matrix separation

Virus extraction from food is the first step for the detection of enteric viruses in food matrices. During this step, viral particles are separated from the food matrix, and the final concentrate must be free of any inhibitors which may be co-extracted or co-concentrated from the sample matrix (Stals *et al.*, 2012). A variety of biological and chemical substances which are present in food matrices or are used during sample processing have been found to act as inhibitors, including polysaccharides, haeme, phenol, and cations (Atmar, 2006). Known PCR inhibitors in shellfish extracts include glycogen and acidic polysaccharides (Schwab *et al.*, 1998).

The extraction method depends to a great extent on the composition of the food. Protocols for concentration of viruses in food samples often start with a washing step (in the case of fresh produce) or a homogenization step

Table 3.1 Molecular methods for detection of the main enteric viruses

Technique	Target	Matrix	Sample size	Limit of detections	Reference
Hepatitis A virus (HAV)					
RT-PCR	External capsid protein-coding region (VP1)	Oysters	Flesh of two oysters (10–20 g of tissue)	3.3 FFU[a] of HAV	Coelho *et al.*, 2003
Long RT-PCR, 5'RACE and 3'RACE	Whole genome (nt pos. 105–7328), 5'NTR, 3'NTR	Serum samples	50 µl of serum		Fujiwara *et al.*, 2001
RT-qPCR	5' noncoding region	Clinical and shellfish samples	10 % (w/v) stool suspension in PBS 150 µl of serum 1.5 g stomach and digestive diverticula homogenates	10 molecules of synthetic ssRNA transcript, 1 viral RNA molecule, and 0.05 infectious virus per reaction	Costafeda *et al.*, 2006
RT-qPCR	5' noncoding region	Serum samples	140 µl of serum	72 copies/ml	Costa-Mattioli *et al.*, 2002
Molecular-Beacon-based RT-qPCR	5' noncoding region	Environmental samples (food or water)	1000 l of groundwater	20 PFU	Abd El Galil *et al.*, 2004
Molecular-Beacon-based real-time NASBA	5' noncoding region	Surface lake water	Surface lake water	10 PFU	Abd El Galil *et al.*, 2005
RT-PCR and nested RT-PCR	5' noncoding region	Vegetables (green onions)		0.2 PFU of HAV	Hu and Arsov, 2009
RT-qPCR	5' noncoding region	Tap, river, mineral and coastal water samples	2.5 l of water	60 ge/ml	Villar *et al.*, 2006
RT-qPCR	5'UTR	Cell culture		$10^{1.8}$ $TCID_{50}$/ml	Houde *et al.*, 2007
RT-qPCR	5'UTR	oysters	25 g	5 $TCID_{50}$/g	Casas *et al.*, 2007

(*Continued*)

Table 3.1 Continued

Technique	Target	Matrix	Sample size	Limit of detections	Reference
RT-qLAMP	5'UTR	Clinical samples (faeces)		0.4–0.8 FFU/reaction	Yoneyama *et al.*, 2007
RT-q PCR	5'UTR	Shellfish (blue mussels)	6 g of digestive glands	5×10^2 $TCID_{50}$/g	Di Pasquale *et al.*, 2010
RT-q PCR	5'UTR	Clinical or environmental samples		0.5 PFU and 40 copies of a synthetic transcript	Jothikumar *et al.*, 2005b
RT-nested PCR	5'UTR VP1-P2A junction (C terminal)	Serum samples		4 HAV particles/ml	Bower *et al.*, 2000
RT-nested PCR	VP1-P2A junction	Serum samples	100 µl of serum		De Paula *et al.*, 2002
RT-qPCR (SYBR Green)	VP3-VP1 regions	Ocean water	4 l of ocean water	90 HAV copies/l	Brooks *et al.*, 2005
RT-PCR	Protease coding region and polymerase coding region of HAV	Vegetables (green onions)	25 g of green onions	1 TCID50/25 g	Guévremont *et al.*, 2006
RT-nested PCR and cell culture+RT-PCR	VP2 and VP4 capsid region	Shellfish (blue mussels and other bivalve molluscs)	75 g of body and liquor homogenate		De Medici *et al.*, 2001
Norovirus (NoV)					
RT-PCR	RNA-dependent RNA polymerase gene (ORF1)	Faecal samples	100 µl of a 10–20% stool suspension		Vennema *et al.*, 2002
RT-PCR and nested PCR	RNA-dependent RNA polymerase gene (ORF1)	Faecal samples			Kawamoto *et al.*, 2001

RT-PCR and nested PCR	RNA-dependent RNA polymerase gene (ORF1)	Seawater, river water, estuarine water, and sewage treatment plant effluents	10 l of seawater, river and estuarine water and 50 ml of sewage samples		La Rosa *et al.*, 2007
		Faecal samples	10% stool suspension in PBS		
RT-PCR + hybridization with membrane-bound oligonucleotides	RNA-dependent RNA polymerase gene (ORF1)	Faecal samples	10% (wt/vol) stool suspension	3–30 RNA particles	Vinjè and Koopmans, 2000
RT-PCR	ORF1/ORF2 junction region (NoV ggI)	Recreational waters	1 l of bathing water concentrated in 140 µl of RNase free water		Silva *et al.*, 2009
RT-qPCR	ORF1/ORF2 junction region	Faecal samples	140 µl 10% (wt/vol) stool suspension	2.0 × 10^4 copies of NoV RNA per gram of stool	Kageyama *et al.*, 2003
RT-qPCR	ORF1/ORF2 junction region (NoV ggI, ggII, ggIII)	Faecal samples and shellfish samples	Clarified PBS extracts of stool samples and 133 µl of shellfish homogenate	<10 copies of viral genome per reaction	Jothikumar *et al.*, 2005c
RT-qPCR	ORF1/ORF2 junction region	Shellfish (oysters)	1.5 g of stomach and digestive diverticula of 2–5 oysters	7 and 70 RT-PCR Units for NVGII and NVGI	Loisy *et al.*, 2005
Duplex RT-qPCR	ORF1/ORF2 junction region	Faecal samples	10% (wt/vol) stool suspension		Ishida *et al.*, 2008
MultiplexRT-qPCR	ORF1/ORF2 junction region	Faecal samples	200 µl of 20% stool suspension	NoV were detected from neat to 10^{-4}	Pang *et al.*, 2005
RT-PCR and nested PCR	ORF1/ORF2 junction region (NoV ggII.4)	Faecal samples	20% (wt/vol) stool suspension		Bull *et al.*, 2006

(*Continued*)

Table 3.1 Continued

Technique	Target	Matrix	Sample size	Limit of detections	Reference
RT-qPCR	ORF1-ORF2 junction region	Faecal samples and oysters	10% stool suspension in PBS Homogenate of digestive diverticulum tissue		Nenonen *et al*., 2009
Real-time NASBA	RNA-dependent RNA polymerase gene (ORF1)	Surface water	12.5 μl of water concentrates (neg membrane filtration/ ultrafiltration of 600 L)		Rutjes *et al*., 2006b
RT-PCR	ORF2	Faecal samples	140 μl 10% (wt/vol) stool suspension in water		Kojima *et al*., 2002
Most probable number-RT-PCR	Capsid region and polymerase region	Faecal samples and shellfish (oysters)	1.5 g stomach and digestive diverticula form 1 to 4 oysters		Le Guyader *et al*., 2003
Hepatitis E virus (HEV)					
RT-PCR and nested PCR	ORF1 and ORF2	Serum samples			Mizuo *et al*., 2002
RT-PCR and nested RT-PCR + Southern blot	ORF1 and ORF2	Stool samples (pig)	100 μl of 10% stool suspension in Hanks balanced salt solution.		Van der Poel *et al*., 2001
RT-PCR and nested PCR	ORF 2	Faecal and serum samples (pig)		31.6 PID_{50}/ml[b] of swine HEV	Huang *et al*., 2002
RT-PCR and nested PCR	ORF 2	Serum samples and tissue homogenates (mesenteric, tracheobronchial, and hepatic lymph nodes)	100 μl of 10% tissue homogenates or serum samples		Williams *et al*., 2001
RT-PCR + Southern blot	ORF2	Serum and faecal samples	140 μl of serum or 10% stool suspension		Herremans *et al*., 2007

RT-qPCR	ORF2	Serum samples			Colson *et al.*, 2007
RT-PCR and nested PCR	ORF2	Wild boar meat and human serum samples			Li *et al.*, 2005
RT-qPCR	ORF2	Serum samples, stool and bile (pig)	200 µl of serum, 10% stool suspension or bile		Kaba *et al.*, 2009
RT-qPCR (SYBR Green)	ORF2	Faecal samples	300 µl of a 10% stool suspension in PBS	10 molecules of HEV cDNA per reaction	Orrú *et al.*, 2004
RT-qPCR	ORF2	Serum samples	140 µl of serum	1.68×10^1 copies	Ahn *et al.*, 2006
RT-qPCR	ORF2	Serum and faecal samples	200 µl of serum samples or 10% stool suspension in PBS		Enouf *et al.*, 2006
RT-qPCR	ORF2	Pig faeces and intestinal contents, liver from a pig, stool and serum samples from humans, stool sample from a Rhesus monkey	Faecal samples and tissues were homogenized in grinding tubes containing beads and then prepared as 5% (w/v) suspensions in TE-buffer (Tris EDTA) pH 7.6	Between 1 and 20 genomic equivalents per reaction for both assays	Gyarmati *et al.*, 2007
RT-PCR and nested PCR	ORF2/ORF3 overlapping region	Serum samples	100 µl of serum	Two or three times more sensitive than Mizuo *et al.*, 2002	Inoue *et al.*, 2006
RT-qPCR	ORF 3	Surface water	10 l of surface water	0.12 PID50/ml of swine HEV	Jothikumar *et al.*, 2006
RT-PCR and semi-nested PCR		Serum and faecal samples (pig)			Seminati *et al.*, 2008

Adapted from Rodríguez-Lázaro *et al.*, 2012.
[a] The virus titre was determined by counting fluorescent cells at the highest virus dilution. Virus quantity is expressed as focus-forming units (FFU) per millilitre.
[b] 50% pig infectious dose.

Table 3.2 Principal characteristics of an ideal procedure used for detection of enteric viruses in food

	Features	Comments
Extraction and concentration method	Generic	Concentrates vast range of foodborne viruses in a small volume of concentrate
	Efficiency	Possesses good virus recovery
	Labor intensity	Not time-consuming, short procedure
	Feasibility	Easy to perform and can be done in suitably equipped food laboratories
	Robustness	Good reproducibility and repeatability
Detection method	Detection limit	Should be as low as possible allowing detection of only a few virus particles in food samples
	Identification of virus type or strain	Should allow identification of virus type or strain without applying additional procedures
	Cost	Inexpensive

Adapted from Rodríguez-Lázaro *et al.*, 2007.

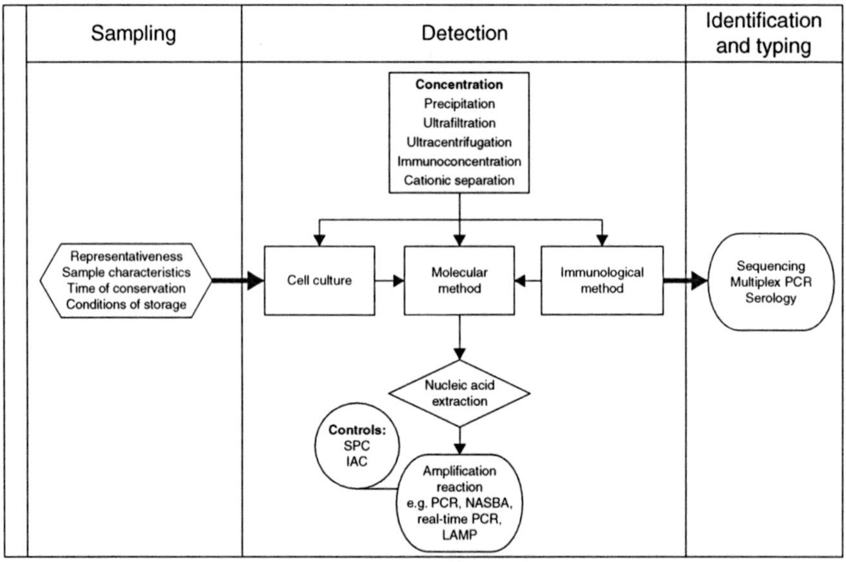

Fig. 3.1 Flowchart of the analytical process of detection and identification of entric viruses in food. (Source: Adapted from Rodríguez-Lázaro *et al.*, 2012.)

(in the case of, for example, shellfish); the virus is concentrated after this first step (Croci *et al.*, 2008; Rodríguez-Lázaro *et al.*, 2007). If appropriate, a minimal volume of a diluent can be added to favour dissociation of the virus from the food matrix if it will not interfere with subsequent virus extraction.

Food matrices can be divided into three categories: (i) carbohydrate and water-based foods like fruits and vegetables; (ii) fat- and protein-based foods like many ready-to-eat products; and (iii) shellfish as filter-feeding organisms which concentrate and accumulate virus particles and other pathogens in the shellfish digestive system (Baert *et al.*, 2008; Stals *et al.*, 2012).

Three main approaches can be followed: elution of the viral particles (whether or not preceded by an acid adsorption step) with subsequent concentration; direct extraction of the viral RNA from the food matrix without the elution-concentration step; and extraction of viruses from the food by proteinase K treatment. After most of the extraction protocols, purification (removal of food debris and inhibitory substances) of the virus eluate/concentrate or extracted RNA is performed by either filtration through cheese cloth or 0.45 and 0.20 μm filters or by treatment of virus suspension with Freon 13, Vertrel®XF or chloroform: butanol (Stals *et al.*, 2012).

Elution-concentration can be used for extraction of foodborne viruses from various matrices like shellfish, carbohydrate/water-based foods and fatty/protein-based foods. The elution-concentration step is based on washing the viral particles from the food surface using an appropriate buffer followed by concentration of the eluted viruses.

Several elution buffers have been used to elute viral agents from food matrices. Alkaline buffer at pH 9 to 10.5 is applied as an alkaline environment allows virus particles to detach from the food solids. The acidic nature of fruits and vegetables can impair virus elution and consequently reduce detection sensitivity (Dubois *et al.*, 2002). A neutral buffer for virus elution is mainly used when anionic exchange or ultracentrifugation are applied to concentrate the eluted viral particles. For foods which contain acidic substances (such as fruits and vegetables) an alkaline Tris-based buffer system is usually used (reviewed in Stals *et al.*, 2012). Phosphate buffer (Bidawid *et al.*, 2000a) or sodium bicarbonate buffer (Kurdziel *et al.*, 2001) have been also used successfully. Elution buffers are frequently combined with beef extract (1–3%) and glycine, which reduce non-specific virus adsorption to the food matrix during their extraction (Dubois *et al.*, 2002; Love *et al.*, 2008) and can therefore increase virus recovery (Butot *et al.*, 2007). However, beef extract might interfere with molecular detection methods (Katayama *et al.*, 2002), and a virus elution purification step can be necessary. When viruses are extracted from fruits and vegetables, pectinase is frequently added to prevent jelly formation in the eluate by breaking the pectin bonds (Dubois *et al.*, 2002). When ultracentrifugation is used as concentration step, soya protein powder can be added in the purification step to facilitate the liberation of viruses from food surfaces and thus improve recovery efficiency (Rzeżutka *et al.*, 2006, 2008).

Direct virus nucleic acid extraction includes treatment of the food with guanidinium isothiocyanate (GITC)/phenol, followed by purification of the viral RNA. It has been used for NoV concentration from foods (Baert *et al.*, 2008; Boxman *et al.*, 2007; Stals *et al.*, 2011b). Direct RNA extraction combined with shredding of the digestive tissue using zirconium beads has successfully recovered NoV from oyster digestive tissue (de Roda Husman *et al.*, 2007).

The combination of proteinase K and heat treatment (usually 65°C) damages the virus capsid and releases virus nucleic acid (Nuanualsuwan and Cliver, 2002). This approach has been used successfully for the detection of NoVs in several foodborne outbreaks related to shellfish (Comelli *et al.*, 2008; Jothikumar *et al.*, 2005c) and has been selected by the *ad hoc* CEN/TC275/WG6/TAG4 working group for the extraction of the most common enteropathogenic viruses from shellfish digestive tissue (Lees and CEN WG6 TAG4, 2010).

3.2.2 Concentration of viruses

The aim of concentrating virus is to collect most of the virus present in the sample in a minimal volume (Cliver, 2008). Concentration methods appropriate for a wide variety of matrices include differential precipitation, ultracentrifugation and ultrafiltration (Rodríguez-Lázaro *et al.*, 2007).

Precipitation with polyethilene glycol (PEG) after elution of viruses from food matrices is frequently used. It allows the precipitation of viruses at neutral pH and at high ionic concentrations without precipitation of other organic material (Baert *et al.*, 2008; Kim *et al.*, 2008; Stals *et al.*, 2011a).If alkaline elution is used, pH is normally firstly adjusted to 7.0–7.4. High NaCl concentrations (0.3–0.525 M) are also required to successfully precipitate virus particles.

Viruses can be precipitated with ultracentrifugation from 100 000 × g to 235 000 × g (Casas *et al.*, 2007; Rutjes *et al.*, 2006a; Rzeżutka *et al.*, 2006, 2008). Virus eluates should be purified prior to ultracentrifugation, either with high speed conventional centrifugation or 0.22–0.45 μm-filtration, to remove food debris and other components which can co-sediment with virus particles during ultracentrifugation (Croci *et al.*, 2008).

Viruses can be also concentrated by ultrafiltration based on molecular weight. While pores in membrane filters permit the passage of liquids and low molecular mass particles (10–100 kDa), viruses are captured by the filter (Croci *et al.*, 2008). Similar to ultracentrifugation, the virus eluates should be purified prior to ultrafiltration, in order to prevent clogging of the filters (Butot *et al.*, 2007; Stals *et al.*, 2012). Treatment of filters with bovine serum albumin (BSA) or sonication of the purified virus eluate can increase virus recovery (Jones *et al.*, 2009). The recoveries using ultrafiltration as a concentration method after alkaline elution of viruses from food matrices varied from 0.1% to 19.6% for different viruses used (reviewed in Stals *et al.*, 2012).

Viruses can also be concentrated by cationic separation or by immunoconcentration. Cationic separation uses positively charged magnetic particles in conjunction with a magnetic capture system to concentrate and purify viral particles from food matrices, as the negatively charged proteins of the virus capsid bind to the positively charged magnetic particles. This approach has been successfully used for concentration of NoVs in lettuce and fresh cheese, providing a recovery efficiency ranging from 5.2% to 56.3% (Fumian *et al.*, 2009). The commercially available automatic cationic separation system Pathatrix™ (Matrix MicroScience, Newmarket, UK) has been used for the concentration of HAV from different types of foods with a recovery efficiency ranging between 17.0% and 81.7% (Papafragkou *et al.*, 2008). However, less promising results with this method were obtained for NoVs (Morton *et al.*, 2009). Immunoconcentration methods can use magnetic beads covered with histo-bloodgroup antigens (HBGA) or porcine gastric mucin for concentration of NoVs (Morton *et al.*, 2009; Tian *et al.*, 2008) or magnetic beads covered by monoclonal HAV antibodies (Bidawid *et al.*, 2000a). This allows specific extraction of foodborne viruses in different categories of foods and efficient removal of PCR inhibitors. However, due to immunogenetic drift, the long-term use of the same antibodies could be problematic (Stals *et al.*, 2012).

3.2.3 Extraction of viral nucleic acids

Genomic nucleic acids must be first extracted from the viral concentrates using in-house or commercial procedures (Mattison and Bidawid, 2009). A wide variety of commercial kits are currently available, most of them based on a method which uses guanidiniumisothyocianate-based lysis followed by capture of nucleic acids on silica (Boom *et al.*, 1990; Haramoto *et al.*, 2005; Le Guyader *et al.*, 2009; Rutjes *et al.*, 2005). Commercial kits offer ease of use, reliability and reproducibility. In recent years, automated nucleic acid extraction platforms have been developed and have been efficiently applied for the analysis of viruses in food samples (Comelli *et al.*, 2008; Perele *et al.*, 2009; Stals *et al.*, 2011b). Other methods for viral nucleic acid extraction include proteinase K treatment followed by phenol-chloroform extraction and ethanol precipitation, sonication and heat treatment (Comelli *et al.*, 2008; Guévremont *et al.*, 2006; Jothikumar *et al.*, 2005c; Le Guyader *et al.*, 2009).

3.2.4 Molecular-based detection

The development of q-PCR represents a significant advance in molecular techniques involving nucleic acids analysis. q-PCR allows monitoring of the synthesis of new amplicon molecules during the PCR (i.e., in real time). Data is therefore collected throughout the PCR process and not just at the end of the reaction (as occurs in conventional PCR) (Heid *et al.*, 1996). The q-PCR results consist of amplification curves that can be used to quantify the initial amounts of template nucleic acid molecules with high precision over a wide range of concentrations (Schmittgen *et al.*, 2000).

Major advantages of q-PCR are the closed-tube format that avoids the risk of carryover contamination, fast and easy to perform analysis, the extremely wide dynamic range of quantification (more than eight orders of magnitude), and the significantly higher reliability and sensitivity of the results compared to conventional PCR (Rodríguez-Lázaro *et al.*, 2007). As with conventional PCR, q-PCR can be used for many different purposes, particularly for quantifying nucleic acids and for genotyping. It combines primer amplification with detection of the amplified product in a single reaction mix. When detection is accomplished due to binding interactions with an extra (third) oligonucleotide, the detection signal also provides confirmation of the identity of the target amplicon (Mattison and Bidawid, 2009). However, q-PCR also suffers from some limitations. The volume used in the amplification reaction is very small; therefore only concentration methods that can deliver a very small volume of the resulting nucleic acid solution from a realistic food sample can be used. In addition, the quality of the nucleic acids is an important factor that directly affects the analytical sensitivity of the assay, and diverse compounds present in samples can inhibit the amplification reaction.

Other molecular techniques used in the detection of enteric viruses are NASBA and loop-mediated isothermal amplification (LAMP). NASBA employs three different enzymes: T7 RNA polymerase, R Nase H and avian myeloblastosis virus (AMV) reverse transcriptase, which act in concert to amplify sequences from an original single-stranded RNA template (Compton, 1991). The reaction also includes two oligonucleotide primers, complementary to the RNA region of interest, and one of the primers also contains a promoter sequence that is recognized by T7 RNA polymerase at the 5'-end. At least four primers are used for the LAMP method, two of which are the loop primers that recognize two regions each in the target genetic sequence. The target sequence is amplified using a strand-displacing DNA-dependent DNA polymerase, and detection is accomplished by measuring an increase in turbidity or the binding of a fluorescent detection reagent; this can be monitored in real time (Fukuda *et al.*, 2007; Yoneyama *et al.*, 2007). The specificity of the reaction in the LAMP system can be increased using additional primers (Mattison and Bidawid, 2009).

3.3 Current issues in molecular detection of viruses in foods

Several major issues are extremely relevant in molecular detection of enteric viruses in food, particularly with the aim of an effective implementation in the food microbiology labs: the representativeness of the sample taken for analysis and the definition of rational and science-based sampling plans; the validity of the results obtained by molecular detection methods; the definition and use of controls to prevent misinterpretations of the analytical results and the implications for public health of the results obtained.

3.3.1 Food sample representativeness and sampling plans

Representativeness is a qualitative factor, which is largely dependent on the appropriate design of the sampling programme and expresses the degree to which sample data accurately and precisely reflect a characteristic or variable at a sampling point. This is essential for the analysis of enteric viruses, which may be present in small quantities and distributed heterogeneously in matrices; there should be a risk-based approach to planning (Andrews and Hammack, 2003; Food Standard Agency, 2004a,b). Consequently, a sample must represent the original matrix, and the sampling process must not alter the condition of the sample, and thus not affect the subsequent analysis (Food Standard Agency, 2004a,b). Other aspects that must also be considered are the characteristics of the food to be analysed (nature – solid, semisolid, viscous or liquid; composition – rich in fat, protein, or plant contents such as tannins; and amount – scarce or abundant), and the subsequent analytical method to be used (cell culture, immunological or molecular). Any inadequacy concerning one of the aspects will affect the validity of the final analytical result.

Various international bodies and national bodies have defined principles and/or standards for the sampling of foods. However, there is no specific mention of sampling for enteric viruses. The CEN/ISO *ad hoc* expert committee for viruses in food 'CEN/TC 275/WG6/TAG4' is currently working on the first international standard for a horizontal method for detection of HAV and NoVs in food. However, the sampling process is not included in this planned standard, and the committee has decided to examine the ISO 6887 series for suitability. Similarly, the US FDA's Bacteriological Analytical Manual (BAM) includes a general protocol for 'Food sampling and preparation of sample homogenate' (Andrews and Hammack, 2003), in which the scientific basis for sampling only uses previously published bacteriological criteria (ICMSF, 1986, 2002), despite the BAM having defined a specific protocol for the detection and quantification of HAV (Goswami, 2001).

Several important lessons can be learnt from the scientific literature (a revision of PCR-based methods in foods with the description of the principal features is shown Table 3.3). First, there is an evident lack of harmonization in the sample size, and therefore a serious risk as to the representativeness of the sampling strategies used. This is most important as most of those studies are related to viral diarrhoeal outbreaks: the consequences may include the true aetiological agent of the gastroenteritis not being found, or the infectious dose being under- or over-estimated. In these studies, sizes of samples used were extremely diverse, ranging from 1.5 to 200 g. Second, with shellfish there is a lack of homogeneity in the selection of the animal tissues or part of the sample tested once the sample is collected. This also can affect the detection of human pathogenic viruses. For example, different shellfish tissues can be tested for human enteric viruses (i.e., the whole shellfish, the mantle, the gills, the stomach, or the digestive diverticula). However, it has been demonstrated that the efficiency of recovery can differ substantially between types of sample and even that the virus may not be detectable in some; the percentages of samples positive in naturally contaminated oysters in a study varied for the whole oyster (0.7%), mantle (2.2%), gills (14.7%), stomach (13.9%), and the digestive diverticula (13.2%), and detection was not possible when the adductor muscles were tested (Wang *et al.*, 2008).

Table 3.3 Sampling methods used for detection of viral hazards in food matrices

Matrix	Sample size	Type of sample	Conservation	Detection method	Reference
Shellfish					
Oysters	1.5 g	Stomach and digestive diverticula	4°C	RT-qPCR and RT-PCR, Sequencing	Le Guyader *et al.*, 2006; 2008
	1.5 g	Digestive tissues	−20°C	q-RT-PCR	Le Guyader *et al.*, 2009
	1.5 g	Digestive tissues (at least 12 oysters from the same bay)	−20°C	RT-PCR, Dot-blot hybridization, Sequencing	Costantini *et al.*, 2006
	1.5 g	Gills, stomach, digestive diverticula, fringe of mantle, adductor muscle and the whole oysters	4°C	RT-PCR, Sequencing	Wang *et al.*, 2008
	5–10 g	Stomach and digestive diverticula (about 20 shellfish)	−20°C	RT-PCR, Sequencing	Boxman *et al.*, 2006
	10–20 g	Oyster tissue (two oysters)	−80°C	RT-PCR	Coelho *et al.*, 2003
	25 g	Whole oysters	−70°C	RT-PCR, nested PCR, sequencing	Shieh *et al.*, 2007
	150 mg	Digestive gland tissue (pool of two oysters)	n.d., processed within 24 h	RT-PCR, sequencing, southern blot hybridization	De Roda Husman *et al.*, 2007
	0.6 g	Stomach and digestive tracts	−80°C	RT-PCR,[1,2] Sequencing,[1] RT-qPCR[1,2]	[1]Nishida *et al.*, 2003; [2]Nishida *et al.*, 2007
	2 g	Hepatopancreas (pool of 6 animals)	4°C	Nested PCR, RT-qPCR	Jothikumar *et al.*, 2005a
Oysters and Mussels	1.5 g	Stomach and digestive diverticula (pools of 5 animals)	4°C[1] −20°C[2,3]	RT-PCR,[1,2] Sequencing,[1,3] Hybridization[2,3]	[1]Beuret *et al.*, 2003; [2]Le Guyader *et al.*, 2003; [3]Le Guyader *et al.*, 2000
	10 g or 30 g	Digestive glands (pool of 20 animals)	4°C	RT-PCR, Sequencing, RT-qPCR, Cell culture	Formiga-Cruz *et al.*, 2002

Mussels	12 g	Digestive tissue (pool of 30 animals)	−20°C	PCR	Karamoko *et al.*, 2005
Mussels and Clams	1.5 g	Stomach and digestive diverticula (pool of 6 animals)	−20°C	RT-PCR Hybridization	Elamri *et al.*, 2006
Mussels, Clams and Cockles	50 g	Stomach and hepatopancreas	n.d.	RT-PCR, Southern blot hybridization	Romalde *et al.*, 2002
Clams	1.5 g	Stomachs and digestive diverticula	−20°C	RT-PCR,[1,2] Sequencing,[1] Southern blot hybridization,[2] RT-qPCR[3]	[1]Sánchez *et al.*, 2002; [2]Bosch *et al.*, 2001; [3]Costafreda *et al.*, 2006
Clams	Approximately 10 g	Stomachs and digestive diverticula with some surrounding tissue (poll of 59 animals)	Frozen	RT-PCR, Sequencing	Kingsley *et al.*, 2002
	5 g	clam tissue	n.d.	RT-PCR, Sequencing (HAV)	Goswami *et al.*, 2002
Mussels	25 g	Whole mussel (pool of at least 29 animals)	4°C	RT-PCR, Semi-nested PCR, Sequencing	Di Pinto *et al.*, 2003
Produce					
Raspberries	10 g	Frozen berries in a sealed package	−20°C	RT-PCR; Nested PCR; Dot-blot hybridization, Sequencing	La Guyader *et al.*, 2004
Blueberries	100 g	Whole frozen blueberries.	−70°C	RT-PCR, Dot-blot hybridization, Sequencing	Calder *et al.*, 2003
Salads	Approximately 20 g		n.d.	RT-PCR (NoV)	Anderson *et al.*, 2001
Lettuce	Approximately 30 g	Salad made on site	4°C	RT-PCR, southern hybridization, nested PCR, sequencing	Daniels *et al.*, 2000

(*Continued*)

Table 3.3 Continued

Matrix	Sample size	Type of sample	Conservation	Detection method	Reference
Meat and ready-to-eat foods					
Ham Turkey Salami	Approximately 20 g	Collected from a university cafeteria during an outbreak investigation	4°C	RT-PCR Southern blot hybridization Sequencing	Schwab *et al.*, 2000
	Approximately 30 g	Sliced products from a delicatessen bar	4°C	RT-PCR Southern hybridization Nested PCR Sequencing	Daniels *et al.*, 2000
Soups, sauces, meat, vegetables Desserts Bread Sausages Cheese	25 g	Collected from a sanatorium kitchen (menu pools and individual samples)	n.d.	RT-semi-nested PCR, Sequencing	Mayr *et al.*, 2009
Dairy, meat, fruits and vegetables, grains	5 g	Outbreak investigation	−80°C	RT-PCR/Southern blot hybridization	Rutjes *et al.*, 2006a
Pork liver	100–200 mg	Several tissue specimens	−80°C	Nested RT-PCR Sequencing	Yazaki *et al.*, 2003
	250 mg	Three parts of inner liver were excised and homogenized manually	−80°C	RT-PCR/Southern blot hybridization Sequencing	Bouwknegt *et al.*, 2007

n.d.: not described. RT-qPCR: RT real-time PCR.

Another important factor is the ambiguous use of individual or pooled samples for foodstuffs. It directly affects both the representativeness and analytical sensitivity of the final results; pooling can affect the final results negatively, and can produce false negative results due to the simple mechanism of reducing the size of each individual sample used in the pool. Conversely, the use of individual samples can also affect the representativeness of the population studied, and a representative number of individual samples must be needed; however this could greatly increase both the cost and the time required for the analyses, and may be unfeasible in the field.

The period of time before starting the analysis, and the conditions of storage of the sample during that period are also other critical aspects in the virus analysis. It is of particular importance if complex foods are analysed, as the stability of the virus may be compromised. However, they are usually not rigorously addressed during sampling, and most studies do not provide the relevant details. Even where this information is provided, the lack of uniformity is again evident. Samples are sometimes stored frozen, refrigerated at 4°C, at room temperature or kept on ice.

3.3.2 Interpretation of test results: validity of molecular detection methods

One of the major differences between the study of the presence and enumeration of bacteria and that of viruses in food and in the environment is the availability of a 'gold standard' method for detection. Classical culture-based techniques are considered the gold standard for the detection of bacteria, but no accepted standard method exists for enteric viruses in foods. Consequently, the reliability of the results produced by molecular techniques is undermined by the lack of standard methods and the wide diversity of viruses, foods, assays and recovery efficiencies described. The lack of a defined and consensus standard method for detection and quantification of viruses is hindering and slowing the adaptation of quantitative viral risk assessment (QVRA) models for food and food environments. Therefore, the establishment and application of a common and validated method for virus detection would make a major contribution to the effective harmonization of QVRA studies.

3.3.3 Control suite for assessing the method performance

Molecular techniques, if used with the appropriate quality controls, could allow substantial progress in the control of viral contamination of foods. If these methods are to be used for monitoring of food supply chains for viruses, then it is vitally necessary that their analytical results can be reliably verified. Many matrices from the food supply chains most prone to virus contamination are complex, and can furthermore contain substances which can inhibit nucleic acid amplification. It is essential therefore that verification includes recognition of analyses where the method has failed to perform correctly, as this may mask the presence of a virus in a sample by a false negative interpretation of the absence of a signal. Incorrect performance can occur during the sample treatment or the assay, and failed methods can be identified by the use of a suite of controls to verify the performance of the complete analytical process. A recent review paper has described them in detail and discussed the significance of the results depending on the signal obtained from the different controls (D'Agostino *et al.*, 2011; see chapter 6 for detailed information). The two controls which are critical for molecular detection of viruses in food are sample process control (SPC) and internal amplification control (IAC), and will be discussed in depth in Chapter 6.

3.3.4 Infectious particles *versus* PCR genome equivalents: implications for public health

Virus infectivity is defined as the capacity of viruses to enter the host cell and exploit its resources to replicate and produce progeny infectious viral particles (Rodríguez *et al.*, 2009), which may lead to infection and subsequent disease in the human host. Therefore, the key information from a public health perspective is the number of virus particles with infective capacity present in food or water. Obviously, cell culture-based methods are the soundest methodologies for the estimation of the number of infective particles. However, as indicated above, there are no available culture models for the most significant enteric viruses. In these cases, only molecular methods are available, but although RT q-PCR is a quantitative and sensitive tool, it cannot distinguish between infective and non-infective viruses (Richards, 1999). This limits its usefulness for public health purposes. The ratio between genome equivalents (GE) and infectious particles has been reported to increase with time, and is strongly dependent upon water and climatic conditions and virus type.

Damage to the virus capsid may result in the loss of its capacity to protect the genome and its ability to replicate in the host. Consequently, the detection of an intact viral genome can be an indication that the virus capsid is still in good condition, protecting the genome from degradation. Determining the relationship between damage to the viral capsid and degradation of the viral genome can provide information that can be used to correlate the detection of the viral genome with the infectivity of the virus. Therefore, several strategies have been developed to adapt PCR to quantify infective virus particles (reviewed in Rodríguez *et al.*, 2009; Table 3.4).Two different approaches have been used: direct RT-PCR (Li *et al.*, 2002; Ma *et al.*, 1994; Simonet and Gantzer, 2006a,b) or the use a pre-PCR sample treatment coupled to a subsequent RT-PCR (Nuanualsuwan and Cliver, 2002, 2003).

In the first approach, several strategies have been followed: targeting the 5' non-translated region (NTR) of the viral genome by PCR or analysing a long target region (LTR) of the viral genome by RT-PCR. The 5' NTR has been reported as the most easily degraded region of the genome of the HAV upon exposure to chlorine and chlorine dioxide, which affect the secondary

Table 3.4 Molecular-based methods used for assessing viral infectivity

Treatment	Detection	Target virus	References
Proteinase and RNase	RT-PCR	*Feline calicivirus*, HAV, murine NoV, Poliovirus 1,	Nuanualsuwan and Cliver, 2002, 2003; Baert *et al.*, 2008
Proteinase and RNase	Real-time NASBA	Human NoV, Feline calicivirus	Lamhoujeb *et al.*, 2008, 2009
RNase protection assay	RT-qPCR	Human NoV, Feline calicivirus	Topping *et al.*, 2009
	5' NTR RT-PCR	HAV	Bhattacharya *et al.*, 2004; Li *et al.*, 2002, 2004
	Long target region (LTR) q-RT-PCR	HAV, Poliovirus 1, F-specific RNA phages	Li *et al.*, 2002; Simonet and Gantzer, 2006a, 2006b
Attachment to cell monolayer	RT-PCR	HAV, Poliovirus 1, Feline calicivirus	Nuanualsuwan and Cliver, 2003
Virus replication in cell culture (ICC: integrated cell culture)	RT-PCR	Human *adenovirus*, *Astrovirus*, *Enterovirus*, *Poliovirus*, *Rotavirus*, HAV, MS2	Blackmer *et al.*, 2000; Chapron *et al.*, 2000; Jiang *et al.*, 2004; Ko *et al.*, 2003, 2005; Lee and Kim, 2002; Lee and Jeong, 2004; Li *et al.*, 2009; Nuanualsuwan and Cliver, 2003; Reynolds *et al.*, 1996; Shieh *et al.*, 2008
Antibody capture	RT-PCR	HAV, Human NoV, *Poliovirus 1*, *Feline calicivirus*	Gilpatrick *et al.*, 2000; Myrmel *et al.*, 2000; Schwab *et al.*, 1996
Immunomagnetic separation	RT-qPCR	HAV	Abd El Galil *et al.*, 2004

Adapted from Rodríguez-Lázaro *et al.*, 2012.

structure of this region (Bhattacharya *et al.*, 2004; Li *et al.*, 2002). The lack of amplification of the 5' NTR is accompanied by the loss of virus infectivity in cell culture. Li *et al.* (2002) also found that the first 600 bases of the HAV genome containing the 5' NTR are more sensitive to chlorine degradation than the rest of the genome. Similar conclusions were reached for *poliovirus* genome degradation during exposureto chlorine dioxide using a q-PCR approach (Simonet and Gantzer, 2006a). Similarly, analyses of longer regions of the genome by RT-PCR can screen for damage in the genome that will eventually reduce its infectivity, as an intact genome is necessary for the virus to remain infectious (Rodríguez *et al.*, 2009). However, even though the longest targeted fragment (6989 bp) disappeared very rapidly after disinfection treatment with chlorine dioxide, reduction of *poliovirus* infectivity was underestimated by this approach (Simonet and Gantzer, 2006a).Similarly, Simonet and Gantzer (2006b) observed that fragment size alone could not be used as the sole judge of RNA damage among different viruses, as bacteriophage MS-2 exhibited greater resistance to UV than *poliovirus* for a similar fragment size. In a study by Pecson *et al.* (2011), a framework was developed to quantify damage to the entire genome based on the q-PCR amplification of smaller targets. Genome regions of MS2 showed heterogeneous sensitivities to UV_{254} treatment. It was demonstrated that the q-PCR-based framework can accurately estimate MS2 infectivity after UV_{254} inactivation if appropriate analyses are used.

In the second approach, an enzymatic treatment (ET) with proteases (mostly proteinase K) and nucleases (RNase or DNase) before PCR is used: if protein capsid integrity is affected, it is more susceptible to degradation by proteinases. Consequently, the virus genome would be released, and be more susceptible to nuclease degradation than capsid-enclosed nucleic acid (Nuanualsuwan and Cliver, 2003). This approach has been successfully used to determine the effectiveness of UV light and chlorine disinfection, and thermal treatment at 72°C in the inactivation of HAV, FCV, and *poliovirus* (Nuanualsuwan and Cliver, 2002). However, some controversy still exists over the completed correlation between the results obtained and infectivity. In a recent study in our lab (Diez-Valcarce *et al.*, 2011), we assessed the use of molecular methods – RT-qPCR and ET coupled to RT-qPCR – to quantify the infectivity of NoV (NoV) after application of various inactivating food-processing technologies. In general, RT-qPCR and ET-RT-qPCR gave significantly different ($p < 0.01$) results concerning the reduction in viral genome counts by all inactivation procedures and conditions used. These findings indicate that the ET prior to RT-qPCR has an effect on the estimation of the reduction of virus genome counts, and may eliminate genomes of affected virus particles. However, no correlation was found between the results obtained by ET-RT-qPCR and those obtained by cell culture. Therefore, the effect is presumably only partial, and not sufficient to allow accurate estimation of virus inactivation. Consequently, the results indicate that the quantification of virus genomes by PCR, regardless of prior ET, is not adequate for establishing virus inactivation and/or infectivity.

3.4 Conclusion

Monitoring the presence of enteric viruses in foods is a challenging task, and efforts to achieve it are growing. However, the principal task must be definitive international standardization to guarantee effective implementation in real-life analytical contexts. Ultimately, there needs to be a focused drive towards taking proven methods from the scientist's laboratory and implementing them in the analyst's laboratory. However, further developments are needed if routine analysis of viruses in food laboratories is expected to be done. Among the main issues that must be addressed are the development of rational and easy-to-use strategies for maximizing virus recovery and the assessment of the infectivity of the detected viruses. However, the absolute prerequisite for successful adoption of a molecular-based diagnostic methodology is international validation and subsequent standardization (Hoorfar and Cook, 2003). A good example of this international validation has been provided in the EU research project 'Integrated monitoring and control of foodborne viruses in European food supply chains' (KBBE 213178; VITAL; www.eurovital.org) (D'Agostino *et al.*, 2012).This work is a landmark in virus detection in foods as it represents the first internationally shared effort to evaluate the robustness of a PCR-based method for detection of an enteric virus in a niche food matrix, that is, raspberries. All the steps were clearly defined and different standard operating procedures (SOPs) were defined, not only for each analytical step (concentration, nucleic acid purification, and PCR) but also for how to prepare the artificial inoculation and use of the blind samples. In addition, the evaluated method incorporated a sample process control and an internal amplification control to verify its correct operation; the significance of the results was interpreted according to the signal obtained from the different controls (D'Agostino *et al.*, 2011).

3.5 References

ABD EL GALIL KH, EL SOKKARY MA, KHEIRA SM, SALAZAR AM, YATES MV, CHEN W and MULCHANDANI A (2004) Combined immunomagnetic separation-molecular beacon-reverse transcription-PCR assay for detection of hepatitis A virus from environmental samples. *Appl Environ Microbiol* **70**: 4371–4.

ABD EL GALIL KH, EL SOKKARY MA, KHEIRA SM, SALAZAR AM, YATES MV, CHEN W and MULCHANDANI A (2005) Real-time nucleic acid sequence-based amplification assay for detection of hepatitis A virus. *Appl Environ Microbiol* **71**: 7113–6.

AHN JM, RAYAMAJHI N, GYUN KANG S and SANG YOO H (2006) Comparison of real-time reverse transcriptase-polymerase chain reaction and nested or commercial reverse transcriptase-polymerase chain reaction for the detection of hepatitis E virus particle in human serum. *Diagn Microbiol Infect Dis* **56**: 269–74. Erratum in: *Diagn Microbiol Infect Dis* (2007) **57**: 353

ANDERSON AD, GARRETT VD, SOBEL J, MONROE SS, FANKHAUSER RL, SCHWAB KJ, BRESEE JS, MEAD PS, HIGGINS C, CAMPANA J and GLASS RI (2001) Outbreak Investigation Team. Multistate outbreak of Norwalk-like virus gastroenteritis associated with a common caterer. *Am J Epidemiol* **154**: 1013–19.

ANDREWS WH and HAMMACK TS (2003) *Food Sampling and Preparation of Sample Homogenate. Bacteriological Analytical Manual (BAM)*. US Food and Drug Administration. US Department of Health and Human Services, Washington, DC.

ATMAR RL (2006) Molecular methods of virus detection in foods. *Viruses in Foods. Food Microbiology and Food Safety Series* (GOYAL SM, ed.), pp. 121–49, Springer, New York.

ATMAR RL and ESTES MK (2001) Diagnosis of noncultivatable gastroenteritis viruses, the human caliciviruses. *Clin Microbiol Rev* **14**: 15–37.

BAERT L, UYTTENDAELE M and DEBEVERE J (2008) Evaluation of viral extraction methods on a broad range of ready-to-eat foods with conventional and real-time RT-PCR for Norovirus GII detection. *Int J Food Microbiol* **123**(1–2): 101–8.

BEURET C, BAUMGARTNER A and SCHLUEP J (2003) Virus-contaminated oysters: a three-month monitoring of oysters imported to Switzerland. *Appl Environ Microbiol* **69**: 2292–7.

BHATTACHARYA SS, KULKA M, LAMPEL KA, CEBULA TA and GOSWAMI BB (2004) Use of reverse transcription and PCR to discriminate between infectious and non-infectious hepatitis A virus. *J Virol Methods* **116**: 181–7.

BIDAWID S, FARBER JM and SATTAR SA (2000a) Rapid concentration and detection of hepatitis A virus from lettuce and strawberries. *J Virol Methods*, **88**(2): 175–85.

BLACKMER F, REYNOLDS KA, GERBA CP and PEPPER IL (2000) Use of integrated cell culture-PCR to evaluate the effectiveness of poliovirus inactivation by chlorine. *Appl Environ Microbiol* **66**: 2267–8.

BOOM R, SOL CJ, SALIMANS MM, JANSEN CL, WERTHEIM-VAN DILLEN PM and VAN DER NOORDAA J (1990) Rapid and simple method for purification of nucleic acids. *J Clin Microbiol* **28**(3): 495–503.

BOSCH A, SÁNCHEZ G, LE GUYADER F, VANACLOCHA H, HAUGARREAU L and PINTÓ RM (2001) Human enteric viruses in Coquina clams associated with a large hepatitis A outbreak. *Water Sci Technol* **43**: 61–5.

BOSCH A, SANCHEZ G, ABBASZADEGAN M, CARDUCCI A, GUIX S, LE GUYADER FS, NETSHIKWETA R, PINTÓ RM, VAN DER POEL W, RUTJES S, SANO D, RODRÍGUEZ-LÁZARO D, KOVAC K, TAYLOR MB, VAN ZYL W and SELLWOOD J (2011) Analytical methods for virus detection in water and food. *Food Anal Methods* **4**: 4–13.

BOUWKNEGT M, LODDER-VERSCHOOR F, VAN DER POEL W, RUTJES SA and DE RODA HUSMAN AM (2007) Hepatitis E virus RNA in commercial porcine livers in the Netherlands. *J Food Prot* **70**: 2889–95.

BOXMAN IL, TILBURG JJ, TE LOEKE NA, VENNEMA H, JONKER K, DE BOER E and KOOPMANS M (2006) Detection of noroviruses in shellfish in the Netherlands. *Int J Food Microbiol* **108**: 391–6.

BOXMAN IL, TILBURG JJ, te LOEKE NA, VENNEMA H, DE BOER E and KOOPMANS M (2007) An efficient and rapid method for recovery of norovirus from food associated with outbreaks of gastroenteritis. *J Food Prot* **70**(2): 504–8.

BOWER WA, NAINAN OV, HAN X and MARGOLIS HS (2000) Duration of viremia in hepatitis A virus infection. *J Infect Dis* **182**: 12–17.

BROOKS HA, GERSBERG RM and DHAR AK (2005) Detection and quantification of hepatitis A virus in seawater via real-time RT-PCR. *J Virol Methods* **127**: 109–18.

BULL RA, TU ET, MCIVER CJ, RAWLINSON WD and WHITE PA (2006) Emergence of a new norovirus genotype II.4 variant associated with global outbreaks of gastroenteritis. *J Clin Microbiol* **44**: 327–33.

BUTOT S, PUTALLAZ T and SÁNCHEZ G (2007) Procedure for rapid concentration and detection of enteric viruses from berries and vegetables. *Appl Environ Microbiol* **73**(1): 186–92.

CALDER L, SIMMONS G, THORNLEY C, TAYLOR P, PRITCHARD K, GREENING G and BISHOP J (2003) An outbreak of hepatitis A associated with consumption of raw blueberries. *Epidemiol Infect* **131**: 745–51.

CASAS N, AMARITA F and DE MARAÑÓN IM (2007) Evaluation of an extracting method for the detection of Hepatitis A virus in shellfish by SYBR-Green real-time RT-PCR. *Int J Food Microbiol* **120**(1–2): 179–85.

CHAPRON CD, BALLESTER NA, FONTAINE JH, FRADES CN and MARGOLIN AB (2000) Detection of astroviruses, enteroviruses, and adenovirus types 40 and 41 in surface waters collected and evaluated by the information collection rule and an integrated cell culture-nested PCR procedure. *Appl Environ Microbiol* **66**: 2520–5.

CLIVER DO (2008) Historic overview of food virology. *Foodborne Viruses: Progress and Challenges* (KOOPMANS MPG, CLIVER DO and BOSCH A, eds). pp. 1–28. ASM press, Washington, DC.

COELHO C, HEINERT AP, SIMÕES CM and BARARDI CR (2003) Hepatitis A virus detection in oysters (*Crassostreagigas*) in Santa Catarina State, Brazil, by reverse transcription-polymerase chain reaction. *J Food Prot* **66**: 507–11.

COLSON P, COZE C, GALLIAN P, HENRY M, DE MICCO P and TAMALET C (2007) Transfusion-associated hepatitis E, France. *Emerg Infect Dis* **13**: 648–9.

COMELLI HL, RIMSTAD E, LARSEN S and MYRMEL M (2008) Detection of norovirus genotype I.3b and II.4 in bioaccumulated blue mussels using different virus recovery methods. *Int J Food Microbiol* **127**(1–2): 53–9.

COMPTON J (1991) Nucleic acid sequence-based amplification. *Nature* **350**: 91–2.

COSTA-MATTIOLI M, MONPOEHO S, NICAND E, ALEMAN MH, BILLAUDEL S and FERRÉ V (2002) Quantification and duration of viraemia during hepatitis A infection as determined by real-time RT-PCR. *J Viral Hepat* **9**: 101–6.

COSTAFREDA MI, BOSCH A and PINTÓ RM (2006) Development, evaluation, and standardization of a real-time TaqMan reverse transcription-PCR assay for quantification of hepatitis A virus in clinical and shellfish samples. *Appl Environ Microbiol* **72**: 3846–55.

COSTANTINI V, LOISY F, JOENS L, LE GUYADER FS and SAIF LJ (2006) Human and animal enteric caliciviruses in oysters from different coastal regions of the United States. *Appl Environ Microbiol* **72**: 1800–9.

CROCI L, DUBOIS E, COOK N, DE MEDICI D, SCHULTZ AC, CHINA B, RUTJES SA, HOORFAR J, VAN DER POEL WHM (2008) Current methods for extraction and concentration of enteric viruses from fresh fruit and vegetables: towards international standards. *Food Anal Methods* **1**(2): 73–84.

D'AGOSTINO M, COOK N, RODRÍGUEZ-LÁZARO D and RUTJES S (2011) Nucleic acid amplification-based methods for detection of enteric viruses: definition of controls and interpretation of results. *Food Environ Virol* **3**: 55–60.

D'AGOSTINO M, COOK C, DI BARTOLO I, RUGGERI FM, MARTELLI F, BANKS M, VASICKOVA P, KRALIK P, PAVLIK I, KOKKINOS P, VANTARAKIS A, SÖDERBERG K, MAUNULA L, VERHAELEN K, RUTJES S, DE RODAHUSMAN AM, HAKZE R, VAN DER POEL W, KOZYRA I, RZEŻUTKA A, PRODANOV J, LAZIC S, PETROVIC T, CARRATALA A, GIRONÉS R, DIEZ-VALCARCE M, HERNÁNDEZ M and RODRÍGUEZ-LÁZARO D (2012) Multicenter collaborative trial evaluation of a method for detection of human adenoviruses in berry fruit. *Food Anal Methods* **5**: 1–7.

DANIELS NA, BERGMIRE-SWEAT DA, SCHWAB KJ, HENDRICKS KA, REDDY S, ROWE SM, FANKHAUSER RL, MONROE SS, ATMAR RL, GLASS RI and MEAD P (2000) A foodborne outbreak of gastroenteritis associated with Norwalk-like viruses: first molecular traceback to deli sandwiches contaminated during preparation. *J Infect Dis* **181**: 1467–70.

DE MEDICI D, CROCI L, DI PASQUALE S, FIORE A and TOTI L (2001) Detecting the presence of infectious hepatitis A virus in molluscs positive to RT-nested-PCR. *Lett App Microbiol* **33**: 362–6.

DE PAULA VS, BAPTISTA ML, LAMPE E, NIEL C and GASPAR AM (2002) Characterization of hepatitis A virus isolates from subgenotypes IA and IB in Rio de Janeiro, Brazil. *J Med Virol* **66**: 22–7.

DE RODA HUSMAN AM, LODDER-VERSCHOOR F, VAN DEN BERG HH, LE GUYADER FS, VAN PELT H, VAN DER POEL WH and RUTJES SA (2007) Rapid virus detection procedure for molecular tracing of shellfish associated with disease outbreaks. *Food Protect*, **70**(4): 967–74.

DI PASQUALE S, PANICONI M, DE MEDICI D, SUFFREDINI E and CROCI L (2010) Duplex real time pcr for the detection of hepatitis a virus in shellfish using feline calicivirus as a process control. *J Virol Methods* **163**: 96–100.

DI PINTO A, FORTE VT, TANTILLO GM, TERIO V and BUONAVOGLIA C (2003) Detection of hepatitis A virus in shellfish (*Mytilus galloprovincialis*) with RT-PCR. *J Food Prot* **66**: 1681–5.

DIEZ-VALCARCE M, KOVAČ K, RODRÍGUEZ-LÁZARO D and HERNÁNDEZ M (2011) Virus genome quantification does not predict norovirus infectivity after application of food inactivation processing technologies. *Food Environ Virol* **3**: 141–6.

DUBOIS E, AGIER C, TRAORÉ O, HENNECHART C, MERLE G, CRUCIÈRE C and LAVERAN H (2002) Modified concentration method for the detection of enteric viruses on fruits and vegetables by reverse transcriptase-polymerase chain reaction or cell culture. *J Food Prot* **65**(12): 1962–9.

DUIZER E, SCHWAB KJ, NEILL FH, ATMAR RL, KOOPMANS MP and ESTES MK (2004) Laboratory efforts to cultivate noroviruses. *J Gen Virol* **85**(Pt 1): 79–87.

ELAMRI DE, AOUNI M, PARNAUDEAU S and LE GUYADER FS (2006) Detection of human enteric viruses in shellfish collected in Tunisia. *Lett Appl Microbiol* **43**: 399–404.

ENOUF V, DOS REIS G, GUTHMANN JP, GUERIN PJ, CARON M, MARECHAL V and NICAND E (2006) Validation of single real-time TaqMan PCR assay for the detection and quantitation of four major genotypes of hepatitis E virus in clinical specimens. *J Med Virol* **78**: 1076–82.

FOOD STANDARDS AGENCY (2004a) *Practical Sampling Guidance for Food Standards and Feeding Stuffs. Part 1: Overall Objectives of Sampling*. Food Standards Agency, United Kingdom.

FOOD STANDARDS AGENCY (2004b) *Practical Sampling Guidance for Food Standards and Feeding Stuffs. Part 2: Food Standards Sampling*. Food Standards Agency, United Kingdom.

FORMIGA-CRUZ M, TOFIÑO-QUESADA G, BOFILL-MAS S, LEES DN, HENSHILWOOD K, ALLARD AK, CONDEN-HANSSON AC, HERNROTH BE, VANTARAKIS A, TSIBOUXI A, PAPAPETROPOULOU M, FURONES MD and GIRONES R (2002) Distribution of human virus contamination in shellfish from different growing areas in Greece, Spain, Sweden, and the United Kingdom. *Appl Environ Microbiol* **68**: 5990–8.

FUJIWARA K, YOKOSUKA O, FUKAI K, IMAZEKI F, SAISHO H and OMATA M (2001) Analysis of full-length hepatitis A virus genome in sera from patients with fulminant and self-limited acute type A hepatitis. *J Hepatol* **35**: 112–19.

FUKUDA S, SASAKI Y, KUWAYAMA M and MIYAZAKI K (2007) Simultaneous detection and genogroup-screening test for norovirus genogroups I and II from fecal specimens in single tube by reverse transcription loop-mediated isothermal amplification assay. *Microbiol Immunol* **51**: 547–50.

FUMIAN TM, LEITE JP, MARIN VA and MIAGOSTOVICH MP (2009) A rapid procedure for detecting noroviruses from cheese and fresh lettuce. *J Virol Methods* **155**(1): 39–43.

GILPATRICK SG, SCHWAB KJ, ESTES MK and ATMAR RL (2000) Development of an immunomagnetic capture reverse transcription-PCR assay for the detection of Norwalk virus. *J Virol Methods* **90**: 69–78.

GOSWAMI WW (2001) *Detection and Quantitation of Hepatitis A Virus in Shellfish by the Polymerase Chain Reaction. Bacteriological Analytical Manual (BAM)*. US Food and Drug Administration. US Department of Health and Human Services.

GOSWAMI BB, KULKA M, NGO D, ISTAFANOS P and CEBULA TA (2002) A polymerase chain reaction-based method for the detection of hepatitis A virus in produce and shellfish. *J Food Prot* **65**: 393–402.

GUÉVREMONT E, BRASSARD J, HOUDE A, SIMARD C and TROTTIER YL (2006) Development of an extraction and concentration procedure and comparison of RT-PCR primer systems for the detection of hepatitis A virus and norovirus GII in green onions. *J Virol Methods* **134**(1–2): 130–5.

GYARMATI P, MOHAMMED N, NORDER H, BLOMBERG J, BELÁK S and WIDÉN F (2007) Universal detection of hepatitis E virus by two real-time PCR assays: TaqMan and Primer-Probe Energy Transfer. *J Virol Methods* **146**: 226–35.

HARAMOTO E, KATAYAMA H, OGUMA K and OHGAKI S (2005) Application of cation-coated filter method to detection of noroviruses, enteroviruses, adenoviruses, and torque teno viruses in the Tamagawa River in Japan. *Appl Environ Microbiol* **71**(5):2403–11.

HEID CA, STEVENS J, LIVAK KJ and WILLIAMS PM (1996) Real time quantitative PCR. *Genome Res* **6**(10): 986–94.

HERREMANS M, VENNEMA H, BAKKER J, VAN DER VEER B, DUIZER E, BENNE CA, WAAR K, HENDRIXKS B, SCHNEEBERGER P, BLAAUW G, KOOIMAN M and KOOPMANS MP (2007) Swine-like hepatitis E viruses are a cause of unexplained hepatitis in the Netherlands. *J Viral Hepat* **14**: 140–6.

HOORFAR J and COOK N (2003) Critical aspects of standardization of PCR. *Methods Mol Biol* **216**: 51–64.

HOUDE A, GUÉVREMONT E, POITRAS E, LEBLANC D, WARD P, SIMARD C and TROTTIER YL (2007) Comparative evaluation of new TaqMan real-time assays for the detection of hepatitis A virus. *J Virol Methods* **140**: 80–9.

HU Y and ARSOV I (2009) Nested real-time PCR for hepatitis A detection. *Lett Appl Microbiol* **49**: 615–19.

HUANG FF, HAQSHENAS G, GUENETTE DK, HALBUR PG, SCHOMMER SK, PIERSON FW, TOTH TE and MENG XJ (2002) Detection by reverse transcription-PCR and genetic characterization of field isolates of swine hepatitis E virus from pigs in different geographic regions of the United States. *J Clin Microbiol* **40**: 1326–32.

ICMSF (1986) *Microorganisms in Foods 2: Sampling for Microbiological Analysis: Principles and Specific Applications*. University of Toronto Press, Toronoto, Canada.

ICMSF (2002) *Microorganisms in Foods 7: Microbiological Testing in Food Safety Management*. Kluwer Academic/Plenum Publishers, New York.

INOUE J, TAKAHASHI M, YAZAKI Y, TSUDA F and OKAMOTO H (2006) Development and validation of an improved RT-PCR assay with nested universal primers for detection of hepatitis E virus strains with significant sequence divergence. *J Virol Methods* **137**: 325–33.

ISHIDA S, YOSHIZUMI S, IKEDA T, MIYOSHI M, OKANO M and OKUI T (2008) Sensitive and rapid detection of norovirus using duplex TaqMan reverse transcription-polymerase chain reaction. *J Med Virol* **80**: 913–20.

JIANG YJ, LIAO GY, ZHAO W, SUN MB, QIAN Y, BIAN CX and JIANG SD (2004) Detection of infectious hepatitis A virus by integrated cell culture/strand-specific reverse transcriptase-polymerase chain reaction. *J Appl Microbiol* **97**: 1105–12.

JONES TH, BRASSARD J, JOHNS MW and GAGNÉ MJ (2009) The effect of pre-treatment and sonication of centrifugal ultrafiltration devices on virus recovery. *J Virol Methods* **161**(2): 199–204.

JOTHIKUMAR N, CROMEANS TL, ROBERTSON BH, MENG XJ and HILL VR (2005a) A broadly reactive one-step real-time RT-PCR assay for rapid and sensitive detection of hepatitis E virus. *J Virol Methods* **131**: 65–71.

JOTHIKUMAR N, CROMEANS TL, SOBSEY MD and ROBERTSON BH (2005b) Development and evaluation of a broadly reactive TaqMan assay for rapid detection of hepatitis A virus. *Appl Environ Microbiol* **71**: 3359–63.

JOTHIKUMAR N, LOWTHER JA, HENSHILWOOD K, LEES DN, HILL VR and VINJÉ J (2005c) Rapid and sensitive detection of noroviruses by using TaqMan-based one-step reverse transcription-PCR assays and application to naturally contaminated shellfish samples. *Appl Environ Microbiol* **71**(4): 1870–75.

KABA M, DAVOUST B, MARIÉ JL, BARTHET M, HENRY M, TAMALET C, RAOULT D and COLSON P (2009) Frequent transmission of hepatitis E virus among piglets in farms in Southern France. *J Med Virol* **81**: 1750–9.

KAGEYAMA T, KOJIMA S, SHINOHARA M, UCHIDA K, FUKUSHI S, HOSHINO FB, TAKEDA N and KATAYAMA K (2003) Broadly reactive and highly sensitive assay for Norwalk-like viruses based on real-time quantitative reverse transcription-PCR. *J Clin Microbiol* **41**: 1548–57.

KARAMOKO Y, IBENYASSINE K, AITMHAND R, IDAOMAR M and ENNAJI MM (2005) Adenovirus detection in shellfish and urban sewage in Morocco (Casablanca region) by the polymerase chain reaction. *J Virol Methods* **126**: 135–7.

KATAYAMA H, SHIMASAKI A and OHGAKI S (2002) Development of a virus concentration method and its application to detection of enterovirus and Norwalk virus from coastal seawater. *Appl Environ Microbiol* **68**(3): 1033–9.

KAWAMOTO H, YAMAZAKI K, UTAGAWA E and OHYAMA T (2001) Nucleotide sequence analysis and development of consensus primers of RT-PCR for detection of Norwalk-like viruses prevailing in Japan. *J Med Virol* **64**: 569–76.

KIM HY, KWAK IS, HWANG IG and KO G (2008) Optimization of methods for detecting norovirus on various fruit. *J Virol Methods* **153**(2): 104–10.

KINGSLEY DH, MEADE GK and RICHARDS GP (2002) Detection of both hepatitis A virus and Norwalk-like virus in imported clams associated with foodborne illness. *Appl Environ Microbiol* **68**: 3914–18.

KO G, CROMEANS TL and SOBSEY MD (2003) Detection of infectious adenovirus in cell culture by mRNA reverse transcription-PCR. *Appl Environ Microbiol* **69**: 7377–84.

KO G, CROMEANS TL and SOBSEY MD (2005) UV inactivation of adenovirus type 41 measured by cell culture mRNA RT-PCR. *Water Res* **39**: 3643–9.

KOJIMA S, KAGEYAMA T, FUKUSHI S, HOSHINO FB, SHINOHARA M, UCHIDA K, NATORI K, TAKEDA N and KATAYAMA K (2002) Genogroup-specific PCR primers for detection of Norwalk-like viruses. *J Virol Methods* **100**: 107–14.

KURDZIEL AS, WILKINSON N, LANGTON S and COOK N (2001) Survival of poliovirus on soft fruit and salad vegetables. *J Food Prot* **64**(5): 706–9.

LAMHOUJEB S, FLISS I, NGAZOA SE and JEAN J (2008) Evaluation of the persistence of infectious human noroviruses on food surfaces by using real-time nucleic acid sequence-based amplification. *Appl Environ Microbiol* **74**: 3349–55.

LAMHOUJEB SFI, NGAZOA SE and JEAN J (2009) Molecular study of the persistence of infectious human norovirus on food-contact surfaces. *Food Environ Virol* **1**: 51–6.

LA ROSA G, FONTANA S, DI GRAZIA A, IACONELLI M, POURSHABAN M and MUSCILLO M (2007) Molecular identification and genetic analysis of Norovirus genogroups I and II in water environments: comparative analysis of different reverse transcription-PCR assays. *Appl Environ Microbiol* **73**: 4152–61. Erratum in: *Appl Environ Microbiol* (2007) **73**: 6329.

LE GUYADER F, HAUGARREAU L, MIOSSEC L, DUBOIS E and POMMEPUY M (2000) Three-year study to assess human enteric viruses in shellfish. *Appl Environ Microbiol* **66**: 3241–8.

LE GUYADER FS, BON F, DEMEDICI D, PARNAUDEAU S, BERTONE A, CRUDELI S, DOYLE A, ZIDANE M, SUFFREDINI E, KOHLI E, MADDALO F, MONINI M, GALLAY A, POMMEPUY M, POTHIER P and RUGGERI FM (2006) Detection of multiple noroviruses associated with an international gastroenteritis outbreak linked to oyster consumption. *J Clin Microbiol* **44**: 3878–82.

LE GUYADER FS, LE SAUX JC, AMBERT-BALAY K, KROL J, SERAIS O, PARNAUDEAU S, GIRAUDON H, DELMAS G, POMMEPUY M, POTHIER P and ATMAR RL (2008) Aichi virus, norovirus, astrovirus, enterovirus, and rotavirus involved in clinical cases from a French oyster-related gastroenteritis outbreak. *J Clin Microbiol* **46**: 4011–17.

LE GUYADER FS, MITTELHOLZER C, HAUGARREAU L, HEDLUND KO, ALSTERLUND R, POMMEPUY M and SVENSSON L (2004) Detection of noroviruses in raspberries associated with a gastroenteritis outbreak. *Int J Food Microbiol* **97**: 179–86.

LE GUYADER FS, NEILL FH, DUBOIS E, BON F, LOISY F, KOHLI E, POMMEPUY M and ATMAR RL (2003) A semiquantitative approach to estimate Norwalk-like virus contamination of oysters implicated in an outbreak. *Int J Food Microbiol* **87**: 107–12.

LE GUYADER FS, PARNAUDEAU S, SCHAEFFER J, BOSCH A, LOISY F, POMMEPUY M and ATMAR RL (2009) Detection and quantification of noroviruses in shellfish. *Appl Environ Microbiol* **75**: 618–24.

LEE HK and JEONG YS (2004) Comparison of total culturable virus assay and multiplex integrated cell culture-PCR for reliability of waterborne virus detection. *Appl Environ Microbiol* **70**: 3632–6.

LEE SH and KIM SJ (2002) Detection of infectious enteroviruses and adenoviruses in tap water in urban areas in Korea. *Water Res* **36**: 248–56.

LEES D and CEN WG6 TAG4 (2010) International standardisation of a method for detection of human pathogenic viruses in molluscan shellfish. *Food Environ Virol* **2**: 146–55.

LI D, GU AZ, HE M, SHI H-C and YANG W (2009) UV inactivation and resistance of rotavirus evaluated by integrated cell culture and real-time RT-PCR assay. *Water Res* **43**: 3261–9.

LI TC, CHIJIWA K, SERA N, ISHIBASHI T, ETOH Y, SHINOHARA Y, KURATA Y, ISHIDA M, SAKAMOTO S, TAKEDA N and MIYAMURA T (2005) Hepatitis E virus transmission from wild boar meat. *Emerg Infect Dis* **11**: 1958–60.

LI JW, XIN ZT, WANG XW, ZHENG JL and CHAO FH (2002) Mechanisms of inactivation of hepatitis A virus by chlorine. *Appl Environ Microbiol* **68**(10): 4951–5.

LI JW, XIN ZT, WANG XW, ZHENG JL and CHAO FH (2004) Mechanisms of inactivation of hepatitis A virus in water by chlorine dioxide. *Water Res* **38**: 1514–19.

LOISY F, ATMAR RL, GUILLON P, LE CANN P, POMMEPUY M and LE GUYADER FS (2005) Real-time RT-PCR for norovirus screening in shellfish. *J Virol Methods* **123**: 1–7.

LOVE DC, CASTEEL MJ, MESCHKE JS and SOBSEY MD (2008) Methods for recovery of hepatitis A virus (HAV) and other viruses from processed foods and detection of HAV by nested RT-PCR and TaqMan RT-PCR. *Int J Food Microbiol* **126**(1–2): 221–6.

MA JF, STRAUB TM, PEPPER IL and GERBA CP (1994) Cell culture and PCR determination of poliovirus inactivation by disinfectants. *Appl Environ Microbiol* **60**(11): 4203–6.

MATTISON K and BIDAWID S (2009) Analytical methods for food and environmental viruses. *Food Environ Virol* **1**(3–4): 107–22.

MAYR C, STROHE G and CONTZEN M (2009) Detection of rotavirus in food associated with a gastroenteritis outbreak in a mother and child sanatorium. *Int J Food Microbiol* **135**: 179–82.

MEDICI MC, MARTINELLI M, RUGGERI FM, ABELLI LA, BOSCO S, ARCANGELETTI MC, PINARDI F, DE CONTO F, CALDERARO A, CHEZZI C and DETTORI G (2005) Broadly reactive nested reverse transcription-PCR using an internal RNA standard control for detection of noroviruses in stool samples. *J Clin Microbiol* **43**: 3772–8.

MIZUO H, SUZUKI K, TAKIKAWA Y, SUGAI Y, TOKITA H, AKAHANE Y, ITOH K, GOTANDA Y, TAKAHASHI M, NISHIZAWA T and OKAMOTO H (2002) Polyphyletic strains of hepatitis E virus are responsible for sporadic cases of acute hepatitis in Japan. *J Clin Microbiol* **40**: 3209–18.

MORTON V, JEAN J, FARBER J and MATTISON K (2009) Detection of noroviruses in ready-to-eat foods by using carbohydrate-coated magnetic beads. *Appl Environ Microbiol* **75**(13): 4641–3.

MYRMEL M, RIMSTAD E and WASTESON Y (2000) Immunomagnetic separation of a Norwalk-like virus (genogroup I) in artificially contaminated environmental water samples. *Int J Food Microbiol* **62**: 17–26.

NENONEN NP, HANNOUN C, OLSSON MB and BERGSTRÖM T (2009) Molecular analysis of an oyster-related norovirus outbreak. *J Clin Virol* **45**: 105–8.

NUANUALSUWAN S and CLIVER DO (2002) Pretreatment to avoid positive RT-PCR results with inactivated viruses. *J Virol Methods* **104**(2): 217–25.

NUANUALSUWAN S and CLIVER DO (2003) Capsid functions of inactivated human picornaviruses and feline calicivirus. *Appl Environ Microbiol* **69**(1): 350–7.

NISHIDA T, KIMURA H, SAITOH M, SHINOHARA M, KATO M, FUKUDA S, MUNEMURA T, MIKAMI T, KAWAMOTO A, AKIYAMA M, KATO Y, NISHI K, KOZAWA K and NISHIO O (2003) Detection, quantitation, and phylogenetic analysis of noroviruses in Japanese oysters. *Appl Microbiol* **69**: 5782–6.

NISHIDA T, NISHIO O, KATO M, CHUMA T, KATO H, IWATA H and KIMURA H (2007) Genotyping and quantitation of noroviruses in oysters from two distinct sea areas in Japan. *Microbiol Immunol* **51**: 177–84.

ORRÙ G, MASIA G, ORRÙ G, ROMANÒ L, PIRAS V and COPPOLA RC (2004) Detection and quantitation of hepatitis E virus in human faeces by real-time quantitative PCR. *J Virol Methods* **118**: 77–82.

PANG XL, PREIKSAITIS JK and LEE B (2005) Multiplex real time RT-PCR for the detection and quantitation of norovirus genogroups I and II in patients with acute gastroenteritis. *J Clin Virol* **33**: 168–71.

PAPAFRAGKOU E, PLANTE M, MATTISON K, BIDAWID S, KARTHIKEYAN K, FARBER JM and JAYKUS LA (2008) Rapid and sensitive detection of hepatitis A virus in representative food matrices. *J Virol Methods* **147**(1): 177–87.

PECSON BM, ACKERMANN M and KOHN T (2011) Framework for using quantitative PCR as a nonculture based method to estimate virus infectivity. *Environ Sci Technol* **45**: 2257–63.

PERELLE S, CAVELLINI L, BURGER C, BLAISE-BOISSEAU S, HENNECHART-COLLETTE C, MERLE G and FACH P (2009) Use of a robotic RNA purification protocol based on the Nucli Senseasy MAG for real-time RT-PCR detection of hepatitis A virus in bottled water. *J Virol Methods* **157**(1): 80–3.

PINTÓ RM and BOSCH A (2008) Rethinking virus detection in food. *Foodborne Viruses: Progress and challenges* (KOOPMANS M, CLIVER DO and BOSCH A eds.) pp. 171–88, ASM Press, Washington, DC, USA.

REYNOLDS KA, GERBA CP and PEPPER IL (1996) Detection of infectious enteroviruses by an integrated cell culture-PCR procedure. *Appl Environ Microbiol* **62**: 1424–7.

RICHARDS GP (1999) Limitations of molecular biological techniques for assessing the virological safety of foods. *J Food Prot* **62**: 691–7.

RODRÍGUEZ RA, PEPPER IL and GERBA CP (2009) Application of PCR-based methods to assess the infectivity of enteric viruses in environmental samples. *Appl Environ Microbiol* **75**(2): 297–307.

RODRÍGUEZ-LÁZARO D, LOMBARD B, SMITH HV, RZEŻUTKA A, D'AGOSTINO M, HELMUTH R, SCHROETER A, MALORNY B, MIKO A, GUERRA B, DAVISON J, KOBILINSKY A, HERNÁNDEZ M, BERTHEAU Y and COOK N (2007) Trends in analytical methodology in food safety and quality: monitoring microorganisms and genetically modified organisms. *Trends Food Sci Technol* **18**: 306–19.

RODRÍGUEZ-LÁZARO D, COOK N, D'AGOSTINO M and HERNANDEZ M (2009) Current challenges in molecular diagnostics in food microbiology. *Global Issues in Food Science and Technology* (BARBOSA-CÁNOVAS G, MORTIMER A, COLONNA P, LINEBACK D, SPIESS W and BUCKLE K eds.) Elsevier, Burlington, USA, 211–228.

RODRÍGUEZ-LÁZARO D, RUGGERI FM, SELLWOOD J, NASSER A, SAO JOSE NASCIMENTO M, D'AGOSTINO M, SANTOS R, SAIZ JC, RZEŻUTKA A, BOSCH A, GIRONÉS R, CARDUCCI A, MUSCILLO M, KOVAC K, DIEZ-VALCARCE M, VANTARAKIS A, COOK N, VON BONSDORFF CH, HERNÁNDEZ M and VAN DER POEL WHM (2012) Virus hazards from food and the environment. *FEMS Microbiol Rev* **36**: 786–814.

ROMALDE JL, AREA E, SÁNCHEZ G, RIBAO C, TORRADO I, ABAD X, PINTÓ RM, BARJA JL and BOSCH A (2002) Prevalence of enterovirus and hepatitis A virus in bivalve molluscs from Galicia (NW Spain): inadequacy of the EU standards of microbiological quality. *Int J Food Microbiol* **74**: 119–30.

RUTJES SA, ITALIAANDER R, VAN DEN BERG HH, LODDER WJ and DE RODA HUSMAN AM (2005) Isolation and detection of enterovirus RNA from large-volume water samples by using the NucliSens mini MAG system and real-time nucleic acid sequence-based amplification. *Appl Environ Microbiol* **71**(7): 3734–40.

RUTJES SA, LODDER-VERSCHOOR F, VAN DER POEL WH, VAN DUIJNHOVEN YT and DE RODA HUSMAN AM (2006a) Detection of noroviruses in foods: a study on virus extraction procedures in foods implicated in outbreaks of human gastroenteritis. *J Food Prot* **69**(8): 1949–56.

RUTJES SA, VAN DEN BERG HH, LODDER WJ and DE RODA HUSMAN AM (2006b) Real-time detection of noroviruses in surface water by a broadly reactive nucleic acid based amplification assay. *Appl Environ Microbiol* **72**: 5349–58.

RZEŻUTKA A, D'AGOSTINO M and COOK N (2006) An ultracentrifugation-based approach to the detection of hepatitis A virus in soft fruits. *Int J Food Microbiol* **108**(3): 315–20.

RZEŻUTKA A, CHROBOCINSKA M, KAUPKE A and MIZAK B (2008) Application of anultracentrifugation-based method for detection of feline calicivirus (a norovirus surrogate) in experimentally contaminated delicatessen meat samples. *Food Anal Methods* **1**(1): 56–60.

SÁNCHEZ G, PINTÓ RM, VANACLOCHA H and BOSCH A (2002) Molecular characterization of hepatitis A virus isolates from a transcontinental shellfish-borne outbreak. *J Clin Microbiol* **40**: 4148–55.

SCHWAB KJ, DELEON R and SOBSEY MD (1996) Immunoaffinity concentration and purification of waterborne enteric viruses for detection by reverse transcriptase PCR. *Appl Environ Microbiol* **62**: 2086–94.

SCHWAB KJ, NEILL FH, ESTES MK, METCALF TG and ATMAR RL (1998) Distribution of Norwalk virus within shellfish following bioaccumulation and subsequent depuration by detection using RT-PCR. *J Food Prot* **61**: 1674–80.

SCHWAB KJ, NEILL FH, FANKHAUSER RL, DANIELS NA, MONROE SS, BERGMIRE-SWEAT DA, ESTES MK and ATMAR RL (2000) Development of methods to detect "Norwalk-like viruses" (NLVs) and hepatitis A virus in delicatessen foods: application to a food-borne NLV outbreak. *Appl Environ Microbiol* **66**: 213–18.

SCHMITTGEN TD, ZAKRAJSEK BA, MILLS AG, GORN V, SINGER MJ and REED MW (2000) Quantitative reverse transcription-polymerase chain reaction to study mRNA decay: comparison of endpoint and real-time methods. *Anal Biochem* **285**: 194–204.

SEMINATI C, MATEU E, PERALTA B, DE DEUS N and MARTIN M (2008) Distribution of hepatitis E virus infection and its prevalence in pigs on commercial farms in Spain. *Vet J* **175**: 130–2.

SHIEH YC, KHUDYAKOV YE, XIA G, GANOVA-RAEVA LM, KHAMBATY FM, WOODS JW, VEAZEY JE, MOTES ML, GLATZER MB, BIALEK SR and FIORE AE (2007) Molecular confirmation of oysters as the vector for hepatitis A in a 2005 multistate outbreak. *J Food Prot* **70**: 145–50.

SHIEH YC, WONG CI, KRANTZ JA and HSU FC (2008) Detection of naturally occurring enteroviruses in waters using direct RT-PCR and integrated cell culture-RT-PCR. *J Virol Methods* **149**: 184–9.

SILVA AM, VIEIRA H, MARTINS N, GRANJA AT, VALE MJ and VALE FF (2009) Viral and bacterial contamination in recreational waters: a case study in the Lisbon bay area. *J Appl Microbiol* **108**: 1023–31.

SIMONET J and GANTZER C (2006a) Degradation of the Poliovirus 1 genome by chlorine dioxide. *J Appl Microbiol* **100**(4): 862–70.

SIMONET J and GANTZER C (2006b) Inactivation of poliovirus 1 and F-specific RNA phages and degradation of their genomes by UV irradiation at 254 nanometers. *Appl Environ Microbiol* **72**(12): 7671–7.

STALS A, BAERT L, VAN COILLIE E and Uyttendaele M (2011a) Evaluation of a norovirus detection methodology for soft red fruits. *Food Microbiol* **28**(1): 52–8.

STALS A, BAERT L, DE KEUCKELAERE A, VAN COILLIE E and UYTTENDAELE M (2011b) Evaluation of a norovirus detection methodology for ready-to-eat foods. *Int J Food Microbiol* **145**(2–3): 420–5.

STALS A, BAERT L, VAN COILLIE E and UYTTENDAELE M (2012) Extraction of foodborne viruses from food samples: a review. *Int J Food Microbiol* **153**(1–2): 1–9.

TIAN P, ENGELBREKTSON A and MANDRELL R (2008) Two-log increase in sensitivity for detection of norovirus in complex samples by concentration with porcine gastric mucin conjugated to magnetic beads. *Appl Environ Microbiol* **74**(14): 4271–6.

TOPPING JR, SCHNERR H, HAINES J, SCOTT M, CARTER MJ, WILLCOCKS MM, BELLAMY K, BROWN DW, GRAY JJ, GALLIMORE CI and KNIGHT AI (2009) Temperature inactivation of Feline calicivirus vaccine strain FCV F-9 in comparison with human noroviruses using an RNA exposure assay and reverse transcribed quantitative real-time polymerase chain reaction-A novel method for predicting virus infectivity. *J Virol Methods* **156**: 89–95.

VAN DER POEL WH, VERSCHOOR F, VAN DER HEIDE R, HERRERA MI, VIVO A, KOOREMAN M and DE RODAHUSMAN AM (2001) Hepatitis E virus sequences in swine related to sequences in humans, The Netherlands. *Emerg Infect Dis* **7**: 970–6.

VENNEMA H, DE BRUIN E and KOOPMANS M (2002) Rational optimization of generic primers used for Norwalk-like virus detection by reverse transcriptase polymerase chain reaction. *J Clin Virol* **25**: 233–5.

VILLAR LM, DE PAULA VS, DINIZ-MENDES L, LAMPE E and GASPAR AM (2006) Evaluation of methods used to concentrate and detect hepatitis A virus in water samples. *J Virol Methods* **137**: 169–76.

VINJÉ J and KOOPMANS MP (2000) Simultaneous detection and genotyping of "Norwalk-like viruses" by oligonucleotide array in a reverse line blot hybridization format. *J Clin Microbiol* **38**: 2595–601.

WANG D, WU Q, YAO L, WEI M, KOU X and ZHANG J (2008) New target tissue for foodborne virus detection in oysters. *Lett Appl Microbiol* **47**: 405–9.

WILLIAMS TP, KASORNDORKBUA C, HALBUR PG, HAQSHENAS G, GUENETTE DK, TOTH TE and MENG XJ (2001) Evidence of extrahepatic sites of replication of the hepatitis E virus in a swine model. *J Clin Microbiol* **39**: 3040–6.

YAZAKI Y, MIZUO H, TAKAHASHI M, NISHIZAWA T, SASAKI N, GOTANDA Y and OKAMOTO H (2003) Sporadic acute or fulminant hepatitis E in Hokkaido, Japan, may be food-borne, as suggested by the presence of hepatitis E virus in pig liver as food. *J Gen Virol* **84**: 2351–7.

YONEYAMA T, KIYOHARA T, SHIMASAKI N, KOBAYASHI G, OTA Y, NOTOMI T, TOTSUKA A and WAKITA T (2007) Rapid and real-time detection of hepatitis A virus by reverse transcription loop-mediated isothermal amplification assay. *J Virol Methods* **145**: 162–8.

4

Sampling strategies for virus detection in foods, food-processing environments, water and air

A. Rzeżutka, National Veterinary Research Institute, Poland and A. Carducci, University of Pisa, Italy

DOI: 10.1533/9780857098870.2.79

Abstract: Currently there are no specific guidelines providing recommendations on how to perform sampling on foods at different phases of their supply chains as well as on environmental matrices when they require testing for virus presence. This chapter considers the significance and the role of sampling during food chain or water monitoring, surveillance and research studies. An overview and discussion of sampling strategies and criteria to be followed during sample selection are presented. Finally, some food- and waterborne outbreaks are described, highlighting the role of efficient sampling strategies in successful virus detection both in the tested samples and the case patients.

Key words: enteric viruses, sampling strategies, food chain monitoring, water monitoring, air, surfaces.

4.1 Introduction

In recent years human enteric viruses have been recognized as important threats to human health. Breaching of current regulations regarding sewage discharge, the use of unclean water for irrigation purposes and poor hygiene practices, often reported among food handlers, have led to viral contamination of food, resulting in outbreaks of disease. It is often the case that the presence of a virus in food cannot be proved, even after testing of the suspected food. Moreover, food testing for viruses is not currently incorporated in any official regulations. For the purposes of an outbreak investigation, a large number of methods for virus detection in different types of food have been published (Croci *et al.*, 2008; Mattison and Bidawid, 2009; Rzeżutka and Cook, 2009). Some of these techniques are specific and can only be used for one type of sample; others can be applied for virus detection in a broad range of matrices. Surprisingly, contrary to the broad range of virus detection methods used for food testing, food sampling methodology seems to be neglected in the literature. Currently there are no specific guidelines addressing food sampling issues for virus analysis and the existing recommendations for food sampling for bacterial pathogens do not necessarily meet virological requirements (Bosch *et al.*, 2011). Furthermore, differences in the sample size and in the initial treatment of the sample can have an influence on the detection limit of the method used. The sampling strategy and methodology are even more crucial when possible environmental sources of food contamination (water, surfaces and air), both for routine safety purposes and for the study of outbreaks, are monitored. Successful virus detection in food and the environment depends not only on the method selected or the capability of the laboratory in testing samples, but also, in most cases, on the sampling strategy applied. In this chapter the significance and the role of sampling in virus monitoring, surveillance and research studies will be discussed.

4.2 Virus monitoring at different levels of the food supply chain

4.2.1 Monitoring the fresh produce supply chain

Food crops are grown in natural conditions and are constantly exposed to different pathogens present in the environment. When crops are grown in virus-polluted soil or irrigated by polluted water, produce originating from these areas can be also contaminated, posing a health risk to consumers. In these cases, low numbers of heterogeneously distributed viruses are usually present on the raw produce or in the end product if the produce is destined for further processing. These factors significantly hamper sampling activity by making selection of the representative food sample(s) for testing difficult. Therefore the testing of environmental samples, for example irrigation or washing water, at a production or a processing stage may provide more relevant results than testing the produce directly.

There are many points at which viruses may enter the food chain so it is important that before the food chain monitoring starts, all sampling points along the chain are identified (Fiore, 2004). This may indicate that for certain types of food, particularly in the case of fresh produce, monitoring may be limited to the production stage. At this stage, the risk of produce contamination is mostly associated with polluted water used for agrarian purposes and a lack of observance in hand hygiene requirements among the field workers. As well as the initial picking of a fruit or vegetable harvest, manual trimming, bunching and packing are all considered risk factors for produce contamination. In these circumstances, it would be more appropriate to undertake virus screening on swabs from the hands of the harvesters, than to test the produce directly. It is known that noroviruses (NoVs) appear to be stable on human hands without significant virus loss (Liu *et al.*, 2009).

What is more, direct contact between food and contaminated surfaces or tools can play an important role in virus transmission. Caliciviruses show a long-term persistence on environmental surfaces (Liu *et al.*, 2009), making it possible to detect them on kitchen surfaces, for example, for at least a few days after food consumption (Boxman *et al.*, 2009). Swabbing of environmental surfaces can therefore be a good alternative to food sampling, providing strong evidence of virus presence in outbreak settings, especially where food samples tested have given negative results.

Apart from during primary production and processing, food can be contaminated at the point of sale (Rzeżutka *et al.*, 2007). Currently published protocols for virus testing in food do not focus on a sampling procedure. Even where procedures are described, the information is usually insufficient and does not provide such details as: (i) how the representative sample was chosen from the batch of the suspected produce; (ii) the frequency and period of sampling in the case of regular food monitoring; and (iii) how a test portion was prepared from the original sample. To find positive produce, at least a dozen sampling occasions should be envisaged. During each visit, several randomly selected samples (e.g., lettuce heads or boxes of fruit) representing different batches need to be collected. Generally, the whole product, that is a whole lettuce head or basket of berries, is sampled.

In the laboratory a test portion is prepared from the original sample which comprises 10 g (Scherer *et al.*, 2010) to 100 g (Dubois *et al.*, 2002) of raspberries or 25 g (Shan *et al.*, 2005) to 90 g of strawberries (Rzeżutka *et al.*, 2006). For leafy vegetables 25 g of green onions (Guévremont *et al.*, 2006; Shan *et al.*, 2005) or 5 g (Rutjes *et al.*, 2006) to 25 g (Dubois *et al.*, 2006) of lettuce can be taken for analysis.

Usually the contamination of fresh produce is external, with viruses present on food surfaces. However, the possibility of internalized contamination of plant tissues cannot be excluded (Urbanucci *et al.*, 2009). Viruses are usually removed from the food surface by washing or homogenization. An alternative approach to food washing can be swabbing of the produce surface (Mäde *et al.*, 2005; Scherer *et al.*, 2009) and this method has been used for detection of NoV and rotavirus (RV) on artificially contaminated apples, cucumbers and pepper (Scherer *et al.*, 2009). The moistened cotton swab was moved across the sample area five times horizontally, vertically and diagonally, whilst simultaneously being turned to expose the whole swab to the contaminated surface. Besides producing virus recovery levels of up to 78%, this method has the added advantage over commonly used virus extraction and concentration procedures of leaving the food matrix intact and reducing the amount of released substances potentially acting as inhibitors. However, it can be only used for solid produce with flat surfaces.

Data on sample storage and transportation to the laboratory is usually sufficiently described in varied sampling protocols. Soft fruit and leafy vegetables should be transported in refrigerated conditions and tested soon after delivery, because long storage times could cause decomposition of produce and significantly hinder virus extraction. However, enteric viruses do have the potential to persist at refrigerated conditions between the time of produce purchase and consumption (De Paula *et al.*, 2010). This should guarantee virus stability for a few days before analysis. In addition, virus survival on different surfaces may be stabilized by the presence of organic matter, coliform bacteria, proteins or fats (Vasickova *et al.*, 2010). Freezing can allow the retention of virus infectivity but frozen transportation or storage is usually only used for berries. Finally, transportation and the sampling process itself should be conducted in such a way that the sample conditions are not negatively affected (Rodríguez-Lázaro *et al.*, 2012). According to the most common indications, transport and storage should occur as quickly as possible, at a controlled temperature (5 ± 3°C). In this temperature range, samples can be stored for up to 48 h. If this cannot be achieved, they should be frozen at −70°C. Samples should never be repeatedly frozen and thawed prior to analysis, as this can affect virus infectivity (Rodríguez-Lázaro *et al.*, 2012).

In summary, sporadic virus occurrence on ready-to-eat produce can lead to negative results, but this will not prove that the implemented surveillance system at a food production or processing level was fully effective. Interpretation of positive results gathered during produce monitoring, especially at the point of sale, is difficult, as it is not possible to identify the stage at which food contamination occurred from such results.

4.2.2 Monitoring the shellfish supply chain

Different types of enteric viruses have been found in shellfish. However, the data on virus prevalence mostly relates to NoV, enterovirus (EV), hepatitis A virus (HAV), astrovirus (AV) and RV (Boxman, 2010). The contamination of shellfish with viruses is expected to be higher than that of fresh produce, as they accumulate viral pathogens during their filter-feeding activities. The level of molluscs contamination ranges from 10^2 to 10^5 virus genome copies per gram of digestive tissues (Butot *et al.*, 2009; Costafreda *et al.*, 2006). Even in shellfish harvested from the same area at the same time, there is still a possibility of non-homogeneous distribution of viral contaminants within the batch. Therefore a negative test result on a single shellfish sample will not guarantee that the whole batch is free from virus contamination (Boxman, 2010). Whether the shellfish are collected from production (growing areas) or point of sale, the sampling procedures may not vary significantly, and differences usually only arise from the type of shellfish sampled. The level of virus accumulation in shellfish may depend on their type, as oyster samples may be less frequently contaminated by viruses than mussels (Le Guyader *et al.*, 2000). In addition, the persistence of viruses has been observed to differ in different shellfish species (Hernroth and Allard, 2007).

Any seasonal occurrence of shellfish contamination should be considered when devising a sampling strategy (Le Guyader *et al.*, 2000; Bosch *et al.*, 2011). It has been shown that shellfish harvested during the winter months contain a higher percentage of human pathogenic viruses (Coelho *et al.*, 2003; Elamri *et al.*, 2006) than those sampled during summer (Formiga-Cruz *et al.*, 2002; Woods and Burkhardt III, 2010). An example of successful shellfish monitoring at primary production was conducted by Le Guyader *et al.* (2000). In this study, samples were collected monthly over a period of three years, with one sample taken from each harvesting area. The single sample was composed of at least 20 oysters or 30 mussels. In the laboratory, the digestive glands were dissected, aliquoted under 1.5 g portions (three aliquots were prepared for each sample) and kept frozen at −20°C until analysis. Based on the example of this and other studies, the following conclusion can be drawn: (i) the shortest monitoring period of shellfish in growing or harvesting areas should be one year and the sampling visits need to be distributed equally during this period; and (ii) several shellfish should be taken per sampling (2–30), preferably from three different locations (sampling points) representing the same harvesting area.

Shellfish are very commonly eaten raw, so upon harvesting they may be temporarily stored or directly marketed. As the shellfish production chain is short, monitoring can be performed either during farming or at retail level. An example of shellfish monitoring covering different retail outlets, such as restaurants, sea food markets and wholesalers, was successfully conducted in the United States (DePaola *et al.*, 2010). The sampling strategy relied on collection of two oyster samples (each consisted of 36 live animals) from these places twice a month in each of nine states. In total, 397 oyster samples were collected representing different lots. The samples were transported to a laboratory under refrigerated conditions (2–10°C) and analysed the day after collection. Before analysis, oysters were washed and shucked and the digestive diverticula from 5 to 10 oysters were removed and homogenized to obtain a tested portion of 25 g. The monitoring resulted in detection of positive NoV and HAV samples.

Despite the use of different schedules in terms of sampling frequency, period of sampling and the number of samples taken per visit, a single sample is usually made up from a pool of several (Beuret *et al.*, 2003; Wang *et al.*, 2008) to a dozen animals (Kingsley *et al.*, 2002). The whole animal body can be used, but an individual sample can also consist of dissected digestive glands from several animals (Kingsley *et al.*, 2002; Sánchez *et al.*, 2002) or homogenates of shellfish bodies (Cromeans *et al.*, 1997; Croci *et al.*, 1999;

Coelho *et al.*, 2003; Elamri *et al.*, 2006). This pooling can affect the sensitivity of the detection method and can produce a false negative result due to reduction of the sample size and the diluting effect of the uncontaminated samples present in the pool. De Roda Husman *et al.* (2007) proved that testing of pooled samples of digestive glands never resulted in a positive signal, whereas testing of the digestive glands individually confirmed the virus presence. Because most viruses are found in the digestive organs, the majority of protocols used for virus extraction employ homogenization of the digestive glands instead of the shellfish bodies (Lees and CEN WG6 TAG4, 2010). The advantage of this approach is increased test sensitivity and elimination of the excess food matrix, which may contain substances inhibitory to molecular detection. This is also the procedure described in the forthcoming CEN 275/WG6/TAG4 horizontal method and recommended by a European Community Reference Laboratory for monitoring bacteriological and viral contamination of bivalve molluscs (Lees and CEN WG6 TAG4, 2010).

Transportation and shellfish storage before analysis should not affect the test results as long as shellfish are kept under refrigerated conditions and processed within 24 h. Viruses inside the shellfish are well protected from temperature changes (Croci *et al.*, 1999). HAV, for example, can persist in shellfish for more than 3 weeks, whilst viral RNA can be detected over 6 weeks after contamination (Kingsley and Richards, 2003).

4.3 The significance of water, air and surface sampling during food chain monitoring

4.3.1 Water

Water can be involved as a source of contamination at various different levels along the food chain: in primary production, through irrigation of crops, used in the growth of shellfish, during food processing, as an ingredient of the food itself or for washing and cleaning, and in food distribution and serving. Furthermore, water is also directly ingested as in the case of bottled or tap water. Recreational waters, although not related to the food chain, can also be a vehicle for virus transmission. There is vast evidence related to the viral contamination of water (Bosch *et al.*, 2011; Rodríguez-Lázaro *et al.*, 2012): a great variety of virus are excreted by faeces, but the most regularly detected are EV, RV, NoV, HAV, HEV and adenovirus (AdV).

Water monitoring along the food chain can involve different types of water, depending on their origin and role in food production. The viral load can differ significantly in different types of water. Typically the fate of any enteric virus excreted by the faecal route involves the contaminated matter coming into sewage that is treated and discharged in surface waters, then collected and treated for drinking, leading to a decrease in viral concentration from 10^{10} particles per gram of faeces to less than one in 100 l of tap water (Gantzer and Schwartzbrod, 2002). Consequently, the sampling volume depends on the water quality, and the use of sampling procedures is related to the volume of the sample. In the reported studies on virological analysis of different types of water, the sample sizes range from 50 μl to 3000 l (Rodríguez-Lázaro *et al.*, 2012), with an evident lack of harmonization. However, the following volumes are the most often indicated: 1 l for raw sewage, 10 l for treated sewage, surface and marine recreational waters, 100 l for ground or drinking water treatment, and 1000 l for tap or mineral water.

Sampling and analytical procedures for the testing of water samples are well documented. Viral recovery and concentration techniques include: ultrafiltration; adsorption-elution using filters or positively or negatively charged membranes; glass wool or glass powder; two-phase separation with polymers; flocculation; and the use of monolithic chromatographic columns (Bosch *et al.*, 2011). The recovery efficiency of these techniques can be affected by the physicochemical quality of the water, including pH, conductivity, turbidity, presence of particulate matter, and organic acids.

For the concentration of large volume samples, two phases may be necessary, such as the use of glass wool, for example, followed by flocculation. The glass wool adsorption-elution procedure for the recovery of enteric viruses from large volumes of water has proved cost-effective. This method can also be adapted for the in-line recovery of viruses from water, thus avoiding transportation of large volumes of potentially highly polluted water. However, no single method can be recognized as superior for all water types and sample volumes: efficiency, constancy of performance, robustness, cost, and complexity are all factors to be considered for each method, and application and performance characteristics must be continually monitored, possibly using process quality controls (Wyn-Jones *et al.*, 2011).

4.3.2 Air and surfaces

Although viral particles are generally at very low concentrations in the air and on surfaces, enteric viruses have been detected in different contexts and buildings, and on a great variety of surfaces and objects (Boone and Gerba, 2007). In most studies the environmental monitoring was related to an outbreak (Cheesbrough *et al.*, 2000; Gallimore *et al.*, 2005; Wu *et al.*, 2005; Jones *et al.*, 2007; Boxman *et al.*, 2009), but in some (Scherer *et al.*, 2009; Boxman *et al.*, 2011; Carducci *et al.*, 2011; Ganime *et al.*, 2011) viral contamination of surfaces and sometimes air was detected in the absence of clinical cases. The settings most studied were hospitals and institutions such as nursing homes and daycare centers. In these settings, the viruses most frequently searched for were RV and NoV. In a recent study based on extensive surface monitoring for NoV in catering companies (Boxman *et al.*, 2011) the prevalence of positive samples in companies without recently reported outbreaks (4.2%) was significantly lower than in outbreak settings (61.1%) and the strains detected were similar to those identified in clinical specimens. The higher prevalence of environmental contamination found during outbreaks suggests that swabs may be used as an adjunct to clinical diagnosis of NoV for the outbreak investigation (Boxman *et al.*, 2009, 2011; Ganime *et al.*, 2011).

A sampling strategy for air and surfaces should have very clear specific aims if reliable results are to be produced and excessive costs avoided. In practice, the detection of viruses in air and on surfaces encounters significant obstacles, largely related to the high dilution factor and the sensitivity limits of the proposed techniques. These limitations are more evident for air monitoring, which, as well as requiring more complex sampling techniques and special expensive devices, produces extremely variable results strongly influenced by particle size and air movements. For this reason air sampling should be limited to particular settings and aims, for example, the diffusion of viral aerosol from a wastewater treatment plant (Carducci *et al.*, 2000), from a toilet flush or from vomit. The sample size is dependent on the level of contamination (generally 1 m^3 of air indoor and 3 m^3 outdoor) and the sampling points can be located close to the source of the aerosol or in areas of possible diffusion, taking air movements into account.

Surface monitoring is simpler and less expensive, and can also provide evidence on the contamination of air through the settling of droplets. It has been most extensively used in health care (Carducci *et al.*, 2011) and food production (Scherer *et al.*, 2009) settings to assess not only viral contamination, but also the efficacy and correct application of disinfection procedures. For enteric virus detection on environmental surfaces, different sampling protocols have been reported and sometimes compared (Wu *et al.*, 2005; Gallimore *et al.*, 2006; Boxman *et al.*, 2009; Scherer *et al.*, 2009; Julian *et al.*, 2011). A definite surface area is sampled using swabs, contact plates or sponge bags (Rodríguez-Lázaro *et al.*, 2012). These devices are all eluted after sampling. The most recommended procedure is based on the swab-rinse technique (Scherer *et al.*, 2009; Rodríguez-Lázaro *et al.*, 2012), but swab material, eluent and rinsing procedures can differ. Comparison studies have shown the highest recovery rate for polyester swabs eluted with PBS (Julian *et al.*, 2011). Electrostatic interactions, van der Waals forces and hydrophobic effects are assumed to be involved in the interactions between virus particles and solid substrates. For these reasons virus type and chemico-physical surface properties may affect recovery efficiency; efficiency is higher when the sample is taken from non-porous and smooth surfaces such as steel, metal, ceramic, glass and high-density polyethylene (Scherer *et al.*, 2009). Further factors affecting recovery are pH, ionic concentration, surface charge and organic matter. This should be taken into account when the aim of monitoring is the control of cleaning procedures (Julian *et al.*, 2011). The swab sampling method revealed a remarkable variability of recovery rates (Scherer *et al.*, 2009) making standardization difficult. Consequently, in the interpretation of swab sample data, positive results indicate surface contamination, implying a potential risk of exposure, whereas the negative ones do not necessarily prove the absence of contamination.

When the surface monitoring along a food chain is aimed at risk assessment, specific pathogens have to be detected, such as NoV and RV. Alternatively, if monitoring is being conducted for safety purposes, a more conservative approach should be used based on the research of indicators. In these cases, because classical bacterial parameters are not representative of viruses, we can choose a viral agent largely spread in healthy people and eliminated via the faecal route. For air and surfaces, AdV and TT virus (Carducci *et al.*, 2011) have been used, but other virus types such as polyomavirus have also been proposed (Foulongne *et al.*, 2011).

The choice of sampling points is also important to guarantee highly sensitive monitoring. The most representative surfaces are the ones subjected to contamination and contact by food handlers, and infrequently cleaned or difficult to clean, for example the flushing chain or knob and the toilet seat, the handle of refrigerators or other kitchen machines, the handles of knives, the cruet set or the soap dispenser. An interesting approach was adopted in a study of catering companies in the Netherlands (Boxman *et al.*, 2011). For each company only three swabs were collected, but each swab was used to sample two or three surfaces in the same context, thus increasing the probability of obtaining positive samples.

Besides the sensitivity of detection methods, the technique and the medium used for virus collection can also affect the infectivity and the persistence of their nucleic acid (Rodríguez-Lázaro *et al.*, 2012). Air samplers based on filtration, impact on solid surfaces, or impingement have all been successfully used for virus detection (Verreault *et al.*, 2008), but all present advantages and disadvantages: filtration samplers are simple to use, but the flow rate, sampling duration and the membrane composition have to be strictly controlled to avoid dehydration; impact samplers are also easy, but dehydration or impact trauma are possible and then flow rate and duration are critical; impinger samplers avoid dehydration but flow rate and collection fluid composition are determinant for virus recovery (Bosch *et al.*, 2011)

4.4 Sampling strategy in relation to food- and waterborne outbreaks

Detection of enteric viruses related to foodborne outbreaks in foodstuffs other than shellfish have rarely been reported, and data on the prevalence of enteric viruses on fresh produce are very scarce. This may reflect either difficulties in the detection of low numbers of viruses on contaminated food, or the fact that monitoring of food for viruses is rarely conducted. So far, only one outbreak linked to the consumption of virus-contaminated produce has been fully documented (Calder *et al.*, 2003). In the many other viral foodborne outbreaks recorded all over the world, including the recently described hepatitis A outbreak associated with semi-dried tomatoes (Petrignani *et al.*, 2010a,b; Gallot *et al.*, 2011), contaminated vegetables have only been identified as putative vehicles of virus transmission. In the food- and waterborne outbreak example discussed below, different sampling strategies as well as the role of epidemiological investigation in identification of the virus source are described. Detailed information on each food sampling strategy and the virus concentration methods applied during the investigation of food-, water- and airborne outbreaks have been presented in the appendix.

4.4.1 Outbreak linked to virus-contaminated fruit

A hepatitis A outbreak associated with the consumption of raw blueberries was reported in 2002 in New Zealand (Calder *et al.*, 2003). An epidemiological study revealed 39 cases of HAV which were directly associated with the consumption of contaminated berries. Clinical samples (faeces, blood or serum) collected from infected persons and samples of frozen blueberries from the cold store were analysed for the presence of HAV. The virus was detected in faecal and food samples, but not in any blood or serum specimens of the patients. HAV- positive samples were confirmed by DNA hybridization and sequencing of PCR products.

Investigation of the berries consumed revealed that the produce may have originated from one source orchard. Based on information obtained from the retailer, the probable source orchard in which the contamination occurred was identified. A sanitary audit conducted in the orchard revealed multiple opportunities for an infected worker to contaminate the product or processing equipment during picking and packing. However, there was no evidence that any of the orchard workers had recently suffered from hepatitis A or revealed symptoms of the disease, and none were tested for HAV-specific antibodies. An unequivocal identification of the mode of contamination of the berries was not possible, although all the evidence supported assumptions that the likely source of the virus was either an infected food handler or faecally polluted ground water. The orchard was not irrigated, but the high rainfall observed during the harvest season may have raised the ground water level, which could easily have been contaminated by field latrine effluents.

The contaminated batch of blueberries contained 36 tonnes of fruit. Fourteen tonnes had been introduced to the market and 22 were still in cold storage. For analysis, only six samples consisting of 100 g portions of frozen blueberries were collected from the cold storage. Among them three gave positive results, suggesting that these samples were highly contaminated. However, the authors concluded that the overall level of fruit contamination was low, or that the viral distribution in the batch was not uniform as, despite over 14 tonnes of berries being sold, only a limited number of cases were recorded.

4.4.2 Outbreak linked to virus-contaminated shellfish

Shellfish is a common, globally recognized food and is usually eaten raw or only slightly cooked. When shellfish are grown in and harvested from polluted water, they may become contaminated by human pathogenic viruses including HAV, which is responsible for the most serious viral infection associated with shellfish consumption (Pintó *et al.*, 2009). Detection of HAV in foodborne outbreaks is rarely successful, as it is hampered by such factors as a lack of efficient sampling methodologies or a low level of virus recovery from shellfish tissues. A good example of a successful HAV detection in shellfish was a multistate foodborne outbreak which occurred in the United States in 2005 (Bialek *et al.*, 2007; Shieh *et al.*, 2007). The outbreak-related oysters were farmed and harvested from two Louisiana harvest areas in the Gulf of Mexico. The oysters were shipped to several restaurants, where they were eaten in a raw state. Only one person among 39 victims admitted eating baked oysters. None of the case patients reported any risk factors related to the infection other than a specific food consumption. Viruses were detected in the oyster meat and HAV RNA sequences derived from the oysters were identical to virus sequences obtained from patients' sera. It was confirmed that all cases were due to infection by a single strain, classified as HAV genotype IA.

This outbreak originated from a common source and it was established that the contamination occurred before oyster distribution, since employees involved in the shipment had minimal physical contact with shellfish and none reported hepatitis symptoms. All implicated oysters had been harvested from approved areas remote from any fixed sources of sewage discharge. Contamination of the shellfish beds by harvesters was unlikely, as they met all hygiene requirements. The probable source of contamination was illegal waste disposal from recreational boats within legal harvest areas, or an illegal harvesting in closed areas. The identification of the virus source was possible due to the epidemiological investigation, which identified contaminated oysters at a wholesaler. They were harvested from the same areas at the same time as oysters served to the case patients.

Two oyster samples were collected for testing, one representing the product as sold in a frozen state, and the second representing the fresh product. Both samples were frozen at the time of collection. In the laboratory, the oysters were thawed and aseptically separated from icy water and adductor muscle. From each sample, three to four 25 g portions of oyster meat were prepared for testing (three to eight oysters were homogenized to obtain a tested portion). HAV RNA was detected in only one 25 g portion of the sample prepared from produce offered in a frozen state. The authors highlighted that maintaining oysters at constant freezing temperatures before sampling prevented viral RNA degradation through enzymes released by the oyster meat. Several elements played a significant role during identification of a foodborne aetiology of this outbreak, including application of molecular epidemiology, an efficient protocol for virus extraction, a proper preservation of food samples and finally a field investigation.

4.4.3 Outbreak linked to virus-contaminated drinking water

The frequency of waterborne outbreaks reported by the existing surveillance systems is increasing, owing to the increasing attention being paid to possible viral origin. The aetiology of these epidemics is generally determined with the detection of viruses in clinical specimens and rarely confirmed by environmental samples (Brunkard *et al.*, 2011).

At the end of July 2006, more than 2800 patients with acute diarrhoea were reported by accident and emergency departments in Taranto, Italy (Martinelli *et al.*, 2007). A field investigation was conducted between July and October 2006, including epidemiological data collection, case–control study and microbiological analysis of both faecal and environmental samples (tap water, sea water and shellfish). The drinking water samples (1 l) were collected at the local waterworks, from major water pipelines and wells, and from tap water in a public bar. The samples were concentrated using the cation-coated filter method reported by Haramoto *et al.* (2005). In July two samples were collected; one produced a positive result for NoV and one a positive result for RV. In August a further six samples were collected, two of which tested positive for RV. In September 14 samples were analysed with two positive results for NoV and four for RV, and in October 22 samples were collected with four positive for RV. Of the total of 44 tap water samples tested, four (9%) were positive for NoV and 11 (25%) for RV.

Twelve sea water samples were also taken and concentrated by the same method cited above: four (33%) were positive for NoV and one (8.3%) for RV. Among 70 faecal samples, 48% tested positive for RV and 40% positive for NoV. No faecal indicator bacteria or endotoxins were detected in the environmental samples of tap water collected in Taranto, and no shellfish samples were positive for bacteria or viruses. The case-control study confirmed tap water as the probable source of the outbreak, supported by the fact that molecular profiles of RV and NoV identified in a number of tap water samples were the same as the ones found in some patients' stool samples. Sequence analysis showed these to be the new NoV strain GGII.4 2006a and RV genotype G9.

4.5 Conclusion

The existing protocols dealing with sample preparation, virus extraction and concentration from food have been described in great detail, yet information on sampling methodology is usually insufficient. So far the necessary food sampling requirements for detection of viruses have not been covered by any standards. Understanding of food- and waterborne viruses has significantly increased during the last ten years as a result of several studies on virus prevalence and persistence in these environments. An increasing number of outbreaks attributed to the consumption of virus-contaminated shellfish have been particularly key, triggering broad research on the development of virus detection methods and studies on virus survival and elimination. The results of these studies are certainly a basis for future shellfish monitoring programmes. However, routine screening of shellfish for the presence of viruses is an expensive, difficult process and may not be fully effective (Pintó *et al.*, 2009, 2010). Testing of shellfish as well as other food types should therefore be considered only as a verification element of the production process. To control the virological quality of shellfish, continuous water monitoring in growing and harvesting areas would be more efficient than direct shellfish testing. Quantitative virological analysis of shellfish could then be performed prior to marketing (Pintó *et al.*, 2009).

The increasing involvement of fresh vegetables in viral outbreaks (EFSA and ECDC, 2011) makes it of vital importance that the virological quality of water used in the production chains is efficiently measured. In addition, assessing the role of food handlers in the contamination of food production environments could help to guarantee food safety in catering (Boxman *et al.*, 2011) and 'ready-to-eat food' companies. Such assessment relies on effective strategies, and one of the main future challenges faced in food virology will be an elaboration of sampling guidelines, especially for fresh produce chain monitoring. The identification and analysis of virus-relevant critical control points within each food supply chain is therefore key to the continued development of effective sampling strategies and the successful application of these strategies for the identification and prevention of possible contamination.

4.6 Sources of further information and advice

The latest developments in virus detection methods and sampling protocols used during virus surveillance or outbreak investigation can be obtained in such scientific journals as *Food and Environmental Virology*, *International Journal of Food Microbiology*, *Applied and Environmental Virology*, *Journal of Food Protection, Water Research, Journal of Virological Methods* and *Journal of Applied Microbiology.* The website of the International Organization for Standardization (www.iso.org) provides information on current operating standards related to sampling techniques and sampling procedures applied in food microbiology.

4.7 References

BEURET CH, BAUMGARTNER A and SCHLUEP J (2003), 'Virus-contaminated oysters: a three – month monitoring of oysters imported to Switzerland', *Appl Environ Microbiol*, **69**, 2292–7.

BIALEK S R, GEORGE P A, XIA G L, GLATZER M B, MOTES M L, VEAZEY J E, HAMMOND R M, JONES T, SHIEH Y C, WAMNES J, VAUGHAN G, KHUDYAKOV Y and FIORE A E (2007), 'Use of molecular epidemiology to confirm a multistate outbreak of hepatitis A caused by consumption of oysters', *Clin Infect Dis*, **44**, 838–40.

BOONE S A and GERBA C P (2007), 'Significance of fomites in the spread of respiratory and enteric viral disease', *Appl Environ Microbiol*, **73**, 1687–96.

BOSCH A, SÁNCHEZ G, ABBASZADEGAN M, CARDUCCI A, GUIX S, LE GUYADER F S, NETSHIKWETA R, PINTÓ R M, VAN DER POEL W H M, RUTJES S, SANO D, TAYLOR M B, VAN ZYL W B, RODRÍGUEZ-LÁZARO D, KOVAČ K and SELLWOOD J (2011), 'Analytical methods for virus detection in water and food', *Food Anal Methods*, **4**, 4–12.

BOXMAN I L A (2010), 'Human enteric viruses occurrence in shellfish from European markets', *Food Environ Virol*, **2**, 156–166.

BOXMAN I L A, DIJKMAN R, TE LOEKE N A J M, HÄGELE G, TILBURG J J H C, VENNEMA H and KOOPMANS M (2009), 'Environmental swabs as a tool in norovirus outbreak investigation, including outbreaks on cruise ships', *J Food Prot*, **72**, 111–19.

BOXMAN I L A, VERHOEF L, DIJKMAN R, HÄGELE G, TE LOEKE N A J M and KOOPMANS M (2011), 'Year-round prevalence of norovirus in the environment of catering companies without a recently reported outbreak of gastroenteritis', *Appl Environ Microbiol*, **77**, 2968–2974.

BRUNKARD J M, AILES E, ROBERTS V A, HILL V, HILBORN E D, CRAUN G F, RAJASINGHAM A, KAHLER A, GARRISON L, HICKS L, CARPENTER J, WADE T J, BEACH M J and YODER J S (2011), 'Surveillance for waterborne disease outbreaks associated with drinking water – United States, 2007–2008', *MMWR*, **60**, 38–68.

BUTOT S, PUTALLAZ T, AMOROSO R and SÁNCHEZ G (2009), 'Inactivation of enteric viruses in minimally processed berries and herbs', *Appl Environ Microbiol*, **75**, 4155–61.

CALDER L, SIMMONS G, THORNLEY C, TAYLOR P, PRITCHARD K, GREENING G and BISHOP J (2003), 'An outbreak of hepatitis A associated with consumption of raw blueberries', *Epidemiol Infect*, **131**, 745–51.

CARDUCCI A, TOZZI E, RUBULOTTA E, CASINI B, CANTIANI L, ROVINI E and PACINI R (2000), 'Assessing airborne biological hazard from urban wastewater treatment', *Water Res*, **34**, 1173–8.

CARDUCCI A, VERANI M, LOMBARDI R, CASINI B and PRIVITERA G (2011), 'Environmental survey to assess viral contamination of air and surfaces in hospital settings', *J Hosp Inf*, **77**, 242–247.

CHEESBROUGH J S, GREEN J, GALLIMORE C I, WRIGHT P A and BROWN D W G (2000), 'Widespread environmental contamination with Norwalk-like viruses (NLV) detected in a prolonged hotel outbreak of gastroenteritis', *Epidemiol Infect*, **125**, 93–8.

COELHO C, HEINERT A P, SIMÕES C M O and BARARDI C R M (2003), 'Hepatitis A virus detection in oysters (*Crassostrea gigas*) in Santa Catarina state, Brazil, by reverse transcription-polymerase chain reaction', *J Food Prot*, **66**, 507–11.

COSTAFREDA M I, BOSCH A and PINTÓ R M (2006), 'Development, evaluation, and standardization of a real-time TaqMan reverse transcription-PCR assay for quantification of hepatitis A virus in clinical and shellfish samples', *Appl Environ Microbiol*, **72**, 3846–5385.

CROCI L, DE MEDICI D, MORACE G, FIORE A, SCALFARO C, BENEDUCE F and TOTI L (1999), 'Detection of hepatitis A virus in shellfish by nested reverse transcription-PCR', *Int J Food Microbiol*, **48**, 67–71.

CROCI L, DUBOIS E, COOK N, DE MEDICI D, SCHULTZ A CH, CHINA B, RUTJES S A, HOORFAR J and VAN DER POEL W H M (2008), 'Current methods for extraction and concentration of enteric viruses from fresh fruit and vegetables: towards international standards', *Food Anal Methods*, **1**, 73–84.

CROMEANS T L, NAINAN O V and MARGOLIS H S (1997), 'Detection of hepatitis A virus RNA in oyster meat', *Appl Environ Microbiol*, **63**, 2460–3.

DEPAOLA A, JONES J L, WOODS J, BURKHARDT W, CALCI K R, KRANTZ J A, BOWERS J C, KASTURI K, BYARS RH, JACOBS E, WILLIAMS-HILL D and NABE K (2010), 'Bacterial and viral pathogens in live oysters: 2007 United States market survey', *Appl Environ Microbiol*, **76**, 2754–68.

DE PAULA V S, GASPAR A M C and VILLAR L M (2010), 'Optimization of methods for detecting hepatitis A virus in food', *Food Environ Virol*, **2**, 47–52.

DE RODA HUSMAN A M, LODDER-VERSCHOOR F, VAN DEN BERG H H, LE GUYADER F S, VAN PELT H, VAN DER POEL W H and RUTJES S A (2007), 'Rapid virus detection procedure for molecular tracing of shellfish associated with disease outbreaks', *J Food Prot*, **70**, 967–74.

DUBOIS E, AGIER C, TRAORÉ O, HENNECHART C, MERLE G, CRUCIERE C and LAVERAN H (2002), 'Modified concentration method for the detection of enteric viruses on fruits and vegetables by reverse transcriptase-polymerase chain reaction or cell culture', *J Food Prot*, **65**, 1962–9.

DUBOIS E, HENNECHART C, DEBOOSÈRE N, MERLE G, LEGEAY O, BURGER C, LE CALVÉ M, LOMBARD B, FERRÉ V and TRAORÉ O (2006), 'Intra-laboratory validation of a concentration method adapted for the enumeration of infectious F-specific RNA coliphage, enterovirus, and hepatitis A virus from inoculated leaves of salad vegetables', *Int J Food Microbiol*, **108**, 164–71.

EFSA and ECDC (2011), 'The European Union summary report on trends and sources of zoonoses, zoonotic agents and food-borne outbreaks in 2009', *EFSA J*, **9**, 2090–2468.

ELAMRI D E, AOUNI M, PARNAUDEAU S and LE GUYADER F S (2006), 'Detection of human enteric viruses in shellfish collected in Tunisia', *Lett Appl Microbiol*, **43**, 399–404.

FIORE A E (2004), 'Hepatitis A transmitted by food', *Clin Infect Dis*, **38**, 705–15.

FORMIGA-CRUZ M, TOFIÑO-QUESADA G, BOFILL-MAS S, LEES D N, HENSHILWOOD K, ALLARD A K, CONDEN-HANSSON A C, HERNROTH B E, VANTARAKIS A, TSIBOUXI A, PAPAPETROPOULOU M, FURONES M D and GIRONES R (2002), 'Distribution of human virus contamination in shellfish from different growing areas in Greece, Spain, Sweden, and the United Kingdom', *Appl Environ Microbiol*, **68**, 5990–8.

FOULONGNE V, COURGNAUD V, CHAMPEAU W and SEGONDY M (2011), 'Detection of merkel cell polyomavirus on environmental surfaces', *J Med Virol*, **83**, 1435–9.

GALLIMORE C I, TAYLOR C, GENNERY A R, CANT A J, GALLOWAY A, LEWIS D and GRAY J J (2005), 'Use of a heminested reverse transcriptase PCR assay for detection of astrovirus in environmental swabs from an outbreak of gastroenteritis in a paediatric primary immunodeficiency unit', *J Clin Micriobiol*, **43**, 3890–4.

GALLIMORE C I, TAYLOR C, GENNERY A R, CANT A J, GALLOWAY A, XERRY J, ADIGWE J and GRAY J J (2006), 'Environmental monitoring for gastroenteric viruses in a paediatric primary immunodeficiency unit', *J Clin Micriobiol*, **44**, 395–9.

GALLOT C, GROUT L, ROQUE-AFONSO A M, COUTURIER E, CARRILLO-SANTISTEVE P, POUEY J, LETORT M J, HOPPE S, CAPDEPON P, SAINT-MARTIN S, DE VALK H and VAILLANT V (2011), 'Hepatitis A associated with semidried tomatoes, France, 2010', *Emerg Infect Dis*, **17**, 566–7.

GANIME A C, CARVALHO-COSTA F A, MENDONÇA M C L, VIEIRA CB, SANTOS M, COSTA FILHO R, MIAGOSTOVICH M P and LEITE J P G (2011), 'Group A rotavirus detection on environmental surfaces in a hospital intensive care unit', *Am J Inf Control*, **30**, 1–4.

GANTZER C and SCHWARTZBROD L (2002), 'Enteroviruses: occurrence and persistence in the environment', in BRITTON, G. (ed.), *Encyclopedia of Environmental Microbiology*, New York, Wiley and Sons Publ, 2338–48.

GUÉVREMONT E, BRASSARD J, HOUDE A, SIMARD C and TROTTIER Y L (2006), 'Development of an extraction and concentration procedure and comparison of RT-PCR primer systems for the detection of hepatitis A virus and norovirus GII in green onions', *J Virol Methods*, **134**, 130–5.

HARAMOTO E, KATAYAMA H, OGUMA K and OHGAKI S (2005), 'Application of cation-coated filter method to detection of noroviruses, enteroviruses, adenoviruses, and torque teno viruses in the Tamagawa River in Japan', *Appl Environ Microbiol*, **71**, 2403–11.

HERNROTH B and ALLARD A (2007), 'The persistence of infectious adenovirus (type 35) in mussels (*Mytilus edulis*) and oysters (*Ostrea edulis*)', *Int J Food Microbiol*, **113**, 296–302.

JONES E L, KRAMER A, GAITHER M and GERBA C P (2007), 'Role of fomite contamination during an outbreak of norovirus on houseboats', *Int J Environ Health Res*, **17**, 123–31.

JULIAN T R, TAMAYO F J, LECKIE J O and BOEHM A B (2011), 'Comparison of surface sampling methods for virus recovery from fomites', *Appl Environ Microbiol*, **77**, 6918–25.

KINGSLEY D H, MEADE G K and RICHARDS G P (2002), 'Detection of both hepatitis A virus and Norwalk-like virus in imported clams associated with food-borne illness', *Appl Environ Microbiol*, **68**, 3914–18.

KINGSLEY D H and RICHARDS G P (2003), 'Persistence of hepatitis A virus in oysters', *J Food Prot*, **66**, 331–4.

LEES D and CEN WG6 TAG4 (2010), 'International standardisation of a method for detection of human pathogenic viruses in molluscan shellfish', *Food Environ Virol*, **2**, 146–55.

LE GUYADER F, HAUGARREAU L, MIOSSEC L, DUBOIS E and POMMEPUY M (2000), 'Three-year study to assess human enteric viruses in shellfish', *Appl Environ Microbiol*, **66**, 3241–8.

LIU P, CHIEN Y, PAPAFRAGKOU E, HSIAO H, JAYKUS L and MOE CH (2009), 'Persistence of human noroviruses on food preparation surfaces and human hands', *Food Environ Virol*, **1**, 141–147.

MÄDE D, KAHLE S TRÜBNER K and STARK R (2005), 'Detection of noroviruses in food and environmental samples by RT-PCR: Application in routine diagnostics', *Archiv Lebensmittelhyg*, **56**, 1–24.

MARTINELLI D, PRATO D, CHIRONNA M, SALLUSTIO A, CAPUTI G, CONVERSANO M, CIOFI DEGLI ATTI M, D'ANCONA F P, GERMINARIO C A and QUARTO M (2007), 'Large outbreak of viral gastroenteritis caused by contaminated drinking water in Apulia, Italy, May – October 2006' *Euro Surveill*, **12**, Available from: http://www.eurosurveillance.org/.

MATTISON K and BIDAWID S (2009), 'Analytical methods for food and environmental viruses', *Food Environ Virol*, **1**, 107–22.

PETRIGNANI M, HARMS M, VERHOEF L, VAN HUNEN R, SWAAN C, VAN STEENBERGEN J, BOXMAN I, PERAN I, SALA R, OBER H, VENNEMA H, KOOPMANS M and VAN PELT W (2010a), 'Update: a food-borne outbreak of hepatitis A in the Netherlands related to semi-dried tomatoes in oil, January – February 2010', *Euro Surveill*, **15**, Available from: http://www.eurosurveillance.org/.

PETRIGNANI M, VERHOEF L, VAN HUNEN R, SWAAN C, VAN STEENBERGEN J, BOXMAN I, OBER H J, VENNEMA H and KOOPMANS M (2010b), 'A possible foodborne outbreak of hepatitis A in the Netherlands, January–February 2010', *Euro Surveill*, **15**, Available from: http://www.eurosurveillance.org/.

PINTÓ R M, COSTAFREDA M I and BOSCH A (2009), 'Risk assessment in shellfish-borne outbreaks of hepatitis A', *Appl Environ Microbiol*, **75**, 7350–5.

PINTÓ R M, COSTAFREDA M I, PÉREZ-RODRIGUEZ F J, D'ANDREA L and BOSCH A (2010), 'Hepatitis A virus: state of the art', *Food Environ Virol*, **2**, 127–35.

RODRÍGUEZ-LÁZARO D, COOK N, RUGGERI F M, SELLWOOD J, NASSER A, NASCIMENTO M S, D'AGOSTINO M, SANTOS R, SAIZ J C, RZEŻUTKA A, BOSCH A, GIRONÉS R, CARDUCCI A, MUSCILLO M, KOVAČ K, DIEZ-VALCARCE M, VANTARAKIS A, VON BONSDORFF CH, DE RODA

HUSMAN A M, HERNÁNDEZ M and van DER POEL W H (2012), 'Virus hazards from food, water and other contaminated environments', *FEMS Microbiol Rev*, **36**, 786–814.

RUTJES S A, LODDER-VERSCHOOR F, VAN DER POEL W H, VAN DUIJNHOVEN Y T and DE RODA HUSMAN A M (2006), 'Detection of noroviruses in foods: a study on virus extraction procedures in foods implicated in outbreaks of human gastroenteritis', *J Food Prot*, **69**, 1949–56.

RZEŻUTKA A and COOK N (2009), 'Review of currently applied methodologies used for detection and typing of foodborne viruses', in BARBOSA-CANOVAS G, MORTIMER A, LINEBACK D, SPIESS W, BUCKLE K, COLONNA P, eds, *Global Issues in Food Science and Technology*, Elsevier Publishing, 229–46.

RZEŻUTKA A, D'AGOSTINO M and COOK N (2006), 'An ultracentrifugation-based approach to the detection of hepatitis A virus in soft fruits', *Int J Food Microbiol*, **108**, 315–20.

RZEŻUTKA A, KOZYRA I, CHROBOCIŃSKA M, KAUPKE A and MIZAK B (2007), 'Do noroviruses in the environment and food constitute a new threat?', *Medycyna Wet*, **63**, 379–83.

SÁNCHEZ G, PINTÓ R M, VANACLOCHA H and BOSCH A (2002), 'Molecular characterization of hepatitis A virus isolates from a transcontinental shellfish-borne outbreak', *J Clin Microbiol*, **40**, 4148–55.

SCHERER K, JOHNE R, SCHRADER C, ELLERBROEK L, SCHULENBURG J and KLEIN G (2010), 'Comparison of two extraction methods for viruses in food and application in a norovirus gastroenteritis outbreak', *J Virol Methods*, **169**, 22–7.

SCHERER K, MÄDE D, ELLERBROEK L, SCHULENBURG J, JOHNE R and KLEIN G (2009), 'Application of a swab sampling method for the detection of norovirus and rotavirus on artificially contaminated food and environmental surfaces', *Food Environ Virol*, **1**, 42–9.

SHAN X C, WOLFFS P and GRIFFITHS M W (2005), 'Rapid and quantitative detection of hepatitis A virus from green onion and strawberry rinses by use of real-time reverse transcription-PCR', *Appl Environ Microbiol*, **71**, 5624–6.

SHIEH Y C, KHUDYAKOV Y E, XIA G, GANOVA-RAEVA L M, KHAMBATY F M, WOODS J W, VEAZEY J E, MOTES M L, GLATZER M B, BIALEK S R and FIORE A E (2007), 'Molecular confirmation of oysters as the vector for hepatitis A in a 2005 multistate outbreak', *J Food Prot*, **70**, 145–50.

URBANUCCI A, MYRMEL M, BERG I, VON BONSDORFF C H and MAUNULA L (2009), 'Potential internalisation of caliciviruses in lettuce', *Int J Food Microbiol*, **135**, 175–8.

VASICKOVA P, PAVLIK I, VERANI M and CARDUCCI A (2010), 'Issues concerning survival of viruses on surfaces', *Food Environ Virol*, **2**, 24–34.

VERREAULT D, MOINEAU S and DUCHAINE C (2008), 'Methods for sampling of airborne viruses', *Microbiol Mol Biol Rev*, **72**, 413–444.

WANG D, WU Q, YAO L, WEI M, KOU X and ZHANG J (2008), 'New target tissue for food-borne virus detection in oysters', *Lett Appl Microbiol*, **47**, 405–9.

WOODS J W and BURKHARDT III W (2010), 'Occurrence of norovirus and hepatitis A virus in U.S. oysters', *Food Environ Virol*, **2**, 176–182.

WU H M, FORNEK M, SCHWAB K J, CHAPIN A R, GIBSON K, SCHWAB E, SPENCER C and HENNING K (2005), 'A norovirus outbreak at a long-term-care facility: the role of environmental surface contamination', *Infect Control Hosp Epidemiol*, **26**, 802–10.

WYN-JONES A P, CARDUCCI A, COOK N, D'AGOSTINO M, DIVIZIA M, FLEISCHER J, GANTZER C, GAWLER A, GIRONES R, HÖLLER C, DE RODA HUSMAN AM, KAY D, KOZYRA I, LOPEZ-PILA J, MUSCILLO M, SÃO JOSÉ NASCIMENTO M, PAPAGEORGIOU G, RUTJES S, SELLWOOD J, SZEWZYK R and WYER M (2011), 'Surveillance of adenoviruses and noroviruses in European recreational waters', *Water Res*, **45**, 1025–38.

4.8 Appendix: sampling from food and air

Virus detection in fruit – reported in Calder *et al.* (2003)

Sampling method

1. Randomly collect 6 samples consisting of minimum 100 g of frozen berries from different locations of suspicious or contaminated batch. Frozen fruit are usually distributed as pre-packed produce, so in this case take 6 bags of fruit. For soft fruit separate aseptically individual fruit pieces up to the minimum weight of 100 g into separate sterile containers.
2. After collection, fruit should be transported in a chilled or frozen state. Do not allow fruit to defrost or become juicy when longer storage is required before analysis.
   NOTE: Samples should be processed within 24 h after collection

Virus detection in shellfish – reported in Shieh *et al.* (2007)

Sampling method

1. At least 8 oysters from suspected or contaminated batch of shellfish should be aseptically collected into sterile containers. If shellfish are packed, make sure that the sampled unit contains the required number of molluscs.
   NOTE: Sampling can be performed on fresh or frozen shellfish
2. Shellfish should be transported to the laboratory as chilled or frozen, depending on their state during collection.
3. In the laboratory, take three to eight whole oysters to obtain a tested portion of 25 g of the shellfish meat (frozen oysters need to be thawed and aseptically separated from icy water and adductor muscle).
4. Tested portions can be stored at −70°C pending analysis.

Virus detection in aerosol – reported in Carducci *et al.* (2011) for detection with cultural methods, and for detection with biomolecular methods.

Sampling method

1. Aerosol samples are collected with an impactor sampler. For virus detection, 1000 l of air (indoor) and 3000 l (outdoor) are sampled on Rodac plates containing Tryptone Soy Agar (TSA).
2. After collection, plates should be transported in a chilled or frozen state. Do not allow plates to dry.
   NOTE: Samples should be processed within 24 h after collection

5

Molecular detection of viruses in water and sewage

G. La Rosa and M. Muscillo, Istituto Superiore di Sanità, Italy

DOI: 10.1533/9780857098870.2.97

Abstract: Enteric viruses are the causes of many sporadic cases and outbreaks originating from contaminated water. Valid and reproducible methods for the detection of waterborne viral pathogens are crucial in order to determine the extent of contamination, the types of pathogens involved and the correlation between viral contamination and environmental factors. Strategies have been developed to overcome the difficulties associated with virological analysis of water samples. Various assays are available for the detection of the major pathogenic viruses potentially present in urban sewage, drinking or recreational waters. Monitoring of sewerage systems provides valuable epidemiological information regarding serotypes circulating in the community.

Key words: viruses, water, sewage, concentration of viruses, molecular detection.

5.1 Introduction

Enteric virus contamination in recreational and drinking waters has been a major health concern worldwide in recent years. Since the 1980s, following advances in environmental virology, enteric viruses have been recognized as the causative agents of many sporadic cases and outbreaks of gastroenteritis and other illness originating from contaminated water. Viral transmission can occur through all types of water. This includes drinking water, recreational (fresh, marine and swimming pool), river and irrigation water. Despite the relatively low concentration of viruses in water, these micro-organisms can still pose a significant health risk, since they often require very low infectious doses.

Despite progress in water and wastewater treatment technology, waterborne diseases continue to have far-reaching public health consequences in both non-industrialized and industrialized countries.

The availability of valid and reproducible methods for the detection of waterborne viral pathogens is crucial in order to determine the extent of contamination, the types of pathogens involved and the correlation between viral contamination and environmental factors. Such information will lead to a better understanding of the associated health risks and to improved methods for control of waterborne viral pathogens, thus reducing their impact on public health.

The first section of the chapter covers methods for processing water samples for virological analysis and strategies developed to overcome the difficulties associated with such analysis.

The second section provides an overview of available assays for the detection of the major pathogenic viruses potentially present in urban sewage, drinking or recreational waters. Historically, the standard method for the isolation of human viruses from environmental water samples was based on the ability of viruses to grow in cell cultures. These assays require considerable resources and expertise and are labor-intensive and time-consuming. In addition, some viruses are impossible or difficult to grow in cell culture systems. Molecular methods, on the other hand, can detect non-culturable viruses and viruses that are difficult to cultivate, saving time, reducing costs, and increasing detection sensitivity. These methods, however, also have shortcomings: their sensitivity is often reduced by inhibitory compounds co-concentrated in water samples, resulting in false negatives. They may also yield ambiguous positive results by detecting non-infectious or inactivated virions. In response to the limitations of both cell cultures and molecular methods, integrated cell culture/polymerase chain reaction (PCR) methods were developed. Most recently, novel molecular methods have also been developed to preclude the detection of non-infectious viruses, as well as tools for the multiplex detection of pathogens in water using microarrays.

Finally the monitoring of sewerage systems is discussed as a valuable source of epidemiological information regarding serotypes circulating in the community which may not be detectable in clinical samples if infections are mild or asymptomatic, and as a potential early warning system for epidemics.

5.2 Sample treatment: adsorption-elution methods

The concentration of enteric viruses in the feces of infected individuals is extremely high (10^5–10^{11} virions per gram of stool). The degree of wastewater inflow contamination depends on the season, the prevalence of viral infections and the characteristics of circulating viruses. Sewage treatment can reduce viral contamination 10–1000-fold, but cannot eliminate it. When sewage effluent is discharged into fresh or marine waters these pathogens can find their way into recreational water, drinking water and agricultural products. Typically, the minimum infectious dose for human enteric viruses (10–100 virions) is lower than that of enteric bacteria, so that even a few viral particles in water can pose health risks.

The detection of viruses in water, as in other environmental samples, presents particular challenges. The basic steps of virological analysis of environmental waters include sampling, virus concentration and purification and detection, whether by cell culture assays or by molecular methods.

The concentration of viruses in water is usually too low to allow detection by direct analysis. A multi-stage process is therefore, required, involving the concentration of the potential virus-containing sample. The sample is processed so as to concentrate viruses into small volumes, usually less than 10 ml. The volume of water to be analysed and the degree of concentration required depends on the likely level of contamination. In groundwater or treated drinking water several thousand-fold concentrations may be needed to allow detection, and 100 l or more will need to be processed. Superficial waters may require a concentration of about 1000-fold and samples of about 10 l of water. In raw sewage, viruses may be detectable in small samples (20–50 ml) without previous concentration.

A good concentration method should be able to concentrate viral particles while at the same time avoiding the co-concentration of inhibitory compounds that could interfere with detection. Ideally, such a method should be easy and quick to implement, have a high virus recovery rate and the ability to concentrate a wide range of viruses, provide a small volume of concentrate, be relatively inexpensive, be capable of processing large volumes of water, and be both repeatable and reproducible (Block and Schwartzbrod, 1989; Wyn-Jones and Sellwood, 2001). No single method fulfills all the above criteria, however. Despite efforts to improve recovery for many waterborne enteric viruses, no method gives consistently high recovery rates. Moreover, the efficiency of viral recovery depends on the type of virus analysed. Three main techniques are used (each with numerous variations): adsorption/elution, ultrafiltration, and ultracentrifugation. These are based on different properties/characteristics of the viruses: ionic charge, particle size, and density and sedimentation coefficient. Methods usually consist of a two-stage concentration process: a first stage (primary concentration) which reduces the initial volume to between 100 and 1000 ml, and a second stage (secondary concentration) which reduces the volume further, to 5–20 ml.

5.2.1 Adsorption

Most methods employed for the concentration of viral particles from environmental water samples depend on adsorption of the virus to a solid matrix, such as a filter, membrane or cartridge, followed by viral elution. The adsorption of any specific virus to the matrix depends on particular conditions of pH and ionic strength. Once the virus is adsorbed, the water is discarded and the virus is then eluted from the matrix. Processing conditions, type of adsorbing matrix and eluting fluids are choices to be made based on the nature of the sample. The class of filters most commonly used for virus collection from large volumes of water is adsorption-elution microporous filters (0.2–0.45 μm), commonly known as VIRADEL. Systems for virus concentration usually consist of a pump, filter housing, and a flow metre. Although the pore sizes of the filters are considerably larger than the diameter of the viruses (up to 90 nm), viruses adsorb by electrostatic and hydrophobic interactions (Gerba, 1984). Two classes of adsorbent filters have been used to concentrate enteric viruses from freshwater and sewage effluents: negatively charged and positively charged filters.

Electronegative filtration

Concentration of viruses in water using negatively charged microporous filters has been practised for many years and there are many variations of the technique. Under ambient conditions, enteric viruses are negatively charged and will adsorb to a negatively charged matrix only in the presence of multivalent cations or, more commonly, under acidic conditions (when their net charge becomes positive). Therefore, the pH of the sample is to be reduced to levels lower than the isoelectric point of the viruses. The first filtration system using negatively charged filters, developed by Wallis and Melnick in 1967 (Wallis and Melnick, 1967), requires the addition of divalent or trivalent salts ($AlCl_3$ or $MgCl_2$) and an adjustment of the water sample to pH 3.5 prior to filtration (Wallis *et al.*, 1972). This method has been employed to concentrate viruses from different types of waters (Beuret, 2003; Fuhrman *et al.*, 2005; Goyal and Gerba, 1983; Haramoto *et al.*, 2004; Katayama *et al.*, 2002; Melnick *et al.*, 1984; Villar *et al.*, 2006). Negatively charged filters, such as nitrocellulose (Millipore HA), fibreglass (Filterite, the most widely used) and cellulose (Whatman) generally have small pore sizes (0.2–0.45 μm). This results in a tendency to clog which limits the volume of sample that can be filtered. To overcome this problem, membranes can be changed more frequently or larger cartridge filters used, where sheets of negatively charged plates are rolled into a cartridge holder (Farrah *et al.*, 1976). Electronegative filters are ideal when concentrating viruses from seawater, high in organic matter and turbidity (Gerba *et al.*, 1978). Katayama *et al.* (2002) developed a method capable of high virus recovery (with minimal inhibitory effect) to concentrate virus from seawater using negatively charged membranes (HA, Millipore) and an acid rinse step to remove cations and other inibitors. Haramoto modified this method to allow recovery of viruses from fresh water, by coating the HA membrane with $AlCl_3$ prior to filtering samples, obtaining high virus recoveries in seeded experiments (Haramoto *et al.*, 2004). In a recent study, Bofill-Mas and colleagues recovered viral particles from seawater samples using a nitrocellulose negatively charged membrane filter-based method and results showed highly variable recoveries ranging between 1.9 and 35.4% (Bofill-Mas *et al.*, 2010a).

Electropositive filtration

Electropositive filters have also been widely used for virus concentration. They can accommodate larger sample volumes due to their greater porosity and extensive surface area. These filters have the advantage that viruses can be concentrated from waters near neutral pH, without prior treatment. Sobsey and Jones first evaluated the positively charged filters on seeded poliovirus in tap water (Sobsey and Glass, 1980; Sobsey and Jones, 1979). These filters were later used to concentrate bacteriophages and viruses in a variety of media in seeded laboratory studies (Chang *et al.*, 1981; Goyal *et al.*, 1980; Logan *et al.*, 1980). Some of the positively charged filters have also been successfully used for the detection of naturally occurring viruses in swimming pools, drinking water and wastewater (Chapron *et al.*, 2000; Ma *et al.*, 1994; Melnick *et al.*, 1984; Pinto *et al.*, 1995; Rose *et al.*, 1984). Electropositive microfilters of different filter materials are available, all designed to capture viruses by electrostatic attraction. The electropositive 1 MDS filter is the most commonly used filter for the concentration of viruses from water, and is recommended by the United States Environmental Protection Agency (USEPA) for the recovery of waterborne human enteric viruses. Thanks to its electrostatic properties, this filter, designed specifically for virus capture and recovery from water, can also effectively capture waterborne bacteria and parasites (Polaczyk *et al.*, 2007). Elution efficiencies and associated recovery efficiencies, however, varied for each organism (Polaczyk *et al.*, 2007).

Advances in membrane technology have resulted in the development of charge-modified nylon membranes for the concentration of viruses from water. Nylon membranes are made in varying pore sizes and carry a positive surface charge, which strongly binds negatively charged virus particles (Wyn-Jones and Sellwood, 2001). A highly sensitive isolation protocol using filtration with a positively charged nylon membrane was developed by Gilgen and colleagues for the concentration and detection of different human pathogenic virus groups from water samples (Gilgen *et al.*, 1997). These filters have been shown to be useful to bind viruses in freshwater samples. Their performance in marine waters, however, is poor, as they tend to clog, since the salinity and alkalinity of seawater result in a low absorption of viruses to the filter (Fong and Lipp, 2005; Lukasik *et al.*, 2000). Indeed, electropositive filters showed lower virus recoveries from marine water and waters of high turbidity than did electronegative filters (Enriquez and Gerba, 1995; Katayama *et al.*, 2002; Lipp *et al.*, 2001; Lukasik *et al.*, 2000).

Recently, new, highly electropositive cartridge filters have been developed, able to adsorb viruses and bacteriophages from water over a wide range of pH, turbidity and salinity conditions (Tepper and Kaledin, 2007). These (NanoCeram, Argonide Corporation, Sanford, FL) are nonwoven pleated cartridge depth filters made of microglass filaments coated with nano alumina fibres derived from boehmite (Ikner *et al.*, 2011). They have an extensive surface area and an effective pore size of approximately 2.0 μm (Ikner *et al.*, 2011). NanoCeram cartridge filters have been tested and found useful as a primary step in the concentration of enteroviruses and noroviruses from source waters, having similar efficiency in the recovery of those viruses to electropositive cartridge filters (Karim *et al.*, 2009). Incidentally, NanoCeram cartridge filters are also less expensive than 1 MDS cartridge filters. Ikner and colleagues successfully used a NanoCeram virus sampler to concentrate seeded tap water; this method appeared to be compatible with both cell culture and PCR assays (Ikner *et al.*, 2011).

Glass wool filtration
Glass wool filtration is an economic and easy-to-implement method of virus concentration. Packed in a column at a defined density, glass wool is an efficient virus adsorbent. Thanks to the presence of both hydrophobic and electropositive sites on their surface, such filters are able to adsorb negatively charged virus particles at near-neutral pH. The technique was first applied to a range of viruses from surface, drinking and wastewaters (Vilagines *et al.*, 1993). Glass wool was subsequently used in virus monitoring studies involving wastewater (Gantzer *et al.*, 1997), seawater (Calgua *et al.*, 2008), drinking water and groundwater (Grabow *et al.*, 2001; van Heerden *et al.*, 2005; Vivier *et al.*, 2004). Recently, this method has been used to recover from water three types of enteric viruses (enterovirus, adenovirus, and norovirus), all on the US Environmental Protection Agency (EPA)'s Contaminant Candidate List. Recovery efficiencies, however, were highly variable (Lambertini *et al.*, 2008). This method has the advantage of diminishing PCR inhibition (Lambertini *et al.*, 2008; van Heerden *et al.*, 2005). Moreover, the fact that it works at or near neutral pH, and without the addition of cations, makes it suitable for viruses sensitive to acids. A recent paper by Bofill-Mas and colleagues describes the recovery of viral particles from fresh water performed by applying a procedure based on the use of glass wool columns for adsorption, followed by elution with glycine-beef extract buffer. Variable recoveries ranging from 6% to 81.5% were obtained (Bofill-Mas *et al.*, 2010a).

5.2.2 Elution

Following binding, viruses are desorbed and eluted from the filter matrix by pressure filtering a small volume (0.5–2 l) of the eluting solution. The eluent is usually a slightly alkaline proteinaceous fluid that commonly incorporates either beef extract, glycine or tryptose phosphate buffer (Brassard *et al.*, 2005; Horman *et al.*, 2004; Katayama *et al.*, 2002; Villar *et al.*, 2006). The most commonly used eluting solution for the recovery of viruses from both electronegative and electropositive filters is beef extract (~pH 9.0–9.5). It may be used at concentrations ranging from 1.5% to 3.0% (Cashdollar and Dahling, 2006; Dahling, 2002; Karim *et al.*, 2009; Lambertini *et al.*, 2008; Ma *et al.*, 1994) with or without glycine (Fout *et al.*, 1996; Karim *et al.*, 2009; Lambertini *et al.*, 2008; Melnick *et al.*, 1984). An important disadvantage of this method, however, is that it inhibits RT-PCR, an effect that is further exacerbated during flocculation procedures for second-step virus concentration (see below) (Abbaszadegan *et al.*, 1993).

Secondary concentration of the eluate

Depending on the filter used, the volume of the eluate can range from 400 ml to more than 1.6 l (Ikner *et al.*, 2011). A second concentration step may, therefore, be necessary to reduce the volume before testing. For this purpose, organic flocculation, polyethylene glycol (PEG) precipitation or celite concentration may be used.

Organic flocculation

Organic flocculation is the most widely used for the secondary concentration step following elution with beef extract. The pH is lowered to below the protein's isoelectric point (usually pH 3.5 ± 0.1), at which the proteins flocculate (Katzenelson *et al.*, 1976). Virus is adsorbed to the floc which is then centrifuged and dissolved in a small volume of neutral buffer. Beef extract can be replaced with skimmed milk (Payment *et al.*, 1983) or inorganic eluting medium (NaOH), the inhibitory effect of which is less pronounced than that of the beef extract. Calgua and colleagues used flocculation as a one-step low-cost procedure to concentrate viruses from seawater samples. Pre-flocculated skimmed milk solution 1% (w/v) was used to concentrate viruses directly from previously acidified (pH 3.5) 10 l seawater samples (Calgua *et al.*, 2008). A single-step concentration of viruses from marine waters by skimmed milk flocculation showed recoveries of 42.5–52.0%, consistent with those found by Calgua and colleagues (Bofill-Mas *et al.*, 2010a). Recently, the procedure based on skimmed milk flocculation has been optimized to be applied to a variety of fresh water samples (Calgua *et al.*, 2010).

Organic flocculation is the method recommended by the Information Collection Rule of the USEPA, and has been used extensively for the secondary concentration of viruses from environmental samples (Guttman-Bass and Armon, 1983; Hurst *et al.*, 1984; Katzenelson *et al.*, 1976; Morris and Waite, 1980; Sobsey and Glass, 1980).

Polyethylene glycol (PEG) precipitation

Concentration and purification of different types of viruses (e.g, enveloped and non-enveloped animal viruses, and bacteriophages) by polyethylene glycol (PEG) precipitation from a variety of aqueous matrices (Lewis and Metcalf, 1988a) has been a common practice in virology laboratories for many years (Huang *et al.*, 2000; Lewis and Metcalf, 1988; Li *et al.*, 1998; McSharry and Benzinger, 1970; Schwab *et al.*, 1996). In this technique, viruses in suspension are precipitated with PEG which essentially removes the water, allowing proteins and viruses to fall out of solution. The virus can then be readily harvested by low speed centrifugation. PEG precipitation has been found to be effective for concentration of enteric viruses (polioviruses, hepatitis A and noroviruses) in water environments (Schwab *et al.*, 1993). Nevertheless, this method has to be optimized for the type of water tested.

Other adsorbents
Other adsorbents can be used to concentrate viruses, for example, magnesium silicate mixed with celite (diatomaceous earth) (Sattar and Ramia, 1979; Sattar and Westwood, 1978). By manipulating pH, viruses are eluted from the celite in small volumes of phosphate buffer (Dahling and Wright, 1986; Dahling and Wright, 1988; Hill *et al.*, 1974; Karim *et al.*, 2009).

One of the major limitations of adsorption-elution methods for virus concentration is interference. Dissolved and colloidal substances in water, especially organic matter, can interfere with virus adsorption by competing with viruses for adsorption sites on filters. The result is a large variation in viral recovery rates due to differences in water chemistry (pH, salinity, presence of proteinaceous materials, humic compounds). In addition, adsorption-elution efficiencies depend on the enteric virus to be concentrated. Sobsey and Glass (1980), for example, reported different virus recovery rates in tap water samples for poliovirus 1, echovirus 1, and reovirus 3 using the electropositive Virosorb 1 MDS filter (Sobsey and Glass, 1980). Despite its limitations, virus adsorption-elution from microporous filters is the method of choice for concentrating enteric viruses from large volumes of water. A determination of recovery efficiency for seeded virus is recommended in parallel with the analysis of water samples.

5.3 Sample treatment: ultrafiltration and ultracentrifugation

The VIRADEL procedure has been the most commonly used method to recover and concentrate enteric viruses from water for decades. Ultrafiltration and ultracentrifugation represents alternatives to the adsorption and elution method for concentrating viral particles.

5.3.1 Ultrafiltration

An alternative method for concentrating viral particles from very large volumes of water samples is ultra filtration (UF). Here, virus is concentrated by virtue of its molecular size rather than by electric charge: molecules smaller than the pore size of the filter, including salts and water, pass through the membrane; larger molecules, including viruses, are withheld. The fluid is circulated through membrane layers (Divizia *et al.*, 1989a, 1989b) or hollow fibre cartridges (Belfort *et al.*, 1982) until the retentate reaches the desired volume. The volume obtained depends on the apparatus; if this is small enough then it may be analysed by detection procedures, or it may have to be further processed by secondary concentration. Ultrafiltration has also been used for second-step concentration (after adsorption/elution) with special micro-apparatuses designed for small volumes (Divizia *et al.*, 1989b). The most frequently used fibre systems have cut-off levels of 30–100 KDa. Some ultafilters employ tangential flow (TFF) or vortex flow (VFF) to reduce clogging (Paul *et al.*, 1991; Wommack *et al.*, 2010). Ultrafiltration methods such as TFF and VFF involve minimal manipulation of the water sample, which can be processed under natural pH, and does not require an elution step. The result is not only a simpler process, but also a more efficient one, since some of the procedures avoided here, as compared to the filtration-elution methods, can cause viral inactivation (i.e., pH adjustment) or PCR inhibition (i.e., beef extract elution and organic flocculation). Ultrafiltration methods, employed since the 1970s (Belfort *et al.*, 1974), have been gaining recognition as effective procedures for the concentration and recovery, from large volumes of water, of microbes in general, and of viruses in particular (Hill *et al.*, 2007). Their very small pore sizes (enough to filter out molecules having molecular weights in the order of 10–100 KDa), allow ultrafiltration membranes to simultaneously concentrate viruses, bacteriophages, bacteria, and parasites. Moreover, recovery efficiencies appear to be higher than those obtained using adsorption/elution techniques (Divizia *et al.*, 1989b; Patti *et al.*, 1996; Soule *et al.*, 2000) although, as with other methods, efficiency varies depending on factors such as sample composition, the process employed, the amount of backpressure used and the operator's skill. Both VFF and TFF are less cost- and time-effective than adsorption-elution because of the high cost of equipment and limitations on the volume of sample that can be concentrated at any one time. Over the last ten years, TFF methods have been used to concentrate viruses from surface and drinking waters (Gibson and Schwab, 2011; Grassi *et al.*, 2010; Jiang *et al.*, 2001; Muscillo *et al.*, 1999; Muscillo *et al.*, 2001; Skraber *et al.*, 2009).

5.3.2 Ultracentrifugation

Ultracentrifugation has the significant advantage of being able to efficiently sediment even the smallest viruses. The relatively small volumes that can be processed and the high cost and lack of portability of the equipment restrict its applicability in primary concentration, but it remains very useful in secondary concentration (Fumian *et al.*, 2010), with the limitation that supplementary stages may be required to remove PCR inhibitors following primary and secondary concentration. Ultracentrifugation has been shown to be an appropriate method for recovering viruses from wastewaters where viruses may be detectable directly in small sample volumes (Fumian *et al.*, 2010; La Rosa *et al.*, 2010b; Nordgren *et al.*, 2009).

5.4 Key assays for virus detection

Virus detection in water samples can be performed with either cell culture methods or molecular techniques. Molecular methods, which target viral genomes, offer speed, sensitivity and specificity. This section will provide an overview of available molecular assays for the detection of the major pathogenic viruses potentially present in urban sewage, drinking or recreational waters.

5.4.1 DNA/RNA extraction and purification

After virus elution or concentration, a variety of protocols may be employed to extract and purify nucleic acids by removing cell debris and inhibitors. All of these involve the lysis of cells followed by the removal of contaminants and DNA/RNA recovery. Removal of proteins is typically achieved by digestion with proteinase K, followed by phenol-chloroform extraction and ethanol precipitation. One of the most widely used methods for viral nucleic acid extraction and purification, known as Boom's method after the scholar who developed it (Boom *et al.*, 1990), is based on guanidinium thiocyanate extraction and the use of silica columns or beads to bind and wash nucleic acids. A wide variety of reliable, reproducible and easy-to-use commercial kits are available for nucleic acid extraction and purification. In recent years, a growing number of automated platforms have been developed, and there are multiple reports of their use in assays for pathogen detection (Albinana-Gimenez *et al.*, 2009; Burgener *et al.*, 2003; Knepp *et al.*, 2003; Kok *et al.*, 2000; Perelle *et al.*, 2009; Rutjes *et al.*, 2005). Most commercial nucleic acid extraction systems that incorporate a spin or vacuum column allow for efficient removal of inhibitors because the nucleic acid trapped in the silica gel membrane can be very effectively washed before elution. Methods incorporating magnetic bead techniques to capture and retain nucleic acid also achieve high recovery and purity rates (Ratcliff *et al.*, 2007).

Manual methods, being less expensive, are generally preferred over kit-based protocols for routine laboratory procedures. The choice of method depends on many factors: the required quantity and molecular weight of the DNA, the level of purity required, and time and expense considerations.

In environmental virology, the presence of PCR inhibitors is a major challenge, since these substances are able to reduce viral nucleic acid extraction efficiency and interfere with cDNA synthesis and/or polymerase activity, leading to an underestimation of target concentrations or to false negative results (Wilson, 1997). To overcome these problems, methodologies have been developed for assessing nucleic acid yields and DNA amplification efficiencies for environmental water samples (see below) using different types of internal controls (Gregory *et al.*, 2011; Hata *et al.*, 2011; La Rosa *et al.*, 2010a; Parshionikar *et al.*, 2004).

5.4.2 Viral detection

Historically, the standard method for the isolation of human viruses from environmental water samples was based on the ability of viruses to produce visible cytopathic effects (CPE) in cell cultures. These assays involve a labor-intensive and time-consuming procedure and, more importantly, are not universally applicable to all viruses. Molecular methods, on the other hand, which target viral genomes, are rapid procedures that can detect all viruses, save time, reduce costs, and increase detection sensitivity and specificity.

Advantages and disadvantages of molecular methods for the detection of viruses from water environments are summarized in Table 5.1.

5.4.3 Molecular methods

PCR, nested/semi-nested PCR

Molecular techniques are based on protocols of nucleic acid amplification, of which PCR is the most commonly used, allowing the detection of the amplified sequence at the reaction end-point. The specificity of PCR can be confirmed either by sequencing the PCR product, or by its hybridization to a labelled internal oligonucleotide probe. Modifications to the basic PCR technique have been used to increase the sensitivity, specificity, and efficiency of virus detection (e.g., by using nested and semi-nested primers), to allow the detection of more than one virus in a single assay (e.g., by using multiplex primers), or to quantify the number of viruses detected (by using real-time PCR). Nested/semi-nested PCR assays have been developed for a variety of enteric viruses and shown increased sensitivities as compared to conventional PCR assays (Allard *et al.*, 1992; Chapron *et al.*, 2000; Van Heerden *et al.*, 2003). Moreover, the two-step amplification mitigates the effect of inhibitory substances on the reaction. Inhibition of nucleic acid amplification is often partial (Gibson and Schwab, 2011; La Rosa *et al.*, 2010a). The first amplification can thus produce a small amount of template that may not be visible after gel electrophoresis of the PCR product, but can nonetheless be successfully amplified in a second round of PCR in which only a small aliquot (about 1/20) of first-cycle mixture is used. Nested PCR has been shown to have a high probability of carryover contamination, however.

Multiplex PCR (M-PCR) is the co-amplification of several nucleic acid sequences, and thus also virus types, in the same assay (Formiga-Cruz *et al.*, 2005; Fout *et al.*, 2003; Lee *et al.*, 2005; Verma and Arankalle, 2010). Its feasibility depends on the compatibility of the PCR primers used in the reaction. A study by Fout and collaborators showed that even under optimal conditions, five target virus groups could not be amplified efficiently (Fout *et al.*, 2003). Green and Lewis successfully detected enterovirus, rotavirus and Hepatitis A virus (HAV) in a single multiplex PCR, but a number of non-specific products were also obtained, requiring a second PCR to confirm results. M-PCR could potentially save considerable time and effort in the laboratory, but it must

Table 5.1 Molecular methods for the detection of enteric viruses from water environments: advantages and disadvantages

Method	Advantages	Disadvantages
PCR/RT-PCR	Rapid; highly sensitive; highly specific if assay is well-designed; less laborious and time-consuming than cell culture.	Non-quantitative; unable to assess infectivity; sensitive to the interference of inhibitors; need to confirm amplicons (e.g., by sequencing).
Nested/semi-nested-PCR	More sensitive and specific than conventional PCR; no need for confirmation step.	Potential risk of carryover contamination.
Multiplex PCR	Saves time and costs; ability to detect several viruses in a single reaction.	Sensitivity varies by target virus; potential non-specific amplification in environmental samples.
Real-time PCR	Can provide quantitative data; no need for confirmation step; reduced risk of contamination thanks to the closed system.	Expensive equipment and materials (labelled probes).
ICC-PCR	More sensitive than cell culture alone; detects viruses that do not – or are slow to – exhibit CPE in cell culture; requires considerably less time than conventional cell culture.	More costly and less time-efficient than PCR alone.
Non-PCR methods: NASBA	Continuous, isothermal process, thus not requiring a thermocycler, with comparable or better sensitivity than RT-PCR.	Reaction temperature cannot exceed 42°C. RNA target should be 120–250 nucleotides long, as shorter or longer sequences are amplified less efficiently.
Microarray	Capable of identifying multiple viral targets simultaneously, in a single assay.	Expensive equipment and materials.
Whole genome sequencing	High speed and throughput able to produce an enormous volume of sequences.	Complex sequence analysis; expensive equipment and materials.

be employed with caution using appropriate controls (internal amplification controls – IACs – see below) and only after prior optimization of the assay. Indeed, this technique requires extensive optimization because primer dimers and other non-specific products may interfere with the amplification of specific ones. The more targets are assayed in a multiplex reaction, the more likely both competition for reagents and inhibition between assays become. Hot Start PCR can reduce primer dimer formation and improve sensitivity. PCR specificity is also influenced by other factors, such as primer design, cycling conditions (temperature and duration of annealing and extension steps), and primer concentration. The time and expense required to optimize and validate multiplex PCR data may ultimately outweigh the potential benefits of the multiplex design.

Non-PCR methods: NASBA

Non-PCR methods have also been developed to amplify nucleic acids; the most widely used of these is nucleic acid sequence base amplification (NASBA). This technique is also known as isothermal amplification, since it does not require the temperature cycling used in conventional PCR methods. NASBA has been reported as useful for the detection of microbial pathogens in food and environmental samples (Cook, 2003; Rutjes *et al.*, 2005, 2006). The technique, reviewed by Cook in 2003 (Cook, 2003), was subsequently used successfully by Rutjes and colleagues for the detection of enteric viruses (enteroviruses and noroviruses) in surface waters (Rutjes *et al.*, 2005, 2006).

Quantitative real-time PCR

Quantitative real-time PCR (qPCR) represents an advance on traditional end-point PCR. It allows the detection of viruses by continuous monitoring ('real-time'), as well as their quantification. All real-time PCR systems rely upon a fluorescent reporter, the signal of which increases in direct proportion to the amount of PCR product in a reaction. The most economical reporter is the double-strand DNA-specific dye SYBR Green. The two most popular alternatives to SYBR Green are TaqMan and Molecular Beacons, both of which are hybridization probes relying on fluorescence resonance energy transfer (FRET) for quantitation (Parida, 2008). TaqMan probes and Molecular Beacons allow multiple DNA species to be measured in a single sample. The method is less time-consuming than traditional PCR because confirmation (e.g., agarose gel electrophoresis, additional hybridization or sequencing) is generally not required. The entire analysis can be done in a closed system, which may reduce the potential for contamination.

A number of home-made and commercial applications are available for the detection of viruses in water environments; these methods allow researchers to rapidly and specifically detect viruses of public health concern. Real-time PCR assays have also been shown to be more sensitive than conventional PCR in several studies (Beuret, 2004; Donaldson *et al.*, 2002; Noble *et al.*, 2003). The high cost of instruments required for qPCR assays is an important drawback that tends to limit the use of this method.

Microarrays

DNA microarrays, developed in the 1990s and significantly improved throughout the 2000s, is a method that uses probes to detect sequences that complement them. These probes, immobilized on solid surfaces or attached to small beads, can interrogate millions of sequences in a single assay. Microarrays have become a well-established strategy for the rapid detection, serotyping and genotyping of microbial pathogens, including foodborne and waterborne viruses, in the fields of food safety, medical and environmental monitoring, and gene expression profiling (Ayodeji *et al.*, 2009; Brinkman and Fout, 2009; Chizhikov *et al.*, 2002; Jaaskelainen and Maunula, 2006; Lemarchand *et al.*, 2004; Wang *et al.*, 2002). Brinkman and Fout (2009) investigated a generic microarray format as a tool to genotype noroviruses in environmental samples. The method was successful in the genotyping of noroviruses seeded into tap and river water samples.

'Next-generation' sequencing (NGS)

Novel DNA sequencing techniques, referred to as 'next-generation' sequencing (NGS), provide high speed and an enormous throughput in terms of sequences. This field is rapidly expanding and new platforms are continuously being developed and released, with many possible research as well as clinical applications. These NGS methods have different underlying biochemistries and differ in sequencing protocol, throughput and sequence lengths. Typical applications of NGS methods in virology include: detection of unknown viral pathogens and discovery of novel viruses; full-length viral genome sequencing; investigation of viral genome variability and characterization of viral quasispecies; epidemiology of viral infections; and viral evolution (Barzon *et al.*, 2011). Viral metagenomics analyses have been applied to environmental samples like fresh water, reused wastewater and ocean water (Angly *et al.*, 2006; Djikeng *et al.*, 2009; Lopez-Bueno *et al.*, 2009; Rosario *et al.*, 2009). These studies have shown that many of the detected viral sequences are unique and unknown. As the costs reduce, these techniques will probably become essential diagnostic tools in clinical routines.

5.5 Advantages and disadvantages of polymerase chain reaction (PCR) and related methods

The advantages of PCR and related methods are considerable. The main benefit is saving time: a result may be obtained after only 12–24 h as compared to several days or weeks for a cell culture assay. The design and synthesis of oligonucleotide primers has been simplified by the availability of computer software for primer design and of rapid commercial services for the synthesis of custom oligonucleotides. Optimization of PCR conditions may also be needed (e.g., annealing temperature, magnesium concentration, primer concentration). Once the optimal conditions for a reaction have been obtained, however, the reaction can be simply repeated and requires less operator skill and training than cell culture. Although PCR is an extremely sensitive tool, a drawback is that false positive results can be obtained from contaminating nucleic acid sequences. Since the same operations are performed repeatedly, the contamination of new tests with amplified products from previous amplifications of the same target sequence is the commonest cause of false positive results. A number of procedures to control PCR contamination are generally recognized and implemented. These include space and time separation of pre- and post-PCR activities, use of physical aids, use of ultraviolet (UV) light, use of aliquoted PCR reagents, incorporation of numerous positive and negative or blank PCRs (water substituted for template), and use of one or more contamination control methods involving chemical and biochemical reactions. The most widely used such technique, based on the bacterial enzyme UNG, relies on the substitution of uracil (dUTP) for thymine (dTTP) during PCR to generate amplification products with distinguishing characteristics relative to the native DNA template. When the bacterial enzyme Uracil N-glycosylase (UNG) is added to the PCR mix, selective hydrolyzation occurs, since only dUTP-containing amplicons are susceptible to hydrolysis by this enzyme, which results in the removal of the contaminating amplicons from the mix (Aslanzadeh, 2004).

PCR assays are also prone to false negative results due to inhibition by a variety of substances, target nucleic acid degradation, sample processing errors, thermal cycler malfunction, and – in reverse- transcription PCR – failure of the reverse transcription step. One method to control false negatives is to test for nucleic acid either intrinsically present in the sample or added as a control (IACs). Such nucleic acid, if co-extracted, can act both as an extraction and an inhibition control (Nolte, 2004). For a realistic assessment of the presence of inhibitors, an amplification control should be present only at levels close to the assay's detection limit. The inhibitory effect can be reduced by serial dilution of the extracted nucleic acid prior to PCR reaction (Brooks *et al.*, 2005; Hamza *et al.*, 2009). This is an excellent method in cases where a gene target is present in high copy numbers, but it can lead to false negatives when targets are diluted below their detection limit.

The testing of water environments for enteric viruses involves numerous steps, however, and most current internal control systems assess only the amplification step of this complex procedure. It is, for example, possible to obtain a positive IAC result in spite of failures in other steps such as recovery of viruses during the concentration step or reduced quality of the extracted RNA. To establish a broader control over the entire sample process from concentration, through elution and extraction to amplification, a sample process control can be used. Such a control, added to each water sample prior to processing, should be similar to the enteric viruses assayed, but not be expected to be present in the analysed matrix. Feline calicivirus (FCV), a respiratory virus that is not associated with human disease (Radford *et al.*, 2007), has been successfully used as a sample process control in the detection of RNA viruses from food and water for some time. It is a non-enveloped, positive-sense RNA virus for which cultivation and molecular detection techniques are readily available. The FCV standard provided a valuable quality control tool when testing potentially contaminated food and water samples (Mattison *et al.*, 2009). Recently, murine norovirus (MNV) has also been proposed as a sample process control (Hata *et al.*, 2011; Lee *et al.*, 2011). One study compared these two viruses as surrogates for the assessment of potential risks to public health associated with the presence in source waters of human enteric viruses that cannot be grown in cell cultures (e.g., noroviruses); MNV showed great promise as a human norovirus surrogate due to its genetic similarity and environmental stability, while FCV was much less stable (Bae and Schwab, 2008).

As noted above, PCR-based molecular tests may yield ambiguous positive results by detecting non-infectious or inactivated viruses. The inability to discriminate between infectious and non-infectious particles is of critical relevance in environmental virology. To overcome this problem, different studies have employed additional processing intended to preclude the detection of non-infectious viruses and increase the potential for detecting infectious viruses. An overview, as well as an evaluation of the strengths and weaknesses of current approaches to overcome the limitations of both traditional cell culture assays and PCR assays used for the detection of human enteric viruses in environmental water samples, can be found in the detailed review by Hamza and colleagues (Hamza *et al.*, 2011). Combination of cell culture with PCR (Integrated Cell Culture-PCR – ICC-PCR) has permitted the detection of slow-growing viruses, or viruses that fail to produce CPE. The method relies on an initial biological amplification of the viral nucleic acids, followed by PCR/RT-PCR or real-time PCR/RT-PCR assays. This method is designed to compensate for some of the disadvantages of cell culture, such as its limited sensitivity and the fact that it is time-consuming. It was first proposed as an alternative method for the detection of waterborne enteric viruses in environmental samples by Reynolds in 1996 (Reynolds *et al.*, 1996) and was later used for a wide variety of human waterborne pathogenic viruses in aquatic environments such as astrovirus (Abad *et al.*, 1997; Chapron *et al.*, 2000), adenoviruses (Chapron *et al.*, 2000; Lee and Jeong, 2004) rotaviruses (Li *et al.*, 2010), hepatitis A virus (Jiang *et al.*, 2004), enteroviruses (Balkin and Margolin, 2010; La Rosa *et al.*, 2010b; Lee and Jeong, 2004; Murrin and Slade, 1997) and reoviruses (Spinner and Di Giovanni, 2001).

Other methods include *in-vitro* specific capture of virus by cellular receptors and immunomagnetic separation to recover enteric viruses from environmental water samples (Casas and Sunen, 2002; Gilpatrick *et al.*, 2000; Jothikumar *et al.*, 1998). Here, antibody-coated paramagnetic beads bind to antigens present on the surface of the target pathogen, thus facilitating the magnetic concentration of the latter (Hamza *et al.*, 2011). The method has the advantage of being able to discriminate between infectious and non-infectious viruses. If the viral capsid is damaged, the antigenic properties of the virus may change, and specific viral antigen–antibody complexes may not form, leading to negative PCR results. Limitations of the method include the fact that the antibody may be unable to target all possible strains of the virus under study, making strain-specific assays necessary, as well as the fact that the antigen-binding affinity depends on several factors, including the amount of colloidal particles and the pH value of the water sample (Hamza *et al.*, 2011).

To test the infectivity of DNA viruses that do not replicate well in cell culture (e.g., enteric adenovirus), a method based on the detection of virus-specific mRNA present in the cell during virus replication may be used (Ko *et al.*, 2003).

An alternative molecular approach to assess virus infectivity is the analysis of long target viral genome regions. The rationale behind this 'long template' RT-PCR strategy is that the integrity of viral genome may correlate with virus infectivity (Allain *et al.*, 2006; Li *et al.*, 2004; Simonet and Gantzer, 2006; Wolf *et al.*, 2009). The fact that viral inactivation does not necessarily compromise genome integrity, however (e.g., when inactivation occurs through the degradation of surface proteins), limits the applicability of this strategy.

In general, molecular techniques are likely to play an increasing role in the detection and characterization of enteric viruses and have also proved to be important in the diagnosis of emerging virus infections. Current techniques, however, need to be further simplified, automated, and standardized, before they can be widely applied to the diagnosis of viruses in clinical and environmental contexts. Validation is a key issue in scientific research, and molecular studies are no exception. The primary goal of method validation is to ensure the accuracy of reported results. A valid method must be repeatable, that is, it must consistently deliver the same result for a given sample when repeated several times and/or in different laboratories. An essential tool for the attainment of repeatable results is the standardization of laboratory methods. In molecular biology today, the multiplicity of assay methods, standard reference materials and targets for molecular diagnostics contribute to inter-laboratory variability in test results. Proficiency testing is the use of inter-laboratory comparisons to determine the performance of individual laboratories for specific tests or measurements, and to monitor a laboratory's performance.

There are many standards and guidelines issued by numerous organizations. Documentary standards from the ISO (International Organization for Standardization) and CEN (the European Committee for Standardization) provide guidance on the use of reference materials and methods. Holden and colleagues (Holden *et al.*, 2011) recently published a comprehensive overview of reference materials, related documentary standards and proficiency testing programs which can be helpful tools for molecular diagnostic laboratories.

5.6 Current applications and results

Over 100 types of pathogenic viruses can be found in human wastes. All these viruses can spread in the environment through water of all kinds (Fong and Lipp, 2005) where sewage is untreated or insufficiently treated. Moreover, several groups of enteric pathogens are resistant to current water treatment processes and may represent potentially emerging waterborne pathogens. These include circoviruses (torque tenovirus and torque tenovirus-like virus), picobirnaviridae, parvoviruses, and polyomaviruses (including JC virus and BK virus).

The prevalence, environmental distribution and health implications of the presence of viruses in water for drinking and recreational use are extensively documented in other parts of the present volume and will thus not be addressed here.

In recent years, more attention has been focused on the virological quality of sewage as well as on the analysis of sewage as an instrument of environmental surveillance and as a proxy for clinical surveillance.

5.6.1 Pathogen surveillance through monitoring of urban sewage

Sewer systems collect pathogens excreted in a range of bodily fluids during active infection from a wide area to a central facility. The monitoring of centralized wastewater enables natural, accidental, or intentional contamination events to be detected (Sinclair *et al.*, 2008). Moreover, untreated wastewater provides a rich matrix in which novel viruses can be identified and virus diversity studied. In the late 1960s Nelson and collaborators first noted the relationship between the occurrence of poliovirus in sewage and the clinical incidence of disease in a community (Nelson *et al.*, 1967). Environmental poliovirus surveillance systems were later adopted, especially in populations with high vaccine coverage. This surveillance is considered an extremely important component of an early warning system. Deshpande and colleagues detected 'silent' wild type poliovirus three months before any clinical cases were observed, demonstrating the great sensitivity of sewage surveillance (Deshpande *et al.*, 2003). Small outbreaks of enteroviruses and adenoviruses have also been predicted by monitoring the sewage of the respective affected communities (Sinclair *et al.*, 2008). Recently, a large Finnish echovirus 30 outbreak was preceded by silent circulation of the same genotype, detected in sewage. It has been suggested that recombination with co-circulating divergent lineages may have empowered the pre-epidemic strains to cause a nationwide outbreak (Savolainen-Kopra *et al.*, 2011).

In the past few years several studies have demonstrated the advantage of environmental surveillance as an additional tool to determine the epidemiology of different viruses circulating in a given community (Barril *et al.*, 2010; Bofill-Mas *et al.*, 2000; Heijnen and Medema, 2011; Iwai *et al.*, 2009; Kokkinos *et al.*, 2011; La Rosa *et al.*, 2010a; Meleg *et al.*, 2008; Pinto *et al.*, 2007). The clinical surveillance of infections requiring hospital visits provides some insight into circulating genotypes, but as a monitoring tool, this method misses mild, asymptomatic or subclinical infections. Genotypes not detected clinically would, however, be shed into local sewage. For example, different studies showed a higher environmental prevalence of noroviruses genogroup I relative to genogroup II (da Silva *et al.*, 2007; La Rosa *et al.*,

2010a,b; Nordgren *et al.,* 2009). This is seemingly at odds with epidemiological data on the occurrence of norovirus outbreaks and sporadic cases, mostly attributable to norovirus genogroup II. This discrepancy, suggesting that GI does circulate in human populations to a significant degree, has prompted different studies to address possible differences in the pathogenic potential of these genogroups, their ability to evade the human immune system, and their stability in water environments.

Richter and collaborators recently studied the circulation of non-polio enteroviruses in the population of Cyprus and assessed the clinical relevance of different serotypes through the analysis of both clinical specimens and sewage (Richter *et al.*, 2011). All serotypes detected in sewage were also present in clinical samples, but the relative frequencies of the different serotypes in the two kinds of samples were significantly different. Phylogenetic analysis revealed that certain enterovirus strains circulate in the population over long periods of time, while others are observed only sporadically and disappear quickly. Another study compared the specimens isolated from Finnish sewage with clinical isolates over a period of 20 years (Hovi *et al.*, 1996). Here, too, sewage and clinical isolates differed in the relative frequency of the serotypes detected. Several explanations for this phenomenon were proposed: (i) differences in the general virulence of different serotypes, (ii) differences in the mean amount of virus excreted by an infected person, or (iii) different rates of inactivation in the environment.

In some cases, data on the prevalence of specific pathogens in untreated wastewaters formed the basis for subsequent clinical studies in the population. For example, GIV noroviruses, a genogroup scarcely known, rarely tested for and only sporadically identified in human patients, were detected in raw sewage samples in Italy. This finding, suggesting that the virus circulates in human populations, later prompted a clinical study and the discovery of GIV norovirus in patients with severe gastroenteritis symptoms (La Rosa *et al.*, 2008). Similarly, environmental studies aimed at investigating the circulation of Hepatitis E viruses (HEV) in industrialized countries, where the acute form of this infection is rarely diagnosed, have demonstrated a sustained circulation of HEV in non-endemic regions (Clemente-Casares *et al.*, 2009; La Fauci *et al.*, 2010; La Rosa *et al.*, 2010c; Rodriguez-Manzano *et al.*, 2010). The later detection of sporadic cases of acute hepatitis E (travel-related and/or of zoonotic origin) in such countries (Cacciola *et al.*, 2011; La Rosa *et al.*, 2011; Veitt *et al.*, 2011) confirmed the above environmental findings.

Bofill-Mas and colleagues recently analysed the presence and characteristics of newly described human polyomaviruses in urban sewage in order to assess the excretion level and the potential role of water as a transmitter of these viruses (Bofill-Mas *et al.*, 2010b). This was the first report on the presence of a virus strongly related to human cancer in sewage and river water samples. The data obtained indicate that the Merkel cell polyomavirus circulates in the population and that it may be disseminated through the fecal/urine contamination of water.

Finally, untreated wastewater provides a rich matrix in which novel viruses can be identified. Cantalupo and colleagues recently explored viral diversity by deep sequencing nucleic acids obtained from virion populations enriched from raw sewage (Cantalupo *et al.*, 2011). They identified 234 known viruses, including 17 that infect humans. Classifiable novel viruses represented 51 different families, but the vast majority of sequences showed little or no relation to known viruses and thus could not be placed into specific taxa. Untreated wastewater, therefore, represents the most diverse viral metagenome examined thus far, and our knowledge of the viral universe with which we are liable to come into contact is limited to a tiny fraction of the viruses that exist.

5.7 References

ABAD, F.X., PINTO, R.M., VILLENA, C., GAJARDO, R., and BOSCH, A. 1997. Astrovirus survival in drinking water. *Appl. Environ. Microbiol.*, **63**, 3119–22.

ABBASZADEGAN, M., HUBER, M.S., GERBA, C.P., and PEPPER, I.L. 1993. Detection of enteroviruses in groundwater with the polymerase chain reaction. *Appl. Environ. Microbiol.*, **59**(5), 1318–24.

ALBINANA-GIMENEZ, N., CLEMENTE-CASARES, P., CALGUA, B., HUGUET, J.M., COURTOIS, S., and GIRONES, R. 2009. Comparison of methods for concentrating human adenoviruses, polyomavirus JC and noroviruses in source waters and drinking water using quantitative PCR. *J. Virol. Methods*, **158**(1–2), 104–109 available from: PM:19428577.

ALLAIN, J.P., HSU, J., PRANMETH, M., HANSON, D., STASSINOPOULOS, A., FISCHETTI, L., CORASH, L., and LIN, L. 2006. Quantification of viral inactivation by photochemical treatment with amotosalen and UV A light, using a novel polymerase chain reaction inhibition method with preamplification. *J. Infect. Dis.*, **194**(12), 1737–44.

ALLARD, A., ALBINSSON, B., and WADELL, G. 1992. Detection of adenoviruses in stools from healthy persons and patients with diarrhea by two-step polymerase chain reaction. *J. Med. Virol.*, **37**(2), 149–157 available from: PM:1629713.

ANGLY, F.E., FELTS, B., BREITBART, M., SALAMON, P., EDWARDS, R.A., CARLSON, C., CHAN, A.M., HAYNES, M., KELLEY, S., LIU, H., MAHAFFY, J.M., MUELLER, J.E., NULTON, J., OLSON, R., PARSONS, R., RAYHAWK, S., SUTTLE, C.A., and ROHWER, F. *et al.* 2006. The marine viromes of four oceanic regions. *PLoS Biol.*, **4(11)**, e368.

ASLANZADEH, J. 2004. Preventing PCR amplification carryover contamination in a clinical laboratory. *Ann. Clin Lab Sci.*, **34**(4), 389–96.

AYODEJI, M., KULKA, M., JACKSON, S.A., PATEL, I., MAMMEL, M., CEBULA, T.A., and GOSWAMI, B.B. 2009. A microarray based approach for the identification of common foodborne viruses. *Open. Virol. J.*, **3**, 7–20.

BAE, J. and SCHWAB, K.J. 2008. Evaluation of murine norovirus, feline calicivirus, poliovirus, and MS2 as surrogates for human norovirus in a model of viral persistence in surface water and groundwater. *Appl. Environ. Microbiol.*, **74**(2), 477–84 available from: PM:18065626.

BALKIN, H.B. and MARGOLIN, A.B. 2010. Detection of poliovirus by ICC/qPCR in concentrated water samples has greater sensitivity and is less costly using BGM cells in suspension as compared to monolayers. *Virol. J.*, **7**, 282.

BARRIL, P.A., GIORDANO, M.O., ISA, M.B., MASACHESSI, G., FERREYRA, L.J., CASTELLO, A.A., GLIKMANN, G., and NATES, S.V. 2010. Correlation between rotavirus A genotypes detected in hospitalized children and sewage samples in 2006, Cordoba, Argentina. *J. Med. Virol.*, **82**(7), 1277–81.

BARZON, L., LAVEZZO, E., MILITELLO, V., TOPPO, S., and PALÙ, G. 2011. Applications of next-generation sequencing technologies to diagnostic virology. *Int. J. Mol. Sci.*, **12**(11), 7861–84.

BELFORT, G., PALUSZEK, A., and STURMAN L.S. 1982. Enterovirus concentration using automated hollow fiber ultrafiltration. *Water Sci. Technol.*, **14**(4–5), 257–72.

BELFORT, G., ROTEM, Y., and KATZENELSON, E. 1974. Virus concentration using hollow fiber membranes. *Wat. Res.*, **9**, 79–85.

BEURET, C. 2003. A simple method for isolation of enteric viruses (noroviruses and enteroviruses) in water. *J. Virol. Methods.*, **107**(1), 1–8.

BEURET, C. 2004. Simultaneous detection of enteric viruses by multiplex real- time RT-PCR. *J. Virol. Methods*, **115**(1), 1–8 available from: ISI:000187795500001.

BLOCK, J.C. and SCHWARTZBROD, L. 1989. *Viruses in Water Systems. Detection and Identification*. New York: VCH Publishers.

BOFILL-MAS, S., CALGUA, B., CLEMENTE-CASARES, P., LA ROSA, G., IACONELLI, M., MUSCILLO, M., RUTJES, S., DE RODA HUSMAN, A.M., GRUNERT, A., GRÄBER, I., VERANI, M., CARDUCCI, A., CALVO, M., WYN-JONES, P., and GIRONES, R. *et al.* 2010a. Quantification of human adenoviruses in European recreational waters. *Food Environ. Virol.*, **2**, 101–9.

BOFILL-MAS, S., PINA, S., and GIRONES, R. 2000. Documenting the epidemiologic patterns of polyomaviruses in human populations by studying their presence in urban sewage. *Appl. Environ. Microbiol.*, **66**(1), 238–45, available from: PM:10618230.

BOFILL-MAS, S., RODRIGUEZ-MANZANO, J., CALGUA, B., CARRATALA, A., and GIRONES, R. 2010b. Newly described human polyomaviruses Merkel cell, KI and WU are present in urban sewage and may represent potential environmental contaminants. *Virol. J.*, **7**, 141.

BOOM, R., SOL, C.J., SALIMANS, M.M., JANSEN, C.L., WERTHEIM-VAN DILLEN, P.M., and VAN DER NOORDAA, J. 1990. Rapid and simple method for purification of nucleic acids. *J. Clin. Microbiol.*, **28**, 495–503.

BRASSARD, J., SEYER, K., HOUDE, A., SIMARD, C., and TROTTIER, Y.L. 2005. Concentration and detection of hepatitis A virus and rotavirus in spring water samples by reverse transcription-PCR. *J. Virol. Methods*, **123**(2), 163–9.

BRINKMAN, N.E. and FOUT, G.S. 2009. Development and evaluation of a generic tag array to detect and genotype noroviruses in water. *J. Virol. Methods.*, **156**(1–2), 8–18.

BROOKS, H.A., GERSBERG, R.M., and DHAR, A.K. 2005. Detection and quantification of hepatitis A virus in seawater via real-time RT-PCR. *J. Virol. Methods*, **127**(2), 109–18.

BURGENER, M., CANDRIAN, U., and GILGEN, M. 2003. Comparative evaluation of four large-volume RNA extraction kits in the isolation of viral RNA from water samples. *J. Virol. Methods*, **108**(2), 165–70.

CACCIOLA, I., MESSINEO, F., CACOPARDO, B., DI, M., V, GALLI, C., SQUADRITO, G., MUSOLINO, C., SAITTA, C., POLLICINO, T., and RAIMONDO, G. 2011. Hepatitis E virus infection as a cause of acute hepatitis in Southern Italy. *Dig. Liver Dis.*, **43**, (12) 996–1000.

CALGUA, B., FUMIAN, T., RUSINOL, M., RODRIGUES, J., BOFILL, S., MIAGOSTOVICH, M., and GIRONES, R. 2010. Detection and quantification of classical and emerging viruses in river water by applying a low cost one-step procedure. *J. Virol. Methods*, **153**, 79–83.

CALGUA, B., MENGEWEIN, A., GRUNERT, A., BOFILL-MAS, S., CLEMENTE-CASARES, P., HUNDESA, A., WYN-JONES, A.P., LOPEZ-PILA, J.M., and GIRONES, R. 2008. Development and application of a one-step low cost procedure to concentrate viruses from seawater samples. *J. Virol. Methods*, **153**(2), 79–83 available from: PM:18765255.

CANTALUPO, P.G., CALGUA, B., ZHAO, G., HUNDESA, A., WIER, A.D., KATZ, J.P., GRABE, M., HENDRIX, R.W., GIRONES, R., WANG, D., and PIPAS, J.M. 2011. Raw sewage harbors diverse viral populations. *MBio.*, **2**(5), e00180–11.

CASAS, N. and SUNEN, E. 2002. Detection of enteroviruses, hepatitis A virus and rotaviruses in sewage by means of an immunomagnetic capture reverse transcription-PCR assay. *Microbiol. Res.*, **157**(3), 169–75, available from: PM:12398285.

CASHDOLLAR, J.L. and DAHLING, D.R. 2006. Evaluation of a method to re-use electropositive cartridge filters for concentrating viruses from tap and river water. *J. Virol. Methods*, **132**, (1–2), 13–17, available from: PM:16194574.

CHANG, L.T., FARRAH, S.R., and BITTON, G. 1981. Positively charged filters for virus recovery from wastewater treatment plant effluents. *Appl. Environ. Microbiol.*, **42**(5), 921–24.

CHAPRON, C.D., BALLESTER, N.A., FONTAINE, J.H., FRADES, C.N., and MARGOLIN, A.B. 2000. Detection of astroviruses, enteroviruses, and adenovirus types 40 and 41 in surface waters collected and evaluated by the information collection rule and an integrated cell culture-nested PCR procedure. *Appl. Environ. Microbiol.*, **66**(6), 2520–5.

CHIZHIKOV, V., WAGNER, M., IVSHINA, A., HOSHINO, Y., KAPIKIAN, A.Z., and CHUMAKOV, K. 2002. Detection and genotyping of human group A rotaviruses by oligonucleotide microarray hybridization. *J. Clin. Microbiol.*, **40**(7), 2398–407.

CLEMENTE-CASARES, P., RODRIGUEZ-MANZANO, J., and GIRONES, R. 2009. Hepatitis E virus genotype 3 and sporadically also genotype 1 circulate in the population of Catalonia, Spain. *J. Water Health*, **7**(4), 664–73, available from: PM:19590134.

COOK, N. 2003. The use of NASBA for the detection of microbial pathogens in food and environmental samples. *J. Microbiol. Methods*, **53**(2), 165–74, available from: PM:12654488.

DA SILVA, A.K., LE SAUX, J.C., PARNAUDEAU, S., POMMEPUY, M., ELIMELECH, M., and LE GUYADER, F. 2007. Evaluation of removal of noroviruses during wastewater treatment, using real-time reverse transcription-PCR: different behaviors of genogroups I and II. *Appl. Environ. Microbiol.*, **73**(24), 7891–7, available from: PM:17933913.

DAHLING, D.R. 2002. An improved filter elution and cell culture assay procedure for evaluating public groundwater systems for culturable enteroviruses. *Water Environ. Res.*, **74**(6), 564–8, available from: PM:12540097.

DAHLING, D.R. and WRIGHT, B.A. 1986. Recovery of viruses from water by a modified flocculation procedure for second-step concentration. *Appl. Environ. Microbiol.*, **51**(6), 1326–31.

DAHLING, D.R. and WRIGHT, B.A. 1988. A comparison of recovery of virus from wastewaters by beef extract-Celite, ferric chloride, and filter concentration procedures. *J. Virol. Methods*, **22**(2–3), 337–46, available from: PM:3220927.

DESHPANDE, J.M., SHETTY, S.J., and SIDDIQUI, Z.A. 2003. Environmental surveillance system to track wild poliovirus transmission. *Appl. Environ. Microbiol.*, **69**(5), 2919–27.

DIVIZIA, M., DE FILIPPIS, P., DI NAPOLI, A., GABRIELI, R., SANTI, A.L., and PANA, A. 1989a. HAV recovery from tap water: evaluation of different types of membranes. *Ann. Ig*, **1**(1–2), 57–64, available from: PM:2483083.

DIVIZIA, M., SANTI, A.L., and PANA, A. 1989b. Ultrafiltration: an efficient second step for hepatitis A virus and poliovirus concentration. *J. Virol. Methods*, **23**(1), 55–62, available from: PM:2536381.

DJIKENG, A., KUZMICKAS, R., ANDERSON, N.G., SPIRO, D.J., and PALÙ, G. 2009. Metagenomic analysis of RNA viruses in a fresh water lake. *PLoS One*, **4(9)**, e7264.

DONALDSON, K.A., GRIFFIN, D.W., and PAUL, J.H. 2002. Detection, quantitation and identification of enteroviruses from surface waters and sponge tissue from the Florida Keys using real-time RT-PCR. *Water Res.*, **36**(10), 2505–14.

ENRIQUEZ, C.E. and GERBA, C.P. 1995. Concentration of enteric adenovirus 40 from tap, sea and waste water. *Water Res.*, **29**, 2554–60.

FARRAH, S.R., GERBA, C.P., WALLIS, C., and MELNICK, J.L. 1976. Concentration of viruses from large volumes of tap water using pleated membrane filters. *Appl. Environ. Microbiol.*, **31**(2), 221–226.

FONG, T.T. and LIPP, E.K. 2005. Enteric viruses of humans and animals in aquatic environments: health risks, detection, and potential water quality assessment tools. *Microbiol. Mol. Biol. R.*, **69**(2), 357–71.

FORMIGA-CRUZ, M., HUNDESA, A., CLEMENTE-CASARES, P., ALBINANA-GIMENEZ, N., ALLARD, A., and GIRONES, R. 2005. Nested multiplex PCR assay for detection of human enteric viruses in shellfish and sewage. *J. Virol. Methods*, **125**(2), 111–18, available from: PM:15794979.

FOUT, G.S., MARTINSON, B.C., MOYER, M.W., and DAHLING, D.R. 2003. A multiplex reverse transcription-PCR method for detection of human enteric viruses in groundwater. *Appl. Environ. Microbiol.*, **69**(6), 3158–64, available from: PM:12788711.

FOUT, G.S., SCHAEFER III, F.W., MESSER, J.W., DAHLING, D.R., and STETLER, R.E. 1996. *Information Collection Rule (ICR) Microbial Laboratory Manual.* EPA/600/R-95/178. U.S. Environmental Protection Agency, Washington, DC.

FUHRMAN, J.A., LIANG, X., and NOBLE, R.T. 2005. Rapid detection of enteroviruses in small volumes of natural waters by real-time quantitative reverse transcriptase PCR. *Appl. Environ. Microbiol.*, **71**(8), 4523–30.

FUMIAN, T.M., LEITE, J.P., CASTELLO, A.A., GAGGERO, A., CAILLOU, M.S., and MIAGOSTOVICH, M.P. 2010. Detection of rotavirus A in sewage samples using multiplex qPCR and an evaluation of the ultracentrifugation and adsorption-elution methods for virus concentration. *J. Virol. Methods.*, **170**(1–2), 42–6.

GANTZER, C., SENOUCI, S., MAUL, A., LEVI, Y., and SCHWARTZBROD, L. 1997. Enterovirus genomes in wastewater: Concentration on glass wool and glass powder and detection by RT-PCR. *J. Virol. Methods*, **65**, 265–71.

GERBA, C.P. 1984. Applied and theoretical aspects of virus adsorption to surfaces. *Adv. Appl. Microbiol.*, **30**, 133–68.

GERBA, C.P., FARRAH, S.R., GOYAL, S.M., WALLIS, C., and MELNICK, J.L. 1978. Concentration of enteroviruses from large volumes of tap water, treated sewage, and seawater. *Appl. Environ. Microbiol.*, **35**(3), 540–8.

GIBSON, K.E. and SCHWAB, K.J. 2011. Tangential-flow ultrafiltration with integrated inhibition detection for recovery of surrogates and human pathogens from large-volume source water and finished drinking water. *Appl. Environ. Microbiol.*, **77**(1), 385–91.

GILGEN, M., GERMANN, D., LUTHY, J., and HUBNER, P. 1997. Three-step isolation method for sensitive detection of enterovirus, rotavirus, hepatitis A virus, and small round structured viruses in water samples. *Int. J. Food Microbiol.*, **37**(2–3), 189–99.

GILPATRICK, S.G., SCHWAB, K.J., ESTES, M.K., and ATMAR, R.L. 2000. Development of an immunomagnetic capture reverse transcription-PCR assay for the detection of Norwalk virus. *J. Virol. Methods*, **90**(1), 69–78.

GOYAL, S.M. and GERBA, C.P. 1983. Viradel method for detection of rotavirus from seawater. *J. Virol. Methods.*, **7**(5–6), 279–85.

GOYAL, S.M., HANSSEN, H., and GERBA, C.P. 1980. Simple method for the concentration of influenza virus from allantoic fluid on microporous filters. *Appl. Environ. Microbiol.*, **39**(3), 500–4.

GRABOW, W.O., TAYLOR, M.B., and DE VILLIERS, J.C. 2001. New methods for the detection of viruses: call for review of drinking water quality guidelines. *Water Sci. Technol.*, **43**(12), 1–8.

GRASSI, T., BAGORDO, F., IDOLO, A., LUGOLI, F., GABUTTI, G., and DE DONNO, A. 2010. Rotavirus detection in environmental water samples by tangential flow ultrafiltration and RT-nested PCR. *Environ. Monit. Assess.*, **164**(1–4), 199–205.

GREGORY, J.B., WEBSTER, L.F., GRIFFITH, J.F., and STEWART, J.R. 2011. Improved detection and quantitation of norovirus from water. *J. Virol. Methods*, **172**(1–2), 38–45.

GUTTMAN-BASS, N. and ARMON, R. 1983. Concentration of simian rotavirus SA-11 from tap water by membrane filtration and organic flocculation. *Appl. Environ. Microbiol.*, **45**(3), 850–5.

HAMZA, I.A., JURZIK, L., STANG, A., SURE, K., UBERLA, K., and WILHELM, M. 2009. Detection of human viruses in rivers of a densly-populated area in Germany using a virus adsorption elution method optimized for PCR analyses. *Water Res.*, **43**(10), 2657–68, available from: PM:19361832.

HAMZA, I.A., JURZIK, L., UBERLA, K., and WILHELM, M. 2011. Methods to detect infectious human enteric viruses in environmental water samples. *Int. J. Hyg. Environ. Health*, **214**(6), 424–36.

HARAMOTO, E., KATAYAMA, H., and OHGAKI, S. 2004. Detection of noroviruses in tap water in Japan by means of a new method for concentrating enteric viruses in large volumes of freshwater. *Appl. Environ. Microbiol.*, **70**(4), 2154–60, available from: PM:15066808.

HATA, A., KATAYAMA, H., KITAJIMA, M., VISVANATHAN, C., NOL, C., and FURUMAI, H. 2011. Validation of internal controls for extraction and amplification of nucleic acids from enteric viruses in water samples. *Appl. Environ. Microbiol.*, **77**(13), 4336–43.

HEIJNEN, L. and MEDEMA, G. 2011. Surveillance of influenza A and the pandemic influenza A (H1N1) 2009 in sewage and surface water in the Netherlands. *J. Water Health*, **9**(3), 434–42.

HILL, V.R., KAHLER, A.M., JOTHIKUMAR, N., JOHNSON, T.B., HAHN, D., and CROMEANS, T.L. 2007. Multistate evaluation of an ultrafiltration-based procedure for simultaneous recovery of enteric microbes in 100-liter tap water samples. *Appl. Environ. Microbiol.*, **73**(13), 4218–25.

HILL Jr, W.F., AKIN, E.W., BENTON, W.H., MAYHEW, C.J., and METCALF, T.G. 1974. Recovery of poliovirus from turbid estuarine water on microporous filters by the use of celite. *Appl. Microbiol.*, **27**(3), 506–12.

HOLDEN, M.J., MADEJ, R.M., MINOR, P., and KALMAN, L.V. 2011. Molecular diagnostics: harmonization through reference materials, documentary standards and proficiency testing. *Expert Rev. Mol. Diagn.*, **11**(7), 741–55.

HORMAN, A., RIMHANEN-FINNE, R., MAUNULA, L., VON BONSDORFF, C.H., TORVELA, N., HEIKINHEIMO, A., and HANNINEN, M.L. 2004. *Campylobacter spp., Giardia spp., Cryptosporidium spp.*, noroviruses, and indicator organisms in surface water in southwestern Finland, 2000–2001. *Appl. Environ. Microbiol.*, **70**(1), 87–95, available from: PM:14711629.

HOVI, T., STENVIK, M., and ROSENLEW, M. 1996. Relative abundance of enterovirus serotypes in sewage differs from that in patients: clinical and epidemiological implications. *Epidemiol. Infect.*, **116**, 91–7.

HUANG, P.W., LABORDE, D., LAND, V.R., MATSON, D.O., SMITH, A.W., and JIANG, X. 2000. Concentration and detection of caliciviruses in water samples by reverse transcription-PCR. *Appl. Environ. Microbiol.*, **66**(10), 4383–8, available from: PM:11010887.

HURST, C.J., DAHLING, D.R., SAFFERMAN, R.S., and GOYKE, T. 1984. Comparison of commercial beef extracts and similar materials for recovering viruses from environmental samples. *Can. J. Microbiol.*, **30**(10), 1253–63, available from: PM:6095985.

IKNER, L.A., SOTO-BELTRAN, M., and BRIGHT, K.R. 2011. New method using a positively charged microporous filter and ultrafiltration for concentration of viruses from tap water. *Appl. Environ. Microbiol.*, **77**(10), 3500–6.

IWAI, M., HASEGAWA, S., OBARA, M., NAKAMURA, K., HORIMOTO, E., TAKIZAWA, T., KURATA, T., SOGEN, S., and SHIRAKI, K. 2009. Continuous presence of noroviruses and sapoviruses in raw sewage reflects infections among inhabitants of Toyama, Japan (2006 to 2008). *Appl. Environ. Microbiol.*, **75**(5), 1264–70.

JAASKELAINEN, A.J. and MAUNULA, L. 2006. Applicability of microarray technique for the detection of noro- and astroviruses. *J. Virol. Methods*, **136**(1–2), 210–16, available from: PM:16781784.

JIANG, S., NOBLE, R., and CHU, W. 2001. Human adenoviruses and coliphages in urban runoff-impacted coastal waters of Southern California. *Appl. Environ. Microbiol.*, **67**(1), 179–184 available from: PM:11133443.

JIANG, Y.J., LIAO, G.Y., ZHAO, W., SUN, M.B., QIAN, Y., BIAN, C.X., and JIANG, S.D. 2004. Detection of infectious hepatitis A virus by integrated cell culture/strand-specific reverse transcriptase-polymerase chain reaction. *J Appl. Microbiol.*, **97**(5), 1105–12.

JOTHIKUMAR, N., CLIVER, D.O., and MARIAM, T.W. 1998. Immunomagnetic capture PCR for rapid concentration and detection of Hepatitis A virus from environmental samples. *Appl. Environ. Microbiol.*, **64**(2), 504–8.

KARIM, M.R., RHODES, E.R., BRINKMAN, N., WYMER, L., and FOUT, G.S. 2009. New electropositive filter for concentrating enteroviruses and noroviruses from large volumes of water. *Appl. Environ. Microbiol.*, **75**(8), 2393–9 available from: PM:19218410.

KATAYAMA, H., SHIMASAKI, A., and OHGAKI, S. 2002. Development of a virus concentration method and its application to detection of enterovirus and norwalk virus from coastal seawater. *Appl. Environ. Microbiol.*, **68**(3), 1033–9, available from: PM:11872447.

KATZENELSON, E., FATTAL, B., and HOSTOVESKY, T. 1976a. Organic flocculation: an efficient second-step concentration method for the detection of viruses in tap water. *Appl. Environ. Microbiol.*, **32**(4), 638–9.

KNEPP, J.H., GEAHR, M.A., FORMAN, M.S., and VALSAMAKIS, A. 2003. Comparison of automated and manual nucleic acid extraction methods for detection of enterovirus RNA. *J. Clin. Microbiol.*, **41**(8), 3532–36.

KO, G., CROMEANS, T.L., and SOBSEY, M.D. 2003. Detection of infectious adenovirus in cell culture by mRNA reverse transcription-PCR. *Appl. Environ. Microbiol.*, **69**(12), 7377–7384 available from: PM:14660388.

KOK, T., WATI, S., BAYLY, B., DEVONSHIRE-GILL, D., and HIGGINS, G. 2000. Comparison of six nucleic acid extraction methods for detection of viral DNA or RNA sequences in four different non-serum specimen types. *J. Clin. Virol.*, **16**(1), 59–63.

KOKKINOS, P.A., ZIROS, P.G., MPALASOPOULOU, A., GALANIS, A., and VANTARAKIS, A. 2011. Molecular detection of multiple viral targets in untreated urban sewage from Greece. *Virol. J.*, **8**:195.

LA FAUCI, V., SINDONI, D., GRILLO, O.C., CALIMERI, S., LO, G.D., and SQUERI, R. 2010. Hepatitis E virus (HEV) in sewage from treatment plants of Messina University Hospital and of Messina City Council. *J. Prev. Med. Hyg.*, **51**(1), 28–30.

LA ROSA, G., IACONELLI, M., POURSHABAN, M., and MUSCILLO, M. 2010a. Detection and molecular characterization of noroviruses from five sewage treatment plants in central Italy. *Water Res.*, **44**(6), 1777–84.

LA ROSA, G., MUSCILLO, M., SPURI, V., GARBUGLIA, A.R., LA, S.P., and CAPOBIANCHI, M.R. 2011. Hepatitis E Virus In Italy: molecular analysis of travel-related and autochthonous cases. *J.Gen. Virol.*, **92**(Pt 7), 1617–1626.

LA ROSA, G., POURSHABAN, M., IACONELLI, M., and MUSCILLO, M. 2008. Detection of genogroup IV noroviruses in environmental and clinical samples and partial sequencing through rapid amplification of cDNA ends. *Arch. Virol.*, **153**(11), 2077–83.

LA ROSA, G., POURSHABAN, M., IACONELLI, M., and MUSCILLO, M. 2010b. Quantitative real-time PCR of enteric viruses in influent and effluent samples from wastewater treatment plants in Italy. *Ann. Ist. Super. Sanita.*, **46**(3), 266–73.

LA ROSA, G., POURSHABAN, M., IACONELLI, M., VENNARUCCI, V.S., and MUSCILLO, M. 2010c. Molecular detection of hepatitis E virus in sewage samples. *Appl. Environ. Microbiol.*, **76**(17), 5870–3.

LAMBERTINI, E., SPENCER, S.K., BERTZ, P.D., LOGE, F.J., KIEKE, B.A., and BORCHARDT, M.A. 2008. Concentration of enteroviruses, adenoviruses, and noroviruses from drinking water by use of glass wool filters. *Appl. Environ. Microbiol.*, **74**(10), 2990–2996.

LEE, H., KIM, M., PAIK, S.Y., LEE, C.H., JHEONG, W.H., KIM, J., and KO, G. 2011. Evaluation of electropositive filtration for recovering norovirus in water. *J. Water Health*, **9**(1), 27–36.

LEE, H.K. and JEONG, Y.S. 2004. Comparison of total culturable virus assay and multiplex integrated cell culture-PCR for reliability of waterborne virus detection. *Appl. Environ. Microbiol.*, **70**(6), 3632–3636.

LEE, S.H., LEE, C., LEE, K.W., CHO, H.B., and KIM, S.J. 2005. The simultaneous detection of both enteroviruses and adenoviruses in environmental water samples including tap water with an integrated cell culture-multiplex-nested PCR procedure. *J. Appl. Microbiol.*, **98**(5), 1020–29.

LEMARCHAND, K., MASSON, L., and BROUSSEAU, R. 2004. Molecular biology and DNA microarray technology for microbial quality monitoring of water. *Crit. Rev. Microbiol.*, **30**(3), 145–72.

LEWIS, G.D. and METCALF, T.G. 1988a. Polyethylene glycol precipitation for recovery of pathogenic viruses, including hepatitis A virus and human rotavirus, from oyster, water, and sediment samples. *Appl. Environ. Microbiol.*, **54**(8), 1983–8, available from: PM:2845860.

LI, D., GU, A.Z., YANG, W., HE, M., HU, X.H., and SHI, H.C. 2010. An integrated cell culture and reverse transcription quantitative PCR assay for detection of infectious rotaviruses in environmental waters. *J. Microbiol. Methods.*, **82**(1), 59–63.

LI, J.W., WANG, X.W., RUI, Q.Y., SONG, N., ZHANG, F.G., OU, Y.C., and CHAO, F.H. 1998. A new and simple method for concentration of enteric viruses from water. *J. Virol. Methods*, **74**(1), 99–108.

LI, J.W., XIN, Z.T., WANG, X.W., ZHENG, J.L., and CHAO, F.H. 2004. Mechanisms of inactivation of hepatitis A virus in water by chlorine dioxide. *Water Res.*, **38**(6), 1514–19.

LIPP, E.K., LUKASIK, J., and ROSE, J.B. 2001. Human enteric viruses and parasites in the marine environment. *Methods Microbiol.*, **30**, 559–88.

LOGAN, K.B., REES, G.E., SEELEY, N.D., and PRIMROSE, S.B. 1980. Rapid concentration of bacteriophages from large volumes of freshwater: evaluation of positively charged, microporous filters. *J. Virol. Methods.*, **1**(2), 87–97.

LOPEZ-BUENO, A., TAMAMES, J., VELAZQUEZ, D., MOYA, A., QUESADA, A., and ALCAMI, A. 2009. High diversity of the viral community from an Antarctic lake. *Science*, **326**, 858–61.

LUKASIK, J., SCOTT, T.M., ANDRYSHAK, D., and FARRAH, S.R. 2000. Influence of salts on virus adsorption to microporous filters. *Appl. Environ. Microbiol.*, **66**(7), 2914–20.

MA, J.F., NARANJO, J., and GERBA, C.P. 1994. Evaluation of MK filters for recovery of enteroviruses from tap water. *Appl. Environ. Microbiol.*, **60**(6), 1974–7.

MATTISON, K., BRASSARD, J., GAGNE, M.J., WARD, P., HOUDE, A., LESSARD, L., SIMARD, C., SHUKLA, A., PAGOTTO, F., JONES, T.H., and TROTTIER, Y.L. 2009. The feline calicivirus as a sample process control for the detection of food and waterborne RNA viruses. *Int. J. Food Microbiol.*, **132**(1), 73–7.

MCSHARRY, J. and BENZINGER, R. 1970. Concentration and purification of vesicular stomatitis virus by polyethylene glycol 'precipitation'. *Virology*, **40**(3), 745–746.

MELEG, E., BANYAI, K., MARTELLA, V., JIANG, B., KOCSIS, B., KISFALI, P., MELEGH, B., and SZUCS, G. 2008. Detection and quantification of group C rotaviruses in communal sewage. *Appl. Environ. Microbiol.*, **74**(11), 3394–9.

MELNICK, J.L., SAFFERMAN, R., RAO, V.C., GOYAL, S., BERG, G., DAHLING, D.R., WRIGHT, B.A., AKIN, E., STETLER, R., SORBER, C., MOORE, B., SOBSEY, M.D., MOORE, R., LEWIS, A. L. and WELLINGS, F.M. 1984. Round robin investigation of methods for the recovery of poliovirus from drinking water. *Appl. Environ. Microbiol.*, **47**(1), 144–50, available from: PM:6320720.

MORRIS, R. and WAITE, W. 1980. Evaluation of procedures for recovery of viruses from water—1 concentration systems. *Wat. Res.*, **14**(7), 791–3.

MURRIN, K. and SLADE, J. 1997. Rapid detection of viable enteroviruses in water by tissue culture and semi-nested polymerase chain reaction. *Water Sci. Technol.*, **35**, 429–32.

MUSCILLO, M., LA ROSA, G., CARDUCCI, A., CANTIANI, L., and MARIANELLI, C. 1999. Molecular and biological characterization of poliovirus 3 strains isolated in Adriatic seawater samples. *Wat. Res.*, **33**(14), 3204–12.

MUSCILLO, M., LA ROSA, G., MARIANELLI, C., ZANIRATTI, S., CAPOBIANCHI, M.R., CANTIANI, L., and CARDUCCI, A. 2001. A new RT-PCR method for the identification of reoviruses in seawater samples. *Wat. Res.*, **35**(2), 548–56, available from: PM:11229010.

NELSON, D.B., CIRCO, R., and EVANS, A.S. 1967. Strategic viral surveillance of sewage during and following an oral poliovirus vaccine campaign. *Am. J. Epidemiol.*, **86**(3), 641–52.

NOBLE, R.T., ALLEN, S.M., BLACKWOOD, A.D., CHU, W., JIANG, S.C., LOVELACE, G.L., SOBSEY, M.D., STEWART, J.R., and WAIT, D.A. 2003. Use of viral pathogens and indicators to differentiate between human and non-human fecal contamination in a microbial source tracking comparison study. *J. Water Health*, **1**(4), 195–207, available from: PM:15382724.

NOLTE, F.S. 2004. Novel internal controls for real-time PCR assays. *Clin. Chem.*, **50**(5), 801–2.

NORDGREN, J., MATUSSEK, A., MATTSSON, A., SVENSSON, A., and LINDGREN, P. 2009. Prevalence of norovirus and factors influencing virus concentrations during one-year in a fullscale wastewater treatment plant. *Wat. Res.*, **43**(4), 1117–25.

PARIDA, M.M. 2008. Rapid and real-time detection technologies for emerging viruses of biomedical importance. *J. Biosci.*, **33**(4), 617–28.

PARSHIONIKAR, S.U., CASHDOLLAR, J., and FOUT, G.S. 2004. Development of homologous viral internal controls for use in RT-PCR assays of waterborne enteric viruses. *J. Virol. Methods.*, **121**(1), 39–48.

PATTI, A.M., AULICINO, F.A., SANTI, A.L., MUSCILLO, M., ORSINI, P., BELLUCCI, C., LA ROSA, G., MASTROENI, I., and VOLTERRA, L. 1996. Enteric virus pollution of tyrrhenian areas. *Wat. Air. Soil Pollut.*, **88**(3–4), 261–7.

PAUL, J.H., JIANG, S.C., and ROSE, J.B. 1991. Concentration of viruses and dissolved DNA from aquatic environments by vortex flow filtration. *Appl. Environ. Microbiol.*, **57**(8), 2197–204.

PAYMENT, P., AYACHE, R., and TRUDEL, M. 1983. A survey of enteric viruses in domestic sewage. *Can. J. Microbiol.*, **29**(1), 111–19.

PERELLE, S., CAVELLINI, L., BURGER, C., BLAISE-BOISSEAU, S., HENNECHART-COLLETTE, C., MERLE, G., and FACH, P. 2009. Use of a robotic RNA purification protocol based on the NucliSens easyMAG for real-time RT-PCR detection of hepatitis A virus in bottled water. *J. Virol. Methods*, **157**(1), 80–83, available from: PM:19114058.

PINTO, R.M., ALEGRE, D., DOMINGUEZ, A., EL-SENOUSY, W.M., SANCHEZ, G., VILLENA, C., COSTAFREDA, M.I., ARAGONES, L., and BOSCH, A. 2007. Hepatitis A virus in urban sewage from two Mediterranean countries. *Epidemiol. Infect.*, **135**(2), 270–73, available from: PM:16817987.

PINTO, R.M., GAJARDO, R., ABAD, F.X., and BOSCH, A. 1995. Detection of fastidious infectious enteric viruses in water. *Environ. Sci. Technol.*, **29**(10), 2636–8.

POLACZYK, A.L., ROBERTS, J.M., and HILL, V.R. 2007. Evaluation of 1MDS electropositive microfilters for simultaneous recovery of multiple microbe classes from tap water. *J. Microbiol. Methods*, **68**(2), 260–6.

RADFORD, A.D., COYNE, K.P., DAWSON, S., PORTER, C.J., and GASKELL, R.M. 2007. Feline calicivirus. *Vet. Res.*, **38**(2), 319–35.

RATCLIFF, R.M., CHANG, G., KOK, T., and SLOOTS, T.P. 2007. Molecular diagnosis of medical viruses. *Curr. Issues Mol. Biol.*, **9**(2), 87–102.

REYNOLDS, K.A., GERBA, C.P., and PEPPER, I.L. 1996. Detection of infectious enteroviruses by an integrated cell culture-PCR procedure. *Appl. Environ. Microbiol.*, **62**(4), 1424–7.

RICHTER, J., TRYFONOS, C., and CHRISTODOULOU, C. 2011. Circulation of enteroviruses in Cyprus assessed by molecular analysis of clinical specimens and sewage isolates. *J. Appl. Microbiol.*, **111**(2), 491–8.

RODRIGUEZ-MANZANO, J., MIAGOSTOVICH, M., HUNDESA, A., CLEMENTE-CASARES, P., CARRATALA, A., BUTI, M., JARDI, R., and GIRONES, R. 2010. Analysis of the evolution in the circulation of HAV and HEV in eastern Spain by testing urban sewage samples. *J. Water Health*, **8**(2), 346–54.

ROSARIO, K., NILSSON, C., LIM, Y.W., RUAN, Y., and BREITBART, M. 2009. Metagenomic analysis of viruses in reclaimed water. *Environ. Microbiol.*, **11**, 2806–20.

ROSE, J.B., SINGH, S.N., GERBA, C.P., and KELLEY, L.M. 1984. Comparison of microporous filters for concentration of viruses from wastewater. *Appl. Environ. Microbiol.*, **47**(5), 989–92.

RUTJES, S.A., ITALIAANDER, R., VAN DEN BERG, H.H., LODDER, W.J., and DE RODA HUSMAN, A.M. 2005. Isolation and detection of enterovirus RNA from large-volume water samples by using the NucliSens miniMAG system and real-time nucleic acid sequence-based amplification. *Appl. Environ. Microbiol.*, **71**(7), 3734–40, available from: PM:16000783.

RUTJES, S.A., VAN DEN BERG, H.H., LODDER, W.J., and DE RODA HUSMAN, A.M. 2006. Real-time detection of noroviruses in surface water by use of a broadly reactive nucleic acid sequence-based amplification assay. *Appl. Environ. Microbiol.*, **72**(8), 5349–58, available from: PM:16885286.

SATTAR, S. and WESTWOOD, J.C. 1978. Viral pollution of surface waters due to chlorinated primary effluents. *Appl. Environ. Microbiol.*, **36**(3), 427–31.

SATTAR, S.A. and RAMIA, S. 1979. Use of talc-celite layers in the concentration of enteroviruses from large volumes of potable waters. *Wat. Res.*, **13**(7), 637–43.

SAVOLAINEN-KOPRA, C., PAANANEN, A., BLOMQVIST, S., KLEMOLA, P., SIMONEN, M.L., LAPPALAINEN, M., VUORINEN, T., KUUSI, M., LEMEY, P., and ROIVAINEN, M. 2011. A large Finnish echovirus 30 outbreak was preceded by silent circulation of the same genotype. *Virus Genes*, **42**(1), 28–36.

SCHWAB, K.J., DE LEON, R., and SOBSEY, M.D. 1993. Development of PCR methods for enteric virus detection in water. *Wat. Sci. Technol.*, **27**(3–4), 211–18.

SCHWAB, K.J., DE LEON, R., and SOBSEY, M.D. 1996. Immunoaffinity concentration and purification of waterborne enteric viruses for detection by reverse transcriptase PCR. *Appl. Environ. Microbiol.*, **62**(6), 2086–94, available from: PM:8787407.

SIMONET, J. and GANTZER, C. 2006. Degradation of the poliovirus 1 genome by chlorine dioxide. *J. Appl. Microbiol.*, **100**(4), 862–70, available from: PM:16553743.

SINCLAIR, R.G., CHOI, C.Y., RILEY, M.R., and GERBA, C.P. 2008. Pathogen surveillance through monitoring of sewer systems. *Adv. Appl. Microbiol.*, **65**, 249–269, available from: PM:19026868.

SKRABER, S., GANTZER, C., HELMI, K., HOFFMANN, L., and CAUCHIE, H.M. 2009. Simultaneous concentration of enteric viruses and protozoan parasites: a protocol based on tangential flow filtration and adapted to large volumes of surface and drinking waters. *Food Environ. Virol.*, **1**(2), 66–76.

SOBSEY, M.D. and GLASS, J.S. 1980. Poliovirus concentration from tap water with electropositive adsorbent filters. *Appl. Environ. Microbiol.*, **40**(2), 201–10, available from: PM:6258472.

SOBSEY, M.D. and JONES, B.L. 1979. Concentration of poliovirus from tap water using positively charged microporous filters. *Appl. Environ. Microbiol.*, **37**(3), 588–95, available from: PM:36844.

SOULE, H., GENOULAZ, O., GRATACAP-CAVALLIER, B., CHEVALLIER, P., LIU, J.-X., and SEIGNEURIN, J.-M. 2000. Ultrafiltration and reverse transcription-polymerase chain reaction: an efficient process for poliovirus, rotavirus and hepatitis A virus detection in water. *Wat. Res.*, **34**(3), 1063–7.

SPINNER, M.L. and DI GIOVANNI, G.D. 2001. Detection and identification of mammalian reoviruses in surface water by combined cell culture and reverse transcription-PCR. *Appl. Environ. Microbiol.*, **67**(7), 3016–20.

TEPPER, F. and KALEDIN, L. 2007. *Virus and Protein Separation Using Nano Alumina Fiber Media.* Available at: http://www.argonide.com/publications/bioseparations.pdf.

VAN HEERDEN, J., EHLERS, M.M., HEIM, A., and GRABOW, W.O. 2005. Prevalence, quantification and typing of adenoviruses detected in river and treated drinking water in South Africa. *J. Appl. Microbiol.*, **99**(2), 234–42, available from: PM:16033453.

VAN HEERDEN, J., EHLERS, M.M., Van ZYL, W.B., and GRABOW, W.O.K. 2003. Incidence of adenoviruses in raw and treated water. *Wat. Res.*, **37**(15), 3704–8.

VEITT, R., REICHARDT, M., WENZEL, J., and JILG, W. 2011. Autochthonous hepatitis E-virus infection as cause of acute hepatitis in Germany – a case report. *Z. Gastroenterol.*, **49**(1), 42–6.

VERMA, V. and ARANKALLE, V.A. 2010. Hepatitis E virus-based evaluation of a virion concentration method and detection of enteric viruses in environmental samples by multiplex nested RT-PCR. *J. Appl. Microbiol.*, **108**(5), 1630–41.

VILAGINES, P., SARRETTE, B., HUSSON, G., and VILAGINES, R. 1993. Glass wool for virus concentration at ambient water pH level. *Wat. Sci. Technol.*, **27**(3–4), 299–306.

VILLAR, L.M., de PAULA, V.S., DINIZ-MENDES, L., LAMPE, E., and GASPAR, A.M. 2006. Evaluation of methods used to concentrate and detect hepatitis A virus in water samples. *J. Virol. Methods.*, **137**(2), 169–76.

VIVIER, J.C., EHLERS, M.M., and GRABOW, W.O. 2004. Detection of enteroviruses in treated drinking water. *Wat. Res.*, **38**(11), 2699–705.

WALLIS, C., HENDERSON, M., and MELNICK, J.L. 1972. Enterovirus concentration on cellulose membranes. *Appl. Microbiol.*, **23**(3), 476–80.

WALLIS, C. and MELNICK, J.L. 1967. Concentration of enteroviruses on membrane filters. *J. Virol.*, **1**(3), 472–477.

WANG, D., COSCOY, L., ZYLBERBERG, M., AVILA, P.C., BOUSHEY, H.A., GANEM, D., and DERISI, J.L. 2002. Microarray-based detection and genotyping of viral pathogens. *Proc. Natl. Acad. Sci. USA*, **99**(24), 15687–92.

WILSON, I.G. 1997. Inhibition and facilitation of nucleic acid amplification. *Appl. Environ. Microbiol.*, **63**(10), 3741–51.

WOLF, S., RIVERA-ABAN, M., and GREENING, G.E. 2009. Long-range reverse transcription as a useful tool to assess the genomic integrity of norovirus. *Food Environ. Virol.*, **1**, 129–36.

WOMMACK, K.E., SIME-NGANDO, T., WINGET, D.M., JAMINDAR, S., and HELTON, R.R. 2010. 'Filtration-based methods for the collection of viral concentrates from large water samples', In Wilhelm, S., Weinbauer, M. and Suttle, C., eds., *Manual of Aquatic Viral Ecology*, Waco, TX, USA: American Society of Limnology and Oceanography, 110–17.

WYN-JONES, A.P. and SELLWOOD, J. 2001. Enteric viruses in the aquatic environment. *J. Appl. Microbiol.*, **91**(6), 945–62, available from: PM:11851802.

6

Quality control in the analytical laboratory: analysing food- and waterborne viruses

M. S. D'Agostino, Food and Environment Research Agency (Fera), UK

DOI: 10.1533/9780857098870.2.126

Abstract: Current ISO standards describe the main controls for the molecular detection of enteric viruses in food and the environment. This chapter clarifies these descriptions and the results which are likely to be obtained, and also suggests ways of rectifying any problems which may be realised when interpreting the results. Additional controls not included in ISO standards are also described. A more comprehensive suite of controls for molecular detection should prove useful to analysts wishing to declare a food/environmental sample target-free.

Key words: sample treatment controls, nucleic acid amplification, amplification controls, pathogen detection, food, environment, enteric viruses.

6.1 Introduction

This chapter outlines a variety of techniques that may be employed for quality control (QC) in an analytical laboratory focusing on virus analysis. Reliable data for analysis of viruses in food and the environment depend on the strict observance of a wide range of operating procedures. Analysts in different laboratories (and in different countries) may use a range of different terms and acronyms, depending on the standards they are using, and this can cause confusion. A list of acronyms used in this chapter, with their definitions, is provided in Table 6.1.

In most cases, analytical laboratories which routinely test food/environmental samples for enteric viruses will already have a comprehensive quality control program in place, covering all stages from sampling to reporting of

Table 6.1 Acronyms

Acronym	Meaning
Cp value	The qPCR cycle at which fluorescence intensity rises above the background
EAC	External amplification control
IAC	Internal amplification control
NAA	Nucleic acid amplification
NEC/EB	Negative extraction control/extraction blank
NNAAC[a]	Negative NAA control[a]
NPC	Negative process control
PCR	Polymerase chain reaction
PNAAC[a]	Positive NAA control[a]
PPC	Positive process control
qPCR[b]	Quantitative polymerase chain reaction
RT-qPCR[b]	Reverse transcription qPCR
SPC	Sample process control
RT-NAA	Reverse transcription nucleic acid amplification
SPC-NC	Sample process control – negative control

[a] In the case of specific NAA-based assays such as PCR, loop-mediated amplification (LAMP) or nucleic acid sequence-based amplification (NASBA), these terms can be changed to reflect the NAA assay used (e.g., PPCRC, NLAMPC, etc.).
[b] The term 'qPCR' is used for quantitative 'real-time' PCR, and 'RT-qPCR' is used for reverse transcription qPCR throughout this article, in accordance with the recommendations of Bustin *et al.* (2009).

results. Since this can be a time-consuming and costly exercise, it is important to ensure that the definitions and requirements are set out clearly and adhered to. The document 'ISO 7218:2007, Microbiology of Food and Animal Feeding Stuffs – General Requirements and Guidance for Microbiological Examinations' will be invaluable in this regard.

Over the years, there has been a great increase in the number of NAA-based assays for foodborne pathogen detection, with analysts using non-proprietorial NAA-based methods, as opposed to commercially available complete NAA detection systems, having a particularly wide choice of methods. A series of International Standards has been developed as a result of progress in PCR-based detection of foodborne bacterial pathogens (see Table 6.2). Molecular detection methods for foodborne viruses require the integration of the principles used for bacterial pathogens found within these standards, but with the addition of specific controls relating to viruses. The detection of viruses in foods poses a greater challenge than that for most foodborne bacteria. Enteric viruses are sub-microscopically small, generally only around thirty millionths of a millimeter in diameter. Due to their low infectious dose they only need to be present as contaminants in very low numbers in a foodstuff to constitute a risk to health. They do not change the appearance or sensory qualities of food, and their presence cannot be detected by sight or smell. They are incapable of growth in food, and unlike

Table 6.2 International Standards relating to PCR-based methods for pathogen detection

ISO 20837:2006	Microbiology of food and animal feedstuffs. Polymerase chain reaction (PCR) for the detection of foodborne pathogens. Requirements for sample detection for qualitative detection. Refers to ISO 22174 for controls.
ISO 20838:2006	Microbiology of food and animal feedstuffs. Polymerase chain reaction (PCR) for the detection of foodborne pathogens. Requirements for amplification and detection for qualitative methods. Refers to ISO 22174 for controls.
ISO 22118:2011	Microbiology of food and animal feedstuffs. Polymerase chain reaction (PCR) for the detection of foodborne pathogens. Performance characteristics. Refers to ISO 22174 for controls. Includes requirements for validation.
ISO 22119:2011	Microbiology of food and animal feedstuffs. Real-time polymerase chain reaction (PCR) for the detection of foodborne pathogens. General requirements and definitions. Refers to ISO 22174 for controls; ISO 22119 specifically mentions the internal amplification control.

bacteria, their numbers in a food sample cannot be increased by enrichment. In addition, they cannot yet be cultured in a laboratory, although research is underway to overcome this issue.

The most effective methods for detection of foodborne viruses are those based on viral nucleic acid amplification (NAA) (Croci *et al.*, 2008). There are two basic processes: sample treatment and detection assay. Sample treatment consists of four stages: (1) removal of viruses from the foodstuff to leave them in suspension; (2) removal of food substances from the virus suspension; (3) concentration of suspended viruses for delivery to the detection assay; and (4) extraction of nucleic acids from the concentrated viruses. As well as the sample preparation stages, the assay itself requires its own special set of controls (Fig. 6.1).

Thus, a complete suite of controls should be employed to verify that the method has performed correctly, and that the results can be interpreted unambiguously. These controls are described in the following sections. For food samples, a positive result obtained from a molecular method should be confirmed as such by the standard ISO method, which, if available, should therefore be useful principally as a reliable screen for negative samples. In the case of viruses, of course, this is not an option, since there are currently no standards available. However the CEN TG275/WG6/TAG4 committee is in the process of developing a standard for detection of norovirus and hepatitis A in foodstuffs, including bivalve molluscs, fruits and vegetables and bottled water.

Guidelines on the Application of General Principles of Food Hygiene to the Control of Viruses in Food, produced by Codex Alimentarius, are available (FAO/WHO, 2012). Prevention is of course the first line of defence, and

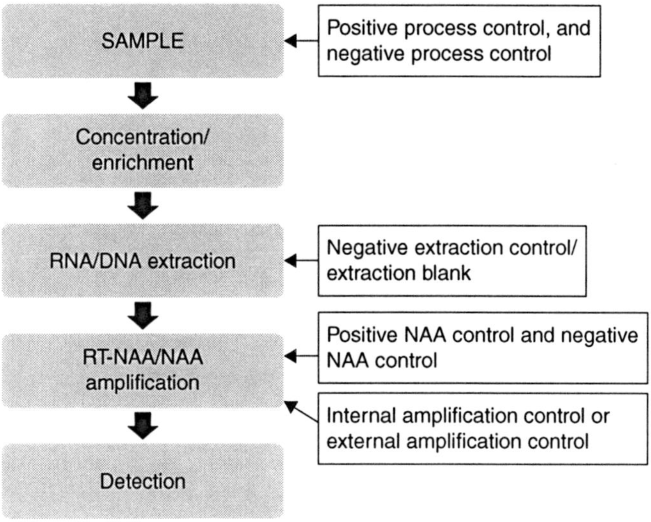

Fig. 6.1 The analytical process and associated controls.

producers, the food industry and distributors should seek to prevent the risk of foodstuffs being contaminated with enteric viruses in the first place using internal industry guidelines and HACCP procedures (see chapters in Part III of this book). Should these procedures fail, it is the analyst's responsibility to perform a reliable test for these viruses on the sample, which is why appropriate controls specifically related to the sample preparation are of particular importance.

6.2 Controls for the sample treatment step

NAA assays are highly efficient, but they can also be negatively affected by the presence of food matrix-derived substances which can interfere or prevent the reaction from performing correctly. This is the case whether a commercially available system or a freely available non-proprietorial published method is being used. Therefore, the use of appropriate and careful sample treatment must be applied to remove these inhibitory substances as far as possible.

The figures referred to in the following descriptions provide a graphical representation of the results which should be observed using a gel-based detection system, and they should be interpreted in conjunction with the descriptions of the IAC, EAC and NEC in each section as appropriate. The principles are transferrable to other detection formats.

The controls that need to be included at the beginning of the actual sample treatment so that both sample treatment and detection can be controlled are the positive process control (PPC) and the negative process control (NPC).

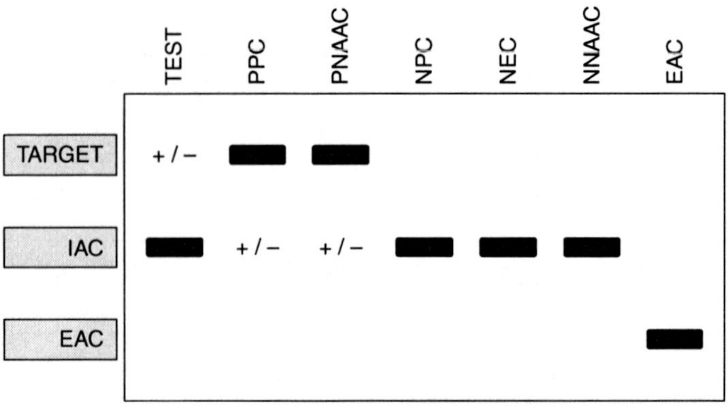

Fig. 6.2 Satisfactory result using the PPC. Key: ■ Positive signal. +/– This means the signal may be present or absent. In the case of the test sample, the IAC signal may be present or absent when a target signal is present, but it must be present in the absence of a target signal.

The PPC, defined in ISO 22174:2005, ensures that a sample is not declared target microorganism-free due to failure of the sample treatment. A separate identical sample spiked with the target microorganism should be treated in the same way as the test samples to verify that the sample treatment has functioned correctly, and identify those samples in which it has failed. Its frequency of use should be determined as part of the QC programme within the laboratory. If no signal is obtained for the PPC, and all test samples are negative, it means that the sample treatment has failed. If the target signal is present, however, then the sample should be reported as positive, provided that all the other controls are correct, for example, the negative control should be free of target signal. Figure 6.2 shows an acceptable set of results in conjunction with other relevant controls.

The purpose of the negative process control (NPC) is to verify the sample treatment reagents and/or equipment are not contaminated with the target and/or its amplicon. It is defined in ISO 22174:2005 and is a 'sample' which does not contain any matrix so, in place of the sample which would have been tested could be the same weight/volume of sterile distilled water. This control is processed exactly the same as a real test sample would be. An NPC must be used with every batch of samples analysed. *Interpretation:* There must be no target signal obtained from the NPC. If there is a signal, then contamination has occurred and the results of the actual test sample cannot be trusted. In this case the whole process should be repeated on the batch of samples and decontamination of equipment should be carried out. If necessary, new reagents should be used. Figure 6.3 shows an acceptable set of results in conjunction with other relevant controls.

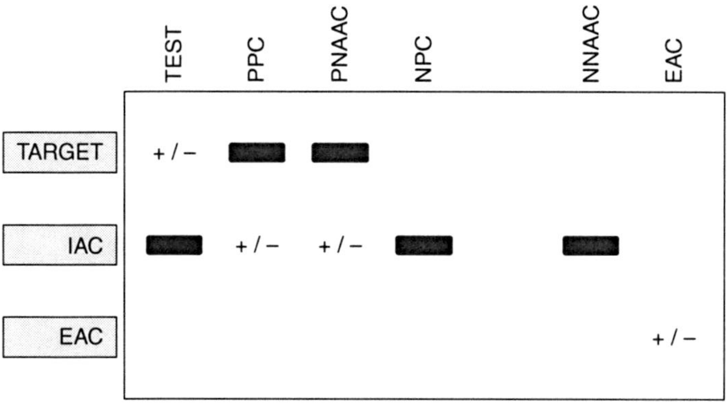

Fig. 6.3 Satisfactory result using the NPC.

6.3 Controls for the nucleic acid extraction step

It is possible that nucleic acid extraction reagents could become contaminated by the target or its amplicon. A control is thus required to ensure this has not happened in the process of extraction.

6.3.1 Negative extraction control (NEC) and extraction blank (EB)

This control is target-free, consisting solely of water which is carried through all steps of the nucleic acid extraction procedure. It is defined in ISO 22174:2005 and is used to ensure that the extraction process has not been contaminated with the target or its amplicon. If the target signal is present in the NEC / EB, it means the extraction reagents or equipment have been contaminated. In this case, the extraction reagents should be replaced and all equipment should be decontaminated. It is not necessary to use this control if an NPC is used. Figure 6.4 shows an acceptable set of results in conjunction with other relevant controls.

6.4 Controls for the amplification step

It is equally important to be able to control the performance of the NAA assays, so controls have to be included in the reagent mixtures, or additional reactions have to be performed. These controls include either an internal (IAC) or external (EAC) amplification control and a positive and negative control (PNAAC and NNAAC).

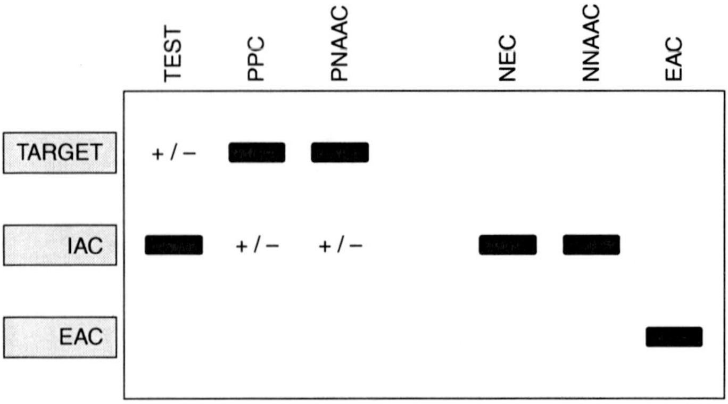

Fig. 6.4 Satisfactory result using the NEC.

6.4.1 Internal amplification control (IAC)

Perhaps one of the most important controls which should be incorporated into an NAA-based assay is an internal amplification control which is a non-target DNA/RNA sequence (in a defined amount or copy number) present in the same reaction as the sample or target nucleic acid extract. If it is successfully amplified to produce a signal, non-production of a target signal in the reaction indicates that the sample did not contain the target pathogen or organism – a true negative. Because the IAC is co-amplified using the same primers as the target, and as long as the concentration of the IAC is carefully optimized (Cook *et al.*, 2013) so as not to out-compete the target signal, the presence of the IAC is not essential when the target signal is present. If, however, the reaction produces neither a signal from the target nor one from the IAC, it signifies that the reaction has failed. This may be due to inhibitory substances or to reaction failure (Hoorfar *et al.*, 2003, 2004). There may be several reasons why reaction failure has occurred, apart from inhibition:

- Human error may have been the cause – was the sample really added?.
- The thermocycler may not be functioning correctly (Anonymous ISO/TS 20836:2005).
- One of the NAA reagents may not have been added.
- The sample may not have been loaded into the well correctly (e.g., when using qPCR).
- The sample may not have been loaded onto the gel correctly (e.g., when using conventional PCR).

Whatever the reason, if a sample does not contain an IAC, there is no way of knowing whether a negative signal response is due to inhibition, is a failed reaction, or is a true negative, and this may have serious consequences when analysing for potentially harmful foodborne pathogens. The amplification should be repeated using new mastermix and/or a dilution of the test sample

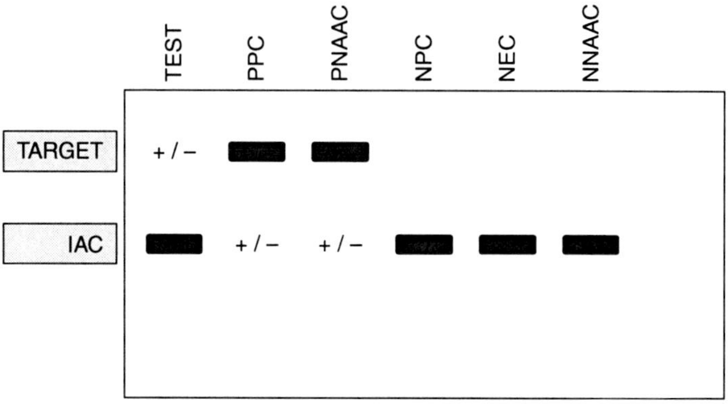

Fig. 6.5 Possible result using the IAC.

nucleic acid should be tested in case inhibitory substances were present in the original extraction (depending on the sensitivity of the assay, you may dilute to beyond its detection limit). Figure 6.5 shows a possible set of results when using an IAC.

6.4.2 External amplification control (EAC)

ISO 22174 includes the option of using an EAC as an alternative to an IAC. Its purpose is to verify reactions which have functioned correctly, and identify those which have been inhibited: in effect the same function as the IAC. The incorporation of an EAC aids in the identification of possible inhibited reactions.

The EAC is not added to the same reaction as the test sample, hence the term 'external', but is included as a separate reaction for each batch of samples tested in an NAA run. It is assumed that the reaction has performed with a similar efficiency to the test sample, so if a negative signal (i.e., none at all) is obtained, it indicates that the nucleic acid extract must contain inhibitory substances, and this is the reason the reaction has failed.

It can be argued, however, that since an EAC involves a completely separate reaction, its use cannot show what has happened in the test sample if it produces no signal – is it a true negative, or has the reaction failed? Since there is always a possibility of human error, it is impossible to unambiguously define the cause of the reaction failure if no signal whatsoever is obtained.

There is some argument as to whether an IAC or an EAC should be used. Some are concerned that because the IAC is contained within the same reaction as the sample, and in most cases it uses the same target primers, competition will occur and sensitivity will be compromised. This could happen if the IAC concentration has not been carefully optimised. Although use of an EAC avoids this potential problem, it does not control for all situations which could occur, such as individual pipetting errors or failure of the individual

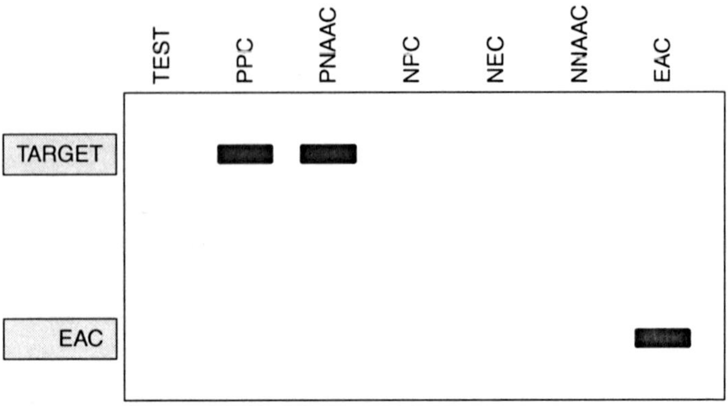

Fig. 6.6 Possible result using the EAC.

reaction. An additional concern with an EAC is that it increases the number of reactions required for each sample, significantly adding to costs. If an IAC signal is obtained, we can at least be certain that the amplification reaction has performed correctly and it must be the sample treatment which has been the problem (unless it is a true negative, which can be deduced if the sample process control (SPC) has correctly performed due to its signal being present). The use of the SPC is described in Section 6.5. When no signal is obtained, as is possible in the case of an EAC, we do not know whether it is due to the sample treatment (because it will not contain an IAC), human error or thermocycler failure. Nor can we rely solely on a positive amplification control to give that answer. Figure 6.6 shows a representation of a possible set of results when using an EAC.

6.4.3 Positive NAA control (PNAAC)

This control is a nucleic acid amplification reaction, into which a known amount of nucleic acid from the target pathogen is included and its role is to verify amplification performance of the target sequence. When an IAC or EAC is used, it can fulfil the role of a positive amplification control; however, when using a reaction involving a probe for the target, as used in real-time PCR, the PNAAC verifies probe performance. If conventional PCR is being used, it need only be included if using an EAC and if so, should be included in every batch of samples in a cycler run. A target signal must be obtained from the PNAAC. Figures 6.2–6.6 represent possible results which include the PNAAC.

6.4.4 Negative NAA control (NNAAC)

The final control to be considered is the NNAAC. It is an amplification reaction in which water takes the place of the nucleic acid extract. It must either include an IAC, or an EAC as a control. The NNAAC verifies that the mastermix reagents are not contaminated with the target, or the PPC, or their amplicons. No target or PPC signal should be obtained from the NNAAC; however, IAC/EAC signals must be obtained. Figures 6.2–6.6 show possible results which include the NNAAC.

6.5 Additional recommended controls

An additional control which is not mentioned within ISO 22174, but is recommended as an addition to the PPC, is termed a sample process control (SPC). In this case, the SPC would be a *non-target* organism added to every test sample, including the NPC, at the start of analysis, and must be detected in every sample to which it has been added. It should *not* be added to its associated 'SPC-NC' (sample process control – negative control), and it requires a separate internal amplification control to be constructed. The principle of an SPC is that if it is detected, then the method was performed correctly. If it is not detected and the target signal is also absent, the method has failed and the foodstuff must be reanalyzed; however, if the target signal is present, the SPC signal is absent and all other controls are correct, then the sample should be reported as positive (the sample treatment must have worked if the target signal is present and all negative controls indicate the target signal is not due to any type of contamination). In addition to this qualitative interpretation of an analytical result, when using a quantitative assay such as qPCR, the SPC can also allow a determination of the recovery efficiency for each individual sample. This can be done by comparing the values of the SPC before and after addition to the sample. This control has already been used successfully in the detection of foodborne viruses (D'Agostino *et al.*, 2012, 2011) and it is being used in research into the detection of foodborne bacteria.

The SPC-NC is equivalent to the NPC and is a sample which does not contain any matrix, that is, no food or environmental material. It should therefore be replaced with the equivalent weight/volume of water. It should also be processed in the same way as a normal sample and should be included in every batch of samples analysed. Its purpose is to verify that the sample treatment reagents and/or equipment are not contaminated with the SPC virus or its amplicon. If, however, the SPC signal is present in the SPC-NC, you can still accept the results of that batch of analyses and record the Cp value of the SPC if performing qPCR. Meanwhile equipment should be decontaminated and fresh sample treatment reagents prepared. The SPC can facilitate determination of the efficiency of detection, therefore if there is widespread contamination with SPC, efficiency can be overestimated. If this occurs, the number of contaminating SPCs can be regarded as background and subtracted from the number detected in the test sample. In this way, the actual number can be calculated, but only if the Cp values of the SPC-NC, which should be

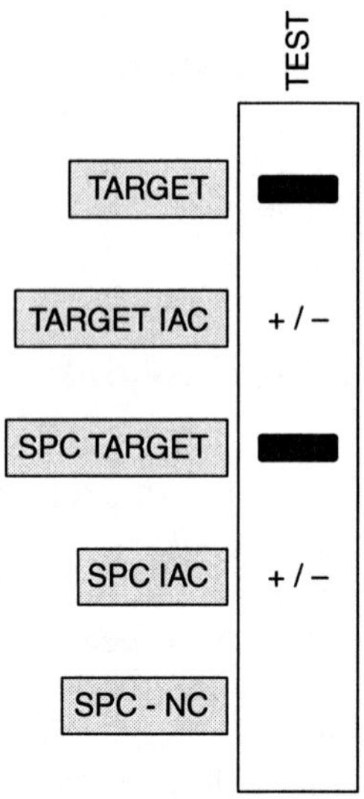

Fig. 6.7 Possible result using the SPC, SPC IAC and SPC-NC.

negative for SPC, are similar. Figure 6.7 shows a representation of a possible set of results when using an SPC, its associated IAC and the SPC-NC.

6.6 Reference materials

Quantified reference materials for NAA assays can be used as positive extraction controls, for method calibration and characterization and as positive controls in semi-quantitative NAA analysis. There are a limited number of reference materials available for quantitative and qualitative determination. Standards for norovirus GI, GII and hepatitis A RT-qPCR are available from the UK Health Protection Agency. More certified reference materials are needed to ensure agreement between the growing number of laboratories analysing food and water samples for viruses, or performing research on these viruses.

6.7 Conclusion

This chapter has focused on controls required for the molecular detection of enteric viruses in food and the environment and has assumed that the analytical laboratory which uses such molecular-based detection methods will include these controls as part of their overall quality assurance/quality control policy. General quality control policies may differ in each lab, but will usually consist of documents covering aspects such as laboratory organisation and responsibilities of staff, calibration procedures and frequencies, internal quality control checks, for example, air quality, water quality etc., data reporting, performance and system audits, preventive maintenance, equipment records and associated servicing reports/schedules, corrective actions etc. These wider sets of controls should of course be adhered to and considered in conjunction with more specific controls such as those set out here, since the results of an analytical procedure may be affected by poor equipment maintenance or poor quality air handling systems giving rise to contamination events.

With regard to the suite of controls described in this chapter, it can be seen that there are several controls required when using an NAA-based method for detection of foodborne viruses. It is important to realise that the results from each of these controls should not be considered in isolation and that a conclusive result must take into account the results of the complete set of controls. For example, if a target signal is obtained but a positive process control signal is not obtained, we should not just assume that the result is acceptable – we should consider that the presence of the target signal may be as a result of contamination, so we should also check the results of the negative process control, the negative extraction control/extraction blank, and the negative NAA control. If any of these show a target positive result, we should not accept the positive test result as it is possible it is due to contamination – a false positive.

This suite of controls, when incorporated into the routine testing of foodstuffs/environmental samples for the presence of viruses, should give the analyst confidence that the results obtained from their NAA analyses are both accurate and unambiguous.

6.8 References

ANONYMOUS ISO/TS 20836:2005 Microbiology of food and animal feeding stuffs – Polymerase chain reaction (PCR) for the detection of food-borne pathogens – Performance testing for thermal cyclers.

BUSTIN, S.A., BENES, V., GARSON, J.A., HELLEMANS, J., HUGGETT, J., KUBISTA, M., MUELLER, R., NOLAN, T., PFAFFL, M.W., SHIPLEY, G.L., VANDESOMPELE, J. and WITTWER, C.T. (2009) The MIQE guidelines: minimum information for publication of quantitative Real-Time PCR experiments. *Clin Chem*, **55**: 611–22.

COOK, N., DE RIDDER, G.A., D'AGOSTINO, M. and TAYLOR, M.B. (2013) Internal amplification controls in real-time polymerase chain reaction-based methods for pathogen detection. In: RODRÍGUEZ-LÁZARO, D. (ed.) *Real-Time PCR in Food Science: Current Technology and Applications*. Caister Academic Press, ISBN: 978-1-908230-15-7.

CROCI, L., DUBOIS, E., COOK, N., DE MEDICI, D., SCHULTZ, A.C., CHINA, B., RUTJES, S.A., HOORFAR, JEFFREY and VAN DER POEL, W.H.M. (2008) Current methods for extraction and concentration of enteric viruses from fresh fruit and vegetables: towards international standards. *Food Anal Methods*, **1**(2): 73–84.

D'AGOSTINO, M., COOK, N., DI BARTOLO, I., RUGGERI, F.M., BERTO, A., MARTELLI, F., BANKS, M., VASICKOVA, P., KRALIK, P., PAVLIK, I., KOKKINOS, P., VANTARAKIS, A., SÖDERBERG, K., MAUNULA, L., VERHAELEN, K., RUTJES, S., DE RODA HUSMAN, A.M., HAKZE, R., VAN DER POEL, W., KAUPKE, A., KOZYRA, I., RZEŻUTKA, A., PRODANOV, J.,

LAZIC, S., PETROVIC, T., CARRATALA, A., GIRONÉS, R., DIEZ-VALCARCE, M., HERNANDEZ, M. and RODRÍGUEZ-LÁZARO, D. (2012) Multicenter collaborative trial evaluation of a method for detection of human adenoviruses in Berry fruit. *Food Anal Methods*, **5**: 1–7.

D'AGOSTINO, M., COOK. N., RODRÍGUEZ-LÁZARO, D. and RUTJES S. (2011) Nucleic acid amplification-based methods for detection of enteric viruses: definition of controls and interpretation of results. *Food Environ Virol*, **3**: 55–60.

FAO/WHO (Food and Agriculture Organization of the United Nations/World Health Organization). (2012) *Guidelines on the Application of General Principles of Food Hygiene to the Control of Viruses in Food* (CAC/GL 79–2012).

HOORFAR, J., COOK, N., MALORNY, B., WAGNER, M., DE MEDICI, D., ABDULMAWJOOD, A. and FACH, P. (2003) Making internal amplification control mandatory for diagnostic PCR: making internal amplification control mandatory. *J Clin Microbiol*, **41**: 5835.

HOORFAR, J., MALORNY, B., ABDULMAWJOOD, A., COOK, N., WAGNER, M. and FACH, P. (2004) Practical considerations in design of internal amplification control for diagnostic PCR assays. *J Clin Microbiol*, **42**(5): 1863–8.

7

Tracing the sources of outbreaks of food- and waterborne viral disease and outbreak investigation using molecular methods

M. B. Taylor, University of Pretoria & National Health Laboratory Service, South Africa

DOI: 10.1533/9780857098870.2.139

Abstract: Tracing sources of human faecal contamination in food and water is essential to ensure food and water safety in developing and industrialised countries. Index organisms, for example, F+ RNA bacteriophages, human adenoviruses and human polyomaviruses, are used as indicators of human viral pollution. Direct monitoring of food and water matrices for enteric viruses, followed by nucleotide sequence-based identification of the virus strains, has successfully been applied to source tracking contamination and transmission routes in outbreaks of food- and waterborne viral disease. Newer methods under development, including metagenomics, will provide new opportunities to expand the repertoire of assays available for multi-taxa molecular source tracking.

Key words: faecal indicator bacteria, microbial source tracking, pathogen source tracking, norovirus, hepatitis A virus, hepatitis E virus, foodborne viruses, waterborne viruses.

7.1 Introduction

Contaminated food and water have been implicated in millions of cases of illnesses per year (Bosch *et al.*, 2008; Grabow, 2007; Hayes *et al.*, 2003; Roslev and Bukh, 2011; Scallan *et al.*, 2011) and are a matter of public health concern in both developing and industrialised countries (Ashbolt, 2004; Widdowson and Vinjé, 2008). Immuno-compromised people are more susceptible to food- and waterborne illnesses which contribute significantly to morbidity and mortality in human immunodeficiency virus-infected individuals (Hayes *et al*., 2003; Okoh *et al*., 2010). Bacteria, viruses, parasites, toxins, metals and prions have all been identified as causal agents of food- and waterborne illness (McCabe-Sellers and Beattie, 2004; Mead *et al*., 1999; Newell *et al*., 2010).

The health risk posed by the faecal contamination of water and food, specifically concerning micro-organisms, has been well documented (Bofill-Mas *et al*., 2011; Field and Samadpour, 2007; Purnell *et al*., 2011; Steele and Odumeru, 2004). The interaction between water, food and human health is complex, as human health can be affected by the ingestion of contaminated water either directly or indirectly through contaminated food (Grabow, 2007; Helmer, 1999). Tap water has been implicated in 14–40% of gastrointestinal illness (Carter, 2005). Globally 1.1 billion people still lack access to clean water and 2.6 billion people lack access to adequate sanitation (Barry and Hughes, 2008; Grabow, 2007). The faecal contamination of water can originate from a wide variety of sources such as human sewage and non-human (urban, agricultural, wildlife) runoff (Purnell *et al*., 2011; Steele and Odumeru, 2004). Fresh produce is vulnerable to contamination by human microbial pathogens at any point along the food chain, from pre-harvest in the field to post-harvest during processing and packaging (Berger *et al*., 2010). Pre-harvest contamination originates from human or animal faecal contamination reaching the crops via contaminated irrigation water, improperly composted manure or wastewater (Baert *et al*., 2008, 2011; Steele and Odumeru, 2004; Wei and Kniel, 2010). Post-harvest contamination can be due to contaminated washing water (Steele and Odumeru, 2004) or infected food handlers during harvest and packaging (Barrabeig *et al*., 2010; Berger *et al*., 2010; Okoh *et al*., 2010; Steele and Odumeru, 2004; Tuan Zainazor *et al*., 2010).

As early as 1999, 67% of foodborne illness in the United States (US) was attributed to viruses (Carter, 2005; Mead *et al*., 1999) and during the period 1990–2006 fresh produce was responsible for the largest number of cases of illness, although seafood was responsible for the highest number of outbreaks (Smith DeWaal *et al*., 2008). Reports of waterborne outbreaks of enteric virus-associated illness are common (Ashbolt, 2004; Carter, 2005; Grabow, 2007; Hoebe *et al*., 2004; Räsänen *et al*., 2010). Although the reported numbers of food- and waterborne viral diseases are considerable and are increasingly recognised, the true global burden of food- and waterborne viral disease is currently unknown (Bosch *et al*., 2008; Grabow, 2007; McCabe-Sellers and Beattie, 2004; Newell *et al*., 2010) and is probably underestimated (Grabow, 2007; O'Brein, 2008), due to lack of environmental epidemiological data, technical limitations in pathogen identification, and difficulties in determining the source (Bosch *et al*., 2008). Person-to-person and food- and waterborne routes of transmission may overlap, which further impacts the estimation of food- and waterborne illness (Koopmans and Duizer, 2004). Tracing or tracking the source of viruses in suspected food- and waterborne outbreaks is essential to facilitate evidence-based intervention strategies for prevention and control of further outbreaks.

7.2 Challenges in food- and waterborne outbreak tracing and investigation

Food or water contaminated with viruses may conform to acceptable bacterial standards (Ebdon *et al.*, 2011; van Heerden *et al.*, 2005; Verhoef *et al.*, 2009) and look, smell and taste normal (Koopmans and Duizer, 2004). One of the challenges in identifying the food item or water source in an outbreak is that individuals have a poor recollection of their ingestion of specific products. In addition, food is often consumed as a mixture of items, for example, salads, making the identification of any one item as the source of infection problematical (Berger *et al.*, 2010; de Roda Husman *et al.*, 2007). This is further compounded by the fact that one of the important foodborne viruses, hepatitis A virus (HAV), has a relatively long incubation period of 15–50 days with a mean of 28–30 days (Amon *et al.*, 2005; Dotzauer, 2008), which results in illness manifesting some time after consumption of the contaminated food item(s). Even if an individual food item is identified as the vehicle of infection it may still, however, be difficult to trace the contamination source due to multiple factors, including inadequate labelling of food items, sketchy record keeping and the global distribution of food (Berger *et al.*, 2010; Newell *et al.*, 2010).

The investigation of virus-associated outbreaks remains challenging, as an outbreak which may have begun as exposure to a single common source of contaminated food or water may be perpetuated by person-to-person spread (Atmar, 2010; Patel *et al.*, 2009). Until recently, direct monitoring for enteric viruses and parasites, which provides indisputable evidence of faecal contamination, was limited by the fact that many of these organisms cannot be cultured and are present in low numbers in environmental sources, and analysis is costly (Ebdon *et al.*, 2012; Field and Samadpour, 2007; Gómez-Doñate *et al.*, 2011).

7.3 Microbial source tracking

A variety of bacteria and viruses of human and non-human sources may be present in faecally contaminated food and water, and because of the human health and economic implications, it is necessary to maintain the microbiological quality and safety of water and food sources. As exposure to contamination of human origin poses a greater risk to human health than exposure to pollution of animal origin, it is essential to identify the source of the faecal contamination (Field and Samadpour, 2007; Grabow, 1996; Meays *et al.*, 2004; Purnell *et al.*, 2011; Scott *et al.*, 2002).

Indicator micro-organisms have been used to predict the presence of pathogens (Ashbolt *et al.*, 2001). As the presence of these indicator micro-organisms may not necessarily demonstrate the presence or absence of pathogens, they are categorised into different groups. General process microbial indicators, such as heterotrophic plate count bacteria and total coliforms, are used to monitor the efficiency of a process such as chlorination, while faecal indicators such as thermotolerant coliforms and *Escherichia coli* demonstrate the presence of faecal contamination. Index or model organisms such as *E. coli* and F-RNA coliphages are used to indicate the presence of *Salmonella* or as models of human enteric viruses, respectively (Ashbolt *et al.*, 2001). Faecal indicator bacteria (FIB), such as *Clostridium perfringens*, *E. coli*, total and thermotolerant (faecal) coliforms and faecal enterococci, used to evaluate the level and potential health risks of faecal contamination in food and water sources (Ashbolt *et al.*, 2001; Meays *et al.*, 2004; Roslev and Bukh, 2011; Scott *et al.*, 2002), do not identify the source of contamination (Field and Samadpour, 2007; Gómez-Doñate *et al.*, 2011; Murugan *et al.*, 2011; Roslev and Bukh, 2011). In addition factors such as precipitation, currents, wind-induced turbulence and host biology may influence the level of faecal pollution in environmental water sources (Roslev and Bukh, 2011).

Faecal source tracking, also referred to as microbial source tracking (MST), is the process whereby a particular source (human, animal or avian) of faecal contamination is identified (Field and Samadpour, 2007; Meays *et al.*, 2004; Roslev and Bukh, 2011; Scott *et al.*, 2002). Various strategies have been applied to MST and include culture-based library-based methods, culture-based library-independent methods, culture-independent library-dependent methods and culture-independent library-independent methods (Field and Samadpour, 2007; Meays *et al.*, 2004; Scott *et al.*, 2002; Stewart *et al.*, 2003).

For library-dependent, also referred to as library-based methods, a library or database of phenotypic or genotypic characteristics, that is, 'fingerprints', is created from a variety and large number of source organisms of human, animal and avian origin. Identification of the faecal source is then achieved by comparison of the characteristics or 'fingerprints' of isolates from food and water sources to the library or database (Field and Samadpour, 2007). The main disadvantage of this approach is that many thousands of patterns are required to capture the bacterial diversity in faeces and environmental sources. It is therefore expensive to create such a database and it has to be updated regularly. In addition, the larger the database, the greater the chance of non-host-specific patterns manifesting.

The culture-based library-based MST methods, each with their own advantages and disadvantages (Meays *et al.*, 2004), include phenotypic methods such as antibiotic resistance analysis, multiple antibiotic resistance analysis and carbon source utilisation profiling, and genotypic approaches such as DNA fingerprinting, ribotyping and related methods (Field and Samadpour,

2007; Scott *et al.*, 2002; Wong *et al.*, 2012). A combination of these assays has been applied to the identification of *E. coli* of human and non-human origin in water from the Tirumanimuttar river in India (Murugan *et al.*, 2011).

Culture-based library-independent microbiological methods, where bacteria and viruses are known to be from specific hosts, include a number of different strategies:

- The determination of the faecal coliform/faecal streptococcus ratio, where a ratio of >4.0 is reportedly indicative of human faecal pollution and ≤0.7 indicates non-human faecal pollution (Scott *et al.*, 2002). This is based on the rationale that human faeces contain higher faecal coliform counts than animal faeces but the method is unreliable due to different bacterial survival rates (Field and Samadpour, 2007) and application difficulties in agricultural settings (Meays *et al.*, 2004).
- The detection of *Bifidobacteium* spp., obligate anaerobic non-spore-forming bacteria, which are more abundant in human faeces than animal faeces and indicate the presence of recent pollution (Scott *et al.*, 2002).
- The detection of *Bacteriodes* bacteriophages, for example, HSP40 which can be indicative of human pollution with the added advantage that their presence has been shown to correlate with the presence of human enteric viruses (Scott *et al.*, 2002), GB-124 which is proposed to be a surrogate for selected human pathogenic viruses in municipal wastewaters (Ebdon *et al.*, 2012) or phages specific for faecal contamination of animal origin (Gómez-Doñate *et al.*, 2011).
- The typing of F+ RNA coliphages. Human and animal faeces contain different types, that is, genogroups I and IV are predominantly of animal origin while genogroups II and III are associated with human faecal contamination (Muniesa *et al.*, 2009; Schaper *et al.*, 2002; Stewart-Pullaro *et al.*, 2006; Wolf *et al.*, 2008). This method can, therefore, be applied to distinguish between faecal contamination of human and non-human origin. In one study this method identified birds as the major source of faecal contamination in a New York reservoir (Stewart *et al.*, 2003).
- The bacteriophage lysis of *Enterococcus* spp. as the bacteriophages display a similar narrow host range to those of *Bacteriodes* spp. (Purnell *et al.*, 2011).

To date the applicability of culture-independent, library-dependent methods, such as community fingerprinting, for the identification of faecal sources in environmental samples, has not been proven (Field and Samadpour, 2007). Culture-independent, library-independent methods can be either chemical or molecular. Chemical indicators such as faecal sterols and bile acids can be applied to identify sources of faecal contamination but their presence may not correlate with the presence of FIB and pathogens (Field and Samadpour, 2007). Chemical markers are affected by a number of different factors including their removal by wastewater treatment (caffeine), their degradation in aerobic or anaerobic environments (sterols), or their settling and storage in sediments (bile acids) (Field and Samadpour, 2007).

7.4 Molecular-based source tracking

The development and application of molecular methods has resulted in radical changes in source tracking approaches and methodology. Host-specific gene markers are unique nucleic acid sequences which can be detected directly from the food or water matrix by molecular-based assays such as the polymerase chain reaction (PCR) (Field and Samadpour, 2007; Fong and Lipp, 2005; Roslev and Bukh, 2011; Santo Domingo *et al.*, 2007; Stewart *et al.*, 2003) or microarrays (Dubinsky *et al.*, 2012; Lauri and Mariani, 2009) without prior amplification or isolation in culture (Field and Samadpour, 2007; Roslev and Bukh, 2011). These methods are not dependent on, or biased by, cultivable microbes and can utilise the genetic diversity in faeces and sewage as molecular markers for source tracking (Cantalupo *et al.*, 2011; Field and Samadpour, 2007; Santo Domingo *et al.*, 2007). The molecular markers could be from prokaryotes, eukaryotes or viruses (Roslev and Bukh, 2011).

Within the prokaryotes, host-specific molecular markers from certain faecal anaerobic bacteria, for example, the *Bacteriodales* and *Bifidobacterium,* can be applied to determine the human or non-human source of faecal contamination while *Rhodococcus coprophilus* markers identify faecal contamination from herbivores (Bernhard and Field, 2000; Field and Samadpour, 2007; Roslev and Bukh, 2011). The targets can either be rRNA or host-specific protein genes with the protein gene targets being more specific. The *Bacteriodales* PCR markers have proved to be most effective for faeces identification and a correlation between *Bacteriodales* markers and FIB has been demonstrated, while the usefulness of bifidobacteria is affected by their limited survival in the environment (Field and Samadpour, 2007). In epidemiological studies, molecular methods have been used for the typing and source tracking of gram-negative foodborne pathogens, for example, *Salmonella* spp. (Foley *et al.*, 2009).

Host-specific mitochondrial (mt) DNA markers are applied to identify eukaryotic host cells rather than micro-organisms from human or animal origin (Roslev and Bukh, 2011). Host blood and intestinal cells are shed in faeces and then excreted into the environment (Field and Samadpour, 2007; Roslev and Bukh, 2011). Quantitative real-time PCR assays have been developed for both human and animal mtDNA marker groups (Roslev and Bukh, 2011), but as the host cells in faeces are sparse, larger water samples may be required to detect them (Field and Samadpour, 2007).

As enteric virus infection is host-specific, enteric viruses are ideal candidates for MST (Wong *et al.*, 2012). Molecular-based techniques has revolutionised the direct detection of enteric viruses in food and water (Girones *et al.*, 2010; Havelaar and Rutjes, 2008). Reverse transcription (RT)-PCR/PCR is now considered the gold standard for virus detection (Bosch *et al.*, 2008; Mattison and Bidawid, 2009) and is the favoured assay for the detection of viruses in outbreaks of suspected viral origin (Parshionikar *et al.*, 2003). The PCR primers can be designed to be group-specific (e.g., enteroviruses), serotype-specific (e.g., polioviruses) or host (human, bovine, etc.) serotype-specific (Fong and Lipp, 2005). Further improvements and advances in technology have led to the development of real-time RT-PCR/PCR assays which apply highly specific probes to detect the amplified molecular marker. Some authors claim real-time RT-PCR/PCR to be the new gold standard for the quantification of enteric viruses (El-Senousy *et al.*, 2007) and the method of choice for the detection of enteric viruses in environmental samples (Rodríguez-Lázaro *et al.*, 2012). Real-time RT-PCR/PCR has the advantage over conventional RT-PCR/PCR of being less time-consuming and more sensitive, and human error is reduced in that the post-amplification interpretation of results is automated (Wong *et al.*, 2012). Contamination is minimal as the reaction is contained in a closed system (Fong and Lipp, 2005; Rodríguez-Lázaro *et al.*, 2012). Furthermore, molecular-based techniques have facilitated advances in the understanding of the genetic variation and evolution of viruses, which has provided the molecular basis for tracing the source of the virus or infection (Lappalainen *et al.*, 2001; Stewart *et al.*, 2003). In food- and waterborne outbreaks nucleotide sequence comparisons of viruses detected in patient specimens with those from the food and/or water sources have been used to link and identify the source of infection (Beaudeau *et al.*, 2008; Lappalainen *et al.*, 2001; Stewart *et al.*, 2003).

Although over 100 types of pathogenic enteric viruses are shed in waste of human and animal origin (Fong and Lipp, 2005; Scott *et al.*, 2002), many of these viruses are often present in low concentrations in the environment and may therefore not be suitable for molecular-based MST (Wong and Xagoraraki, 2011). Indicator viruses are therefore used to predict the possible presence of human virus contamination (Carter, 2005). Prior to the poliovirus eradication initiative, when vaccine-derived polioviruses were still abundant in water sources, enteroviruses were used for assessing viral pollution in water as they could easily be isolated and quantified in cell culture (Carter, 2005; Fong and Lipp, 2005; Pina *et al.*, 1998).

As early as 1998, human adenoviruses (AdVs) were proposed as an indicator of human viral pollution (Pina *et al.*, 1998) as they are highly host-specific (Jiang, 2006), persist for longer and are present in higher numbers in polluted water sources and are therefore a more suitable predictor of pathogenic human viruses than enteroviruses (Fong and Lipp, 2005; Pina *et al.*, 1998). Together with human AdVs, another DNA virus, human polyomavirus (PyV) (JCPyV and BKPyV), have been proposed as an index of human viral contamination as, like human AdVs, they are widespread in human populations (Bofill-Mas *et al.*, 2000) and are present in high titres and stable in environmental water, wastewater and sewage (Bofill-Mas *et al.*, 2006; McQuaig *et al.*, 2006; Rodríguez-Lázaro *et al.*, 2012). It has therefore been suggested that the detection and quantification of human AdVs and JCPyV can reliably be applied to assessing the virological quality of water in water treatment plants (Albinana-Gimenez *et al.*, 2009; Bofill-Mas *et al.*, 2006).

As AdVs and PyVs are host-specific, qPCR methods have been developed for the detection of bovine and porcine AdVs and bovine PyVs for MST of faecal contamination of animal origin (Bofill-Maas *et al.*, 2011; Maluquer de Motes *et al.*, 2004; Wong and Xagoraraki, 2011). Real-time RT-PCR assays have been developed for the detection of F+ RNA coliphages (Ogorzaly and Gantzer, 2006). These assays have been shown to be more reliable than nucleic acid hybridisation for the molecular identification of the origin of faecal contamination in water sources (Stewart *et al.*, 2006), and have successfully been applied to tracking the source of faecal contamination in wastewater and faecal samples in Asia (Lee *et al.*, 2009), river water (Ogorzaly *et al.*, 2009; Wolf *et al.*, 2008); and shellfish (Wolf *et al.*, 2008). Besides being applied to identify the source of faecal contamination F+ RNA coliphages are considered to be indicators of enteric viral contamination in water and shellfish (Wolf *et al.*, 2008), with F+ RNA levels being a predictor of enteric virus concentration (Yee *et al.*, 2006). Although a positive correlation between the F+ RNA genogroup II concentration, bacterial indicators and human AdVs has been observed in river water (Ogorzaly *et al.*, 2009), the detection of F+ RNA coliphages does not necessarily prove the presence of human enteric viruses (Carter, 2005).

7.5 Molecular tracing in outbreaks

Most of the viruses associated with foodborne disease are of human origin (Greening, 2006). With the globalisation of the food market, which can result in geographically disseminated foodborne outbreaks (Le Guyader *et al.*, 2006; Verhoef *et al.*, 2010), food- and waterborne outbreaks of viral infection have significant direct and indirect economic implications (Grabow, 2007; Lopman *et al.*, 2012; Mayet *et al.*, 2011; Newell *et al.*, 2010). The prevention of outbreaks is therefore essential and relies on identification of the source and mode of transmission. Nucleotide sequence-based identification has proven an effective tool for the tracking of pathogenic organisms ('pathogen source tracking') (Roslev and Bukh, 2011) in the environment for source tracking contamination and transmission routes in food- and waterborne outbreaks. This is essential to discriminate between multiple introductions and prolonged outbreaks with sustained transmission of the pathogen (Kirby and Iturriza-Gómara, 2012).

7.5.1 Noroviruses

Noroviruses (NoVs) are considered to be a common cause of foodborne (Atmar, 2010) and waterborne (Rodríguez-Lázaro *et al.*, 2012) outbreaks of gastroenteritis, with secondary person-to-person transmission following point-source food or water exposure (Centers for Disease Control and Prevention (CDC), 2011). NoV GI has been more commonly associated with traveller's diarrhoea and shellfish-associated disease, while NoV GII is associated with sporadic cases and outbreaks of gastroenteritis (Atmar and Estes, 2009). Since 2001 genotype GII.4 has been associated with the majority of gastroenteritis outbreaks worldwide (CDC, 2011).

The molecular detection and genetic characterisation of NoVs in outbreak-associated stool specimens and food and water samples has identified the source of infection and routes of transmission in a large number of food- and waterborne outbreaks of gastroenteritis worldwide. Detection and partial sequence analysis of the RNA polymerase region have been applied to a number of waterborne outbreak investigations (Hewitt *et al.*, 2007; Maunula *et al.*, 2005; Parshionikar *et al.*, 2005). In Finland, NoVs were identified in a number of paired samples from water and patients in outbreaks in small communities.

Norovirus GII.4 was the most frequent genotype aetiologically associated with the outbreaks, with NoV GI.3 also evident (Maunula *et al.*, 2005). Inadequate disinfection was identified as the underlying reason for one of the outbreaks. During the same period, NoV GI.3 was identified as the source of infection in waterborne outbreaks of gastroenteritis in the United States (Parshionikar *et al.*, 2003) and the Netherlands (Hoebe *et al.*, 2004). More recently a contaminated public water network was identified as the source of infection in a NoV GI.3-associated outbreak of gastroenteritis in Sweden (Riera-Montes *et al.*, 2011). In contrast, NoV GI.5 was identified in a faecally contaminated water supply and linked stool specimens implicated in a gastroenteritis outbreak in a ski resort in New Zealand (Hewitt *et al.*, 2007). In an epidemiological and virological investigation of a gastroenteritis outbreak in two factories and a construction site in China which shared a common water supply, NoV GII.4 with >99% nucleotide sequence identity was identified in the water and stool samples. A breakage in the water pipe was identified as the point of contamination (He *et al.*, 2010). In larger community-wide outbreaks more than one NoV type may be identified in stool specimens, which provides strong evidence of faecally contaminated water as the common source in the absence of other identified point sources of infection (Nenonen *et al.*, 2012).

Shellfish, which are often eaten raw or inadequately cooked, have been implicated in a number of NoV-associated outbreaks of gastroenteritis, as depuration does not effectively remove the NoVs (Patel *et al.*, 2009). In a large outbreak of NoV-associated gastroenteritis involving two countries, France and Italy, oysters were identified as a common source of infection as they had been harvested from the same area in France. Five different NoV strains (GII.4, GII.b, GI.4), with one stool containing two strains, were identified in the French patients, while GII.4, GII.8, GI.6 and GI.4 were evident in the stool specimens from Italy. Analysis of the oysters yielded GI.4, GII.8 and GII.4 in individual oysters, and heavy rains and sewage treatment failures were considered responsible for the contamination of the oyster bed. Sewage contamination is implicated in food sources containing multiple NoV strains and shellfish are often implicated as the source in outbreaks where a number of NoV strains are identified (Le Guyader *et al.*, 2006).

Outbreaks of NoV-associated gastroenteritis in which approximately 1100 people from Denmark, Sweden and France were affected were associated with the consumption of frozen imported raspberries. The raspberries were imported from Poland from different small-scale farms. The source of contamination could not be determined although irrigation water was suspected since it would give the most plausible answer as to how different NoV strains were found to be the cause of infection in individuals (Cotterelle *et al.*, 2005; Falkenhorst *et al.*, 2005; Korsager *et al.*, 2005). In Finland, imported raspberries were linked to a large number of NoV-associated outbreaks affecting 900 people over an extended period in 2009. The raspberries, of Polish origin, were implicated in 12/13 outbreaks. Norovirus GII were identified in 10 outbreaks and NoV GI in one outbreak, with NoV GII.4 most commonly detected. In one outbreak NoV GI.4 was identified in stool specimens as the causal agent (Sarvikivi *et al.*, 2012), but only detected in Polish raspberry samples associated with outbreaks of gastroenteritis later in the same year (Maunula *et al.*, 2009). As the batch of raspberries had originated from 62 different farms it could not be established whether the original source of contamination was irrigation water or food handler hygiene (Sarvikivi *et al.*, 2012). In Sweden NoV GII.8 strains were identified in two outbreaks of gastroenteritis where raspberries imported from China were implicated (Lysén *et al.*, 2009).

The extent and enormity of the shellfish and raspberry outbreaks, with severe public health, trade and economic implications, serve to highlight the applicability of molecular pathogen source tracking where the culture-based library-independent methods are used more often for quality management.

As there can be a high diversity of NoVs detected in food- and waterborne outbreaks (Le Guyader *et al.*, 2004, 2006; Lysén *et al.*, 2009), sensitive broad reaction methods are required for the detection of NoVs in outbreak investigations (Lysén *et al.*, 2009; Wolf *et al.*, 2007), while more specific assays are required for tracking and linking outbreak strains (Xerry *et al.*, 2008, 2010). The nucleotide sequence diversity of a 459 bp region of the protruding P2 domain has been shown to provide an appropriate region for analysis for more specific discriminatory source tracking of NoV GI and NoV GII strains in outbreaks of gastroenteritis (Xerry *et al.*, 2008, 2010). The rationale for this approach was that the P2 domain exhibits the greatest sequence variation and can be applied to separate or link patients more reliably.

Strains with 100% sequence identity were considered to have originated from a common source (Xerry *et al.*, 2008, 2010). Using this methodology catering staff were identified as the source of NoV GII.2 in a foodborne outbreak in two schools, while NoV GII.3 outbreaks linked to oyster consumption in a chain of seafood restaurants were considered to be unrelated. Identical NoV GII.4s were detected in passengers on five cruise ships over a 10-week period, and on the basis of epidemiological and virological evidence, it was concluded that there was a point-source outbreak, followed by person-to-person spread (Xerry *et al.*, 2009).

7.5.2 Hepatitis A virus

Clusters of HAV strains within genotypes and subgenotypes dominate in certain geographical regions (Faber *et al.*, 2009; Hollinger and Emerson, 2007; Robertson *et al.*, 1992). Despite genetic heterogeneity at the nucleotide level, immunological evidence has demonstrated the existence of a single serotype of HAV (Hollinger and Emerson, 2007) with two phylogenetically distinct biotypes (Knowles *et al.*, 2012; Stanway *et al.*, 2005). Molecular-based nucleic acid detection techniques can be applied to detect HAV in samples of different origin (Spradling *et al.*, 2009), but are also not generally used for diagnostic purposes. Assays such as RT-PCR and real-time RT-PCR, which are currently the most sensitive widely used assays for the detection of HAV RNA, are used in epidemiological studies to identify infection sources and transmission patterns (Hollinger and Emerson, 2007). Nucleic acid sequence analysis of selected genomic regions can be used for the determination of the genetic relatedness of isolates (Hollinger and Emerson, 2007) and for tracing the origin of outbreaks (Pintó *et al.*, 2010).

Waterborne outbreaks of hepatitis A have been reported (Bloch *et al.*, 1990; Bosch *et al.*, 1991; Morse *et al.*, 1972), but since the introduction of effective drinking water treatment, waterborne outbreaks of hepatitis A are less common (Fiore, 2004; Pintó and Saiz, 2007). A number of different types of food, including shellfish, cold foods and fresh produce, have been implicated in foodborne outbreaks of hepatitis A (Fiore, 2004). Soft fruits and vegetables are being increasingly identified as sources of foodborne viral illness (Croci *et al.*, 2008) and it is estimated that 5% of HAV infections are due to the consumption of contaminated foods (Le Guyader and Atmar, 2008).

The first foodborne outbreak of infectious hepatitis was described in Sweden in 1955, when 629 cases were associated with the consumption of raw oysters. The largest outbreak of HAV, affecting almost 300 000 individuals, occurred in Shanghai, China in 1988, and was associated with consumption of clams harvested from a sewage-polluted area (Pintó *et al.*, 2010; Richards, 2006). Shellfish have been implicated in a number of other outbreaks of hepatitis A where viral genetic analysis was applied to identify the source of contamination (Bosch *et al.*, 2001; Kaul *et al.*, 2000; Pintó *et al.*, 2009; Richards, 2006). Genetic analysis has also been applied to the investigation of outbreaks of hepatitis A associated with green onions (Dentinger *et al.*, 2001; Wheeler *et al.*, 2005), strawberries (Hutin *et al.*, 1999) and lettuce (Cuthbert, 2001). Epidemiological investigations from the outbreaks due to contaminated green onions and strawberries showed that these items required extensive handling during harvesting which may have been the original source of contamination (Fiore, 2004). The ingestion of raw blueberries was identified as the cause of an outbreak of hepatitis A in New Zealand (Calder *et al.*, 2003), and raspberries have been linked to outbreaks of hepatitis A (Croci *et al.*, 2008).

In an outbreak of hepatitis A in Belgium, an unexpected food item, that is, contaminated raw beef, was implicated as the vehicle of transmission with an infected food handler as the source of contamination (Robesyn *et al.*, 2009). As noted with NoVs, molecular-based pathogen source tracking has facilitated the identification of transcontinental outbreaks (Bosch *et al.*, 2001; Pintó *et al.*, 2009; Sánchez *et al.*, 2002) and intracontinental outbreaks of HAV infection (Amon *et al.*, 2005; Dentinger *et al.*, 2001; Hutin *et al.*, 1999).

7.5.3 Hepatitis E virus

HEV is recognised as a zoonotic disease which is transmitted primarily via the faecal–oral route. Contaminated water is the main source of hepatitis E epidemics, while shellfish, contaminated animal meat and contact with infected animals have been associated with sporadic cases (Meng, 2010; Miyamura, 2011). In developing countries waterborne outbreaks, affecting large numbers of people, are usually linked to faecally contaminated drinking water or flooding (Aggarwal, 2011; Miyamura, 2011). In industrialised countries such as Japan, zoonotic transmission of HEV has been linked to the consumption of contaminated meat (Pavio *et al.*, 2010). Using viral genetic analysis, HEV genotype 3 transmission from wild boar meat to humans has been identified (Li *et al.*, 2005). The consumption of raw wild Sika deer meat was implicated in an outbreak of HEV infection involving two families when HEV RNA sequences from the patient specimens and frozen Sika meat were shown to be identical (Meng, 2011; Miyamura, 2011; Pavio *et al.*, 2010). The far-reaching implications of zoonotic foodborne HEV were further elaborated when a case of transfusion-transmitted hepatitis E was linked to a blood donor who had acquired a foodborne HEV infection (Matsubayashi *et al.*, 2008). The latter linkage would not have been possible without molecular pathogen source tracking.

7.6 Conclusion

The microbiological criteria for food and water quality rely on bacterial standards. As FIB, for example, *E. coli* and enterococci, do not necessarily correlate with the presence or absence of potentially pathogenic human enteric viruses, more specific advanced virus detection and characterisation techniques may replace FIB culture-based library-dependent and culture-based library-independent methods for source tracking viruses in food and waterborne outbreaks.

Improvements in recovery techniques and molecular methods have increased the opportunities for the direct detection and characterisation of viruses in clinical specimens, and in food and water matrices. However, as no one method will be applicable to every situation, a multi-tiered approach should be applied to identify the pathogen source. For successful public health interventions in outbreak situations, rapid techniques for pathogen monitoring and tracking need to be applied. Sensitive and highly specific virus molecular-based detection assays are required for the monitoring of food and water sources, and these assays are susceptible to inhibitors in environmental samples. In addition, prior knowledge of the suspected target virus is required to ensure the application of appropriate detection assays. Unknown or untargeted viruses will not, however, be detected by this approach. Despite the challenges encountered with the detection of viruses in the environment, molecular pathogen source tracking by nucleotide sequence analysis has added a valuable new dimension to the MST toolbox. Newer culture-independent, library-dependent molecular-based methods under development and validation, for example, metagenomics, which facilitate simultaneous comparison of many genomes from multiple taxa, will provide new opportunities to expand the repertoire of assays available for broad-based molecular source tracking.

7.7 References

AGGARWAL, R. 2011. Hepatitis E: Historical, contemporary and future perspectives. *J Gastroenterol Hepatol* **26**(Suppl 1), 72–82.

ALBINANA-GIMENEZ, N., MIAGOSTOVICH, M.P. CALGU., HUGUET, J.M., MATIA, L. and GIRONES, R. 2009. Analysis of adenoviruses and polyomaviruses quantified by qPCR as indicators of water quality in source and drinking-water treatment plants. *Wat Res* **43**, 2011–19.

AMON, J.J., DEVASIA, R., XIA, G., NAINAN, O.V., HALL, S., LAWSON, B., WOLTHUIS, J.S., MACDONALD, P.D.M., SHEPARD, C,W., WILLIAMS, I.T., ARMSTRONG, G.L., GABEL, J.A., ERWIN, P., SHEELER, L., KUHNERT, W., PATEL, P., VAUGHAN, G., WELTMAN, A., CRAIG, A.S., BELL, B.P. and FIORE, A. 2005. Molecular epidemiology of foodborne hepatitis A outbreaks in the United States, 2003. *J Inf Dis* **192**, 1323–30.

ASHBOLT, N.J. 2004. Microbial contamination of drinking water and disease outcomes in developing regions. *Toxicol* **198**, 229–38.

ASHBOLT, N.J., GRABOW, W.O.K and SNOZZI, M. 2001. Indicators of microbial water quality. In: FEWTRELL, L., BARTRAM, J. editors. *Water Quality: Guidelines, Standards and Health*, London: IWA Publishing, pp. 289–315.

ATMAR, R.L. 2010. Noroviruses: State of the art. *Food Environ Virol* **2**, 117–26.

ATMAR, R. L. and ESTES, M.K. 2009. Human caliciviruses. In: RICHMAN, D.D., WHITLEY, R.J., HAYDEN. F.G. editors. *Clinical Virology*. 3rd ed. Washington: ASM Press, pp.1109–26.

BAERT, L., MATTISON, K., LOISY-HAMON, F., HARLOW, J., MARTYRES, A., LEBEAU, B., STALS, A., VAN COILLIE, E., HERMAN, L. and UYTTENDAELE, M. 2011. Review: norovirus prevalence in Belgian, Canadian and French fresh produce: a threat to human health. *Int J Food Microbiol* **151**, 261–9.

BAERT, L., UYTTENDAELE, M., VERMEERSCH, M., VAN COILLIE, E. and DEBEVERE, J. 2008. Survival and transfer of murine norovirus 1, a surrogate for human noroviruses, during the production process of deep-frozen onions and spinach. *J Food Protect* **71**, 1590–7.

BARRABEIG, I., ROVIRA, A., BUESA, J., BARTOLOMÉ, R., PINTÓ, R., PRELLEZO, H. and DOMINGUEZ, A. 2010. Foodborne norovirus outbreak: the role of an asymptomatic food handler. *BMC Infect Dis* **10**, 269.

BARRY, M. and HUGHES, J.M. 2008. Talking dirty – the politics of clean water and sanitation. *NEJM* **359**, 784–7.

BEAUDEAU P., DE VALK, H., VAILLANT, V., MANNSCHOTT, C., TILLIER, C., MOULY, D. LEDRANS, M. 2008. Lessons learned from ten investigations of waterborne gastroenteritis outbreaks, France, 1998–2006. *J Water Health* **6**, 491–503.

BERGER, C.N., SODHA, S.V., SHAW, R.K., GRIFFIN, P.M., PINK, D., HAND, P. and FRANKEL, G. 2010. Fresh fruit and vegetables as vehicles for the transmission of human pathogens. *Environ Microbiol* **12**, 2385–97.

BERNHARD, A.E. and FIELD, K.G. 2000. A PCR assay to discriminate human and ruminant feces on the basis of host differences in *Bacteriodes-Prevotella* genes encoding 16S rRNA. *Appl Environ Microbiol* **66**, 4571–4.

BLOCH, A.B., STRAMER, S.L., SMITH, J.D., MARGOLIS, H.S., FIELDS, H.A., MCKINLEY, T.W., GERBA, C.P., MAYNARD, J.E. and SIKES, R.K. 1990. Recovery of hepatitis A virus from a water supply responsible for a common source of outbreak of hepatitis A. *Am J Pub Health* **80**, 428–30.

BOFILL-MAS, S., ALBINANA-GIMENEZ, N., CLEMENTE-CASARES, P., HUNDESA, A., RODRIGUEZ-MAZANO, J., ALLARD, A., CALCO, M. and GIRONES, R. 2006. Quantification and stability of human adenoviruses and polyomavirus JCPyV in wastewater matrices. *Appl Environ Microbiol* **72**, 7894–6.

BOFILL-MAS, S., HUNDESA, A., CALGUA, B., RUSIÑOL, M., MALUQUER DE MOTES, C. and GIRONES, R. 2011. Cost-effective method for microbial source tracking using specific human and animal viruses. *J Vis Exp* **58**, e2820, DOI: 10:3791/2820.

BOFILL-MAS, S., PINA, S. and GIRONES, R. 2000. Documenting the epidemiologic patters of polyomaviruses in human populations by studying their presence in urban sewage. *Appl Environ Microbiol* **66**, 238–45.

BOSCH, A., GUIX, S., SANO, D and PINTÓ, R.M. 2008. New tools for study and direct surveillance of viral pathogens in water. *Curr Opin Biotechnol* **19**, 295–301.

BOSCH, A., LECENA, F., DÍEZ, J.M., GAJARDO, R., BLASI, M. and JOFRE, J. 1991. Waterborne viruses associated with hepatitis outbreak. *J Am Wat Works Ass* **83**, 80–3.

BOSCH, A., SÁNCHEZ, G., LE GUYADER, F., VANACLOCHA, H., HAUGARREAU, L. and PINTÓ, R.M. 2001. Human enteric viruses in Coquina clams associated with a large outbreak of hepatitis A. *Wat Sci Technol* **43**, 61–5.

CALDER, L., SIMMONS, G., THORNLEY, C., TAYLOR, P., PRITCHARD, K., GREENING, G. and BISHOP, J. 2003. An outbreak of hepatitis A associated with consumption of raw blueberries. *Epidemiol Infect* **131**, 745–1.

CANTALUPO, P.G., CALGUA, B., ZHAO, G., HUNDESA, A., WIER, A.D., KATZ, J.P., GRABE, M., HENDRIX, R.W., GIRONES, R., WAND, D. and PIPAS, J. M. 2011. Raw sewage harbours diverse viral populations. *mBio* **2**, 1–11.

CARTER, M.J. 2005. Enterically infecting viruses: Pathogenicity, transmission and significance for food and waterborne infection. *J Appl Microbiol* **98**, 1354–80.

CENTERS FOR DISEASE CONTROL and PREVENTION. 2011. Updated norovirus outbreak management and diseases prevention guidelines. *Morbidity and Mortality Weekly Report* **60**, 1–14.

COTTERELLE, B., DROUGARD, C., ROLLAND, J, BECAMEL, M, BOUDON, M., PINEDE, S., TRAORÉ, O., BALAY, K., POTHIER, P. and ESPIÉ, E. 2005. Outbreak of norovirus infection associated with the consumption of frozen raspberries, France, March 2005. Euro Surveill **10**(17), 2690.

CROCI, L., DUBOIS, E., COOK, N., de MEDICI, D., SCHULTZ, A.C., CHINA, B., RUTJES, S.A., HOORFAR, J. and VAN DER POEL, W.H.M. 2008. Current methods for extraction and concentration of enteric viruses from fresh fruit and vegetables; towards international standard. *Food Anal Methods* **1**, 73–84.

CUTHBERT, J.A. 2001. Hepatitis A: old and new. *Clin Microbiol Rev* **14**, 38–58.

DENTINGER, C., BOWER, W.A., NAINAN, O.V., COTTER, S.M., MYERS, G., DUBUSKY, L.M., FOWLER, S., SALEHI, E.D. and BELL, B.P. 2001. An outbreak of hepatitis A associated with green onions. *J Infect Dis* **183**, 1273–6.

DE RODA HUSMAN, A.M., LODDER-VERSCHOOR, F., VAN DEN BERG, H.H.J.L., LE GUYADER, F.S., VAN PELT, H., VAN DER POEL, W.H.M. and RUTJES, S. 2007. Rapid virus detection procedure for molecular tracing of shellfish associated with disease outbreaks. *J Food Prot* **70**, 967–74.

DOTZAUER, A. 2008. Hepatitis A virus. In: MAHY, B.W.J., VAN REGENMORTEL, M.V. editors. *Encyclopedia of Virology*, 3rd ed. Slovenia: Academic Press, pp. 343–50.

DUBINSKY, E.A., ESMAILI, L., HULLS, J.R., CAO, Y., GRIFFITH, J.F. and ANDERSON, G.L. 2012. Application of pylogenetic microarray analysis to discriminate soures of faecal pollution. *Environ Sci Tecnol* **46**, 4340–7.

EBDON, J.E., SELLWOOD, J., SHORE, J. and TAYLOR, H.D. 2012. Phages of *Bacteriodes* (GB-124): a novel tool for viral waterborne disease control. *Environ Sci Technol* **46**, 1163–9.

EL-SENOUSY, W.M., GUIX, S., ABID, I., PINTÓ, R.M. and BOSCH, A. 2007. Removal of astrovirus from water and sewage treatment plants, evaluated by a competitive reverse transcription-PCR. *Appl Environ Microbiol* **7**, 164–7.

FABER, M.S., STARK, K., BEHNKE, S.C., SCHREIER, E. and FRANK, C. 2009. Epidemiology of hepatitis A infection, Germany, 2007–2008. *Emerg Infect Dis* **15**, 1760–8.

FALKENHORST, G., KRUSELL, L., LISBY, M., MADSEN, S.B., BÖTTIGER, B. and MØLBAK, K. 2005. Imported frozen raspberries cause a series of norovirus outbreaks in Denmark, 2005. *Eurosurveillance* **10**(38), 2795.

FIELD, F. G and SAMADPOUR, M. 2007. Fecal source tracking, the indicator paradigm, and managing water quality. *Wat Res* **41**, 3517–38.

FIORE, A. 2004. Hepatitis A transmission by food. *Clin Infect Dis* **38**, 705–15.

FOLEY, S.L., LYNNE, A.M. and NAYAK, R. 2009. Molecular typing methodologies for microbial source tracking and epidemiological investigations of gram-negative bacterial foodborne pathogens. *Infect Genet Evol* **9**, 430–40.

FONG, T-T. and LIPP, E.K. 2005. Enteric virus of humans and animals in aquatic environments: Health risks, detection, and potential water quality assessment tools. *Microbiol Mol Biol Rev* **69**, 357–71.

GIRONES, R., FERRÙS A.A., ALONSO., J.L., RODRIGUEZ-MANZANO, J., CALGUA, B., DE ABREU CORRÊA, A., HUNDESA, A., CARRATALA, A. and BOFILL-MAS, S. 2010. Molecular detection of pathogens in water – The pros and cons of molecular techniques. *Wat Res* **44**, 4325–39.

GRABOW, W.O.K. 1996. Waterborne diseases: update on water quality assessment and control. *Water SA* **22**, 193–202.

GRABOW, W.O.K. 2007. Overview of health-related water virology. In: BOSCH, A. editor. *Human Viruses in Water*. 1st ed. Amsterdam: Elsevier, pp. 1–25.

GREENING, GE. 2006. Human and animal viruses in food (including taxonomy of enteric viruses). In: GOYAL, S.M. editor. *Viruses in Foods*. New York: Springer Science and Business Media, pp. 5–42.

GÓMEZ-DOÑATE, M., PAYÁN, A., CORTÉS, I., BLANCH, AR., LUCERNA, F., JOFRE, J. and MUNIESA, M. 2011. Isolation of bacteriophage host strains of *Bacteriodes* species suitable for tracking sources of animal faecal pollution in water. *Environ Microbiol* **13**, 1622–31.

HAVELAAR, A.H. and RUTJES, S.A. 2008. Risk assessment of viruses in food: opportunities and challenges. In: KOOPMANS, M.P.G., CLIVER, D.O., BOSCH, A. editors. *Food-Borne Viruses Progress and Challenges*. Washington, DC: ASM Press, pp. 221–36.

HAYES, C., ELLIOT, E, KRALES, E. and DOWNER G. 2003. Food and water safety for persons infected with human immunodeficiency virus. *Clin Infect Dis* **36**, S106–S109.

HE, Y-Q., MA, H-W., YAO, X-J., BU, Q-L., YANG H., HUANG, W., DUAN, Y-X., ZHANG, H-L., DAI., C-H., LI, W-K. and YU, S-Y. 2010. Norovirus gastroenteritis outbreak is associated with contaminated drinking water in South China. *Food Environ Virol* **2**, 207–10.

HELMER, R. 1999. Water quality and human health. *The Environmentalist* **19**, 11–16.

HEWITT, J., BELL, D., SIMMONS, G.C., RIVERA-ABAN, M., WOLF, S. and GREENING, G.E. 2007. Gastroenteritis outbreak caused by waterborne norovirus at a New Zealand ski resort. *Appl Environ Microbiol* **73**, 7853–7.

HOEBE, C.J.P.A., VENNEMA, H., DE RODA HUSMAN, A.M. and VAN DUYNHOVEN, Y.T.P.H. 2004. Norovirus outbreak among primary school children who played in a recreational water fountain. *J Inf Dis* **189**, 699–705.

HOLLINGER, F.B. and EMERSON, S.U. 2007. Hepatitis A virus. In: KNIPE, D.M., GRIFFIN, D.E., LAMB, R.A., STRAUS, S.E., HOWLEY, P.M., MARTIN, M.A., ROIZMAN, B., editors. *Fields Virology*. 5th ed. Philadelphia: Lippincott Williams & Wilkins, pp. 911–47.

HUTIN, Y.J., POOL, V., CRAMER, E.H., NAINAN, O.V., WETH, J., WILLIAM, I.T., GOLDSTEIN, S.T., GENSHEIMER, K.F., BELL, B.P., SHAPIRO, C.N., ALTER, M.J. and MARGOLIS, H.S. 1999. A multistate, foodborne outbreak of hepatitis A. National hepatitis A investigation team. *NEJM* **340**, 595–602.

JIANG, S.C. 2006. Human adenoviruses in water: occurrence and health implications: a critical review. *Environ Sci Technol* **40**, 7132–40.

KAUL, A., CHAVES, R.L., EISSELE, R., MATZDORFF, P., NORMANN, A. and FLEHMIG, B. 2000. Analysis of a foodborne hepatitis A epidemic. *Biotest Bull* **6**, 217–224.

KIRBY, A. and ITURRIZA-GÓMARA, M. 2012. Norovirus diagnostics. Options, applications and interpretations. *Expert Rev Anti Infect Ther* **10**, 423–33.

KNOWLES, N.J. HOVI, T., HYYPIÄ, T., KING, A.M.Q., LINDBERG, A.M., PALLANSCH, M.A., PALMENBERG, A.C., SIMMONDS, P., SKERN, T., STANWAY, G., YAMASHITA, T. and ZELL, R. 2012. Family Picornaviridae. In: KING, A.M.Q., ADAMS, M.J., CARSTENS, E.B. AND LEFKOWITZ, E.J. editors, *Virus Taxonomy. Classification and Nomenclature of Viruses. Ninth Report of the International Committee on Taxonomy of Viruses*. San Diego: Academic Press, pp. 855–80.

KOOPMANS, M. and DUIZER, E. 2004. Foodborne viruses: an emerging problem. *Int J Food Microbiol* **90**, 23–41.

KORSAGER, B., HEDE, S., BØGGILD, H., BÖTTIGER, B. and MØLBAK, K. 2005. Two outbreaks of norovirus infections associated with the consumption of imported frozen raspberries, Denmark, May–June 2005. *Euro Surveill* **10**, E050623.1.

LAPPALAINEN, M., CHEN, R.W. and MAUNULA, L. 2001. Molecular epidemiology of viral pathogens and tracing of transmission routes: hepatitis-, calici- and hantaviruses. *J Clin Virol* **21**, 177–85.

LAURI, A. and MARIANI, P.O. 2009. Potential and limitations of molecular diagnostic methods in food safety. *Genes Nutr* **4**, 1–12.

LEE, J.E., LIM, M.Y., KIM., S.Y., LEE, S., LEE, H., OH, H., HUR, H. and KO, G. 2009. Molecular characterization of bacteriophages for microbial source tracking in Korea. *Appl Environ Microbiol* **75**, 7107–14.

LE GUYADER, F.S. and ATMAR, R.L. 2008. Binding and inactivation of viruses on and in food, with a focus on the role of the matrix. In: KOOPMANS, M.P.G., CLIVER, D.O., BOSCH, A. editors. *Food-Borne Viruses: Progress and Challenges*. Washington DC: ASM Press, pp. 189–208.

LE GUYADER, F.S., BON F., DEMEDICI, D., PARNAUDAU, S., BERTONE, A., CRUDELI, S., DOYLE, A., ZIDANE, M., SUFFREDINI, E., KOHLI, E., MADDALO, F., MONINI, A., GALLAY, A., POMMEPUY, M., POTHIER, P. and RUGGERI, F.M. 2006. Detection of multiple noroviruses associated with an international gastroenteritis outbreak linked to oyster consumption. *J Clin Microbiol* **44**, 3878–82.

LE GUYADER, F.S., MITTELHOLZER, C., HAUGARREAU, L., HEDLUND, K.-O., ALSTERLUND, R., POMMEPUY, M. and SVENSSON, L. 2004. Detection of noroviruses in raspberries associated with a gastroenteritis outbreak. *Int J Food Microbiol* **97**, 179–86.

LI, T-C, CHIJIWA, K., SERA, N., ISHIBASHI, T., ETOH., Y., SHINOHARA, Y., KURATA, Y., ISHIDA, M., SAKAMOTO., S., TAKEDA, N. and MIYAMURA, T. 2005. Hepatitis E virus transmission from wild boar meat. *Emerg Infect Dis* **11**, 1958–60.

LOPMAN, B., GASTAÑADUY, P., PARK., G.W., HALL, A.J., PARASHAR, U.D. and VINJÉ, J. 2012. Environmental transmission of norovirus gastroenteritis. *Curr Opin Virol* **2**, 96–102.

LYSÉN, M., THORHAGEN, M., BRYTTING, M., HJERTQVIST, M., ANDERSSON, Y. and HEDLUND, K.-O. 2009. Genetic diversity among food-borne and waterborne norovirus strains causing outbreaks in Sweden. *J Clin Microbiol* **47**, 2411–18.

MALUQUER DE MOTES, C., CLEMENTE-CASARES, P., HUNDESA, A., MARTÍN, M. and GRONES, R. 2004. Detection of bovine and porcine adenoviruses for tracing the source of fecal contamination. *Appl Environ Microbiol* **70**, 1448–54.

MATSUBAYASHI, K., KANG, J-H., SAKATA, H., TAKAHASHI, K., SHINDO, M., KATO, M., SATO, S., KATO, T., NISHIMORI, H, TSUJI, K., MAGUCHI, H., YOSHIDA, J-I., MAEKUBO, H, MISHIRO, S. and IKEDA, H. 2008. A case of transfusion transmitted hepatitis E caused by blood from a donor infected with hepatitis E virus via zoonotic food-borne route. *Transfusion* **48**, 1368–75.

MATTISON, K. and BIDAWID, S. 2009. Analytical methods for food and environmental viruses. *Food Environ Virol* **1**, 107–22.

MAUNULA, L., MIETTINEN, I,T. and VON BONSDORFF, C-H. 2005. Norovirus outbreaks from drinking water. *Emerg Infect Dis* **11**, 1716–21.

MAUNULA, L., ROIVAINEN, M., KERÄNEN, M., MÄKELÄ, S., SÖDDERBERG, K., SUMMA, M., von BONSDORFF, C.H., LAPPALAINEN, M., KORHONEN, T., KUUSI, M. and NISKANEN, T. 2009. Detection of human norovirus from frozen raspberries in a cluster of gastroenteritis outbreaks. *Euro Surveill* **14**(49), pii=19435.

MAYET, A., ANDRÉO, V., BÉDUBOURG, G., VICTORION, S., PLANTEC, J.Y., SOULLIÉ, B., MEYNARD, J.B., DEDIEU, J.J., POLVÉCHE, P.Y. and MIGLIANI, R. 2011. Food-borne outbreak of norovirus infection in a French military parachuting unit, April 2011. *Euro Surveill* **16**(30), pii=19930.

MCCABE-SELLERS, B.J. and BEATTIE, S.E. 2004. Food safety: Emerging trends in food-borne illness surveillance and prevention. *J Am Diet Assoc* **104**, 1708–17.

MCQUAIG, S.M., SCOTT, T.M., HARWOOD, V.J., FARRAH, S.R. and LUKASIK, J.O. 2006. Detection of human-derived fecal pollution in environmental waters bu use of a PCR-based human polyomavirus assay. *Appl Environ Microbiol* **72**, 7567–74.

MEAD, P.S., SLUTSKER, L., DIETZ, V., MCCAIG, L. F., BRESEE, J.S., SHAPIRO, C., GRIFFIN, P.M. and TAUXE, R.V. 1999. Food-related illness and death in the United States. *Emerg Infect Dis* **5**, 607–25.

MEAYS, C.L., BROERSMA, K., NORDIN, R. and MAZUMDER, A. 2004. Source tracking fecal bacteria in water: a critical review of current methods. *J Environ Manage* **73**, 71–9.

MENG, X-J. 2010. Recent advances in hepatitis E virus. *J Viral Hepat* **17**, 153–61.

MENG, X-J. 2011. From barnyard to food table: the omnipresence of hepatitis E virus and risk for zoonotic infection and food safety. *Virus Res* **161**, 23–30.

MIYAMURA, T. 2011. Hepatitis E virus infection in developed countries. *Virus Res* **161**, 40–6.

MORSE, L.J., BRYAN, J.A., HURLEY, J.P., MURPHY, J.F., O'BRIEN, T.F. and WACKER, W.E. 1972. The Holy Cross college football team hepatitis outbreak. *JAMA* **219**, 702–8.

MUNIESA, M., PAYAN, A., MOCE-LLIVINA, L., BLANCH, A.R. and JOFRE, J. 2009. Differential persistence of F-specific RNA phage subgroups hinders their use as single tracers for faecal source tracking in surface water. *Wat Res* **43**, 1559–64.

MURUGAN, K., PRABHAKARAN, P., AL-SOHAIBANI, S. and SEKAR, K. 2011. Identification of the source of faecal pollution of Tirumanimuttar River, Tamilnadu, India using microbial source tracking. *Environ Monit Assess* (available on line DOI: 10.1007/s10661-011-2398-7).

NENONEN, N.P., HANNOUN, C., LARSSON, C.U. and BERGSTRÖM, T. 2012. Marked genetic diversity of norovirus genogroup I strains in a waterborne outbreak. *Appl Environ Microbiol* **78**, 1846–52.

NEWELL, D.G., KOOPMANS, M., VERHOEF, L., DUIZER, E., AIDARA-KANE, A., SPRONG, H., OPSTEEGH, M., LANGELAAR, M., THREFALL, J., SCHEUTZ, F., VAN DER GIESSEN, J. and KRUSE, H. 2010. Food-borne diseases – The challenges of 20 years ago still persist while new ones continue to emerge. *Int J Food Microbiol* **139**, S3–S15.

O'BREIN, S J. 2008 The challenge of estimating the burden of an underreported disease. In: KOOPMANS, M.P.G., CLIVER, D.O. and BOSCH, A. editors. *Food-borne Viruses Progress and Challenges*. Washington, DC: ASM Press, pp. 87–115.

OGORZALY, L. and GANTZER, G. 2006. Development of real-time RT-PCR methods for specific detection of F-specific RNA bacteriophage genogroups: application to urban raw wastewater. *J Virol Methods* **138**, 131–9.

OGORZALY, L., TISSIER, A., BERTRAND, I., MAUL, A. and GANTZER C. 2009. Relationship between F-specific RNA phage genogroups, faecal pollution indicators and human adenoviruses in river water. *Wat Res* **43**, 1257–64.

OKOH, A.I., SIBANDA, T. and GUSHA, S.S. 2010. Inadequately treated wastewater as a source of human enteric viruses in the environment. *Int J Environ Res Public Health* **7**, 2620–37.

PARSHIONIKAR, S.U., WILLIAN-TRUE, S., FOUT, G. S, ROBBINS, D. E, SEYS, S.A., CASSADY, J.D. and HARRIS, R. 2003. Waterborne outbreak of gastroenteritis associated with a norovirus. *Appl Environ Microbiol* **69**, 5263–8.

PATEL, M.M., HALL, A.J., VINJÉ, J. and PARASHAR, U.D. 2009. Noroviruses: a comprehensive review. *J Clin Virol* **44**, 1–8.

PAVIO, N., MENG, X-J. and RENOU, C. 2010. Zoonotic hepatitis E: animal reservoirs and emerging risks. *Vet Res* **41**, 46.

PINA, S., PUIG, M., LUCENA, F., JOFRE, J. and GIRONES, R. 1998. Viral pollution in the environment and in shellfish: Human adenovirus detection by PCR as an index of human viruses. *Appl Environ Microbiol* **64**, 3376–82.

PINTÓ, R.M., COSTAFREDA, M.I. and BOSCH, A. 2009. Risk assessment in shellfish-borne outbreaks of hepatitis A. *Appl Environ Microbiol* **75**, 7350–5.

PINTÓ, R.M., COSTAFREDA, M,I., PÉREZ-RODRIGUEZ, F.J., D'ANDREA, L. and BOSCH, A. 2010. Hepatitis A virus: State of art. *Food Environl Virol* **2**, 127–35.

PINTÓ, R.M. and SAIZ, J-C. 2007. Enteric hepatitis viruses. In: BOSCH, A. editor. *Human Viruses in Water*. Amsterdam: Elsevier B.V., pp. 39–67.

PURNELL, S.E., EBDON, J.E. and TAYLOR, H.D. 2011. Bacteriophage lysis of *Enterococcus* host strains: A tool for microbial source tracking. *Environ Sci Technol* **45**, 10699–705.

RÄSÄNEN, S., LAPPALAINEN, S., KAIKKONEN, S., HÄMÄLÄINEN, M., SALMINEN, M. and VESIKARI, T. 2010. Mixed viral infections causing acute gastroenteritis in children in a waterborne outbreak. *Epidemiol Infect* **138**, 1227–34.

RICHARDS, G.P. 2006. Shellfish-associated viral diseases outbreaks. In: GOYAL, S.M. editor. *Viruses in Foods*. New York: Springer, p. 223–34.

RIERA-MONTES, M., SJÖLANDER, K.B., ALLESTAM, G., HALLIN, E., HEDLUND, K.-O. and LÖFDAHL, M. 2011. Waterborne norovirus outbreak in a municipal drinking-water supply in Sweden. *Epidemiol Infect* **139**, 1928–35.

ROBERTSON, B.H., JANSEN, R.W., KHANNA, B., TOTSUKA, A., NAINAN, O.V., SIEGL, G., WIDELL, A., MARGOLIS, H.S., ISOMURA, S., ITO, K., ISHIZU, T., MORITSUGU, Y. and LEMON, S.M. 1992. Genetic relatedness of hepatitis A virus strains recovered from different geographical regions. *J Gen Virol* **73**, 1365–77.

ROBESYN, E., DE SCHRIJVER, K., WOLLANTS, E., TOP, G., VERBEECK, J. and VAN RANST, M. 2009. An outbreak of hepatitis A associated with the consumption of raw beef. *J Clin Virol* **44**, 207–10.

RODRÍGUEZ-LÁZARO, D., COOK, N., RUGGERI, F.M., SELLWOOD, J., NASSER, A., NASCIMENTO, M.S.J., D'AGOSTINO, M., SANTOS, R., SAIZ, J.C., RZEŻUTKA, A., BOSCH, A., GIRONÉS, R., CARDUCCI, A., MUSCILLO, M., KOVAČ, K., DIEZ-VALCARCE, M., VANTAKARIS, A, VON BONSDORFF, C-H., DE RODA HUSMAN, A.M, HERNÁNDEZ, M. and VAN DER POEL, W.H.M. 2012. Virus hazards from food, water and other contaminated environments. *FEMS Microbiol Rev* **36**, 786–814.

ROSLEV, P. and BUKH, A.S. 2011. State of the art molecular markers for fecal pollution source tracking in water. *Appl Microbiol Biotechnol* **89**, 1341–55.

SÁNCHEZ, G., PINTÓ, R.M., VANACLOCHA, H. and BOSCH, A. 2002. Molecular characterization of hepatitis A virus isolates from a transcontinental shellfish-borne outbreak. *J Clin Microbiol* **40**, 4148–55.

SANTO DOMINGO, J.W., BAMBIC, D.G., EDGE, T.A. and WUERTZ, S. 2007. Quo vadis source tracking? Towards a strategic framework for environmental monitoring of fecal pollution. *Wat Res* **41**, 3539–52.

SARVIKIVI, E., ROIVAINEN, M., MAUNULA, L., NOSKANEN, T, KORHONEN, T., LAPPALAINEN, M. and KUUSI, M. 2012. Multiple norovirus outbreaks linked to imported frozen raspberries. *Epidemiol Infect* **140**, 260–67.

SCALLAN, E., GRIFFIN, P.M., ANGULO, F.J., TAUXE, R.V. and HOEKSTRA, R.M. 2011. Foodborne illness acquired in the United States – Unspecified agents. *Emerg Infect Dis* **17**, 16–22.

SCHAPER, M., JOFRE, J., UYS, M. and GRABOW, W.O. 2002. Distribution of genotypes of F-specific RNA bacteriophages in human and non-human sources of faecal pollution in South Africa and Spain. *J Appl Micro* **92**, 657–67.

SCOTT, T.M., ROSE, J.B., JENKINS, T.M., FARRAH, S.R. and LUKASIK, J. 2002. Microbial source tracking: Current methodology and future directions. *Appl Environ Microbiol* **68**, 5796–803.

SMITH DEWAAL, C., TIAN, X.A. and BHUIYA, F. 2008. Outbreak Alert 2008 report-final. Centre for Science in the Public Interest 10.

SPRADLING, P.R., MARTIN, A. and FEINSTONE, S.M. 2009. HEPATITIS A VIRUS. In: RICHARDS, D.D., WHITLEY, R.J., HAYDEN, F.G., editors. *Clinical Virology*. 3rd ed. Washington, DC: ASM Press, pp. 1083–1108.

STANWAY, G., BROWN, F., CHRISTIAN, P., HOVI, T., HYYPIÄ, T., KING, A.M.Q., KNOWLES, N.J., LEMON, S.M., MINOR, P.D. PALLANSCH, M.A., PALMENBERG, A.C. and SKERN, T. 2005. Family Picornaviridae. In: FAUQUET, C.M., MAYO, M.A., MANILOFF, J., DESSELBERGER, U. and BALL, L.A. editors, *Virus Taxonomy. Eighth Report of the International Committee on Taxonomy of Viruses*. San Diego: Academic Press, pp. 757–78.

STEELE, M. and ODUMERU, J. 2004. Irrigation water as source of foodborne pathogens on fruit and vegetables. *J Food Protect* **67**, 2839–49.

STEWART, J.R., ELLENDER, R.D., GOOCH, J.A., JIANG, S., MYODA, S.P. and WEISBERG, S.B. 2003. Recommendations for microbial source tracking: lessons from a methods comparison. *J Water Health* **1**, 225–31.

STEWART, J.R., VINJÉ, J., OUDEJANS, S.J.G., SCOTT, G.I. and SOBSEY, M. 2006. Sequence variation among group III F-specific RNA coliphages from water samples and swine lagoons. *Appl Environ Microbiol* **72**, 1226–30.

STEWART-PULLARO, J., DAUGOMAH, J.W., CHESTNUT, D.E., GRAVES, D.A., SOBSEY, M.D. and SCOTT, G.I. 2006. F+ RNA coliphage typing for microbial source tracking in surface waters. *J Appl Microbiol* **101**, 1015–26.

TUAN ZAINAZOR, C., HIDAYAH, M.S.N., CHAI, L.C., TUNUNG, R., GHAZALI, F.M. and SON, R. 2010. The scenario of norovirus contamination in food and food handlers. *J Microbiol Biotechnol* **20**, 229–37.

VAN HEERDEN, J., EHLERS, M.M., HEIM, A. and GRABOW, W.O.K. 2005. Prevalence, quantification and typing of adenoviruses detected in river and treated drinking water in South Africa. *J Appl Microbiol* **99**, 234–42.

VERHOEF, L., VENNEMA, H., VAN PELT, W., LEES, D., BOSHUIZEN, H., HENSHILWOOD, K. and KOOPMANS, M. 2010. Use of norovirus genotypes profiles to differentiate origins of foodborne outbreaks. *Emerg Infect Dis* **16**, 617–29.

VERHOEF, L.P.B., KRONEMAN, A., VAN DUYNHOVEN, Y., BOSHUIZEN, H., VAN PELT, W and KOOPMANS, M. 2009. Selection tool for foodborne norovirus outbreaks. *Emerg Infect Dis* **15**, 31–8.

WEI, J. and KNIEL, K.E. 2010. Pre-harvest viral contamination of crops originating from fecal matter. *Food Environ Virol* **2**, 195–206.

WHEELER, C., VOGT, T.M, ARMSTRONG, G.L., VAUGHAN, G., WELTMAN, A., NAINAN, O.V., DATO, V., XIA, G., WALLER, K., AMON, J.A., LEE, T.M., HIGHBAUGH-BATTLE, A., HEMBREE, C., EVENSON, S., RUTA, M.A., WILLIAMS, I.T., FIORE, A.E. and BELL, B.P. 2005. An outbreak of hepatitis A associated with green onions. *NEJM* **353**, 890–7.

WIDDOWSON, M-A. and VINJÉ, J. 2008. Food-borne viruses – state of the art. In: KOOPMANS, M.P.G., CLIVER, D.O. and BOSCH, A. editors. *Food-Borne Viruses: Progress and Challenges*. Washington, DC: ASM Press, pp. 29–64.

WOLF, S., HEWITT J., RIVERA-ABAN, M. and GREENING, G.E. 2008. Detection and characterization of F+RNA bacteriophages in water and shellfish: application of a multiplex real-time reverse transcription PCR. *J Virol Methods* **149**, 123–8.

WOLF, S., WILLIAMSON, W.M., HEWITT, J., RIVERA-ABAN, M., LIN, S., BALL, A., SCHOLES, P. and GREENING, G.E. 2007. Sensitive multiplex real-time reverse transcription-PCR assay for the detection of human and animal noroviruses in clinical and environmental samples. *Appl Environ Microbiol* **73**, 5464–70.

WONG, K., FONG, T-T., BIBBY, K. and MOLINA, M. 2012. Application of enteric viruses for fecal pollution source tracking in environmental samples. *Environ Int* **45**, 151–64.

WONG, K. and XAGORARAKI, I. 2011. Evaluating the prevalence and genetic diversity of adenovirus and polyomavirus in bovine waste for microbial source tracking. *Appl Microbiol Biotechnol* **90**, 1521–6.

XERRY, J., GALLIMORE, C.I., ITURRIZA-GÓMARA, M., ALLEN, D.J. and GRAY, J.J. 2008. Transmission events within outbreaks of gastroenteritis determined through analysis of nucleotide sequences of the P2 domain of genogroup II noroviruses. *J Clin Microbiol* **46**, 947–53.

XERRY, J., GALLIMORE, C.I., ITURRIZA-GÓMARA, M and GRAY, J.J. 2009. Tracking the transmission routes of genogroup II noroviruses in suspected food-borne or environmental outbreaks of gastroenteritis through sequence analysis of the P2 domain. *J Med Virol* **81**, 1298–1304.

XERRY, J., GALLIMORE, C.I., ITURRIZA-GÓMARA, M. and GRAY, J.J. 2010. Genetic characterization of genogroup I norovirus in outbreaks of gastroenteritititis. *J Clin Microbiol* **48**, 2560–2.

YEE, S.Y.F., FONG, N.Y., FONG, G.T., TAK, O.J., HUI, G.T. and MING, Y.S. 2006. Male-specific RNA coliphages detected by plaque assay and RT-PCR in tropical river waters and animal fecal matter. *Int J Environ Health Res* **16**, 59–68.

8

Quantitative risk assessment for food- and waterborne viruses

A. M. De Roda Husman and M. Bouwknegt,
National Institute for Public Health and the Environment (RIVM), The Netherlands

DOI: 10.1533/9780857098870.2.159

Abstract: Quantitative microbiological risk assessment has hitherto been predominantly used for risks from exposure to bacteria. This chapter examines the requirements for full quantitative risk assessments for food- and waterborne viruses, and the interpretation and extrapolation of risk assessment outcomes. Future trends in quantitative risk assessment research for viruses are considered.

Key words: quantitative microbiological risk assessment, viruses, food, water, health risk.

8.1 Introduction

One of the first applications of risk assessment was by NASA in the aftermath of the 1967 Apollo fire. Risk assessment is now widely used in different scientific disciplines including nuclear science, finance, industrial processing and infectious diseases. Charles Haas used risk assessment in the latter field for the first time to mathematically describe a dose–response relationship for adverse public health events due to virus ingestion (Haas, 1983). The methodology was subsequently extended, with an exposure assessment to estimate the exposure dose and a risk quantification for the presence of viruses in drinking water (Haas *et al.*, 1993). Despite this initial focus on viruses, quantitative microbiological risk assessments for food and water have so far focused predominantly on bacteria. Recently completed assessments include models that describe the entire food production pathway from production of the raw material until the moment of consumption (often described as 'farm-to-fork' models). Examples of this include the assessment of public health risks associated with *E. coli* O157:H7 in ground beef hamburgers (Cassin *et al.*, 1998) with *Listeria monocytogenes* in ready-to-eat foods (Rocourt *et al.*, 2003) and with *Campylobacter* on broiler meat (Nauta *et al.*, 2009). Such farm-to-fork models are now being developed for viruses in food for the first time as part of the European project 'Integrated monitoring and control of foodborne viruses in European food supply chains' ('VITAL').

Public health risk assessment comprises several stages: identifying and defining the risk, assessing the level of exposure and the exposure–response relationship, and assessing the associated risk for humans. This process can be done either qualitatively (providing a 'yes' or 'no' answer about a risk), semi-quantitatively (weighting or scoring each contributory component of the overall risk and adding them together to calculate the final risk), or quantitatively (using probability distributions to provide mean estimates of the risk and 95% intervals). This chapter will focus on quantitative risk assessments only and will not consider qualitative or semi-quantitative risk assessments.

Health outcomes are based on information concerning exposure to viruses in foods and water and on dose–response relationships. Several frameworks for assessing microbiological risks have been proposed (ILSI, 1996; Haas *et al.*, 1999), that are designed to structure and harmonize the approach to risk assessment. These frameworks use the following steps for the assessment of food- and waterborne viruses:

- Identification of the pathogenic virus(es) that may be present in a particular type of food or water sample and that are capable of causing infection, illness or death in humans. In addition, the identification of the exposure route or routes; ingestion or inhalation, for example.
- Estimation of the virus dose to which individuals are exposed per exposure route.
- Quantitative evaluation of the nature and probability of adverse health effects associated with a certain intake of the pathogenic agent.
- Integration of the estimated exposure dose and the probability of the adverse health event, given that dose. This aspect thus consists of the joining of the results from the previous two steps.

Risk assessment is preferably an iterative process of systematic and objective evaluation of all available information pertaining to a given hazard – in this particular instance, viruses in foods and water.

Some risk profiles, a term which references the first part of quantitative microbial risk assessment (QMRA) as described above, have been reported for foodborne viruses. These include risk profiles of the Norwalk-like virus in raw molluscs in New Zealand (Greening *et al.*, 2003), foodborne norovirus infections (HPA, 2004) and hepatitis E virus (HEV) (Bouwknegt *et al.*, 2009). Factors such as the persistence of virus infectivity on foods, the role of irrigation water and food handlers' hands in virus transmission and the effectiveness of existing virus control points in food harvesting, processing and handling have had limited representation in the available data. It was therefore concluded at an international meeting of experts in 2007 that undertaking a full quantitative risk assessment for foodborne virus may be premature (FAO/WHO, 2008a). However, increased interest in food- and waterborne viruses and the development of accessible and rapid diagnostic methods in recent years has increased the amount of available data related to viruses in foods and water, with the result that quantitative models could now start to be designed.

8.2 Quantitative microbiological risk assessments (QMRAs) and their outcomes

Studies employing QMRA to assess adverse public health events associated with viruses in food and water are presented in Table 8.1, with aspects of particular relevance to QMRA for viruses highlighted later.

The viruses selected in the quantitative risk assessment models described include the relatively larger DNA adenoviruses (Crabtree *et al.*, 1997; van Heerden *et al.*, 2005) and the small RNA enteroviruses (Regli *et al.*, 1991; Mena *et al.*, 2003) and rotaviruses (Regli *et al.*, 1991; Haas *et al.*, 1993; Gerba *et al.*, 1996) The enteroviruses included coxsackieviruses, (Mena *et al.*, 2003) polioviruses type 1 and 3, and echovirus 12 (Regli *et al.*, 1991). Some studies refer to viruses in general (Petterson and Ashbolt, 2001; Petterson *et al.*, 2001; Hamilton *et al.*, 2006).

Three of the papers mentioned in Table 8.1 describe quantitative risk assessment models for fresh produce (Petterson and Ashbolt, 2001; Petterson *et al.*, 2001; Hamilton *et al.*, 2006). One is a methodological study to examine the implications of over- dispersion in virus concentration on crops (Petterson and Ashbolt, 2001). Hamilton *et al.* (2006) provide a simplified model for estimating the number of viruses on an item of produce as a result of spray irrigation. Masago *et al.* (2006) estimate the risk of a norovirus infection due to the consumption of drinking water and describe an approach for estimating virus concentrations based on presence/absence data. The remaining studies describe the estimation of risks associated with drinking and/or coming into recreational contact with potentially contaminated waters.

The choice of statistical distribution used and the assumptions made in quantitative risk assessments affect the estimated risks. Regli *et al.* (1991) and Haas *et al.* (1993) present a theoretical background for the estimation of infection risks due to consumption of drinking water which can also be useful for estimates related to the consumption of other matrices such as food. Regli *et al.* (1991) describe the assumptions which must be made in virological risk assessment and go on to evaluate different dose–response models

Table 8.1 Quantitative risk assessments for viruses in the environment

Matrix	Virus	Title publication	Reference
Drinking water	Rotavirus, Poliovirus 1, Poliovirus 3, Echovirus 12	Modeling the risk from Giardia and viruses in drinking water	Regli *et al.*, 1991
Drinking water	Rotavirus	Risk assessment of virus in drinking water	Haas *et al.*, 1993
Drinking and recreational waters	Rotavirus	Waterborne rotavirus: a risk assessment	Gerba *et al.*, 1996
Drinking and recreational waters	Adenovirus	Waterborne adenovirus: a risk assessment	Crabtree *et al.*, 1997
Salad crops	Enteroviruses	Viral risks associated with wastewater reuse: modeling virus persistence on wastewater irrigated salad crops	Petterson and Ashbolt, 2001
Salad crops	Viruses	Microbial risks from wastewater irrigation of salad crops: a screening-level risk assessment	Petterson *et al.*, 2001
Drinking and recreational waters	Coxsackie virus	Risk assessment of waterborne coxsackie virus	Mena *et al.*, 2003
Drinking water; recreational water	Adenovirus	Risk assessment of adenoviruses detected in treated drinking water and recreational water.	van Heerden *et al.*, 2005
Raw vegetables	Viruses	Quantitative microbial risk assessment models for consumption of raw vegetables irrigated with reclaimed water	Hamilton *et al.*, 2006
Drinking water	Norovirus	Quantitative risk assessment of noroviruses in drinking water based on qualitative data in Japan	Masago *et al.*, 2006

(exponential and Beta-Poisson). Haas *et al.* (1993) provide an approach to including uncertainty and variability in risk assessments. These studies demonstrate the importance of understanding (1) the theoretical backgrounds of the distributions used; (2) the randomness or lack thereof in processes leading to viral contamination; (3) the essence of the dose–response models used (the measured response, the unit of the exposure dose); and (4) that uncertainty and variability are essential to the proper conduct of quantitative viral risk assessments.

The three models particularly focused on fresh produce centre on the use of irrigation water for salad crops (Petterson and Ashbolt, 2001; Petterson *et al.*, 2001; Hamilton *et al.*, 2006). Petterson *et al.* (2001) modelled the clinging of viruses to lettuce crops through sprayed irrigation water. The volume of retained water was taken from a study in which lettuce heads were immersed fully in water (Shuval *et al.*, 1997). The estimated volume amounted to approximately 10 ml per 100 g of lettuce, an estimate which was also later used in relation to lettuce by Hamilton *et al.* (2006). In the absence of better data, this volume was considered to represent a worst-case situation and provided an acceptable starting point for risk assessment. However, full immersion of crops might not accurately represent a worst-case scenario, as the volume of water on shrubs is reported to be larger after a simulated rainfall event than after immersion (Garcia-Estringana *et al.*, 2010). Ideally the volume of water retained after single and multiple irrigation events should therefore be examined experimentally. Such an experiment was conducted by Hamilton *et al.* (2006) for broccoli, grand slam cabbage, savoy cabbage and winter head cabbage. Their results gave estimated mean water retentions of 0.02, 0.04, 0.04 and 0.09 ml per g of produce, respectively, after a single irrigation event of 20 min. Probability distributions that may be used in a quantitative risk assessment are provided in the paper.

For estimation of the exposure dose, data on the amount of food or water consumed are required, with the data preferably representative of the general population. The available papers on food consumption lack specific monitoring data, relying instead on such valid approaches as the hypothetical consumption of 100 g of a food product (Petterson *et al.*, 2001) or the amount of food consumed being a function of body weight (Hamilton *et al.*, 2006).

The outcome of any risk assessment depends on the research aim and this is usually decided in consultation with risk managers. Examples of risk assessment outcomes include identification of exposure doses, risk of infection, likelihood of illness and possible resultant levels of mortality. An important aspect to consider in the translation of an exposure dose to an adverse health event is the dose–response model used, and how this relates to the subsequent modelled response. Depending on the data set underlying the dose–response model, the initial translation from dose to risk represents the risk of the specific response used in that particular dose–response model. For instance, the model to estimate norovirus infection risks was based on volunteers showing faecal excretion of virus and seroconversion (Teunis *et al.*, 2008). In the same study, the dose–response model for illness conditional on infection was based on a response including diarrhoea and/or vomiting combined with other symptoms such as abdominal pain, myalgia, fatigue, chills and headache more than 8 h after the infection. Thus, employing either the first or both models estimates the respective response observed in the volunteers.

Similarly, a dose–response model for hepatitis A virus (HAV) is based on data from Ward *et al.* (1958), whose observed response was the development of jaundice in institutionalized individuals, the majority of whom were children.

Hence, the latter dose–response model predicts the probability of developing jaundice for a particular ingested HAV dose in instances where the exposed individual is a child. Due to the lack of more appropriate data, such estimates are often extrapolated to other (sub)populations or to the general population in order to proclaim a risk of adverse health events. The imperfection of this approach should be considered when interpreting the estimated risk and possibly using it for setting health-based targets for public health protection.

The published public health risk estimates summarized in Table 8.1 are mainly presented as infection probabilities, with some estimates alternatively translated into probability of disease or mortality. An additional measure used to represent adverse health effects for humans is an estimate of the disability adjusted life years (DALY) (Murray, 1994). This measure is a combination of the years of life lost due to premature mortality and the period of time spent in a suboptimal health status as a consequence of infection. A DALY estimate therefore includes the translation from infection to actual disease. Such an estimate may give greater insight into the adverse health effects of pathogens in the environment than an infection risk, as infection does not necessarily produce illness. The WHO use DALY to measure the global burden of disease (Lopez *et al.*, 2006), and also base their targets for public health protection on DALYs rather than on infection or illness risks. However, the progression from infection to disease can be highly variable between individual humans and depends on multiple host and pathogen characteristics. Depending on the data underlying the dose–response model, such progression cannot always be estimated.

8.3 Data gaps and needs

We conclude that quantitative risk assessment for food- and waterborne viruses may be very useful for the estimation of public health risks and for evaluating risk-reducing interventions. However, several data gaps and needs exist with respect to hazard identification, exposure assessment and dose-response relations which encompass the different steps in risk assessment.

8.3.1 Hazard identification

Viruses in foods that have been identified as being of highest priority are norovirus (NoV), hepatitis A (HAV) and E (HEV) viruses (Table 8.2). Other hazardous foodborne viruses include SARS coronavirus, avian influenza viruses and tick-borne encephalitis virus (Duizer and Koopmans, 2008). Furthermore, Duizer and Koopmans (2008) conclude that most emerging viruses cannot easily be excluded from being foodborne.

Food items implicated in human viral infection and disease include shellfish products, fresh produce and meat products. Contamination of these food items may occur during production, for example due to filter feeding by

Table 8.2 Priority viruses in foods and examples of implicated food items

Foodborne viruses	Implicated food items
Norovirus	Shellfish, raspberries, drinking water
Hepatitis A virus (HAV)	Bivalve molluscs, particularly oysters, clams and mussels, salad crops, as lettuce, green onions and other greens, and soft fruits, such as raspberries and strawberries
Hepatitis E virus (HEV)	Pork pies, liver pate, wild boar, under-cooked or raw pork, home-made sausages, meat (in general), unpasteurized milk, shellfish and ethnic foods
SARS coronavirus	Food of animal origin
Tick-borne encephalitis virus	(Raw) milk, yoghurt, butter and cheese
Nipah virus	Fruits
Avian influenza viruses	Duck blood

shellfish or irrigation of fresh produce, during processing as a result of contact with contaminated hands and utensils, and at point of sale due to contact with contaminated hands of store personnel and customers (Table 8.2). With respect to fresh produce, the FAO/WHO expert meeting on the microbiological hazards in fresh fruits and vegetables recommended that leafy green vegetables should be considered the highest global priority in terms of fresh produce safety (FAO/WHO, 2008b). With respect to foodborne virus contamination of meat, the presence of HEV in raw pig meat has been demonstrated (Deest *et al.*, 2007; Matsubayashi *et al.*, 2008). Meat products that are consumed raw or moderately heated, such as locally produced sausages (Colson *et al.*, 2010) and sashimi (Tei *et al.*, 2003), are of particular concern in this respect.

8.3.2 Exposure assessment

Data on the prevalence of enteric viruses for reservoirs and sources that may relate to potential contamination during food production have been published (Van den Berg *et al.*, 2005; Cheong *et al.*, 2009; Gentry *et al.*, 2009; Boxman *et al.*, 2011). For shellfish, numerous data on viruses in harvesting waters are available, whilst for fresh produce, studies are available on foodborne viruses in irrigation water and manure. With regard to food handling environments, a number of foodborne viruses in human and animal faeces have been determined. Nevertheless, such virus presence has largely been determined qualitatively (presence or absence) rather than quantitatively (enumerated). Sufficient quantitative data (with respect to sampling size and numbers of viral particles) for the performance of quantitative risk assessments is therefore still largely unavailable. Furthermore, the infectivity status of the viruses detected often remains unknown, as molecular methods detect nucleic acids from both infectious and non-infectious viruses, and the non-infectious viruses do not contribute to the risk of a public health event. These data issues therefore enforce an integrated approach at present for quantitative risk assessments – an approach which includes proper data collection by means of statistically sound sampling plans and proper data analysis using appropriate diagnostic controls and statistical controls for quantification.

Another aspect that needs to be considered in quantitative risk assessment is the transfer of viruses to and from foods due to handling and processing. Some data are available for transfer between surfaces and hands (Sattar *et al.*, 2000; Bidawid *et al.*, 2004). and these have been used to model, for example, norovirus transfer (Mokhtari and Jaykus, 2009). However, robust estimates are not possible given the limited data available. Details of virus transfer between food products, hands and utensils are therefore largely unquantified and uncertain. Viral transfer from carriers (humans, animals) or environmental sources to foods, as well as between foods, are largely based on assumptions rather than on experimental or field data. It is deemed important to perform virus transfer experiments under controlled settings following specified protocols. It was shown that 1% (0.2–10%), 0.05% (0.02–0.3%) and 9% (4–15%) of infectious MNV-1 was transferred from fingertips to raspberries, strawberries and lettuce, respectively (Verhaelen *et al.*, in preparation). The transfer fractions of MNV-1 from raspberries and lettuce to fingertips was 3% (<0.01–5%) and 4% (2–6%), respectively. Using the estimated mean transfer rate of MNV-1, it was estimated that a single food handler with an initial virus contamination level of 10^4 infectious virus particles on his fingertips could contaminate more than 2 kg of raspberries with a minimum of one virus particle per raspberry. This study exemplifies the substantial role a single food handler can have in the transmission of hNoV, considering hNoV is the most infectious agent described, with an infection probability of about 0.5 per single hNoV particle (Teunis *et al.*, 2008).

An important requirement for any exposure assessment is data on the amount of a particular food item that has been consumed and the methods used to prepare it. Regarding the amount of food consumed, several national studies have been collected and assembled by the European Food Safety Authority (EFSA) in a database (Comprehensive European Food Consumption Database) (EFSA, 2011a). This database specifies consumption data on aggregated levels (for example 'total consumption of vegetables per day, including mushrooms and other fungi'). Specific data for a particular product, such as 'lettuce' or, more specifically 'butterhead lettuce', are not retrievable from the database. Data for the consumption of drinking water are similarly available from a Dutch national food consumption survey (Teunis *et al.*, 1997), which used as a mean estimated volume of unboiled drinking water 0.28 L day^{-1} (described by a lognormal distribution with $\mu = -1.86$ and $\sigma = 1.07$).

Specific information and population frequencies for food and beverage preparation methods used in homes and restaurants prior to consumption are scarce but essential for quantitative risk assessments. Heating or cross-contamination can affect the virus concentration in and on food items, and therefore alter the exposure dose. Consumer habits and preferences regarding the state of a food item at consumption are highly diverse within and across different countries, meaning region- specific data is often required. Ignoring any possible effects of preparation on virus presence/infectivity can result in either under- or overestimation of the risk, depending on the efficacy of the disregarded process.

For high-risk foods such as soft fruits and salad vegetables, irrigation water may be one of the sources of contamination, making it important that this process is modelled. To estimate virus concentration on fresh produce due to irrigation, it is important to assess the volume of retained water on such products, as a function of the duration of irrigation. Furthermore, the clinging of viruses to food products and their wash-off during prolonged or subsequent irrigation events needs to be determined. The lack of available data currently prevents the possibility of such estimation.

The persistence of viruses on sources is another important aspect to consider in QMRA. On surfaces, hands and other environments, as well as once transferred onto food products or into water, viruses may persist for prolonged periods of time (De Roda Husman *et al.*, 2009). Experiments to assess the stability of viruses make use of cell culture systems. By growing *in vitro* cells that are susceptible to infection with the virus under study, and assessing the infectivity before and after treatment, parameters on virus stability can be estimated and used in risk assessment studies. However, no robust cell culture systems for the detection of infectious human noroviruses, HEV and HAV are available. Information on the persistence of infectious particles of these viruses in the environment is therefore limited.

However, several studies have been conducted using a surrogate virus, especially with regard to norovirus. At present, the most promising surrogate is the culturable murine norovirus due to its genetic similarity and environmental stability (Bae and Schwab, 2008). In general, infectivity reduction rates of surrogates were shown to be higher at higher temperatures (>25°C) and room temperature than at 4°C, for matrices such as surfaces of stainless steel, lettuce, berries, deli ham, surface and ground waters (Cannon *et al.*, 2006; D'Souza *et al.*, 2006; Bae and Schwab, 2008; Butot *et al.*, 2008). In addition, the relative humidity is an important determinant for survival in the environment (Stine *et al.*, 2005; Cannon *et al.*, 2006). Data obtained for the stability of norovirus-like particles, as well as surrogate viruses, demonstrated stability over a pH range of 3–7 and up to 55°C (Duizer *et al.*, 2004; Ausar *et al.*, 2006; Cannon *et al.*, 2006).

Viruses on foods are challenged by applied or natural production conditions, such as storage temperature, storage humidity and exposure to sunlight. Furthermore, the food matrix itself can induce virus inactivation due to, for example, the effects of pH or the presence of proteases. For most intact fresh produce, no recommended storage temperature is provided by legislation. In general, low temperatures and high relative humidity are applied in the fresh produce industry to prolong shelf-life and maintain produce quality. Unlike for bacteria, these conditions generally promote viral persistence. The usual storage temperature of lettuce, for example, is about 4°C with a relative humidity (RH) of about 80%. The shelf-life of lettuce is strongly dependent on storage conditions: a shelf-life of 21–28 days can be expected at 0°C with >95% RH, whilst at 5°C a shelf-life of 14 days can be expected. At point of sale, whole lettuces are usually stored at ambient temperature. Persistence of feline calicivirus (FCV) on lettuce in commonly applied storage conditions was studied by Mattison *et al.* (2007). After 4 days of storage at 21°C the virus was not detectable, which is equivalent to a reduction of about 2.5 $\log_{10}$-units. Infectious FCV was reduced by about 2 $\log_{10}$-units after seven days of storage at 4°C. Murine norovirus (MNV-1) was found to be persistent on raspberries and strawberries at 4 and 10°C, meaning that the D-values (first 1 $\log_{10}$ unit reduction) reached or exceeded the 7-day shelf-life of the berries. However, MNV-1 infectivity dropped by about 1.5 $\log_{10}$-units on strawberries after just one day of storage at room temperature, whereas no virus decay was observed on raspberries in this period. A 1 $\log_{10}$ unit decrease in MNV-1 infectivity only occurred on raspberries after 3 days at room temperature, yet in practice, raspberries are rarely stored over such a long period. Please see Chapter 13 for further information on the natural persistence of food- and waterborne viruses.

Foodborne viruses may be reduced by a diverse range of food treatment processes, each displaying a different estimated efficiency in infectivity reduction (EFSA, 2011b). The effectiveness of treatment processes for virus reduction has mostly been assessed by use of indicator viruses such as MNV (murine norovirus), FCV and phages. In order to conduct a quantitative viral risk assessment, it is important to first know which decontamination practices are used in food production, whilst acknowledging that lab experiments may produce different outcomes to standard virus reduction due to current practices in the field. Secondly, process-specific infectivity rates for each virus, or an appropriate surrogate thereof, should ideally be included in the study. Unfortunately, the lack of cell culture systems for certain viruses, including NoV and HEV, have hampered such studies in the past. Alternative approaches, such as the use of animal models (Feagins *et al.*, 2008) or cell infection experiments (Emerson *et al.*, 2005) have been developed for HEV. The results of both experiments indicated that HEV was still viable when heated at 56°C for an hour. In an *in vivo* animal experiment, one of the livers used as inoculum was stir-fried at 191°C (internal temperature of 71°C) for 5 min and infection was not observed after inoculation. Furthermore, recent progress in HEV culture has been made in the European project 'VITAL', by using a 3D cell culture system. Initial experiments with this system have shown that HEV is inactivated when heated at 100°C for 15 s (Berto *et al.*, 2012). Hence, HEV seems to be heat-intolerant above a certain temperature, but the temperature-dependent inactivation rate is yet to be determined.

8.3.3 Dose–response relation

Limited information on human dose–response relationships is currently available for quantitative viral risk assessment. The dose–response relationship for norovirus is based on ingestion of an inoculum by volunteers (Teunis *et al.*, 2008), and the exposure dose was quantified using RT-PCR detection. The currently available dose–response model for HAV is based on inoculation of institutionalized individuals with faeces from a patient (Haas *et al.*, 1999), with the original dose quantified as grams of faeces ingested (Ward *et al.*, 1958). In a subsequent study, this dose was adjusted to PCR detectable units using a maximum likelihood estimation to increase usability (the ingested amount of faeces is not frequently the outcome of an exposure assessment) (Bouwknegt *et al.*, in preparation). The dose–response model currently available for HEV is based on intravenous inoculation of pigs, and corrected to represent oral ingestion (Bouwknegt *et al.*, 2011). This dose was quantified using RT-PCR.

These available dose–response models need to be applied with particular care in quantitative microbiological risk assessment. The ratio between infectious and non-infectious viruses can differ by several orders of magnitude between samples (De Roda Husman *et al.*, 2009). This ratio can therefore differ significantly between the samples used for the dose–response modelling and those used for quantifying viruses at the risk assessment contamination points. The estimated risk can consequently over- or underestimate the actual risk by several orders of magnitude. As indicated before, the availability of an efficient cell culture system might allow the inclusion of a correction factor in the dose–response model, although this factor is expected to be highly variable.

In addition, dose–response relations are often established based on foodborne viruses in spiked water, whilst no dose relation for viruses on food is determined. However, very recently, a study was published in which it was demonstrated, via consumption of treated shellfish by human volunteers, that high hydrostatic pressure was effective for reduction of Norwalk virus in oysters (Leon *et al.*, 2011).

In comparison to the general population, vulnerable subpopulations exposed to foodborne viruses may experience higher disease incidence. However, the details of this are at present largely unknown (Gerba *et al.*, 1996). Exposure to foodborne viruses may result in more severe disease outcomes for vulnerable subpopulations, such as immunodeficient transplant recipients, those living with HIV/AIDS, children, the elderly and pregnant women. NoV infection, for example, is common in all age groups but the incidence is highest in young children (<5 years). In the case of HAV, severe infections among adults are rare in high endemic areas, due to induced life-long immunity upon exposure as a child, whereas in low endemic areas the disease mainly occurs in adulthood with an increased likelihood of severe symptomatic illness developing. Hepatitis E (at least genotype 1) poses a particular threat to pregnant women, who are at high risk of developing severe hepatitis, resulting in mortality in up to 25% of cases for this category. Similarly, individuals with underlying diseases, such as chronic liver disease, liver cirrhosis or a history of high alcohol consumption, are at a higher risk of developing hepatitis E, whilst immunosuppressed transplant patients are at risk of developing chronic hepatitis E.

8.4 Future trends

One of the predominant challenges in quantitative viral risk assessment is absolute quantification of infectious viruses. When cell culture systems are available, the number of cytopathologic effects in cell culture, or plaques in a plaque assay, can be interpreted as an indicator of the number of infectious viruses per unit of sample examined. At present, where such systems are not routinely available, as is the case for NoV, HAV and HEV, one is restricted to the use of indirect methods such as genome detection by (RT-) PCR. The methods for correct use of qPCR data in QMRA are lacking, which leads to the currently unavoidable ignorance of two aspects: (1) distinction between genomes originating from infectious (viable) and non-infectious (non-viable) micro-organisms; and (2) uncertainty and bias in quantification of genomes. Distinction between infectious and non-infectious micro-organisms cannot be made with qPCR directly, because the detected genomes can originate from both types of micro-organism. Approaches to distinguish between the two by using, for example, enzymatic pre-treatment are being explored at present (Schielke *et al.*, 2011).

Furthermore, results generated by qPCR are, when quantified, translated into point estimates of genome quantities according to a standard curve, based on samples with supposedly known concentrations of targets (i.e., the standards). However, points often neglected in such quantification are the efficiency of isolation of genomic material from samples, differences in amplification efficiency between the standard and targeted micro-organisms, uncertainty around the estimated concentration of the standards and target viruses, and measurement error of the apparatus. For results of molecular methods to be useful in quantitative risk assessment, new procedures are needed to provide correct quantification, including uncertainty, of infectious micro-organisms detected by qPCR.

Another important challenge for QMRA for viruses is to retrieve the required information from published results. The majority of published studies on the contamination of products with enteric viruses, for instance, detail the percentage of detected positive samples (so-called prevalence studies). The efficiency of such prevalence studies correlates directly to the efficiency of the method used, thus yielding method-dependent data. Factors that affect the data quality include the limit of the detection assay, the inefficiency in amplification, the RNA isolation efficiency, and the procedure used for quantification.

Approaches have been published to account for several of these factors in the detection protocol (see, e.g., Costafreda *et al.*, 2006). Nevertheless, when proper controls were included in the analyses, results described in published papers were found to be lacking sufficiently detailed descriptions for the data to be included in risk assessment studies. A solution to this problem can be to share the raw laboratory data with risk analysts. Alternatively, the science community could strive to ensure raw data are included with their publications in journals, and/or submitted to global databases maintained by networks of scientists, such as the COST network for Food and Environmental Virologists or the Food and Water Department of the responsible Ministry, Inspectorate or research institute. When made available, virus concentrations, including details of any uncertainty/variation may be assessed and included in risk estimation.

Once conceptual mathematical models for estimating virus contamination per contamination source have been developed, validated and made available, such models can be reprogrammed as stand-alone software tools. This allows the access and application of data by a wide range of scientists and specialists, including those who have not necessarily received specialist training in mathematics and statistics (e.g., Schijven *et al.*, 2011; Schijven *et al.*, submitted). Future trends might include the use of such tools for real-time monitoring, for instance when the models are linked to rapid methods for virus detection in potential sources.

8.5 Conclusion

Quantitative microbiological risk assessment is a method by which available quantitative data for a microbiological hazard can be utilized to estimate human exposure and resultant levels of morbidity or mortality. The results of applying this method can be more informative for public health risk management than the presence/absence and pathogen prevalence data often presented. Furthermore, by identifying the available and the required data for a risk assessment, data gaps can be identified. This information can subsequently be used to allocate research resources more effectively. The development of statistically sound sampling plans for proper data collection that target these data gaps can extend the effective use of resources still further.

The continuing development of risk assessment models and their possible distribution in risk assessment tools might increase the use of QMRA to assist public health policy makers. However, it is important that the limitations of risk assessments are considered when interpreting the results, which differ per hazard examined. These, usually data-driven, limitations should be included in the development or analysis of any product based on risk assessment outcomes, as should the assumptions made during assessment.

8.6 References

AUSAR SF, FOUBERT TR, HUDSON MH, VEDVICK TS and MIDDAUGH CR. Conformational stability and disassembly of Norwalk virus-like particles. Effect of pH and temperature. *J Biol Chem*. 2006; **281**(28): 19478–19488.

BAE J and SCHWAB KJ. Evaluation of murine norovirus, feline calicivirus, poliovirus, and MS2 as surrogates for human norovirus in a model of viral persistence in surface water and groundwater. *Appl Environ Microbiol*. 2008; **74**(2): 477–484.

BERTO A, VAN DER POEL WHM, HAKZE-VAN DER HONING R, MARTELLI F, LA RAGIONE RM, INGLESE N, *et al.* Replication of hepatitis E virus in three-dimensional cell culture. *J Virol Methods*. 2012; pii: S0166-0934(12)00381-3.

BIDAWID S, MALIK N, ADEGBUNRIN O, SATTAR SA and FARBER JM. Norovirus cross-contamination during food handling and interruption of virus transfer by hand antisepsis: experiments with feline calicivirus as a surrogate. *J Food Prot*. 2004; **67**(1): 103–109.

BOUWKNEGT M, RUTJES SA, DE RODA HUSMAN AM. *Hepatitis E Virus Risk Profile: Identifying Potential Animal, Food and Water Sources for Human Infection.* Bilthoven: National Institute for Public Health and the Environment 2009. Report No.: 330291001.

BOUWKNEGT M, TEUNIS PF, FRANKENA K, DE JONG MC and DE RODA HUSMAN AM. Estimation of the likelihood of fecal-oral HEV transmission among pigs. *Risk Anal*. 2011; **31**(6): 940–950.

BOUWKNEGT M, VERHAELEN K, RZEZUTKA A, MAUNULA L, KOKKINOS P, PETROVIC T, *et al.* Farm-to-fork risk assessment for norovirus and hepatitis A virus in European soft fruit and salad vegetables supply chains. In preparation.

BOXMAN IL, VERHOEF L, DIJKMAN R, HAGELE G, TE LOEKE NA and KOOPMANS M. Year-round prevalence of norovirus in the environment of catering companies without a recently reported outbreak of gastroenteritis. *Appl Environ Microbiol*. 2011; **77**(9): 2968–2974.

BUTOT S, PUTALLAZ T and SANCHEZ G. Effects of sanitation, freezing and frozen storage on enteric viruses in berries and herbs. *Int J Food Microbiol*. 2008; **126**(1–2): 30–35.

CANNON JL, PAPAFRAGKOU E, PARK GW, OSBORNE J, JAYKUS LA and VINJE J. Surrogates for the study of norovirus stability and inactivation in the environment: a comparison of murine norovirus and feline calicivirus. *J Food Prot*. 2006; **69**(11): 2761–2765.

CASSIN MH, LAMMERDING AM, TODD EC, ROSS W and MCCOLL RS. Quantitative risk assessment for *Escherichia coli* O157:H7 in ground beef hamburgers. *Int J Food Microbiol*. 1998; **41**: 21–44.

CHEONG S, LEE C, SONG SW, CHOI WC, LEE CH and KIM SJ. Enteric viruses in raw vegetables and groundwater used for irrigation in South Korea. *Appl Environ Microbiol*. 2009; **75**(24): 7745–7751.

COSTAFREDA MI, BOSCH A and PINTO RM. Development, evaluation, and standardization of a real-time TaqMan reverse transcription-PCR assay for quantification of hepatitis A virus in clinical and shellfish samples. *Appl Environ Microbiol*. 2006; **72**(6): 3846–3855.

CRABTREE KD, GERBA CP, ROSE JB and HAAS CN. Waterborne adenovirus: a risk assessment. *Wat Sci Tech*. 1997; **35**(11): 1–6.

DEEST G, ZEHNER L, NICAND E, GAUDY-GRAFFIN C, GOUDEAU A and BACQ Y. Autochthonous hepatitis E in France and consumption of raw pig meat. *Gastroenterol Clin Biol*. 2007; **31**(12): 1095–1097.

DE RODA HUSMAN AM, LODDER WJ, RUTJES SA, SCHIJVEN JF and TEUNIS PF. Long-term inactivation study of three enteroviruses in artificial surface and groundwaters, using PCR and cell culture. *Appl Environ Microbiol*. 2009; **75**(4): 1050–1057.

D'SOUZA DH, SAIR A, WILLIAMS K, PAPAFRAGKOU E, JEAN J, MOORE C, Jaykus L. Persistence of caliciviruses on environmental surfaces and their transfer to food. *Int J Food Microbiol.* 2006; **108**(1): 84–91.

DUIZER E, BIJKERK P, ROCKX B, DE GROOT A, TWISK F and KOOPMANS M. Inactivation of caliciviruses. *Appl Environ Microbiol.* 2004; **70**(8): 4538–4543.

DUIZER E and KOOPMANS MPG. Emerging food-borne viral diseases. In: KOOPMANS MPG, BOSCH A, CLIVER DO, editors. *Food-Borne Viruses: Progress and Challenges.* Washington, DC, USA: ASM Press, 2008, pp. 117–156.

EFSA. Use of the comprehensive European food consumption database in exposure assessment. *EFSA J.* 2011a; **9**(3): 2097.

EFSA. Scientific Opinion on an update on the present knowledge on the occurrence and control of foodborne viruses. *EFSA J.* 2011b; **9**(7): 2190.

EMERSON SU, ARANKALLE VA and PURCELL RH. Thermal stability of hepatitis E virus. *J Infect Dis.* 2005; **192**(5): 930–933.

FAO/WHO. Microbiological risk assessment series. Viruses in food: scientific advice to support risk management activities. Meeting report. 2008a.

FAO/WHO. Microbiological risk assessment series. Microbiological hazards in fresh fruits and vegetables. Meeting report. 2008b.

FEAGINS AR, OPRIESSNIG T, GUENETTE DK, HALBUR PG and MENG XJ. Inactivation of infectious hepatitis E virus present in commercial pig livers sold in local grocery stores in the United States. *Int J Food Microbiol.* 2008; **123**(1–2): 32–37.

GARCIA-ESTRINGANA P, ALONSO-BLÁZQUEZ N and ALEGRE J. Water storage capacity, stemflow and water funneling in Mediterranean shrubs. *J Hydrol.* 2010; **389**: 363–372.

GENTRY J, VINJE J, GUADAGNOLI D and LIPP EK. Norovirus distribution within an estuarine environment. *Appl Environ Microbiol.* 2009; **75**(17): 5474–5480.

GERBA CP, ROSE JB, HAAS CN and CRABTREE KD. Waterborne rotavirus: a risk assessment. *Wat Res.* 1996; **30**(12): 2929–2940.

GERBA CP, ROSE JB and HAAS CN. Sensitive populations: who is at the greatest risk? *Int J Food Microbiol.* 1996; **30**(1–2): 113–123.

GREENING GE, LAKE R, HUDSON JA, CRESSEY P, NORTJE G. Risk profile: Norwalk-like virus in mollusca (raw). 2003. Client Report FW0312 prepared for NZ Food Safety Authority.

HAAS CN. Estimation of risk due to low doses of microorganisms: a comparison of alternative methodologies. *Am J Epidemiol.* 1983; **118**(4): 573–582.

HAAS CN, ROSE JB, GERBA C and REGLI S. Risk assessment of virus in drinking water. *Risk Anal.* 1993; **13**(5): 545–552.

HAAS CN, ROSE JB and GERBA CP. *Quantitative Microbiological Risk Assessment.* New York: Wiley and Sons, Inc., 1999.

HAMILTON AJ, STAGNITTI F, PREMIER R, BOLAND AM and HALE G. Quantitative microbial risk assessment models for consumption of raw vegetables irrigated with reclaimed water. *Appl Environ Microbiol.* 2006; **72**(5): 3284–3290.

HPA. Microbiological risk assessment for Norovirus infection – contribution to the overall burden: Health Protection Agency 2004.

ILSI. A conceptual framework to assess the risks of human disease following exposure to pathogens. ILSI Risk Science Institute Pathogen Risk Assessment Working Group. *Risk Anal.* 1996; **16**(6): 841–848.

LEON JS, KINGSLEY DH, MONTES JS, RICHARDS GP, LYON GM, ABDULHAFID GM, SEITZ SR, FERNANDEZ ML, TEUNIS PF, FLICK GJ and MOE CL. Randomized, double-blinded clinical trial for human norovirus inactivation in oysters by high hydrostatic pressure processing. *Appl Environ Microbiol.* 2011; **77**(15): 5476–5482.

LOPEZ AD, MATHERS CD, EZZATI M, JAMISON DT and MURRAY CJL. *Global Burden of Disease.* New York: Oxford University Press, 2006.

MASAGO Y, KATAYAMA H, WATANABE T, HARAMOTO E, HASHIMOTO A, OMURA T, HIRATA T and OHGAKI S. Quantitative risk assessment of noroviruses in drinking water based on qualitative data in Japan. *Environ Sci Technol.* 2006; **40**(23): 7428–7433.

MATSUBAYASHI K, KANG JH, SAKATA H, TAKAHASHI K, SHINDO M, KATO M, NISHIMORI H, TSUJI K, MAGUCHI H, YOSHIDA J, MAEKUBO H, MISHIRO S and IKEDA H. A case of transfusion-transmitted hepatitis E caused by blood from a donor infected with hepatitis E virus via zoonotic food-borne route. *Transfusion.* 2008; **48**(7): 1368–1375.

MATTISON K, KARTHIKEYAN K, ABEBE M, MALIK N, SATTAR SA, FARBER JM, BIDAWID S. Survival of calicivirus in foods and on surfaces: experiments with feline calicivirus as a surrogate for norovirus. *J Food Prot.* 2007; **70**(2): 500–503.

MENA KD, GERBA CP, HAAS CN and ROSE JB. Risk assessment of waterborne coxsackievirus. *J Am Wat Works Assoc.* 2003; **95**(7): 122–131.

MOKHTARI A and JAYKUS LA. Quantitative exposure model for the transmission of norovirus in retail food preparation. *Int J Food Microbiol.* 2009; **133**(1–2): 38–47.

MURRAY CJL. Global burden of disease. *WHO Bull OMS.* 1994; **72**: 427–445.

NAUTA M, HILL A, ROSENQUIST H, BRYNESTAD S, FETSCH A, VAN DER LOGT P, FAZIL A, CHRISTENSEN B, KATSMA E, BORCK B and HAVELAAR A. A comparison of risk assessments on Campylobacter in broiler meat. *Int J Food Microbiol.* 2009; **129**(2): 107–123.

PETTERSON SR and ASHBOLT NJ. Viral risks associated with wastewater reuse: modeling virus persistence on wastewater irrigated salad crops. *Water Sci Technol.* 2001; **43**(12): 23–26.

PETTERSON SR, ASHBOLT NJ and SHARMA A. Microbial risks from wastewater irrigation of salad crops: a screening-level risk assessment. *Water Environ Res.* 2001; **73**(6): 667–672.

REGLI S, ROSE JB, HAAS CN and GERBA CP. Modeling the risk from Giardia and viruses in drinking water. *J Am Wat Works Assoc.* 1991; **83**: 76–84.

ROCOURT J, BENEMBAREK P, TOYOFUKU H and SCHLUNDT J. Quantitative risk assessment of *Listeria monocytogenes* in ready-to-eat foods: the FAO/WHO approach. *FEMS Immunol Med Microbiol.* 2003; **35**(3): 263–267.

SATTAR SA, JASON T, BIDAWID S and FARBER J. Foodborne spread of hepatitis A: recent studies on virus survival, transfer and inactivation. *Can J Infect Dis.* 2000; **11**(3): 159–163.

SCHIELKE A, FILTER M, APPEL B and JOHNE R. Thermal stability of hepatitis E virus assessed by a molecular biological approach. *Virol J.* 2011; **8**: 487.

SCHIJVEN JF, BOUWKNEGT M, DE RODA HUSMAN AM, RUTJES SA, SUK J, SUDRE B, *et al.* A decision support tool to compare waterborne and foodborne infection and/or illness risks associated with climate change. Submitted.

SCHIJVEN JF, TEUNIS PF, RUTJES SA, BOUWKNEGT M and DE RODA HUSMAN AM. QMRA spot: a tool for Quantitative Microbial Risk Assessment from surface water to potable water. *Wat Res.* 2011; **45**(17): 5564–5576.

SHUVAL H, LAMPERT Y and FATTAL B. Development of a risk assessment model for evaluating wastwater reuse standards for agriculture. *Wat Sci Tech.* 1997; **35**(11–12): 15–20.

STINE SW, SONG I, CHOI CY and GERBA CP. Effect of relative humidity on preharvest survival of bacterial and viral pathogens on the surface of cantaloupe, lettuce, and bell peppers. *J Food Prot.* 2005; **68**(7): 1352–1358.

TEI S, KITAJIMA N, TAKAHASHI K and MISHIRO S. Zoonotic transmission of hepatitis E virus from deer to human beings. *Lancet.* 2003; **362**(9381): 371–373.

TEUNIS PF, MEDEMA GJ, KRUIDENIER L and HAVELAAR AH. Assessment of the risk of infection by *Cryptosporidium* or *Giardia* in drinking water from a surface water source. *Wat Res.* 1997; **31**: 1333–1346.

TEUNIS PF, MOE CL, LIU P, MILLER SE, LINDESMITH L, BARIC RS, LE PENDU J and CALDERON RL. Norwalk virus: how infectious is it? *J Med Virol.* 2008; **80**(8): 1468–1476.

VAN DEN BERG HH, LODDER WJ, VAN DER POEL WH, VENNEMA H and DE RODA HUSMAN AM. Genetic diversity of noroviruses in raw and treated sewage water. *Res Microbiol.* 2005; **156**(4): 532–540.

VAN HEERDEN J, EHLERS MM and GRABOW WO. Detection and risk assessment of adenoviruses in swimming pool water. *J Appl Microbiol.* 2005; **99**(5): 1256–1264.

VERHAELEN K, BOUWKNEGT M, CARRATALA A, LODDER-VERSCHOOR F, DIEZ-VALCARCE M, RUTJES SA, *et al.* Viral transfer rates between fingertips, soft berries, and lettuce and associated health risks. In preparation.

WARD R, KRUGMAN S, GILES JP, JACOBS AM and BODANSKY O. Infectious hepatitis; studies of its natural history and prevention. *N Engl J Med.* 1958; **258**(9): 407–416.

Part III

Virus transmission routes and control of food and water contamination

9

Natural persistence of food- and waterborne viruses

P. Vasickova and K. Kovarcik, Veterinary Research Institute, Czech Republic

DOI: 10.1533/9780857098870.3.179

Abstract: This chapter summarises data on the persistence of food- and waterborne viruses in the natural environment and discusses the different factors which can affect this persistence. Conventional and alternative methods by which persistence can be studied are described, and the natural factors influencing virus persistence outside the host organism are discussed. Available data concerning virus persistence in water, soil, on surfaces and in food products are reviewed.

Key words: food and waterborne viruses, methods for study of virus persistence, factors affecting virus persistence, water, soil, surfaces, food products.

9.1 Introduction

Data on the persistence of food- and waterborne viruses in the environment outside the host organism are very important to the understanding of viral ecology and thus in determining the risk to the human population represented by these viruses. This chapter is focused mainly on the existing data regarding virus persistence in the most important matrices associated with the spread of food- and waterborne viruses (i.e., water, soil, food-related surfaces and food products). As virus persistence outside the host organism is affected by a combination of biological (presence of envelope, type of virus genome and presence of other micro-organisms), physical (temperature, relative humidity and UV) and chemical (pH, presence of salts and adsorption state) factors, the chapter contains a special section which reviews these factors and their general influence on virus survival.

To obtain information about the presence and persistence of viruses in the environment, appropriate methods are needed; that is, methods which are able to distinguish between infectious and non-infectious particles and are thus suitable for use to determine the real risk of infection. Therefore, common (cell culture and polymerase chain reactions) and new alternative methods (e.g., a special pre-treatment of samples) available for such studies are also discussed. Virus transmission via food and the environment is now a well-recognised problem. This awareness of foodborne and waterborne viruses emphasises the need for data regarding the persistence of viruses in the environment and the effects of preservation methods upon viruses. During recent years, there has been an increase in the number of survival and inactivation studies. Despite this, there is still a lack of general information about the risk to the human population represented by these viruses, which remains to be fully determined. The important foodborne and waterborne viruses (e.g. norovirus, hepatitis E virus and several strains of hepatitis A virus) cannot be commonly cultivated in the laboratory, which hinders the study of their stability in food and in the water environment. To address this problem, cultivable surrogate viruses, which are genetically related to the strains which infect humans, are used as substitutes for these viruses in order to obtain preliminary information about their behaviour in the environment. However, the use of such surrogates is questionable due to the differences in susceptibility to environmental factors even between different surrogates. At present, novel cultivation methods and approaches based on molecular methods are being developed. Although these methods are promising, none are commonly applicable to all viruses and thus further development is needed. In addition, the study of virus persistence in the environment would benefit greatly from the standardisation of experimental protocols, which would allow the generation of complete and comparable data about the natural persistence of viruses. Therefore, standardised methods applicable for the detection of infectious viral particles in different kinds of matrices and for the subsequent statistical analysis of results need to be developed.

Environmental and especially food virology are relatively new subject areas. To date, most experiments dealing with virus persistence in the environment have been conducted at the national level. At the international level, only a limited number of projects focused on waterborne and foodborne viruses have been implemented: for example, Virobathe (http://www.virobathe.org) which evaluated methods for detecting noroviruses and adenoviruses in recreational waters; VITAL (Integrated Monitoring and Control of Foodborne Viruses in European Food Supply Chains; http://eurovital.csl.gov.uk) which focused on foodborne viruses within selected food supply chains from farm to market; and COST 929 ENVIRONET (http://www.cost929-environet.org/index.html) as an international network for environmental and food virology. These projects have contributed to our understanding of foodborne and waterborne viruses and enabled the risks associated with the transmission of these viruses to be partially determined. Notably, the research group CEN/

WG6/TC275/TAG4 'Detection of viruses in foods' has been established to investigate the occurrence of foodborne viruses and introduce standard methods for their detection. Current information about virus persistence and appropriate methodologies is available in the form of research articles and several reviews; for example, data concerning viral persistence, particularly in fresh food, were summarised by Seymour and Appleton (2001) and Rzezutka and Cook (2004); overviews of data with respect to reduction/inactivation of viruses by food preservation methods were published by Baert *et al.* (2009) and Hirneisen *et al.* (2010) and methods for assessing the infectivity of enteric viruses in environmental samples were discussed by Rodriguez *et al.* (2009).

9.2 Methods for studying persistence

Studies of virus persistence and the determination of a virus' ability to retain infectivity are usually performed according to basic principles. Generally, tested samples are artificially contaminated by a virus suspension containing a determined concentration of infectious particles, and the samples are then stored or processed under defined conditions. Subsequently, the viruses are extracted from the sample and the infectious units are quantified. The comparison between the number of infectious viruses isolated from a tested sample and the number which was originally introduced provides data about the persistence of virus infectivity (Rzezutka and Cook, 2004).

Studies can be performed under laboratory or under natural conditions; both approaches have their advantages and disadvantages. The advantages of studies carried out under laboratory conditions are the possibilities for precisely defined and stable conditions (e.g., temperature, relative humidity and pH). Virus persistence is influenced by combinations of biological, physical and chemical factors, the various permutations of which it is not feasible to successfully recreate under laboratory conditions. Thus, laboratory experiments can be used only to define the effect of individual factors on virus persistence. In contrast, experiments performed under natural conditions provide compact data about virus persistence in the environment. However, problems could arise with regard to the definition of specific conditions influencing virus persistence.

Successful detection of infectious viral particles in the environment is complicated by several factors, such as virus size; the wide variation among and within viral genera; low concentration; the presence of substances which can interfere with analysis procedures; the limits of detection of different techniques; and the absence of reliable controls (Vasickova *et al.*, 2010). Various methods have been developed for the isolation of infectious viral particles from different kinds of matrices (e.g., water, soil, food and surfaces). Their basic principles and aims are similar; they involve the separation of viral particles (elution, washing or filtration) from the sample and their subsequent concentration to an amount which is suitable for proper detection. Successful detection depends on both the extraction method and the detection techniques. The standard methods for the detection of infectious viruses and thus viral survival under different conditions still involve cell culture. Susceptible cell lines in which the viruses propagate are required. The advantages are the direct detection of infectious viral particles and their sensitivity; theoretically, these methods can detect a single viable viral particle (Reynolds *et al.*, 2001). The quantification of infectious virions can be achieved through the use of cell culture in a quantitative format, for example, plaque assay. Routinely cultivable viruses include poliovirus and related viruses from the family *Picornaviridae*, astroviruses, rotaviruses, and cell culture-adapted strains of hepatitis A virus (HAV) (Richards, 2012). However, each virus type or even strain has different capabilities and thus may require different conditions for effective propagation in cell culture; for example, not all enteroviruses are able to propagate on one cell line (Dahling, 1991). Cell culture methods can also be time-consuming: the time required varying between 4 and 30 days depending on the virus. Furthermore, detection is problematic in the case of those viruses which cannot be grown in conventional cell culture: for example, human norovirus (NoV) and hepatitis E virus (HEV) (Rodriguez *et al.*, 2009; Vasickova *et al.*, 2010). Although NoVs have been reported growing in highly differentiated 3D cell cultures (Straub *et al.*, 2007), these systems require specialised equipment and extensive experience, and have proven difficult to successfully reproduce (Parshionikar *et al.*, 2010).

The lack of a suitable cell culture method for the detection of infectious particles of non-cultivable viruses such as HAV or NoV has led to the use of surrogate viruses, which provides at least predictive data about the survival of non-cultivable viruses in the environment. The selection of a proper surrogate virus is usually based on its ability to propagate in cell cultures, and its genetic, biological, physical and chemical relatedness to the virus which is to be isolated. Although several surrogates have been used specifically for NoV (Table 9.1; Richards, 2012), differences have been found between the inactivation of NoV and these viruses. It was also shown that the susceptibility of different NoV surrogates to temperature, environmental and food-processing conditions or disinfectants differs dramatically between these viruses (Cannon *et al.*, 2006). Moreover, several studies reported that HAV strains that have adapted to cell culture have diverse sensitivities to heat and high pressure (Shimasaki *et al.*, 2009). Differing modes of inactivation can be anticipated also among other kinds of viruses. However, differences between strains of viruses belonging to the same genus may not be as pronounced as those between the non-cultivable viruses and their surrogates. Therefore, data obtained by the use of surrogate viruses should be evaluated and presented carefully as presumptive evidence of how pathogens may respond to different treatments. Thus, the use of surrogate viruses and extrapolation from the persistence of surrogates to the persistence of non-cultivable viruses is questionable (Richards, 2012).

Table 9.1 Summary of surrogate viruses and their characteristic

Surrogate virus	Classification (family/genus)	Genome	Host	Infection	Represented virus	References
Feline calicivirus (FCV)	*Caliciviridae*/*Vesivirus*	+ssRNA	Cat	Respiratory infection	Norovirus	Duizer *et al.*, 2004
Canine calicivirus	*Caliciviridae*/*Vesivirus*	+ssRNA	Dog	Glossitis, enteritis	Norovirus	Duizer *et al.*, 2004
Murine Norovirus (MNV)	*Caliciviridae*/*Norovirus*	+ssRNA	Mouse	Enteritis	Norovirus	Takahashi *et al.*, 2011
Attenuated hepatitis A virus (HAV)	*Picornaviridae*/*Hepatovirus*	+ssRNA	Human		Hepatitis A virus	Hewitt and Greening, 2004; Kingsley *et al.*, 2005
Attenuated poliovirus	*Picornaviridae*/*Enterovirus*	+ssRNA	Human		Poliovirus	Alvarez *et al.*, 2000
Rotavirus SA-11	*Reoviridae*/*Rotavirus*	dsRNA	Monkeys		Rotavirus	Kingsley *et al.*, 2005; Raphael *et al.*, 1985
Phage MS2	*Leviviridae*/*Levivirus*	+ssRNA	*Escherichia coli*		Norovirus, HAV, enterovirus, rotavirus	Dawson *et al.*, 2005; Helmi *et al.*, 2008
Phage ΦX174	*Microviridae*/*Microvirus*	ssDNA	*Escherichia coli*		Norovirus, HAV, enterovirus, rotavirus	Helm *et al.*, 2008; Dawson *et al.*, 2005
Virus-like particles					Norovirus, rotavirus	Kingsley *et al.*, 2005; Caballero *et al.*, 2004

ss single-stranded; ds double-stranded; + positive sense.
Source: Supplemented and adapted from Baert *et al.* (2009).

Molecular techniques are an alternative method, particularly for the detection of viral genomes. These methods, based on polymerase chain reaction (PCR) (Malorny *et al.*, 2003; Shimasaki *et al.*, 2009), nucleic acid sequence-based amplification (NASBA) (Cook, 2003; Casper *et al.*, 2005) or their quantitative format (qPCR, qNASBA), represent highly sensitive and specific assays. Molecular techniques can be used for all types of viruses (in the case of RNA viruses it is necessary to run a reverse transcription reaction prior to PCR), can determine the presence of different agents in the same sample, allow the identification of non-cultivable viruses, are rapid and can be used to quantify the viral load in the sample. Besides, additional sequencing of the amplicons allows the establishment of epidemiologic associations. The disadvantage of molecular methods is that, when used alone, they are not able to distinguish between infectious and non-infectious viruses. This is due to the ability of the viral capsid to protect nucleic acid even in non-infectious particles in many instances (Ogorzaly *et al.*, 2010). Therefore, the use of PCR or NASBA has limited application for persistence studies.

In the case of commonly cultivable viruses, the integration of cell culture and PCR (i.e., integrated cell culture-polymerase chain reaction; ICC-PCR) can allow the detection of infectious viruses in the space of hours or days compared with the days or weeks necessary with cell culture alone (Reynolds *et al.*, 2001; Gallagher and Margolin, 2007). The detection of enteroviruses in water samples, for example, is reduced to 5 days. The assay is based on an initial replication of viral particles using an appropriate cell culture for a short period, followed by PCR amplification of a specific part of the viral genome. The sensitivity of this method is comparable to that obtained using a second passage of cell culture (Rodriguez *et al.*, 2009). So far the use of ICC-PCR has been described for the detection of enteroviruses, HAV, enteric adenoviruses, reovirus and astroviruses (Shoeib *et al.*, 2009; Rigotto *et al.*, 2010; Schlindwein *et al.*, 2010). Despite the major advantages of this method, namely that it overcomes the limitations of cell culture and PCR methodologies when used alone, a system capable of detecting infectious non-cultivable viruses is still lacking.

To address the limitations of PCR-based methods several new approaches have been developed recently (Gilpatrick *et al.*, 2000; Nuanualsuwan and Cliver, 2002; Parshionikar *et al.*, 2010; Li *et al.*, 2011). Generally, these include additional sample pre-treatment steps prior to nucleic acid isolation or PCR. These steps utilize the essential properties of the viral capsid, which are associated with the loss of infectivity of viruses (Cliver, 2009a). However, these methods have been successfully used to prevent the transcription and amplification only of certain kinds of inactivated viruses, and serious limitations have arisen during their application (Table 9.2). Despite this, with further development these assays have the potential to provide more information about virus persistence.

Generally, for public health protection, the usefulness of a method is determined by its applicability to all cultivable and non-cultivable strains

Table 9.2 Summary of pre-treatment approaches prior to PCR to distinguish between infectious and non-infectious viruses

Pre-treatment prior to PCR	Rationale	Advantages	Disadvantages	References
Proteinase and RNase treatment	The capsid of infectious particles is resistant to proteinase digestion and thus protects the genome against RNases.	Usable for non-cultivable viruses Practical and easy pre-treatment	Works only in defined conditions (inactivation at 72°C or by hypochlorite)	Nuanualsuwan and Cliver, 2002; Nuanualsuwan and Cliver, 2003
Antibody capture of the virus	The capsid of inactivated virus might alter its antigenic properties (changes in protein conformation) thus resulting in the virus losing the ability of being recognised by specific antibodies.	Isolation of infectious viruses from large volume samples (e.g., water) Usable for non-cultivable viruses	Detection depends on antigenic properties of viral capsid Effective only if the antigenic properties of the viral capsid are defined	Gilpatrick *et al.*, 2000; Schwab *et al.*, 1996
Attachment of virus to cell monolayer	The capsid of inactivated viruses changes its antigenic properties (changes in protein conformation) and thus the virus is unable to attach to receptors of the monolayer.	Able to distinguish between infectious and non-infectious viruses (in defined conditions)	Applicable only for cell-cultivable viruses	Li *et al.*, 2011; Nuanualsuwan and Cliver, 2003
Intercalating dyes treatment (Propidium monoazid; PMA)	The PMA penetrates the damaged or compromised capsid of the inactivated virus and binds covalently to the genome upon exposure to visible light, which prevents detection of the genome using PCR.	Usable for non-cultivable viruses	Works only in defined conditions (inactivation at 72 and 37°C or by hypochlorite)	Fittipaldi *et al.*, 2010; Parshionikar *et al.*, 2010

of viruses, its detection of viral infectivity and the rapidity with which the results are obtained (Parshionikar *et al.*, 2010). A combination of molecular techniques and cell culture methods should be used (Cliver, 2009b), but this does not solve the problem of non-cultivable viruses such as NoV and HEV. The ideal solution would involve a single and simple pre-treatment of any kind of sample that would quickly preclude nucleic acid amplification from non-infectious viruses (Cliver, 2009b). None of the pre-treatment procedures is applicable for all viruses and further optimization is needed.

9.3 General factors affecting the natural persistence of viruses

The potential for viral spread depends in large part on the ability of viruses to persist in the environment (Boone and Gerba, 2007). The infectious viral particles of non-enveloped viruses consist of two major components: the viral genome and the genome-protecting capsid. Infectivity requires the functional integrity of both of these components; that is, infection will occur only if the viral genome (either DNA or RNA) has retained its functional integrity and if the undamaged capsid is able to attach to the host cell receptor and initiate the process by which the nucleic acid enters the host cell (Cliver, 2009a). Many viruses have an envelope that covers the capsid and helps viruses to enter the host cell. Thus, the persistence (preservation of infectivity) of enveloped viruses is also strongly dependent on the integrity of the envelope. Viruses cannot replicate outside the host organism. Since specific and living host cells are not present in the environment, the number of viral particles cannot increase and the amount of any contaminating viruses should decline over the storage time. However, foodborne and waterborne viruses are infectious in very low doses (Carter, 2005; Vasickova *et al.*, 2005).

Virus survival outside the host organism (i.e., the integrity of the viral genome, capsid and envelope) is affected by a combination of various environmental conditions and biological, physical and chemical factors (Carter, 2005; Fig. 9.1).

The persistence of viral particles can be predicted primarily according to the virus classification. However, variations in virus survival occur within a given virus family or even genus. Based on its similar classification and thus similar genomic organisation and physicochemical characterisation, feline calicivirus (FCV) is used as a surrogate virus for non-cultivable NoV. It has been found, however, that human NoVs are more resistant than FCV to low and high pH and to other environmental factors (Duizer *et al.*, 2004; Hewitt and Greening, 2004). The coronaviruses OC43 and 229E represent an example of variation within the same genus. The infectivity of these two viruses differs temporally: coronavirus 229E was detectable after 3 h on various surfaces, while coronavirus OC43 persisted for only 1 h or less (Sizun *et al.*, 2000).

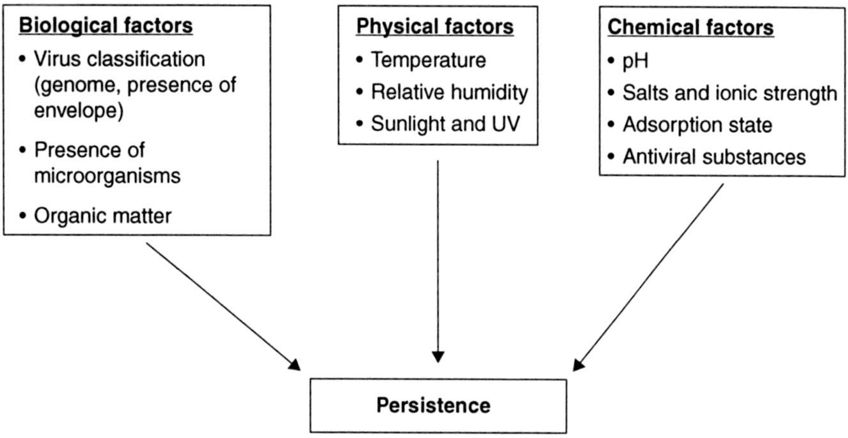

Fig. 9.1 General factors affecting virus persistence in the environment.

It is generally known that non-enveloped viruses have a higher resistance to desiccation and thus are spread more easily than viruses with an envelope, which corresponds also with their mode of transmission. Enteric viruses produced in either the intestines or livers of infected humans or animals are relatively stable outside the host organism. They are able to withstand a pH as low as 3 for a certain period, which allows them to pass through the stomach and cause infection in the small intestine or liver. In contrast, enveloped non-enteric viruses such as respiratory viruses are less often transmitted via contaminated food or surfaces and are generally less stable in the environment (Duizer *et al.*, 2004; Howie *et al.*, 2008).

The biological factors affecting viral persistence also include other micro-organisms present in the environment. Bacteria or microscopic fungi can attack and inactivate viral particles by direct or indirect actions. For example, micro-organisms produce metabolites such as proteolytic enzymes, which adversely affect viral capsid. In fact, some environmental isolates of bacteria can use viral proteins and nucleic acids as a nutrient source. Also, certain bacteria can produce substances with antiviral activity, but whose molecular weight is so low that they cannot act enzymatically (Cliver and Hermann, 1972; Deng and Cliver, 1992; Deng and Cliver, 1995). In contrast, an increasing number of microbes have been described that can protect viruses from desiccation or disinfection (Storey and Ashbolt, 2001, 2003). Various studies have suggested that infectious viral particles may be trapped and thus accumulate in biofilms. Although the rate of contamination of biofilms with pathogenic viruses could be low and even though this attachment was observed only experimentally, biofilms should be considered as protective reservoirs of pathogenic viruses and could be implicated in numerous persistent viral infections (Lacroix-Gueu *et al.*, 2005).

Notwithstanding all of the above, the degree of virus persistence is mostly affected by temperature, with relative humidity (RH) also being an influencing factor. However, infectious viral particles are able to persist for days or even months over a range of temperatures and RH. Although viruses are destroyed by boiling, the thermal stability of some is remarkable (Carter, 2005). Temperatures of above 90°C are required for a 2 $\log_{10}$ inactivation of HAV in shellfish extract (Millard *et al.*, 1987), and a heat treatment of 100°C for 2 min has been recommended (Croci *et al.*, 1999) for elimination of infectious HAV in shellfish. On the other hand, viruses are preserved by refrigeration or freezing, which could cause problems, particularly with regard to food. The effect of RH is variable within virus types. It is believed that viruses with higher lipid contents (enveloped viruses) tend to be more resistant to lower RH and non-enveloped viruses are more stable at higher RH; that is, the survival of enveloped viruses is better when the RH is lower than 50%, while an RH level of higher than 80% is more beneficial for the persistence of non-enveloped viruses. However, several exceptions exist. In general, reducing the temperature promotes virus survival, but viruses might respond differently to RH (Mbithi *et al.*, 1991).

Sunlight and ultraviolet radiation (UV) are also important factors promoting the inactivation of viruses (Sagripanti and Lytle, 2007). UV light primarily not only targets viral nucleic acids but can also modify proteins of the viral capsid. Virus resistance to UV is influenced by virus type. Viruses with double-stranded genomes (dsRNA or dsDNA) are more resistant against UV inactivation than viruses with genomes consisting of single-stranded nucleic acid (ssRNA or ssDNA). Viral resistance to UV also depends on the size of the genome; it was found that the rate of nucleic acid degradation increased linearly with increasing fragment size of nucleic acid (Gerba *et al.*, 2002; Hijnen *et al.*, 2006; Tseng and Li, 2007). According to Hijnen *et al.* (2006), adenoviruses are the most resistant to UV. The sensitivities of FCV, rotaviruses, poliovirus and coxsackievirus were found to be similar, and HAV was the most sensitive virus (Hijnen *et al.*, 2006). The effect of UV-B was less pronounced on surrogate caliciviruses (enteric canine calicivirus and respiratory FCV) than on bacteria (De Roda Husman *et al.*, 2004) and enteroviruses (Gerba *et al.*, 2002) and it was more effective against phage MS2 (De Roda Husman *et al.*, 2004), adenoviruses (Gerba *et al.*, 2002) and *Bacillus subtilis* spores (Chang *et al.*, 1985).

Due to the mode of infection, foodborne and waterborne viruses are able to survive the extremes of the gastrointestinal tract. In general, viruses are most stable at pH values close to 7 and prefer low pHs (3–5) rather than alkaline pHs; 9–12 (Vasickova *et al.*, 2010). The pH can also indirectly affect virus persistence in the environment when persistence is influenced by the virus adsorption on solid particles or surfaces (Gerba, 1984). Persistence is increased while the viral particles are immobilised (adsorbed) and viruses are able to keep their infectivity after desorption. The interactions that take place between the viral particles and the adsorbent surface are determined by their characteristics and involve electrostatic and hydrophobic interactions and ionic strength. Due to the fact that the surface charge of viral particles is related to the pH of the environment, any disruption of such interactions is connected with pH changes (Gerba, 1984). Viral adsorption is theoretically better at high ionic strengths. Therefore, salt solutions are commonly used to promote adsorption. Viruses can also be stabilised and protected by dissolved, colloidal and solid organic matter including faecal and humic material. Organic matter has a low isoelectric point and thus carries a negative surface charge at most pH levels (Boone and Gerba, 2007). Besides, some viruses have more than one isoelectric point (Michen and Graule, 2010).

Comprehensive information regarding the influence of the environment on all viruses and their stability under external conditions does not exist. Most studies have used only a few target viruses or their surrogates and have not considered the effects of a combination of treatments or factors. Since experimental conditions or methods vary and studies performed to date have yielded contradictory results, it is difficult to draw conclusions from these experiments (Carter, 2005). Because the majority of foodborne or waterborne viruses which infect humans cannot be cultivated routinely, survival data and inactivation rates are sparse.

9.4 Persistence in aquatic environments

Foodborne and waterborne viruses can be present in any kind of untreated water, due to contamination caused by faecal material of human or animal origin. Despite the possible dilution of faecal contamination, the evidence suggests that viruses can persist in water in sufficient amounts to cause disease (Seymour and Appleton, 2001). In addition, it was found that these viruses might survive wastewater treatment and thus pose a threat to recreational users, consumers of shellfish or consumers of fresh fruit and vegetables (Carter, 2005). Viruses present in untreated water are inactivated slowly by a combination of the biological, physical and chemical effects mentioned above. It appears that the temperature of contaminated water and virus type are the most important factors affecting persistence (Seymour and Appleton, 2001). It was found that outbreaks caused by NoV are much more prevalent in the winter than in the summer, which is possibly due to lower temperatures (Doultree *et al.*, 1999; Mattison *et al.*, 2007) promoting survival.

Virus persistence in water can vary widely; the time required for a reduction of 1 log titre of enteroviruses ranges between 14 and 288 h (Chung and Sobsey, 1993; Callahan *et al.*, 1995). These variations could reflect the different conditions under which the experiments and studies were performed. Generally, viruses are capable of persisting for weeks or months at environmental temperatures and when sheltered from UV in combination with low temperature can even survive for years (Carter, 2005). HAV and poliovirus were shown to persist for more than 1 year in mineral water stored at 4°C while they remained infectious for 90 days at 10°C in wastewater and ground water (Biziagos *et al.*, 1988). According to Sobsey *et al.* (1988), HAV can survive in fresh or salt water for up to a year. No significant loss of infectious rotavirus particles was observed after 64 days at 4°C in raw water, treated tap water or filtered water, but a 99% reduction of titre was observed after 10 days at 20°C (Raphael *et al.*, 1985). Further studies have demonstrated the persistence of rotaviruses and poliovirus for more than 1 year (Biziagos *et al*, 1988) and of adenoviruses 40 and 41 for 300 days in artificially contaminated water (Enriquez *et al.*, 1995). Infectious astroviruses have been detected in drinking water after 90 days at 4°C (Abad *et al.*, 1997).

The rate of virus inactivation at or below 30°C is dependent on the pH and the ionic composition of the water environment (Salo and Cliver, 1976). It was previously thought that inactivation is associated mostly with genome degradation (Dimmock, 1967 as quoted by Cliver, 2009b), but subsequent studies demonstrated that viruses can be inactivated due to the loss of the ability to attach to host cell receptors, which implies a subtle denaturation of capsid proteins (Nuanualsuwan and Cliver, 2003). The pH and presence of salts do not appear to have a significant direct effect on virus persistence in natural waters, but instead influence the interaction between viruses and solid particles present in water (Seymour and Appleton, 2001). Several studies have suggested that adsorption of viral particles to particulate matter and sediments can result in substantial protection against inactivation procedures in the water environment (Mandel, 1971 as quoted by Cliver, 2009b). It was found that enteric viruses are destabilised and subsequently inactivated in water which is poor in salt ions, for example, Mg^{2+}. In contrast, increased concentrations of salts (NaCl) could be virucidal for several kinds of viruses (Vasickova *et al.*, 2010).

Viruses have been shown to persist better in sewage-polluted water than in non-polluted water environments, probably due to the presence of organic matter or solid particles. Alvarez *et al.* (2000) reported that the inactivation of bacteriophage MS2 and poliovirus in pre-filtered ground water was faster than in raw ground water. On the other hand, micro-organisms normally present in fresh and sea water can play an important role in the inactivation of viruses. The study by Gordon and Toze (2003) showed reduced inactivation rates in ground water in the absence of bacteria. Bosch (1995) demonstrated antiviral activity of bacteria present in sea water. However, bacteria can also have protective effects on viruses. Recent studies have revealed that viral particles are able to penetrate biofilms and in this way benefit from protection against environmental stress such as desiccation or other effects of antimicrobial agents. In addition, during biofilm erosion or sloughing, protected immobilised viral particles may be released into the environment, subsequently contacting their target host organism (Lehtola *et al.*, 2007; Briandet *et al.*, 2008; Helmi *et al.*, 2008). Although the protective effects of biofilms were observed only experimentally and natural biofilms can be contaminated with only a very low dose of viruses, biofilms should be considered as a protective reservoir for pathogenic viruses and facilitate numerous persistent viral infections (Lacroix-Gueu *et al.*, 2005).

As mentioned above, enteric viruses may survive the treatment of wastewater. Most enteric viruses (e.g., HAV, NoV, adenoviruses and HEV), have been found in wastewater and subsequently in treated water using PCR, and the detection of viruses in sewage systems has been conducted using the cultivation of some viruses on cell lines (Gantzer *et al.*, 1998; Matsuura *et al.*, 2000; Sedmak *et al.*, 2005). Astroviruses have been found in sewage treatment plant inlet and effluent waters. A reduction of approximately 2 log of viruses was detected in response to waste water processing and 10^5 copies of the astrovirus genome (per 1 l) were found in effluent water (Le Cann *et al.*, 2004). In addition, intact enteroviruses and HAV were detected in sludge originating from a wastewater plant (Albert and Schwarzbrod, 1991; Graff *et al.*, 1993).

Enteric viruses have also been detected in drinking water (Payment, 1989), which may in large part be due to contamination of water sources, failures in the treatment process (i.e., pressure failure, insufficient disinfection or exceptionally high concentration of pathogenic viruses), or contamination of an already treated water source (Carter, 2005). Filtration together with subsequent disinfection (e.g., chlorination, ozone or UV treatment) of raw water can achieve up to a 10 000-fold reduction in contaminating agents. During the disinfection procedure, turbidity has the greatest effect on virus survival. It was found that increasing the turbidity could decrease the effect of free chlorine, shield the viruses from UV and promote virus aggregation. It is assumed that the procedures used commonly for drinking water treatment might not destroy all viruses (Le Chevallier *et al.*, 1981). The study by Gofti-Laroche *et al.* (2003) showed a correlation between the presence of astrovirus RNA in drinking water and increased risk of intestinal disease. The results of a volunteer study suggested that NoVs could survive some water chlorination (Keswick *et al.*, 1985). In contrast, rotavirus is inactivated efficiently by chlorine (Carter, 2005).

Water plays an important role not only in the spread of human pathogenic viruses. Waterborne outbreaks of enteric viruses are common and thus interest has recently been focused on virus survival in drinking water and wastewater. The contamination of water sources is also a crucial step determining the potential of viruses to contaminate soil or crops. Although a number of experiments have been performed to investigate this issue, there is still a lack of information regarding the persistence of these viruses in the water environment and thus about their ecology. Furthermore, variations in the results of studies of virus survival in water emphasise the need for experiments to be performed under comparable conditions using standardised methods.

9.5 Persistence in soils

Predictably, viruses and bacteria are more abundant in a diverse range of moist soil types compared to dry and arid soil (Srinivasiah *et al.*, 2008).

Foodborne or waterborne viruses may contaminate soil via land disposal of sewage sludge or already contaminated water. Prolonged persistence has also been shown in such environments. Hurst *et al.* (1980) investigated the effects of several environmental conditions on virus persistence in soil. Based on the results, temperature, soil moisture content, degree of virus adsorption to soil, soil levels of resin-extractable phosphorus, exchangeable aluminium and the pH of the soil were found to influence virus persistence. Temperature and virus adsorption appear to be the most important factors affecting virus persistence in soil (Hurst *et al.*, 1980).

Infectious poliovirus was detected in spray-irrigated soil after 96 days during the winter season, while a maximum survival period of 11 days was demonstrated during the summer time (Tierney *et al.*, 1977). Oron *et al.* (1995) found that a relatively high soil temperature (30°C) together with low moisture content hindered poliovirus survival. Infectious particles of poliovirus and echovirus were recovered from loamy soil (pH 7.5) after 110–130 days at 3–10°C, while at 18–23°C the duration of persistence fell to 40–90 days (Bagdasaryan, 1964 as quoted by Rzezutka and Cook, 2004). Damgaard-Larsen *et al.* (1977) studied the persistence of enteroviruses in sludge-amended soil, where the temperature of the environment varied between −12°C and 26°C. A loss of 0.5–1 $\log^{10}$ of viral titre was observed per month, and viruses were still detected after 6 months of monitoring. In general, viruses persist for longer periods of time at low temperatures (4–8°C) than higher temperatures (20–37°C).

Wet conditions are usually associated with low soil temperatures. Poliovirus persistence was found to increase as more liquid was added to soil beyond the saturation point and then decrease as the soil moisture content increased up to the soil saturation point; that is, the inactivation rate for poliovirus increased when the water content of sandy soil increased from 5% to 15% and subsequently decreased when the water content further increased to 25% (Hurst *et al.*, 1980). It was also shown that virus survival is apparently prolonged in anaerobic conditions. The effect of soil water content on virus inactivation is dependent on soil type (Zhao *et al.*, 2008). Yeager and O'Brien (1979) found that viruses are able to persist for at least 180 days in saturated sandy loam or sandy soil, while no infectious viral particles were detected in dried soil regardless of soil type after 25 days.

Phosphorus, aluminium and pH have indirect effects on virus survival via their influence on the adsorption state. Whilst the presence of aluminium increases the virus adsorption rate, the level of resin-extractable phosphorus (phosphate anions) results in the elution of the adsorbed viral particles from soil. Virus adsorption on soil particles is increased when the pH of soil decreases and higher pH values result in the release of virus from soil particles (Zhao *et al.*, 2008). Virus survival is likely to be highest in types of soil that would be most effective in preventing ground water contamination. The study of Sobsey *et al.* (1980) compared the interactions of different soil materials and two different virus types (poliovirus type 1 and reovirus type 3). The behaviour of both viruses was found to be similar, which is in contrast to the studies of Goyal and Gerba (1979) and Landry *et al.* (1979), who found that adsorption varies with virus type and even strain. Generally, clay materials efficiently adsorbed viruses from waste water over a range of pH values, while sands and organic soil materials were poorer adsorbents; their ability to adsorb viruses increased only at low pHs together with the addition of total dissolved solids or divalent ions (Sobsey *et al.*, 1980). A further study confirmed that the presence of clay mineral enhances the persistence of viruses (Vettori *et al.*, 2000). Sobsey *et al.* (1980) also reported that even under unsaturated conditions viruses could still be washed from sandy soil and were able to contaminate water sources during heavy rainfall.

Virus survival in soil also appears to be generally greater under sterile than non-sterile conditions, suggesting the influence of other micro-organisms on virus survival in soil (Nasser *et al.*, 2002). The presence of aerobic bacteria could decrease virus survival due to the production of proteolytic substances. However, a study by Hurst *et al.* (1980) did not confirm this hypothesis: virus survival was not significantly affected by the addition of sewage effluent.

Due to the increasing land application of wastewater, it is important to evaluate the influence of different factors on virus survival in soil and thus evaluate the risk of resultant human illness. In addition, virus adsorption and inactivation in soil are crucial steps determining the potential of viruses to contaminate water resources (Zhao *et al.*, 2008). Viruses from sewage do not bind readily with soil particles and thus they can easily enter ground waters and in this way contaminate water sources (Seymour and Appleton, 2001). Furthermore, studies with poliovirus suggested that viral particles can infiltrate the roots and body of tomato plants (leaves) from the soil (Oron *et al.*, 1995). However, there is no evidence of illness from this source. To date, only a few studies have examined the biological factors affecting virus persistence in soil. Therefore, further studies are needed to evaluate the complex interactions between viral particles, soil and autochthonous micro-organisms; due to the restrictions of laboratory conditions the study of viruses under natural conditions, that is, in the field, should be emphasised.

9.6 Persistence on food-related surfaces

Surfaces can be contaminated directly through contact with body secretions and fluids containing infectious viral particles or indirectly via the aerosol or other contaminated fomites. Once a surface is contaminated, it may serve as a source of infectious viral particles for animate and inanimate subjects; for example, contaminated door handles or hands were found to be an efficient vector of viruses. It has been reported that at least 14 persons could be infected or their hands contaminated by touching a polluted door handle. Successive transmission of infectious viral particles from one person to another could be followed for up to six contacts via contaminated hands (von Rheinbaben *et al.*, 2000). The main factors influencing such kinds of transmission are temperature, RH, adsorption state and the character of the surface (Vasickova *et al.*, 2010).

Studies based on surrogate viruses indicate that NoV can persist for prolonged periods at low temperatures and that it can then be transmitted by different environmental matrices such as surfaces (Mattison *et al.*, 2007). It was found that FCV dried onto a glass surface and stored at 4°C displayed a 4.75 log reduction over 56 days, while the virus titre declined to undetectable levels at room temperature over 21–28 days and the infectious virions were not detectable at 37°C after 1 day (Doultree *et al.*, 1999). Studies focused on other viruses showed similar results. The infectivity of rotaviruses decreases more rapidly at 37°C than at 20°C or 4°C. Approximately 10% of rotavirus particles remain infectious at 4°C (RH 25–50%) after 10 days, while less than 1% of infectious virions persist at 20°C after 2 days (Moe and Shirley, 1982). A study by Abad *et al.* (1994) found that enteric viruses, including HAV and rotavirus, can persist for extended periods (up to 30 days) on fomites, and virus survival was prolonged at 4°C compared to 20°C. Astroviruses also showed a faster rate of inactivation at 20°C compared to 4°C (Abad *et al.*, 2001). Overall, astroviruses can persist longer than poliovirus (as a representative of enteroviruses) and adenoviruses; however, they show less robust survival than HAV or rotavirus (Mbithi *et al.*, 1991).

When the effect of RH on the survival of viruses was studied, it was found that RH had little effect on virus persistence at 5°C, while at 20°C viruses were able to survive for longer periods in low RH (Sattar *et al.*, 1987; Bidawid *et al.*, 2000). Mbithi *et al.* (1991) demonstrated that 34% and 52% of HAV particles remained infectious at high RH (80%) and low RH (25%) after 4 h at 20°C on a non-porous surface, respectively. In contrast, Abad *et al.* (1994) reported that HAV persistence was enhanced at high RH (90%) in comparison with moderate RH (50%) on a non-porous material after 60 days of storage at 20°C. Enhanced survival of rotavirus was observed at high RH on porous surfaces (Abad *et al.*, 1994), while Sattar *et al.* (1986) reported better rotavirus survival in low and medium RH on non-porous ones. Based on these results, virus persistence on surfaces is mainly related to virus strain, type of surface and temperature. The results with regard to the effect of RH are contradictory.

Foodborne viruses are potentially resistant to drying: 7% of rotavirus and 16–30% of HAV infectious particles persisted on finger pads after drying at room temperature for 4 h, although 68% of virions lost their infectivity within the first hour of the experiment (Ansari *et al.*, 1988; Mbithi *et al.*, 1992). It was also demonstrated that NoVs could be transferred from a contaminated surface to clean hands and via contaminated hands could cross-contaminate a series of seven surfaces (Barker *et al.*, 2004). The infectivity of NoVs is complicated by their problematic cultivation. A protective effect of organic material was demonstrated by Lee *et al.* (2008): a 2.7 log^{10} reduction of murine norovirus (MNV; surrogate of NoV) was detected at 18°C in a stool suspension, compared to 5.3 log^{10} on gauze or diaper surface. On non-porous surfaces, poliovirus and adenovirus persisted better in the presence of stool than did HAV and rotavirus. In contrast, the presence of stool material had a negative influence on the survival of poliovirus and adenovirus on porous fomites. The presence of food residues on steel material increased the persistence of MNV; a decline of 6.2 log and 1.4 log was observed at day 30 in residue-free and residua-present fomites, respectively (Takahashi *et al.*, 2011). The bacteria present in organic matter produce certain virucidal substances and thus are able to inactivate viruses (see section 9.3). This kind of inactivation is also dependent on temperature: the lower the temperature, the lower the activity of other micro-organisms and the longer the virus is able to maintain infectivity (Deng and Cliver, 1995).On the other hand, protective effects of bacterial reactions on viruses and bacterial biofilms have been reported and micro-organisms could have protective effects on virus survival due to their production of biofilms (see section 9.4).

The relationship between virus persistence and adsorption state is influenced by characteristics of both the virus and the type of surface. The majority of viruses remain viable for a longer period on non-porous materials (Tiwari *et al.*, 2006; Boone and Gerba, 2007; Lamhoujeb *et al.*, 2008); however, there are also several exceptions. Astroviruses are able to remain infectious at 4°C (RH 90 ± 5%) for 60 days when adsorbed on non-porous and for 90 days on porous surfaces; poliovirus and adenovirus persist also longer on porous (paper and cotton towel) than on non-porous material (aluminium, china, glazed tile, latex and polystyrene). Obviously, the physical properties of the surface may inhibit the recovery of viral particles from surfaces. Viruses can be trapped within the matrix, especially within the porous surface, and thus the results of virus persistence could be misinterpreted. Furthermore, several studies indicate virucidal activity of surfaces such as aluminium or copper. Adenovirus, poliovirus and the B40-8 phage showed lower persistence on aluminium than on other non-porous material (Thurman and Gerba, 1988; Abad *et al.*, 1997). The antimicrobial properties of copper and copper-based surfaces were also demonstrated (Faundez *et al.*, 2004; Noyce *et al.*, 2007).

The transmission of pathogenic viruses via contaminated surfaces is clear, but comparable data concerning the persistence of viral particles on surfaces is still lacking and studies performed to date have yielded contradictory results. Virological monitoring as well as studies of virus persistence on surfaces, using comparable methods and natural conditions, would be very useful in assessing the risks of virus spread via contaminated surfaces. In addition, information about the presence of non-cultivable viruses (NoV or HEV) on surfaces and hence their transmission via contaminated surfaces are still based on surrogate viruses (von Rheinbaben *et al.*, 2000; Boone and Gerba, 2007).

9.7 Persistence in food

If food cannot be reliably decontaminated during production, adequate food preparation such as cooking becomes critical (Cliver, 2009b). The consumption of food which is processed only minimally before consumption or served raw such as fruit and vegetables, shellfish or some traditional meat specialities represents a considerable risk (Vasickova *et al.*, 2005). Many studies have shown that the washing of already contaminated fruit and vegetables is not sufficient and that the depuration of virus-contaminated shellfish is highly unreliable (Croci *et al.*, 1999; Dawson *et al.*, 2005). Most of these studies were focused on predicting the persistence of foodborne viruses or their surrogates in cases of non-cultivability. Despite the variability in experimental conditions, several conclusions can be drawn. Virus persistence in fresh food is primarily influenced by the surrounding temperature, RH and the characteristics of the food and surrounding environment. Studies indicate that viruses are able to persist well on chilled, acidified, frozen foods and foods packed under modified atmosphere or in dry conditions even when preservation methods such as high hydrostatic pressure processing and irradiation were used. The virus viability was found to usually exceed the shelf-life of fresh food. Further, the decontamination of fresh produce can cause a reduction of at most 1 to 3 log in infectious viral particles even when chlorine and peroxyacetic acid solutions are used (Gulati *et al.*, 2001; Baert *et al.*, 2009).

Although viruses can be rapidly diluted in water, virus concentration can greatly increase in shellfish due to their filter-feeding; levels can be 100–1000-fold higher than in the surrounding water (Carter, 2005). According to DiGirolamo *et al.* (1970) and Tierney *et al.* (1982), viruses can persist well in shellfish; no loss of viral infectivity was observed over a month's refrigerated storage or 4 months when frozen. Ueki *et al.* (2007) compared the persistence of MNV and NoV in the digestive tissues of oysters and after depuration for 10 days. FCV was completely depleted, while NoV still persisted. Hewitt and Greening (2004) inoculated commercially prepared marinated mussels with HAV and FCV. A 1.7 log reduction in HAV was observed after 4 weeks of storage at 4°C and a 7 log decrease in MNV was seen after 1 week of storage under the same conditions.

The persistence of enteric viruses has been determined in a range of different fruits and vegetables (Seymour and Appleton, 2001). Konowalchuk *et al.* (1974) found no significant loss in coxsackievirus titre in lettuce stored for 16 days under moist conditions at 4°C, but inactivation was observed during storage in dry conditions. Badawy *et al.* (1985) reported that a cultivable strain of rotavirus (SA-11) is able to survive on lettuce, carrot and radishes for up to 30 days at 4°C. However, virus inactivation was greater at room temperature, and it was found that the virus still remained infectious after 25 days. These results contradict the documented persistence of HAV (Stine *et al.*, 2005). It was found that HAV survived longer on lettuce in medium RH (45.1–48.4%), compared to high RH (85.7–90.3%) (Abad *et al.*, 1994). In general, chilled storage (2–8°C) retards respiration, senescence, product browning, moisture loss and microbial growth in fruit and vegetables, but contributes to virus persistence (Seymour and Appleton, 2001). Besides, based on the inactivation rates calculated in the study by Stine *et al.* (2005), a 99.9% reduction in HAV-infectious particles could require as much as 822 days in pre-harvest lettuce.

According to Vega *et al.* (2008), it seems that electrostatic forces play the major role in virus adsorption to lettuce. Mattison *et al.* (2007) suggested that smooth surfaces such as those of lettuce might provide less protection to the viral particles compared to ham; ham is rich in proteins and fats, which might protect virions against dryness and other factors.

Several studies indicate that on plant surfaces viruses can be exposed to potentially virucidal substances, such as organic acids, phenols, ethanol or acetaldehyde, which could accelerate the inactivation of virions (Lamhoujeb *et al.*, 2008). The presence of such substances was described in several kinds of fruits and vegetables. Significant differences in viral recovery were found for strawberries, cherries and peaches kept in a humid atmosphere at 4°C; virus survival was lower than on other studied fruit (Konowalchuk and Speirs, 1975). Kurdziel *et al.* (2001) reported 1 log reduction of poliovirus after 11.6 days of storage for lettuce and after 14.2 days for white cabbage, while no significant decline in virus titre was observed on green onions after 15 days. A longer survival of HAV was detected on lettuce than on fennel and carrots (Baert *et al.*, 2009). A faster decline of HAV was also observed on fennel and carrots, which was reported to be due to the antimicrobial activity of carrot extracts (Babic *et al.*, 1994). Similar effects regarding poliovirus inactivation were also found for grape juice, apple juice and tea (Konowalchuk and Speirs, 1976). These findings could also be connected with the low pH of such food products. Despite this, outbreaks of hepatitis A have been associated with several kinds of fruit and fruit juices (Seymour and Appleton, 2001).

Freezing has a minimal effect on enteric viruses in berries (strawberries, blueberries and raspberries) and herbs (parsley and basil). The infectivity of NoV, HAV and rotavirus was not reduced significantly after freezing for 3 months; the number of infectious FCV decreased in frozen raspberries and strawberries, however (Butot *et al.*, 2008). A reduction ($<2 \log^{10}$) in infectious poliovirus on frozen strawberries was also reported after 15 days of storage (Kurdziel *et al.*, 2001), which could be explained by the low pH values of this fruit or the presence of virucidal substances. Deep-frozen storage of onions and spinach for 6 months had no effect on MNV survival (Baert *et al.*, 2008). On the other hand, FCV and canine calicivirus showed declines in infectivity of 0.34 log and 0.44 log after 5 cycles of freeze-thawing (Duizer *et al.*, 2004).

Despite the number of studies regarding the persistence of viruses in food, much remains to be learned (Cliver, 2009b), especially because of the absence of methods for the detection of all types of infectious viral particles. To resolve this problem predictive models have been proposed for persistence in shellfish which were based on surrogate viruses such as MNV and a cell culture-adapted HAV strain (Buckow *et al.*, 2008; Grove *et al.*, 2009). However, owing to the problems with surrogates, such models are not likely to portray the inactivation of pathogenic viruses in food accurately and the data obtained should be presented only as preliminary evidence of how pathogens might respond to different conditions (Richards, 2012).

9.8 Acknowledgement

This work was supported by the Ministry of Agriculture (No. MZE0002716202) and the Ministry of Education, Youth and Sports of the Czech Republic (AdmireVet CZ 1.05/2.1.00/01.0006-ED0006/01/01).

9.9 References

ABAD FX, PINTO RM and BOSCH A (1994), 'Survival of enteric viruses on environmental fomites', *Appl Environ Microbiol*, **60**, 3704–10.

ABAD FX, PINTO RM, VILLENA C, GAJARDO R and BOSCH A (1997), 'Astrovirus survival in drinking water', *Appl Environ Microbiol*, **63**, 3119–22.

ABAD FX, VILLENA C, GUIX S, CABALLERO S, PINTÓ RM and BOSCH A (2001), 'Potential role of fomites in the vehicular transmission of human astroviruses', *Appl Environ Microbiol*, **67**, 3904–7.

ALBERT M and SCHWARZBROD L (1991), 'Recovery of enterovirus from primary sludge using three elution concentration procedures', *Water Sci Technol*, **24**, 225–8.

ALVAREZ ME, AGUILAR M, FOUNTAIN A, GONZALEZ N, RASCON O and SAENZ D (2000), 'Inactivation of MS-2 phage and poliovirus in groundwater', *Can J Microbiol*, **46**, 159–65.

ANSARI SA, SATTAR SA, SPRINGTHORPE VS, WELLS GA and TOSTOWARYK W (1988), 'Rotavirus survival on human hands and transfer of infectious virus to animate and nonporous inanimate surfaces', *J Clin Microbiol*, **26**, 1513–18.

BABIC I, NGUYEN-THE C, AMIOT MJ and AUBERT S (1994), 'Antimicrobial activity of shredded carrot extracts on food-borne bacteria and yeast', *J Appl Bacteriol*, **76**, 135–41.

BADAWY AS, GERBA CP and KELLY LM (1985), 'Survival of rotavirus SA-11 on vegetables', *Food Microbiol*, **2**, 261–4.

BAERT L, UYTTENDAELE M, VERMEERSCH M, VAN COILLIE E and DEBEVEREI J (2008), 'Survival and transfer of murine norovirus 1, a surrogate for human noroviruses, during the production process of deep-frozen onions and spinach', *J Food Protect*, **71**, 1590–7.

BAERT L, DEBEVERE J and UYTTENDAELE M (2009), 'The efficacy of preservation methods to inactivate foodborne viruses', *Int J Food Microbiol*, **131**, 83–94.

BAGDASARYAN GA (1964), 'Survival of viruses of the enterovirus group (poliomyelitis, echo, coxsackie) in soil and on vegetables', *Hyg Epid Microbiol Immunol*, **8**, 497–505.

BARKER J, VIPOND IB and BLOOMFIELD SF (2004), 'Effects of cleaning and disinfection in reducing the spread of Norovirus contamination via environmental surfaces', *J Hosp Infect*, **58**, 42–9.

BIDAWID S, FARBER JM and SATTAR SA (2000), 'Contamination of foods by food handlers: Experiments on hepatitis A virus transfer to food and its interruption', *Appl Environ Microbiol*, **66**, 2759–63.

BIZIAGOS E, PASSAGOT J, CRANCE JM and DELOINCE R (1988), 'Long-term survival of hepatitis A virus and poliovirus type 1 in mineral water', *Appl Environ Microbiol*, **54**, 2705–10.

BOONE SA and GERBA CP (2007), 'Significance of fomites in the spread of respiratory and enteric viral disease', *Appl Environ Microbiol*, **73**, 1687–96.

BOSCH A (1995), 'The survival of enteric viruses in the water environment', *Microbiologia*, **11**, 393–6.

BRIANDET R, LACROIX-GUEU P, RENAULT M, LECART S, MEYLHEUC T, BIDNENKO E, STEENKESTE K, BELLON-FONTAINE MN and FONTAINE-AUPART MP (2008), 'Fluorescence correlation spectroscopy to study diffusion and reaction of bacteriophages inside biofilms', *Appl Environ Microbiol*, **74**, 2135–43.

BUCKOW R, ISBARN S, KNORR D, HEINZ V and LEHMACHER A (2008), 'Predictive model for inactivation of feline calicivirus, a norovirus surrogate, by heat and high hydrostatic pressure', *Appl Environ Microbiol*, **74**, 1030–8.

BUTOT S, PUTALLAZ T and SANCHEZ G (2008), 'Effects of sanitation, freezing and frozen storage on enteric viruses in berries and herbs', *Int J Food Microbiol*, **126**, 30–5.

CABALLERO S, ABAD FX, LOISY F, LE GUYADER FS, COHEN J, PINTÓ RM and BOSCH A (2004), 'Rotavirus virus-like particles as surrogates in environmental persistence and inactivation studies', *Appl Environ Microbiol*, **70**, 3904–9.

CALLAHAN KM, TAYLOR DJ and SOBSEY MD (1995), 'Comparative survival of hepatitis A virus, poliovirus and indicator viruses in geographically diverse seawaters', *Water Sci Technol*, **31**, 189–93.

CANNON JL, PAPAFRAGKOU E, PARK GW, OSBORNE J, JAYKUS LA and VINJE J (2006), 'Surrogates for the study of norovirus stability and inactivation in the environment: A comparison of murine norovirus and feline calicivirus', *J Food Prot*, **69**, 2761–5.

CARTER MJ (2005), 'Enterically infecting viruses: pathogenicity, transmission and significance for food and waterborne infection', *J Appl Microbiol*, **98**, 1354–80.

CASPER ET, PATTERSON SS, SMITH MC and PAUL JH (2005), 'Development and evaluation of a method to detect and quantify enteroviruses using NASBA and internal control RNA (IC-NASBA)', *J Virol Methods*, **124**, 149–55.

CHANG JC, OSSOFF SF, LOBE DC, DORFMAN MH, DUMAIS CM, QUALLS RG and JOHNSON JD (1985), 'UV inactivation of pathogenic and indicator microorganisms', *Appl Environ Microbiol*, **49**, 1361–5.

CHUNG H and SOBSEY MD (1993), 'Comparative survival of indicator viruses and enteric viruses in seawater and sediments', *Water Sci Technol*, **27**, 425–8.

CLIVER DO and HERMANN JE (1972), 'Proteolytic and microbial inactivation of enteroviruses', *Water Res*, **6**, 338–53.

CLIVER DO (2009a), 'Capsid and infectivity in virus detection', *Food Environ Virol*, **1**, 123–8.

CLIVER DO (2009b), 'Control of viral contamination of food and environment', *Food Environ Virol*, **1**, 3–9.

COOK N (2003), 'The use of NASBA for the detection of microbial pathogens in food and environmental samples', *J Microbiol Methods*, **53**, 165–74.

CROCI L, DE MEDICI D, MORACE G, FIORE A, SCALFARO C, BENEDUCE F and TOTI L (1999), 'Detection of hepatitis A virus in shellfish by nested reverse transcription-PCR', *Int J Food Microbiol*, **48**, 67–71.

DAHLING D (1991), 'Detection and enumeration of enteric viruses in cell culture', *CRC Rev Environ Contam*, **21**, 237–63.

DAMGAARD-LARSEN S, JENSEN KO, LUND E and NISSEN B (1977), 'Survival and movement of enterovirus in connection with land disposal of sludges', *Water Res*, **11**, 503–8.

DAWSON DJ, PAISH A, STAFFELL LM, SEYMOUR IJ and APPLETON H (2005), 'Survival of viruses on fresh produce, using MS2 as a surrogate for norovirus', *J Appl Microbiol*, **98**, 203–9.

DENG MY and CLIVER DO (1992), 'Inactivation of poliovirus type-1 in mixed human and swine wastes and by bacteria from swine manure', *Appl Environ Microbiol*, **58**, 2016–21.

DENG MY and CLIVER DO (1995), 'Antiviral effects of bacteria isolated from manure', *Microb Ecol*, **30**, 43–54.

DE RODA HUSMAN AM, BIJKERK P, LODDER W, VAN DEN BERG H, PRIBIL W, CABAJ A, GEHRINGER P, SOMMER R and DUIZER E (2004), 'Calicivirus inactivation by nonionizing (253.7-nanometer-wavelength (UV)) and ionizing (Gamma) radiation', *Appl Environ Microbiol*, **70**, 5089–93.

DIGIROLAMO R, LISTON J and MATCHES JR (1970), 'Survival of virus in chilled, frozen and processed oysters', *Appl Microbiol*, **20**, 58–63.

DIMMOCK NJ (1967), 'Differences between the thermal inactivation of picornaviruses at "high" and "low" temperatures', *Virology*, **31**, 338–53.

DOULTREE JC, DRUCE JD, BIRCH CJ, BOWDEN DS and MARSHALL JA (1999), 'Inactivation of feline calicivirus, a Norwalk virus surrogate', *J Hosp Infect*, **41**, 51–7.

DUIZER E, BIJKERK P, ROCKX B, DE GROOT A, TWISK F and KOOPMANS M (2004), 'Inactivation of caliciviruses', *Appl Environ Microbiol*, **70**, 4538–43.

ENRIQUEZ CE, HURST CJ and GERBA CP (1995), 'Survival of the enteric adenoviruses 40 and 41 in tap, sea and wastewater', *Water Res*, **29**, 2548–53.

FAUNDEZ G, TRONCOSO M, NAVARRETE P and FIGUEROA G (2004), 'Antimicrobial activity of copper surfaces against suspensions of *Salmonella enterica* and *Campylobacter jejuni*', *BMC Microbiol*, **4**, 4–19.

FITTIPALDI M, RODRIGUEZ NJP, CODONY F, ADRADOS B, PENUELA GA and MORATO J (2010), 'Discrimination of infectious bacteriophage T4 virus by propidium monoazide real-time PCR', *J Virol Methods*, **168**, 228–32.

GALLAGHER EM and MARGOLIN AB (2007), 'Development of an integrated cell culture – real-time RT-PCR assay for detection of reovirus in biosolids', *J Virol Methods*, **139**, 195–202.

GANTZER C, MAUL A, AUDIC JM and SCHWARTZBROD L (1998), 'Detection of infectious enteroviruses, enterovirus genomes, somatic coliphages, and *Bacteroides fragilis* phages in treated wastewater', *Appl Environ Microbiol*, **64**, 4307–12.

GERBA CP (1984), 'Applied and theoretical aspects of virus adsorption to surfaces', *Adv Appl Microbiol*, **30**, 133–68.

GERBA CP, GRAMOS DM and NWACHUKU N (2002), 'Comparative inactivation of enteroviruses and adenovirus 2 by UV light', *Appl Environ Microbiol*, **68**, 5167–9.

GILPATRICK SG, SCHWAB KJ, ESTES MK and ATMAR RL (2000), 'Development of an immunomagnetic capture reverse transcription-PCR assay for the detection of Norwalk virus', *J Virol Methods*, **90**, 69–78.

GOFTI-LAROCHE L, GRATACAP-CAVALLIER B, DEMANSE D, GENOULAZ O, SEIGNEURIN JM and ZMIROU D (2003), 'Are waterborne astrovirus implicated in acute digestive morbidity (E.MI.R.A. study)?', *J Clin Virol*, **27**, 74–82.

GORDON C and TOZE S (2003), 'Influence of groundwater characteristics on the survival of enteric viruses', *J Appl Microbiol*, **95**, 536–44.

GOYAL SM and GERBA CP (1979), 'Comparative adsorption of human enteroviruses, simian rotavirus, and selected bacteriophages to soils', *Appl Environ Microbiol*, **38**, 241–7.

GRAFF J, TICEHURST J and FLEHMIG B (1993), 'Detection of hepatitis A virus in sewage sludge by antigen capture polymerase chain reaction', *Appl Environ Microbiol*, **59**, 3165–70.

GROVE SF, LEE A, STEWART CM and ROSS T (2009), 'Development of a high pressure processing inactivation model for hepatitis A virus', *J Food Prot*, **72**, 1434–42.

GULATI BR, ALLWOOD PB, HEDBERG CW and GOYAL SM (2001), 'Efficacy of commonly used disinfectants for the inactivation of calicivirus on strawberry, lettuce, and a food-contact surface', *J Food Prot*, **64**, 1430–4.

HELMI K, SKRABER S, GANTZER C, WILLAME R, HOFFMANN L and CAUCHIE HM (2008), 'Interactions of *Cryptosporidium parvum*, *Giardia lamblia*, vaccinal poliovirus type 1, and bacteriophages phi X174 and MS2 with a drinking water biofilm and a wastewater biofilm', *Appl Environ Microbiol*, **74**, 2079–88.

HEWITT J and GREENING GE (2004), 'Survival and persistence of norovirus, hepatitis A virus, and feline calicivirus in marinated mussels', *J Food Prot*, **67**, 1743–50.

HIJNEN WA, BEERENDONK E and MEDEMA GJ (2006), 'Inactivation credit of UV radiation for viruses, bacteria and protozoan (oo)cysts in water: A review', *Water Res*, **40**, 3–22.

HIRNEISEN KA, BLACK EP, CASCARINO JL, FINO VR, HOOVER DG and KNIEL KE (2010), 'Viral inactivation in foods: A review of traditional and novel food-processing technologies', *Comp Rev Food Sci Food Safety*, **9**, 3–20.

HOWIE R, ALFA MJ and COOMBS K (2008), 'Survival of enveloped and non-enveloped viruses on surfaces compared with other micro-organisms and impact of suboptimal disinfectant exposure', *J Hosp Infect*, **69**, 368–76.

HURST CJ, GERBA CP and CECH I (1980), 'Effects of environmental variables and soil characteristics on virus survival in soil', *Appl Environ Microbiol*, **40**, 1067–79.

KESWICK BH, SATTERWHITE TK, JOHNSON PC, DUPONT HL, SECOR SL, BITSURA JA, GARY GW and HOFF JC (1985), 'Inactivation of Norwalk virus in drinking water by chlorine', *Appl Environ Microbiol*, **50**, 261–4.

KINGSLEY DH, GUAN D and HOOVER DG (2005), 'Pressure inactivation of hepatitis A virus in strawberry puree and sliced green onions', *J Food Prot*, **68**, 1748–51.

KONOWALCHUK J, SPEIRS JL, PONTEFRACT RD and BERGERON G (1974), 'Concentration of enteric viruses from water with lettuce extract', *Appl Microbiol*, **28**, 717–19.

KONOWALCHUK J and SPEIRS JI (1975), 'Survival of enteric viruses on fresh fruit', *J Milk Food Technol*, **38**, 598–600.

KONOWALCHUK J and SPEIRS JI (1976), 'Virus inactivation by grapes and wines', *Appl Environ Microbiol*, **32**, 757–63.

KURDZIEL AS, WILKINSON N, LANGTON S and COOK N (2001), 'Survival of poliovirus on soft fruit and salad vegetables', *J Food Prot*, **64**, 706–9.

LACROIX-GUEU P, BRIANDET R, LEVEQUE-FORT S, BELLON-FONTAINE MN and FONTAINE-AUPART MP (2005), '*In situ* measurements of viral particles diffusion inside mucoid biofilms', *C R Biol*, **328**, 1065–72.

LAMHOUJEB S, FLISS I, NGAZOA SE and JEAN J (2008), 'Evaluation of the persistence of infectious human noroviruses on food surfaces by using real-time nucleic acid sequence-based amplification', *Appl Environ Microbiol*, **74**, 3349–55.

LANDRY EF, VAUGHN JM, THOMAS MZ and BECKWITH CA (1979), 'Adsorption of enteroviruses to soil cores and their subsequent elution by artificial rainwater', *Appl Environ Microbiol*, **38**, 680–7.

LE CANN P, RANARIJAONA S, MONPOEHO S, LE GUYADER F and FERRE V (2004), 'Quantification of human astroviruses in sewage using real-time RT-PCR', *Res Microbiol*, **155**, 11–15.

LECHEVALLIER MW, EVANS TM and SEIDLER RJ (1981), 'Effect of turbidity on chlorination efficiency and bacterial persistence in drinking water', *Appl Environ Microbiol*, **42**, 159–67.

LEE J, ZOH K and KO G (2008), 'Inactivation and UV disinfection of murine norovirus with TiO2 under various environmental conditions', *Appl Environ Microbiol*, **74**, 2111–17.

LEHTOLA MJ, TORVINEN E, KUSNETSOV J, PITKÄNEN T, MAUNULA L, VON BONSDORFF CH, MARTIKAINEN PJ, WILKS SA, KEEVIL CW and MIETTINEN IT (2007), 'Survival of *Mycobacterium avium, Legionella pneumophila*, *Escherichia coli*, and caliciviruses in drinking water-associated biofilms grown under high-shear turbulent flow', *Appl Environ Microbiol*, **73**, 2854–9.

LI D, BAERT L, VAN COILLIE E and UYTTENDAELE M (2011), 'Critical studies on binding-based RT-PCR detection of infectious noroviruses', *J Virol Methods*, **177**, 153–9.

MALORNY B, TASSIOS PT, RADSTROM P, COOK N, WAGNER M and HOORFAR J (2003), 'Standardization of diagnostic PCR for the detection of foodborne pathogens', *Int J Food Microbiol*, **83**, 39–48.

MANDEL B (1971), 'Characterisation of type 1 poliovirus by electrophoretic analysis', *Virology*, **44**, 554–68.

MATSUURA K, ISHIKURA M, YOSHIDA H, NAKAYAMA T, HASEGAWA S, ANDO S, HORIE H, MIYAMURA T and KITAMURA T (2000), 'Assessment of poliovirus eradication in Japan: genomic analysis of polioviruses isolated from river water and sewage in toyama prefecture', *Appl Environ Microbiol*, **66**, 5087–91.

MATTISON K, KARTHIKEYAN K, ABEBE M, MALIK N, SATTAR SA, FARBER JM and BIDAWID S (2007), 'Survival of calicivirus in foods and on surfaces: Experiments with feline calicivirus as a surrogate for norovirus', *J Food Prot*, **70**, 500–3.

MBITHI JN, SPRINGTHORPE VS and SATTAR SA (1991), 'Effect of relative-humidity and air-temperature on survival of hepatitis A virus on environmental surfaces', *Appl Environ Microbiol*, **57**, 1394–9.

MBITHI JN, SPRINGTHORPE VS, BOULET JR and SATTAR SA (1992), 'Survival of hepatitis A virus on human hands and its transfer on contact with animate and inanimate surfaces', *J Clin Microbiol*, **30**, 757–63.

MICHEN B and GRAULE T (2010), 'Isoelectric points of viruses', *J Appl Microbiol*, **109**, 388–97.

MILLARD J, APPLETON H and PARRY JV (1987), 'Studies on heat inactivation of hepatitis A virus with special reference to shellfish. Part 1. Procedures for infection and recovery of virus from laboratory-maintained cockles', *Epidemiol Infect*, **98**, 397–414.

MOE K and SHIRLEY JA (1982), 'The effects of relative-humidity and temperature on the survival of human rotavirus in feces', *Arch Virol*, **72**, 179–86.

NASSER AM, GLOZMAN R and NITZAN Y (2002), 'Contribution of microbial activity to virus reduction in saturated soil', *Water Res*, **36**, 2589–95.

NOYCE JO, MICHELS H and KEEVIL CW (2007), 'Inactivation of influenza A virus on copper versus stainless steel surfaces', *Appl Environl Microbiol*, **73**, 2748–50.

NUANUALSUWAN S and CLIVER DO (2002), 'Pretreatment to avoid positive RT-PCR results with inactivated viruses', *J Virol Methods*, **104**, 217–25.

NUANUALSUWAN S and CLIVER DO (2003), 'Capsid functions of inactivated human picornaviruses and feline calicivirus', *Appl Environ Microbiol*, **69**, 350–7.

OGORZALY L, BERTRAND I, PARIS M, MAUL A, GANTZER C. (2010), 'Occurrence, survival, and persistence of human adenoviruses and F-specific RNA phages in raw groundwater', *Appl Environ Microbiol*, **76**, 8019–25.

ORON G, GOEMANS M, MANOR Y and FEYEN J (1995), 'Poliovirus distribution in the soil-plant system under reuse of secondary wastewater', *Water Res*, **29**, 1069–78.

PARSHIONIKAR S, LASEKE I and FOUT GS (2010), 'Use of propidium monoazide in reverse transcriptase PCR to distinguish between infectious and noninfectious enteric viruses in water samples', *Appl Environ Microbiol*, **76**, 4318–26.

PAYMENT P (1989), 'Elimination of viruses and bacteria during drinking water treatment: review of 10 years of data from the Montréal metropolitan area', In: *Biohazards of Drinking Water Treatment*, LARSON RA (ed.), Lewis Publishers, Chelsea, MI, pp. 59–65.

RAPHAEL RA, SATTAR SA and SPRINGTHORPE VS (1985), 'Long-term survival of human rotavirus in raw and treated river water', *Can J Microbiol*, **31**, 124–8.

REYNOLDS KA, GERBA CP, ABBASZADEGAN M and PEPPER LL (2001), 'ICC/PCR detection of enteroviruses and hepatitis A virus in environmental samples', *Can J Microbiol*, **47**, 153–7.

RICHARDS GP (2012), 'Critical review of norovirus surrogates in food safety research: rationale for considering volunteer studies', *Food Environ Virol*, **4**, 6–13.

RIGOTTO C, VICTORIA M, MORESCO V, KOLESNIKOVAS CK, CORRÊA AA, SOUZA DS, MIAGOSTOVICH MP, SIMÕES CM and BARARDI CR (2010), 'Assessment of adenovirus, hepatitis A virus and rotavirus presence in environmental samples in Florianopolis, South Brazil', *J Appl Microbiol*, **109**, 1979–87.

RODRIGUEZ RA, PEPPER IL and GERBA CP (2009), 'Application of PCR-based methods to assess the infectivity of enteric viruses in environmental samples', *Appl Environ Microbiol*, **75**, 297–307.

RZEZUTKA A and COOK N (2004), 'Survival of human enteric viruses in the environment and food', *FEMS Microbiol Rev*, **28**, 441–53.

SAGRIPANTI JL and LYTLE CD (2007), 'Inactivation of influenza virus by solar radiation', *Photochem Photobiol*, **83**, 1278–82.

SALO RJ and CLIVER DO (1976), 'Effect of acid pH, salts, and temperature on the integrity and physical integrity of enteroviruses', *Arch Virol*, **52**, 269–82.

SATTAR SA, LLOYD-EVANS N, SPRINGTHORPE VS and NAIR RC (1986), 'Institutional outbreaks of rotavirus diarrhoea: potential role of fomites and environmental surfaces as vehicles for virus transmission', *J Hyg (Lond)*, **96**, 277–89.

SATTAR SA, KARIM YG, SPRINGTHORPE VS and JOHNSONLUSSENBURG CM (1987), 'Survival of human rhinovirus type-14 dried onto nonporous inanimate surfaces – effect of relative-humidity and suspending medium', *Can J Microbiol*, **33**, 802–6.

SCHLINDWEIN AD, RIGOTTO C, SIMOES CM and BARARDI CR (2010), 'Detection of enteric viruses in sewage sludge and treated wastewater effluent', *Water Sci Technol*, **61**, 537–44.

SCHWAB KJ, DE LEON R and SOBSEY MD (1996), 'Immunoaffinity concentration and purification of waterborne enteric viruses for detection by reverse transcriptase PCR', *Appl Environ Microbiol*, **62**, 2086–94.

SEDMAK G, BINA D, MACDONALD J and COUILLARD L (2005), 'Nine-year study of the occurrence of culturable viruses in source water for two drinking water treatment plants and the influent and effluent of a wastewater treatment plant in Milwaukee, Wisconsin (August 1994 through July 2003)', *Appl Environ Microbiol*, **71**, 1042–50.

SEYMOUR IJ and APPLETON H (2001), 'Foodborne viruses and fresh produce', *J Appl Microbiol*, **91**, 759–73.

SHIMASAKI N, KIYOHARA T, TOTSUKA A, NOJIMA K, OKADA Y, YAMAGUCHI K, KAJIOKA J, WAKITA T and YONEYAMA T (2009), 'Inactivation of hepatitis A virus by heat and high hydrostatic pressure: variation among laboratory strains', *Vox Sang*, **96**, 14–19.

SHOEIB AR, ABD EL MAKSOUD SN, BARAKAT AB, SHOMAN SA and EL ESNAWY NA (2009), 'Comparative assessment of mammalian reoviruses versus enteroviruses as indicator for viral water pollution', *J Egypt Public Health Assoc*, **84**, 181–96.

SIZUN J, YU MWN and TALBOT PJ (2000), 'Survival of human coronaviruses 229E and OC43 in suspension and after drying on surfaces: a possible source of hospital-acquired infections', *J Hosp Infect*, **46**, 55–60.

SOBSEY MD, DEAN CH, KNUCKLES ME and WAGNER RA (1980), 'Interactions and survival of enteric viruses in soil materials', *Appl Environ Microbiol*, **40**, 92–101.

SOBSEY MD, SHIELDS PA, HAUCHMAN FS, DAVIS AL, RULLMAN VA and BOSH A (1988), 'Survival and persistence of hepatitis A virus in environmental samples', In: *Viral Hepatitis and Liver Diseases*, Zuckerman AJ (ed.), Alan R Liss, New York, pp. 121–4.

SRINIVASIAH S, BHAVSAR J, THAPAR K, LILES M, SCHOENFELD T and WOMMACK KE (2008), 'Phages across the biosphere: contrasts of viruses in soil and aquatic environments', *Res Microbiol*, **159**, 349–57.

STINE SW, SONG I, CHOI CY and GERBA CP (2005), 'Effect of relative humidity on preharvest survival of bacterial and viral pathogens on the surface of cantaloupe, lettuce, and bell peppers', *J Food Prot*, **68**, 1352–8.

STOREY MV and ASHBOLT NJ (2001), 'Persistence of two model enteric viruses (B40–8 and MS-2 bacteriophages) in water distribution pipe biofilms', *Water Sci Technol*, **43**, 133–8.

STOREY MV and ASHBOLT NJ (2003), 'Enteric virions and microbial biofilms – a secondary source of public health concern?', *Water Sci Technol*, **48**, 97–104.

STRAUB TM, HÖNER ZU BENTRUP K, OROSZ-COGHLAN P, DOHNALKOVA A, MAYER BK, BARTHOLOMEW RA, VALDEZ CO, BRUCKNER-LEA CJ, GERBA CP, ABBASZADEGAN M and NICKERSON CA (2007), '*In vitro* cell culture infectivity assay for human noroviruses', *Emerg Infect Dis*, **13**, 396–403.

TAKAHASHI H, OHUCHI A, MIYA S, IZAWA Y and KIMURA B (2011), 'Effect of food residues on norovirus survival on stainless steel surfaces', *PLoS One*, **6**, e21951.

THURMAN RB and GERBA CP (1988), 'Characterization of the effect of aluminium metal on poliovirus', *J Indust Microbiol*, **3**, 33–8.

TIERNEY JT, SULLIVAN R and LARKIN EP (1977), 'Persistence of poliovirus 1 in soil and on vegetables grown in soil previously flooded with inoculated sewage sludge or effluent', *Appl Environ Microbiol*, **33**, 109–13.

TIERNEY JT, SULLIVAN R, PEELER JT and LARKIN EP (1982), 'Persistence of polioviruses in shell stock and shucked oysters stored at refrigeration temperature', *J Food Protect*, **45**, 1135–7.

TIWARI A, PATNAYAK DP, CHANDER Y, PARSAD M and GOYAL SM (2006), 'Survival of two avian respiratory viruses on porous and nonporous surfaces', *Avian Dis*, **50**, 284–7.

TSENG CC and LI CS (2007), 'Inactivation of viruses on surfaces by ultraviolet germicidal irradiation', *J Occup Environ Hyg*, **4**, 400–5.

UEKI Y, SHOJI M, SUTO A, TANABE T, OKIMURA Y, KIKUCHI Y, SAITO N, SANO D and OMURA T (2007), Persistence of caliciviruses in artificially contaminated oysters during depuration', *Appl Environ Microbiol*, **73**, 5698–5701.

VASICKOVA P, DVORSKA L, LORENCOVA A and PAVLIK I (2005), 'Viruses as a cause of foodborne diseases: a review of the literature', *Veterinarni Medicina*, **50**, 89–104.

VASICKOVA P, PAVLIK P, VERANI M and CARDUCCI A (2010), 'Issues concerning survival of viruses on surfaces', *Food Environ Virol*, **2**, 24–34.

VEGA E, GARLAND J and PILLAI SD (2008), 'Electrostatic forces control nonspecific virus attachment to lettuce', *J Food Prot*, **71**, 522–9.

VETTORI C, GALLORI E and STOTZKY G (2000), 'Clay minerals protect bacteriophage PBS1 of *Bacillus subtilis* against inactivation and loss of transducing ability by UV radiation', *Can J Microbiol*, **46**, 770–3.

VON RHEINBABEN F, SCHUNEMANN S, GROSS T and WOLFF MH (2000), 'Transmission of viruses via contact in a household setting: experiments using bacteriophage phi X174 as a model virus', *J Hosp Infect*, **46**, 61–6.

YEAGER JG and O'BRIEN RT (1979), 'Structural changes associated with poliovirus inactivation in soil', *Appl Environ Microbiol*, **38**, 702–9.

ZHAO B, ZHANG H, ZHANG J and JIN Y (2008), 'Virus adsorption and inactivation in soil as influenced by autochthonous microorganisms and water content', *Soil Biol Biochem*, **40**, 649–59.

10

Occurrence and transmission of food- and waterborne viruses by fomites

C. P. Gerba, University of Arizona, USA

DOI: 10.1533/9780857098870.3.205

Abstract: This chapter discusses occurrence and survival of viruses on fomites (inanimate objects). The significance of transfer of viruses from fomites to hands and other objects is discussed and existing data reviewed. The application of transmission models to better quantify the extent of fomite transmission and the success of interventions are discussed.

Key words: fomites, transmission of viruses, virus survival, modeling transmission.

10.1 Introduction: the role of fomites in virus transmission

Fomites are defined as inanimate objects that can be involved in the transmission of infectious disease. Fomites usually become contaminated by exposure to bodily fluids (mucus, salvia, feces, vomit, urine, blood) or tissues (skin, organs) from infected individuals. Transmission occurs when the contaminated fomite comes into contract with the skin, portal of entry (mouth, nose, eye, wound), or tissue. Transmission of blood-borne viruses can occur from the re-use of needles, tattooing, sharing of towels or other fomites which may come into contact with open wounds. Respiratory viruses can be transmitted by contact of the fingers with fomites contaminated with respiratory secretions and placement of the finger in the nose or eyes. Enteric viruses, similarly, can be transmitted by contact with fecally contaminated surfaces and hand or object movement to the mouth.

Contact with virus-contaminated inanimate objects or fomites has long been known as a route of potential transmission. Blankets from smallpox-infected individuals were provided to Native Americans in hopes of spreading the infection and blankets transported from India to England were suspected of being involved in cases of smallpox. The involvement of fabrics and cotton was suspected more than 100 years ago during a smallpox outbreak in England traced to raw cotton contaminated with variola virus crusts and scabs (Fenner and White, 1970). Variola virus can survive for several months in fabrics (England, 1982). Studies conducted on the transmission of the common cold clearly demonstrated the importance of this route in its transmission. In more recent years the importance of fomites is seen with the rapid spread of noroviruses in large facilities such as hotels, cruise ships and institutions.

The significance of fomites in the transmission of respiratory and enteric infections is currently probably greater than at any time in the history of the developed world. Most of us today work in an office and spend more time indoors than any generation in history (~80%). We go to ever larger malls, stores, cruise ships, airports and schools. In addition we have more buttons to push than any other generation in history. This allows contact with more surfaces that increasing numbers of other individuals have come into contact with previously. Every time an individual touches a surface with his hands there is the potential for them to contaminate the surface with viruses. When the surface is touched by another individual a number of these viruses will be transferred to that person's hand.

Human activity results in the rapid contamination of fomites and hands by viruses. Placement of coliphage ΦX174 on a household doorknob was transferred to 14 successive people by contact and then to an additional six successive people by the shaking of hands (Rheinbaben *et al.*, 2000). Plotkin (2012) placed coliphage MS-2 on the entrance door push plate to an office of 80 persons and within 4 h 44–56% of the fomites in communal areas (e.g. coffee break rooms) were contaminated with the phage.

10.2 Occurrence and survival of viruses on fomites

Until the development of molecular methods, such as the polymerase chain reaction, data on the occurrence of viruses on fomites was limited by the cost of gathering the information. Infectious viruses have been detected on a wide variety of fomites (Table 10.1). It has been demonstrated that respiratory viruses are common on fomites when infected individuals are present. For example, it was found that in households with two children with influenza infections, the virus could be detected on more than half of commonly touched fomites such as phones, TV remotes, faucets, computer equipment and doorknobs (Boone and Gerba, 2005). Rhinovirus has been detected on 40–90% of the hands of adults with colds, and from 6% to 25% of selected fomites in rooms with persons with colds. In a study of offices, parainfluenza virus could be isolated in one-third of the offices tested across the United States (Boone and Gerba, 2010). Noroviruses are commonly detected on desk tops in elementary school classrooms (Bright *et al.*, 2010) and during outbreaks (Jones *et al.*, 2007; Wu *et al.*, 2005).

Table 10.1 Detection viruses on fomites

Fomite	Virus
Household	
Toys	*Coxsackie*[*] B6, hepatitis B
Door knob	*Rhinovirus,* influenza virus
Living room and kitchen floors	*Poliovirus* 2 and 3
Faucet	*Rhinovirus*
Commercial	
Drinking glasses in coffee shops	*Adenovirus* types 3 and 4; *echovirus* 11; *herpesvirus*
Daycare centers and schools	
Toys	*Cytomegalovirus*, rotavirus, rhinovirus
Toilets, door handles, tables, play mats	Rotavirus, influenza, noro, astro
Offices	
Desk tops, computer equipment, phones	Parainfluenza
Hotels, cruise ships	Noro
Hospital and other health care facilities	Noro, respiratory syncytial virus, hepatitis A, parainfluenza, rota, corona

[*]Italics indicate infectious virus detected.
Sources: Boone and Gerba, 2007; Sattar and Springthorpe, 1996; England, 1982; Fox and Hall, 1980.

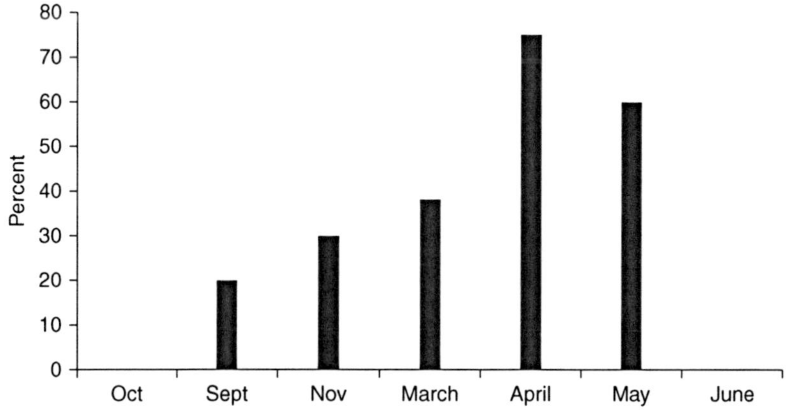

Fig. 10.1 Seasonal variation in the occurrence of influenza viruses on fomites in day care centers (percent of fomites on which virus detected).

The occurrence of viruses on fomites may vary seasonally. Influenza viruses were only found on fomites in day care centers during the influenza season, with the number of positive fomites increasing with the number of infections (Boone and Gerba, 2005, Fig. 10.1).

Table 10.2 Factors influencing the survival of viruses on fomites

Factor	Remark
Type of virus	Usually enteric viruses survive longer than respiratory viruses on fomites
Suspension media	Longer survival in feces and bodily fluids
Drying rate	Generally the faster the drying rate, the more rapid the inactivation of the virus
Relative humidity	Some survive better at low relative humidity and others at high relative humidity
Type of fomite	Longer survival on surfaces that hold moisture longer, like clothes
Inoculum titer	Slower inactivation rate with higher titer

The survival of viruses on fomites is depended on a number of factors, as listed in Table 10.2. The type of virus is probably the most significant factor, as survival can vary with the type and strain. There are no standardized procedures for assessing the survival of viruses on fomites and so only broad generalizations on survival of viruses on fomites can be made. The relative importance of environmental factors varies depending upon the virus. Thus, lipid-containing viruses such as influenza virus tend to survive better at lower relative humidity (RH), while the non-lipid-containing reoviruses survive better at high RH. Still others, such as poliovirus, survive best at intermediate RH. As a general rule non-lipid-containing enteric viruses survive longer on fomites than respiratory viruses (Table 10.3). The survival of respiratory viruses is usually a matter of hours to days, whereas with enteric viruses it ranges from days to weeks. Usually virus die-off on surfaces is biphasic with greater inactivation occurring during drying of the suspension material followed by a slower inactivation rate. The drying rate is governed by the RH and the suspending media. In most indoor buildings with air-conditioning systems, the RH ranges from 40% to 60%. It is usually less in the winter because of indoor heating, which may favor the survival of some respiratory viruses such as influenza.

Fomite type is an important and highly variable influence on viral inactivation. Fomites vary in porosity, hydrophobicity, surface charge, micro-topography and antimicrobial properties. These same properties may also influence the ability to recovery the viruses from the fomite over time. Influenza A and B were found to be inactivated more rapidly on porous than non-porous surfaces (Bean *et al.*, 1982). Similarly avian influenza virus and avian metapneumonia virus survived longer on non-porous surfaces than porous surfaces (Tiwari *et al.*, 2005). These results suggest that related viruses exhibit similar inactivation patterns on porous *vs* non-porous surfaces. However, Abad *et al.* (2001) found human astrovirus 4 was inactivated more rapidly on non-porous surfaces than porous surfaces. In contrast hepatitis A

Table 10.3 Inactivation rates of viruses on fomites

Virus	Inactivation rate (log_{10}/hour)
Respiratory viruses	
Avian influenza	0.0138
Influenza A	0.0278
Influenza B	0.0417
Cornavirus-229E	0.167
Cornavirus-OC43	0.2
Rhinovirus 14	0.25
Parainfluenza	0.5
Respiratory syncytial virus (RSV)	0.625
Enteric viruses	
Astrovirus 4	0.0025
Hepatitis A virus	0.00278
Adenovirus 40	0.011
Poliovirus type 1	0.01
Coliphage	
P22	0.0379
MS-2	0.0062

Data from Bonne and Gerba, 2007; Henley, 2008.

virus was inactivated more rapidly on porous surfaces. Some metal fomites such as brass, copper and silver surfaces are antiviral (Noyce *et al.*, 2007; Silvestry-Rodriguez *et al.*, 2007). Murray and Laband (1979) found that poliovirus was very stable on silicon dioxide surfaces (glass) but degraded on CuO, MnO_2 and Al_2O_3 surfaces.

The medium in which virus is suspended can have a significant effect on survival. Sattar *et al.* (1987) found suspension in bovine mucin and nasal discharge decreased infectivity of human rhinovirus type 14, while suspension in tryptose phosphate broth had no effect. Bovine mucin and discharge were found to substantially reduce human rhinovirus 14 infectivity during the drying processes, with 82% and 89% of infectivity loss, respectively. In contrast only 3% of the viral infectivity was lost during drying in tryptose phosphate broth suspensions. Feces also appear to be protective. Inactivation of hepatitis A virus and human rotavirus was unaffected by the presence of fecal matter (Abad *et al.*, 1994). However, poliovirus type 1, enteric adenovirus and human astrovirus type 4 exhibited varying inactivation rates in the presence of fecal material. Poliovirus 1 and enteric adenovirus decreased in titer at a slower rate on non-porous surfaces in the presence of fecal material; however, the opposite was observed on porous surfaces. Human astrovirus 14 had different inactivation rates in the presence of fecal material, depending on both fomite type and temperature (Abad *et al.*, 2001).

10.3 Virus transfer and modeling transmission

Modeling of transfer is useful in predicting the probability of infection and the potential success of interventions.

10.3.1 Virus transfer via fomites

Transfer of pathogens can occur by direct contact of the fomite with an animate surface where infection can be initiated. Such would be the case with a contaminated needle or medical device used in the examination of a patient. Wounds inflicted by contact with a contaminated surface are another potential route. Risk from these can be reduced by proper disinfection of devices or treatment of the wounds with antimicrobials. However, the most common infections transmitted by fomites are respiratory, enteric and dermal. Respiratory and enteric infections are readily transmitted when the pathogen is placed in the eyes, mouth or nose by movement of the hands to the face. This occurs on a regular basis during normal everyday activity, and in small children it occurs an average of 81 times per hour (Table 10.4). Studies have shown that hands easily become contaminated when fomites with infectious virus on them are touched. The reverse is also true. Rhinovirus was found to be transferred after 1–3 h on a plastic surface after being touched (Hendley *et al.*, 1973). Persons infected with rhinovirus have been shown to readily contaminate fomites which they touched (Reed, 1975), and infectious virus could be recovered from objects previously touched by infected persons (Pancic *et al.*, 1980). Hand transfer from fomites of other respiratory viruses, such as influenza (Bean *et al.*, 1982) and respiratory syncytial virus (RSV) (Hall *et al.*, 1980) has been demonstrated.

Hand contact with fabrics that had been dried was shown to be capable of transferring polio and vaccinia virus to the hands (Sidwell *et al.*, 1970). Rotavirus can be readily transferred to hands by touching metal surfaces (Ansari *et al.*, 1989). Transfer to metal surfaces and finger-to-finger transfer was also demonstrated. Transfer efficiency is dependent on a number of factors including the pressure applied to the surface (Ansari *et al.*, 1989), the nature of the contact (e.g. turning a door knob *vs.* pushing a button) (Rusin *et al.*, 2002), the type of virus, suspending material, relative humidity and the composition of the fomite. Generally, the smoother and the easier it is to clean a surface, the more virus can be transferred (Fig. 10.2). Thus, ~68% of coliphage MS-2 was transferred from an acrylic plastic surface to the finger *vs* less than 1% from paper money when the relative humidity was 40–60% (Lopez, 2012). The transfer from the acrylic was reduced to 21% when the relative humidity was 15–30%. Julian *et al.* (2010) reported that coliphage fr transfer from fomites was significantly greater than that of MS-2 and ΦX174 and transfer was less to recently washed hands.

The pressure of the hand applied in a handshake or to a doorknob when opening a door has been estimated at 1 kg/cm^2 (Ansari *et al.*, 1988). Mbithi

Table 10.4 Frequency of face touching

Age	Average touching frequency per hour	Reference
Adults (alone)		Nicas and Best, 2008
Total	15.7	
Lips	8	
Nostrils	5.3	
Eyes	2.4	
Adults (public-church)		Henley *et al.*, 1973
Nostrils	0.33	
Eye	0.37	
Children (indoor)		Tulve *et al.*, 2002
<2 years (mouth)	81	
2–5 years (mouth)	42	
Children (indoor)		Xue *et al.*, 2007
3 months to 11 years, mouth	6.7–28	
Children (outdoor)		
7–12 years, mouth	12.6	Beamer *et al.*, 2012

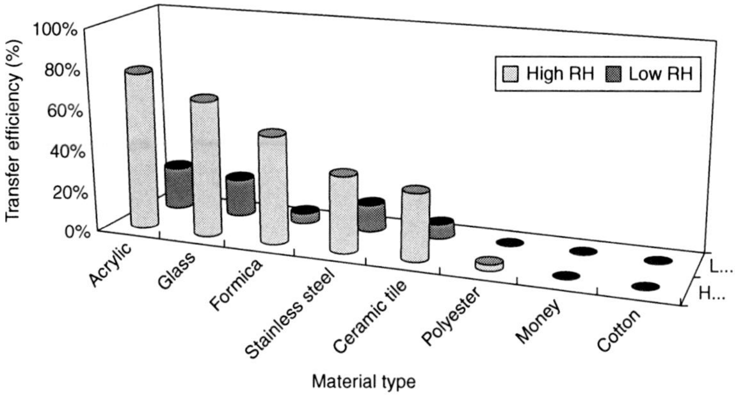

Fig. 10.2 Transfer efficiency of viruses from various surfaces.

et al. (1992) and Julian *et al.* (2009) used 0.2 kg/cm^2 and 25 kPa (kilo Pascals) respectively. Mbithi *et al.* (1992) found that the amount of pressure and friction applied made a significant difference to the amount of HAV transferred. Ansari *et al.* (1991) found transfer efficiency for rotavirus and parainfluenza virus 3 to be 0.67% and 1.48%, respectively, from stainless steel disks to fingers using a 5-s contact time. The difference between the rotavirus transfer rates can possibly be attributed to the decrease in contact time. Using 0.2 kg/cm^2

of pressure with a 10-s contact time Mbithi *et al.* (1992) found a 22% transfer rate for hepatitis A virus from stainless steel to fingers. Rusin *et al.* (2002) found a transfer rate for PRD1 phage to be 33.5% and 65.8% from faucet handle and phone receiver, respectively. They used a 30-s contact time, and both the palms of hands and fingers were tested for virus.

10.3.2 Modeling virus transmission via fomites

Modeling of virus transmission in indoor environments is seen as a way of better understanding the importance of different routes of transmission (inhalation *vs* finger contact to the nose) and assessment of different interventions in the reduction of transmission (Spicknall *et al.*, 2010). For example, the relative importance of inhalation of droplets and finger contact to the face in the transmission of influenza virus is still debated (Nicas and Jones, 2009). In the case of influenza, models can help us understand if face mask wearing would be more successful than hand sanitizers in reducing transmission. A number of factors have to be considered in the development of models as shown in Table 10.5. Data are limited at the current time on concentration of viruses on fomites and degree of exposure.. The significance of fomite transmission increases with the density of the population in an indoor environment, but is not as dependent upon population density as aerosol transmission, if the virus has a reasonable survival time on the fomite. Nicas and Sun (2006) used an integrated model with a discrete-time Markov chain model to represent

Table 10.5 Factors to consider in modeling virus transmission in an indoor environment

Shedding
Shedding rate
Shedding magnitude (~volume)
Virus concentration
Shedding Distribution
Size of the droplets
Target membrane surface area
Environment
Inactivation rate on the fomite
Host/Environment
Touch rate
Transfer efficiency (skin/surface area)
Inactivation rate on the skin
Rate of changing location (movement from one surface to the next)
Host density
Exposure
Self-inoculation rate
Transfer efficiency to the skin-membrane
Infectivity (dose response)

Modified from Spicknall *et al.*, 2010.

physical elements in various source-receptor pathways as 'states' in a health care environment. The various pathways considered movement of the organism from hands to various fomites (such as bed sheets) to resuspension in the air and movement from one fomite to another by the hands. By using these models to change the die-off rate of the virus on a surface, reflecting treatment with an antimicrobial, the reduction in the probability of infection can be determined. Thus, Nicas and Sun (2006) determined that the use of textiles in health care facilities treated with an antimicrobial would result in a 50% reduction in risk of infection. By application of a model to estimate probability of rotavirus infection in small children Julian *et al.* (2009) found that direct mouthing of fomites was a more significant route than hand-to-fomite contact. Similar approaches could be used to assess other interventions such as hand washing, hand sanitizers and self-sanitizing surfaces.

10.4 Disinfection and other interventions to prevent fomite transmission

Successful interruption of virus transmission by fomites has been documented by numerous studies on hand washing and the use of hand sanitizers (Aiello *et al.*, 2007). Interventions with the use of disinfectants and self-sanitizing surfaces are not as well documented, and data on reduction of illness for specific viruses is generally lacking. The use of virucidal iodine solution on the fingers was found to reduce the infection rate of rhinovirus among mothers by 66% relative to a control group (Hendley *et al.*, 1988). Instruction in hand hygiene and encouragement of the use of hand sanitizers was found to reduce the number of laboratory-confirmed influenza cases by 52% and total absenteeism by 26% (Stebbins *et al.*, 2011). Norovirus is more resistant to inactivation by commonly used alcohol gel sanitizer than many other common enteric viruses, and its preferential use over hand washing was found to result in increased risk of norovirus infection (Blaney *et al.*, 2011). A specially formulated alcohol gel sanitizer has been developed which is more effective against norovirus (Macinga *et al.*, 2008).

Ward *et al.* (1991) assessed the ability of a disinfectant spray to prevent the transmission of rotavirus from contaminated surfaces using human volunteers. Volunteers who licked a contaminated surface or touched the surface with their finger and then licked the finger became infected. Prior treatment of the surface with a disinfectant prevented transmission. In similar studies it was shown that rhinovirus transmission could be prevented by treatment of stainless steel surfaces with a disinfectant spray (quat) or bleach (Sattar and Springthrope, 1996). Combined use of hand sanitizer and a disinfectant wipe was shown to reduce both school absenteeism and gastroenteritis. Significantly fewer noroviruses were detected on fomites (desk tops, water fountain) in classrooms using the wipes and hand sanitizers (Sandora *et al.*, 2008).

10.5 Future trends

The role of fomites in virus transmission has probably increased in recent years because of changes in work, play, healthcare and lifestyle which bring us into more contact with more surfaces that have the potential to be contaminated with infectious virus. Integration of exposure models with quantitative microbial risk assessment will help us better understand the importance of fomites in virus transmission and provide tools for understanding which interventions may be most successful in reducing exposure. The greatest amount of uncertainty in risk models is created by the lack of quantitative data on exposure (Haas *et al.*, 1999). Data on the concentration and occurrence of viruses on fomites is needed to reduce the amount of uncertainty in risk estimates. Knowledge of the infectivity of viruses on fomites is also needed. Hopefully, advances in molecular biology will allow us to provide a low-cost method of assessing virus infectivity on fomites. This would also enable assessment of the effectiveness of disinfectants or self-sanitizing surfaces in actual indoor environments in occupied buildings.

10.6 References

ABAD FX, PINTO RM and BOSCH A (1994) 'Survival of enteric viruses on environmental fomites', *Appl Environ Microbiol*, **60**, 3704–10.

ABAD FX, VILLENA C, GUIX S, CABALLERO S, PINTO RM and BOSCH A (2001) 'Potential role of fomites in the vehicular transmission of human astroviruses', *Appl Environ Microbiol*, **67**, 3904–7.

AIELLO AE, LARSON EL and SEDLAK R (2007) *Against disease. The impact of hygiene and cleanliness on health*. Washington, DC: Soap and Detergent Association.

ANSARI SA, SATTAR SA, SPRINGTHORPE VS, WELLS GA and TOSTOWARYK W (1988) 'Rotavirus survival on human hands and transfer of infectious virus to animate and nonporous inanimate surfaces', *J Clin Microbiol*, **26**, 1513–8.

ANSARI SA, SATTAR SA, SPRINGTHORPE VS, WELLS GA and TOSTOWARYK W (1989) '*In vivo* protocol for testing efficacy of hand washing agents against viruses and bacteria: experiments with rotavirus and *Escherichia coli*', *Appl Environ Microbiol*, **55**, 3113–8.

ANSARI SA, SPRINGTHORPE VS, SATTAR SA, RIVARD S and RAHMAN M (1991) 'Potential role of hands in the spread of respiratory viral infections: studies with human parainfluenza virus 3 and rhinovirus 14', *J Clin Microbiol*, **29**, 2115–19.

BEAMER PI, LUIK CE, CANALES RA and LECKIE JO (2012) 'Quantified outdoor micro-activity data for children aged 7–12 years old', *J Exposure Sci Environ Epidemiol*, **22**, 82–92.

BEAN B, MOORE BM, STERNER B, PETERSON, LR, GERDING DN and BELFOUR JR. HH (1982) 'Survival of influenza viruses on environmental surfaces', *J Infect Dis*, **146**, 47–51.

BLANEY DD, DALY ER, KIRKLAND KB, TONGREN JE, KELSO and TALBOT EA (2011) 'Use of alcohol-based hand sanitizers as a risk factor for norovirus outbreaks in long-term care facilities in northern New England: December 2006 to March 2007', *Am J Infect Contr*, **39**, 296–301.

BOONE SA and GERBA CP (2005) 'The occurrence of influenza A virus on household and day care center fomites', *J Infect*, **73**, 1687–96.

BOONE SA and GERBA CP (2007) 'Significance of fomites in the spread of respiratory and enteric disease', *Appl Environ Microbiol*, **73**, 1687–96.
BOONE SA and GERBA CP (2010) 'The prevalence of human parainfluenza virus 1 on indoor office fomites', *Food Environ Virol*, **2**, 41–6.
BRIGHT KR, BOONE SA and GERBA CP (2010) 'Occurrence of bacteria and viruses on elementary classroom surfaces and the potential of classroom hygiene in the spread of infectious diseases', *J Sch Nurs*, **26**, 33–41.
ENGLAND BL (1982) 'Detection of viruses on fomites', in GERBA CP and GOYAL SM. *Methods in environmental virology*. New York: Marcel Dekker, 179–220.
FENNER FJ and WHITE DO (1970) *Medical virology*. New York: Academic Press.
FOX JP and HALL CE (1980) *Viruses in families*. Littleton, MA: PSG Publishing.
HAAS CN, ROSE JB and GERBA CP (1999) *Quantitative microbial risk assessment*. New York: Wiley.
HALL CB, DOUGLAS RG and GEIMAN JM (1980) 'Possible transmission by fomites of respiratory syncytial virus', *J Infect Dis*, **141**, 98–102.
HENDLEY JO and GWALTNEY JM (1988) 'Mechanisms of transmission of rhinovirus transmission', *Epid Rev*, **10**, 241–58.
HENDLEY JO, WENZEL RP and GWALTNEY JM (1973) 'Transmission of rhinovirus colds by self-inoculation', *New Eng J Med*, **288**, 1361–4.
HENLEY JB (2008) *Determination of inactivation rates of viruses on indoor surfaces*. Master Thesis, Tucson, AZ: University of Arizona.
JONES EL, KRAMER A, GAITHER M and GERBA CP (2007) 'Role of fomite contamination during an outbreak of norovirus on houseboats', *Int J Environ Health Res*, **17**, 123–31.
JULIAN TR, CANALES R A, LECKIE JO and BOEHM AB (2009) 'A model of exposure to rotavirus from nondietary ingestion iterated by simulated intermittent contacts', *Risk Anal*, **29**, 617–32.
JULIAN TR, LECKIE JO and BOEHM AB (2010) 'Virus transfer between fingerpads and fomites', *J Appl Microbiol*, **109**, 1868–74.
LOPEZ GU (2012) *Comparative fomite-to-finger transfer efficiency of bacteria, spores and coliphage from environmental nonporous and porous surfaces*. Master Thesis, Tucson, AZ: University of Arizona.
MACINGA DR, SATTAR SA, JAYKUS LA and ARBOGAST JW (2008) 'Improved inactivation of nonenveloped enteric viruses and their surrogates by a novel alcohol-based hand sanitizer', *Appl Environ Microbiol*, **74**, 5047–52.
MBITHI JN, SPRINGTHORPE VS, BOULET JR and SATTAR SA (1992) 'Survival of hepatitis A virus on human hands and its transfer on contact with animate and inanimate surfaces', *J ClinMicrobiol*, **30**, 757–63.
MURRAY JP and LABAND SJ (1979) 'Degradation of poliovirus by adsorption on inorganic surfaces', *Appl Environ Microbiol*, **37**, 480–6.
NICAS M and BEST D (2008) 'A study quantifying the hand-to-face contact rate and its potential application to predicting respiratory tract infection', *J Occup Environ Hyg*, **5**, 347–52.
NICAS M and JONES RM (2009) 'Relative contributions of four exposure pathways to influenza infection risk', *Risk Anal*, **29**, 1292–1303.
NICAS M and SUN G (2006) 'An integrated model of infection risk in a health care environment', *Risk Anal*, **26**, 1085–96.
NOYCE JO, MICHELS H and KEEVIL CW (2007) 'Inactivation of influenza A virus on copper versus stainless steel surfaces', *Appl Environ Microbiol*, **73**, 2748–2750.
PANCIC F, CARPENTIER DC and CAME PE (1980) 'Role of infectious secretions in the transmission of rhinovirus', *J Clin Microbiol*, **12**, 567–71.
PLOTKIN KR (2012) *Risk modeling of human viruses on fomites and the impact of a healthy workplace intervention*. Master Thesis, Tucson, AZ: University of Arizona.

REED S (1975) 'An investigation of the possible transmission of rhinovirus colds through direct contact', *J Hyg (Camb)*, **75**, 249–58.

RHEINBABEN F, SCHÜNERMANN S, GROSS T and WOLFF MH (2000) 'Transmission of viruses via contact in a household setting: experiments using bacteriophage ΦX174 as a model virus', *J Hosp Infect*, **46**, 61–6.

RUSIN P, MAXWELL S and GERBA C (2002) 'Comparative surface-to-hand and fingertip-to-mouth transfer efficiency of gram-positive bacteria, gram-negative bacteria, and phage', *J Appl Microbiol*, **93**, 585–92.

SANDORA TJ, SHIH MC and GOLDMAN DA (2008) 'Reducing absentism from gastrointestinal and respiratory illness in elementary school students: a randomized, controlled trial of an infection-control intervention', *Pediatrics*, **121**, 1555–62.

SATTAR, SA, KARIM YG, SPRINGTHORPHE VS and JOHNSON-LUSSENBURG CM (1987) 'Survival of human rhinovirus type 14 dried onto nonporous inanimate surfaces: effect of relative humidity and suspending medium', *Canad J Microbiol*, **33**, 802–6.

SATTAR SA and SPRINGTHORPE VS (1996) 'Transmission of viral infections through animate and inanimate surfaces and infection control through chemical disinfection', in C. HURST. *Modeling disease transmission and its prevention by disinfection.* Cambridge: Cambridge University Press, 224–57.

SIDWELL RW, DIXON GJ, WESTBROOK L and FORZIATI FH (1970) 'Quantitative studies on fabrics as disseminators of viruses', *Appl Environ Microbiol*, **19**, 950–4.

SILVESTRY-RODRIGUEZ N, SICAIROS-RUELAS EE, GERBA CP and KELLY KR (2007) 'Silver as a disinfectant', *Rev Environ Contam Toxicol*, **191**, 23–45.

SPICKNALL IH, KOOPMAN JS, NICAS M, PUJOL JM, SHENG L and EISENBERG JNS (2010) 'Informing optimal environmental influenza interventions: how the host, agent, and environment alter dominant routes of transmission', *PLoS Computation Biol*, **6**(10), e1000969, 1–11.

STEBBINS S, CUMMINGS DA, STARK JH, VUKOTICH C, MITRUKA K, THOMPSON W, RINAIDO C, ROTH L, WAGNER M, WISNIEWSKI SR, DATO V, ENG H and BURKE DS (2011) 'Reduction in the incidence of influenza A but not influenza B associated with the use of hand sanitizer and cough hygiene in schools: a randomized controlled trial', *Pediatr Infect Dis*, **30**, 921–6.

TIWARI A, PATNAYAK DP, CHANDER Y, PARSAD M and GOYAL SM (2005) 'Survival of two avian respiratory viruses on porous and nonporous surfaces', *Avian Dis*, **50**, 284–7.

TULVE NS, SUGGS JC, MCCURDY T, COHEN-HUBAL EA and MOYA J (2002) 'Frequency of mouthing behavior in young children', *J Expo Anal Environ Epidemiol*, **12**, 259–64.

WARD RL, BERNSTEIN DI, KNOWLTON DR, SHERWOOD JR, YOUNG EC, CUSACK TM, RUBINO JR and SCHIFF GM (1991) 'Prevention of surface-to-surface human transmission of rotavirus by treatment with disinfectant spray', *J Clin Microbiol*, **29**, 1991–6.

WU HM, FOMEK M, SCHWAB KJ, CHAPIN AR, GIBSON K, SCHWAB E, SPENCER C and HENNING K (2005) 'A norovirus outbreak at a long-term-care facility: the role of environmental surface contamination', *Infect Contr Hosp Epidemiol*, **26**, 802–10.

XUE J, ZARTARIAN V, MOYA J, FREEMAN N, BEAMER P, TULVE N and SHALAT S (2007) 'A meta-analysis of children's hand-to-mouth frequency data for estimating nondietary ingestion exposure', *Risk Anal*, **27**, 411–20.

11

Viral contamination by food handlers and recommended procedural controls

I. L. A. Boxman, Food and Consumer Product Safety Authority (NVWA), The Netherlands

DOI: 10.1533/9780857098870.3.217

Abstract: Virus contamination as a consequence of human handling can occur at any stage of food production or processing. This chapter reviews evidence of the role of food handlers in the transmission of norovirus and hepatitis A virus, their current knowledge of and compliance with food hygiene practices. The new Codex Guidelines for the Application of the General Principles of Food Hygiene to the Control of Viruses in Food are summarized. The possible application of risk analysis of critical control points to control of viruses is discussed.

Key words: norovirus, hepatitis A virus, virus transmission, food safety, HACCP.

11.1 Introduction

Norovirus (NoV) and hepatitis A virus (HAV) are the human enteric viruses most frequently reported as involved in foodborne outbreaks. Estimates of the proportion of viral illness attributed to food are in the range of around 5% for HAV and 12–47% for NoV (FAO/WHO, 2008). Food handlers play an important role in transmission of these viruses at the end of the food chain, for example, in food establishments, but also further back in the production chain. During production, harvest, packaging and preparation, food can become contaminated with viruses by food handlers or after contact with virus-contaminated water or surfaces.

This chapter summarizes evidence that has been collected in recent years on the role of food handlers in the transmission of foodborne viruses. It also reviews studies of food handlers' knowledge of and compliance with food hygiene rules, with particular reference to foodborne viruses and to hand washing practices. The provision of effective food hygiene training is essential for food handlers and for everyone employed in the food business. The Codex Alimentarius Commission has adopted the new Guidelines for the Application of the General Principles of Food Hygiene to the Control of Viruses in Food (FAO/WHO, 2012b).

11.2 Role of food handlers in virus transmission

Food handlers play an important role in transmission of foodborne pathogens, including foodborne viruses, to food. Virus contamination as a consequence of human handling can occur at virtually any stage during harvest, packing, distribution, at retail establishments or at home. Alternatively, food can become contaminated with foodborne viruses by contaminated water or surfaces (WHO/FAO, 2008; EFSA, 2011). Not practising good hand hygiene after toilet use may lead to the spread of enteric pathogens to food handling areas and to food. In food service operations, Kassa *et al.* (2001) reported that enteric bacteria, as indicators of faecal contamination, were present on 57% of the cooler and freezer door handles tested in the food processing area, and spreading on this scale may also occur with enteric viruses. Bidawid *et al.* (2000) were the first to demonstrate transmission of enteric viruses from artificially contaminated finger pads to food or inanimate surfaces; this study was followed by others (Bidawid *et al.*, 2004; D'Souza *et al.*, 2006). Whereas transmission of viruses to food is a direct risk for consumers, transmission to food contact surfaces in food handling areas may serve as a secondary source of contamination (Boone and Gerba, 2007; Lopman *et al.*, 2011). From the analyses of more than 800 outbreak reports, it was concluded that the most frequently reported factor associated with the involvement of the infected worker was bare hand contact with the food (Todd *et al.*, 2007). Manual handling of food by food handlers seems therefore to be crucial in the transmission of viruses to food.

Foodborne viral outbreaks of NoV or HAV are often associated with ill food handlers on the premises (Koopmans *et al.*, 2002; Fiore, 2004). In early outbreak reports, evidence was mainly based on epidemiological data. The development of molecular techniques enabled the epidemiological link between the suspected food or setting and cases to be identified through detection of NoV and HAV in faecal samples of patients and food handlers. In some studies, further evidence was obtained in the form of identical viral sequences in faecal specimens of both cases and (asymptomatic) food handlers (Barrabeig *et al.*, 2010; Nicolay *et al.*, 2011), or viral RNA on food samples prepared by the food handler (Schwab *et al.*, 2000; Boxman *et al.*, 2007, 2009a). Detection of viral RNA on food samples is, however, difficult, and absence of faecal samples or leftovers for analysis could hamper outbreak investigations.

Environmental swabs taken from surfaces in outbreak settings in institutions and hotels (Dalling, 2004; Wu *et al.*, 2005) and in the catering industry (Boxman *et al.*, 2009a) have recently become an additional tool in NoV-associated outbreak investigation, especially when leftover food or faecal specimens are not available. Particularly when viral RNA is detected from swabs taken from surfaces in the food handling area, positive samples indicate that a contagious person has been present. Boxman *et al.* (2009a) showed that where clinical symptoms consistent with viral gastroenteritis were present and the period between consumption of suspected food and onset of symptoms was at least 12 h, NoV RNA was present in 48 of 119 (40%) environmental swab specimens taken from 14 of 27 outbreaks (52%), with sequences matching positive clinical samples in seven outbreaks. Further evidence for the feasibility of transmission of NoV by a food handler to food was the presence of NoV RNA on a food handler's hands in an outbreak setting (Boxman *et al.*, 2009b). In addition, a year-round prevalence study showed that the presence of NoV in catering companies seemed to mirror the presence of NoV in the population. Sequence analysis of NoV-positive environmental swabs showed that NoV strains in catering companies were interspersed with NoV strains found in outbreaks of illness in humans. The prevalence of NoV on surfaces in catering companies associated with recently reported outbreaks (61%) was, however, significantly higher than in randomly selected catering companies with no such association (4%) (Boxman *et al.*, 2011).

This raises the question of who is shedding viruses in food handling areas. It is known that NoV- and HAV-infected persons shed large amounts of virus in their stools while symptoms are present, and that infectious particles are also present in vomit (Atmar *et al.*, 2008; FAO/WHO, 2008). For both NoV and HAV infections the infectious dose is only a few particles (Teunis *et al.*, 2008). For these reasons, ill persons should be excluded from the premises. The timing of exclusion (e.g., for 48 h after symptoms have ceased) may, however, vary in practice. Sometimes employees continue working or return immediately after symptoms have ceased. Shedding of viruses is, however, not restricted to the symptomatic period. It can occur before the onset of symptoms, and may continue on average for at least two weeks post-symptomatically (FAO/WHO, 2008), which further underlines the importance of strict hand hygiene after toilet use at all times. In addition, prolonged shedding has been noticed in immuno-suppressed persons and small children (Murata *et al.*, 2007; Kirkwood and Streitberg, 2008; Henke-Gendo *et al.*, 2009). Food handlers should therefore be particularly careful to practice good hand hygiene before returning to the food preparation area when known to be immuno-suppressed or after caring for young children or being in contact with soiled nappies. Daniels *et al.* (2000) were the first to report an outbreak most likely caused by a food handler caring for a child with watery diarrhoea before returning to the food premises to prepare deli sandwiches.

Some NoV and HAV infections occur without symptoms (Atmar *et al.*, 2008; FAO/WHO, 2008). Recently, the age-adjusted prevalence of asymptomatic

NoV infection in the UK has been shown to be 12%; it is at its highest in children under five, especially in winter (Phillips *et al.*, 2010). It can be assumed that the proportion of food handlers shedding without symptoms is similar to that in the general population. Ozawa *et al.* (2007) reported that NoV was detected in 73% of the stool specimens of symptomatic food handlers and 7% of the stool specimens of asymptomatic food handlers associated with outbreak- and sporadic NoV-associated gastroenteritis. Even in non-outbreak facilities, NoV was detected in 12% of the stool samples derived from asymptomatic food handlers (Okabayashi *et al.*, 2008). Good hand hygiene should therefore always be practiced as it is impossible for food handlers to know when they are contagious and shedding viruses.

11.3 Current knowledge and hygiene practices among food handlers

A major barrier to the spread of disease in food production is the practice of good hand hygiene before handling food, but also after using the toilet, or being in contact with faecal material (e.g., after changing nappies or cleaning toilets) or vomit (FAO/WHO, 2008). Unfortunately, the level of compliance with hand washing practices is low and where it does occur, it is not always performed effectively. Among the 800 establishments in a study by the US Food and Drug Administration (FDA), failure to wash hands properly and adequately was observed in every type of facility investigated, but was noticed in up to 76% of the full service (i.e., non-fast food) restaurants (FDA, 2009). By law, hand washing by food workers is required after many activities, for example, after touching raw food of animal origin, non-food surfaces or the worker's own body parts such as nose or mouth. Precise and detailed observational studies indicate that the number of times hand washing is required, according to the national FDA Food Code (USA), ranges from 9 to 29 times per hour. This number depends greatly on the variety of work activities and the organizational structure within the food establishment. There will be a higher number of hand washing requirements, for example, where food handlers are undertaking many different activities than when handling of money and food is performed by separate workers (Green *et al.*, 2006; Strohbehn *et al.*, 2008). The observational studies (Green *et al.*, 2006; Strohbehn *et al.*, 2008; Lubran *et al.*, 2010), unfortunately, focused on hand hygiene practices within the food handling area only and did not collect data on food employees' behavior outside the food area, that is, hand washing after toilet use. A survey about food hygiene knowledge and behavior in the workplace among catering managers and staff in the United Kingdom (FSA, 2002) showed that more than a third of those questioned (39%) did not wash their hands after visiting the toilet at work, and as it was a self-reporting study, the true proportion may be even higher. Moreover, hand washing methods may be inadequate, for example, omitting the use of soap or omitting to dry hands (Todd *et al.*, 2010a). Reasons given by employees for not washing hands at appropriate times included laziness, time pressure, inadequate facilities and supplies, lack of accountability, risk of skin damage or dermatitis, lack of involvement of managers and co-workers, and organizations that were not supportive of hand washing (Pragle *et al.*, 2007; Todd *et al.*, 2010b). The findings of Green *et al.* (2007) support this list of reasons. In their study, hand washing was more likely to occur in restaurants whose food workers received food safety training, with more than one hand sink available, and with a hand sink in the worker's sight.

Apart from good hand hygiene, hygiene practice in the workplace also includes the exclusion of ill food handlers and efficient sanitation and disinfection. A food worker who has an infectious disease that can be transmitted through food, for example, diarrhoea, should be excluded from food handling or from the premises. Five per cent of the respondents in the Environmental Health Specialists Network telephone survey among food service workers in restaurants indicated that they had worked while sick with vomiting or diarrhoea (Green *et al.*, 2005). In a recent questionnaire in the Netherlands, 20% of the food handlers working in catering companies and 11% of those working in institutional settings responded that they would continue working while sick with diarrhoea (Verhoef, submitted). In the same study, 10% of those working in catering companies and 5% of those working in institutional settings responded that they would continue while sick with vomiting in the future (Verhoef, submitted) despite the sickness benefits that apply in the Netherlands. It is thought that company policies, such as not allowing ill food workers sick leave, result in food workers staying at the premises instead of reporting their illness to the manager (Moe, 2009). In order to examine how obligatory exclusion of ill food handlers works in practice, the FDA looked at whether establishments have written policies on when to exclude food employees based on illnesses or symptoms, when to remove such exclusions (e.g., length of time without symptoms of diarrhoea and vomiting after which employees are allowed to return to work), and also on responsibilities of food employees to report illnesses and symptoms. Formal written employee health policies were most likely to be found in hospital food service settings (46%) and least likely in full service restaurants (13%), but were missing in all types of facility investigated (FDA, 2009).

Further differences between various food service settings were noted, such as the educational level of food workers, management involvement and hand hygiene facilities. Institutional food service facilities placed a higher value on formal food safety training than did restaurants (Strohbehn *et al.*, 2008) and committed significantly fewer violations than restaurants (Kassa *et al.*, 2010). Given the high turnover of largely part-time workers, lack of training in food safety in restaurants is not surprising. Any training provided will generally be on the job.

In order to improve food safety practices in kitchens, accredited training and testing programs have been set up for certification of kitchen managers

(CKM). In a systematic environmental evaluation of food safety in restaurants, the presence of a certified kitchen manager (CKM) was shown to be the major difference between outbreak and non-outbreak restaurants. Thirty-two percent of outbreak restaurants had a CKM against 71% of non-outbreak restaurants (OR 0.2; 95% CI 0.1–0.5). In addition, the presence of a CKM was associated with fewer NoV outbreaks (Hedberg *et al.*, 2006). In an FDA study, the presence of a CKM during regular inspections was shown to positively correlate with higher percentage of observed occurences being in compliance with the appropriate FDA Food Code provision for the data item in full service restaurants and retail food centres for deli food, seafood or produce (FDA, 2009). No significant difference was observed with or without the presence of a CKM for institutional settings or meat and poultry retail stores, probably because these food service types already had a higher IN compliance percentage (FDA, 2009). One of the risk factors for which the presence of a certified food protection manager had the most positive correlation in restaurants and deli retail stores was poor personal hygiene. Other studies support these results by showing that restaurants with a CKM present were less likely to have critical violations than those without a CKM (Cates *et al.*, 2009; Kassa *et al.*, 2010).

Knowledge among food handlers of food safety issues and transmission routes of specific foodborne pathogens has been assessed using questionnaires (Angelillo *et al.*, 2000; DeBess *et al.*, 2009) that targeted common food safety issues but included one question about HAV. In a study by DeBess *et al.* (2009), 47% of the food handlers in Oregon, USA, knew that HAV can be spread through uncooked food via faeces transferred from unwashed hands. In the Angelillo *et al.* study (2000), the proportion of food handlers in Italy who were able to specify a food vehicle for transmission of HAV was significantly higher for those with a higher educational level and with longer food handling experience. Data on knowledge of NoV among food handlers is limited. A recent questionnaire study among food handlers in the Netherlands (Verhoef, submitted), showed that in catering industries 60% of the staff members were aware of the existence of a virus that causes gastroenteritis, but only 20% had ever heard of NoV. In institutional settings, such as hospitals and care homes, awareness among food handlers was much higher – 90% for a virus causing gastroenteritis, and 84% for NoV specifically. Regardless of whether they were working in catering or institutional settings, however, 20% of all participants incorrectly answered that shedding of NoV was restricted to 1–2 days after symptoms have subsided, and 12% incorrectly stated that vomit was not infectious.

11.4 Guidance documents on food hygiene

It is clear that guidance documents on food hygiene and on food transmittable diseases are essential for the training of food handlers as well as companies in the food business. In the early 1960s, the Codex Alimentarius Commission was created by FAO and WHO to develop food standards, guidelines and related publications under the Joint FAO/WHO Food Standards Programme; The main aims of the programme are to protect consumer health, ensure fair practices in the food trade and promote the coordination of all food standards work undertaken by international governmental and non-governmental organizations (FAO/WHO, 2011), and its publications are based on the work of eminent individuals and organizations in food and related fields. Countries have responded by introducing food legislation and Codex-based standards and by establishing or expanding the role of food control agencies to monitor compliance with such regulations (FAO/WHO, 2006).

The basic document for food hygiene is the Codex document 'General Principles of Food Hygiene', CAC/RCP 1-1969, which was adopted in 1969, and has since been revised (FAO/WHO, 2003). It defines hygiene practices during the production, processing, manufacturing, transport and storage of foods that are considered essential for safety and suitability of food for consumption. The document describes measures to be taken at different stages along the chain, and these are summarized below.

The 'General Principles of Food Hygiene' (CAC/RCP 1-1969) (FAO/WHO, 2003) comprises ten sections referred to as Sections I to X. Sections I and II describe the objectives and scope of the document and how it should be used. Section III states that primary production should be managed in a way that ensures that food is safe and suitable for its intended use, including, where necessary, avoiding the use of areas where the environment poses a threat to the safety of food, and controlling contaminants, pests and diseases of animals and plants so as not to pose a threat to food safety. In addition, primary production should adopt practices and measures to ensure food is produced under appropriately hygienic conditions (FAO/WHO, 2003).

Section IV targets premises, equipment and facilities (FAO/WHO, 2003). These should be located, designed and constructed in such a way as to minimize contamination of food and permit appropriate maintenance, cleaning and disinfection. Surfaces and materials, in particular those in contact with food, should be non-toxic. Temperature and humidity should be controlled where appropriate, and there should be effective protection against pests. Section V recommends the control of food hazards through systems such as Hazard Analysis and Critical Control Points (HACCP). The HACCP food safety system itself and guidelines for its application are described in detail in the Annex to the General Principles of Food Hygiene (CAC/RCP 1-1969) (FAO, 1997; FAO/WHO, 2003).Effective control procedures should be implemented and monitored in any parts of a food business where food safety is critical. Periodically, and at times of change, the control procedures should be reviewed.

In addition, each establishment should have effective systems for adequate and appropriate maintenance and cleaning, pest control, waste management and monitoring of effective maintenance and sanitation procedures (Section VI). Food workers themselves should maintain an appropriate degree of personal cleanliness (Section VII) to avoid contamination of food and transmission of illness to consumers. People known or suspected to be suffering from, or to be a carrier of a disease or illness likely to be transmitted through food, should not be allowed to enter any food handling area if there is a likelihood of their contaminating food. Any person so affected should immediately report their illness or symptoms to the management. Examples of symptoms of illnesses given in the document are jaundice, diarrhoea and vomiting (FAO/WHO, 2003).

During transportation (Section VIII) (FAO/WHO, 2003) of food, measures should be taken where necessary to protect food from potential sources of contamination or damage likely to render it unsuitable for consumption, and to provide an environment which effectively controls the growth of pathogenic or spoilage micro-organisms and the production of toxins in food. Products should be labelled (Section IX) (FAO/WHO, 2003) to ensure that adequate and accessible information is available to the next person in the food chain to enable them to handle, store, process, prepare and display the product safely and correctly. Labelling also enables easy identification and recall of a lot or batch if necessary. Finally those engaged in food operations who come directly or indirectly into contact with food should be trained (Section X), and/or instructed in food hygiene to a level appropriate to the operations they are to perform (FAO/WHO, 2003).

11.5 Guidelines on the application of general principles of food hygiene to the control of viruses in food

Following adoption of the 'General Principles of Food Hygiene' (CAC/RCP 1-1969) (FAO/WHO, 2003) codes and standards for specific products or specific pathogen-product combinations have been developed. These are all available on the Codex Alimentarius Commission website (FAO/WHO, 2012a). However since foodborne viral infections remain very common in many parts of the world, the Commission decided to develop specific guidelines for the control of viruses in food, and an Expert Meeting on Viruses in Food was convened by FAO and WHO in May 2007 (FAO/WHO, 2008). The meeting concluded that NoV and HAV are the viruses of greatest concern from a food safety perspective, whereas the virus-commodities combinations of greatest public health concern are NoV and HAV in ready-to-eat-food, bivalve shellfish or fresh produce. Two Physical Working Groups (2009, 2010) and an Electronic Working Group (2011) led by the Netherlands, but with contributions from many other countries, resulted in the Codex 'Proposed Draft Guidelines on the Application of General Principles of Food Hygiene to the Control of Viruses in Food' (FAO/WHO, 2012b), which was agreed on at the 43rd Session of Codex Committee on Food Hygiene in Miami, USA,

December 2011, and at time of writing has been sent for adoption to the Codex Alimentarius Commission.

The introduction to the new Codex guidelines provides background information on foodborne viruses, but also emphasizes that management strategies regarding foodborne viruses and associated illnesses should be different from those for bacterial pathogens. The Codex guidelines for the control of viruses in food do not replace the General Principles of Food Hygiene (CAC/RCP 1-1969), but should be used in conjunction with it, together with the Code of Hygienic Practice for Precooked and Cooked Foods in Mass Catering (CAC/RCP 39-1993), the Code of Practice for Fish and Fishery Products (CAC/RCP 52-2003) and Code of Hygienic Practice for Fresh Fruits and Vegetables (CAC/RCP 53-2003) (FAO/WHO, 2012a). Two supplementary annexes provide specific virus-commodity recommendations on the Control of Hepatitis A Virus (HAV) and Norovirus (NoV) in Bivalve Molluscs (Annex I) and on the Control of Hepatitis A Virus (HAV) and Norovirus (NoV) in Fresh Produce (Annex II) (FAO/WHO, 2012b). References in brackets below are to the specified sections of the new document.

The primary purpose of the Codex guidelines for the control of viruses in food is to give guidance on how to prevent or minimize the presence of human enteric viruses in food, and more specifically of NoV and HAV in food. The guidelines are applicable to all foods, with a focus on ready-to-eat food, from primary production through consumption. The new guidelines should be used as a main reference (FAO/WHO, 2012b) to describe procedural controls to prevent viral contamination of food, but especially by food handlers. Below, references are made to the to the specified sections of the new document in brackets.

Section III identifies different aspects of primary production processes that should be controlled to reduce the chance of viral contamination of food. Sources of contamination can be either water, soil, harvesting containers or utensils contaminated by faeces or vomit or by food handlers. Prior to starting production activities, such potential sources of viral contamination of the environment should be identified. In areas where the presence of viruses may lead to the viral contamination of food, primary food production should consequently not be started or carried on. During growing and harvesting of fresh produce only water that does not compromise food safety should be used, and aquaculture should not be established in areas susceptible to sewage contamination. Hygiene and health requirements for food handlers (described in Section VII, see below) (FAO/WHO, 2012b) should also be followed in primary production to ensure that those who come directly into contact with food during production do not contaminate the product with faecal material or vomit.

The establishment of appropriately designed and adequate hygiene facilities is a prerequisite for hygienic production of food (Section IV) (FAO/WHO, 2012b). Equipment and facilities should be designed, constructed and laid out to ensure that surfaces can be cleaned and disinfected if needed.

Sufficient hygiene and sanitary facilities should be available to food workers to ensure that an appropriate and acceptable degree of personal hygiene can be maintained. These sanitary facilities should be sufficient to cover any seasonal influx of workers. They should also be located in close proximity to the production area, but without direct access to it. There should be adequate facilities for washing and drying hands which can be appropriately cleaned and disinfected and, when feasible, separate facilities should be provided for staff and visitors.

All hand washing facilities should be supplied with hand cleanser (soap) and, where possible, have non-hand operable taps and single-use towels. Hand washing and drying instructions should be displayed for all users of these facilities. Hand washing and drying facilities should be suitably located in food preparation or production areas to ensure food handlers have ready access to them. There should also be hand washing facilities in close proximity to the toilets and positioned so that staff must pass by them before returning to the food handling area.

Widespread contamination of food production premises is likely to occur after vomiting and diarrhoea events and when individuals are shedding viruses. Each establishment should keep a record of regular cleaning and disinfection and have procedures and equipment readily available. As viruses are more resistant than bacteria, specific guidance on disinfectant methods is provided in Section VI of the document (FAO/WHO, 2012b). Cleaning and disinfection should take place immediately after each vomiting event in premises or rooms, but also after reported symptoms of gastroenteritis or of hepatitis in any personnel. All food handling in the same area(s) should be stopped. Disinfection should include all surfaces suspected to be contaminated with viruses, both in the hygiene facilities and toilets and (as a preventive measure) in food production areas (e.g., equipment, utensils, telephones, keyboards, door handles, etc.), as viruses in vomit, aerosols and faecal matter are persistent and can remain infectious for a long period.

Surfaces should always be cleaned prior to disinfection. For surface disinfection, (preferably fresh) solutions of ≥1000 ppm free chlorine should be applied for 5–10 min at room temperature. Alternatively, chlorine dioxide solutions at concentrations of 200 ppm may be used. These solutions are corrosive, and need to be thoroughly rinsed from food contact surfaces afterwards. Alternative disinfection procedures include vaporized hydrogen peroxide (VHP) treatment at >100 ppm for 1 h, which has been shown to be effective against a range of viruses or UV irradiation at >40 mWs/cm^2 (=mJ/cm^2). Most other surface disinfectants are not effective against enteric viruses at manufacturer's recommended concentrations and exposure times (FAO/WHO, 2012b).

Ideally, cleaning and disinfection is performed by a person who is trained in cleaning up infectious material, wearing disposable materials such as gloves, facemasks and aprons. Absorbent material such as paper towels and tissues may be used to limit the spread of contaminated fluids, but should then be properly disposed of, for example, in closed plastic bags, to avoid further contamination of foods, surfaces or personnel.

Food handlers should perform strict personal hygiene, particularly in relation to the prevention of NoV and HAV contamination (Section VII) (FAO/WHO, 2012b). Food handlers with clinical symptoms of gastroenteritis or with symptoms of acute hepatitis should be excluded from food production areas. Worker(s) should leave the food handling area, if possible, before the onset of vomiting or any diarrhoea event and in any case directly after these events. Any person with symptoms of acute hepatitis should seek medical advice.

An employee who has had gastroenteritis should only be allowed to return to work after a period without symptoms of diarrhoea and vomiting. No agreement has been reached on the precise length of this period, but 48 h without symptoms is generally recommended (Moe, 2009). A person who has had hepatitis should only be allowed to return to work after disappearance of jaundice – this is not actually a very strict criterion. More importantly, all staff should receive training on the infectivity, transmission and disinfection of foodborne viruses, and the importance of following strict hand hygiene instructions at all times as shedding of viruses, such as NoV or HAV, may continue for several weeks after symptoms have subsided (FAO/WHO, 2012b).

Moreover, when an employee has symptoms of gastroenteritis or hepatitis, other staff members may be or become (asymptomatically) infected. Similarly, when an employee's family member has symptoms of gastroenteritis or hepatitis, that employee may be (asymptomatically) infected, and/or act as a carrier of infectious virus. Compliance with strict hand hygiene measures is particularly important in these situations to reduce the risk of further spread of the illness. Vaccination of food handlers against hepatitis A should be recommended where necessary to reduce the risk of viral contamination of the food, taking into account the epidemiological situation and/or immune status of the local population. As visitors may introduce viruses into the food handling areas, non-authorized persons and, as far as possible, children should not be present in areas where food is grown, harvested, stored or prepared (FAO/WHO, 2012b).

The most effective way of preventing the spread of viruses is thorough hand washing. Hands should be lathered with soap and then washed with clean running water. If gloves are used, a procedure for glove use should be developed and followed, for example, when gloved hands have been in contact with potentially contaminated items new gloves should be put on before preparing food. Clothes of food handlers who have been infected, or suspected to have been infected, should be washed. It has been shown that conventional household washing detergents have a good virucidal efficiency at 40°C (FAO/WHO, 2012b).

Food handlers were shown to be unaware of or less familiar with controls specific to enteric viruses. Food handlers who come directly or indirectly into contact with foods should therefore receive specific training on this subject at a level appropriate to their job (Section X) (FAO/WHO, 2012b). Training programmes should include the following information:

- The potential for food to be a vehicle of virus transmission.
- The potential sources and routes of transmission of human enteric viruses.
- The potential for persistence of infectious virus in/or contaminated food and food production settings.
- The incubation periods of foodborne viruses, specifically NoV and HAV.
- The duration of virus shedding during and even after recovery from clinical symptoms and the possibility of pre- and post-symptomatic shedding.
- The infectivity of vomit.
- Procedures for cleaning and disinfection of contaminated surfaces; proper hand washing practices and the importance of strict compliance with hand washing instructions at all times, particularly after being in contact with faecal or vomit matter.
- Proper hand washing instructions given to each new employee.
- The possibility that if one staff member or household member has a viral illness, other staff members or household members may also be infected.
- The need for children, as far as possible, to stay away from food growing fields and food preparation areas in HAV-endemic areas (since in endemic areas children are a primary source of the virus).
- Procedures for the disposal of contaminated food items.

Managers are responsible for educating and training their staff, and for the proper operation of both cleaning and disinfection programmes. Employers and management are responsible for monitoring to ensure that employees are undertaking good hygiene practices; this should include regular observation of hand washing prior to entering food handling areas. Employees are responsible for informing the supervisor or employer if they are ill with diarrhoea or vomiting, or have symptoms indicative of hepatitis or gastrointestinal illnesses. It is also the responsibility of all employees to adhere to strict hand washing instructions after returning from the toilet or after being in contact with faecal or vomit matter. Extensive training and instructions should be given to all new staff, but also to inspectors or other relevant authorities who inspect fields, post-harvest processing plants, and eating facilities. Incorporation of these instructions into National Codes of Hygienic Practice would be advisable.

The control of human enteric viruses such as NoV and HAV in food (Section V) (FAO/WHO, 2012b) typically requires hygiene control systems, such as good hygiene practice and standard sanitation operating procedures, to be strictly applied. Table 11.1 summarizes the key aspects of the hygiene control systems recommended in the guidelines.

Table 11.1 Key aspects of hygiene control systems to control viruses in food

Risks posed by food possibly contaminated by viruses

- Any food possibly contaminated by vomit particles or by aerosols containing vomit particles should be disposed of.
- Any food handled by an ill person should be evaluated to determine the need to dispose of it, having the incubation periods for NoV (1–2 days) and HAV (at least 2 weeks) in mind.
- If an outbreak has been traced back to an establishment, the necessary steps should be taken to find the source of contamination, to eliminate the virus, and to avoid future outbreaks.
- Based on the determined level of risk associated with the presence of viruses in a given food product, a decision may be taken to recall the contaminated product from the market. The need for public information and communicated warnings should be considered.

Effects of food processing steps on viral infectivity

- Cooling and freezing processes should not be considered suitable for the control of foodborne viruses.
- The effects of heat treatment on virus infectivity in foods are highly dependent on virus (sub)-type, food matrix and the initial level of viral contaminants. Cooking procedures in which an internal temperature of the food reaches at least 90 °C for 90 s are considered adequate treatments. Conventional pasteurization may not adequately inactivate viruses in a contaminated food. Commercial canning is considered an adequate treatment.
- Other food processing (combinations) should be subject to rigorous validation to ensure consumer protection.
- Various types of packaging are not effective against human viruses.

Management responsibilities to prevent contamination of food by viruses

- Preferably use raw ingredients from suppliers or production plants with an adequate food safety management system as raw ingredients contaminated with viruses may lead to contamination of food handlers' hands, other foods, or food contact surfaces.
- Managers and supervisors should understand the importance of applying good hygiene practices, including the availability of adequate hygiene facilities; compliance with hand washing instructions; and how to clean and disinfect surfaces.
- Exclusion from the premises of food handlers or any persons, including children, with symptoms of gastroenteritis or acute hepatitis or those recovering from these infections.
- Control procedures used for viruses should be monitored to ensure their continuing effectiveness.

11.6 Designing Hazard Analysis and Critical Control Points (HACCP) with the viruses NoV and HAV in mind

The HACCP system identifies specific hazards and measures for their control to ensure the safety of food (FAO, 1997). Any HACCP system is capable of accommodating change, such as advances in equipment design or processing procedures, or technological developments. HACCP can be applied throughout the food chain from primary production to final consumption and its implementation should be guided by scientific evidence of risks to human health. HACCP is a tool to assess hazards and establish control systems that focus on prevention rather than relying mainly on end-product testing.

Although international standard detection methods for NoV and HAV RNA in shellfish and fresh produce, and on food surfaces for end products are expected in 2013 (Anonymous, 2013a, b), end-product testing is both time-consuming and expensive. Moreover, to date no critical or safe limits have been set internationally, although the issue is under discussion (EFSA, 2012). A further complication is that the commonly used molecular detection methods do not discriminate between the presence of infectious and non-infectious virus particles in food, and NoV and HAV cannot be reliably cultured *in vitro*. This makes it hard to state whether food that tests positive for NoV or HAV RNA poses a risk to human health (although the presence of NoV and HAV in the food chain is clearly undesirable). The absence of methods for assessing the level of inactivation of viruses in food has made it impossible to directly evaluate the effectiveness of (new) food processing steps in reducing virus infectivity. This has led to the use of surrogate viruses, for example, feline calicivirus or murine calicivirus (see Chapter 12). Surrogates, however, will not always mimic the resistance of the subject foodborne viruses. With these uncertainties, it is particularly challenging to design an HACCP system that focuses on prevention of viral contamination of food.

Each HACCP system should follow seven Principles (FAO, 1997). The first two principles describe hazard analysis and determination of the critical control points. In this case, the most important hazards are NoV and HAV in food (FAO/WHO, 2008). Zoonotic transmission of hepatitis E virus (HEV) may occur after consumption of raw or undercooked products of animal origin (EFSA, 2011); the risks of HEV will be discussed in Chapter 19. Hazard analysis (Principle 1) is the process of collecting and evaluating information on NoV and HAV in food and conditions leading to their presence. With this information one can decide which conditions are significant for food safety and which therefore should be addressed in the HACCP plan. This process leads to determination of critical control points (CCPs) (Principle 2). CCPs are defined as stages at which control can be applied and which are essential to prevent or eliminate a food safety hazard or reduce the hazard to an acceptable level. For foodborne viruses there are several stages at which food can become contaminated and at which prevention therefore should take place. However, once contaminated, elimination of the hazard is almost impossible except for a treatment requiring the internal temperature of the food to reach at least 90°C for 90 s (FAO/WHO, 2008). Verification of hazard reduction to an acceptable level, as is possible for bacterial pathogens, is hard to achieve for NoV and HAV, due to the lack of infectivity assays and because acceptable levels have not yet been established. Currently, the HACCP system for the control of viruses can only identify the stages where contamination of food should be prevented and regard these stages as CCPs. Examples may be the exclusive use of water that does not compromise food safety in primary production; the exclusion of ill food handlers from farm to fork; and good hand hygiene practices at all times, but particularly after being in contact with faecal material or vomit, or when returning to work after illness.

The next three principles of a HACCP system deal with the control loop, that is, the critical limit(s) (Principle 3); the system for monitoring CCPs (Principle 4); and corrective actions if monitoring indicates that a particular CCP is not under control (Principle 5). A critical limit in the HACCP system is defined as a criterion that separates acceptability from unacceptability (Principle 3). Not being able to refer to acceptable levels of NoV and HAV in food at this moment other than being absent, we can only refer to acceptable processes that prevent food becoming contaminated. In the examples described above, production of food can only continue if water used for production of food is not contaminated with human excreta, all sick members of staff report their illness and stay away from the premises, and good hygiene practices are used at all times. Under the monitoring system, water would be tested for its quality, employees' knowledge of foodborne illnesses would be checked, and compliance with good hand hygiene would be monitored. If, in the examples above, the water is not safe for use, a sick employee is reported to be present on the premises or hand hygiene is not being satisfactorily practiced, the management must take corrective actions (Principle 5). To increase the control of hand hygiene compliance, some food industries have, for example, incorporated hand hygiene stations that only allow staff to enter the food processing area if their hands have been washed sufficiently.

The last two principles of a HACCP system relate to documentation and verification, that is, verification procedures to confirm that the HACCP system is working effectively (Principle 6) and to establish documentation concerning all procedures and records appropriate to these principles and their application (Principle 7).

11.7 Conclusion and future trends

The main requirements for hygienic production of food were described several decades ago in the General Principles of Food Hygiene (CAC/RCP 1-1969) (FAO/WHO, 2003). Recently a Codex document on the Application of the General Principles of Food Hygiene to the Control of Viruses in Food (FAO/WHO, 2012b) has been added. Studies reveal a lack of knowledge on the former document and also non-compliance with hand hygiene practices among food handlers., Ensuring familiarity among food handlers and their managers, but also food branches, with the contents of the new Codex document and compliance with these new guidelines (FAO/WHO, 2012b) will be challenging.

Reasons given by food handlers for not complying with good hand hygiene frequently include lack of time and skin irritation (Pragle *et al.*, 2007). If all staff members adhered to good hand hygiene practice after each visit to the toilet – only a few on a working day – viral spread in food preparation areas would be greatly reduced. Moreover, setting the same priority to the cleaning of frequently touched sites, such as refrigerator and freezer handles, or kitchen equipment, as to food contact surfaces might offer a useful adjunct to the current preoccupation with hand hygiene and lead to a reduction in the transmission of foodborne viruses through contaminated surfaces. This would be in line with a recently suggested approach to reduce hospital-acquired infections of prioritizing the cleaning of hand-touch sites near patients, such as bed rails, bedside lockers, or nurse-call buttons, in addition to floors and toilets (Dancer, 2009). Warnings of any sudden rise in national NoV activity issued by public health authorities to the food industry, may reduce foodborne illnesses by drawing attention to the need for compliance with good hand hygiene and the exclusion of ill members of staff. HAV vaccination should be mandatory for non-immune employees planning to travel to HAV- endemic areas. Reducing the potential for food handlers to contaminate foods would be a significant step towards reducing the public health challenge of foodborne viruses.

11.8 Acknowledgement

I thank the Food and Agriculture Organization of the United Nations for their permission to reproduce and summarize parts of the recommended international code of practice, General Principles of Food Hygiene, CAC/RCP 1 and of the Draft Guidelines on the Application of General Principles of Food Hygiene to the Control of Viruses in Food.

11.9 References

ANGELILLO I F, VIGGIANI N M, RIZZO L, BIANCO A, (2000), 'Food handlers and foodborne diseases: knowledge, attitudes, and reported behavior in Italy', *J Food Prot*, **63**, 381–5.

ANONYMOUS, (2013a), 'Microbiology of food and animal feed – Horizontal method for determination of hepatitis A virus and norovirus in food using real-time RT-PCR – part 1: method for quantification', *CEN ISO/TS 15216–1 (draft).*

ANONYMOUS, (2013b), 'Microbiology of food and animal feed – horizontal method for determination of hepatitis A virus and norovirus in food using real-time RT-PCR – part 2: method for qualitative detection', *CEN ISO/TS 15216–2 (draft).*

ATMAR R L, OPEKUN A R, GILGER M A, ESTES M K, CRAWFORD S E, NEILL F H and GRAHAM D Y, (2008), 'Norwalk virus shedding after experimental human infection', *Emerg Infect Dis*, **14**, 1553–7.

BARRABEIG I, ROVIRA A, BUESA J, BARTOLOMÉ R, PINTÓ R, PRELLEZO H and DOMÍNGUEZ A, (2010), Foodborne norovirus outbreak: the role of an asymptomatic food handler', *BMC Infect Dis*, **10**, 269. Doi: 10.1186/1471-2334-10-269.

BIDAWID S, FARBER J M and SATTAR S A, (2000), 'Contamination of foods by food handlers: experiments on hepatitis A virus transfer to food and its interruption', *Appl Environ Microbiol*, **66**, 2759–63.

BIDAWID S, MALIK N, ADEGBUNRIN O, SATTAR S A and FARBER J M, (2004), 'Norovirus cross-contamination during food handling and interruption of virus transfer by hand antisepsis: experiments with feline calicivirus as a surrogate', *J Food Prot*, **67**, 103–9.

BOONE S A and GERBA CP, (2007), 'Significance of fomites in the spread of respiratory and enteric viral disease', *Appl Environ Microbiol*, **73**, 1687–96.

BOXMAN I L, TILBURG J J, TE LOEKE N A, VENNEMA H, DE BOER E and KOOPMANS M, (2007), 'An efficient and rapid method for recovery of norovirus from food associated with outbreaks of gastroenteritis', *J Food Prot*, **70**, 504–8.

BOXMAN I L, DIJKMAN R, TE LOEKE N A, HÄGELE G, TILBURG J J, VENNEMA H and KOOPMANS M, (2009a), 'Environmental swabs as a tool in norovirus outbreak investigation, including outbreaks on cruise ships', *J Food Prot*, **72**, 111–19.

BOXMAN I, DIJKMAN R, VERHOEF L, MAAT A, VAN DIJK G, VENNEMA H and KOOPMANS M, (2009b), 'Norovirus on swabs taken from hands illustrate route of transmission: a case study', *J Food Prot*, **72**, 1753–5.

BOXMAN I L, VERHOEF L, DIJKMAN R, HÄGELE G, TE LOEKE N A and KOOPMANS M, (2011), 'Year-round prevalence of norovirus in the environment of catering companies without a recently reported outbreak of gastroenteritis', *Appl Environ Microbiol*, **77**, 2968–74.

CATES S C, MUTH M K, KARNS S A, PENNE M A, STONE C N, HARRISON J E and RADKE V J, (2009), 'Certified kitchen managers: do they improve restaurant inspection outcomes?', *J Food Prot*, **72**, 384–91.

DALLING J, (2004), 'A review of environmental contamination during outbreaks of Norwalk-like virus', *Br J Infect Contr*, **5**, 9–13.

DANCER S J, (2009), 'The role of environmental cleaning in the control of hospital-acquired infection', *J Hosp Infect*, **73**, 378–85.

DANIELS N A, BERGMIRE-SWEAT D A, SCHWAB K J, HENDRICKS K A, REDDY S, ROWE S M, FANKHAUSER R L, MONROE S S, ATMAR R L, GLASS R I and MEAD P, (2000), 'A foodborne outbreak of gastroenteritis associated with Norwalk-like viruses: first molecular traceback to deli sandwiches contaminated during preparation', *J Infect Dis*, **181**, 1467–70.

DEBESS E E, PIPPERT E, ANGULO F J and CIESLAK P R, (2009), 'Food handler assessment in Oregon', *Foodborne Pathog Dis*, **6**, 329–35.

D'SOUZA D H, SAIR A, WILLIAMS K, PAPAFRAGKOU E, JEAN J, MOORE C and JAYKUS L, (2006), 'Persistence of caliciviruses on environmental surfaces and their transfer to food', *Int J Food Microbiol*, **108**, 84–91.

EFSA (EUROPEAN FOOD SAFETY AUTHORITY), (2011) 'Scientific Opinion on an update on the present knowledge on the occurrence and control of foodborne viruses', *EFSA J*, **9**, 2190 (96 pp). Doi: 10.2903/j.efsa.2011.2190. Available from http://www.efsa.europa.eu/en/efsajournal/doc/2190.pdf (Accessed 17 February 2012).

EFSA (European Food Safety Authority), Panel on Biological Hazards (BIOHAZ), (2012), 'Scientific opinion on Norovirus (NoV) in oysters: methods, limits and control options', *EFSA J*, **10**, 2500. (39 pp) Doi: 10.2903/j.efsa.2012.2500. Available from http://www.efsa.europa.eu/en/efsajournal/pub/2500.htm (Accessed 17 February 2012).

FAO (Food and Agriculture Organization of the United Nations), (1997), Hazard analysis and critical control point (HACCP) system and guidelines for its application.

Annex to CAC/RCP 1-1969'. Available from http://www.fao.org/docrep/005/y1579e/y1579e03.htm (Accessed 17 February 2012).

FAO/WHO (FOOD AND AGRICULTURE ORGANIZATION OF THE UNITED NATIONS/WORLD HEALTH ORGANIZATION), (2003), 'General Principles of Food Hygiene, CAC/RCP-1'. Available from http://www.codexalimentarius.net/web/more_info.jsp?id_sta=23 (Accessed 17 February 2012).

FAO/WHO (FOOD AND AGRICULTURE ORGANIZATION OF THE UNITED NATIONS/WORLD HEALTH ORGANIZATION), (2006), 'Understanding the Codex Alimentarius'. ISBN: 978-92-5-105614-1. Available from ftp://ftp.fao.org/codex/Publications/understanding/Understanding_EN.pdf (Accessed 17 February 2012).

FAO/WHO (FOOD AND AGRICULTURE ORGANIZATION OF THE UNITED NATIONS/WORLD HEALTH ORGANIZATION), (2008), 'Viruses in food: scientific advice to support risk management activities: meeting report', *Microbiological Risk Assessment Series*, No. 13. Available from http://www.who.int/foodsafety/publications/micro/mra13/en/index.html (Accessed 17 February 2012).

FAO/WHO (FOOD AND AGRICULTURE ORGANIZATION OF THE UNITED NATIONS/WORLD HEALTH ORGANIZATION), (2011), 'FAO/WHO Food standards Codex Alimentarius'. Available from http://www.codexalimentarius.net/web/index_en.jsp (Accessed 17 February 2012).

FAO/WHO (FOOD AND AGRICULTURE ORGANIZATION OF THE UNITED NATIONS/WORLD HEALTH ORGANIZATION), (2012a), 'FAO/WHO Codex Alimentarius International Food Standards'. Available from http://www.codexalimentarius.org/standards/list-of-standards (Accessed 3 May 2012).

FAO/WHO (FOOD AND AGRICULTURE ORGANIZATION OF THE UNITED NATIONS/WORLD HEALTH ORGANIZATION), (2012b), 'Guidelines on the Application of General Principles of Food Hygiene to the Control of Viruses in Food (CAC/GL 79–2012)'.

FDA (FOOD AND DRUG ADMINISTRATION) NATIONAL RETAIL FOOD TEAM, (2009), 'FDA Report on the Occurrence of Foodborne Illness Risk Factors in Selected Institutional Foodservice, Restaurant, and Retail Food Store Facility Types'. Available from: http://www.fda.gov/downloads/Food/FoodSafety/RetailFoodProtection/FoodborneIllnessandRiskFactorReduction/RetailFoodRiskFactorStudies/UCM224682.pdf (Accessed 17 February 2012).

FIORE A, (2004), 'Hepatitis A transmitted by food', *Clin Inf Dis*, **38**, 705–15.

FSA (FOOD STANDARDS AGENCY UK), (2002), 'Why are you focusing on handwashing after using the toilet?'. Available from http://www.food.gov.uk/multimedia/faq/fhc-2qanda/ (Accessed 17 February 2012).

GREEN L, SELMAN C, BANERJEE A, MARCUS R, MEDUS C, ANGULO FJ, RADKE V, BUCHANAN S and EHS-NET WORKING GROUP, (2005), 'Food service workers' self-reported food preparation practices: an EHS-Net study', *Int J Hyg Environ Health*, **208**, 27–35.

GREEN L R, SELMAN C A, RADKE V, RIPLEY D, MACK JC, REIMANN DW, STIGGER T, MOTSINGER M and BUSHNELL L, (2006), 'Food worker hand washing practices: an observation study', *J Food Prot*, **69**, 2417–23.

GREEN L R, RADKE V, MASON R, BUSHNELL L, REIMANN D W, MACK J C, MOTSINGER M D, STIGGER T and SELMAN CA, (2007), 'Factors related to food worker hand hygiene practices', *J Food Prot*, **70**, 661–6.

HEDBERG C W, SMITH S J, KIRKLAND E, RADKE V, JONES T F, SELMAN C A and EHS-NET WORKING GROUP, (2006), 'Systematic environmental evaluations to identify food safety differences between outbreak and nonoutbreak restaurants', *J Food Prot*, **69**, 2697–702.

HENKE-GENDO C, HARSTE G, JUERGENS-SAATHOFF B, MATTNER F, DEPPE H and HEIM A, (2009), 'New real-time PCR detects prolonged norovirus excretion in highly immunosuppressed patients and children', *J Clin Microbiol*, **47**, 2855–62.

KASSA H, HARRINGTON B, BISESI M and KHUDER S, (2001), 'Comparison of microbiological evaluation of selected kitchen areas with visual inspection for preventing potential risk of food borne outbreaks in food service operations', *J Food Prot*, **64**, 509–13.

KASSA H, SILVERMAN G S and BAROUDI K (2010), 'Effect of a manager training and certification program on food safety and hygiene in food service operations', *Environ Health Insights*, **4**, 13–20.

KIRKWOOD C D and STREITBERG R, (2008), 'Calicivirus shedding in children after recovery from diarrhoeal disease', *J Clin Virol*, **43**, 346–8.

KOOPMANS M, VON BONSDORFF CH, VINJÉ J, DE MEDICI D and MONROE S, (2002), 'Foodborne viruses', *FEMS Microbiol Rev*, **26**, 187–205.

LOPMAN B, GASTAÑADUY P, WOO PARK G, HALL AJ, PARASHAR UD and VINJÉ J, (2011), 'Environmental transmission of norovirus gastroenteritis', *Curr Opin Virol*, **2**, 1–7. Doi: 10.1016/j.coviro.2011.11.005; Doi: 10.1016/j.coviro.2011.11.005.

LUBRAN M B, POUILLOT R, BOHM S, CALVEY E M, MENG J and DENNIS S, (2010), 'Observational study of food safety practices in retail deli departments', *J Food Prot*, **73**, 1849–57.

MOE C L, (2009), 'Preventing norovirus transmission: how should we handle food handlers?', *Clin Infect Dis*, **48**, 38–40.

MURATA T, KATSUSHIMA N, MIZUTA K, MURAKI Y, HONGO S and MATSUZAKI Y, (2007), 'Prolonged norovirus shedding in infants <or=6 months of age with gastroenteritis', *Pediatr Infect Dis J*, **26**, 46–9.

NICOLAY N, MCDERMOTT, R, KELLY M, GORBY M, PRENDERGAST T, TUITE G, COUGHLAN S, MCKEOWN P and SAYERS G, (2011) 'Potential role of asymptomatic kitchen food handlers during a food-borne outbreak of norovirus infection, Dublin, Ireland, march 2009', *Euro Surveill*, **16**(30), pii=19931.

OKABAYASHI T, YOKOTA S, OHKOSHI Y, OHUCHI H, YOSHIDA Y, KIKUCHI M, YANO K and FUJII N, (2008), 'Occurrence of norovirus infections unrelated to norovirus outbreaks in an asymptomatic food handler population', *J Clin Microbiol*, **46**, 1985–8.

OZAWA K, OKA T, TAKEDA N and HANSMAN G S, (2007), 'Norovirus infections in symptomatic and asymptomatic food handlers in Japan', *J Clin Microbiol*, **45**, 3996–4005.

PHILLIPS G, TAM CC, RODRIGUES LC and LOPMAN B, (2010), 'Risk factors for symptomatic and asymptomatic norovirus infection in the community', *Epidemiol Infect*, **139**, 1676–86.

PRAGLE A S, HARDING A K and MACK J C, (2007), 'Food workers' perspectives on handwashing behaviors and barriers in the restaurant environment', *J Environ Health*, **69**, 27–32.

SCHWAB K J, NEILL F H, FANKHAUSER R L, DANIELS N A, MONROE S S, BERGMIRE-SWEAT D A, ESTES MK and ATMAR R L, (2000), 'Development of methods to detect "Norwalk-like viruses" (NLVs) and hepatitis A virus in delicatessen foods: application to a food-borne NLV outbreak', *Appl Environ Microbiol*, **66**, 213–18.

STROHBEHN C, SNEED J, PAEZ P and MEYER J, (2008), 'Hand washing frequencies and procedures used in retail food services', *J Food Prot*, **71**, 1641–50.

TEUNIS P F, MOE C L, LIU P, MILLER S E, LINDESMITH L, BARIC R S, LE PENDU J and CALDERON R L, (2008), 'Norwalk virus: how infectious is it?', *J Med Virol*, **80**, 1468–76.

TODD E C, GREIG J D, BARTLESON C A and MICHAELS B S, (2007), 'Outbreaks where food workers have been implicated in the spread of foodborne disease. Part 3. Factors contributing to outbreaks and description of outbreak categories', *J Food Prot*, **70**, 2199–217.

TODD E C, MICHAELS B S, SMITH D, GREIG J D and BARTLESON C A, (2010a), 'Outbreaks where food workers have been implicated in the spread of foodborne disease. Part 9. Washing and drying of hands to reduce microbial contamination', *J Food Prot*, **73**, 1937–55.

TODD E C, GREIG J D, MICHAELS B S, BARTLESON C A, SMITH D and HOLAH J, (2010b), 'Outbreaks where food workers have been implicated in the spread of foodborne disease. Part 11. Use of antiseptics and sanitizers in community settings and issues of hand hygiene compliance in health care and food industries', *J Food Prot*, **73**, 2306–20.

WU H M, FORNEK M, SCHWAB K J, CHAPIN AR, GIBSON K, SCHWAB E, SPENCER C and HENNING K, (2005), 'A norovirus outbreak at a long-term care facility: the role of environmental surface contamination', *Infect Control Hosp Epidemiol*, **26**, 802–10.

12

Foodborne virus inactivation by thermal and non-thermal processes

L. Baert, Ghent University, Belgium

DOI: 10.1533/9780857098870.3.237

Abstract: Heating, high hydrostatic pressure processing and irradiation are preservation methods that can be used to establish microbial inactivation. The efficacy of these techniques to inactivate foodborne viruses is discussed in this chapter. Cultivable viral strains and representative surrogate viruses generally used to estimate the behavior of their related human-infecting strains are summarized and their appropriateness discussed.

Key words: inactivation of foodborne viruses, heating, irradiation, high hydrostatic pressure, surrogates.

12.1 Introduction

In contrast to most microbiological agents, viruses cannot grow on food and thus the contamination level cannot increase during processing or storage, but due to their low infectious dose their survival should be considered a risk (Carter, 2005; Koopmans and Duizer, 2004). Several studies indicate that chilling or freezing fresh produce (Baert *et al.*, 2008b; Butot *et al.*, 2008; Dawson *et al.*, 2005; Hewitt and Greening, 2004) and other foods (Mattison *et al.*, 2007) can allow survival of a considerable number of viruses before deterioration of the food. Enteric viruses are also reported to be able to persist at acid conditions (pH 2) (Cannon *et al.*, 2006) or modified atmosphere packaging (Bidawid *et al.*, 2001).

Food preservation methods which have a greater impact on micro-organisms are required due to (i) the long-term survival of viruses under these conditions and (ii) the low infectious dose. Heating, high hydrostatic pressure processing and irradiation are preservation methods that can be used to establish microbial inactivation. The efficacy of these techniques to inactivate viruses is discussed in this chapter.

Because the majority of the human-infecting viruses present on food cannot be cultivated in the laboratory, inactivation rates are mainly obtained by the use of surrogate viruses. Cultivable viral strains and representative surrogate viruses generally used to estimate the behavior of their related human-infecting strains are summarized and their appropriateness is discussed.

12.2 Thermal processes

The inactivating effect of thermal processes on foodborne viruses is presented in Table 12.1. Duizer *et al.* (2004) observed similar inactivation rates of feline calicivirus (FCV) and canine calicivirus (CaCV) at temperatures ranging from 37°C to 100°C. Similar thermal inactivation rates at 63°C and 72°C were noted for FCV and murine norovirus 1 (MNV-1) (Cannon *et al.*, 2006). Different reductions of FCV for the same time-temperature combination (Table 12.1) were achieved by Doultree *et al.* (1999) and Buckow *et al.* (2008). The experimental set-up, 100 μL virus suspension heated in Sarstedt tubes (Doultree *et al.*, 1999) or 10 μL virus suspension added to 90 μL preheated medium in 0.2 ml reaction tubes (Buckow *et al.*, 2008), was likely to be responsible for the differences in heat inactivation rates.

Bidawid *et al.* (2000b) studied heat inactivation of hepatitis A virus (HAV) in sterile skimmed milk (0% fat), homogenized milk (3.5% fat) and table cream (18% fat). At 71°C, exposure of 0.16, 0.18 and 0.52 min were needed respectively in skimmed milk, homogenized milk and cream to reduce HAV by 1 log, whereas 4 log reduction required 6.55 (skim), 8.31 (homogenized) and 12.67 (cream) min (Table 12.1). Longer heat treatment was needed in cream to achieve similar inactivation of HAV compared to milk. The high fat content presumably protected HAV towards heat. In addition, the loss of infectivity of poliovirus was lower in milk after 0.25 min exposure to 72°C than in water, proving the protective effect of milk constituents (Strazynski *et al.*, 2002). Furthermore, a protective effect of the mussel matrix was indicated by Croci *et al.* (1999). A 2 log reduction of HAV was found in a mussel homogenate after 10 min exposure at 60°C or after 3 min at 80°C, whereas at least 4.6 log reduction was induced after the same treatments in cell culture medium (Croci *et al.*, 1999).

Bidawid *et al.* (2000b) noted a nonlinear decline of HAV in milk between 65°C and 75°C. The observed rapid inactivation in the initial phase could be explained by freely suspended virus particles along with virus particles present at the outside of viral aggregates. It is assumed that a subpopulation of a viral aggregate may be more resistant as presumably all virions must be inactivated before the aggregate is non-infective as a whole (Grant, 1995). Virus aggregation is influenced by the ionic composition of the medium, pH

Table 12.1 The efficacy of thermal processes in the inactivation of foodborne viruses

Virus		Heat treatment	Matrix	Log_{10} reduction[a]	Reference
Reoviridae	Rotavirus	60°C 10 min	Cell culture medium	7	Mahony *et al.*, 2000
Picornaviridae	HAV[b]	85°C < 0.5 min	Milk	5	Bidawid *et al.*, 2000b
		71°C 6.55 min (skimmed); 8.31 min (homogenized); 12.67 min (cream)	Milk	4; 4; 4	
	HAV	85°C 0.96 min (28°Brix); 4.98 min (52°Brix)	1 g mashed strawberry	1; 1	Deboosere *et al.*, 2004
		80°C 8.94 min (52°Brix)		1	
	HAV	95°C 2.5 min; 75°C 2.5 min	15 g basil	> 3; 1.87	Butot *et al.*, 2009
	HAV	60°C 10 min; 80°C 3 min	4 ml virus suspension	> 4.6; > 4.6	Croci *et al.*, 1999
		60°C 10 min; 80°C 3 min	4 ml shellfish homogenate	2; 2	
	HAV	60°C 74.6 min	5 ml PBS	1	Gibson and Schwab, 2011
	HAV	90°C 3 min	Clam digestive gland	4.75	Sow *et al.*, 2011
	HAV	62.8°C 30 min; 71.6°C 0.25 min	Milk	3; 2	Parry and Mortimer, 1984
	HAV	Steaming 6 min	Mussels	> 3	Harlow *et al.*, 2011
	HAV	63°C 1 min; 72°C 1 min	Milk	1.60; 2.22	Hewitt *et al.*, 2009
	HAV	63°C 30 min	Milk	3	Mariam and Cliver, 2000b
	Poliovirus	72°C 0.25 min; 72°C 0.5 min	Milk	0.56; >5	Strazynski *et al.*, 2002
	Poliovirus	Steaming 30 min	Oysters	2	Di Girolamo *et al.*, 1970

(*Continued*)

Table 12.1 Continued

Virus		Heat treatment	Matrix	Log_{10} reduction[a]	Reference
Caliciviridae	FCV[c], CaCV[d]	71.3°C 1 min	Cell culture medium	3	Duizer *et al.*, 2004
	FCV, CaCV	37°C 24 h; 56°C 8 min	Cell culture medium	3; 3	
	FCV	0.5 min immersion of 6–8 cockles in boiling water	Cockles	1.7	Slomka and Appleton, 1998
	FCV	56°C 3 min; 56°C 60 min	Cell culture medium	No red[g]; 7.5	Doultree *et al.*, 1999
		70°C 1 min; 3 min; 5 min	Cell culture medium	3; 6.5; 7.5	
		Boiling 1 min	Cell culture medium	7.5	
	FCV	70°C 1.5 min	Cell culture medium	6	Buckow *et al.*, 2008
	FCV	63°C 0.41 min; 72°C 0.12 min	Cell culture medium	1; 1	Cannon *et al.*, 2006
	FCV	60°C 14.1 min	5 ml PBS	1	Gibson *et al.*, 2011
	FCV	95°C 2.5 min; 75°C 2.5 min	15 g basil	>4; 3.98	Butot *et al.*, 2009
	NoV[e]	60°C 30 min		Incomplete	Dolin *et al.*, 1972
	MNV-1[f]	80°C 2.5 min	Cell culture medium	6.5	Baert *et al.*, 2008c
	MNV-1	65°C 0.5 min; 75°C 0.25 min	Raspberry puree (9.2°Brix)	1.86; 2.81	Baert *et al.*, 2008a
	MNV-1	80°C 1 min	50 g spinach leaves	> 2.4	Baert *et al.*, 2008b
	MNV-1	63°C 1 min; 72°C 1 min	Milk	1.35; >3.5	Hewitt *et al.*, 2009
	MNV-1	63°C 0.44 min; 72°C 0.17 min	Cell culture medium	1; 1	Cannon *et al.*, 2006
	MNV-1	60°C 13.7 min	5 ml PBS	1	Gibson *et al.*, 2011
	MNV-1	90°C 1.5 min	Clam digestive gland	5.47	Sow *et al.*, 2011

Table redrafted from Baert *et al.* (2009).
[a] Log10 reduction represents the reduction in infectivity; [b] HAV: hepatitis A virus; [c] FCV: feline calicivirus; [d] CaCV: canine calicivirus; [e] NoV: norovirus; [f] MNV-1: murine norovirus 1; [g] No red: no reduction; [h] FCS: fetal calf serum.

and the isoelectric point of the virus. Conditions inducing aggregation will consequently be influenced by the food matrix and virus type.

Thermoresistance of HAV inoculated in synthetic media mimicking chemical characteristics of strawberry mashes was investigated by Deboosere *et al.* (2004). These experiments showed that sucrose concentration (indicated as Brix value) and pH affected HAV heat inactivation. In 1 g strawberry mash, HAV was lowered by 1 log after a heat treatment consisting of 2 min to reach 85°C followed by 0.96 min at 85°C (Deboosere *et al.*, 2004). It appeared that a pH increase (from 2.5 to 3.3) resulted in increased heat resistance (Deboosere *et al.*, 2010). MNV-1 showed a reduction of 2.81 log after exposure to 75°C for 0.25 min in 10 g of preheated raspberry puree (Baert *et al.*, 2008a). Although HAV seems to be more resistant than MNV-1, the composition of strawberries – their higher sucrose concentration (indicated by Brix in Table 12.1) and possibly other protective components – the initial warming-up period of 2 min and the higher virus inoculum (10^7 PFU/ml HAV *vs* 10^5 PFU/ml MNV-1), could be partially responsible for this difference in inactivation rate.

Slomka and Appleton (1998) investigated the inactivation of FCV by immersion of FCV-contaminated cockles in boiling water for 0.5 min and found 1.7 log reduction of the virus. At that time, the internal temperature of the cockles reached approximately 60°C. After 1 min, the internal temperature reached 78°C and FCV (initially 4.5 log TCID50/g present) could no longer be detected. Di Girolamo *et al.* (1970) demonstrated the survival (7%) of poliovirus in oysters after steaming for 30 min, when a temperature above 88°C was reached after 25 min. Millard *et al.* (1987) reported that when the internal temperature of the cockle meat was raised to 85–90°C and maintained for 1 min, HAV was inactivated. In contrast with the observations in oysters reported by Di Girolamo *et al.* (1970), Millard *et al.* (1987) noted that an internal temperature of more than 88°C was reached after 2 min steaming of cockles. At this point infectious HAV (initially 10^5 infectious units) was no longer detected. Harlow *et al.* (2011) demonstrated that 6 min steaming of mussels induced >3 log reduction of HAV. HAV was completely inactivated (>5.47 log) after heating clams (spiked digestive gland surrounded by clam meat) for 180 s at 90°C (Sow *et al.*, 2011). Without the clam meat, HAV (spiked in digestive gland without the surrounded clam meat) was not completely inactivated at this time-temperature combination (Table 12.1), because the time needed to reach the temperature of interest was much longer in the presence of clam meat. Since the preparation of shellfish usually involves heating until the shells are opened, which in clams occurs at 70°C after 47 ± 5 s (Koff and Sear, 1967), this heating process would be insufficient to inactivate viruses. Remarkably, Dolin *et al.* (1972) reported remaining infectivity of norovirus (NoV) in stool filtrate exposed to 60°C for 30 min. The latter study showed contradictory results compared to studies using NoV surrogates (Table 12.1) and should be further investigated.

The mechanism of inactivation above 65°C might be caused by large irreversible structural changes presumably due to extensive protein unfolding (Volkin *et al.*, 1997). Nuanualsuwan and Cliver (2003) suggested that heat treatment did not cause loss of RNA infectivity of poliovirus but that the presumed target might be the viral capsid. In addition, the quaternary structure of NoVLPs (NoV virus-like particles) was found to be unaffected up to 60°C. However, above 65°C particles were irregular in shape with significant disruption of their icosahedral structure (Ausar *et al.*, 2006).

Because heat affects primarily the viral capsid, a discrepancy between virus infectivity and RT-PCR detection can be observed. MNV-1 exposed to 80°C for 2.5 min showed 6.5 log reduction of PFU, while the RT-PCR signal, representing the number of genomic copies, was not reduced even after 1 h heating at this temperature (Baert *et al.*, 2008c). Accordingly, genomic copies of MNV-1 were detected on blanched spinach by real-time RT-PCR while no infective MNV-1 particles were found by plaque assay (Baert *et al.*, 2008c). Discrepancies between the reduction in the number of infectious viruses (HAV and FCV) and the number of genomes detected by real-time RT-PCR were observed in heat-treated herbs (Butot *et al.*, 2009). RT-PCR signals obtained from heat-treated samples should therefore be interpreted with caution because an overestimation of viral infectivity and consequently viral risk would be possible. However, similar log reductions were observed for HAV by plaque assay and real-time RT-PCR in clam digestive gland surrounded by clam meat (Sow *et al.*, 2011). The effect of heating foods upon NoV GII.3 was studied by real-time RT-PCR (Mormann *et al.*, 2010). After heat treatment, NoV was extracted from the artificially inoculated food and subjected to an RNase treatment before detection by real-time RT-PCR. More than 4 log reduction was obtained by baking frozen pizza (200°C 12 min), while no significant decrease was observed when pasteurizing (72–74°C 1 min) spiced tomato sauce. Due to the lack of a culturing system, it is difficult to evaluate the correlation with the viral infectivity. Topping *et al.* (2009) predicted by extrapolation that for a 2 min treatment the temperature of 76.6°C obtained a maximal RNA exposure (and thus reduction) of NoV GII.4 in diluted clinical specimen. These authors investigated the heat resistance of NoV GII by a combination of RNase treatment and RT-PCR in low copy number samples.

12.3 Non-thermal processes

Besides thermal processes or heat treatments, non-thermal processes are of interest for more perishable foods such as shellfish and fresh produce. Milder treatments are preferred in order to maintain nutritional and sensorial properties of the food. High pressure processing and irradiation are two alternative non-thermal treatments which are discussed in this chapter.

12.3.1 High hydrostatic pressure processing

High hydrostatic pressure processing (HPP) has been applied to raw bivalve shellfish, fruit juices, cider, jams and jellies, drinkable yoghurt, smoothies, avocado products, chopped onions and ready-to-eat meat products (Kingsley *et al.*, 2005). In contrast to heat, HPP does not disrupt covalent bonds, thus maintaining the primary structure of proteins and retaining appearance, flavor, texture and nutritional qualities of the unprocessed product (Murchie *et al.*, 2005). Changes in the tertiary structure, maintained by hydrophobic and ionic interactions, are in general usually observed for proteins above 200 MPa (Balny and Masson, 1993). It was shown for MNV-1 that the viral structure was disrupted by HPP without degradation of the viral capsid protein or the antigenicity (Lou *et al.*, 2011). The efficacy of HPP to inactivate foodborne viruses is shown in Table 12.2.

Kingsley *et al.* (2005) studied the persistence of HAV in mashed raspberries and sliced green onions. HAV exposed to pressures of 375 MPa at 21°C for 5 min was reduced by respectively 4.3 and 4.7 log in strawberry puree and on sliced green onions (Table 12.2). Structural and organoleptic changes were observed for treated whole green onions and strawberries, although sliced green onions or strawberry puree might be accepted by consumers and can be used as flavor enhancers or as ingredients for cream, jams, juices or smoothies. HPP was used to treat oysters with a pressure of 400 MPa for 1 min (9.0°C) and induced 3 log reduction of HAV (Calci *et al.*, 2005), whereas MNV-1 was reduced by 4 log (5°C) (Kingsley *et al.*, 2007).

The effect of HPP on picornaviruses other than HAV was investigated in cell culture medium by Kingsley *et al.* (2004). Aichivirus and coxsackievirus B5 remained fully infectious if 600 MPa was applied for 5 min at ambient temperature, whereas coxsackievirus A9 was reduced by 7.6 log under the same conditions. Similarly poliovirus was found to be resistant to 600 MPa for 1 h (Wilkinson *et al.*, 2001).

It can be concluded that sensitivity towards HPP can vary even between genetically related taxonomic groups or strains. A possible explanation could be the difference in protein sequence and structure. Members of the genus *Enterovirus*, such as poliovirus, are characterized by large capsid proteins that may protect these virions against HPP (Grove *et al.*, 2008). The aberrant behavior of coxsackievirus A9 by HPP (high sensitivity towards HPP compared to other viruses within the genus *Enterovirus*) may be explained by an atypical protein capsid. Herrmann and Cliver (1973) indicated a distinct viral capsid of coxsackievirus A9 as this virus was sensitive towards protease activity in contrast to the majority of enteroviruses. The mechanism by which HPP inactivates viruses (HAV) is believed to be due to alterations in the proteins of the viral capsid (Kingsley *et al.*, 2002).

The efficacy of HPP might be influenced by ionic strength. Kingsley *et al.* (2005) reported an ionic strength (salt concentration) of 3.2‰, 2.1‰, 4.1‰ and 5–20‰ for respectively strawberry mash, onion extracts, cell culture media and oysters. An inverse correlation between ionic strength and viral inactivation induced by HPP (Table 12.2) was observed. Moreover, an increasing concentration of sucrose (0–40%) or NaCl (0–12%) had an increasing protective effect upon FCV in cell culture medium regarding HPP (Kingsley

Table 12.2 The efficacy of the non-thermal processes high hydrostatic pressure processing and irradiation in the inactivation of foodborne viruses

Virus		Inactivation treatment	Matrix	Log_{10} reduction[a]	Reference
High hydrostatic pressure processing					
Reoviridae	Rotavirus	300 MPa, 25°C, 2 min	Cell culture medium	8	Khadre and Yousef, 2002
Picornaviridae	HAV[b]	450 MPa, ambient temp[g], 5 min	Cell culture medium	>6	Kingsley *et al.*, 2002
	HAV	400 MPa, ambient temp, 10 min	Cell culture medium	2	Grove *et al.*, 2009
	HAV	400 MPa, ambient temp, 10 min	Cell culture medium + marine salt mixture	< 0.5	Grove *et al.*, 2009
	HAV	400 MPa, 9°C, 1 min	Oysters	3	Calci *et al.*, 2005
	HAV	400 MPa, 18–22°C, 5 min	Blue mussels	2.1	Terio *et al.*, 2010
	HAV	375 MPa, 21°C, 5 min	Mashed strawberries; sliced green onions	4.3; 4.7	Kingsley *et al.*, 2005
	HAV	500 MPa, 4°C, 5 min	Sausages	3.23	Sharma *et al.*, 2008
	Poliovirus	600 MPa, ambient temp, 5 min	Cell culture medium	No red[f]	Kingsley *et al.*, 2004
		600 MPa, 20°C, 60 min		No red	Wilkinson *et al.*, 2001
	Poliovirus	600 MPa, ambient temp, 5 min	Cell culture medium	No red	Grove *et al.*, 2008
	Aichivirus	600 MPa, ambient temp, 5 min	Cell culture medium	No red	Kingsley *et al.*, 2004
	Coxsackievirus B5			No red	
	Coxsackievirus A9			7.6	
Caliciviridae	FCV[c]	275 MPa, ambient temp, 5 min	Cell culture medium	>6	Kingsley *et al.*, 2002
	FCV	200 MPa, −10°C or 20°C, 4 min	Cell culture medium	5 or 0.3	Chen *et al.*, 2005
	FCV	300 MPa, ambient temp, 3 min	Cell culture medium	5	Grove *et al.*, 2008
	FCV	500 MPa, 4°C, 5 min	Sausages	2.89	Sharma *et al.*, 2008

	MNV-1	400 MPa, 5°C, 5 min	Oyster tissue	4	Kingsley *et al.*, 2007
	MNV-1	450 MPa, 20°C, 5 min	Cell culture medium	6.85	Kingsley *et al.*, 2007
	MNV-1	350 MPa, 4°C, 2 min	Cell culture medium	8.1	Lou *et al.*, 2011
	MNV-1	350 MPa, 4°C, 2 min	Cut lettuce; strawberries	2.4; 2.2	Lou *et al.*, 2011
	MNV-1	400 MPa, 0°C, 5 min	Oysters	>4	Li *et al.*, 2009
Leviviridae	MS2	600 MPa, 21°C, 10 min	Cell culture medium	3.5	Guan *et al.*, 2006
	MS2	500 MPa, 4°C, 5 min	Sausages	1.47	Sharma *et al.*, 2008
Irradiation					
Reoviridae	Rotavirus	2.4 kGy	Oysters; clams	1; 1	Mallett *et al.*, 1991
Picornaviridae	HAV	UV dose: 40 mW s/cm^2	Lettuce; green onions; strawberries	4.3; 4.2; 1.3	Fino and Kniel, 2008
		UV dose: 120 mW s/cm^2	Lettuce; green onions; strawberries	4.5; 5.3; 1.8	
	HAV	Pulsed UV light: 50 mW s/cm^2 6 pulses	PBS + 5% FCS^h; PBS	2.4; 4.8	Jean *et al.*, 2011
	HAV	3 kGy	Lettuce; strawberries	1; 1	Bidawid *et al.*, 2000a
	HAV	2.0 kGy	Oysters; clams	1; 1	Mallett *et al.*, 1991
	HAV	High intensity broad spectrum pulsed light 1 J/cm^2	PBS + 5% FCS^h; PBS	4.1; > 5.7	Roberts and Hope, 2003
	Aichivirus	UV dose: 40 mW s/cm^2	Lettuce; green onions; strawberries	4.0; 2.4; 1.5	Fino and Kniel, 2008
		UV dose: 120 mW s/cm^2	Lettuce; green onions; strawberries	4.4; 3.7; 1.6	
	Poliovirus	High intensity broad spectrum pulsed light 1 J/cm^2	PBS + 5% FCS; PBS	3.2; > 6.7	Roberts and Hope, 2003
	Coxsackievirus B2	7 kGy	Ground beef	1	Sullivan *et al.*, 1973

(*Continued*)

Table 12.2 Continued

Virus		Inactivation treatment	Matrix	Log_{10} reduction[a]	Reference
Caliciviridae	FCV	UV dose: 12 mW s/cm^2; 200 Gy	Virus suspension with low protein content	3; 1.6	De Roda Husman *et al.*, 2004
	CaCV[d]	UV dose: 20 mW s/cm^2; 200 Gy		3; 2.4	
	FCV	UV dose: 40 mW s/cm^2	Lettuce; green onions, strawberries	3.5; 2.5; 1.1	Fino and Kniel, 2008
		UV dose: 120 mW s/cm2	Lettuce; green onions; strawberries	3.8; 3.9; 1.6	
	FCV	UV dose: 25 mW s/cm^2	Cell culture medium (filtrated for dispersion)	4	Park *et al.*, 2011
	MNV[e]	UV dose: 29 mW s/cm^2	Cell culture medium (filtrated for dispersion)	4	Park *et al.*, 2011
	MNV	2. 8; 5.6 kGy	strawberries	1.3; 2.4	Fung *et al.*, 2011
	MNV-1	Pulsed UV light: 50 mW s/cm^2 6 pulses	PBS + 5% FCS[h]; PBS	2.8; > 5	Jean *et al.*, 2011
	MNV	e-beam: 4 kGy	Cut strawberries	< 1	Sanglay *et al.*, 2011
Leviviridae	MS2	UV dose: 65 mW s/cm^2, 200 Gy	Virus suspension with low protein content	3; 7	De Roda Husman *et al.*, 2004
	MS2	UV dose: 70 mW s/cm^2	Cell culture medium (filtrated for dispersion)	4	Park *et al.*, 2011

Table redrafted from Baert *et al.* (2009).
[a]Log_{10} reduction represents the reduction in infectivity; [b]HAV: hepatitis A virus; [c]FCV: feline calicivirus; [d]CaCV: canine calicivirus; [e]MNV: murine norovirus; [f]No red: no reduction; [g]temp: temperature; [h]FCS: fetal calf serum.

and Chen, 2008). Water activity did not necessarily play a role because FCV suspended in a sucrose solution decreased to a lesser extent compared to FCV suspended in a NaCl solution having equivalent water activity. A baroprotective effect was also observed for HAV (suspended in cell culture medium + 10% fetal bovine serum) in the presence of 1–6% NaCl (Kingsley and Chen, 2009). Accordingly, Grove *et al.* (2009) noticed protection of HAV against HPP by the presence of salt (Table 12.2). A protective effect of $CaCl_2$ was observed against HPP for MNV-1 (Sanchez *et al.*, 2011). MNV-1 infectious titer was only reduced marginally by a pressure of 400 MPa at 25°C or 45°C when 10 mM $CaCl_2$ was added to the test suspension. Additionally, a baroprotective effect of some components in oysters was reported for FCV and HAV (Kingsley and Chen, 2009; Murchie *et al.*, 2007). Lou *et al.* (2011) investigated the inactivation of MNV-1 in different purees (strawberry, tomato, carrot, lemon) and observed the effects of both the pH and the food matrix. Observations from the studies mentioned have thus assumed protection of viruses against HPP by components of the food matrix.

Temperature can also affect HPP efficacy. FCV was reduced by 4 to 5 log at low temperatures (−10°C) when treated with a pressure of 200 MPa (4 min); however the same treatment at 20°C only reduced the titer by 0.3 log (Chen *et al.*, 2005). Also Kingsley *et al.* (2007) found only 1.15 log reduction when MNV-1 was treated with a dose of 350 MPa (5 min) in propagation medium at 30°C, while a reduction of 5.56 log was observed at 5°C. On the other hand, temperatures above 30°C can enhance pressure inactivation for particular virus strains as >5 log reduction of MNV-1 was obtained when a pressure of 400 MPa at 45°C was applied (Sánchez *et al.*, 2011).

A possible disadvantage of HPP is that viral strains might develop resistance to this technology, as was assumed by Smiddy *et al.* (2006). They noticed altered plaques of Qß (ss-RNA coliphage) after HPP treatment when the altered shaped plaques persisted by sub-culturing. These phages with unusual plaque morphology might be more pressure-resistant but this hypothesis was not further investigated. An extensive overview of the effect of HPP upon foodborne viruses is described by Kovac *et al.* (2010).

12.3.2 Irradiation

Irradiation is another non-thermal approach to inactivate foodborne viruses; results of studies on its efficacy are presented in Table 12.2. UV light treatment of lettuce at a dose of 40 mW s/cm² achieved 4.3, 4.0 and 3.5 log reduction of respectively HAV, aichivirus and FCV (Fino and Kniel, 2008). Inactivation was greatest on lettuce and least effective on strawberries. Viruses were likely sheltered from UV light by the strawberry matrix. A 3 log reduction was achieved for FCV, CaCV and MS2 in tenfold-diluted cell culture medium after exposure to UV at a dose of, respectively, 12, 20 and 65 mW s/cm² (de Roda Husman *et al.*, 2004). The protein load present in the viral suspension did not influence inactivation obtained by UV treatment,although cell-associated echovirus 12 (harvested by washing infected FRhK-4 cells 3 times with PBS followed by centrifugation) had an approximately threefold higher resistance to UV light than did the viruses obtained by freeze-thaw cell lysis and subsequent serial filtration (0.2 and 0.08 μm pore polycarbonate filters) (Park *et al.*, 2011). It might be that large organic particulates (>1 μm) would influence virus inactivation by UV irradiation (Templeton *et al.*, 2005).

UV inactivation of micro-organisms is probably due to the absorption of UV by nucleic acids causing dimerization of thymine in DNA or uracil in RNA (Nuanualsuwan and Cliver, 2003; Sommer *et al.*, 2001). At higher doses (≥1000 mW s/cm^2) UV light can also affect the capsid proteins. The combined effect of size/type of the virion and nucleic acids is thought to be factors determining the resistance/sensitivity of viruses towards UV (Sommer *et al.*, 2001). Surprisingly, Ma *et al.* (1994) found 4.5 log reduction of infective poliovirus particles by plaque assay after treatment with a UV dose of 22 mW s/cm^2,while only 1 log reduction of the genomic copies, detected by RT-PCR, was observed since UV was thought to break down the genome. Damage of the genomic material after UV treatment might be not sufficient to break down the small target RNA fragment needed for RT-PCR detection. As with heat-treated samples, RT-PCR signals obtained from UV-treated samples should be interpreted with caution.

In contrast to UV treatment, high intensity broad spectrum pulsed light covers wavelengths between 200 and 1100 nm. The latter treatment involves short pulses of high intensity (at least 1000 times that of conventional UV light (Roberts and Hope, 2003)). A dose of 1 J/cm^2 (corresponding with 1 W s/cm^2) reduced HAV and poliovirus by at least 5.7 and 6.7 log in PBS. Jean *et al.* (2011) observed a 4.8 log reduction of HAV in PBS when treated with pulsed UV light at a dose of 0.5 W s/cm^2. In the presence of proteins, in this case, 5% fetal calf serum, virus inactivation was less effective (Table 12.2). Inactivation of viruses in food matrices and the effect of (high intensity broad spectrum) pulsed light upon the organoleptic and structural properties of foods should be further examined.

Gamma irradiation at doses of 2–4 kGy is often used for bacterial control in particular food products, and its effect on foodborne viruses has been reported by several investigators. The US FDA approved doses up to 4.0 kGy to treat fresh iceberg lettuce and spinach (Feng *et al.*, 2011). Due to their small size and genome, enteric viruses are more resistant towards exposure to ionizing irradiation than bacteria, parasites and fungi (Hirneisen *et al.*, 2010). Bidawid *et al.* (2000a) found that 3 kGy was needed in order to achieve 1 log reduction of HAV on lettuce or strawberries. Mallett *et al.* (1991) reported that 2.0 kGy was able to reduce HAV by 1 log in oysters and clams but 2.4 kGy was needed to achieve this for rotavirus. Coxsackievirus B2 was reduced by 1 log in ground beef when treated with 7 kGy (Sullivan *et al.*, 1973). A dose of 200 Gy reduced CaCV and FCV, respectively, by 2.4 and 1.6 log (De Roda Husman *et al.*, 2004). The same conditions induced 7 log reduction of MS2. Gamma irradiation was found to be greatly affected by the presence of proteins (De Roda Husman *et al.*, 2004). The authors assumed that free OH radicals induced by gamma irradiation, which normally interact with nucleic acids and the virus coat, were scavenged and therefore induced less inactivation. The influence of the food matrix in protecting MNV from gamma irradiation was also suggested by Feng *et al.* (2011). These authors showed that gamma irradiation produced physical distortion of the virion geometry, degradation of the capsid protein and genomic RNA of MNV-1. Additionally, the degradation kinetics of human norovirus capsid, studied by using norovirus virus-like particles (VLPs), by gamma irradiation was shown to be similar to that of MNV-1 (Feng *et al.*, 2011).

12.4 Appropriateness of surrogates

The effect of thermal and non-thermal processes are evaluated upon infectivity in Tables 12.1 and 12.2. Consequently, only cultivatable viruses where the reduction in infectivity can be measured are presented. Due to the difficulties or the inability to culture foodborne viruses causing illness in humans, surrogate viruses are used to evaluate the effect of food processes. A point of view on the use of different surrogate viruses is given.

12.4.1 Why use surrogates?

In general, foodborne viruses cannot be cultivated, or difficulties are encountered in the culturing assay. A culture system for human NoV strains is not yet established. Although a 3D cell culture model for NoV was described (Straub *et al.*, 2007), it has not been successfully reproduced. Cell culture propagation of HAV is hampered due to the non-cytopathic replication of wild-type strains, and it may take several weeks before detection based on HAV antigen can be established (Cromeans *et al.*, 1989). Growth of environmental wild-type rotavirus (RoV) strains is also laborious and difficult (Rutjes *et al.*, 2009). Molecular assays have been developed for these viruses to facilitate detection. The major drawback is the lack of the link with viability because detection is based on genomic copies. It has been shown that there is no straightforward correlation between detection of genomic copies and cell culture assay after heat treatment of MNV-1 (Baert *et al.*, 2008c). As a result, the effect of thermal and non-thermal processes upon viruses is mainly studied by the use of surrogate viruses. Busta *et al.* (2003) defined criteria for surrogates used in fresh produce processes although these criteria can also be valuable for the validation of other food-processing steps. Ideally, surrogates should be non-pathogenic, behave similarly to the target micro-organism, have inactivation characteristics and kinetics that can be used to predict those of the target micro-organism, be easily prepared and enumerated using rapid, inexpensive detection methods, be easily differentiated from other microflora and be genetically stable (Busta *et al.*, 2003). An overview of viral strains that have been used as surrogates for foodborne viruses is given in Table 12.3. Below, various proposed surrogates for foodborne viruses such as caliciviruses, bacteriophages, cell culture-adapted strains, VLPs, etc. are briefly discussed.

12.4.2 Caliciviruses

Surrogates used to represent human NoV comprise several strains of the *Caliciviridae* family, selected for their ability to be cultivated in cell lines. Feline calicivirus strain F-9 was introduced as surrogate by Slomka and Appleton (1998). The USEPA recommends the use of FCV as surrogate in virucidal tests for products to disinfect hard non-porous surfaces against NoV (USEPA, http://www.epa.gov/oppad001/pdf_files/initial_virucidal_test.pdf). FCV belongs to the same family as human NoV but is part of the genus *Vesivirus*. It is readily cultured in established cell lines and a $TCID_{50}$ assay (Doultree *et al.*, 1999) and plaque assay is available, allowing enumeration of the virus (Bidawid *et al.*, 2003). FCV cause a severe respiratory illness in cats, which calls into questions the reliability of using this virus as model for NoV, which produces an enteric illness. To a lesser extent canine calicivirus (CaCV), isolated from dogs, has been used to represent human NoV (De Roda Husman *et al.*, 2004; Duizer *et al.*, 2004). In 2004, MNV-1 was identified. This virus is biochemically, pathologically and molecularly more related to human NoV than other surrogates (Wobus *et al.*, 2006), and is also cultivable in vitro. MNV-1 appears to be more stable than FCV, or at least equally persistent towards tested conditions of pH and temperature except at 56°C (Cannon *et al.*, 2006).

An increasing number of other caliciviruses are being characterized and might offer alternative ways to study NoV. Recently, Tulane virus was isolated from rhesus macaques and showed to be genetically more closely related to *Norovirus* than other caliciviruses (Farkas *et al.*, 2008, 2010). Tulane virus probably replicates in the intestine of these animals, but it is not known whether infection is associated with clinical illness, such as diarrhea. Tulane virus has been cultivated successfully in LLC-MK2 rhesus monkey kidney cells, making it a potential model for the study of human caliciviruses (Wei *et al.*, 2008).

12.4.3 Bacteriophages

Bacteriophages are proposed as surrogate for foodborne viruses because they are easily cultured in the laboratory without the need of mammalian cell lines and thus cell culture facilities. MS2, a coliphage (virus that infects *E. coli*), is mostly used (Table 12.2). It is a non-enveloped virus, having a similar virion size to most foodborne viruses and its genome likewise consists of ss-RNA. MS2 belongs to the family *Leviviridae* that infect only F+ male hosts through the F pili. Coliphages are not pathogenic to humans and cultivation assays are inexpensive. These factors contribute to the interest in using bacteriophages as potential surrogates.

Table 12.3 Characteristics of representative viral strains used in studies investigating the stability of foodborne viruses

Viral strains	Genus	Family	Genome	Host	Mode of transmission	Surrogate for viruses infecting humans
Feline calicivirus	*Vesivirus*	*Caliciviridae*	ss-RNA positive	Cat	Respiratory	Norovirus
Canine calicivirus				Dog	Fecal-oral	Norovirus
Murine norovirus 1	*Norovirus*			Mouse	Fecal-oral, respiratory	Norovirus
Attenuated HAV HM-175	*Hepatovirus*	*Picornaviridae*		Human	Fecal-oral	Hepatitis A virus
Attenuated poliovirus (Sabin, Lsc-2ab)	*Enterovirus*			Human	Fecal-oral	Poliovirus
Coxsackieviruses				Human	Fecal-oral	
Aichivirus	*Kobuvirus*			Human	Fecal-oral	
MS2	*Levivirus*	*Leviviridae*		E. coli	Fecal	Enteric viruses[a]
Rotavirus SA11	*Rotavirus*	*Reoviridae*	ds-RNA segmented	Primate	Fecal-oral	Rotavirus

Redrafted from Baert *et al.* (2009).

[a] Human enteric viruses such as noroviruses, hepatitis A virus, enteroviruses and rotaviruses.

12.4.4 Other

NoV VLPs have been tested as potential surrogates for human NoV. The expression of VP1 protein in insect cells results in self-assembly of VLPs that are structurally and antigenetically similar to native virions (Jiang *et al.*, 1992). A similar degradation of the capsid proteins of MNV-1 and human NoV was observed after exposure to gamma irradiation (Feng *et al.*, 2011). RoV VLPs established by coexpression of VP2 and VP6 in a baculovirus expression system have been used to investigate the effect of UV light irradiation (Caballero *et al.*, 2004) and persistence in shellfish (Loisy *et al.*, 2005). It should be noted that detection of VLPs is based on ELISA, flow cytometry or electron microscopy; these assays have a high detection limit requiring the use of high concentrations of VLPs to evaluate treatments.

Although wild-type strains of HAV and RoV are fastidious in growth, cell culture-adapted strains are available. These cell culture-adapted strains are frequently relied upon to study virus survival and the effect of thermal and non-thermal processes. The HAV HM-175 strain can be grown in FRhK-4 cells and plaque assays are described (Mbithi *et al.*, 1991). The simian RoV SA11 is used as model virus because it is easily grown (Estes *et al.*, 1979). Other tissue culture-adapted RoV strains exist, such as the strains Wa and DS1 that can be cultured in the African Green Monkey Kidney cell line MA104 and CaCo-2 cell line (Mahony *et al.*, 2000). Previously, poliovirus was often used as the prototype virus representing enteric viruses. The Sabin attenuated poliovirus type 1 strain (LSc/2ab) was frequently used as a surrogate in studies investigating the stability of enteric viruses. Aichivirus, part of the family *Picornaviridae*, genus *Kobuvirus,* and Coxsackieviruses A9 and B5, family *Picornaviridae,* genus *Enterovirus,* have been investigated regarding their persistence against high hydrostatic pressure by plaque assay and $TCID_{50}$, respectively (Kingsley *et al.*, 2004).

The mengovirus vMC_0 can potentially be used as a surrogate. It can be cultivated, is not pathogenic and belongs to the same family as HAV (Costafreda *et al.*, 2006).

12.4.5 Criticism about the use of surrogates

Surrogates are needed because of the lack of, or the difficulties involved with, infectivity assays for foodborne viruses. Depending on the choice of surrogate, inactivation profiles differ. Differences in susceptibility of the surrogates may lead to different interpretations of the effectiveness of thermal or non-thermal processes. FCV and MNV showed distinct inactivation profiles towards different alcohols (Park *et al.*, 2010). Some tested conditions, such as low pH, indicated that the infectivity of FCV reduced significantly faster than that of MNV (Cannon *et al.*, 2006). No significant difference between MNV-1 and HAV (HM-175) was observed after heat exposure (63°C and 72°C up to 10 min), indicating that MNV-1 is a suitable surrogate (Hewitt *et al.*, 2009). However, log reductions in real-time RT-PCR titre were significantly different between MNV-1, HAV, human NoV GI.2 and GII.4, calling into question the appropriateness of these surrogates for human NoV. It is therefore not appropriate at present to claim that a particular surrogate would be superior to other surrogates. For HAV, no surrogate is used but most experiments are based on the cell culture-adapted HM-175 strain. Wild-type HAV strains are difficult to culture in vitro, hampering the evaluation of their resistance towards processing techniques. However, the extent of the similarity between cell culture-adapted and circulating wild-type HAV strains is not known.

Instead of relying upon surrogates, the use of molecular assays to follow up reduction in nucleic acids of the virus of interest might be meaningful. The heat resistance of NoV GII.4 was investigated by including a RNase treatment prior to real-time RT-PCR detection, and it proved more heat-resistant than FCV (Topping *et al.*, 2009). A theoretical framework was developed to estimate virus infectivity of MS2 after UV treatment, and this might be applicable to nonculturable viruses such as NoV (Pecson *et al.*, 2011). The applicability of these alternative approaches was evaluated only for limited test conditions and should thus be further explored.

Richards (2012) reviewed the uncertainties and drawbacks of the use of surrogates and stated that human clinical studies would provide the definitive answer to what methods are needed to inactivate NoV. Indeed surrogate studies have not answered the question regarding NoV inactivation (Richards, 2012). Differences in the choice of surrogate, matrix and treatment parameters resulted in variability regarding the inactivation outcome. Nevertheless, taking critical account of all the available data could at least enable a reasonable suggestion of treatment conditions per matrix and inactivation method that is likely to be effective to inactivate/reduce the NoV load to some extent Validating the effectiveness of the treatment will require human challenge studies, but due to the costs associated with volunteer studies, data obtained by using surrogates provides valuable initial information. A considerable number of issues will remain when clinical trials of this nature are carried out, because the outcome will only be valid for one specific condition, inactivation method and matrix. Additionally, there will be issues around not only the selection of NoV strains to be investigated to determine the inactivation profile against a particular treatment, but also the choice of volunteers, due to differences in susceptibility in the population.

12.5 Future trends

Studies of foodborne viral inactivation have increased exponentially during the past decade. Treatments frequently applied in the food industry and disinfection strategies with a proven antibacterial effect are increasingly being examined for their effect on enteric viruses. Discrepancies in inactivation profiles that are observed in different studies may be partially due to differences in the experimental approach (initial viral titre, matrix composition, etc.). Harmonization and standardization of test conditions and surrogates is urgently needed.

Tailing of viral inactivation curves was observed in several studies using different inactivation approaches. The reason for tailing is not clear, which is a disadvantage in the use of inactivation models. It is therefore recommended that the inactivation method of interest is validated using the food on which its use is intended. In addition the experimental set-up of the validation needs as far as possible to reflect industry practice in order to obtain a reliable and useful outcome.

The reductions induced by inactivation treatments are described but explanations of the observed effects are lacking at present. The mechanism of viral inactivation needs to be explored in order to understand the strong and the weak points responsible for their stability/sensitivity. This will provide new ideas for potential strategies of viral inactivation.

Food preservation techniques are often a combination of growth-reducing factors and inactivation procedures used either in tandem or sequentially and referred to as hurdle technology. It would be interesting to evaluate the effect of the hurdle approach on foodborne viruses, since they seem to be relatively resistant to the single inactivation strategies used to control bacteria. Studies combining inactivation techniques to examine if a synergistic effect on foodborne viruses can be observed are therefore of interest.

12.6 Sources of further information and advice

Data with regard to survival of viruses on foods and particularly on fresh produce is reviewed by Rzezutka and Cook (2004) and Seymour and Appleton (2001). An extensive overview of the effect of preservation methods and food-processing technologies on inactive foodborne viruses is described by Baert *et al.* (2009) and Hirneisen *et al.* (2010). In particular, the effect of HPP is detailed by Kovac *et al.* (2010). Additional information can be found in *Viruses in food* (ed. Goyal, 2006) and *Foodborne Viruses: Progress and Challenges* (eds Koopmans *et al.*, 2008).

12.7 References

AUSAR, S.F., FOUBERT, T.R., HUDSON, M.H., VEDVICK, T.S. and MIDDAUGH, C.R. 2006. Conformational stability and disassembly of Norwalk virus-like particles – effect of pH and temperature. *Journal of Biological Chemistry* **281**, 19478–88.

BAERT, L., DEBEVERE, J. and UYTTENDAELE, M. 2009. The efficacy of preservation methods to inactivate foodborne viruses. *International Journal of Food Microbiology* **131**, 83–94.

BAERT, L., UYTTENDAELE, M., Van COILLIE, E. and DEBEVERE, J., 2008a. The reduction of murine norovirus 1, B-fragilis HSP40 infecting phage B40–8 and *E-coli* after a mild thermal pasteurization process of raspberry puree. *Food Microbiology* **25**, 871–4.

BAERT, L., UYTTENDAELE, M., VERMEERSCH, M., VAN COILLIE, E. and DEBEVERE, J. 2008b. Survival and transfer of murine norovirus 1, a surrogate for human noroviruses, during the production process of deep-frozen onions and spinach. *Journal of Food Protection* **71**, 1590–7.

BAERT, L., WOBUS, C.E., VAN COILLIE, E., THACKRAY, L.B., DEBEVERE, J. and UYTTENDAELE, M. 2008c. Detection of murine norovirus 1 by using plaque assay, transfection assay, and real-time reverse transcription-PCR before and after heat exposure. *Applied and Environmental Microbiology* **74**, 543–6.

BALNY, C. and MASSON, P. 1993. Effects of high pressure on proteins. *Food Reviews International* **9**, 611–28.

BIDAWID, S., FARBER, J.M. and SATTAR, S.A. 2000a. Inactivation of hepatitis A virus (HAV) in fruits and vegetables by gamma irradiation. *International Journal of Food Microbiology* **57**, 91–7.

BIDAWID, S., FARBER, J.M. and SATTAR, S.A. 2001. Survival of hepatitis A virus on modified atmosphere-packaged (MAP) lettuce. *Food Microbiology* **18**, 95–102.

BIDAWID, S., FARBER, J.M., SATTAR, S.A. and HAYWARD, S. 2000b. Heat inactivation of hepatitis A virus in dairy foods. *Journal of Food Protection* **63**, 522–8.

BIDAWID, S., MALIK, N., ADGBURIN, O., SATTAR, S.A. and FARBER, J.M. 2003. A feline kidney cell line based plaque assay for FCV, a surrogate for Norwalk virus. *Journal of Virological Methods* **107**, 163–7.

BUCKOW, R., ISBARN, S., KNORR, D., HEINZ, V. and LEHMACHER, A. 2008. Predictive model for inactivation of feline Calicivirus, a norovirus surrogate, by heat and high hydrostatic pressure. *Applied and Environmental Microbiology* **74**, 1030–8.

BUSTA, F.F., SUSLOW, T.V., PARISH, M.E., BEUCHAT, L., FARBER, J.M., GARRETT, E.H. and HARRIS, J. 2003. The use of indicators and surrogate microorganisms for the evaluation of pathogens in fresh and fresh-cut produce. *Critical Reviews in Food Science and Nutrition* **2**(S1), 179–85.

BUTOT, S., PUTALLAZ, T., AMOROSO, R. and SANCHEZ, G., 2009. Inactivation of enteric viruses in minimally processed berries and herbs. *Applied and Environmental Microbiology* **75**, 4155–61.

BUTOT, S., PUTALLAZ, T. and SANCHEZ, G. 2008. Effects of sanitation, freezing and frozen storage on enteric viruses in berries and herbs. *International Journal of Food Microbiology* **126**, 30–5.

CABALLERO, S., ABAD, F.X., LOISY, F., LE GUYADER, F.S., COHEN, J., PINTO, R.M. and BOSCH, A. 2004. Rotavirus virus-like particles as surrogates in environmental persistence and inactivation studies. *Applied and Environmental Microbiology* **70**, 3904–9.

CALCI, K.R., MEADE, G.K., TEZLOFF, R.C. and KINGSLEY, D.H. 2005. High-pressure inactivation of hepatitis A virus within oysters. *Applied and Environmental Microbiology* **71**, 339–43.

CANNON, J.L., PAPAFRAGKOU, E., PARK, G.W., OSBORNE, J., JAYKUS, L.A. and VINJE, J., 2006. Surrogates for the study of norovirus stability and inactivation in the environment: a comparison of murine norovirus and feline calicivirus. *Journal of Food Protection* **69**, 2761–5.

CARTER, M.J. 2005. Enterically infecting viruses: pathogenicity, transmission and significance for food and waterborne infection. *Journal of Applied Microbiology* **98**, 1354–80.

CHEN, H.Q., HOOVER, D.G. and KINGSLEY, D.H. 2005. Temperature and treatment time influence high hydrostatic pressure inactivation of feline calicivirus, a norovirus surrogate. *Journal of Food Protection* **68**, 2389–94.

COSTAFREDA, M.I., BOSCH, A. and PINTO, R.M. 2006. Development, evaluation, and standardization of a real-time TaqMan reverse transcription-PCR assay for quantification of hepatitis A virus in clinical and shellfish samples. *Applied and Environmental Microbiology* **72**, 3846–55.

CROCI, L., CICCOZZI, M., DE MEDICI, D., Di PASQUALE, S., FIORE, A., MELE, A. and TOTI, L., 1999. Inactivation of hepatitis A virus in heat-treated mussels. *Journal of Applied Microbiology* **87**, 884–8.

CROMEANS, T., FIELDS, H.A. and SOBSEY, M.D. (1989). Replication kinetics and cytopathic effect of hepatitis-A virus. *Journal of General Virology* **70**, 2051–62.

DAWSON, D.J., PAISH, A., STAFFELL, L.M., SEYMOUR, I.J. and APPLETON, H., 2005. Survival of viruses on fresh produce, using MS2 as a surrogate for norovirus. *Journal of Applied Microbiology* **98**, 203–9.

DEBOOSERE, N., LEGEAY, O., CAUDRELIER, Y. and LANGE, M., 2004. Modelling effect of physical and chemical parameters on heat inactivation kinetics of hepatitis A virus in a fruit model system. *International Journal of Food Microbiology* **93**, 73–85.

DEBOOSERE, N., PINON, A., DELOBEL, A., TEMMAM, S., MORIN, T., MERLE, G., BLAISE-BOISSEAU, S., PERELLE, S. and VIALETTE, M. 2010. A predictive microbiology approach for thermal inactivation of hepatitis A virus in acidified berries. *Food Microbiology* **27**, 962–7.

DE RODA HUSMAN, A.M.D., BIJKERK, P., LODDER, W., VAN DEN BERG, H., PRIBIL, W., CABAJ, A., GEHRINGER, P., SOMMER, R. and DUIZER, E. 2004. Calicivirus inactivation by nonionizing (253.7-nanometer-wavelength (UV)) and ionizing (Gamma) radiation. *Applied and Environmental Microbiology* **70**, 5089–93.

DI GIROLAMO, R., LISTON, J. and MATCHES, J.R. 1970. Survival of virus in chilled, frozen, and processed oysters. *Applied Microbiology* **20**, 58–63.

DOLIN, R., DUPONT, H., WYATT, R.G., HORNICK, R., BUSCHO, R.F., CHANOCK, R.M., BLACKLOW, N.R. and KASEL, J.A. 1972. Biological properties of norwalk agent of acute infectious nonbacterial gastroenteritis. *Proceedings of the Society for Experimental Biology and Medicine* **140**, 578–83.

DOULTREE, J.C., DRUCE, J.D., BIRCH, C.J., BOWDEN, D.S. and MARSHALL, J.A. 1999. Inactivation of feline calicivirus, a Norwalk virus surrogate. *Journal of Hospital Infection* **41**, 51–7.

DUIZER, E., BIJKERK, P., ROCKX, B., de GROOT, A., TWISK, F. and KOOPMANS, M. 2004. Inactivation of caliciviruses. *Applied and Environmental Microbiology* **70**, 4538–43.

ESTES, M.K., GRAHAM, D.Y., GERBA, C.P. and SMITH, E.M. 1979. Simian rotavirus sa11 replication in cell cultures. *Journal of Virology* **31**, 810–15.

FARKAS, T., CROSS, R.W., HARGITT, E., LERCHE, N.W., MORROW, A.L. and SESTAK, K. 2010. Genetic diversity and histo-blood group antigen interactions of rhesus enteric caliciviruses. *Journal of Virology* **84**, 8617–25.

FARKAS, T., SESTAK, K., WEI, C. and JIANG, X. 2008. Characterization of a rhesus monkey calicivirus representing a new genus of Calicivitidae. *Journal of Virology* **82**, 5408–16.

FENG, K., DIVERS, E., MA, Y.M. and LI, J.R. 2011. Inactivation of a human norovirus surrogate, human norovirus virus-like particles, and vesicular stomatitis virus by gamma irradiation. *Applied and Environmental Microbiology* **77**, 3507–17.

FINO, V.R. and KNIEL, K.E. 2008. UV light inactivation of hepatitis A virus, Aichi virus, and feline calicivirus on strawberries, green onions, and lettuce. *Journal of Food Protection* **71**, 908–13.

GIBSON, K.E. and SCHWAB, K.J. (2011). Thermal inactivation of human norovirus surrogates. *Food and Environmental Virology* **3**, 74–7.

GOYAL, S.M. (ed.) 2006. *Viruses in food.* Springer Science+Business Media, New York, ISBN: 978-0-387-28935-9.

GRANT, S.B. (1995). Inactivation kinetics of viral aggregates. *Journal of Environmental Engineering*, **121**, 311–20.

GROVE, S.F., FORSYTH, S., WAN, J., COVENTRY, J., COLE, M., STEWART, C.M., LEWIS, T., ROSS, T. and LEE, A. 2008. Inactivation of hepatitis A virus, poliovirus and a norovirus surrogate by high pressure processing. *Innovative Food Science and Emerging Technologies* **9**, 206–10.

GUAN, D., KNIEL, K., CALCI, K.R., HICKS, D.T., PIVARNIK, L.F. and HOOVER, D.G. 2006. Response of four types of coliphages to high hydrostatic pressure. *Food Microbiology* **23**, 546–51.

HARLOW, J., OUDIT, D., HUGHES, A. and MATTISON, K. 2011. Heat inactivation of hepatitis A virus in shellfish using steam. *Food and Environmental Virology* **3**, 3134.

HERRMANN, J.E. and CLIVER, D.O. 1973. Degradation of Coxsackie virus type-A9 by proteolytic enzymes. *Infection and Immunity* **7**, 513–17.

HEWITT, J. and GREENING, G.E. 2004. Survival and persistence of norovirus, hepatitis A virus, and feline calicivirus in marinated mussels. *Journal of Food Protection* **67**, 1743–50.

HEWITT, J., RIVERA-ABAN, M. and GREENING, G.E. 2009. Evaluation of murine norovirus as a surrogate for human norovirus and hepatitis A virus in heat inactivation studies. *Journal of Applied Microbiology* **107**, 65–71.

HIRNEISEN, K.A., BLACK, E.P., CASCARINO, J.L., FINO, V.R., HOOVER, D.G. and KNIEL, K.E. 2010. Viral Inactivation in foods: a review of traditional and novel food-processing technologies. *Comprehensive Reviews in Food Science and Food Safety* **9**, 3–20.

JEAN, J., MORALES-RAYAS, R., ANOMAN, M.N. and LAMHOUJEB, S. 2011. Inactivation of hepatitis A virus and norovirus surrogate in suspension and on food-contact surfaces using pulsed UV light (pulsed light inactivation of food-borne viruses). *Food Microbiology* **28**, 568–72.

JIANG, X., WANG, J.X., GRAHAM, D.Y. and ESTES, M.K. 1992. Detection of Norwalk virus in stool by polymerase chain reaction. *Journal of Clinical Microbiology* **30**, 2529–34.

KHADRE, M.A. and YOUSEF, A.E. 2002. Susceptibility of human rotavirus to ozone, high pressure, and pulsed electric field. *Journal of Food Protection* **65**, 1441–46.

KINGSLEY, D.H. and CHEN, H.Q. 2008. Aqueous matrix compositions and pH influence feline calicivirus inactivation by high pressure processing. *Journal of Food Protection* **71**, 1598–603.

KINGSLEY, D.H. and CHEN, H.Q. 2009. Influence of pH, salt, and temperature on pressure inactivation of hepatitis A virus. *International Journal of Food Microbiology* **130**, 61–4.

KINGSLEY, D.H., CHEN, H.Q. and HOOVER, D.G. 2004. Inactivation of selected picornaviruses by high hydrostatic pressure. *Virus Research* **102**, 221–4.

KINGSLEY, D.H., GUAN, D.S. and HOOVER, D.G. 2005. Pressure inactivation of hepatitis A virus in strawberry puree and sliced green onions. *Journal of Food Protection* **68**, 1748–51.

KINGSLEY, D.H., HOLLINIAN, D.R., CALCI, K.R., CHEN, H.Q. and FLICK, G.J. 2007. Inactivation of a norovirus by high-pressure processing. *Applied and Environmental Microbiology* **73**, 581–5.

KINGSLEY, D.H., HOOVER, D.G., PAPAFRAGKOU, E. and RICHARDS, G.P., 2002. Inactivation of hepatitis A virus and a calicivirus by high hydrostatic pressure. *Journal of Food Protection* **65**, 1605–9.

KOFF, R.S. and SEAR, H.S. 1967. Internal temperature of steamed clams. *New England Journal of Medicine* **276**, 737–9.

KOOPMANS, M. and DUIZER, E. 2004. Foodborne viruses: an emerging problem. *International Journal of Food Microbiology* **90**, 23–41.

KOOPMANS, M.P.G., O. CLIVER, D. and BOSCH, A. (eds) 2008. *Foodborne Viruses: Progress and Challenges*. Washington, DC: ASM Press, ISBN: 978-1-55581-464-9.

KOVAC, K., DIEZ-VALCARCE, M., HERNANDEZ, M., RASPOR, P. and RODRIGUEZ-LAZARO, D. 2010. High hydrostatic pressure as emergent technology for the elimination of foodborne viruses. *Trends in Food Science and Technology* **21**, 558–68.

LI, D., TANG, Q.J., WANG, J.F., WANG, Y.M., ZHAO, Q. and XUE, C.H. 2009. Effects of high-pressure processing on murine norovirus-1 in oysters (Crassostrea gigas) in situ. *Food Control* **20**, 992–6.

LOISY, F., ATMAR, R.L., LE SAUX, J.C., COHEN, J., CAPRAIS, M.P., POMMEPUY, M. and LE GUYADER, F.S. 2005. Use of rotavirus virus-like particles as surrogates to evaluate virus persistence in shellfish. *Applied and Environmental Microbiology* **71**, 6049–53.

LOU, F.F., NEETOO, H., CHEN, H.Q. and LI, J.R. 2011. Inactivation of a human norovirus surrogate by high-pressure processing: effectiveness, mechanism, and potential application in the fresh produce industry. *Applied and Environmental Microbiology* **77**, 1862–71.

MA, J.F., STRAUB, T.M., PEPPER, I.L. and GERBA, C.P. 1994. Cell-culture and PCR determination of poliovirus inactivation by disinfectants. *Applied and Environmental Microbiology* **60**, 4203–6.

MAHONY, J.O., DONOGHUE, M.O., MORGAN, J.G. and HILL, C., 2000. Rotavirus survival and stability in foods as determined by an optimised plaque assay procedure. *International Journal of Food Microbiology* **61**, 177–85.

MALLETT, J.C., BEGHIAN, L.E., METCALF, T.G. and KAYLOR, J.D. 1991. Potential of irradiation technology for improved shellfish sanitation. *Journal of Food Safety* **11**, 231–45.

MARIAM, T.W. and CLIVER, D.O. 2000b. Small round coliphages as surrogates for human viruses in process assessment. *Dairy, Food and Environmental Sanitation* **20**, 684–9.

MATTISON, K., KARTHIKEYAN, K., ABEBE, M., MALIK, N., SATTAR, S.A., FARBER, J.M. and BIDAWID, S. 2007. Survival of calicivirus in foods and on surfaces: experiments with feline calicivirus as a surrogate for norovirus. *Journal of Food Protection* **70**, 500–3.

MBITHI, J.N., SPRINGTHORPE, V.S. and SATTAR, S.A. 1991. Effect of relative humidity and air temperature on survival of hepatitis A virus on environmental surfaces. *Applied and Environmental Microbiology* **57**, 1394–9.

MILLARD, J., APPLETON, H. and PARRY, J.V. 1987. Studies on heat inactivation of hepatitis A virus with special reference to shellfish. Procedures for infection and recovery of virus from laboratory-maintained cockles. *Epidemiology and Infection* **98**, 397–414.

MORMANN, S., DABISCH, M. and BECKER, B. 2010. Effects of technological processes on the tenacity and inactivation of norovirus genogroup II in experimentally contaminated foods. *Applied and Environmental Microbiology* **76**, 536–45.

MURCHIE, L.W., CRUZ-ROMERO, M., KERRY, J.P., LINTON, M., PATTERSON, M.F., SMIDDY, M., KELLY, A.L. 2005. High pressure processing of shellfish: A review of microbiological and other quality aspects. *Innovative Food Science and Emerging Technologies* **6**, 257–70.

MURCHIE, L.W., KELLY, A.L., WILEY, M., ADAIR, B.M. and PATTERSON, M. 2007. Inactivation of a calicivirus and enterovirus in shellfish by high pressure. *Innovative Food Science and Emerging Technologies* **8**, 213–17.

NUANUALSUWAN, S. and CLIVER, D.O. 2003. Infectivity of RNA from inactivated poliovirus. *Applied and Environmental Microbiology* **69**, 1629–32.

PARK, G.W., BARCLAY, L., MACINGA, D., CHARBONNEAU, D., PETTIGREW, C.A. and VINJE, J. 2010. Comparative efficacy of seven hand sanitizers against murine norovirus, feline calicivirus, and GII.4 norovirus. *Journal of Food Protection* **73**, 2232–8.

PARK, G.W., LINDEN, K.G. and SOBSEY, M.D. 2011. Inactivation of murine norovirus, feline calicivirus and echovirus 12 as surrogates for human norovirus (NoV) and coliphage (F^+) MS2 by ultraviolet light (254 nm) and the effect of cell association on UV inactivation. *Letters in Applied Microbiology* **52**, 162–7.

PARRY, J.V. and MORTIMER, P.P. 1984. The heat sensitivity of hepatitis-A virus determined by a simple tissue-culture method. *Journal of Medical Virology* **14**, 277–83.

PECSON, B.M., ACKERMANN, M. and KOHN, T. 2011. Framework for using quantitative PCR as a nonculture based method to estimate virus infectivity. *Environmental Science and Technology* **45**, 2257–63.

RICHARDS, G.P. 2012. Critical review of norovirus surrogates in food safety research: rationale for considering volunteer studies. *Food and Environmental Virology* **4**(1), 6–13.

ROBERTS, P. and HOPE, A. 2003. Virus inactivation by high intensity broad spectrum pulsed light. *Journal of Virological Methods* **110**, 61–5.

RUTJES, S.A., LODDER, W.J., van LEEUWEN, A.D. and HUSMAN, A.M.D. 2009. Detection of infectious rotavirus in naturally contaminated source waters for drinking water production. *Journal of Applied Microbiolog* **107**, 97–105.

RZEZUTKA, A. and COOK, N. 2004. Survival of human enteric viruses in the environment and food. *Fems Microbiology Reviews* **28**, 441–53.

SANCHEZ, G., AZNAR, R., MARTINEZ, A. and RODRIGO, D. 2011. Inactivation of human and murine norovirus by high-pressure processing. *Foodborne Pathogens and Disease* **8**, 249–53.

SANGLAY, G.C., LI, J., URIBE, R. M. and LEE, K., 2011. Electron-beam inactivation of a norovirus surrogate in fresh produce and model systems. *Journal of Food Protection* **74**, 1155–60.

SEYMOUR, I.J. and APPLETON, H. 2001. Foodborne viruses and fresh produce. *Journal of Applied Microbiology* **91**, 759–73.

SHARMA, M., SHEARER, A.E.H., HOOVER, D.G., LIU, M.N., SOLOMON, M.B. and KNIEL, K.E. 2008. Comparison of hydrostatic and hydrodynamic pressure to inactivate food-borne viruses. *Innovative Food Science and Emerging Technologies* **9**, 418–22.

SLOMKA, M.J. and APPLETON, H., 1998. Feline calicivirus as a model system for heat inactivation studies of small round structured viruses in shellfish. *Epidemiology and Infection* **121**, 401–7.

SMIDDY, M., KELLY, A.L., PATTERSON, M.F. and HILL, C., 2006. High pressure-induced inactivation of Q beta coliphage and c2 phage in oysters and in culture media. *International Journal of Food Microbiology* **106**, 105–10.

SOMMER, R., PRIBIL, W., APPELT, S., GEHRINGER, P., ESCHWEILER, H., LETH, H., CABAJ, A. and HAIDER, T. 2001. Inactivation of bacteriophages in water by means of non-ionizing (UV-253.7 nm) and ionizing (gamma) radiation: A comparative approach. *Water Research* **35**, 3109–16.

SOW, H., DESBIENS, M., MORALES-RAYAS, R., NGAZOA, S.E. and JEAN, J. 2011. Heat inactivation of hepatitis A virus and a norovirus surrogate in soft-shell clams (Mya arenaria). *Foodborne Pathogens and Disease* **8**, 387–93.

STRAUB, T.M., BENTRUP, K.H.Z., OROSZ-COGHLAN, P., DOHNALKOVA, A., MAYER, B.K., BARTHOLOMEW, R.A., VALDEZ, C.O., BRUCKNER-LEA, C.J., GERBA, C.P., ABBASZADEGAN, M. and NICKERSON, C.A. 2007. *In vitro* cell culture infectivity assay for human noroviruses. *Emerging Infectious Diseases*, **13**, 396–403.

STRAZYNSKI, M., KRAMER, J. and BECKER, B. 2002. Thermal inactivation of poliovirus type 1 in water, milk and yoghurt. *International Journal of Food Microbiology* **74**, 73–78.

SULLIVAN, R., SCARPINO, P.V., FASSOLITIS, A.C., LARKIN, E.P. and PEELER, J.T. 1973. Gamma-radiation inactivation of coxsackie-virus-B2. *Applied Microbiology* **26**, 14–17.

TEMPLETON, M.R., ANDREWS, R.C. and HOFMANN, R. 2005. Inactivation of particle-associated viral surrogates by ultraviolet light. *Water Research* **39**, 3487–500.

TERIO, V., MARTELLA, V., MOSCHIDOU, P., DI PINTO, P., TANTILLO, G. and BUONAVOGLIA, C. 2010. Norovirus in retail shellfish. *Food Microbiology* **27**, 29–32.

TOPPING, J.R., SCHNERR, H., HAINES, J., SCOTT, M., CARTER, M.J., WILLCOCKS, M.M., BELLAMY, K., BROWN, D.W., GRAY, J.J., GALLIMORE, C.I. and KNIGHT, A.I. 2009. Temperature inactivation of Feline calicivirus vaccine strain FCV F-9 in comparison with human noroviruses using an RNA exposure assay and reverse transcribed quantitative real-time polymerase chain reaction – A novel method for predicting virus infectivity. *Journal of Virological Methods* **156**, 89–95.

VOLKIN, D.B., BURKE, C.J., MARFIA, K.E., OSWALD, C.B., WOLANSKI, B. and MIDDAUGH, C.R. 1997. Size and conformational stability of the hepatitis A virus used to prepare VAQTA, a highly purified inactivated vaccine. *Journal of Pharmaceutical Sciences* **86**, 666–673.

WEI, C., FARKAS, T., SESTAK, K. and JIANG, X. 2008. Recovery of infectious virus by transfection of in vitro-generated RNA from tulane calicivirus cDNA. *Journal of Virology* **82**, 11429–36.

WILKINSON, N., KURDZIEL, A.S., LANGTON, S., NEEDS, E. and COOK, N. 2001. Resistance of poliovirus to inactivation by high hydrostatic pressures. *Innovative Food Science and Emerging Technologies* **2**, 95–8.

WOBUS, C.E., THACKRAY, L.B. and VIRGIN, H.W. 2006. Murine norovirus: a model system to study norovirus biology and pathogenesis. *Journal of Virology* **80**, 5104–12.

13

Preventing and controlling viral contamination of fresh produce

S. Bidawid, Health Canada, Canada

DOI: 10.1533/9780857098870.3.261

Abstract: With mounting evidence indicating an increase in produce (fruits, vegetables and ready-to-eat food) contamination along the food continuum, food safety authorities and the scientific community are examining the best options to prevent contamination of food (pre- and post-harvest) before it reaches the consumer, as well as devising intervention strategies to combat and contain any inadvertent post-harvest food contamination.

Key words: foodborne viruses, norovirus, hepatitis A virus (HAV), food contamination, outbreaks, prevention, guidelines.

13.1 Introduction: why food contamination occurs

It has been clearly evident in recent years that foodborne viruses are a leading cause of major foodborne outbreaks and illness which have a significant impact on public health and inflict economic losses globally. As a result, more focus has been placed on increasing our knowledge of the characteristics of these viruses, developing more efficient and rapid methods for their isolation and detection in various food matrices, and developing strategies to prevent food contamination and/or inactivating viral contaminants in foods before they reach the consumer. The widespread impact of these foodborne viruses on public health worldwide prompted the World Health Organization and the Food and Agriculture Organization of the United Nations to convene a panel of international experts to assess the risk associated with viruses in foods, identifying, among others, noroviruses (NoV) and hepatitis A virus (HAV) as having the most prominent impact (WHO/FAO, 2008). Many types of foods, such as shellfish, fresh produce, and ready-to-eat foods have been implicated in foodborne outbreaks. Norovirus is now considered the leading non-bacterial causative agent of gastroenteritis in the USA and Europe, and is associated with the greatest burden of illness (WHO/FAO, 2008; EFSA, 2011; Gould *et al.*, 2011). Aspects of the different food- and waterborne viruses, their physical and biological characteristics, and detection methodology have already been detailed in various chapters in this book. Therefore, this chapter will focus on the prevalence and survival of viruses in produce-type foods and environments, mechanisms of virus-food adsorption and internalization, and major outbreaks associated with NoV and HAV, as well as addressing selected control and intervention strategies to mitigate and/or prevent contamination of food and environmental settings.

Enteric viruses are prevalent in environmental soil and water, sewage, and sewage-contaminated irrigation water. These viruses are hardy and survive well for long periods of time in their environmental niche, as well as other matrices such food, on inanimate surfaces, on hands and in faeces (Sattar *et al.*, 2000; Stine *et al.*, 2005b; Ethelberg *et al.*, 2010; Mattison *et al.*, 2010). Enteric viruses may contaminate soil through the land disposal of sewage sludge and dirty irrigation water, which in turn contaminate produce. Mounting evidence suggests that viruses can survive long enough and in high enough numbers to cause human diseases through direct contact with polluted water or contaminated foods (Seymour and Appleton, 2001). Astrovirus survived in drinking water after 90 days at 4°C (Abad *et al.*, 1997), HAV and poliovirus can survive in fresh or salt water for up to a year (Sobsey *et al.*, 1988) and in wastewater and groundwater for 90 days or more at 10°C. HAV, rotavirus and other enteric viruses persisted for extended periods (>30 days) on several types of porous and nonporous surfaces (Abad *et al.*, 1997; Sattar *et al.*, 2000). Viruses survived for more than 12 days on lettuce (Bidawid *et al.*, 2000a; Mattison *et al.*, 2007), 30 days on radishes and carrots, and up to 60 days in roots, closed leaves and internal fruit parts (Smith, 1982). Studies by Fallahi and Mattison (2011) showed a one $\log_{10}$ reduction in Murine Norovirus-MNV (a surrogate for norovirus) infectivity after 29 days in water, 4 days on lettuce, 12 days on soil, and 15 days on stainless steel disks. However, MNV genomes were not significantly reduced for up to 42 days, suggesting that genomic detection is not a reliable indicator of viability. Seitz *et al.* (2011) demonstrated that NoV spiked in groundwater remained infectious, when given to human volunteers, after storage at room temperature in the dark for 61 days, and NoV RNA within intact capsids was detected in groundwater for up to 1266 days with no significant reduction. This study demonstrates that NoV in groundwater can remain detectable for over 3 years and can remain infectious for at least 61 days. Similar studies conducted in Japan showed that 100% influent and 50% effluent samples were positive for NoV genomes belonging to a total of 152 different NoV strains, suggesting that genetically diverse NoV strains are co-circulating in aquatic environments and human populations (Kitajima *et al.*, 2012).

Viruses have also been found in produce. A survey of salad vegetables conducted in the Slovak Republic in spring 2008 found that five out of 60 retail produce samples (of lettuce, leeks, spring onions and mixed vegetables) were contaminated with NoV (EFSA, 2011). In a study conducted in Canada in 2009, Mattison *et al.* (2010) surveyed 275 samples of packaged leafy greens for the presence of NoV and rotavirus. Of the samples tested, 148 (54%) were positive for NoV and one (0.4%) for RoV group A by RT-PCR. However, the detection of these viruses was not linked to illness or outbreaks. In a survey conducted in Belgium in 2009, Stals *et al.* (2011) reported ten of the 29 (34.5%) soft red fruit samples, seven of eight (87.5%) samples of cherry tomatoes and one of two (50%) fruit salads tested were positive for NoV by real-time RT-PCR, but were not confirmed by sequence analysis. These findings were not associated with illness or outbreaks. Based on these and other studies, it is obvious that foodborne viruses are prevalent in our surroundings and that foods are vulnerable to contamination along the food continuum from farm to fork.

13.2 Contamination of produce

Contamination can occur at source in the growth and harvesting area from contact with polluted water and inadequately treated or untreated sewage sludge used for irrigation and fertilization. Pre-harvest contamination of produce by foodborne viruses can occur through a variety of routes, including animal faeces/manures, contaminated soil and irrigation water, animals, and human handling. Problems of contamination are magnified by potential countrywide distribution and the possibility of cross-contamination during transport. A primary contamination arises when food materials are already contaminated before they are harvested (e.g., shellfish grown in contaminated waters or soft fruits and vegetables irrigated or sprayed with contaminated water) (Le Guyader *et al.*, 2004; Hjertqvist *et al.*, 2006; Butot *et al.*, 2007; Vivancos *et al.*, 2009). A secondary contamination occurs at harvest or during processing, emphasizing the role of the food handler where produce can become contaminated with foodborne viruses during food handling and preparation, or through the use of polluted water, contaminated surfaces, utensils or materials in processing. This is mostly attributed to infected food handlers not adhering to hygiene standards and regulations (Calder *et al.*, 2003; Ozawa *et al.*, 2007). The possibility that viruses may be mechanically transmitted to fruits and vegetables during irrigation and harvest was suggested by Sadovski *et al.* (1978) who observed the persistence of inoculated poliovirus in drip irrigation pipes and soil. Enteric viruses are the most likely human pathogens to contaminate groundwater. Their extremely small size allows them to infiltrate soils from contamination sources such as broken sewage pipes and septic tanks, eventually reaching aquifers. Viruses can move considerable distances in the subsurface environment with penetration as great as 67 m and horizontal migration as far as 408 m (Borchardt *et al.*, 2003). In a study in the USA, 72% of groundwater sites were positive for human enteric viruses (Fout *et al.*, 2003). Contaminated soil and irrigation water play an important role in pre-harvest contamination of produce (Mara and Sleigh, 2010). Poliovirus injected 10 cm below the surface of the soil was later detected in tomato plant leaves but not in the tomato fruit itself, suggesting that poliovirus can penetrate into plant tissue through the root system (Oron *et al.*, 1995; Hirneisen *et al.*, 2012). Song *et al.* (2006) indicated that significantly greater virus contamination of lettuce was observed in furrow-irrigated plots than in subsurface drip-irrigated plots, suggesting that the virus was transported through the soil matrix via irrigation water movement. When tomato and cucumber plants were irrigated with wastewater effluent, using surface drip irrigation or subsurface drip irrigation methods, Alum *et al.* (2011) demonstrated that viral contamination of the plant stems was greater than plant roots when surface drip irrigation was applied, whereas greater contamination of plant roots occurred when subsurface drip irrigation was applied. Other studies on contamination of produce (iceberg lettuce, cantaloupe, radishes, bell peppers) with viruses (HAV, coliphage, poliovirus, NoV, adenovirus) have also suggested contamination due to the use of contaminated water as diluents of pesticides for spray application (Ward and Irving, 1987; Stine *et al.*, 2011) and flooding plots with wastewater or discharge of wastewater into feeding rivers (Tierney *et al.*, 1977; Hernandez *et al.*, 1997; Petterson and Ashbolt, 2001; Stine *et al.*, 2005a). Potential contamination of foods with viruses has also been shown to occur during post-harvest processing of produce, which may involve spraying, washing, or immersion into water with disinfectants. Alternatively, fruits and/or vegetables can become contaminated when handled by an infected person(s) who can transmit the virus to these commodities which, in turn, become vehicles for virus transmission to consumers, resulting in subsequent infection and illness (Bidawid *et al.*, 2000b, 2004; Seymour and Appleton, 2001; Sumathi *et al.*, 2004; Verhoef *et al.*, 2008). With increased globalization, immigration and increased consumer demand for a wide variety of foods, a greater influx of non-traditional foods has flooded ethnic and domestic markets, particularly in the developed regions of the world. These foods usually originate from countries where good agricultural practice (GAP) and good manufacturing practice (GMP) may be sub-standard and, thus, pose a greater likelihood of food contamination with various microbial pathogens, including viruses. Recent reports have shown that, in addition to traditional food (e.g., leafy greens, fruits) contamination, norovirus is the leading viral pathogen to have caused outbreaks associated with consumption of tropical fruits, whether consumed as whole, fresh cut, dried, juice blends, frozen, pulp, or nectars in markets around the world (Strawn *et al.*, 2011). At the post-harvest stage, infected food handlers not complying with hygiene regulations play a prominent role in contaminating foods which are handled and not intended to be heated before consumption (Widdowson *et al.*, 2005; Ozawa *et al.*, 2007; Marvin *et al.*, 2009). Epidemiological and/or laboratory-confirmed data of

NoV-related food- and waterborne outbreaks from 2000 to 2007 showed that in 42.5% of the cases food handlers were implicated in these outbreaks, followed by water (27.5%), bivalve shellfish (17.5%) and raspberries (10.0%). Studies in Japan indicate that 73% of symptomatic food handlers and 7% of asymptomatic food handlers were positive for norovirus, suggesting that asymptomatic infections are common and contribute to the spread of the infection in outbreak areas (Ozawa *et al.*, 2007). This is further supported by evidence from an investigation of a 2009 norovirus foodborne outbreak in Ireland suggesting a potential role of asymptomatic kitchen food handlers (Nicolay *et al.*, 2011).

13.2.1 Foodborne outbreaks

Although many types of fruit and vegetable have been implicated, salad, lettuce, juice, melon, sprouts, green onion, tomatoes, berries, and ready-to-eat foods have been most often reported as causing a large number of major outbreaks and illness (Mayet *et al.*, 2011; Smith *et al.*, 2011; Donnan *et al.*, 2012; Sarvikivi *et al.*, 2012). According to EFSA (2011), foodborne viruses are the second most common cause of outbreaks in Europe, second only to the ubiquitous Salmonella bacterium. In 2009, they were responsible for 19% of all outbreaks in the EU, causing over 1000 outbreaks and affecting more than 8700 citizens. The total number of outbreaks caused by viruses has been increasing since 2007, with norovirus having contributed to almost 10.2% of the foodborne outbreaks in Europe (Tuan Zainazor *et al.*, 2010; EFSA, 2011). Data analysis by Gould *et al.* (2011) showed that during 2008, 1034 foodborne disease outbreaks were reported in the USA, resulting in 23,152 cases of illness, 1276 hospitalizations and 22 deaths. Among the 479 outbreaks with a laboratory-confirmed single etiological agent reported, norovirus was the most common, accounting for 49% of outbreaks and 46% of illnesses. A recent CDC survey estimated that 58.3% of all foodborne disease outbreaks associated with leafy greens and with confirmed etiologies between 1973 and 2006 were caused by NoV (Herman *et al.*, 2008). Many outbreaks result from contamination of food during preparation and service via unwashed or improperly washed hands of food handlers who are norovirus shedders, resulting in contamination of more than one food item which transmit the virus to consumers who, in turn, extend the spread of the virus through contact with others (Gould *et al.*, 2011). A norovirus outbreak of acute gastroenteritis involving 147 military staff, in which pasta and some raw vegetables were tested positive for norovirus by PCR, as was a cook who prepared the meals, was also reported by Mayet *et al.* (2011). Likewise, Smith *et al.* (2011) reported norovirus outbreaks afflicting 240 persons who had eaten oysters, passion fruit jelly and lavender dish at a gourmet restaurant over a period of 7 weeks in 2009 in England. It was suggested the virus was persistent in oysters but was spread via infected food handlers or the restaurant environment. Data analysis of outbreaks in 2009 in Finland suggested that at least 6 of the 13 norovirus outbreaks which had affected about 900 people could be linked to imported frozen raspberries (Sarvikivi *et al.*, 2012).

With respect to HAV, surveillance data suggest that foodborne outbreaks account for nearly 5% of reported cases of hepatitis A in the USA (CDC, 1996). Most outbreaks occur in a single food establishment and result from contamination of uncooked or previously cooked food by an infected food handler (Carl *et al.*, 1983; Cliver, 2009). Occasionally, more widespread foodborne outbreaks are associated with uncooked or fresh food contaminated before distribution, including shellfish, lettuce, frozen raspberries and strawberries (Ramsay and Upton, 1989; Rosenblum *et al.*, 1990; Desenclose *et al.*, 1991; Niu *et al.*, 1992). In approximately 40% of reported cases of hepatitis A, the source of infection could not be identified (Astridge *et al.*, 2011). However, HAV was linked to some major outbreaks such as that in which HAV was associated with green onions (Hutin *et al.*, 1999). More recently, a number of hepatitis A cases associated with the consumption of contaminated semi-dried tomatoes, solely or as ingredients in salad, were reported in England, France, the Netherlands and Australia. These cases were epidemiologically linked implicating semi-dried tomatoes to be the potential source of this outbreak; no common source could be identified (Carvalho *et al.*, 2011; Donnan *et al.*, 2012; Fournet *et al.*, 2012).

Foodborne outbreaks are often linked to food handlers who acquire the viral infection as a result of environmental exposure or being in contact with ill family members, including children, before handling food. Consequently, sick food handlers contaminate foods which are eaten raw or not further processed (ready-to-eat (RTE) foods) prior to consumption (Anderson *et al.*, 2001; Lederer *et al.*, 2005; de Wit *et al.*, 2007; Schmid *et al.*, 2007). Food handlers can contaminate food either with particles from vomit (e.g., NoV) or from faeces (NoV/HAV) when practicing sub-standard personal hygiene, especially when shedding viruses themselves, for example, after using the toilet, but also after caring for infected persons (e.g., changing of diapers) or cleaning toilet areas used by infected persons (Rizzo *et al.*, 2007; Todd *et al.*, 2007; Noda *et al.*, 2008). Food handlers can transfer viruses from contaminated surfaces and inanimate objects such as utensils to their hands during preparation of ready-to-eat food or from contaminated food to other ready-to-eat foods and surfaces (Bidawid *et al.*, 2000b, 2004; Boxman *et al.*, 2009). It has also been suggested that food handlers who themselves are not sick but may have been in contact with sick family members can pick up and transmit the virus to the workplace. An example of a foodborne NoV outbreak related to a food handler, which was described by Kuo *et al.* (2009), occurred in Austria in 2007, when 21 out of 63 persons became ill after a pre-Christmas celebration among a group of local people. The outbreak was attributed to ham roll which most likely became contaminated with NoV during preparation by a disease-free kitchen assistant, whose infant became sick with laboratory-confirmed NoV gastroenteritis 2 days before the party. Outbreaks of infection with HAV associated with a food handler have also been reported (Chironna *et al.*, 2004; Greig *et al.*, 2007; Todd *et al.*, 2007; Boxman *et al.*, 2011).

13.3 Attachment, adsorption and internalization

A considerable amount of research has focused on the attachment of bacteria to produce such as lettuce, cabbage and sprouts (Elhariry, 2011; Gutiérrez-Rodriguez *et al.*, 2011; Kroupitski *et al.*, 2011; Saldaña *et al.*, 2011). However, NoV attachment to shellfish tissue has also been reported in some studies. Most of these studies seem to identify specific ligand receptors in shellfish tissue to which NoV binds (Maalouf *et al.*, 2010). Studies on VLP (virus-like particle) binding and bioaccumulation have suggested that specific glycan ligands impact bioaccumulation efficiency and that binding to the sialic acid-containing ligand present in all tissues would contribute to retention of virus particles in the gills or mantle (Le Guyader *et al.*, 2006; Maalouf *et al.*, 2011). In contrast, few studies have been conducted to assess viral-produce attachment, adsorption and internalization. Electrostatic and hydrophobic interactions are thought to contribute towards virus adsorption and are controlled by the characteristics of the soil (Gerba, 1983) and the texture of food matrices and environmental influences such as pH, moisture, and surface receptors (Vega *et al.*, 2005, 2008; Esseili *et al.*, 2012). The protein coats of most viruses give the viral particles a net charge due to the presence of amino acids such as glutamic acid, aspartic acid, histidine and tyrosine that contain ionized carboxylic and amino groups. Most enteric viruses have a net negative charge at a pH above five and a net positive charge below pH 5 (Olson *et al.*, 2005). The adsorptive interaction between the virus particle and the adsorbents is a function of the isoelectric point of the virus, as well as that of the adsorbent particle, and also its hydrophobicity. A number of studies have been conducted to elucidate factors influencing virus adsorption to matter. Vega *et al.*, (2005, 2008) investigated (i) 1% Tween 80, (ii) 1 M NaCl, and (iii) 1% Tween 80 with 1 M NaCl to determine the role of hydrophobic, electrostatic, and combined hydrophobic and electrostatic forces, respectively. Echovirus 11 appeared to exhibit reversible attachment above its pI, while exhibiting strong adsorption below its pI. In contrast, FCV demonstrated maximum adsorption above its pI. Attachment above the pI of FCV and echovirus 11 was reduced or eliminated in the presence of NaCl, indicating an electrostatic interaction between the viruses and lettuce. Overall, 1 M NaCl was the most effective treatment in desorbing viruses from the surface of lettuce at pH 7.0 and 8.0. The results imply that electrostatic forces play a major role in controlling virus adsorption to lettuce, and further indicate that incorporating 1 M NaCl in a washing solution would improve the removal or elution of unenveloped viruses from lettuce (Vega *et al.*, 2005, 2008).

Histo-blood group antigens (HBGA), identified as receptors on human gastrointestinal cells to which NoV virions bind (Marionneau *et al.*, 2002;

Huang *et al.*, 2003; Hutson *et al.*, 2003), are also present on the surfaces of oyster gastrointestinal cells. It is postulated, therefore, that during the course of filter feeding, oysters bioaccumulate NoV particles through HBGA binding (Le Guyader *et al.*, 2006; Tian *et al.*, 2007). Since human NoVs have been shown to use carbohydrates of histo-blood group antigens as receptors/coreceptors, Esseili *et al.* (2012) examined the role of carbohydrates in the attachment of NoV to lettuce leaves by using VLPs of a human NoV/GII.4 strain. Immunofluorescence analysis showed that the VLPs attached to the leaf surface, especially to cut edges, stomata, and along minor veins, suggesting that virus binding to cell wall material (CWM) of older leaves was significantly ($P < 0.05$) higher (1.5–2-fold) than that to CWM of younger leaves. Disrupting the carbohydrates of CWM by using 100 mM sodium periodate significantly decreased the binding by up to 43% in older leaves. These results indicate that NoV VLPs bind to lettuce CWM by utilizing multiple carbohydrate moieties. This binding may enhance virus persistence on the leaf surface and prevent effective decontamination. Likewise, using appropriate carbohydrate-disruptive agents may aid in the removal of virus from produce. In a recent study, Gandhi *et al.* (2010) observed that recombinant Norwalk virus-like particles (rNVLP) applied to the surface of romaine lettuce localized as large clusters primarily on the leaf veins. They indicated that extracts of romaine lettuce leaves (RE) bound rNVLP in a dose-dependent manner. However, RE did not bind rNVLP by histo-blood group antigens (HBGA) nor was RE competitive with rNVLP binding to porcine gastric mucin. These results suggest that non-HBGA molecules in RE bind rNVLP by a binding site(s) that is different from the defined binding pocket on the virion. Extracts of cilantro, iceberg lettuce, spinach, and celery also bound rNVLP. Samples of each of the vegetables spiked with rNVLP and tested with anti-NVLP antibody revealed, by confocal microscopy, the presence of rNVLP not only on the veins of cilantro but also throughout the surface of iceberg lettuce. Therefore, it appears that there are some distinct receptors that promote virus binding to different types of produce. However, these data remain scant, and further research on virus-binding receptors is needed.

13.4 Prevention

Most Hazard Analysis and Critical Control Point (HACCP) plans developed by the food industry focus on identifying Critical Control Points (CCPs) aimed at decreasing the amount of bacteria present in food. These plans, however, may not be entirely applicable when attempting to address issues of viral contamination of foods (Koopmans and Duizer, 2004; WHO/FAO, 2008, 2011). Therefore, food manufacturers should consider enteric viruses as a major public health risk in their HACCP plans, particularly when dealing with raw or ready-to-eat food. Effective measures to control the spread of foodborne viruses should focus on preventing contamination at all levels of production rather than on trying to remove or inactivate these viruses from contaminated food (EFSA, 2011). Various proactive approaches can be applied along the food continuum to reduce or prevent viral contamination of produce. Such measures include GAP and GMP, and also good hygienic and handling practices (GHP), good processing practices (GPP), HACCP, monitoring and surveillance. However, intervention measures should also be planned in case of inadvertent contamination of foods, water, food preparation surfaces, and environmental settings. Key recommendations geared towards developing a HACCP system and advice to prevent virus contamination of produce and ready-to-eat foods have been proposed by various investigators and food safety authorities (Koopmans and Duizer, 2004; Cliver, 2009; Doyle, 2010; Havelaar *et al.*, 2010). Furthermore, other sources are available, such as HACCP plan microbiological criteria set by EU 2073/2005, the Codex Alimentarius code of hygiene practice for fresh fruits and vegetables (CAC/RCP 53-2003), and the Recommended International Code of Practice – General Principles of Food Hygiene (CAC/RCP 1-1969) which aims at controlling microbial hazards associated with all stages of the production of fresh fruits and vegetables from primary production to packing. The code focuses on the water quality to be used in primary production (such as irrigation), the use of manure, biosolids and other natural fertilizers, workers (growers, pickers), and transport (open transportation may provide contamination opportunities) from the field to the packing and processing units. In addressing future challenges in food safety, a more proactive, science-based approach is required, starting with predicting where problems might arise by applying the risk analysis framework. The development of effective surveillance networks for early detection of international common-source outbreaks of foodborne viruses is needed to track potential links between outbreaks (Duizer *et al.*, 2008; Koopmans *et al.*, 2008; Havelaar *et al.*, 2010).

13.5 Recommendations

The following guidelines could help reduce or prevent contamination of produce with food viruses.

- Primary and secondary wastewater treatment processes typically reduce virus loads by 1–2 $\log_{10}$ (Verschoor *et al.*, 2005) and can be rendered more efficient by using novel processes such as a membrane bioreactor (MBR) which can achieve a 2–5 $\log_{10}$ decrease in viral load (Zhang and Farahbakhsh, 2007; Kuo *et al.*, 2009; Hirani *et al.*, 2010).
- Animals (wild or domestic) should be kept away and fenced off from produce-growing lots, irrigation water, river streams, or surface water, in case they become a dumping ground for animal excreta.
- Irrigation water should be monitored for any potential contamination with sewage, and corrective measures should be applied immediately.

Consideration should be given to the disposal or treatment of any produce that may have been exposed to sewage material.

- Adequate and clean hygiene and hand washing facilities are essential and should be located in close proximity to the fields, easily accessible, and well equipped with clean potable water and cleaning supplies such as soap, disinfectants, preferably disposable paper hand towels, etc. (Koopmans and Duizer, 2004).
- Hand washing with soap and running water for at least 20 s is the most effective way to reduce or remove viruses from hands, whereas 70% alcohol-based hand sanitizers might serve as an effective adjunct between proper hand washings, but should not be considered a substitute for soap and water hand washing (Bidawid *et al.*, 2000b; Macinga *et al.*, 2008; Liu *et al.*, 2010; Park *et al.*, 2010). As an additional preventive strategy, it is recommended that there be no bare-hand contact with ready-to-eat foods (Hall *et al.*, 2011).
- Food handlers with clinical symptoms of gastroenteritis (diarrhoea and/or vomiting) or with symptoms of acute hepatitis (fever, headache, fatigue combined with dark urine and light stools, or jaundice) should be excluded from food handling and should not be present in the primary production area. Considering the highly infectious nature of norovirus and HAV, and the shedding of viruses in stool for long periods of time (2–8 weeks), exclusion and isolation of infected persons to minimize contact with other people are often the most practical means of interrupting transmission of virus and limiting contamination of the environment (FDA, 2010; Harris *et al.*, 2010).
- Vaccination is effective against HAV. However, it can be very costly and requires two doses at least 6 months apart; also, worker turnover in some segments of the food industry is too great to ensure that the two-shot series will be completed (Cliver, 2009).
- Food industry operators producing or harvesting plant products should keep work areas, equipment, utensils, transport vehicles, containers and storage areas clean and should use appropriate disinfection strategies to eliminate any risk of contamination.
- Randomized and unannounced inspections of work areas should be undertaken to ensure compliance with best practice in support of the safe production of foods.
- Detailed logs and records of inspection, testing, identified areas of concern and corrective actions taken must be maintained. This will help rectify potential problems in a timely manner.
- Workers in the food industry should be trained on various aspects of GMP, GHP, GPP, and HACCP to increase their awareness of potential areas of contamination risk. They should also be made aware of the importance of personal hygiene.
- Intervention plans and strategies should be in place for immediate application in the event of inadvertent workplace contamination, such as by

a vomiting episode. Action protocols and cleaning procedures should be immediately available so that appropriate corrective decontamination measures can be taken.

- Chlorine is the most widely used sanitizer in the food industry. A level of decontamination can be achieved by incorporating 20–200 ppm free chlorine (if the washing water is not highly contaminated/polluted/of high chlorine demand). It has been shown that these amounts of chlorine may result in a reduction of 2–3 $\log_{10}$ in virus titre as compared to washing produce with water only. Since washing occurs at the beginning of the process in the plants, it is very important to refrain from bare-hand contact after the washing step(s). A possible alternative to chlorine is peracitic acid (PAA), which is a mixture of acetic acid and H_2O_2 in an aqueous solution. It outranges the oxidation potential of chlorine and is hardly influenced by organic compounds present in produce wash water. The use of 150–250 mg/L PAA can induce at least an additional one $\log_{10}$ reduction compared to washing produce in water alone (Fraisse *et al.*, 2011).
- Heat treatment at 90°C for 90 s, or pasteurization at 70°C for 15 min to reduce post-packaging contamination, may also be used (Sattar *et al.*, 2003; Baert *et al.*, 2008).
- Food preparation surfaces and equipment should be routinely cleaned with sanitizers and disinfectants to remove food residues and inactivate pathogens (Lages *et al.*, 2008; Baert *et al.*, 2009; Eterpi *et al.*, 2009; Hall *et al.*, 2011). The use of chemical disinfectants is important to interrupt virus spread from contaminated surfaces. Chlorine, as a gas or calcium/sodium hypochlorite, is probably the most commonly used sanitizing agent. Exposure to a range between 20 and 200 ppm, ≥1000 ppm, and 5000 ppm, for 5–10 min, can reduce virus titres by 1, 3 and 4 $\log_{10}$, respectively (Park *et al.*, 2006; EPA, 2001; Hall *et al.*, 2011; Park and Sobsey, 2011). Other sanitizers include: oxidizing agents, such as peroxyacetic acid and hydrogen peroxide (Fraisse *et al.*, 2011); quaternary ammonium compounds; phenolic compounds and aldehydes (Jimenez and Chiang, 2006; Hudson *et al.*, 2007; Park *et al.*, 2007; Pottage *et al.*, 2010). The USEPA-approved product list is available at http://www.epa.gov/oppad001/list_g_norovirus.pdf.
- Any spillage or contamination with faeces or vomit should be cleaned and disinfected immediately, and food handling in the same area(s) should be stopped.

13.6 Additional intervention strategies

Irradiation may be useful in controlling bacteria and parasites, but the high irradiation doses required to significantly reduce viral titres may adversely affect organoleptic properties of foods. A dose of 3 kGy is required to achieve one $\log_{10}$ reduction of HAV on lettuce or strawberries (Bidawid *et al.*, 2000c). Approved maximum limits for irradiating foods in the USA are: spinach and lettuce, 4.0 kGy; spices and herbs, 10.0 kGy; seeds for sprouting, 8.0 kGy (FDA, 2011).

High hydrostatic pressure effectively inactivates some viruses under certain conditions (Grove *et al.*, 2006, 2009; Kingsley *et al.*, 2006; Kingsley and Chen, 2008, 2009; Hirneisen *et al.*, 2010). Exposure to 500 MPa at 4°C for 5 min reduced titres of HAV/FCV/rotavirus by 2.89–3.23 $\log_{10}$ in sausage, green onions, and strawberries (Kingsley *et al.*, 2005).

UV light can be used for disinfection of water and for inactivating microbes on environmental and food surfaces (Nuanualsuwan *et al.*, 2002). Inactivation of HAV, FCV, and aichi virus on lettuce by UV was 4.5–4.6 $\log_{10}$, as compared to only 1.9–2.6 $\log_{10}$ inactivated on strawberries (Fino and Kniel, 2008).

Ozone gas is a strong oxidizing agent that can effectively kill microbes. However, there are limitations to its use in foods. Organic compounds in foods react with and consume ozone, thereby decreasing the amount available to inactivate microbes. Furthermore, ozone reacts with food components and may alter flavors and colors. Ozone is most stable in solution at pH 5 and starts to decompose as pH increases (Hirneisen *et al.*, 2011).

Vacuum freeze-drying is used to manufacture high-quality dehydrated products. This process inactivated HAV and NoV on berries and herbs by 0.29–3.5 $\log_{10}$ (Butot *et al.*, 2009).

13.7 Future trends

More studies are needed in the following areas to determine the different receptors and mechanisms for virus attachment and internalization:

- The development of more efficient and rapid technologies for the isolation of virus particles from different food matrices.
- The development of more sensitive, rapid and field-deployable methodology for virus detection and genotyping in fresh produce.
- Harmonization of methodology for international standardization to enhance source tracking, risk analysis and real-time intervention strategies for viruses in fresh produce. This will also help in establishing integrated international networks for epidemiological monitoring and surveillance.

13.8 Sources of further information and advice

http://www.fao.org/fileadmin/user_upload/agns/pdf/HAV_Tomatoes.pdf (HACCP)

WHO Guidelines for the safe use of wastewater, excreta and groundwater. ISBN 924 154686

http://www.codexalimentarius.net/web/index_en.jsp

http://www.efsa.europa.eu/en/efsajournal/doc/2190.pdf

http://ec.europa.eu/food/food/biosafety/hygienelegislation/legisl_en.htm (EU 2073/2005)

http://www.codexalimentarius.org/standards/list-of-standards/en/ (Codex Alimentarius code of hygienic practice for fresh fruits and vegetables (CAC/RCP 53-2003), and the Recommended International Code of Practice – General Principles of Food Hygiene (CAC/RCP 1-1969).

CDC (Centers for Disease Control and Prevention) (2006). *Norovirus in Healthcare Facilities Fact Sheet*. Atlanta, GA: US Department of Health and Human Services, http://www.cdc.gov/ncidod/dvrd/revb/gastro/downloads/noro-hc-facilities-fs-508.pdf.

13.9 References

ABAD FX, PINTO RM and BOSCH A (1997), Disinfection of human enteric viruses on fomites. *FEMS Microbiol Lett*, **156**, 107–11.

ALUM A, ENRIQUEZ C and GERBA CP (2011), Impact of drip irrigation method, soil, and virus type on tomato and cucumber contamination. *Food Environ Virol*, **4**,78–85.

ASTRIDGE K H, MCPHERSON M, KIRK M D, KNOPE K, GREGORY J, KARDAMANIDIS K and BELL R (2011), Foodborne disease outbreaks in Australia 2001–2009. *Food Australia*, **63**, 44–50.

ANDERSON A D, GARRETT V D, SOBEL J, MONROE S S, FANKHAUSER R L and SCHWAB K J (2001), Multistate outbreak of Norwalk-like virus gastroenteritis associated with a common caterer. *Am J Epidemiol*, **154**, 1013–19.

BAERT L, UYTTENDAELE M, VAN COILLIE E and DEBEVERE J (2008), The reduction of murine norovirus 1, B. fragilis HSP40 infecting phage B40-8 and *E. coli* after a mild thermal pasteurization process of raspberry puree. *Food Microbilol*, **5**, 871–4.

BAERT L, UYTTENDAELE M, STALS A E, DIERICK K, DEBEVERE J and BOTTELDOORN N (2009), Reported foodborne outbreaks due to noroviruses in Belgium: the link between food and patient investigations in an international context. *Epidemiol Infect*, **137**, 316–25.

BIDAWID S, FARBER J M and SATTAR S A (2000a), Survival of hepatitis A virus on modified atmosphere-packaged (MAP) lettuce. *Food Microbiol*, 2001, **18**, 95–102 doi:10.1006/fmic.2000.0380, Available online at http://www.idealibrary.com.

BIDAWID S, FARBER J M and SATTAR S A (2000b), Contamination of foods by food handlers: experiments on hepatitis A virus transfer to food and its interruption. *Appl Environ Microbiol*, **66**, 2759–63.

BIDAWID S, FARBER J M and SATTAR SA (2000c), Inactivation of hepatitis A virus (HAV) by gamma irradiation. *Int J Food Microbiol*, **57**, 91–7.

BIDAWID S, MALIK N, ADEGBUNRIN O, SATTAR S A and FARBER J M (2004), Norovirus cross-contamination during food handling and interruption of virus transfer by hand antisepsis: experiments with feline calicivirus as a surrogate. *J Food Prot*, **67**, 103–9.

BORCHARDT M A, BERTZ P D, SPENCER S K and BATTIGELLI D A (2003), Incidence of enteric viruses in groundwater from household wells in Wisconsin. *Appl Environ Microbiol*, **69**, 1172–80.

BOXMAN I, DIJKMAN R, VERHOEF L, MAAT A, VAN DIJK G, VENNEMA H and KOOPMANS M (2009), Norovirus on swabs taken from hands illustrate route of transmission: a case study. *J Food Prot*, **72**, 1753–5.

BOXMAN I L, VERHOEF L, DIJKMAN R, HAGELE G, TE LOEKE N A and KOOPMANS M (2011), Year-round prevalence of norovirus in the environment of catering companies without a recently reported outbreak of gastroenteritis. *Appl Environ Microbiol*, **77**, 2968–74.

BUTOT S, PUTALLAZ T and SANCHEZ G (2007), Procedure for rapid concentration and detection of enteric viruses from berries and vegetables. *Appl Environ Microbiol*, **73**, 186–92.

BUTOT S, PUTALLAZ T, AMOROSO R and SÁNCHEZ G (2009), Inactivation of enteric viruses in minimally processed berries and herbs. *Appl Environ Microbiol*, **75**, 4155–61.

CALDER L, SIMMONS G, THORNLEY C, TAYLOR P, PRITCHARD K, GREENING G and BISHOP J (2003), An outbreak of hepatitis A associated with consumption of raw blueberries. *Epidemiol Infect*, **131**, 745–51.

CARL M, FRANCIS DP and MAYNARD JE (1983), Food-borne hepatitis A: recommendations for control. *J Infect Dis*, **148**, 1133–5.

CARVALHO C, THOMAS H L, BALOGUN K, TEDDER R, PEBODY R, RAMSAY M and NGUI S L (2011), A possible outbreak of hepatitis A associated with semi-dried tomatoes, England, July. *Euro Surveill*, **17(6)**, pii=20083.

CDC (CENTERS FOR DISEASE CONTROL AND PREVENTION) (1996), Hepatitis surveillance report no. 56. Atlanta: 25.

CHIRONNA M, LOPALCO P, PRATO R, GERMINARIO C, BARBUTI S and QUARTO M (2004), Outbreak of infection with hepatitis A virus (HAV) associated with a foodhandler and confirmed by sequence analysis reveals a new HAV genotype IB variant. *J Clin Microbiol*, **42**, 2825–28.

CLIVER D O (2009), Disinfection of animal manures, food safety and policy. *Bioresour Technol*, **100**, 5392–4.

DESENCLOSE J C, KLONTZ K C, WILDER M H, NAINAN O V, MARGOLIS H S and GUNN R A (1991), A multistate outbreak of hepatitis A caused by the consumption of raw oysters. *Am J Public Health*, **81**, 1268–72.

DE WIT M A, WIDDOWSON M A, VENNEMA H, DE BRUIN E, FERNANDES T and KOOPMANS M (2007), Large outbreak of norovirus: the baker who should have known better. *J Infect*, **55**, 188–93.

DONNAN E J, FIELDING J E, GREGORY J E, LALOR K, ROWE S, GOLDSMITH P, ANTONIOU M, FULLERTON K E, KNOPE K, COPLAND J G, BOWDEN D S, TRACY S L, HOGG G G, TAN A, Adamopoulos, J, Gaston, J and Vally, H (2012), A multistate outbreak of hepatitis A associated with semidried tomatoes in Australia, *Clin Infect Dis*, **54**, 775–81.

DOYLE E (2010), *White Paper on Effectiveness of Existing Interventions on Virus Inactivation in Meat and Poultry Products*. Food Research Institute, University of Wisconsin–Madison, Madison WI 53706 http://fri.wisc.edu/docs/pdf/FRI_Brief_Virus_Inactivation_inFood_6_10.pdf.

DUIZER E, KRONEMAN A, SIEBENGA J, VERHOEF L, VENNEMA H and KOOPMANS M (2008), Typing database for noroviruses, *Euro Surveill*, **13**(19), pii=18867. Available online: http://www.eurosurveillance.org/ViewArticle.aspx?ArticleId=18867.

EFSA PANEL ON BIOLOGICAL HAZARDS (BIOHAZ) (2011), Scientific Opinion on An update on the present knowledge on the occurrence and control of foodborne viruses. Prevention not inactivation key to tackling foodborne viruses. *EFSA J*, **9**, 2190. Doi: 10.2903/j.efsa.2011.2190. Available online: www.efsa.europa.eu/efsajournal.

ELHARIRY H M (2011), Attachment strength and biofilm forming ability of *Bacillus cereus* on green leafy vegetables: cabbage and lettuce. *Food Microbiol*, **28**, 1266–74.

EPA (ENVIRONMENTAL PROTECTION AGENCY) (2001), *The Effectiveness of Disinfectant Residuals in the Distribution System*. www.epa.gov/ogwdw/disinfection/tcr/.../issue-paper_effectiveness.pdf.

ESSEILI M A, WANG Q and SAIF L J (2012), Binding of human GII.4 norovirus virus-like particles to carbohydrates of romaine lettuce leaf cell wall materials. *Appl Environ Microbiol*, **78**, 786–94.

ETERPI M, MCDONNELL G and THOMAS V (2009), Disinfection efficacy against parvoviruses compared with reference viruses. *J Hosp Infect*, **73**, 64–70.

ETHELBERG S, LISBY M, BOTTIGER B. SCHULTZ A C, VILLIF A, JENSEN T, OLSEN K E and SCHEUTZ F, KJELSO, C. and MULLER L (2010), Outbreaks of gastroenteritis linked to lettuce, Denmark. *Euro Surveill*, **15**(6), pii=19484. Available online: http://www.eurosurveillance.org/ViewArticle.aspx?ArticleId=19484.

FALLAHI S and MATTISON K (2011), Evaluation of murine norovirus persistence in environments relevant to food production and processing. *J Food Prot*, **74**, 1847–51.

FDA (FOOD AND DRUG ADMINISTRATION) (2010), *Food code 2009*. College Park, MD: Food and Drug Administration, http://www.fda.gov/Food/FoodSafety/RetailFoodProtection/FoodCode/FoodCode2009/default.htm

FDA (FOOD AND DRUG ADMINISTRATION) (2011), http://www.accessdata.fda.gov/scripts/cdrh/cfdocs/cfcfr/CFRSearch.cfm?fr=179.26.

FINO V R and KNIEL K E (2008), UV light inactivation of hepatitis A virus, Aichi virus, and feline calicivirus on strawberries, green onions, and lettuce. *J Food Prot*, **71**, 908–13.

FOURNET N, BAAS D, VAN PELT W, SWAAN C, OBER H, ISKEN, L, CREMER J, FRIESEMA I. VENNEMA H, BOXMAN I, KOOPMANS M and VERHOEF L (2012), Another possible food-borne outbreak of hepatitis A in the Netherlands indicated by two closely related molecular sequences, July to October 2011. *Euro Surveill*, **17**(6), pii=20079. Available online: http://www.eurosurveillance.org/ViewArticle.aspx?ArticleId=20079.

FOUT G S, MARTINSON B C, MOYER M W and Dahling D R (2003), A multiplex reverse transcription-PCR method for detection of human enteric viruses in groundwater. *Appl Environ Microbiol*, **69**, 3158–64.

FRAISSE A, TEMMAM S, DEBOOSERE N, GUILLIER L, DELOBEL A, MARIS P, VIALETTE M, MORIN T AND Perelle S (2011), Comparison of chlorine and peroxyacetic-based disinfectant to inactivate Feline calicivirus, Murine norovirus and Hepatitis A virus on lettuce. *Int J Food Microbiol*, **151**, 98–104.

GANDHI K M, MANDRELL R E and TIAN P (2010), Binding of virus-like particles of Norwalk virus to romaine lettuce veins. *Appl Environ Microbiol*, **76**, 7997–8003.

GERBA C P (1983), Methods for recovering viruses from the water environment. In G. BERG (ed.), Viral Pollution of the Environment, Boca Raton, FL: CRC Press, pp. 19–35.

GOULD L H, NISLER A L, HERMAN K M, COLE D J, WILLIAMS I T, MAHON B. E, GRIFFIN P M and HALL A J (2011), Surveillance for foodborne disease outbreaks – United States, 2008. *Morbidity and Mortality Weekly Report*, **60:35**, 1197–202.

GREIG J D, TODD E C, BARTLESON C A and MICHAELS B S (2007), Outbreaks where food workers have been implicated in the spread of foodborne disease. Part 1. Description of the problem, methods, and agents involved. *J Food Prot*, **70**, 1752–61.

GROVE S F, LEE A, LEWIS T, Stewar, C M, CHEN H and HOOVER D G (2006), Inactivation of foodborne viruses of significance by high pressure and other processes. *J Food Prot*, **69**, 957–68.

GROVE S F, LEE A, STEWART C M and ROSS T (2009), Development of a high pressure processing inactivation model for hepatitis A virus. *J Food Prot*, **72**, 1434–42.

GUTIÉRREZ-RODRIGUEZ A, GUNDERSEN A, SBODIO O and SUSLOW T V (2011), Variable agronomic practices, cultivar, strain source and initial contamination dose differentially affect survival of Escherichia coli on spinach. *J Appl Microbiol*, **112**, 109–18.

HALL A J, VINJÉ J, LOPMAN B, PARK G W, YEN C, GREGORICUS N and PARASHAR U (2011), Updated norovirus outbreak management and disease prevention guidelines. *Morbidity and Mortality Weekly Report (MMWR)*, **60**(RR03), 1–15. Available online: http://www.cdc.gov/mmwr/preview/mmwrhtml/rr6003a1.htm

HARRIS J P, LOPMAN B A and O'BRIEN S J (2010), Infection control measures for norovirus: a systematic review of outbreaks in semi-enclosed settings. *J Hosp Infect*, **74**, 1–9.

HAVELAAR A H, BRUL S, DE JONG A, DE JONGE R, ZWIETERING M H and TER KUILE B H (2010), Future challenges to microbial food safety. *Int J Food Microbiol*, **139**(Supplement), S79–S94.

HERMAN K M, AYERS T L and LYNCH M (2008), Foodborne disease outbreaks associated with leafy greens, 1973–2006, p. 27. *Abstr. Int. Conf. Emerg. Infect. Dis.*, Atlanta, GA, 16 to 19 March 2008.

HERNANDEZ F, MONGE R, JIMENEZ C and TAYLOR L (1997), Rotavirus and hepatitis A virus in market lettuce (*Latuca sativa*) in Costa Rica. *Int J Food Microbiol*, **37**, 221–3.

HIRANI Z M, DECAROLIS J F, ADHAM S S and JACANGELO J G (2010), Peak flux performance and microbial removal by selected membrane bioreactor systems. *Water Res*, **44**, 2431–40.

HIRNEISEN K A, BLACK E P, CASCARINO J L, FINO V R, HOOVER D G and KNIEL K E, (2010), Viral inactivation in foods: a review of traditional and novel food-processing technologies. *Comp Rev Food Sci Food Safety*, **9**, 3–20.

HIRNEISEN K A, MARKLAND S M and KNIEL K E (2011), Ozone inactivation of norovirus surrogates on fresh produce. *J Food Prot*, **74**, 836–9.

HIRNEISEN K A, SHARMA M and KNIEL K E (2012), Human enteric pathogen internalization by root uptake into food crops. *Foodborne Pathog Dis*, **9**, 396–405.

HJERTQVIST M, JOHANSSON A, SVENSSON N, ABOM P E, MAGNUSSON C, OLSSON M, HEDLUND K O and ANDERSSON Y (2006), Four outbreaks of norovirus gastroenteritis after consuming raspberries, Sweden, June–August 2006. *Euro Surveill*, **11**: E060907.1. Available at: http://www.eurosurveillance.org/ew/2005/050623.asp#1.

HUANG P, FARKAS T, MARIONNEAU S, ZHONG W, RUVOEN-CLOUET,N, MORROW A L, ALTAYE M, PICKERING L K, NEWBURG D S, LEPENDU J and JIANG X (2003), Noroviruses bind to human ABO, Lewis, and secretor histo-blood group antigens: identification of 4 distinct strain-specific patterns. *J Infect Dis*, **188**, 19–31.

HUDSON J B, SHARMA M and PETRIC M (2007), Inactivation of norovirus by ozone gas in conditions relevant to healthcare. *J Hosp Infect*, **66**, 40–45.

HUTIN Y J, POOL V, CRAMER E H, NAINAN O V, WETH J, WILLIAMS I T, GOLDSTEIN S T, GENSHEIMER K F, BELL B P, SHAPIRO C N, ALTER M J and MARGOLIS H S (1999), A multistate, foodborne outbreak of hepatitis A. National Hepatitis A Investigation Team. *N Engl J Med*, **340**, 595–602.

HUTSON A M, ATMAR R L, MARCUS D M and ESTES M K (2003), Norwalk virus-like particle hemagglutination by binding to histo-blood group antigens. *J Virol*, **77**, 405–15.

JIMENEZ L and CHIANG M (2006), Virucidal activity of a quaternary ammonium compound disinfectant against feline calicivirus: a surrogate for norovirus. *Am J Infect Control*, **34**, 269–73.

KINGSLEY D H, GUAN D and HOOVER D G (2005), Pressure inactivation of Hepatitis A virus in strawberry puree and sliced green onions. *J Food Prot*, **68**, 1748–51.

KINGSLEY D H, GUAN D, HOOVER D G and CHEN H (2006), Inactivation of hepatitis A virus by high-pressure processing: the role of temperature and pressure oscillation. *J Food Prot*, **69**, 2454–9.

KINGSLEY D H and CHEN H (2008), Aqueous matrix compositions and pH influence feline calicivirus inactivation by high pressure processing. *J Food Prot*, **71**, 1598–603.

KINGSLEY D H and CHEN H (2009), Influence of pH, salt, and temperature on pressure inactivation of hepatitis A virus. *Int J Food Microbiol*, **130**, 61–4.

KITAJIMA M, HARAMOTO E, PHANUWAN C, KATAYAMA H and FURUMAI H (2012), Molecular detection and genotyping of human noroviruses in influent and effluent water at a wastewater treatment plant in Japan. *J Appl Microbiol*, **112**, 605–13.

KOOPMANS M and DUIZER E (2004), Foodborne viruses: an emerging problem. *Int J Food Microbiol*, **90**, 23–41.

KOOPMANS M, VENNEMA H, HEERSMA H, van STRIEN E, van DUYNHOVEN Y, BROWN D, REACHER M, LOPMAN B and for the European Consortium on Foodborne Viruses

(2008), Early identification of common-source foodborne virus outbreaks in Europe. *J Public Health*, **30**, 82–90. Doi: 10.1093/pubmed/fdm080.

KROUPITSKI Y, PINTO R, BELAUSOV E and SELA S (2011), Distribution of *Salmonella typhimuriurm* in romaine lettuce leaves. *Food Microbiol*, **28**, 990–7.

KUO H W, SCHMID D, SCHWARZ K, PICHLER A M, KLEIN H, KONIG C, DE MARTIN A and ALLERBERGER F (2009), A non-foodborne norovirus outbreak among school children during a skiing holiday, Austria, 2007. *Wien Klin Wochenschr*, **121**, 120–4.

LAGES S L, RAMAKRISHNAN M A and GOYAL S M (2008), *In-vivo* efficacy of hand sanitisers against feline calicivirus: a surrogate for norovirus. *J Hosp Infect*, **68**, 159–63.

LEDERER I, SCHMID D, PICHLER A M, DAPRA R, KRALE, P, BLASSNIG A, LUCKNER-HORNISCHER A, BERGHOLD C and ALLERBERGER F (2005), Outbreak Of norovirus infections associated with consuming food from a catering company, Austria, September 2005. *Euro Surveill*, **10**, E051020 7.

LE GUYADER F S, MITTELHOLZER C, HAUGARREAU L, HEDLUND K O, ALSTERLUND R, POMMEPUY M and SVENSSON L (2004), Detection of noroviruses in raspberries associated with a gastroenteritis outbreak. *Int J Food Microbiol*, **97**, 179–86.

LE GUYADER F S, BON F, DEMEDICI D, PARNAUDEAU S. BERTONE A, CRUDELI S, DOYLE A, ZIDANE M, SUFFREDINI E, KOHLI E, MADDALO F, MONINI M, GALLAY A, POMMEPUY M, POTHIER P and RUGGERI F M (2006), Norwalk virus specific binding to oyster digestive tissues. *Emerg Infect Dis*, **12**, 931–6.

LIU P, YUEN Y, HSIAO H M, JAYKUS L A and MOE C (2010), Effectiveness of liquid soap and hand sanitizer against Norwalk virus on contaminated hands. *Appl Environ Microbiol*, **76**, 394–9.

MAALOUF H, ZAKOUR M, LE PENDU J, LE SAUX J-C, ATMAR R and LE GUYADER F S (2010), Norovirus genogroup I and II ligands in oysters: tissue distribution and seasonal variations. *Appl Environ Microbiol*, **76**, 5621–30.

MAALOUF H, SCHAEFFER J, PARNAUDEAU S, LE PENDU J, ATMAR RL, CRAWFORD S E and LE GUYADER F S (2011), Strain-dependent norovirus bioaccumulation in oysters. *Appl Environ Microbiol*, **77**, 3189–96.

MACINGA D R, SATTAR S A, JAYKUS L A and ARBOGAST W (2008), Improved inactivation of nonenveloped enteric viruses and their surrogates by a novel alcohol-based hand sanitizer. *Appl Environ Microbiol*, **74**, 5047–52.

MARA D and SLEIGH A (2010), Estimation of norovirus infection risks to consumers of wastewater-irrigated food crops eaten raw. *J Water Health*, **8**, 39–43.

MARIONNEAU S, RUVOEN N, LE MOULLAC-VAIDYE B, CLEMENT, M, CAILLEAU-THOMAS A, RUIZ-PALACOIS G, HUANG P, JIANG X and LE PENDU J (2002), Norwalk virus binds to histo-blood group antigens present on gastroduodenal epithelial cells of secretor individuals. *Gastroenterology*, **122**, 1967–77.

MARVIN H J, KLETE, G A, PRANDINI A, DEKKER S and BOLTON D J (2009), Early identification systems for emerging foodborne hazards. *Food Chem Toxicol*, **47**, 915–26.

MATTISON K, KARTHIKEYAN K, Abeb, M, MALIK N, SATTAR S A, FARBER J M and BIDAWID S (2007), Survival of calicivirus in foods and on surfaces: experiments with feline calicivirus as a surrogate for norovirus. *J Food Prot*, **70**, 500–3.

MATTISON K, HARLOW J, MORTON V, COO A, POLLAR F, BIDAWID S and FARBER J M (2010), Enteric viruses in ready-to-eat packaged leafy greens. *Emerg Infect Dis*, **16**, 1815–17, discussion 1817.

MAYET A, ANDREO V, BEDUBOURG G, VICTORION S, PLANTEC J, SOULLIE B, MEYNARD J, DEDIEU J, POLVECHE P and MIGLIANI R (2011), Food-borne outbreak of norovirus infection in a French military parachuting unit, *Euro Surveill*, **16**(30), pii=19930.

NICOLAY N, MCDERMOTT R, KELLY M, GORBY M, PRENDERGAST T, TUITE G, COUGHLAN S, MCKEOWN P and SAYERS G (2011), Potential role of asymptomatic kitchen food handlers during a food-borne outbreak of norovirus infection, Dublin, Ireland, March 2009. *Euro Surveill*, **16**(30), pii=19931. Available online: http://www.eurosurveillance.org/ViewArticle.aspx?ArticleId=19931.

NIU M T, POLISH L B, ROBERTSON B H, KHANNA B K, WOODRUFF B A, SHAPIRO C N, MILLER M A, SMITH J D, GEDROSE J K, ALTER M J and MARGOLIS H S (1992), Multistate outbreak of hepatitis A associated with frozen strawberries. *J Infect Dis*, **166**, 518–24 Doi: 10.1093/infdis/166.3.518.

NODA M, FUKUDA S and NISHIO O (2008), Statistical analysis of attack rate in norovirus foodborne outbreaks. *Int J Food Microbiol*, **122**, 216–20.

NUANUALSUWAN S, MARIAM T, HIMATHONGKHAM S and CLIVER D O (2002), Ultraviolet inactivation of feline calicivirus, human enteric viruses and coliphages. *Photochem Photobiol*, **76**, 406–10.

OLSON M R, AXLER R P, HICKS R E, HENNECK J R and MCCARTHY B J (2005), Seasonal virus removal by alternative onsite wastewater treatment systems. *J Water Health*, **3**, 139–55.

ORON G, GOEMANS M, MANOR Y and FEYEN J (1995), Poliovirus distribution in the soil-plant system under reuse of secondary wastewater. *Water Res*, **29**, 1069–78.

OZAWA K, OKA T, TAKEDA N and HANSMAN G S (2007), Norovirus infections in symptomatic and asymptomatic food handlers in Japan. *J Clin Microbiol*, **45**, 3996–4005.

PARK G W, VINJÉ J and SOBSEY M D (2006), The comparison of UV and chlorine disinfection profiles of murine norovirus (MNV-1), feline calicivirus (FCV), and coliphage MS2 (Presentation), American Society for Microbiology (ASM), Orlando, Florida, May 21–25.

PARK G W, BOSTON D M, KASE J A, SAMPSON M N and SOBSEY M D (2007), Evaluation of liquid- and fog-based application of Sterilox hypochlorous acid solution for surface inactivation of human norovirus. *Appl Environ Microbiol*, **73**, 4463–8.

PARK G W, BARCLAY L, MACINGA D, CHARBONNEAU D, PETTIGREW CA and VINJE J (2010), Comparative efficacy of seven hand sanitizers against murine norovirus, feline calicivirus, and GII.4 norovirus. *J Food Prot*, **73**, 2232–8.

PARK G W and SOBSEY M D (2011), Simultaneous comparison of murine norovirus, feline calicivirus, coliphage MS2, and GII.4 norovirus to evaluate the efficacy of sodium hypochlorite against human norovirus on a fecally soiled stainless steel surface. *Foodborne Pathogens Dis*, **8**, 1005–10.

PETTERSON S R and ASHBOLT N J (2001), Viral risks associated with wastewater reuse: modeling virus persistence on wastewater irrigated salad crops. *Water Sci Technol*, **43**, 23–6.

POTTAGE T, RICHARDSON C, PARKS S, WALKER J T and BENNETT A M (2010), Evaluation of hydrogen peroxide gaseous disinfection systems to decontaminate viruses. *J Hosp Infect*, **74**, 55–61.

RAMSAY C N and UPTON P A (1989), Hepatitis A and frozen raspberries. *Lancet*, **i**, 43–4.

RIZZO C, DI BARTOLO I, SANTANTONIO M, COSCIA M F, MONNO R, DE VITO D, RUGGERI F M and RIZZO G (2007), Epidemiological and virological investigation of a Norovirus outbreak in a resort in Puglia, Italy. *BMC Infect Dis*, **7**, 135.

ROSENBLUM L S, MIRKIN I R, ALLEN D T, SAFFORD S and HADLER S C (1990), A multifocal outbreak of hepatitis A traced to commercially distributed lettuce. *Am J Public Health*, **80**, 1075–9.

SADOVSKI A Y, FATTAL B, GOLDBERG D, KATZENELSON E and SHUVAL H I (1978), High levels of microbial contamination of vegetables irrigated with wastewater by the drip method. *Appl Environ Microbiol*, **36**, 824–30.

SALDAÑA Z, SÁNCHEZ E, XICOHTENCATL-CORTES J, PUENTE J L and GIRÓN J A (2011), Surface structures involved in plant stomata and leaf colonization by Shiga-toxigenic *Escherichia coli* O157:H7. *Frontiers Microbiol*, **2**, 119. Doi: 10.3389/fmicb.2011.00119.

SARVIKIVI E, ROIVAINEN M, MAUNULA L, NISKANEN T, KORHONEN T, LAPPALAINEN M and KUUSI, M (2012), Multiple norovirus outbreaks linked to imported frozen raspberries. *Epidemiol Infect*, **140**, 260–7.

SATTAR S A, JASON T, BIDAWID S and FARBER J (2000), Foodborne spread of hepatitis A: recent studies on virus survival, transfer and inactivation. *Can J Infect Dis*, **11**, 159–63.

SATTAR S A, SPRINGTHORPE VS, ADEGBUNRIN O, ZAFER A A and BUSA M A (2003), Disc-based quantitative carrier test method to assess the virucidal activity of chemical germicides. *J Virol Methods*, **112**, 3–12.

SCHMID D, STUGER H P, LEDERER I, PICHLER A M, KAINZ-ARNFELSER G, SCHREIER E and ALLERBERGER F (2007), A foodborne norovirus outbreak due to manually prepared salad, Austria 2006. *Infection*, **35**, 232–9.

SEITZ S R, LEON J S, SCHWAB K J, LYON G M, DOWD M, MCDANIELS M, ABDULHAFID G, FERNandEZ M L, LINDESMITH L C, BARIC R S and MOE C L (2011), Norovirus infectivity in humans and persistence in water. *Appl Environ Microbiol*, **77**, 6884–8.

SEYMOUR I J and APPLETON H (2001), Foodborne viruses and fresh produce. *J Appl Microbiol*, **91**, 759–73.

SMITH M A (1982), Retention of bacteria, viruses and heavy metals on crops irrigated with reclaimed water. *Canberra: Australian Water Resource Council*, pp. **308**.

SMITH A J, MCCARTHY N, SALDANA L, IHEKWEAZU C, MCPHEDRAN K, ADAK G. K, ITURRIZA-GOMARA M, BICKLE G and O'MOORE E (2011), A large foodborne outbreak of norovirus in diners at a restaurant in England between January and February 2009. *Epidemiol Infect*, **140**(9), 1695–701.

SOBSEY M D, FUJI T and SHIELDS P A (1988), Inactivation of hepatitis A virus and model viruses in water by free chlorine and monochloramine. *Water Sci Technol*, **20**, 385–91.

SONG I, STINE S W, CHOI C Y and GERBA C P (2006), Comparison of crop contamination by microorganisms during subsurface drip and furrow irrigation. *J Environ Engr*, **132**, 1243–8.

STALS A, BAERT L, DE KEUCKELAERE A, VAN COILLIE E and UYTTENDAELE M (2011), Evaluation of a norovirus detection methodology for ready-to-eat foods. *Int J Food Microbiol*, **145**, 420–5.

STINE S, SONG I, CHOI C and GERBA C (2011), Application of pesticide sprays to fresh produce: a risk assessment for hepatitis A and *Salmonella*. *Food Environ Virol*, **3**, 86–91.

STINE S W, SON I, CHOI C Y and GERBA C P (2005a), Application of microbial risk assessment to the development of standards for enteric pathogens in water used to irrigate fresh produce. *J Food Prot*, **68**, 913–18.

STINE S W, SONG I, CHOI C Y and GERBA C P (2005b), Effect of relative humidity on preharvest survival of bacterial and viral pathogens on the surface of cantaloupe, lettuce, and bell peppers. *J Food Prot*, **68**, 1352–8.

STRAWN L K, SCHNEIDE, K R and DANYLUK M D (2011), Microbial safety of tropical fruits. *Crit Rev Food Sci Nutr*, **51**, 132–45.

SUMATHI S, CINDY R F, LINDA C and ROBERT V T (2004), Fresh produce: a growing cause of outbreaks of foodborne illness in the United States, 1973 through 1997. *J Food Prot*, **67**, 2342–53.

TIAN P, ENGELBREKTSON A L, JIANG X, ZHONG W and MANDRELL R E (2007), Norovirus recognizes histo-blood group antigens on gastrointestinal cells of clams, mussels, and oysters: a possible mechanism of bioaccumulation. *J Food Prot*, **70**, 2140–7.

TIERNEY J T, SULLIVAN R and LARKIN E P (1977), Persistence of poliovirus 1 in soil and on vegetables grown in soil previously flooded with inoculated sewage sludge or effluent. *Appl Environ Microbiol*, **33**, 109–13.

TODD E C, GREIG J D, BARTLESON C A and MICHAELS B S (2007), Outbreaks where food workers have been implicated in the spread of foodborne disease. Part 4. Infective doses and pathogen carriage. *J Food Prot*, **71**, 2339–73.

TUAN ZAINAZOR C, HIDAYAH M S, CHAI L C, TUNUNG R, GHAZALI F M and SON R (2010), The scenario of norovirus contamination in food and food handlers. *J Microbiol Biotechnol*, **20**, 229–37.

VEGA E, SMITH J, GARLAND J, MATOS A and Pillaii S D (2005), Variability of virus attachment patterns to butterhead lettuce. *J Food Prot*, **68**, 2112–7.

VEGA E, GARLAND J and PILLAI S D (2008), Electrostatic forces control nonspecific virus attachment to lettuce. *J Food Prot*, **71**, 522–9.

VERHOEF L, BOXMAN I L and KOOPMANS M (2008), CAB Reviews: Perspectives in Agriculture, Veterinary Science, *Nutrition and Natural Resources* **3**, No. 078.

VERSCHOOR L F, DE RODA HUSMAN D R, VAN DEN BERG A M, STEIN H H, VAN PELT-HEERSCHAP A M and VAN DER POEL W H (2005), Year-round screening of noncommercial and commercial oysters for the presence of human pathogenic viruses. *J Food Prot*, **68**, 1853–9.

VIVANCOS R, SHROUFI A, SILLIS M, AIRD H, GALLIMORE C I, MYERS L, MAHGOUB H and NAIR P (2009), Food-related norovirus outbreak among people attending two barbeques: epidemiological, virological, and environmental investigation. *Int J Infect Dis*, **13**, 629–35.

WARD B K and IRVING L G (1987), Virus survival on vegetables spray-irrigated with wastewater. *Water Res*, **21**, 57–63.

WIDDOWSON M A, SULKA A, BULENS S N, BEARD R S, CHAVES S S, HAMMOND R, SALEHI E D, SWANSON E, TOTARO J, WORON R, MEAD P S, BRESEE J S, MONROE S S and GLASS R I (2005), Norovirus and foodborne disease, United States, 1991–2000. *Emerg Infect Dis*, **11**, 95–102.

WHO/FAO (2008), Viruses in food: scientific advice to support risk management activities, microbiological risk assessment series #13, ISBN: 1726-5274.

WHO/FAO (2011), Prevention and control of hepatitis A virus (HAV) and norovirus (NoV) in ready-to-eat semi-dried products. Available at: http://www.fao.org/filead-min/user_upload/agns/pdf/HAV_Tomatoes.pdf.

ZHANG K and FARAHBAKHSH K (2007), Removal of native coliphages and coliform bacteria from municipal wastewater by various wastewater treatment processes: implications to water reuse. *Water Res*, **41**, 2816–24.

14

Preventing and controlling viral contamination of shellfish

J. W. Woods and W. Burkhardt III, US Food and Drug Administration, USA

DOI: 10.1533/9780857098870.3.281

Abstract: A significant proportion of infectious foodborne illnesses worldwide can be attributed to human enteric viruses. Bivalve molluscan shellfish can act as vehicles for transmission of enteric viruses. Viruses accumulate in shellfish digestive diverticula and can infect humans upon ingestion of raw or undercooked shellfish meat. The use of indicator bacteria as an index of fecal contamination in shellfish growing areas reduced bacterial gastrointestinal infections but this practice is believed to have limited predictive value for enteric viral pathogen contamination. Mitigation strategies such as depuration or relaying exist to reduce or eliminate potential enteric virus hazards associated with consumption of raw or undercooked shellfish.

Key words: enteric viruses, shellfish, norovirus, detection, mitigation.

14.1 Introduction

According to epidemiological evidence, enteric viruses are the leading cause of foodborne infectious diseases in the United States (Mead *et al.*, 1999; Scallan *et al.*, 2011). Of the 9.4 million annual estimated cases of known foodborne illnesses, 58% are attributed to human noroviruses (NoV). Virtually any food may be implicated in norovirus transmission, but bivalved molluscan shellfish present a relatively high risk because of their ability to concentrate viruses from contaminated waters. Human NoV and hepatitis A virus (HAV) are the agents of greatest concern from the consumption of shellfish and NoV is the leading cause of non-bacterial shellfish-associated illnesses in the US (Katz *et al.*, 2012). Since most enteric viruses are robust and environmentally stable, they may withstand current wastewater treatment methods and pollute shellfish harvested for human consumption. The efficiency of wastewater treatment is generally assessed on the removal/inactivation of fecal coliform bacteria. However, since the relationship between indicator organisms and levels of enteric viruses is known to vary, measurement of bacterial levels provides little reassurance of the risks imposed by enteric virus levels in water.

The National Shellfish Sanitation Program (NSSP) and its Model Ordinance (MO) provide information on establishing the sanitary classification of estuarine waters to produce safe shellfish and establishing safeguards to insure minimal pollution impact (NSSP, 2009). The shellfish growing waters in the United States are classified by sanitary surveys and can be categorized as approved, conditionally approved, conditionally restricted, restricted, or prohibited. Shellfish harvested from approved and conditionally approved areas can be directly sold to market without any post-harvest processing and those harvested from restricted areas require relay or depuration. Shellfish from restricted areas are also suitable for low acid canning. While these guidelines are rigorously adhered to by State shellfish control authorities, viral gastroenteritis attributed to shellfish consumption still occurs.

14.2 Human enteric viruses in the environment

Human enteric viruses pose a significant health threat in the aquatic environment since they are transmitted via the fecal oral route. Human activities such as faulty septic systems, agricultural runoff, urban runoff, sewage outfall, and direct illegal wastewater discharge from vessels are ways enteric viruses are introduced into the environment. There are approximately one hundred and forty enteric viruses found in human waste and at least 10% of the population can shed these viruses at any given time (Griffin *et al.*, 2003). Enteric viruses can be transported throughout the environment by attaching to particulates in groundwater, estuarine water, seawater and rivers, estuaries, and by aerosols emitted from sewage treatment plants (Bosch, 1998). The fate of these enteric viruses can take many routes such as rivers, lakes, sewage, land runoff, estuaries, and ground water. Humans can be exposed to enteric viruses through various routes: crops grown in land irrigated with wastewater or fertilized with sewage, shellfish grown in contaminated water, sewage-polluted recreational waters and contaminated drinking water. In a waterborne disease outbreak study between 1946 and 1980, water system deficiencies that contributed to these outbreaks were categorized under five major headings: (i) use of contaminated untreated surface water, (ii) use of contaminated untreated groundwater, (iii) inadequate or interrupted treatment (iv) distribution network problems, and (v) miscellaneous (Lippy and Waltrip, 1984). Deficiencies in treatment and distribution of water contributed to more than 80% of the outbreaks.

Human enteric viruses can be transmitted by water, food, fomites, and by person-to-person contact. They typically have a low infectious dose, which makes them an immediate public health concern. In some instances, such as norovirus infections, the infectious dose can be a little as one to ten virions with a secondary attack rate of 50% (Koopmans and Duizer, 2004; Teunis *et al.*, 2008). The risk for infection when consuming viral contaminated water is at least tenfold greater than that for pathogenic bacteria with similar exposures (Haas *et al.*, 1993).

Enteric viruses in water are of particular concern because of the potential for contamination from a variety of sources. Because significant advances have been made in the area of environmental virology, enteric viruses have now been recognized as the causative agents in many non-bacterial gastroenteritis cases and outbreaks previously identified as of unknown etiological origin (Bosch, 1998). Enteric viruses encompass a diverse group of organisms and have been detected and linked to many outbreaks from contaminated water and foods (Beuret *et al.*, 2002; Daniels *et al.*, 2000; Munnoch *et al.*, 2004). Since it is not practical to monitor all pathogenic viruses and indicator bacteria have not been shown to be effective viral surrogates, an indicator of viral contamination should be a human enteric virus or bacteriophage.

14.3 Enteric viruses in sewage and shellfish

Molluscan bivalves are shellfish that have two shell halves that hinge together. Commercial types commonly harvested and sold in the United States are the Pacific oyster (*Crassostrea gigas*), Eastern oyster (*Crassostrea virginica*), Quahog clam (*Mercenaria mercenaria*), and blue mussel (*Mytilus edulis*) (NOAA, 2007). These animals attach a substrate or bury themselves in the water floor. Molluscan bivalves vary in their characteristics and habitat. When out of the water, most animals close their shell tight to retain a marine environment around their internal parts (Lees, 2000). Most shellfish can survive weeks out of the water under refrigeration, but their quality declines over time.

Individuals infected with enteric viruses transmitted by the fecal oral route can shed billions of viral particles in their feces. Sewage treatment plants remove the majority of viruses and other micro-organisms but removal efficiency, which can range between 87% and 99%, varies between groups of organisms and the type of treatment (Burkhardt *et al.*, 2005). Although enteric viruses may be present in low concentrations after treatment, it may only take one virion to cause disease. Once in the environment, enteric viruses can survive for weeks to months either in the water column or by attaching to particulate matter and accumulating in sediments (Bosch, 1998). In the process of filter feeding, bivalve shellfish concentrate and retain human pathogens from their surrounding water, thus making microbial contamination levels in shellfish tissue significantly greater than those in overlying water (Cabelli, 1988; LaBelle and Gerba, 1979). The risks posed by bioaccumulation of pathogenic micro-organisms are exacerbated by the traditional consumption of shellfish raw or lightly cooked. This circumstance is unique to bivalved shellfish and it represents a special case among microbial hazards associated with food that dates back to the 1800s (Rippey, 1994). Recent epidemiological evidence suggests that human enteric viruses are the most common etiological agent implicated in the transmission of infectious disease due to the consumption of contaminated shellfish (Katz *et al.*, 2012). As these viruses are retained in the shellfish, they do not increase in number because they are obligate intracellular parasites and require human cells in which to replicate. Current microbiological indicators serve as a predictor of fecal contamination in shellfish growing areas and have succeeded to some extent in preventing shellfish-associated bacterial gastrointestinal infections but are considered to have limited predictive value for viral enteric pathogen contamination in shellfish (Goyal, 2006; Pina *et al.*, 1998; Rippey, 1994).

In recent years the incidence of gastroenteritis caused by enteric viruses has not significantly increased while advances in research and technology have allowed for better detection methods and understanding of these viruses. While most enteric viruses are found more commonly in the winter or during colder temperatures, the ability of the shellfish to accumulate viruses coupled with increased community illnesses in colder climates increases the risk of gastrointestinal illnesses associated with shellfish consumption (Mounts *et al.*, 2000). This phenomenon has been documented in several outbreaks which occurred during cold times of the year (DePaola *et al.*, 2010; Iritani *et al.*, 2010; Maalouf *et al.*, 2010; Woods and Burkhardt, 2010). Cold storage or immediate freezing of shellfish after harvest can also represent ideal conditions for maintaining enteric viruses in shellfish. In outbreaks occurring in 2006 and 2011, shellfish that had been imported were flash frozen immediately upon harvesting. These shellfish were later consumed and subsequently implicated in a shellfish outbreak of gastroenteritis cause by norovirus (Woods and Burkhardt, 2010).

14.4 Survival of enteric viruses in the environment

Dissemination of enteric viruses is dependent on their interaction not only with a host, but also with the environment. Viruses possess no inherent metabolism outside the host and may be thought of as inert particles that do not require nutrients to persist outside the host. They somehow possess a level of toughness that allows them to remain infectious during various conditions in the environment as they are transferred from one host to another. The sheer number of enteric viral diseases transmitted by fecal oral route in the environment demonstrates their robustness (Rippey, 1994; Scallan *et al.*, 2011).

Enteric viruses increase their chance of transmission the longer they can survive outside the host. Various environmental conditions and other factors such as heat, moisture and pH will affect their chances of survival (Bosch, 1998). The presence and extent of these conditions will vary among different environments. If the risks that enteric pathogens pose are to be fully understood, knowledge of enteric virus survival in the environment and the factors that influence it are important.

Most studies used to determine the potential for survival of enteric viruses have been conducted using basic principles. For example, a known number of infectious viruses have been artificially introduced into a sample of water, food, soil etc., and the sample stored under conditions relevant to those in the environment. After a specified time, the viruses are extracted and enumerated. There are varying methods for extraction of viruses from the environment and from foods, including, typically, inorganic extraction, ultracentrifugation, and proteinase K digestion, coupled with detection using molecular techniques or cell culture. Using cell culture plaque assays for culturable viruses, along with molecular methods for detection, allows comparison of infectious particles remaining in the sample with the amount of virus that was introduced into the sample. Statistical analysis can be used to determine the significance of the results.

Survival of enteric viruses in environmental water has been extensively investigated. Utilizing simulated natural conditions, Loh *et al.* (1979) inoculated poliovirus into samples of coastal water from the plume of a sewage outlet and samples of water from 6.4 km away from the plume. Samples were mixed continuously and incubated at 24°C for 4 days. The virus titer had dropped slightly after 1 day and a complete inactivation was observed after 72–120 h. In this study, there was data to suggest that a virus-inactivating component of a biological nature was present in the sewage-polluted water and the water retrieved miles away from the plume. In 1980, Fujioka *et al.* (1980) substantiated this evidence in a study where the antiviral activity of the seawater samples was lost when it was challenged with filtration, boiling, or autoclaving. A study conducted by Hurst *et al.* in 1989 analyzed the long-term survival of species of enteroviruses in surface freshwater. Over a period of 12 weeks, temperatures of −20, 1, and 22° C were shown to have virus inactivation levels of 0.4–0.8, 4–5, and 6.5–7.0 log reduction, respectively. In this study, many physical and chemical parameters appeared crucial to virus survival, including turbidity and suspended solids.

Although the majority of foodborne viral outbreaks can be traced to food contaminated by infected food handlers (Koopmans and Duizer, 2004), survival and persistence of enteric viruses in shellfish represents a unique challenge because of the nature of these filter feeding animals. Molluscan shellfish accumulate the virus from contaminated harvest water and when contaminated shellfish are consumed raw or slightly cooked, there is a potential for infection to occur. There are several recorded outbreaks of gastroenteritis where shellfish contaminated with enteric viruses was implicated as the vehicle of transmission (Berg *et al.*, 2000; Butt *et al.*, 2004; Gallimore *et al.*, 2005; Woods *et al.*, 2010). There have also been several studies demonstrating the survival and persistence of enteric viruses in shellfish. A 1970 study conducted by DiGirolamo *et al.* examined the survival of poliovirus in chilled, frozen, and processed Pacific (*Crassostrea gigas*) and Olympia (*Ostrea lurdia*) oysters. After 15 days of storage at 5°C, infectious virus in the Olympia oyster was reduced by 60%, and 13% remained infectious after 30 days. In the frozen

Pacific oyster, infectious virus was reduced by less than 10% in 4 weeks, and by 12 weeks only 10% of the infectious virus remained. In 2004, Hewitt and Greening studied the survival and persistence of NoV, HAV, and feline calicivirus (FCV) in marinated mussels. Marinated green-lipped mussels (*Perna canaliculus*) and marinade liquid were inoculated with NoV, HAV, and FCV and held at 4°C for up to 4 weeks. Survival of HAV and FCV was determined by $TCID_{50}$ and persistence of the non-culturable NoV was determined by RT-PCR assay. Over 4 weeks, HAV survived exposure to the marinade at a low pH (3.75). There was a 1.7 log reduction in HAV $TCID_{50}$ titer but no reduction in the NoV or HAV RT-PCR titer after 4 weeks.

Persistence and survival of enteric viruses in shellfish pose a challenge. Further work is required to gain a better understanding of enteric viruses causing illnesses. Development of robust and reliable detection methods for recovery of these viruses will provide additional information necessary to recover clinically significant enteric viruses in the environment and foods.

14.5 Mitigation strategies and depuration

Mitigation strategies, such as relaying or depuration, are post-harvest treatments used to reduce or eliminate infectious micro-organisms in shellfish. Depuration is a form of post-harvest processing where shellfish are placed in tanks with clean estuarine water and allowed to depurate or purge; the concept dates back to the 1900s (Liu *et al.*, 1967). Relaying of shellfish focuses on harvesting shellfish from contaminated water and relocating them to clean water where shellfish will naturally purge contaminants. These approaches reduce the levels of micro-organisms present in the shellfish in an effort to decrease the potential of ingesting infectious micro-organisms.

The majority of current research on mitigation strategies in relation to viruses has been carried out on surrogates because the most common agents identified as causing shellfish-associated outbreaks cannot be propagated using traditional cell culture techniques.

Many studies on the depuration dynamics of shellfish have consistently demonstrated that bacteria levels can be effectively reduced within 24–48 h, but enteric viruses, such as norovirus and hepatitis A, have depuration rates from 96 h up to 3 weeks (Dore and Lees, 1995; Kingsley and Richards, 2003; Love *et al.*, 2010). Metcalf *et al.*, (1979) demonstrated an 80–88% elimination of poliovirus from soft-shelled clams after 48 h and 98–99% elimination after 144 h. Their study also demonstrated that bioaccumulation rates of viruses vary according to size, with the larger shellfish having higher rates.

Since it has been suggested that male-specific bacteriophage (MSB) would be an ideal indicator for enteric virus contamination, studies have demonstrated that their depuration rates vary depending on the type and species of shellfish (Dore *et al.*, 2000; Humphrey and Martin, 1993; Lasobras *et al.*, 1999). In Love *et al.* (2010), MSB was shown to depurate at much faster rates in oysters (*Crassostrea virginica*) than clams (*Mercenaria mercenaria*), though both shellfish types demonstrate a slower rate of depuration for NoV and HAV than MSB. Despite mitigations strategies, outbreaks due to shellfish contaminated with enteric viruses still occur. In 1983, oysters contaminated with norovirus were implicated an outbreak that occurred in Tower Hamlets, London (Gill *et al.*, 1983). The investigation demonstrated that the oysters were depurated for at least 72 h and there was no evidence of fecal contamination. In 2007 an oyster-associated outbreak occurred in France, where 111 cases of HAV infection were identified (Guillois-Becel *et al.*, 2009).The source of the shellfish contamination was unknown but it was thought that the contamination occurred in the storage depuration tanks. In 2008, Le Guyader *et al.*, reported on a shellfish-associated outbreak where five different enteric viruses (Aichi virus, norovirus, astrovirus, enterovirus, and rotavirus) were detected in oysters. The investigation showed that heavy rains led to increased enteric virus contamination in the shellfish growing area. The French Institute of Research for the Exploitation of the Sea (IFREMER) and the Department Directorate of Health and Social Affairs (DDASS) recommended closure of shellfish harvesting, due to the heavy rainfall. Instead, shellfish were depurated for additional days before being released for sale. In the US, shellfish growing areas are most frequently closed due to increased levels of fecal coliforms that often occur after a significant rainfall event. Reopening is based upon the length of time after the event has occurred and is verified by post-event water monitoring for fecal coliform levels. Due to the limited approved growing areas for harvest of shellfish and the cost of depuration, availability of depurated shellfish in the United States is very limited.

Despite current guidance for the control of shellfish safety, the US does experience occasional shellfish-associated outbreaks due to enteric viruses, although a recent survey demonstrated NoV and HAV levels of 3.8% and 4.2% respectively in US retail market oysters (DePaola *et al.*, 2010).

14.6 Current regulations

In the United States, the NSSP and its Guide identifies the requirements necessary to regulate interstate commerce of molluscan shellfish to assure sale of safe shellfish (NSSP, 2009). The US Food and Drug Administration and shellfish-producing States administer this program. The NSSP and its precursor Committees were established in the 1920s to prevent shellfish-borne *Salmonella typhi* illnesses. Initial controls were premised upon acceptable levels of coliforms (total and fecal coliforms) in the water where the shellfish were harvested. Additional controls to minimize the potential impact of anthropogenic pollution on the sanitary quality were instituted through comprehensive sanitary surveys. These surveys are based upon the identification of pollution sources, their magnitude, and their potential impact on the shellfish growing area. NSSP requirements and guidance are set out in FDA's *Guide for the Control of Molluscan Shellfish*, which is published on the FDA website at: http://wcms.fda.gov/FDAgov/Food/FoodSafety/Product-SpecificInformation/Seafood/federalStatePrograms/NationalShellfishSanitationProgram/ucm046353.htm. In addition to physical and administrative controls, the NSSP requires that State authorities educate all licensed harvesters about the hazards associated with overboard and direct vessel discharge of human body waste into shellfish areas. These are explicitly prohibited by the NSSP and all shellfish industry vessels must be equipped with receptacles on board to contain and segregate body wastes from harvested shellfish.

While the utility of coliform organisms has proven effective in preventing shellfish-borne bacterial illnesses, their effectiveness in preventing enteric viral illnesses has been questioned. The inability of coliforms to index viral pathogens in shellfish and water is likely due to several reasons including differential susceptibility to wastewater treatment and disinfection, environmental stability, differences in bioaccumulation, and elimination by shellfish. With the recognition of this disparity between coliforms (including total coliforms, fecal coliforms, and *E. coli)* and enteric viruses, the identification and utility of alternative indicator micro-organisms have been sought. Currently MSB which have similar physio-chemical properties to those viruses of greatest concern to shellfish consumers (notably norovirus and hepatitis A virus) have been identified as an indicator group that can be used to augment the existing bacterial indicator system.

In 2009 the Interstate Shellfish Sanitation Conference (ISSC) adopted for inclusion into the NSSP's Model Ordinance (MO) the use of MSB to reopen shellfish harvest areas that were closed as an emergency measure due to a discharge of raw untreated sewage from a large community sewage collection system or wastewater treatment plant. The MO requires that: (i) the water quality, as assessed by coliform bacteria, achieve suitable quality, and (ii) the male-specific coliphage (MSC) levels in the shellfish 'not exceed background levels or a level of 50 MSC/100 grams'. This reopening criterion is not currently applicable to conditional closures of the harvest areas due to insufficient scientific support.

In instances when shellfish from a particular harvest area have been implicated in illness, the State Control Authority (SCA) is required to immediately investigate the source of contamination in the implicated shellfish area, close the area to further shellfish harvesting, and remediate the cause. The SCA must also initiate a recall of all shellfish by harvest dates. When viral etiology is indicated by epidemiology, the identified area must remain closed for at least 21 days after the cause has been remedied. In addition, SCAs are required to close classified shellfish areas when adverse events affecting shellfish resources may cause contamination. Such closures may be either precautionary in advance of forecast major storms and distant spills, or in response to accidental discharges and treatment failures.

When an epidemiological link is established between illness and shellfish, the SCA most often issues public notification and institutes a recall of implicated shellfish. Shellfish implicated in illnesses cannot be reconditioned for human consumption, and such products are usually destroyed. A product recall may not be appropriate when the product is no longer available in the market. When the source of illness is attributed to lapses in the sanitary distribution and processing system, the product is immediately recalled (and destroyed) and the cause requires immediate correction.

The NSSP currently does not recommend virus monitoring of shellfish or their waters due to the technical complexity, time requirement, cost, and limitations of technology to detect and recover viruses from shellfish. However, with the advent of more rapid and efficient technologies for extraction and detection of enteric RNA viruses and the adherence to robust validation criterion, these methods may be incorporated into regulatory compliance programs and applied routinely in support of outbreak investigations.

The public health risk associated with the detection of virus genome without knowing viral infectivity is a regulatory challenge. Little progress has been made in propagation of human NoV *in vitro* and therefore it is not possible to determine infectivity of norovirus extracted from shellfish. In an effort to determine relative risk associated with findings of molecular tests to partially overcome this obstacle, investigators have begun to enumerate the levels of HAV and NoV in shellfish that have caused illness to determine if an acceptable threshold can be established (Lowther *et al.*, 2012; Woods and Burkhardt, 2010). This approach, however, has not been adopted by federal or State regulatory agencies.

14.7 Conclusion

Enteric viruses in the aquatic environment have been thoroughly studied over several years. Most enteric viruses can survive under many environmental stressors; effective wastewater treatment is therefore paramount in minimizing the discharge of potentially pathogenic micro-organisms into receiving waters. The use of bacterial indicators as a predictor of fecal contamination in shellfish growing areas is believed to have limited predictive value for viral enteric pathogen contamination. Development of molecular techniques has allowed detection of these viral pathogens, especially those that cannot be readily propagated utilizing cell culture techniques. As the scientific community continues to make advances in understanding the pathogenic nature of these viruses, we can be instrumental in preventing disease outbreaks and improving public health.

14.8 References

BERG, D.E., M.A. KOHN, T.A. FARLEY, and L.M. MCFARLAND. 2000. Multistate outbreaks of acute gastroenteritis traced to fecal-contaminated oysters harvested in Louisiana. *J. Infect. Dis.* **181**: S381–S386.

BEURET, C., D. KOHLER, A. BAUMGARTNER, and T.M. LUTHI. 2002. Norwalk-like virus sequences in mineral waters: one year monitoring of three brands. *Appl. Environ. Microbiol.* **68**: 1925–31.

BOSCH, A. 1998. Human enteric viruses in the water environment: a minireview. *Int. Microbiol.* **1**(3): 191–6.

BURKHARDT, W., J.W. WOODS, and K.R. CALCI. 2005. Evaluation of wastewater treatment plants' efficiency to reduce viral loading using real-time RT-PCR. ASM Abstract. 105 General Meeting.

BUTT, A.A., K.E. ALDRIDGE, and C.B. SANDERS. 2004. Infections related to the ingestion of seafood Part II: viral and bacterial infections. *Lancet.* **4**: 201–12.

CABELLI, V.J. 1988. Microbial indicator levels in shellfish, water and sediment from upper Narragansett Bay conditional shellfish growing area. Report to the Narragansett Bay Project. Providence, RI, Narragansett Bay Project.

DANIELS, N.A., BERGMIRE-SWEAT, D.A., SCHWAB, K.J., HENDRICKS, K.A., REDDY, S., MONROE, S.A., FANKHAUSER, R.L., MONROE, S.S., ATMAR, R.L., GLASS, R.I., and MEAD, P. 2000. A foodborne outbreak of gastroenteritis associated with Norwalk-like viruses: first molecular traceback to deli sandwiches contaminated during preparation. *J. Infect. Dis.* **181**: 1467–70.

DEPAOLA, A., JONES, J.L., WOODS, J., BURKHARDT, W. III, CALCI, K.R., KRANTZ, J.A., BOWERS, J.C., KASTURI, K., BYARS, R.H., JACOBS, E., WILLIAMS-HILL, D., and NABE, K. 2010. Bacterial and viral pathogens in live oysters: 2007 United States market survey. *Appl. Environ. Microbiol.* **76**(9): 2754–68.

DORE, W.J., HENSHILWOOD, K., and LEES, D.N. 2000. Evaluation of F-specific RNA bacteriophage as a candidate human enteric virus indicator for bivalve molluscan shellfish. *Appl. Environ. Microbiol.* **66**(4): 1280–85.

DORE, W.J. and LEES, D.N. 1995. Behavior of *Escherichia coli* and male-specific bacteriophage in environmentally contaminated bivalve molluscs before and after depuration. *Appl. Environ. Microbiol.* **61**(8): 2830–34.

FUJIOKA, R.S., LOH, P.C., and LAU, L.S. 1980. Survival of human enteroviruses in the Hawaiian ocean environment: evidence for virus-inactivating microorganisms. *Appl. Environ. Microbiol.* **39**(6): 1105–10.

GALLIMORE, C.I., PIPKIN, C., SHRIMPTON, H., GREEN, A.D., PICKFORD, Y., MCCARTNEY, C., SUTHERLAND, G., BROWN, D.W., and GRAY, J. J. 2005. Detection of multiple enteric virus strains within a foodborne outbreak of gastroenteritis: an indication of the source of contamination. *Epidemiol.Infect.* **133**(1): 41–7.

GILL, O.N., CUBITT, W.D., MCSWIGGAN, D.A., WATNEY, B.M., and BARTLETT, C.L. 1983. Epidemic of gastroenteritis caused by oysters contaminated with small round structured viruses. *Br. Med. J.(Clin.Res.Ed)* **287**(6404): 1532–34.

GOYAL, S. M. 2006. *Viruses in Food.* Springer Sciences, New York.

GRIFFIN, D.W., DONALDSON, K.A., PAUL, J.H., and ROSE, J.B. 2003. Pathogenic human viruses in coastal waters. *Clin. Microbiol. Rev.* **16**(1): 129–43.

GUILLOIS-BECEL, Y., COUTURIER, E., LE SAUX, J.C., ROQUE-AFONSO, A.M., LE GUYADER, F.S., LE, G.A., PERNES, J., LE, B.S., BRIAND, A., ROBERT, C., DUSSAIX, E., POMMEPUY, M., and VAILLANT, V. 2009. An oyster-associated hepatitis A outbreak in France in 2007. *Euro. Surveill.* **14**(10): 19144.

HAAS, C.N., J.B. ROSE, C.P. GERBA, and R. REGLI. 1993. Risk assessment of viruses in drinking water. *Risk Anal.* **13**: 545–52.

HEWITT, J. and G. E. GREENING. 2004. Survival and persistence of norovirus, hepatitis A virus, and feline calicivirus in marinated mussels. *J. Food Prot.* **67**: 1743–50.

HUMPHREY, T.J. and MARTIN, K. 1993. Bacteriophage as models for virus removal from Pacific oysters (Crassostrea gigas) during re-laying. *Epidemiol. Infect.* **111**(2): 325–35.

IRITANI, N., KAIDA, A., KUBO, H., ABE, N., GOTO, K., OGURA, H., and SETO, Y. 2010. Molecular epidemiology of noroviruses detected in seasonal outbreaks of acute nonbacterial gastroenteritis in Osaka City, Japan, from 1996–1997 to 2008–2009. *J. Med. Virol.* **82**(12): 2097–2105.

KATZ, M. S., HOFFMAN, S. and MORRIS JR., J.G. 2012. Ranking the disease burden of 14 pathogens in food sources in the United States attribution data from outbreak investigations and expert elicitation. *J. Food Prot.* **75**(7): 1278–91.

KINGSLEY, D.H. and RICHARDS, G.P. 2003. Persistence of hepatitis A virus in oysters. *J. Food Prot.* **66**(2): 331–4.

KOOPMANS, M. and E. DUIZER. 2004. Foodborne viruses: an emerging problem. *Int. J. Food Microbiol.* **90**(1): 23–41.

LABELLE, R.L. and C.P. GERBA. 1979. Influence of pH, salinity, and organic matter on the adsorption of enteric viruses to estuarine sediment. *Appl. Environ. Microbiol.* **38**(1): 93–101.

LASOBRAS, J., J. DELLUNDE, J. JOFRE, and F. LUCENA. 1999. Occurrence and levels of phages proposed as surrogate indicators of enteric viruses in different types of sludges. *J. Appl. Microbiol.* **86**(4): 723–9.

LEES, D. 2000. Viruses and bivalve shellfish. *Int. J. Food Microbiol.* **59**(1–2): 81–116.

LE GUYADER, F.S., J.C. LE SAUX, K. AMBERT-BALAY, J. KROL, O. SERAIS, S. PARNAUDEAU, H. GIRAUDON, G. DELMAS, M. POMMEPUY, P. POTHIER, and R.L. ATMAR. 2008. Aichi virus, norovirus, astrovirus, enterovirus, and rotavirus involved in clinical cases from a French oyster-related gastroenteritis outbreak. *J. Clin. Microbiol.* **46**(12): 4011–17.

LIPPY, E. C. and S. L. WALTRIP. 1984. Waterborne disease outbreaks 1946–1980: a thirty-five year prospective. *J. Amer. Water Works Assoc.* **76**: 60–7.

LIU, O.C., H.R. SERAICHEKAS, and B.L. MURPHY. 1967. Viral depuration of the Northern Quahaug. *Appl. Microbiol.* **15**(2): 307–15.

LOH, P. C., R. S. FUJIOKA, and S. LAU. 1979. Recovery, survival and dissimination of human enteric viruses in ocean waters receiving sewage in Hawaii. *Water, Air, Soil Pollut.* **12**: 197–217.

LOVE, D.C., G.L. LOVELACE, and SOBSEY, M.D. 2010. Removal of *Escherichia coli, Enterococcus fecalis*, coliphage MS2, poliovirus, and hepatitis A virus from oysters (*Crassostrea virginica*) and hard shell clams (*Mercinaria mercinaria*) by depuration. *Int. J. Food Microbiol.* **143**(3): 211–17.

LOWTHER, J.A., GUSTAR, N.E., HARTNELL, R.E., and LEES, D.N. 2012. Comparison of norovirus RNA levels in outbreak-related oysters with background environmental levels. *J. Food Prot.* **75**(2): 389–93.

MAALOUF, H., ZAKHOUR, M., LE, P.J., LE SAUX, J.C., ATMAR, R.L., and LE GUYADER, F.S. 2010. Distribution in tissue and seasonal variation of norovirus genogroup I and II ligands in oysters. *Appl. Environ. Microbiol.* **76**(16): 5621–30.

MEAD, P.S., SLUTSKER, L., DIETZ, V., MCCAIG, L.F., BRESEE, J.S., SHAPIRO, C., GRIFFIN, P.M., and TAUXE, R.V. 1999. Food-related illness and death in the United States. *Emerg. Infect. Dis.* **5**(5): 607–25.

METCALF, T.G., MULLIN, B., ECKERSON, D., MOULTON, E., and LARKIN, E.P. 1979. Bioaccumulation and depuration of enteroviruses by the soft-shelled clam, Mya arenaria. *Appl. Environ. Microbiol.* **38**(2): 275–82.

MOUNTS, J. W., ANDO, T.A., KOOPMANS, M., BRESEE, J.S., INOUYE, S., NOEL, J., and GLASS, R.I. 2000. Cold weather seasonality of gastroenteritis associated with Norwalk-like viruses. *J. Infect. Dis.* **181**: 284–7.

MUNNOCH, S., ASHBOLT, R., COLEMAN, D.J., WALTON, N., BEERS-DEBBLE, M.Y., and TAYLOR, R. 2004. A multi-jurisdictional outbreak of hepatitis A related to a youth camp – implications for catering operations and mass gatherings. *Commun. Dis. Intell.* **28**: 521–7.

NATIONAL SHELLFISH SANITATION PROGRAM. 2009. *Guide for the Control of Molluscan Shellfish.* http://wcms.fda.gov/FDAgov/Food/FoodSafety/Product-SpecificInformation/Seafood/federalStatePrograms/NationalShellfishSanitationProgram/ucm046353.htm.

PINA, S., PUIG, M., LUCENA, F., JOFRE, J., and GIRONES, R. 1998. Viral pollution in the environment and in shellfish human adenovirus detection by PCR as an index of human viruses. *Appl. Environ. Microbiol.* **64**: 3376–82.

RIPPEY, S.R. 1994. Infectious diseases associated with molluscan shellfish consumption. *Clin. Microbiol. Rev.* **7**(4): 419–25.

SCALLAN, E., HOEKSTRA, R.M., ANGULO, F.J., TAUXE, R.V., WIDDOWSON, M.A., ROY, S.L., JONES, J.L., and GRIFFIN, P.M. 2011. Foodborne illness acquired in the United States – major pathogens. *Emerg. Infect. Dis.* **17**(1): 7–15.

TEUNIS, P.F., MOE, C.L., LIU, P., MILLER, E., LINDESMITH, L., BARIC, R.S., LE, P.J. and CALDERON, R.L. 2008. Norwalk virus: How infectious is it? *J. Med. Virol.* **80**(8): 1468–76.

WOODS, J.W. and BURKHARDT III, W. 2010. Occurrence of norovirus and hepatitis A virus in U.S. oysters. *Food Environ. Microbiol.* **2**: 176–82.

15

Viral presence in waste water and sewage and control methods

C. P. Gerba, M. Kitajima and B. Iker, University of Arizona, USA

DOI: 10.1533/9780857098870.3.293

Abstract: This chapter discusses the occurrence of different virus types in wastewaters (sewage) and their removal by various treatment processes. Factors that influence the occurrence and concentration of viruses in wastewater are reviewed and the mechanisms involved in their removal discussed.

Key words: wastewater, sewage, treatment, activated sludge, soil aquifer treatment, wetlands, disinfection.

15.1 Introduction: virus occurrence in wastewater

Various types of viruses known to infect humans have been identified in sewage, and most of them can be transmitted through water. Since large quantities of enteric viruses are shed in the feces of infected individuals (up to more than 10^{11} virus/g-feces), raw sewage, which gathers domestic wastewater including toilet flushes, contains viruses at high concentrations. Viruses in raw sewage may be reduced in numbers by certain sewage treatments, but it is very difficult to achieve complete removal with conventional wastewater treatment processes, and it usually does not occur. Thus, viruses can be expected to be present in sewage discharges. The treated sewage and combined sewer overflows (CSOs) are obviously the major sources of environmental pollution with human enteric viruses in surface waters.

Considering that much of the sewage discharged into the world's aquatic environment is untreated, more detailed information on the number and types of viruses associated with raw sewage would be most useful in understanding the fate of these agents after discharge into the environment and the associated risk to public health. The occurrence and concentration of viruses in raw sewage is determined by the incidence of infection in the population, season, and water usage. Higher concentrations of viruses are usually seen in developing countries and arid regions where per capita water use is lower than other regions. Some viral infections are seasonal and seasonal changes in the concentration of the viruses are seen throughout the year. Thus, noroviruses are usually at greater concentrations in sewage in the winter and enteroviruses in the summer or early fall months in temperate climates. The numbers are also affected by the concentration and purification procedures used to isolate the viruses, and in the case of infectivity assays, the cell lines which are used.

Traditionally (until the 1990s), viruses in wastewater were mainly identified by cell-culture assay and the use of specific antisera. Viruses identified were limited to cultivable viruses, such as enteroviruses, adenoviruses, rotaviruses, and reoviruses (Table 15.1). This was made more difficult because of the need for multiple cell lines, since all types of viruses will not grow in the same cell line. Usually greater numbers of viruses are detected by the use of primary cell lines than continuous cell lines (Table 15.1). The highest concentration of viruses in raw sewage ever reported was by Nupen *et al.* (1974) who found 463 500 infectious viruses per liter in raw sewage in South Africa using primary rhesus monkey kidney cells.

The advent of the polymerase chain reaction (PCR) enabled the detection of viral genomes in wastewater, including those of non-cultivable viruses such as human noroviruses and sapoviruses, based on their nucleotide sequence information. Initial attempts were made to determine the concentration of viral genomes with semi-quantitative PCR (based on regular PCR and end-point dilution). Recently, real-time quantitative PCR (qPCR) assays have been applied to detect viral genomes in wastewater, allowing us to determine the concentration of viruses including non-cultivable types. As a result, data on the concentration of viruses in wastewater has rapidly accumulated in recent years (Table 15.2).

However, it should be noted that not all techniques used to detect and quantify viruses in sewage are easily comparable. Methods used for virus recovery, extraction, and quantification may be drastically different from one study to another, making it difficult to compare results from different studies. In addition, inhibition of PCR caused by various substances (such as humic acids and heavy metals) in wastewater can be problematic for molecular quantification. Therefore it has recently been recommended that internal process controls should be included to monitor efficiency of viral nucleic acid extraction and RT-PCR (da Silva *et al.*, 2007). Some researchers back-calculate the actual concentration using the raw qPCR data and extraction-RT-PCR efficiency (called, e.g. 'corrected' copy number) (Sano *et al.*, 2011; Pérez-Sautu *et al.*, 2012), while others report the raw qPCR data (even if they included an internal process control) and use the extraction-RT-PCR efficiency data to ensure the reliability of the qPCR data, not for back-calculation (da Silva *et al.*, 2007; Sima *et al.*, 2011; Skraber *et al.*, 2011).

Table 15.1 Concentration of infectious viruses in wastewater, as determined by cell culture-based assays

Sample	Virus	Country	Max (unit/L)	Unit	Cells/methods	Reference
Raw wastewater	Enteroviruses	US	1×10^3	MPN	BGM	Harwood *et al.*, 2005
	Infectious virus	Israel	4.5×10^4	PFU	MK_2 and Vero	Buras, 1974
		Netherlands	8.3×10^2	PFU	BGM	Lodder and de Roda Husman, 2005
	Rotaviruses	US	2.5×10^2	IF	MA104	Gerba, 1981
	Reoviruses	Netherlands	2.1×10^3	PFU	BGM	Lodder and de Roda Husman, 2005
		US	2.0×10^4	$TCID_{50}$	Primary monkey kidney	England, 1972
	Reo- and enteroviruses	Namibia	7.2×10^4	$TCID_{50}$	Primary monkey kidney	Nupen *et al.*, 1974
	Astrovirus genogroup A	Egypt	3.3×10^3	CC-RT-PCR U	Caco-2/CC-RT-PCR	Morsy El-Senousy *et al.*, 2007
	Astrovirus genogroup B	Egypt	3.3×10^5	CC-RT-PCR U	Caco-2/CC-RT-PCR	Morsy El-Senousy *et al.*, 2007
	Infectious viruses	US	5.0×10^4	MPN	BGM	Simmons and Xagoraraki, 2011
	Adenoviruses	Spain	6.0×10^4	IF	A529 and A293	Calgua *et al.*, 2011
		Greece	70–3200	$TCID_{50}$	Hep	Krilkelis *et al.*, 1985
Treated wastewater	Enteroviruses	US	1×10^{-3}	MPN	BGM	Harwood *et al.*, 2005
		Netherlands	39	PFU	BGM	Lodder and de Roda Husman, 2005
	Rotaviruses	US	75	IU	MA104	Gerba, 1981
	Reoviruses	Netherlands	92	PFU	BGM	Lodder and de Roda Husman, 2005
	Astrovirus genogroup A	Egypt	3.3	CC-RT-PCR U	Caco-2/CC-RT-PCR	Morsy El-Senousy *et al.*, 2007
	Astrovirus genogroup B	Egypt	330	CC-RT-PCR U	Caco-2/CC-RT-PCR	Morsy El-Senousy *et al.*, 2007
	Infectious viruses	US	40	MPN	BGM	Simmons and Xagoraraki, 2011
	Enteroviruses, Adenoviruses	US	140	MPN	PLC/PRF/5	Rodríguez *et al.*, 2009

PFU, plaque forming units; MPN, most probable number; IU, infectious units; CC, cell culture; IF, infectious foci.

Table 15.2 Concentration of viruses in wastewater determined by quantitative or semi-quantitative PCR

Sample	Virus	Country	Max (unit/L)	Unit	Method	Reference
Raw wastewater	Noroviruses	Netherlands	8.5×10^5	PDU	PCR/hybridization	Lodder and de Roda Husman, 2005
		Netherlands	1.0×10^6	PDU	PCR/hybridization	van den Berg *et al.*, 2005
		Sweden	4.5×10^3	MPN	RT-PCR	Ottoson *et al.*, 2006
	Norovirus genogroup I	Japan	1.5×10^6	copy	Real-time PCR	Kitajima *et al.*, 2012a
		Japan	2.6×10^5	PDU	Real-time PCR	Haramoto *et al.*, 2006
		Japan	5.5×10^6	RT-PCR unit	Real-time PCR	Katayama *et al.*, 2008
		Japan	2.3×10^6	Copy	Real-time PCR	Iwai *et al.*, 2009
		France	1×10^9	Copy	Real-time PCR	da Silva *et al.*, 2007
		US	1.2×10^6	Copy	Real-time PCR	Kitajima *et al.*, 2012b
		Spain	10^8–10^9	Copy-corrected	Real-time PCR	Pérez-Sautu *et al.*, 2012
		Sweden	2×10^6	Genomes	Real-time PCR	Nordgren *et al.*, 2009
	Norovirus genogroup II	Japan	9.1×10^5	Copy	Real-time PCR	Kitajima *et al.*, 2012a
		Japan	1.9×10^6	PDU	Real-time PCR	Haramoto *et al.*, 2006
		Japan	5.5×10^6	RT-PCR unit	Real-time PCR	Katayama *et al.*, 2008
		Japan	7.1×10^7	Copy	Real-time PCR	Iwai *et al.*, 2009
		France	6×10^7	Copy	Real-time PCR	da Silva *et al.*, 2007
		US	4.3×10^5	Copy	Real-time PCR	Kitajima *et al.*, 2012b
		US	2.9×10^4	Virus	Real-time PCR	Simmons and Xagoraraki, 2011
		Spain	10^9–10^{10}	Copy-corrected	Real-time PCR	Pérez-Sautu *et al.*, 2012
		Sweden	1×10^7	Genomes	Real-time PCR	Nordgren *et al.*, 2009

Norovirus genogroup IV	Japan	6.9×10^4	Copy	Real-time PCR	Kitajima *et al.*, 2009
	US	3.5×10^4	Copy	Real-time PCR	Kitajima *et al.*, 2012b
Adenovirus	Japan	1.8×10^7	PDU	Real-time PCR	Haramoto *et al.*, 2007
	Japan	6.4×10^6	RT-PCR unit	Real-time PCR	Katayama *et al.*, 2008
	US	6.7×10^8	Virus	Real-time PCR	Simmons and Xagoraraki, 2011
	Spain	3.5×10^6	Copy	Real-time PCR	Calgua *et al.*, 2011
Enterovirus	Japan	5.5×10^5	RT-PCR unit	Real-time PCR	Katayama *et al.*, 2008
	Sweden	9.0×10^5	MPN	RT-PCR	Ottoson *et al.*,, 2006
	US	1.1×10^6	Virus	Real-time PCR	Simmons *et al.*, 2011
Rotavirus	Netherlands	5.5×10^4	PDU	PCR/hybridization	Lodder and de Roda Husman, 2005
Sapovirus	Japan	9.3×10^4	Copy	Real-time PCR	Kitajima *et al.*, 2011
	Japan	1.3×10^5	Copy	Real-time PCR	Haramoto *et al.*, 2008a
	US	3.2×10^6	Copy	Real-time PCR	Kitajima *et al.*, 2012b
	Spain	10^8–10^9	Copy-corrected	Real-time PCR	Sano *et al.*, 2011
Astroviruses	France	3.1×10^8	Genome	Real-time PCR	Le Cann *et al.*, 2004
Astrovirus genogroup A	Egypt	5.6×10^6	Copy	Competitive PCR	Morsy El-Senousy *et al.*, 2007
Astrovirus genogroup B	Egypt	8.7×10^6	Copy	Competitive PCR	Morsy El-Senousy *et al.*, 2007
Aichi viruses	Japan	2.3×10^7	Copy	Real-time PCR	Kitajima, 2011
	US	2.9×10^6	Copy	Real-time PCR	Kitajima *et al.*, 2012b
Torque teno virus	Japan	4.8×10^4	Copy	Real-time PCR	Haramoto *et al.*, 2008b
Influenza A virus	Netherlands	2.6×10^5	Copy	Real-time PCR	Heijnen and Medema, 2011

(*Continued*)

Table 15.2 Continued

Sample	Virus	Country	Max (unit/L)	Unit	Method	Reference
Treated wastewater	Noroviruses	Netherlands	7.5×10^3	PDU	PCR/hybridization	Lodder and de Roda Husman, 2005
		Netherlands	1.0×10^4	PDU	PCR/hybridization	van den Berg *et al.*, 2005
		Sweden	1.5×10^2	MPN	RT-PCR	Ottoson *et al.*,, 2006
	Norovirus GI	Japan	5.6×10^3	Copy	Real-time PCR	Kitajima *et al.*, 2012a
		Japan	7.7×10^2	PDU	Real-time PCR	Haramoto *et al.*, 2006
		Japan	2.9×10^4	RT-PCR unit	Real-time PCR	Katayama *et al.*, 2008
		France	6×10^6	Copy	Real-time PCR	da Silva *et al.*, 2007
		US	5.8×10^3	Copy	Real-time PCR	Kitajima *et al.*, 2012b
		US	2.9×10^4	Virus	Real-time PCR	Simmons and Xagoraraki, 2011
		Spain	10^6–10^7	Copy-corrected	Real-time PCR	Pérez-Sautu *et al.*, 2012
		Sweden	7.9×10^5	Genomes	Real-time PCR	Nordgren *et al.*, 2009
	Norovirus GII	Japan	3.8×10^3	Copy	Real-time PCR	Kitajima *et al.*, 2012a
		Japan	1.5×10^3	PDU	Real-time PCR	Haramoto *et al.*, 2006
		Japan	5.5×10^4	RT-PCR unit	Real-time PCR	Katayama *et al.*, 2008
		France	3×10^6	Copy	Real-time PCR	da Silva *et al.*, 2007
		US	6.22×10^3	Copy	Real-time PCR	Kitajima *et al.*, 2012b
		US	9.4×10^2	Virus	Real-time PCR	Simmons and Xagoraraki, 2011
		Spain	10^6–10^7	Copy-corrected	Real-time PCR	Pérez-Sautu *et al.*, 2012
		Sweden	5.1×10^5	Genomes	Real-time PCR	Nordgren *et al.*, 2009
	Norovirus GIV	Japan	4.8×10^3	Copy	Real-time PCR	Kitajima *et al.*, 2009
		US	2.7×10^4	Copy	Real-time PCR	Kitajima *et al.*, 2012b

Adenovirus	Japan	5.4×10^4	PDU	Real-time PCR	Haramoto *et al.*, 2007
	Japan	3.4×10^5	RT-PCR unit	Real-time PCR	Katayama *et al.*, 2008
	US	2.9×10^4	virus	Real-time PCR	Simmons and Xagoraraki, 2011
	Finland	3.0×10^4	PCR units	Real-time PCR	Manunula *et al.*, 2012
Enterovirus	Japan	1.3×10^4	RT-PCR unit	Real-time PCR	Katayama *et al.*, 2008
	Sweden	3.0×10^4	MPN	RT-PCR	Ottoson *et al.*,, 2006
	US	9.4×10^2	Virus	Real-time PCR	Simmons and Xagoraraki, 2011
Rotavirus	Netherlands	2.9×10^4	PDU	PCR/hybridization	Lodder and de Roda Husman, 2005
Sapovirus	Japan	8.1×10^3	Copy	Real-time PCR	Kitajima *et al.*, 2011
	Japan	5.2×10^2	Copy	Real-time PCR	Haramoto *et al.*, 2008a
	US	1.1×10^5	Copy	Real-time PCR	Kitajima *et al.*, 2012b
	Spain	10^5–10^6	Copy-corrected	Real-time PCR	Sano *et al.*, 2011
Astroviruses	France	3.6×10^5	Copy	Real-time PCR	Le Cann *et al.*, 2004
Astrovirus genogroup A	Egypt	1.1×10^4	Copy	Competitive PCR	Morsy El-Senousy *et al.*, 2007
Astrovirus genogroup B	Egypt	6.2×10^5	Copy	Competitive PCR	Morsy El-Senousy *et al.*, 2007
Aichi viruses	Japan	1.8×10^4	Copy	Real-time PCR	Kitajima, 2011
	US	5.5×10^5	Copy	Real-time PCR	Kitajima *et al.*, 2012b
Torque teno virus	Japan	5.5×10^3	Copy	Real-time PCR	Haramoto *et al.*, 2008b
Influenza A virus	Netherlands	ND	Copy	Real-time PCR	Heijnen and Medema, 2011

PFU, plaque forming units; MPN, most probable number; PDU, PCR detectable units; ND, not detected.

Human caliciviruses, noroviruses and sapoviruses, typically show a strong seasonal trend: their concentration is generally higher in the winter, which is an epidemic period of human caliciviruses in many countries (Haramoto *et al.*, 2006b; Iwai *et al.*, 2009; van den Berg *et al.*, 2005). A study conducted in Sweden, interestingly, found that the concentration of GI NoV in sewage increased in the summer season (Nordgren *et al.*, 2009). On the other hand, concentrations of other commonly found enteric viruses like adenoviruses and Aichi viruses are relatively constant over a year and do not show a clear seasonal trend (Katayama *et al.*, 2008; Kitajima *et al.*, 2012b).

Other potential factors impacting on virus occurrence in raw sewage are population size, structure, and geographical location (Hewitt *et al.*, 2011). The relative abundance and occurrence of a particular viral pathogen greatly depends on how well it is transferred or spread throughout a community, the community's herd, innate, or acquired immunity to said virus, and the rate of introduction of viral pathogens into a population. Although there is no direct evidence that population size has a significant effect on virus occurrence, it is useful to think about the presence or absence of viruses from an epidemiological perspective (Yorke *et al.*, 1979). Community structure and geographical location of a sewage service area may provide a better estimation of virus occurrence in sewage than population size alone. Populations comprising relatively high numbers of immuno-compromised patients, areas of poor socio-economic status, or high rates of immigration and emigration may provide more opportunity for viruses to spread, affecting occurrence in sewage.

Several types of newly recognized viruses, cardioviruses, cosaviruses, bocaviruses, circoviruses, and picobirnaviruses, have also been identified from wastewater, suggesting that these viruses potentially could be transmitted through water (Blinkova *et al.*, 2009; Hamza *et al.*, 2011). Recently, viral metagenomic analyses have been performed on raw sewage and reclaimed water, which demonstrated the high diversity of viral types in wastewater (Cantalupo *et al.*, 2011; Rosario *et al.*, 2009).

15.2 Natural treatment systems

Natural systems for treatment of wastewater usually involve land treatment, wetlands, and aquaculture. Virus removal in these systems is a combination of die-off and attachment to plants, soil, and sediments. Land application systems can be divided into low rate irrigation, overland flow, and high rate infiltration. In developed countries land application systems are usually used as a tertiary treatment for further improvement of water quality after activated sludge treatment. In developing countries application to farmland may occur with no, minimal, or low cost technology (e.g. stabilization ponds) treatment. In irrigation systems the wastewater may be recovered by drain tiles under the farmland or by return flow (e.g. overland flow). The most studied land application systems for virus removal are soil aquifer systems (also referred to as soil aquifer treatment) groundwater, recharge, or rapid infiltration extraction. In these systems the wastewater is allowed to percolate through the soil and into the groundwater where it may be later recovered. Wetlands and aquaculture systems are also used to treat raw wastewater or to further improve water quality. There are two types of wetlands in general use: free water surface (FWS) and subsurface flow systems (SFS). An FWS is a wetland similar to a natural marsh with the water surface being exposed to the atmosphere. Both floating and submerged plants may be used. In an SFS wetland the water is never exposed to the atmosphere and flows underground, usually in a pit or in tanks filled with gravel. Aquaculture systems may be used for production of feed for animals or fish.

15.2.1 Virus removal by activated sludge

All sewage treatment processes usually remove or inactivate most viruses to some degree, but none is likely to remove all of the viruses present. Viruses can be removed during sewage treatment physically and by inactivation of the virus. Physical removal can take place by adsorption to the biological floc that forms during treatment or enzymatic attack, or by antiviral substances produced by the organisms that grow during sewage treatment.

The three most widely used biological processes for secondary sewage treatment are trickling filtration, activated sludge, and oxidation ponds. All three generate large amounts of microbial biomass to which viruses readily adsorb. In fact, the degree of virus removal which occurs during these processes appears to largely depend on virus adsorption to the solids (Berg, 1973; Balluz and Butler, 1978; Sobsey and Cooper, 1973), although biological antagonism may also be a major factor in the eventual inactivation of the virus. Studies on the detection of enteric viruses in activated sludge aeration basins indicate that 83–99% of the indigenous enteroviruses may be solids-associated. Generally, trickling filtration is less effective than activated sludge because of the lower biomass available for viral adsorption, and the relatively short contact time between the wastewater and the biological growth on the filter medium (Sorber, 1983).

Typically, removal rates are around 1–2 $\log_{10}$ or 90–99% by activated sludge treatment (Table 15.3). Recent evidence has suggested that some viruses are resistant to removal by the activated sludge process. For example Aichi virus has been found in high titers in raw sewage, but little removal of at least the genome appears to take place during wastewater treatment (Kitajima, 2011; Kitajima *et al.*, 2012b).

15.2.2 Die-off of viruses in natural systems

The major mechanisms of virus removal by natural systems are die-off (inactivation) and retention by soil or other substrates (e.g. plants). Many factors

Table 15.3 Efficiency of virus removal by activated sludge

Treatment	Virus	Country	Log_{10} removal	Method	Reference
Settlement-activated sludge	Norovirus genogroup I	Japan	1.82 ± 0.61	Real-time PCR	Haramoto *et al.*, 2006
		Sweden	0.7 ± 0.3	Real-time PCR	Nordgren *et al.*, 2009
	Norovirus genogroup II	Japan	2.74 ± 1.19	Real-time PCR	Haramoto *et al.*, 2006
		Sweden	1.2 ± 0.3	Real-time PCR	Nordgren *et al.*, 2009
	Adenoviruses	US	Mean: 2.2	Real-time PCR	Simmons and Xagoraraki, 2011
	Enteroviruses	US	2.0 to 3.7 (Mean: 2.9)	Real-time PCR	Simmons and Xagoraraki, 2011
	Astroviruses	France	1.30 to >4.70 (Mean: 2.69)	Real-time PCR	Le Cann *et al.*, 2004
	Astrovirus genogroup A	Egypt	5.1 ± 1.6	Competitive PCR	Morsy El-Senousy *et al.*, 2007
		Egypt	2.5 ± 1.0	Cell-culture PCR	Morsy El-Senousy *et al.*, 2007
	Astrovirus genogroup B	Egypt	3.6 ± 2.2	Competitive PCR	Morsy El-Senousy *et al.*, 2007
		Egypt	2.5 ± 0.8	Cell-culture PCR	Morsy El-Senousy *et al.*, 2007
	Infectious viruses	US	3.2	Cell-culture PCR	Simmons and Xagoraraki, 2011
Settlement-activated sludge-phosphorous/ nitrogen removal	Noroviruses	Netherlands	0.2 to 2.1	RT-PCR	Lodder and de Roda Husman, 2005
		Netherlands	Mean: 2.0 (Plant A) 2.7 (Plant B)	Semi-quantitative PCR	van den Berg *et al.*, 2005
		Sweden	0.95 ± 0.84	Semi-quantitative PCR	Ottoson *et al.*, 2006

	Norovirus genogroup I	Spain	2.2 ± 0.4	Real-time PCR	Pérez-Sautu *et al.*, 2012
	Norovirus genogroup II	Spain	3.1 ± 0.3	Real-time PCR	Pérez-Sautu *et al.*, 2012
	Rotaviruses	Netherlands	−1.8 to 1.1	Semi-quantitative PCR	Lodder and de Roda Husman, 2005
	Enteroviruses	Netherlands	0.7 to 1.8	Cell culture	Lodder and de Roda Husman, 2005
		Sweden	1.67 ± 0.52	Semi-quantitative PCR	Ottoson *et al.*, 2006
	Reoviruses	Netherlands	0.9 to 1.4	Cell culture	Lodder and de Roda Husman, 2005
Settlement-activated sludge-chlorination	Norovirus genogroup I	Japan	2.27 ± 0.67	Real-time PCR	Haramoto *et al.*, 2006
	Norovirus genogroup II	Japan	3.69 ± 1.21	Real-time PCR	Haramoto *et al.*, 2006
	Astrovirus genogroup B	Egypt	3.8 ± 1.2	Competitive PCR	Morsy El-Senousy *et al.*, 2007
		Egypt	2.5 ± 0.8	Cell-culture PCR	Morsy El-Senousy *et al.*, 2007
	Aichi viruses		2.41 ± 0.42 (Plant A) 2.96 ± 0.40 (Plant B)	Real-time PCR	Kitajima, 2011
	Infectious viruses	US	1.2	Cell-culture	Simmons and Xagoraraki, 2011

Table 15.4 Inactivation rates of viruses in ground water

Organism	Temperature (°C)	Mean inactivation rate (log/day)	Inactivation rate range (log/day)
Poliovirus	0–10	0.02	0.005–0.05
	11–15	0.08	0.03–0.2
	16–20	0.1	0.03–0.2
	26–30	0.08	0.006–1.4
Hepatitis A virus	0–10	0.02	0–0.08
	20–30	0.04	0.009–0.1
Echovirus	11–15	0.1	0.05–0.2
	16–20	0.1	0.05–0.2
	21–25	0.2	0.06–0.6
Coxsackievirus	8–20	0.06	0.002–0.2
	25–30	0.1	0.007–0.3
Rotavirus	3–15	0.4	One study
	23.2	0.03	One study
Adenovirus	4	0.0076	One study
	12–22	0.028	0.01–0.047

Data from John and Rose, 2009; Sidhu and Toze, 2012; Sidhu *et al.*, 2010.

may influence virus die-off in natural systems, but the most predictive factor is temperature since all chemical and biological activities are influenced by temperature (Gerba *et al.,* 1991). Persistence of viruses in the soil also depends upon location, with greater inactivation occurring near the soil surface and in the vadose zone (the water unsaturated region of the subsurface). Temperatures will vary more in this region, moisture content is more variable, and drying during non-flooding events will result in desiccation. Hurst *et al.* (1980) found that poliovirus 1 and echovirus 1 were inactivated at 0.04 to 0.15 log/day when attached to soil during flooding of infiltration basins and 0.11 to 0.52 log per day during drying of the basins. The greater microbial activity in this region can also aid in virus inactivation. Thus, viruses held near the soil surface can be expected to be inactivated at a greater rate than those viruses which reach the saturated zone.

Temperature and the presence of native microflora also influence the persistence of virus detection by PCR (Bae and Schwab, 2008; de Roda Husman *et al.*, 2009; Ogorzaly *et al.*, 2010). A review of the literature on the survival of viruses in groundwater by John and Rose (2005) found that hepatitis A virus and the coliphage PRD-1 were the viruses with the slowest decay rates (Table 15.4). Recent data on the survival of enteric viruses in groundwater suggest that adenoviruses and coliphages ΦX-174and PRD-1 are the longest surviving viruses in groundwater (Table 15.4).

Predicting long-term persistence in the environment is difficult because of the usual bi-phasic nature of die-off, where a rapid inactivation occurs for a few days, followed by a much slower rate. The persistence fraction is most likely simply due to variation in the virus population, where some small fraction of viruses is more thermally stable or more resistant to other factors that influence survival. The nucleic acid of an organism will persist for a much longer time than its infectivity, but it will eventually decay after the organisms become inactivated. The ability to detect viruses by PCR is also dependent on temperature, native microflora (whose enzymes result in the degradation of the nucleic acid), and the type of virus. Ogorzaly *et al.* (2010) found that ability to detect adenovirus 2 by PCR declined at about half the rate of the decline of infectivity in groundwater at 20°C, and no loss occurred after 400 days at 4°C.

15.2.3 Soil aquifer treatment systems

A major parameter in assessing virus removal by soil aquifer treatment systems is their retention during transport through the soil. Factors involved in transport are complex and site-specific and require a good understanding of the nature of the substrata. As with pathogen die-off, removal is different near the soil surface, in the vadose zone and in the saturated aquifer. The greatest removal occurs near the soil surface with decreasing removal at greater distances. It is also important to recognize that viruses are not permanently immobilized on soil particles and can become remobilized, especially if changes in water quality occur. Pang (2009) concluded that virus removal determined from laboratory column studies could be from two to three orders of magnitude greater than that determined from field studies. This is because the greatest removal of viruses usually occurs within the first meter of the soil surface. Thus, removal rates for viruses of less than one meter should not be exploited to rates beyond one meter (Pang, 2009). The removal of viruses is a power function (or hyper-exponentially) decreasing with distance (Pang *et al.*, 2009). Pang (2009) found that removal rates for enteroviruses are lower than MS2 coliphage, other bacteriophages (except in some cases for PRD1), and other human viruses.

15.2.4 Wetlands

A summary of virus removals by wetlands is shown in Table 15.5. Most of these wetlands had a retention time of approximately five days, which is usually considered the minimum time required to ensure wastewater quality. Most studies have involved the reduction of naturally occurring coliphages and added tracer coliphage. Few studies have been done on the reduction of human enteric viruses. As can be seen in Table 15.5, a five-day retention time results in a ~95–99% reduction in viruses if plants are present. Longer retention times have not been studied in detail. Virus removal is increased in the presence of plants (Table 15.5) and is related to plant density (Gersberg *et al.*, 1987; Nokes *et al.*,, 2007). Viruses adsorb onto plant surfaces within hours of entering the wetlands (Quinonoez-Dias *et al.*, 2001). Subsurface flow wetlands appear to result in greater removal of viruses than surface flow wetlands

Table 15.5 Virus removal by wetlands

Location	Vegetation	Virus	% Virus removal
Free water surface			
Glendale, AZ	Typha, Scripus	Coliphage	95
Oxelosunf, Sweden	Typha	Coliphage	94.7
Duplin County, NC	Typha, Sparganium	Somatic coliphage	84–92
		F-specific coliphage	84–92
Houghton Lake, MI	Typha, Carex	Reovirus, polio 1	90
Waldo, FL	Cypress	Coliphage	85.5–99.8
Tucson, AZ	Scripus, Typha	PRD-1	96.4
Subsurface flow			
Santee, CA	Bullrush	MS-2	>99
	Bullrush	Polio 1	>99.9
	Bullrush	F-specific phage	99
	Not planted	F-specific phage	94.5
Duluth, MN	Bullrush	Somatic phage	90 (winter)
	Bullrush	Somatic phage	98.4 (summer)
Abu Arwa, Egypt	Phragmites	Coliphage	99.7
Budds Farms, UK	Phragmites	Coliphage	97.3
	Phragmites	Enteroviruses	99.3
Etenbuttel, Germany	Phragmites	Coliphage	96.7
Wiedersburg, Germany	Phragmites	Coliphage	99.9
Barcelona, Spain	Salix attocinera	Coliphage	94.9
Tucson, AZ	Mixed species, trees, bushes, reeds	PRD-1 Coliphage	98.8 95.2
Wickenburg, AZ	No plants	Coliphage	61.5
Sahuarita, AZ	Cottonwood and Willow	Coliphage	86.6

Data from Kadlec, 2002; Vadles *et al.*, 2003, 2006; Nokes *et al.*, 2003; Reinoso *et al.*, 2008; Thurston *et al.*, 2001.

(Table 15.5), perhaps because of greater contact with roots, surface biofilms or granular material. Removals are also greater in the summer than the winter, probably due to lower temperatures increasing virus survival and lowered plant activity.

15.3 Disinfection of wastewaters

Disinfection of wastewater is required in some locations and countries. Chlorine is the most commonly used disinfectant, but ultraviolet light is becoming more common, because it has less environmental impact.

15.3.1 Ultraviolet light

Ultraviolet (UV) treatment has become a widely accepted method for water and wastewater disinfection in recent years in the United States. UV is highly effective at inactivating most waterborne pathogens and, unlike conventional chloride or ozone disinfection methods, it is generally not thought to create potentially harmful disinfection by-products. Most studies with viruses have been conducted with low-pressure (LP) UV light sources. In addition to LP UV, medium-pressure (MP), pulsed UV lamps and UV light-emitting diodes (LEDs) are also being used for water disinfection. Differences in the source lamp characteristics of these types of UV result in radiation with different spectral outputs, and photon densities that vary in their action on micro-organisms. LP UV lamps used in disinfection contain mercury vapor at a pressure of 0.001–0.01 torr and produce largely monochromatic UV light at 253.7 nm. MP UV lamps contain mercury vapor at 100–10 000 torr and emit polychromatic UV light ranging from 200 nm to as high as 1400 nm with several peaks between 185 and 300 nm, which is generally considered to be the practically applicable germicidal range (US Environmental Protection Agency, 2003). Standard pulsed UV lamps differ from LP and MP lamps in that they contain xenon vapor rather than mercury and emit intense pulses of light at high photon densities rather than the continuous, lower-density wave of LP and MP UV. Pulsed UV lamps emit polychromatic UV ranging from 185 nm to about 800 nm. Pulsed UV units may have an advantage in that they emit light of the highest intensity: the power flux for pulsed UV systems can be 500 000 times greater than that for continuous-wave LP and MP systems (Eischeid *et al.*, 2011). Pulsed UV can therefore be better adapted to deliver higher UV doses than LP or MP; intensity itself may also be a factor in the effectiveness of UV disinfection. Weber and Scheble (2003) estimated that a commercially available pulsed UV unit can deliver a UV dose of 650–1000 mJ/cm^2 in water with the relatively high transmittances generally found in drinking water. The spectral output of pulsed UV lamps can be designed to match a specific need by changing the energy input (McDonald *et al.*, 2000). The differences in emission spectra between LP, MP, and pulsed UV are thought to be primarily responsible for the differences in their action on micro-organisms.

Since LP UV emits very near the 260 nm absorbance maximum for DNA, it inactivates pathogens by damaging their DNA, rendering them incapable of replicating in hosts (Eischeid and Linden, 2011). The primary DNA photoproduct formed during LP UV irradiation of micro-organisms is the cyclobutane pyrimidine dimer (CPD), and it is thought to constitute 90% of DNA photoproducts (Harm, 1980). MP UV and pulsed UV emit other wavelengths – primarily those between 250 and 300 nm – which can damage additional cellular components such as proteins, amino acids, lipids, and small molecules such as carboxylic acid and ketone compounds (Harm, 1980). Wavelengths below approximately 210 nm are absorbed by the phosphate

Table 15.6 Dose needed to inactivate enteric viruses and phage by 90% (low-pressure UV lamps)

Virus	Dose mJs/cm^2
Adeno 1	34.5
2	40
3	34
5	28
6	38.5
40	54.3
Coxsackie	11.0–15.6
Echo	10.8–12.1
Polio	5–12
HAV	3.7–7.3
Rota SA-11	8–9
Coliphage MS-2	18.6

Data from Gerba *et al.*, 2002; Gerba 2009; Nwachuku *et al.*, 2005; Thurston-Enriquez *et al.*, 2003.

backbone of DNA, and those below 240 nm are absorbed by the peptide bonds of proteins. While these lower wavelengths are highly effective at inactivating viruses, they do not travel through water as well as higher wavelengths. However, the additional wavelengths emitted by MP and pulsed UV systems do seem to be advantageous in that they prevent reactivation of microbes; damage to photo repair capabilities has been shown for *Escherichia coli* and *Cryptosporidium parvum* (Oguma *et al.*, 2002). This is probably because damage to DNA caused by LP UV lamps can be repaired but the more widespread damage inflicted by MP or pulsed UV cannot; such damage can be expected to prevent host cell reactivation of the DNA-containing adenovirus as well. Recent work has shown that greater inactivation of adenovirus in cell culture is achieved by both MP and pulsed UV than by LP UV (Eischeid *et al.*, 2009; Guo *et al.*, 2010; Linden *et al.*, 2007); over 4-log inactivation of adenovirus can be achieved using approximately 40 mJ/cm^2 of MP or pulsed UV, and this is in much closer agreement with the LP UV doses required to inactivate other viruses.

The dose needed to inactivate enteric viruses by low-pressure lamps is shown in Table 15.6.

15.3.2 Chlorine and chloramines

Chlorine is the disinfectant most commonly used to disinfect wastewater. Unless highly treated to remove organic matter and nitrogen, most of the chlorine is largely converted to chloramines, which are much slower in the inactivation of viruses. In water applications the effectiveness of a chemical disinfectant is expressed as Ct, where C is the chlorine or chloramine

Table 15.7 Ct values for chlorine inactivation of enteric viruses in water (99% inactivation at 5°C)

Virus	pH	Ct
Polio 1	6.0	1.7
Echo 1	6.0	0.24
	7.8	0.56
	10.0	47.0
Echo 11	7.0	0.54
Echo 12	7.5	2.1
	9.0	8.4
Coxsackie B5	7.8	2.16
	10.0	33.0
Adeno 2	7.0	0.04
Adeno 40	7.0	0.15
Adeno 41	7.0	0.005
Murine noro	7.0 and 8.0	<0.02

Data from Gerba *et al.*, 2003; Gerba, 2009; Black *et al.*, 2009; Cromeans *et al.*, 2010.

concentration and t the time required to inactivate a certain percentage of virus at a specific pH and temperature. This approach allows for proper design of the chlorine dose and contact time needed for drinking and waste water treatment. Because of the often rapid conversion of chlorine to chloramines and presence of suspended particles which interfere with disinfection, infectious enteric viruses can usually be detected in activated sludge-treated wastewater after disinfection. Tertiary treatment in the form of flocculation, filtration or membrane treatment is needed to increase the effectiveness of disinfection by reducing the amount of suspended soluble organic matter.

Hypochlorous acid (HOCl) forms when chlorine is added to water and is the main form of chlorine responsible for its disinfecting properties. The amount of HOCl formed is greater at neutral and lower pH levels, resulting in greater disinfection ability at these pH levels (Table 15.7). Chlorine as HOCl or OCl^- is free available chlorine. In sewage the HOCl combines with ammonia and organic compounds to produce combined chlorine in the form of chloramines. Enteric viruses vary greatly in their resistance to chlorine and chloramines even among the same genus (Tables 15.7 and 15. 8). The greater resistance of echo 11 and 12, and Coxsackie B5 is believed to be due to their ability to easily clump compared to the other enteroviruses (Jensen *et al.*, 1980). To assess the Ct value likely for the human norovirus for which an infectivity assay was not available, Shin and Sobsey (2008) developed a Ct based on RT-PCR detection using short template and long template, and compared it to infectivity results and RT-PCR with poliovirus type 1 and coliphage MS-2. They concluded that despite underestimation of virus inactivation by chlorine based on RT-PCR assays, the predicted Ct values for

Table 15.8 Ct values for chloramines in water (99% inactivation at 5°C)

Virus	pH	Ct
Polio 1	9.0	1420
Coxsackie B3	8.0	240
Coxsackie B5	8.0	670
Echo 1	8.0	8
Echo 11	8.0	880
Hepatitis A	8.0	592
Rota SA-11	8.0	4034
Adeno 2	7.0	600
	8.0	990
Adeno 40	8.0	360
Adeno 41	8.0	190
Mouse Noro	8.0	36
Coliphage MS-2	8.0	2100

Data from Sobsey 1989; Cromeans *et al.*, 2010.

human norovirus based on RT-PCR are lower than the ones for most other enteric viruses.

Chlorine may inactivate viruses by damage of either the viral capsid proteins or nucleic acid (Thurman and Gerba, 1988). The site of action may also depend on the concentration of free chlorine and the type of virus. Alvarez and O'Brien (1982) found that at a free chlorine concentration of less than 0.8 mg/L inactivation of poliovirus RNA occurs without any structure changes, but chlorine concentrations greater than 0.8 mg/L resulted in damage to the viral RNA and protein capsid. The protein coat for the double-stranded RNA rotavirus is the target responsible for inactivation (Vaughn and Novotny, 1991).

15.4 Future trends

As evident from studies over the last decade, almost any virus which infects man can be detected in wastewater by molecular methods. The potential for these viruses to be transmitted via water or food is not currently understood. Many cause illnesses not traditionally associated with waterborne disease, such as respiratory infections, meningitis, skin infections, etc. Linking detection of viruses by molecular methods and risk assessment is needed to better understand the risks of the wide variety of viruses detected in wastewater. Development of molecular methods capable of assessing infectivity would greatly increase our understanding of the survival of viruses which are difficult to culture or cannot be cultured in the laboratory.

Data is surprising limited on the ability of wetland treatment systems to remove naturally occurring enteric viruses. Most of the studies have been done with coliphages. Studies are needed on the removal of enteric viruses by both conventional and molecular methods to get a more quantitative estimate of the performance of these systems over longer retention times than that which has been obtained to date. Data on the removal of adenoviruses in both wetland treatment and aquifer recharge is also lacking. Since adenoviruses appear to be one of the longer surviving viruses in the environment, such information would provide better assurance that these systems are designed to reduce all types of enteric viruses present in wastewater.

The use of UV light systems for disinfection of treated wastewater is a rapidly increasing replacement for chlorine in the United States and elsewhere. Recent studies indicate that this is resulting in increased release of the highly UV light-resistant adenoviruses into surface waters (Rodríguez *et al.,* 2008). UV systems are primarily designed to meet bacterial standards in sewage discharges, which require UV light doses far below these required to inactivate viruses. Studies are needed to assess if this is increasing the risk of water transmission of adenoviruses. UV light technology is rapidly changing and a better understanding of the molecular mechanisms underlying wavelength sensitivity of enteric viruses will aid in the better design of such systems.

Although development of PCR has allowed us to detect any known virus, including non-cultivable viruses such as noroviruses and sapoviruses, PCR is still database-dependent and able to detect only previously identified viruses. Recently, various viruses putatively associated with human gastroenteritis have been newly identified through a metagenomic approach (Rosario *et al.*, 2009). Application of the metagenomic analysis utilizing pyrosequencing, which is a database-independent, unbiased, and high-throughput technique, should provide a better understanding of the viral community in wastewater and the molecular epidemiology of viruses among human populations.

15.5 References

ALVAREZ ME and O'BRIEN RT (1982) 'Effects of chlorine concentration on the structure of poliovirus', *Appl Environ Microbiol*, **43**, 237–9.

BAE KJ and SCHWAB, KB (2008) 'Evaluation of murine norovirus, feline calicivirus, poliovirus, and MS2 as surrogates for human norovirus in a model of viral persistence in surface water and groundwater', *Appl Environ Microbiol*, **74**, 477–84.

BALLUZ SA, BUTLER M and JONES HH (1978) 'The behavior of F2 coliphage in activated sludge treatment', *J Hyg*, **80**, 237–42.

BERG G (1973) Removal of viruses from sewage, effluents, and waters: A review. *Bull World Health Organ*, **49**, 451–60.

BLACK S, THURSTON JA and GERBA CP (2009) 'Determination of Ct values for chlorine resistant enteroviruses', *J Environ Sci Hlth Part A*, **44**, 336–9.

BLINKOVA O, KAPOOR A, VICTORIA J, JONES M, WOLFE N, NAEEM A, SHAUKAT S, SHARIF S, ALAM MM, ANGEZ M, ZAIDI S and DELWART EL (2009) 'Cardioviruses are genetically diverse and cause common enteric infections in South Asian children', *J Virol Methods*, **83**, 4631–41.

BURAS N (1974) 'Recovery of viruses from waste-water and effluent by the direct inoculation method', *Water Res*, **8**, 19–22.

CALGUA B, BARARDI CRM, BOFILL-MAS S, RODRIGUEZ-MANZANO J and GIRONES R (2011) 'Detection and quantitation of infectious human adenoviruses and JC polyomaviruses by immunofluorescence assay', *J Virol Methods*, **171**, 1–7.

CANTALUPO PG, CALGUA B, ZHAO G, HUNDESA A, WIER AD, KATZ JP, GRABE M, HENDRIX RW, GIRONES R, WANG D and PIPAS JM (2011) 'Raw sewage harbors diverse viral populations', *mBio*, **2**(5):e00180–11. Doi: 10.1128/mBio.00180–11.

CROMEANS TL, KAHLER AM and HILL VR (2010) 'Inactivation of adenoviruses, enteroviruses, and murine norovirus in water by free chlorine and monochloroamine', *Appl Environ Microbiol*, **76**, 1028–33.

DA SILVA AK, LE SAUX JC, PARNAUDEAU S, POMMEPUY M, ELIMELECH M and LE GUYADER FS (2007) 'Evaluation of removal of noroviruses during wastewater treatment, using real-time reverse transcription-PCR: different behaviors of genogroups I and II', *Appl Environ Microbiol*, **73**, 7891–7.

EISCHEID AC and LINDEN KG (2011) 'Molecular indications of protein damage in adenoviruses after UV disinfection', *Appl Environ Microbiol*, **77**, 1145–7.

EISCHEID AC, MEYER JN and LINDEN KG (2009) 'UV disinfection of adenovirus: molecular indications of DNA damage efficiency', *Appl Environ Microbiol*, **75**, 23–8.

EISCHEID AC, THURSTON JA and LINDEN KG (2011) 'UV disinfection of adenovirus: present state of the research and future directions', *CRC Crit Environ Sci Technol*, **41**, 1375–96.

GERBA CP (1981) 'Virus survival in wastewater treatment', in GODDARD M and BUTLER M. *Viruses and wastewater treatment*, Pergamon Press, Oxford, UK, pp. 39–48.

GERBA CP (2009) 'Disinfection', in MAIER RM, PEPPER IL and GERBA CP. *Environmental microbiology*, Academic Press, San Diego, CA, pp. 539–52.

GERBA CP, GRAMOS DM and NWACHUKU N (2002) 'Comparative inactivation of enteroviruses and adenovirus type 2 by UV light', *Appl Environ Microbiol*, **68**, 5167–9.

GERBA CP, NWACHUKU N and RILEY KR (2003) 'Disinfection resistance of waterborne pathogens on the United States Environmental Protection Agency's Contaminate Candidate List (CCL)', *J Water Supply Res Technol*, **52**, 81–94.

GERBA CP, YATES MV and YATES SR (1991) 'Quantitation of factors controlling viral and bacterial transport in the subsurface', in HURST CJ. *Modeling the environmental fate of microorganisms*, ASM Press, Washington DC, pp. 77–88.

GERSBERG RM, LYON SR, BRENNER R and ELKINS BV (1987) 'Fate of viruses in artificial wetlands', *Appl Environ Microbiol*, **53**, 731–6.

GUO H, CHU X and HU J (2010) 'Effect of host cells on low- and medium-pressure UV inactivation of adenoviruses', *Appl Environ Microbiol*, **76**, 7068–75.

HAMZA IA, JURZIK L, UBERLA K and WHILHELM M (2011) 'Evaluation of pepper mild mottle virus, human picobirnavirus and Torque teno virus as indicators of fecal contamination in river water', *Water Res*, **45**, 1358–68.

HARAMOTO E, KATAYAMA H and OHGAKI S (2008b) 'Quantification and genotyping of torque teno virus at a wastewater treatment plant in Japan', *Appl Environ Microbiol*, **74**, 7434–6.

HARAMOTO E, KATAYAMA H, OGUMA K and OHGAKI S (2007) 'Quantitative analysis of human enteric adenoviruses in aquatic environments', *J Appl Microbiol*, **103**, 2153–9.

HARAMOTO E, KATAYAMA H, OGUMA K, YAMASHITA H, TAJIMA A, NAKAJIMA H and OHGAKI S (2006) 'Seasonal profiles of human noroviruses and indicator bacteria in a wastewater treatment plant in Tokyo, Japan', *Water Sci Technol*, **54**, 301–8.

HARAMOTO E, KATAYAMA H, PHANUWAN C and Ohgaki S (2008a) 'Quantitative detection of sapoviruses in wastewater and river water in Japan', *Lett Appl Microbiol*, **46**, 408–13.

HARM W (1980) *Biological effects of ultraviolet radiation*. Cambridge University Press, Cambridge, MA.

HARWOOD VJ, LEVINE AD, SCOTT TM, CHIVUKULA V, LUKASIK J, FARRAH SR and ROSE JB (2005) 'Validity of the indicator organism paradigm for pathogen reduction in reclaimed water and public health protection', *Appl Environ Microbiol*, **71**, 3163–70.

HEIJNEN L and MEDEMA G (2011) 'Surveillance of influenza A and the pandemic influenza A (H1N1) 2009 in sewage and surface water in the Netherlands', *J Water Health*, **9**, 434–42.

DE RODA HUSMAN AM, LODDER WJ, RUTJES SA, SCHIJVEN, JF and TEUNIS PFM (2009) 'Long-term inactivation study of three enteroviruses in artificial surface and groundwaters, using PCR and cell culture', *Appl Environ Microbiol*, **75**, 1050–7.

HURST, CJ, GERBA CP and CECH I (1980) 'Effects of environmental variables and soil characteristics on virus survival in soil', *Appl Environ Microbiol*, **40**, 1067–79.

IWAI M, HASEGAWA S, OBARA M, NAKAMURA K, HORIMOTO E, TAKIZAWA T, KURATA T, SOGEN S and SHIRAKI K (2009) 'Continuous presence of noroviruses and sapoviruses in raw sewage reflects infections among inhabitants of Toyama, Japan (2006 to 2008)', *Appl Environ Microbiol*, **75**, 1264–70.

JENSEN H, THOMAS K and SHARP DG (1980) 'Inactivation of coxsackieviruses B3 and B5 in water by chlorine', *Appl Environ Microbiol*, **40**, 633–40.

JOHN JE and ROSE JB (2005) 'Review of factors affecting microbial survival in groundwater', *Environ Sci Technol*, **39**, 7345–56.

KADLEC RH (2001) 'Wastewater treatment-Wetlands and reedbeds', in BITTON, G. *Encyclopedia of Environmental Microbiology*, Wiley, New York, pp. 3389–3401.

KATAYAMA H, HARAMOTO E, OGUMA K, YAMASHITA H, TAJIMA A, NAKAJIMA H and OHGAKI S (2008) 'One-year monthly quantitative survey of noroviruses, enteroviruses, and adenoviruses in wastewater collected from six plants in Japan', *Water Res*, **42**, 1441–8.

KITAJIMA M (2011) *Molecular epidemiological analysis of pathogenic viruses in water environments and risk assessment*, PhD dissertation, University of Tokyo.

KITAJIMA M, HARAMOTO E, PHANUWAN C, KATAYAMA H and FURUMAI H (2012a) 'Molecular detection and genotyping of human noroviruses in influent and effluent water at a wastewater treatment plant in Japan', *J Appl Microbiol*, **112**, 605–13.

KITAJIMA M, HARAMOTO E, PHANUWAN C, KATAYAMA H and OHGAKI S (2009) 'Detection of genogroup IV norovirus in wastewater and river water in Japan', *Lett Appl Microbiol*, **49**, 655–8.

KITAJIMA M, IKER BC, PEPPER IL and GERBA CP (2012b) 'Molecular detection and genetic analysis of human gastroenteritis viruses in wastewater in southern Arizona', *112th General Meeting of American Society for Microbiology*. Q-2016.

KRIKELIS V, SPYROU N, MARKOULATOS P and SERIE C (1985) 'Seasonal distribution of enteroviruses and adenoviruses in domestic sewage', *Can J Microbiol*, **31**, 24–5.

LE CANN P, RANARIJAONA S, MONPOEHO S, LE GUYADER F and FERRÉ V (2004) 'Quantification of human astroviruses in sewage using real-time RT-PCR', *Res Microbiol*, **155**, 11–15.

LINDEN KG, THURSTON J, SCHAEFER R and MALLEY JP (2007) 'Enhanced UV inactivation of Adenoviruses under polychromatic UV lamps', *Appl Environ Microbiol*, **73**, 7571–4.

LODDER WJ and DE RODA HUSMAN AM (2005) 'Presence of noroviruses and other enteric viruses in sewage and surface waters in the Netherlands', *Appl Environ Microbiol*, **71**, 1453–61.

MCDONALD KF, CURRY RD, CLEVENGER TE, UNKLESBAY K, EISENSTARK A, GOLDEN, J and MORGAN RD (2000) 'A comparison of pulsed and continuous ultraviolet light sources for decontamination of surfaces', *IEEE Trans Plasma Sci* **28**, 1581–7.

MORSY EL-SENOUSY W, GUIX S, ABID I, PINTÓ RM and BOSCH A (2007) 'Removal of astrovirus from water and sewage treatment plants, evaluated by a competitive reverse transcription-PCR', *Appl Environ Microbiol*, **73**, 164–7.

NOKES R L, GERBA C P and KARPISCAK MM (2003) 'Microbial water quality improvement by small scale on-site subsurface wetland treatment', *J Environ Sci Hlth*, **A38**, 1849–55.

NORDGREN, J, MATUSSEK A, MATTSSON A, SVENSSON AP and LINDGREN E (2009) 'Prevalence of norovirus and factors influencing virus concentrations during one year in a full-scale wastewater treatment plant', *Water Res*, **43**, 1117–25.

NUPEN EM, BATEMAN BW and MCKENNY NC (1974) 'The reduction of viruses by various unit processes used in the reclamation of sewage to portable waters', in Malina JF and Sagik BP. *Virus survival in water and wastewater systems*, University of Texas, Austin, pp. 105–14.

NWACHUKU N, GERBA CP, OSWALD A and MASHADI FD (2005) 'Comparative inactivation of adenovirus serotypes by UV light disinfection', *Appl Environm Microbiol*, **71**, 5633–6.

OGORZALY L, BERTRAND I, PARIS M, MAUL A and GANTZER C (2010) 'Occurrence, survival, persistence of human adenoviruses and F-Specific RNA phages in raw groundwater', *Appl Environ Microbiol*, **76**, 8019–25.

OGUMA K, KATAYAMA H and OHGAKI S (2002) 'Photoreactivation of *Escherichia coli* after low- or medium-pressure UV disinfection determined by an endonuclease-sensitive site assay', *Appl Environ Microbiol*, **68**, 6029–35.

OTTOSON J, HANSEN A, BJÖRLENIUS B, NORDER H and STENSTRÖM T (2006) 'Removal of viruses, parasitic protozoa and microbial indicators in conventional and membrane processes in a wastewater pilot plant', *Water Res*, **40**, 1449–57.

PANG L (2009) 'Microbial removal rates in subsurface media estimated from published studies of field experiments and large intact soil cores', *J Environ Qual*, **38**, 1531–59.

PÉREZ-SAUTU U, SANO D, GUIX S, KASIMIR G, PINTÓ RM and BOSCH A (2012) 'Human norovirus occurrence and diversity in the Llobregat river catchment, Spain', *Environ Microbiol*, **14**, 494–502.

QUINONOEZ-DIAZ MJ, KARPISCAK MM, ELLMAN ED and GERBA CP (2001) 'Removal of pathogenic and indicator microorganism by constructed a wetland receiving untreated domestic wastewater', *J Environ Sci Hlth*, **36**, 1311–20.

REINOSO R, TORRES AL and BECARES E (2008) 'Efficiency of natural systems for removal of bacteria and pathogenic parasites from wastewater', *Sci Total Environ*, **395**, 80–6.

RODRIGUEZ RA, GUNDY PM and GERBA CP (2008) 'Comparison of BGM and PLC/PRC/5 cell lines for total culturable viral assay of treated sewage', *Appl Environ Microbiol*, **74**, 2583–7.

ROSARIO K, NILSSON C, LIM YW, RUAN Y and BREITBART M (2009) 'Metagenomics analysis of viruses in reclaimed water', *Environ Microbiol*, **11**, 2806–20.

SANO D, PÉREZ-SAUTU U, GUIX S, PINTÓ RM, MIURA T, OKABE S and BOSCH A (2011) 'Quantification and genotyping of human sapoviruses in the Llobregat river catchment, Spain', *Appl Environ Microbiol*, **77**, 1111–4.

SHIN GA and SOBSEY MD (2008) 'Inactivation of norovirus by chlorine disinfection of water', *Water Res*, **42**, 4562–8.

SIDHU JPS and TOZE S (2012) 'Assessment of pathogen survival potential during managed aquifer recharge with diffusion chambers', *J Appl Microbiol*, **113**, 693–700.

SIDHU JPS, TOZE S, HODGERS L, SHACKELTON M, BARRY K and PAGE D (2010) 'Pathogen inactivation during passage of stormwater through a constructed reedbed and aquifer transfer, storage and recovery', *Water Sci Technol*, **62**, 1190–7.

SIMA LC, SCHAEFFER J, LE SAUX JC, PARNAUDEAU S, ELIMELECH M and LE GUYADER FS (2011) 'Calicivirus removal in a membrane bioreactor wastewater treatment plant', *Appl Environ Microbiol*, **77**, 5170–7.

SIMMONS FJ and XAGORARAKI I (2011) 'Release of infectious human enteric viruses by full-scale wastewater utilities', *Water Res*, **45**, 3590–8.

SKRABER S, LANGLET J, KREMER JR, MOSSONG J, DE LANDTSHEER S, EVEN J, MULLER CP, HOFFMANN L AND Cauchie HM (2011) 'Concentration and diversity of noroviruses detected in Luxembourg wastewaters in 2008–2009', *Appl Environ Microbiol*, **77**, 5566–8.

SORBER CA (1983) Removal of viruses from wastewater and efficient by treatment processes. In G. BERG. *Viral Pollution of the Environment*, CRC Press, Boca Raton, Florida, pp. 39–75.

SOBSEY MD (1989) 'Inactivation of health-related microorganisms in water by disinfection processes', *Water Sci Techol*, **21**, 179–95.

SOBSEY MD and COOPER RC (1973) 'Enteric virus survival in algal-bacterial wastewater treatment systems: laboratory studies', *Water Res*, **1**, 669–85.

THURMAN RB and GERBA CP (1988) 'Molecular mechanisms of viral inactivation by water disinfection', *Adv Appl Environ Microbiol*, **33**, 75–105.

THURSTON JA, GERBA CP, FOSTER KF and KARPISCAK MM (2001) 'Fate of indicator microorganisms, *Giardia* and *Cryptosporidium* in subsurface flow constructed wetlands', *Water Res*, **35**, 1547–51.

THURSTON-ENRIQUEZ JA, HAAS CN, JACANGELO J, RILEY K and GERBA CP (2003) 'Inactivation of feline calicivirus and adenovirus type 40 by UV irradiation', *Appl Environ Microbiol*, **69**, 577–82.

US ENVIRONMENTAL PROTECTION AGENCY (2003) *UV disinfection guidance manual.* EPA 815-D-03-007. US Environmental Protection Agency, Washington, DC.

VADLES J A, GERBA CP and KARPISCAK MM (2003) 'Virus removal from wastewater in a multispecies subsurface-flow constructed wetland', *Water Environ Res*, **75**, 238–45.

VADLES, JA, GERBA CP and KARPISCAK MM (2006) 'Transport of coliphage in a surface flow constructed wetland', *Water Environ Res*, **78**, 2253–60.

VAN DEN BERG H, LODDER W, VAN DER POEL W, VENNEMA H and DE RODA HUSMAN AM (2005) 'Genetic diversity of noroviruses in raw and treated sewage water', *Res Microbiol*, **156**, 532–40.

VAUGHN JM and NOVOTNY JF (1991) 'Virus inactivation by disinfectants', in HURST CJ. *Modeling the environmental fate of microorganisms*, ASM Press, Washington, DC, pp. 217–41.

WEBER E and SCHEIBLE K (2003) 'Pulsed-UV unit may inactivate biological agents', *J Am Water Works Assoc*, **95**(6), 34–46.

Part IV

Particular pathogens and future directions

16

Advances in understanding of norovirus as a food- and waterborne pathogen and progress with vaccine development

D. J. Allen, Health Protection Agency, UK,
M. Iturriza-Gómara, University of Liverpool, UK and
D. W. G. Brown, Health Protection Agency, UK

DOI: 10.1533/9780857098870.4.319

Abstract: Noroviruses are the commonest cause of infectious intestinal disease, and are frequently associated with outbreaks of gastroenteritis, mainly in healthcare-associated settings, but also in outbreaks associated with contaminated food and/or water. The contamination of foods can occur during production, preparation, and/or service, or, more rarely by contamination of water supply. Contamination of water supply with norovirus is rare, and usually occurs as a consequence of leakage of sewage or as a result of leaching after heavy rainfall. Outbreaks of norovirus gastroenteritis associated with contaminated food and water can have high impact as a large number of individuals can become affected quickly over a large geographical area, with a high number of secondary cases. However, adequate capture of both epidemiological and laboratory data of norovirus outbreaks remains a major challenge, as many outbreaks fail to be identified and/ or followed up and so the incidence of norovirus-associated foodborne outbreaks is not well defined. Measures for preventing norovirus contamination are centred on good hand hygiene and environmental cleaning practices in healthcare settings, food establishments and on board cruise ships. Several guidelines for responding to outbreaks in food preparation premises are available, and there is a wide range of generic legislation for food processing and handling. There is currently no licenced vaccine or antiviral drug for prophylaxis or treatment of norovirus. However, the first trial demonstrating homologous protection against illness and infection using a norovirus VLP (virus-like particle) was reported recently. Whilst promising, the vaccine is monovalent and the evidence suggests that there is little cross-protection between norovirus strains, a multivalent vaccine is likely to be the only viable option for future vaccine development.

Key words: norovirus, epidemiology, outbreaks, gastroenteritis, vaccine.

16.1 Introduction

Noroviruses are the commonest cause of infectious intestinal disease. These viruses are frequently associated with outbreaks of illness, and a high genetic mutation rate drives the periodic emergence of new norovirus strains, both of which can lead to a significant impact on public health. In this chapter we focus on the epidemiology of norovirus and in particular the public health impact of outbreaks associated with contaminated food and water. We discuss the control measures in place to limit norovirus outbreaks, and the current progress made towards a norovirus vaccine.

16.1.1 History

The identification of norovirus, and its association as an aetiological agent of infectious gastroenteritis, was made in 1972 following electron microscopy (EM) and serology studies on a panel of specimens collected during an outbreak of non-bacterial gastroenteritis from Norwalk, Ohio (Kapikian *et al.*, 1972). Viral particles were visualised by EM in faecal specimens, and patients infected, either during the original outbreak or experimentally, were shown to develop serological markers of infection (Kapikian *et al.*, 1972). The agent became known as the Norwalk agent, and later part of a group of viruses termed the Norwalk-like viruses (NLVs) or the small round structured viruses (SRSVs). Almost 30 years after the first description, the classification of these viruses was reorganised. The terms NLV and SRSV were replaced by an updated description of the virus family *Caliciviridae* (from the Latin *calyx*, meaning cup, after the depressions on the virus surface as viewed by EM) (Mayo, 2002). Within this family, four virus genera were described, of which the genus Norovirus replaced the NLV group (Mayo, 2002).

16.1.2 Impact of norovirus gastrointestinal illness

Infectious gastrointestinal (diarrhoeal) disease remains a major public health challenge worldwide. Morbidity and mortality rates remain high in developing countries, largely due to inadequate treatment of diarrhoeal disease. In developed countries, mortality rates are generally low; however, each year around one in five people suffer from infectious gastrointestinal illness (Tam *et al.*, 2011). The list of aetiological agents of infectious gastroenteritis is extensive, with bacterial, parasite and viral agents all associated with the disease. The association between bacterial agents and foodborne and waterborne infection is well established. The link between many viral agents and foodborne and waterborne infection is less clear; nonetheless, outbreaks of gastroenteritis associated with contaminated food or water, where norovirus has been identified as the aetiological agent, are regularly reported.

16.2 Norovirus virology and clinical manifestations

Studies into the replication and pathogenesis of human norovirus remain limited by the lack of a laboratory cell culture system or suitable small animal model. Detailed information about the structure of the norovirus particle has been obtained through study of recombinant virus-like particles (VLPs), and as genome sequencing has become more accessible, the genetic diversity and evolution of the norovirus genome has been studied in detail.

16.2.1 Classification of the virus

Early studies of the taxonomy of norovirus strains using sequence diversity were confusing, with different laboratories using different regions of the viral genome for strain detection and classification. The high level of genetic diversity among norovirus strains led to data sets that were not comparable, and results between laboratories were conflicting. A classification of norovirus strains that accurately reflected the genetic diversity found among these viruses and allowed straightforward comparison of sequence data was proposed in 2006 (Zheng *et al.*, 2006). Here the authors compared complete amino acid sequences of entire capsid proteins to reveal phylogenetic relationships between norovirus strains. This meant that the norovirus genus was classified into five genogroups, and within these numerous genetic clusters were delineated (Zheng *et al.*, 2006). The majority of human strains are classified within genogroup-I (GI) or genogroup II (GII) strains, with 8 and 17 genetic clusters/genotypes identified among GI and GII noroviruses, respectively (Zheng *et al.*, 2006).

The remaining genogroups (GIII, GIV and GV) contain norovirus strains that have been isolated from mice (Karst *et al.*, 2003), cows (Oliver *et al.*, 2003) and macaques (Farkas *et al.*, 2008), respectively. Despite the diverse range of animal hosts for norovirus strains, there is no evidence of zoonotic transmission of norovirus to date.

16.2.2 Virus structure

The norovirus particle is spherical with protrusions that extend from the strict and local two-fold axes to an outer radius of 190 Å (32 nm) (Prasad *et al.*, 1994; Venkataram Prasad *et al.*, 1999). The virus capsid is formed of 90 homodimers of the major capsid protein (VP1), arranged in an icosahedral formation ($T = 3$ symmetry). The VP1 protein is highly organised into functional domains: N-terminal domain (N), shell domain (S), N-terminal protruding domain (P1–1), hypervariable domain (P2) and a C-terminal protruding domain (P1–2) (Venkataram Prasad *et al.*, 1999). The VP1-S domain forms the internal core, whilst the VP1-P domain extends from the particle surface forming spike-like protrusions. Receptor binding functions and immunoreactivity have been associated with the surface-exposed and hypervariable P2 domain.

The only complete human norovirus capsid structure that has been elucidated is that of the GI-1 norovirus prototype strain Norwalk virus, which was published in 1994 following cryo-EM studies on virus-like particles (VLPs) (Prasad *et al.*, 1994). This led to the detailed X-ray crystallographic structure of this virus (Venkataram Prasad *et al.*, 1999), as well as other norovirus strains and members of the *Caliciviridae* family (Bertolotti-Ciarlet *et al.*, 2002; Chen *et al.*, 2004, 2006; Shanker *et al.*, 2011). The first description of a GII norovirus structure was published in 2007 by Cao *et al.* who detailed the VP1 P domain from baculovirus-expressed VLPs (Cao *et al.*, 2007). The only available structure of a native norovirus particle is that of the cell culture-adapted murine norovirus strain CW.1 (Katpally *et al.*, 2007, 2010).

16.2.3 Virus evolution

Genetic diversity in viral genomes is generated and maintained through several mechanisms: reassortment/recombination, appearance of deletions and/or insertions and point mutations. Norovirus recombination between strains within the same genogroup has been reported (Bull *et al.*, 2007; Rohayem *et al.*, 2005). For recombination to occur co-infection with two or more norovirus strains is a prerequisite and this is often facilitated through primary sewage contamination of water and/or foodstuffs such as bivalve molluscs, and subsequent human exposure through consumption or contact with the contaminated water/shellfish, which may contain more than one norovirus strain. The role of recombination in the emergence of epidemiological important strains is, however, limited, given that such an event is limited to involving strains which are relatively closely related, and in the absence of interspecies transmission of noroviruses, recombination does not result in the emergence of significantly novel strains in the human population.

In contrast, mutation has a major impact on norovirus evolution. Driven by selective pressure from the individual host (Nilsson *et al.*, 2003) and herd immunity (Lindesmith *et al.*, 2011), the norovirus genome depends on its high mutation rate to generate novel genetic variants that maintain viral fitness. The genetic diversity observed among circulating norovirus strains (Gallimore *et al.*, 2007), the observed patterns of evolution of these viruses (Allen *et al.*, 2008; Lindesmith *et al.*, 2008; Siebenga *et al.*, 2007), and the impact of the emergence of novel strains in the epidemiology of norovirus infections and epidemics (Adamson *et al.*, 2007; Lopman *et al.*, 2004a) indicate that mutation is the major contributor to the persistence of noroviruses in the human population.

16.2.4 Features relevant to food, water and environmental transmission of norovirus

Although the majority of transmission events for norovirus are person-to-person, there are properties of the virus that enable its transmission via food, water and through the environment.

The norovirus particle demonstrates exceptional stability under a range of chemical and physical conditions (Ausar *et al.*, 2006; Cuellar *et al.*, 2010; Shoemaker *et al.*, 2010), making these viruses highly stable in the environment. Studies have found virus contamination on environmental surfaces even after deep cleaning (Morter *et al.*, 2011; Xerry *et al.*, 2008). A surrogate virolysis study found the norovirus capsid to be resistant to disruption using common disinfection reagents (Nowak *et al.*, 2011).

This stability means that foodstuffs and water sources at risk of primary contamination must be carefully treated in order to avoid infection following consumption. High-risk foods are those that are eaten raw such as oysters (Le Guyader *et al.*, 2006) and soft fruits (Maunula *et al.*, 2009), and are at risk of primary sewage contamination. In addition, contamination of food by infected food handlers can also be an important source of norovirus outbreaks and often involves foods that may be cooked, but that are prepared and handled without subsequent cooking such as salads, buffets (de Wit *et al.*, 2007; Nicolay *et al.*, 2011; Vivancos *et al.*, 2009; Wadl *et al.*, 2010).

The number of particles required to establish infection in an individual is probably very low; studies have indicated that fewer than 100 particles (Glass *et al.*, 2000; Teunis *et al.*, 2008) may be sufficient to cause symptomatic gastroenteritis in a susceptible individual. This contributes to the ease with which these viruses transmit, and to the high norovirus attack rates reported, particularly in closed or semi-closed environments (Harris *et al.*, 2010; Noda *et al.*, 2008; Ter Waarbeek *et al.*, 2010; Werber *et al.*, 2009).

16.2.5 Clinical presentations and treatment

Noroviruses have long been recognised as a cause of relatively mild, self-limiting gastroenteritis, which was first described as winter vomiting disease in 1929 (Zahorsky, 1929).

Typically norovirus infection has a short incubation period of 24 h (12–48 h). Illness is characterised by diarrhoea and vomiting, which in a proportion of cases is described as 'projectile'. Onset is sudden with no prodromal symptoms and other symptoms reported include nausea, abdominal cramps and flu-like symptoms.

Norovirus illness generally lasts 12–60 h (Kaplan, 1982), however, illness is prolonged by 24 h in the elderly (Lopman *et al.*, 2004c), and infection in severely immunocompromised/immunodeficient patients is associated with severe and persistent symptoms. There are few studies of mortality due to norovirus; in one study using a modelling approach a few deaths linked to norovirus were identified each year in the UK (Harris *et al.*, 2008). Recent research has suggested that norovirus infection may play a role in post-infectious irritable bowel syndrome (PI-IBS), and that the number of individuals developing PI-IBS may be as high as 13% (Zanini *et al.*, 2012).

Community-based studies show that most norovirus infections cause mild disease and are managed without intervention from healthcare services (Pang *et al.*, 2000). Only a minority of cases cause severe gastroenteritis as defined by the Vesikari scale, a scoring system used to measure the severity of gastroenteritis symptoms between 0 and 20 (low to high, respectively) (Ruuska and Vesikari, 1990). The low severity of cases is reflected in the relative lower numbers of paediatric hospital admissions associated with norovirus infection in comparison to rotavirus infections. Dehydration is the main concern and this is generally treatable with oral rehydration therapy (ORT). Few cases require intravenous rehydration.

It is noteworthy that norovirus infection does not always lead to symptomatic illness. Around 16% of infections will lead to asymptomatic carriage of the virus (Amar *et al.*, 2007; Phillips *et al.*, 2010b).

16.2.6 Pathogenesis

Histopathology studies have shown that norovirus infects the villus enterocytes of the duodenum and jejunum (small intestine), where it establishes a cytolytic infection. The infection-induced cell death causes blunting of the villi through cellular atrophy and crypt hyperplasia (Agus *et al.*, 1973). The mucosa and epithelium remain intact (Meeroff *et al.*, 1980).

Noroviruses cause secretary diarrhoea with malabsoption and enzymic dysfunction in the small intestinal brush border. Vomiting is due to changes in gastric motility and delays in gastric emptying (Meeroff *et al.*, 1980).

16.3 Susceptibility, immunity and diagnosis

16.3.1 Volunteer studies

Understanding of the pathogenesis of norovirus remains poor, due to the lack of a laboratory cell culture model (Duizer *et al.*, 2004) and the lack of a suitable animal model in which to study these viruses. Available information on the pathophysiology of norovirus infection has largely been derived from histological and pathological information collected during volunteer studies with the virus.

Soon after the identification of norovirus as an agent of gastroenteritis, it was demonstrated that a specific antibody response to the virus was generated following infection, and protection against subsequent re-infection with an homologous strain was observed (Dolin *et al.*, 1972). Subsequently, a first study with volunteers showed that primary infection with one of three different norovirus strains caused illness in some but not all individuals (Wyatt *et al.*, 1974). Individuals who became symptomatic following primary challenge were immune when re-challenged with the homologous virus 6–14 weeks later. However, heterologous challenge gave a spectrum of results that suggested cross-protection between norovirus strains was limited (Wyatt *et al.*, 1974). Other subsequent volunteer studies have presented conflicting data, demonstrating that some individuals with pre-existing high anti-NV antibodies were more susceptible following virus challenge than others with low or no anti-NV antibodies (Baron *et al.*, 1984; Parrino *et al.*, 1977). The duration of long-term immunity to norovirus following experimental infection is unclear, as studies found volunteers experimentally challenged 27–42 months after original challenge were not necessarily protected from symptomatic infection. When a proportion of the volunteers displaying clinical symptoms after two experimental infections were challenged a third time 4–8 weeks after the second challenge, one quarter remained susceptible to symptomatic infection (Parrino *et al.*, 1977). In a separate study, 77% of volunteers with pre-existing immunity to norovirus became ill following experimental infection, whilst only 32% of volunteers without pre-existing immunity to norovirus became ill (Blacklow *et al.*, 1979).

More recent attempts to evaluate homotypic and heterotypic responses to norovirus infection (Iritani *et al.*, 2007; Parker *et al.*, 1995; Rockx *et al.*, 2005a) have yielded equally mixed results. Results from a seroepidemiological study revealed a slight one-way cross-reactivity (Parker *et al.*, 1995). The authors note that the cross-reactive antibodies could be derived from an immunological memory to a previous infection; this is one of the main problems encountered when studying immunity to noroviruses in the population. Norovirus infection is common in the community: it has been demonstrated that seroprevalence of norovirus in the population is around 24% in infants, and rises to almost 90% in persons >60 years old (Gray *et al.*, 1993), therefore the presence of immunological memory to norovirus infection must be considered in all serological studies. Studies of homotypic and heterotypic antibody responses to natural norovirus infection in a population cohort found a higher degree of intra-genogroup cross-reactivity compared to inter-genogroup cross-reactivity, and that antibody avidity could not distinguish between homotypic and heterotypic antibody responses (Rockx *et al.*, 2005a). A recent study of the humoral response to norovirus infection in children showed that a strong serum IgG response was mounted to infection that remained high more than two years following infection, and that whilst both serum IgM and IgA could be detected following the infection, the IgM response was short-lived and the IgA response was maintained but at a low level (Iritani *et al.*, 2007).

16.3.2 Antigenic drift

Data from volunteer studies seem to indicate a limited role for serum antibody in protection from norovirus infection. The presence of norovirus-specific serum antibody does not correlate with protection against either symptomatic or asymptomatic infection. This is in line with the norovirus strict tropism, infecting only the gut mucosa without further dissemination in the host, meaning that local antibody responses will be important in protection from infection and/or symptoms, and serum antibody is unlikely to significantly impact on the progression of infection. Any immunity generated during a norovirus infection is likely to be limited in cross-protection against different norovirus genotypes (Matsui and Greenberg, 2000). In fact, data suggest that immune response to a particular genotype may be specific to the infecting strain, and may not be reactive against a different strain of the same genotype (Allen *et al.*, 2009). An individual's historical immunity to norovirus infection is further complicated by the rapid mutation rate of some strains, particularly the common GII-4 norovirus strains. These norovirus strains evolve rapidly, and frequently acquire genetic changes that can manifest as phenotypic differences which may confer a selective advantage to the virus, making it 'fitter' in the population (Allen *et al.*, 2008; Lindesmith *et al.*, 2008; Siebenga *et al.*, 2007). If such mutations in the genome give rise to antibody-escape mutants capable of evading existing immunity, then an individual can potentially be re-infected if the virus has sufficiently altered its antigenic profile to be unrecognised by the immunological memory from the first infection.

16.3.3 Virus binding to blood group antigens

The conflicting data from volunteer studies indicates that host susceptibility to norovirus infection is complex, and may not be dependent wholly on immune status of the individual. Some individuals may have a degree of natural resistance to particular norovirus genotypes, conferred by a genetically determined phenotype. Such a link was proposed between susceptibility to infection and the ABO blood group status of the individual (Hutson *et al.*, 2002). In their study, Hutson *et al.* showed that individuals with type O antigen exhibited a higher susceptibility to infection, whereas type B individuals had a decreased risk of infection and did not develop symptoms (Hutson *et al.*, 2002). In humans, HBGAs are expressed on different tissues with particular restrictions: ABH and Lewis antigens are expressed on the epithelia of tissues that are in contact with the external environment, such as the respiratory and gastrointestinal tracts. HBGA molecules are also secreted into the saliva of some individuals. The secretor phenotype is controlled by a single genetic locus; approximately 80% of Caucasian people are HBGA secretors.

Early serological studies showed that different norovirus genotypes displayed different HBGA-binding patterns (Harrington *et al.*, 2002; Huang *et al.*, 2003). Whilst GI-1 norovirus bound to saliva of all secretor but no non-secretor individuals, GII-2 norovirus bound only to that of B/AB secretor individuals, and a third genotype (GII-1) did not bind any of the types (Harrington *et al.*, 2002). Since these initial observations, a number of studies have sought to define binding relationships between the different norovirus genotypes and HBGAs (Huang *et al.*, 2005; Tan *et al.*, 2003, 2006; Tan and Jiang, 2005). These studies suggest a complicated relationship between noroviruses and HBGAs, with no robust binding pattern defined to date. The data certainly indicate a role for HBGAs as a susceptibility factor for norovirus infection (Hutson *et al.*, 2002; Lindesmith *et al.*, 2003; Rockx *et al.*, 2005b), with a strong but not exclusive link demonstrated between non-secretor status and resistance to infection (Lindesmith *et al.*, 2003). Multiple binding patterns have been observed between HBGAs and the ubiquitous GII-4 norovirus strains isolated at different time points (Lindesmith *et al.*, 2008).

Efforts have been made to elucidate the relationship between norovirus and HBGA at a structural level (Bu *et al.*, 2008; Cao *et al.*, 2007; Choi *et al.*, 2008; Shanker *et al.*, 2011; Tan *et al.*, 2003, 2008). The crystallographic structural data presented in these studies support the idea that interactions between norovirus and HBGAs is strain-specific, and demonstrate that the structure of the GI and GII norovirus capsid, especially in the hypervariable P2 domain where it is predicted that the HBGA-binding site is located, are very different (Bu *et al.*, 2008). There is no evidence that HBGAs are the primary receptors for noroviruses; given the systemic distribution of HBGAs, but lack of systemic norovirus infection/disease, it is more likely that HBGAs are co-receptors for efficient virus entry, in parallel to an as yet unidentified primary receptor (Bucardo *et al.*, 2010; Le Pendu *et al.*, 2006; Nordgren *et al.*, 2010).

16.3.4 Diagnosis of norovirus infection

The diagnosis of individual cases of viral gastroenteritis requires specific laboratory investigation but clinical diagnosis of norovirus-associated outbreaks of gastroenteritis is based on the 'Kaplan criteria' and is specific although a little insensitive in discriminating between outbreaks of gastroenteritis caused by norovirus from other aetiological agents (Turcios *et al.*, 2006). Four criteria are used: (i) bacterial/parasitic pathogen cannot be detected; (ii) >50% cases show vomiting; (iii) illness lasts 12–60 h; and (iv) incubation period is 24–48 h (Kaplan *et al.*, 1982a,b). The availability of laboratory-based diagnostic tests for norovirus have evolved dramatically, moving from EM, to immunoassays, reverse transcriptase-polymerase chain reaction (RT-PCR) and gel analysis, and most recently to real-time RT-PCR.

Initially, the only laboratory method for norovirus diagnosis was electron microscopy (EM), a technique used by many laboratories in the routine diagnosis of gastroenteritis until the advent of molecular methods. EM can be a powerful method, and since it is unbiased it is a true 'capture-all' method. However, its effective use in clinical diagnostics suffered from limited sensitivity and the high cost of a labour-intensive method requiring highly skilled workforce. At best the lower limit of detection for EM is one million virus particles per gram faeces, roughly the same as the peak levels of shedding during a norovirus infection. The sensitivity of EM can be increased through concentration of the virus in the clinical sample by ultracentrifugation and/or through the use of immune EM (IEM) (Kapikian *et al.*, 1972).

Following cloning and sequencing of the prototype norovirus strain Norwalk virus (Jiang *et al.*, 1990), VLPs were produced (Jiang *et al.*, 1992b) which could be used as antigen in immunoassays and as immunogens for producing antibodies in laboratory animals. This led to production of norovirus VLPs for representative strains of the different genogroups and genotypes, which in turn allowed the development of enzyme-linked immunosorbent assays (ELISAs) based on type-specific monoclonal antibodies. These assays quickly became adopted in the clinical setting for the detection of norovirus antigen in clinical samples (Jiang *et al.*, 2000). The relatively low sensitivity of these assays limits their use to the diagnosis of outbreaks of gastroenteritis, and only when several samples are available from the same outbreak. These methods are insufficiently sensitive and specific for the diagnosis of sporadic cases and their use may be further limited with time as the viruses evolve.

Also following the cloning and sequencing of norovirus prototype genomes (Jiang *et al.*, 1990; Lambden *et al.*, 1993), molecular assays for the detection of norovirus were developed, concurrent with the increasing use of polymerase chain reaction (PCR) methods in diagnostic laboratories (Ando *et al.*, 1995; Green *et al.*, 1993b; Willcocks *et al.*, 1993). RT-PCR methods for the detection of noroviruses have overtaken other methods for routine detection of norovirus in clinical specimens (Green *et al.*, 1993a; Jiang *et al.*, 1992a).

The majority of assays now used in the clinical settings are real-time RT-PCR based on the methods of Kageyama (Amar *et al.*, 2005; Gunson and Carman, 2005; Kageyama *et al.*, 2003; Richards *et al.*, 2004). With such sensitive methods of detection available, it becomes important to understand when positive results are indeed clinically relevant. Given that a high level of asymptomatic carriage of norovirus has been found in the community, and the relatively high incidence of mixed infections with more than one pathogen, it is increasingly important to understand the limitations of the diagnostic methods in order to assign causality between symptoms/disease and the detection of a pathogen in a clinical sample. To address this, over 300 samples collected from patients in the UK were tested by real-time RT-PCR for detection of norovirus, half of which were from individuals with symptomatic gastroenteritis, and half of which had no symptoms nor recent history of gastroenteritis (Phillips *et al.*, 2010a). Asymptomatic individuals were found to have lower norovirus loads correlating with Ct (cycle threshold) values ≥31: whilst technically positive by definition in the assay, these were not deemed to be clinically relevant, and so defined a cut-off for clinical interpretation of real-time RT-PCR detection of norovirus (Phillips *et al.*, 2010a).

16.4 Epidemiology of norovirus gastroenteritis associated with food, water and the environment

Many countries now run national surveillance programmes with the aim of understanding the burden of disease attributable to norovirus. However, due to underreporting, accurate figures are difficult to determine. The pattern of disease observed for norovirus is affected by a number of factors, and there have now been several studies to understand the impact of both endemic and epidemic norovirus-associated disease. Whilst mostly associated with person-to-person transmission, the environmental stability of norovirus means that it can also be transmitted via contaminated food, water and environmental vehicles.

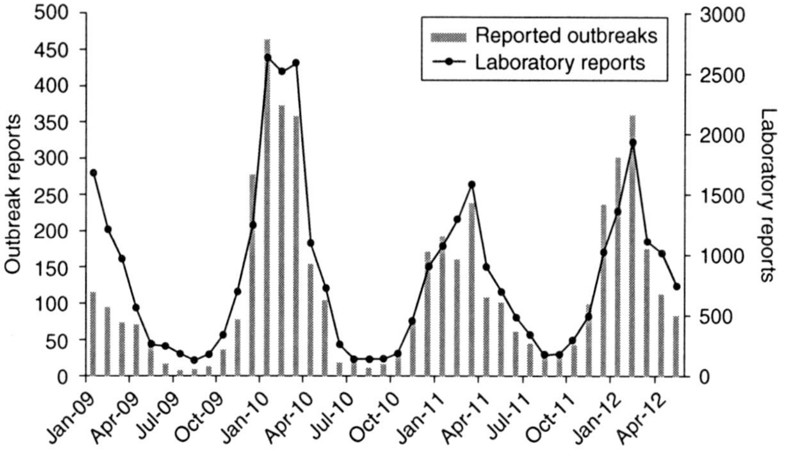

Fig. 16.1 Seasonality of norovirus outbreaks and laboratory confirmed cases in England and Wales, reported to HPA between January 2009 and May 2012. This represents the standard surveillance output, which illustrates the scale of underreporting.

16.4.1 Norovirus epidemiology

Many countries have surveillance programmes for norovirus, but because most cases are not laboratory investigated, and depending upon healthcare system structure the sufferer may or may not seek medical attention (Tam *et al.*, 2011), accurate figures are difficult to determine (Dedman *et al.*, 1998). In the USA, norovirus is estimated to cause about 21 million cases of acute gastroenteritis and 70 000 hospitalisations per year (Centers for Disease Control and Prevention, 2012a).

Food-related norovirus outbreaks resulting from primary sewage contamination of shellfish, soft fruit or salads, and from food contaminated by infected food handlers after cooking or during preparation have been described (Dedman *et al.*, 1998). However, these outbreaks are often not recognised or reported, and so the incidence of foodborne outbreaks is not well defined. In the USA, it is estimated that 50% of foodborne outbreaks due to known agents are related to noroviruses (Centers for Disease Control and Prevention, 2012a). In the UK, it has been estimated that 700 000 (of 3 million) cases annually may be food-related (Adak *et al.*, 2002), but the precise burden of foodborne norovirus illness is not known.

Seasonality

Noroviruses have a worldwide distribution. The range of strains seen in different geographical regions overlap, suggesting global circulation. Some GII-4 strains have been shown to spread rapidly worldwide. In developed countries, norovirus infections exhibit a clear winter peak, with outbreaks occurring predominantly from the autumn through to the spring (see Fig. 16.1). Infections and outbreaks continue to occur throughout the year at a lower rate in temperate countries (Lopman *et al.*, 2003). Foodborne outbreaks reflect this pattern in the community, with the majority of outbreaks occurring in the winter season. This may be a result of both peak incidence of infection leading to more environmental contamination and subsequent contamination of foodstuffs at primary production, and colder temperatures reducing bivalve mollusc activity with reduced rates of viral clearance. The temporal pattern of infection in developing countries is less pronounced, with infections occurring throughout the year.

Age

Studies of norovirus incidence and prevalence all show the peak age of infection to be children under 5 years (Tam *et al.*, 2011). However in contrast to other viral gastroenteritis agents, infection and outbreaks occur throughout the age range due to the extreme viral diversity, rapid evolution and lack of long-term immunity.

Endemic disease

Noroviruses are the commonest cause of infectious intestinal disease (IID). The most comprehensive data on the burden of gastrointestinal disease attributable to norovirus comes from specific studies such as the infectious intestinal disease studies (IID studies) (Phillips *et al.*, 2010a; Tam *et al.*, 2011, 2012) and the Sensor Study in the Netherlands (de Wit *et al.*, 2001). Data from the second IID study estimated that there are 3 million community cases of norovirus in the UK each year, but only 130 000 (4.3%) of these cases result in affected persons accessing healthcare (Phillips *et al.*, 2010a; Tam *et al.*, 2011).

Outbreaks of norovirus gastroenteritis

Norovirus is the major aetiological agent of outbreaks of viral gastroenteritis, and its propensity to cause large outbreaks is related to the high infectivity, broad diversity and lack of long-term immunity of the virus. There is evidence that a proportion of the population excrete norovirus asymptomatically (Amar *et al.*, 2007; Phillips *et al.*, 2010b), and with its low infectious dose and wide dissemination throughout the population, norovirus is easily transmitted (Lopman *et al.*, 2004b). The most significant impact of norovirus-associated outbreaks is found in healthcare settings. Estimates from the Health Protection Agency's Norovirus Outbreak Reporting scheme suggest that in England over a 1-year period, there are more than 2000 outbreaks of norovirus in hospitals, affecting over 24 000 patients and resulting in >15 000 days of ward closure and >47 000 bed-days lost (Harris, 2010; Harris *et al.*, 2011). This is a common problem in hospitals in Europe and the USA. Much smaller numbers of point-source outbreaks in association with food consumption are reported annually (10–20). Food-related outbreaks can be large, potentially resulting in widespread outbreaks. In addition, secondary person-to-person transmission after a point-source outbreak (Alfano-Sobsey *et al.*, 2012) complicates investigation and appropriate ascertainment of food-related norovirus outbreaks and cases.

Epidemic disease

The pattern and burden of norovirus infection varies from year to year and thus has been associated with genotype evolution. In particular the genogroup II-genotype 4 (GII-4) group of norovirus strains have been periodically associated with epidemic waves of gastroenteritis (Adamson *et al.*, 2007; Allen *et al.*, 2008; Lopman *et al.*, 2004a), linked to the appearance in the population of a novel strain that has acquired a new phenotype (Allen *et al.*, 2008). This is suggested to be a change in the antigenic profile of the virus allowing it to evade existing herd immunity, and so cause increased epidemic levels of disease following emergence (Allen *et al.*, 2009). The different norovirus genotypes are detected with varying frequencies; however circulating strains worldwide have been dominated by GII-4 since the early 1990s (Hale *et al.*, 2000). These strains are most frequently associated with outbreaks of gastroenteritis, particularly in healthcare settings, but also other semi-closed environments and even outbreaks in the community. The success of GII-4 norovirus strains is due, at least in part, to their high degree of genetic variation. Over a winter season (September–March) a genetically diverse population of GII-4 norovirus strains circulate, and this diversity gives rise to successful GII-4 strains which circulate for a time and are succeeded by different genetic variant strains (Gallimore *et al.*, 2007). The reservoir of genetic diversity among the GII-4 norovirus population has been used to model the evolution of these viruses (Allen *et al.*, 2008; Lindesmith *et al.*, 2008; Siebenga *et al.*, 2007). It is predicted that there is population-mediated selection exerted on these variants, likely through herd immunity, which gives rise to viruses with novel phenotypes that may have a selective advantage in the population, such as antibody-escape mutants (Allen *et al.*, 2009). This pattern of evolution among GII-4 norovirus strains is supported by epidemiological data collected since the early 2000s. A novel GII-4 norovirus strain was detected in 2002, which went on to cause an increase in outbreaks during the summer and autumn of 2002, resulting in a 77% increase in outbreaks compared to the previous high in 1995 (Lopman *et al.*, 2004a). In the same year, this strain emerged across Europe and caused the same increase in outbreaks in all countries (Lopman *et al.*, 2004a). This pattern was repeated in 2006 when a different, novel, GII-4 norovirus strain emerged, causing a higher than normal number of outbreaks in the summer months, followed by epidemic numbers of outbreaks in the subsequent winter season (Adamson *et al.*, 2007). Since then, an increase in outbreak numbers was reported in the winter of 2009/2010 (Harris *et al.*, 2011), and it is predicted that a change in the virus phenotype will once again lead to the emergence of a novel virus strain resulting in higher numbers of infections and outbreaks of norovirus gastroenteritis.

16.4.2 Transmission routes

There is an extensive literature documenting various causes of outbreaks of norovirus gastroenteritis in a variety of settings. Primary transmission

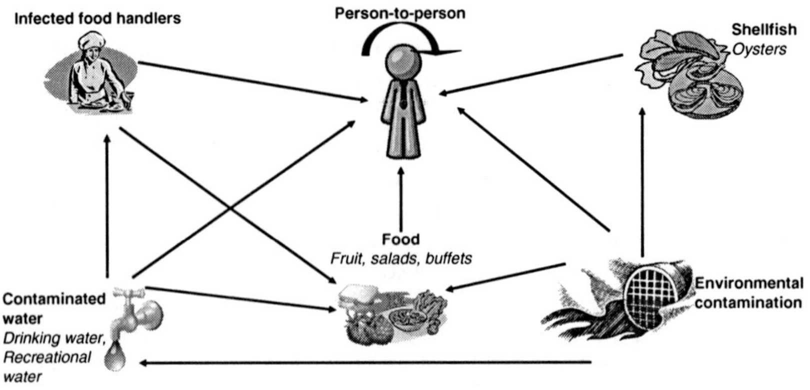

Fig. 16.2 Norovirus transmission routes, illustrating the role of food and environmental contamination in the virus infection cycle. The overall burden of infection is predominantly person-to-person. The significance of foodborne transmission is yet to be defined.

of norovirus occurs via the faecal-oral route. The most common transmission route for norovirus is directly from person to person. Transmission can also occur via foodstuffs contaminated by an infected food handler. Contamination of the environment with faecal material has also been reported associated with outbreaks of norovirus gastroenteritis (Kimura *et al.*, 2010; Morter *et al.*, 2011). See Fig. 16.2.

Person-to-person transmission

Person-to-person transmission of norovirus is the most significant route of transmission, as humans are the only known reservoir of the human-pathogenic strains. During symptomatic infection, the virus is shed in faeces and vomit, both of which can contaminate the environment and/or become aerosolised leading to transmission of the virus. In addition, following infection norovirus can be shed before the onset of clinical symptoms (prodromal excretion), and after clinical symptoms have resolved (Goller *et al.*, 2004), and can also be shed entirely asymptomatically (Phillips *et al.*, 2010b). Although tentative links between asymptomatic carriage of norovirus and transmission in outbreaks have been made in the literature (Barrabeig *et al.*, 2010; Nicolay *et al.*, 2011; Schmid *et al.*, 2011; Sukhrie *et al.*, 2012), the role of asymptomatic cases in transmission has yet to be established. Outbreaks of norovirus gastroenteritis in healthcare settings have been investigated using high-resolution strain tracking methods (Xerry *et al.*, 2008). These methods show that individual strains can be tracked during the course of an outbreak, and demonstrate that transmission events must have occurred between individuals in the hospital (Morter *et al.*, 2011; Xerry *et al.*, 2008). However, to understand the dynamics of transmission in such a setting would require both the high-resolution strain tracking virological data and detailed epidemiological information on patient and staff movements and infection control measures implemented during the outbreak. As well as direct contact between infected and uninfected individuals, the virus can be transmitted through aerosols generated from faeces and vomit contamination of the environment. Virus in vomit aerosols have been linked to transmission during outbreaks in hospitals (Sawyer *et al.*, 1988), cruise ships (Ho *et al.*, 1989), schools and following events at hotels (Marks *et al.*, 2000) and concerts (Evans *et al.*, 2002) and sports venues (Repp and Keene, 2012).

Food, water and environmental transmission

Individuals excreting the virus through faecal material and/or vomit can also contaminate environmental surfaces, foods and water, all of which can then act as vehicles for transmission of the virus. The low infectious dose (Glass *et al.*, 2000; Teunis *et al.*, 2008) and environmental stability (D'Souza *et al.*, 2006; Nowak *et al.*, 2011) of norovirus means that contamination of food and water with human faecal material can also be the cause of large outbreaks. The foods most commonly associated with norovirus gastroenteritis are contaminated shellfish (particularly oysters) (Baker *et al.*, 2010; Doyle *et al.*, 2004; Gallimore *et al.*, 2005; Le Guyader *et al.*, 2006), soft fruits (Falkenhorst *et al.*, 2005; Le Guyader *et al.*, 2004; Maunula *et al.*, 2009), salads (Wadl *et al.*, 2010), and meat products (Malek *et al.*, 2009; Vivancos *et al.*, 2009). In terms of waterborne norovirus illness, both contaminated drinking water (Kvitsand and Fiksdal, 2010; Nenonen *et al.*, 2012; Riera-Montes *et al.*, 2011; Scarcella *et al.*, 2009) and recreational water (Hoebe *et al.*, 2004) have been implicated in outbreaks of gastroenteritis.

16.4.3 Epidemiology of food-, water- and environment-associated norovirus

Contamination of an intermediate vehicle with norovirus can cause outbreaks to become protracted and more widespread as the exposure rates are increased. The impact of transmission via food, water and the environment are discussed in more detail below.

Norovirus food outbreaks

Contamination of foods can occur at the point of production (such as sewage contamination of oysters), the point of preparation (such as food-handler contamination of salads), or at the point of service (such as contact contamination of buffet foods by other diners). One of the major impacts of food-associated norovirus gastroenteritis is that a large number of individuals can become affected quickly, as the number of exposed people can be large and can cover a large geographical area, and these cases can lead to a high number of secondary cases (Alfano-Sobsey *et al.*, 2012). Outbreaks of norovirus gastroenteritis have been reported in association with contaminated oysters (Baker *et al.*, 2010; Lamden *et al.*, 2004; Nenonen *et al.*,

2009; Westrell *et al.*, 2010). Contamination of oysters with micro-organisms occurs when the seawater in oyster beds becomes contaminated with sewage (Pommepuy *et al.*, 2004). Oysters are natural filter-feeder organisms, and as such can concentrate material in their tissues up to 99-fold of that in their surrounding environment (Burkhardt and Calci, 2000). Other foods, particularly fruits and salads, have been implicated as vehicles of transmission for norovirus (de Wit *et al.*, 2007; Malek *et al.*, 2009; Sala *et al.*, 2005; Schmid *et al.*, 2007). Ready-to-eat foods such as these can become contaminated at the point of origin with human sewage. As these foods are eaten uncooked, improper washing or preparation will not remove the virus. Food can also become contaminated during preparation via the food handler (Greig *et al.*, 2007), and a number of outbreaks of norovirus gastroenteritis originating in eating establishments or at catered events have been associated with contamination of the food by food handlers (Barrabeig *et al.*, 2010; Boxman *et al.*, 2009; de Wit *et al.*, 2007; Schmid *et al.*, 2007).

Norovirus water outbreaks

Water supplies in developed countries are predominantly clean. Contamination occurs rarely, usually as a consequence of leakage of sewage from civic basins or septic tanks, or as a result from flooding or runoff associated with heavy rainfall. All of these can leach faecal-contaminated water into water supplies for consumption or recreational use. Contamination of both drinking water (Gallay *et al.*, 2006; Maunula, 2005; Nygard *et al.*, 2004; Ter Waarbeek *et al.*, 2010) and recreational water (Hoebe *et al.*, 2004; Maunula *et al.*, 2004; Podewils *et al.*, 2006; Sartorius *et al.*, 2007) with sewage or faecal material has been association with large outbreaks of norovirus gastroenteritis. Waterborne outbreaks, particularly those involving contamination of drinking water supply chains, have the potential to affect very large numbers of people in a short period of time, as the exposure to the contaminated water source can be high (Maunula, 2005; Riera-Montes *et al.*, 2011; Scarcella *et al.*, 2009).

Norovirus environmental outbreaks

Contamination of the environment can have an important role in the dynamics of an outbreak. The majority of available data on this topic has been collected during outbreaks occurring in healthcare settings. Following outbreaks on hospital wards, environmental contamination has been detected in patient areas (bedside furniture, bathrooms, etc.) and on equipment (blood pressure monitors, keyboards, etc.) (Morter *et al.*, 2011). Contaminated environments can act as temporary reservoirs of the virus that lead to re-seeding of the outbreak if the environmental cleaning is inefficient. In this situation, despite implementation of infection control measures, the outbreak can persist and become protracted. Additionally, if contaminated equipment is shared, then the outbreak can spread geographically (e.g. from one ward into other wards or parts of a hospital). This highlights the need for stringent cleaning protocols during outbreaks and to ensure routine cleaning is efficient and as effective as possible. Fomite contamination can also lead to subsequent foodborne contamination (Repp and Keene, 2012).

16.5 Prevention and control

The stability of the norovirus virion under a wide range of conditions has made establishing and evaluating prevention and control measures challenging. Most routinely used prevention and control methods focus on strict staff training and establishing good hygiene and cleaning practices. Several guidelines for responding to outbreaks in food preparation premises are available, and there is a wide range of generic legislation for food processing and handling. Whilst there remains no licenced vaccine against norovirus, data from the first clinical trials using a virus-like particle vaccine has recently been reported.

16.5.1 Prevention

Detailed discussion on the prevention of foodborne infections can be found elsewhere in this book. Measures for preventing norovirus contamination are centred on good hand hygiene and environmental cleaning practices in healthcare settings, food establishments and on board cruise ships. Comprehensive training of staff in catering establishments focusing on good hand hygiene is a priority for infection control, as the virus is spread along a faecal–oral route. Encouraging passengers on board cruise ships and at catered events to wash their hands before handling buffet foods can significantly reduce the risk of cross-contamination of the foods. Alcohol wipes which are widely used for hand cleaning are not effective for norovirus. Similarly, stringent environmental cleaning protocols must be followed on hospital wards and in kitchens, particularly during outbreaks, to limit the onward spread of contamination, and to remove temporary reservoirs of the virus in the environment that may re-seed the outbreak even after cleaning. At food handling premises, maintenance of clean and hygienic food preparation surfaces and equipment must be maintained, alongside separation of unwashed and washed foods, uncooked and cooked foods, etc. to avoid cross-contamination.

16.5.2 Control

Several guidelines for responding to outbreaks in food preparation premises are available (Centers for Disease Control and Prevention, 2012b; European Food Safety Authority, 2012). They are all based on staff training, hand hygiene, segregation of risk foods from foods to be served, exclusion of symptomatic food handlers for up to 72 h after recovery from symptoms, and systematic environmental cleaning.

Food

There is a wide range of generic legislation for food processing and handling. For shellfish-related norovirus, several approaches have been adopted. In Europe, growing waters for bivalve molluscs (oysters, mussels) are monitored for contamination, and sale of these products is controlled depending upon water quality. This is an imprecise control method and the recent description of a reliable molecular test for detecting noroviruses in shellfish may enable more effective controls to be established. The link between shellfish, particularly oysters, and outbreaks of norovirus gastroenteritis has been well documented. Oysters are treated by a depuration process before sale; however, this process is not efficient at removing micro-organisms from oyster tissues (Savini *et al.*, 2009). As the oyster is normally eaten raw, there is no further opportunity to inactivate any contaminating micro-organism in the tissues before consumption, and even at low numbers, this may be sufficient to establish infection in the individual.

Water

Treatment of water for consumption or recreational use relies mainly on chemical treatments such as chlorination, and/or filtration. When treatment procedures fail, large outbreaks can result from faecal contamination of the water (Werber *et al.*, 2009). It is important to continue to monitor water quality and the effectiveness of water treatment methods. Both chlorination and sand-filtration of water have been shown to reduce norovirus contamination in water supplies (Kitajima *et al.*, 2010; Shirasaki *et al.*, 2010); however, these measurements are based on surrogate systems and do not account for viable virus particles – an important indicator that remains unavailable in norovirus research due to the lack of a cell culture model (Duizer *et al.*, 2004).

Environment

A study looking at the structural stability of the norovirus capsid under disinfection conditions indicated that only a combination of heat and alkali was suitable for breaking down norovirus capsids as a proxy for disinfection. Commonly used disinfectants (e.g. alcohols, quaternary ammonium compounds, and chlorine dioxide) were ineffective at promoting virolysis of the norovirus capsid (Nowak *et al.*, 2011). This highlights the possible challenges faced by infection control teams when faced with an outbreak of norovirus gastroenteritis.

16.5.3 Vaccine

There is currently no licenced vaccine or antiviral drug for prophylaxis or treatment of norovirus. Following the cloning of a norovirus genome in the early 1990s (Jiang *et al.*, 1990), the viral capsid protein was expressed in a recombinant protein expression system and found to self-assemble into antigenically native but replication incompetent VLPs (Jiang *et al.*, 1992b). The effective use of norovirus VLPs as a potential vaccine (Ball *et al.*, 1996) led to a number of studies and trials in both humans and mice (El-Kamary *et al.*, 2010; LoBue *et al.*, 2006, 2009; Parra *et al.*, 2012), and more recently chimpanzees (Bok *et al.*, 2011). With a number of other viral pathogen VLP vaccines now either in development or in use, it is likely that this technology will be the leading choice for future vaccine development (Roy and Noad, 2009). As well as direct administration of VLPs as an oral or intranasal vaccine, there has been research into administration systems involving: transgenic plants including tobacco, potatoes and tomatoes (Mason *et al.*, 1996; Santi *et al.*, 2008; Tacket *et al.*, 2000; Zhang *et al.*, 2006); adenovirus-based systems (Guo *et al.*, 2008); and also yeast systems (Xia *et al.*, 2007). Early studies in mice demonstrated that immunisation with VLPs induced intestinal IgA 24 h after immunisation, and mucosal IgG 9 days later (Ball *et al.*, 1996). Subsequently, a phase I study found that one 250 μg dose of VLP was sufficient to induce a ≥4-fold increase in serum IgG with no side effects (Ball *et al.*, 1999). Subsequent studies using VLPs as human vaccines confirmed the immunological responses they induce in humans (Guerrero *et al.*, 2001; Tacket *et al.*, 2003). El-Kamary and colleagues reported results from a phase I clinical trial in which an intranasal vaccine composed of an adjuvanted VLP was shown to be immunogenic and was safely delivered without serious side effects (El-Kamary *et al.*, 2010). Recently, Atmar et al (2011) reported the first trial demonstrating homologous protection against illness and infection using a norovirus VLP. The study in 47 recipients showed a norovirus-specific IgA seroresponse in 70% of vaccine recipients. Recipients had a significantly reduced frequency of gastroenteritis (37%) vs placebo recipients (69%). Whilst promising, the vaccine is monovalent and with evidence to suggest that there is little cross-protection between norovirus strains, a multivalent vaccine is likely to be the only viable option for vaccine development. The high mutation rate of the virus, coupled with the strain replacement events observed among circulating virus strains, means that any vaccine developed would possibly require regular re-development as the circulating strains change, similarly to that for the seasonal influenza vaccine.

16.6 Conclusion

Noroviruses are an important enteric pathogen and recent findings on viral diversity, human susceptibility and immunity have improved our understanding of the pattern of infection. The transmission of noroviruses via contaminated food and water is well documented; however, the number of food- and/or waterborne outbreaks of gastrointestinal illness attributable to norovirus is difficult to determine accurately and thus the true burden of foodborne norovirus is poorly defined. Recent data from the CDC estimated that in the USA, 49% of the foodborne outbreaks with a single confirmed aetiological agent were associated with norovirus (Centers for Disease Control and Prevention,

2011). Adequate capture of both epidemiological and laboratory data of norovirus outbreaks remains a major challenge, as many outbreaks fail to be identified and/or followed up. Improving structured surveillance and reporting systems would enhance understanding of the epidemiology of noroviruses in outbreaks, and help improve the implementation of timely public health interventions and control measures. In addition laboratory data including strain characterisation from samples obtained from the affected individuals can help understand whether outbreaks are likely to be associated with sewage contamination at the point of production (as these are often associated with multiple norovirus strains), or whether the outbreak is likely to be associated with single point-source contamination, for example, food handler-associated (more likely to be associated with a single strain shared among all affected, including the food handler). Good hand hygiene practices, exclusion of affected staff and strict guidelines in food preparation and catering premises remains the focus for foodborne infection prevention strategies. Methods to control transmission by food and/or water contaminated at the point of production, such as controls on oyster farming, are in place, but these have been proved to be less than completely efficient due to the levels of contamination found in European waters, and thus remain a major challenge.

16.7 References

ADAK, G. K., LONG, S. M. and O'BRIEN, S. J. (2002). Trends in indigenous foodborne disease and deaths, England and Wales: 1992 to 2000. *Gut* **51**, 832–41.

ADAMSON, W. E., GUNSON, R. N., MACLEAN, A. and CARMAN, W. F. (2007). Emergence of a new norovirus variant in Scotland in 2006. *J Clin Microbiol* **45**, 4058–60.

AGUS, S. G., DOLIN, R., WYATT, R. G., TOUSIMIS, A. J. and NORTHRUP, R. S. (1973). Acute infectious nonbacterial gastroenteritis: intestinal histopathology. Histologic and enzymatic alterations during illness produced by the Norwalk agent in man. *Ann Intern Med* **79**, 18–25.

ALFANO-SOBSEY, E., SWEAT, D., HALL, A., BREEDLOVE, F., RODRIGUEZ, R., GREENE, S., PIERCE, A., SOBSEY, M., DAVIES, M. and LEDFORD, S. L. (2012). Norovirus outbreak associated with undercooked oysters and secondary household transmission. *Epidemiol Infect* **140**, 276–82.

ALLEN, D. J., GRAY, J. J., GALLIMORE, C. I., XERRY, J. and ITURRIZA-GOMARA, M. (2008). Analysis of amino acid variation in the P2 domain of norovirus VP1 protein reveals putative variant-specific epitopes. *PLOS One* **3**, 1–9.

ALLEN, D. J., NOAD, R., SAMUEL, D., GRAY, J. J., ROY, P. and ITURRIZA-GÓMARA, M. (2009). Characterisation of a GII-4 norovirus variant-specific surface-exposed site involved in antibody binding. *Virol J* **6**, 150.

AMAR, C. F., EAST, C. L., GRANT, K. A., GRAY, J., ITURRIZA-GÓMARA, M., MACLURE, E. A. and MCLAUCHLIN, J. (2005). Detection of viral, bacterial, and parasitological RNA or DNA of nine intestinal pathogens in fecal samples archived as part of the English infectious intestinal disease study: assessment of the stability of target nucleic Acid. *Diagn Mol Pathol* **14**, 90–6.

AMAR, C. F., EAST, C. L., GRAY, J., ITURRIZA-GÓMARA, M., MACLURE, E. A. and MCLAUCHLIN, J. (2007). Detection by PCR of eight groups of enteric pathogens in 4627 faecal samples: re-examination of the English case-control Infectious Intestinal Disease Study (1993–1996). *Eur J Clin Microbiol Infect Dis* **26**, 311–23.

ANDO, T., MONROE, S. S., GENTSCH, J. R., JIN, Q., LEWIS, D. C. and GLASS, R. I. (1995). Detection and differentiation of antigenically distinct small round structured viruses (Norwalk-like viruses) by reverse transcription PCR and Southern hybridization. *J Clin Microbiol* **33**, 64–71.

ATMAR, R. L., BERNSTEIN, D. I., HARRO, C. D., AL-IBRAHIM, M. S., CHEN, W. H., FERREIRA, J., ESTES, M. K., GRAHAM, D. Y., OPEKUN, A. R., RICHARDSON, C. and MENDELMAN, P. M. (2011). Norovirus vaccine against experimental human Norwalk Virus illness. *N Engl J Med* **365**, 2178–87.

AUSAR, S. F., FOUBERT, T. R., HUDSON, M. H., VEDVICK, T. S. and MIDDAUGH, C. R. (2006). Conformational stability and disassembly of Norwalk virus-like particles. Effect of pH and temperature. *J Biol Chem* **281**, 19478–88.

BAKER, K., MORRIS, J., MCCARTHY, N., SALDANA, L., LOWTHER, J., COLLINSON, A. and YOUNG, M. (2010). An outbreak of norovirus infection linked to oyster consumption at a UK restaurant, February 2010. *J Public Health (Oxf)* **33**, 205–11.

BALL, J. M., ESTES, M. K., HARDY, M. E., CONNER, M. E., OPEKUN, A. R. and GRAHAM, D. Y. (1996). Recombinant Norwalk virus-like particles as an oral vaccine. *Arch Virol* **S12**, 243–9.

BALL, J. M., GRAHAM, D. Y., OPEKUN, A. R., GILGER, M. A., GUERRERO, R. A. and ESTES, M. K. (1999). Recombinant norwalk virus-like particles given orally to volunteers: phase I study. *Gastroenterology* **117**, 40–8.

BARON, R. C., GREENBERG, H. B., CUKOR, G. and BLACKLOW, N. R. (1984). Serological responses among teenagers after natural exposure to Norwalk virus. *J Infect Dis* **150**, 531–4.

BARRABEIG, I., ROVIRA, A., BUESA, J., BARTOLOME, R., PINTO, R., PRELLEZO, H. and DOMINGUEZ, A. (2010). Foodborne norovirus outbreak: the role of an asymptomatic food handler. *BMC Infect Dis* **10**, 269.

BERTOLOTTI-CIARLET, A., WHITE, L. J., CHEN, R., PRASAD, B. V. and ESTES, M. K. (2002). Structural requirements for the assembly of norwalk virus-like particles. *J Virol* **76**, 4044–55.

BLACKLOW, N. R., CUKOR, G., BEDIGIAN, M. K., ECHEVERRIA, P., GREENBERG, H. B., SCHREIBER, D. S. and TRIER, J. S. (1979). Immune response and prevalence of antibody to Norwalk enteritis virus as determined by radioimmunoassay. *J Clin Microbiol* **10**, 903–9.

BOK, K., PARRA, G. I., MITRA, T., ABENTE, E., SHAVER, C. K., BOON, D., ENGLE, R., YU, C., KAPIKIAN, A. Z., SOSNOVTSEV, S. V., PURCELL, R. H. and GREEN, K. Y. (2011). Chimpanzees as an animal model for human norovirus infection and vaccine development. *Proc Natl Acad Sci U S A* **108**, 325–30.

BOXMAN, I., DIJKMAN, R., VERHOEF, L., MAAT, A., VAN DIJK, G., VENNEMA, H. and KOOPMANS, M. (2009). Norovirus on swabs taken from hands illustrate route of transmission: a case study. *J Food Prot* **72**, 1753–5.

BU, W., MAMEDOVA, A., TAN, M., XIA, M., JIANG, X. and HEGDE, R. S. (2008). Structural basis for the receptor binding specificity of Norwalk virus. *J Virol* **82**, 5340–7.

BUCARDO, F., NORDGREN, J., CARLSSON, B., KINDBERG, E., PANIAGUA, M., MOLLBY, R. and SVENSSON, L. (2010). Asymptomatic Norovirus Infections in Nicaraguan Children and its Association With Viral Properties and Histo-Blood Group Antigens. *Pediatr Infect Dis J* **29**, 934–9.

BULL, R. A., TANAKA, M. M. and WHITE, P. A. (2007). Norovirus recombination. *J Gen Virol* **88**, 3347–59.

BURKHARDT, W., III and CALCI, K. R. (2000). Selective accumulation may account for shellfish-associated viral illness. *Appl Environ Microbiol* **66**, 1375–8.

CAO, S., LOU, Z., TAN, M., CHEN, Y., LIU, Y., ZHANG, Z., ZHANG, X. C., JIANG, X., LI, X. and RAO, Z. (2007). Structural Basis for the Recognition of Blood Group Trisaccharides by Norovirus. *J Virol* **81**, 5949–57.

CENTERS FOR DISEASE CONTROL AND PREVENTION. (2011). Surveillance for foodborne disease outbreaks – United States, 2008. *Morbidity and Mortality Weekly Report* **60**, 1197–1202.

CENTERS FOR DISEASE CONTROL AND PREVENTION. (2012a). Burden of norovirus illness and outbreaks, http://www.cdc.gov/norovirus.

CENTERS FOR DISEASE CONTROL AND PREVENTION. (2012b). Updated norovirus outbreak management and disease prevention guidelines. In *Morbidity and Mortality Weekly Report*, Recommendations and Reports, Vol. 60, No. 3, March 4, 2011.

CHEN, R., NEILL, J. D., ESTES, M. K. and PRASAD, B. V. (2006). X-ray structure of a native calicivirus: structural insights into antigenic diversity and host specificity. *Proc Natl Acad Sci U S A* **103**, 8048–53.

CHEN, R., NEILL, J. D., NOEL, J. S., HUTSON, A. M., GLASS, R. I., ESTES, M. K. and PRASAD, B. V. (2004). Inter- and intragenus structural variations in caliciviruses and their functional implications. *J Virol* **78**, 6469–79.

CHOI, J. M., HUTSON, A. M., ESTES, M. K. and PRASAD, B. V. (2008). Atomic resolution structural characterization of recognition of histo-blood group antigens by Norwalk virus. *Proc Natl Acad Sci U S A* **105**, 9175–80.

CUELLAR, J. L., MEINHOEVEL, F., HOEHNE, M. and DONATH, E. (2010). Size and mechanical stability of norovirus capsids depend on pH: a nanoindentation study. *J Gen Virol* **91**, 2449–56.

DEDMAN, D., LAURICHESSE, H., CAUL, E. O. and WALL, P. G. (1998). Surveillance of small round structured virus (SRSV) infection in England and Wales, 1990–95. *Epidemiol Infect* **121**, 139–49.

DE WIT, M. A., KOOPMANS, M. P., KORTBEEK, L. M., WANNET, W. J., VINJE, J., VAN LEUSDEN, F., BARTELDS, A. I. and VAN DUYNHOVEN, Y. T. (2001). Sensor, a population-based cohort study on gastroenteritis in the Netherlands: incidence and etiology. *Am J Epidemiol* **154**, 666–74.

DE WIT, M. A., WIDDOWSON, M. A., VENNEMA, H., DE BRUIN, E., FERNANDES, T. and KOOPMANS, M. (2007). Large outbreak of norovirus: The baker who should have known better. *J Infect* **55**, 188–93.

DOLIN, R., BLACKLOW, N. R., DUPONT, H., BUSCHO, R. F., WYATT, R. G., KASEL, J. A., HORNICK, R. and CHANOCK, R. M. (1972). Biological properties of Norwalk agent of acute infectious nonbacterial gastroenteritis. *Proc Soc Exp Biol Med* **140**, 578–83.

DOYLE, A., BARATAUD, D., GALLAY, A., THIOLET, J. M., LE GUYAGUER, S., KOHLI, E. and VAILLANT, V. (2004). Norovirus foodborne outbreaks associated with the consumption of oysters from the Etang de Thau, France, December 2002. *Euro Surveill* **9**, 24–6.

D'SOUZA, D. H., SAIR, A., WILLIAMS, K., PAPAFRAGKOU, E., JEAN, J., MOORE, C. & JAYKUS, L. (2006). Persistence of caliciviruses on environmental surfaces and their transfer to food. *Int J Food Microbiol* **108**, 84–91.

DUIZER, E., SCHWAB, K. J., NEILL, F. H., ATMAR, R. L., KOOPMANS, M. P. and ESTES, M. K. (2004). Laboratory efforts to cultivate noroviruses. *J Gen Virol* **85**, 79–87.

EL-KAMARY, S. S., PASETTI, M. F., MENDELMAN, P. M., FREY, S. E., BERNSTEIN, D. I., TREANOR, J. J., FERREIRA, J., CHEN, W. H., SUBLETT, R., RICHARDSON, C., BARGATZE, R. F., SZTEIN, M. B. and TACKET, C. O. (2010). Adjuvanted intranasal Norwalk virus-like particle vaccine elicits antibodies and antibody-secreting cells that express homing receptors for mucosal and peripheral lymphoid tissues. *J Infect Dis* **202**, 1649–58.

EUROPEAN FOOD SAFETY AUTHORITY. (2012). Manual for reporting of food-borne outbreaks in accordance with Directive 2003/99/EC from the year 2011.

EVANS, M. R., MELDRUM, R., LANE, W., GARDNER, D., RIBEIRO, C. D., GALLIMORE, C. I. and WESTMORELAND, D. (2002). An outbreak of viral gastroenteritis following environmental contamination at a concert hall. *Epidemiol Infect* **129**, 355–60.

FALKENHORST, G., KRUSELL, L., LISBY, M., MADSEN, S. B., BOTTIGER, B. and MOLBAK, K. (2005). Imported frozen raspberries cause a series of norovirus outbreaks in Denmark, 2005. *Euro Surveill* **10**, E050922 050922.

FARKAS, T., SESTAK, K., WEI, C. and JIANG, X. (2008). Characterization of a Rhesus Monkey Calicivirus Representing a New Genus of Caliciviridae. *J Virol* **82**, 5408–16.

GALLAY, A., DE VALK, H., COURNOT, M., LADEUIL, B., HEMERY, C., CASTOR, C., BON, F., MEGRAUD, F., LE CANN, P. and DESENCLOS, J. C. (2006). A large multi-pathogen waterborne community outbreak linked to faecal contamination of a groundwater system, France, 2000. *Clin Microbiol Infect* **12**, 561–70.

GALLIMORE, C. I., CHEESBROUGH, J. S., LAMDEN, K., BINGHAM, C. and GRAY, J. J. (2005). Multiple norovirus genotypes characterised from an oyster-associated outbreak of gastroenteritis *Int JFood Microbiol* **103**, 323–30.

GALLIMORE, C. I., ITURRIZA-GÓMARA, M., XERRY, J., ADIGWE, J. and GRAY, J., J. (2007). Inter-Seasonal Diversity of Norovirus Genotypes: Emergence and Selection of Virus Variants. *Arch Virol* **152**, 1295–1303.

GLASS, R. I., NOEL, J., ANDO, T., FANKHAUSER, R., BELLIOT, G., MOUNTS, A., PARASHAR, U. D., BRESEE, J. S. and MONROE, S. S. (2000). The epidemiology of enteric caliciviruses from humans: a reassessment using new diagnostics. *J Infect Dis* **181**, S254–S261.

GOLLER, J. L., DIMITRIADIS, A., TAN, A., KELLY, H. and MARSHALL, J. A. (2004). Long-term features of norovirus gastroenteritis in the elderly. *J Hosp Infect* **58**, 286–91.

GRAY, J. J., JIANG, X., MORGAN-CAPNER, P., DESSELBERGER, U. and ESTES, M. K. (1993). Prevalence of antibodies to Norwalk virus in England: Detection by enzyme-linked immunosorbent assay using baculovirus-expressed Norwalk virus capsid antigen. *J Clin Microbiol* **31**, 1022–5.

GREEN, J., NORCOTT, J. P., LEWIS, D., ARNOLD, C. and BROWN, D. W. (1993a). Norwalk-like viruses: demonstration of genomic diversity by polymerase chain reaction. *J Clin Microbiol* **31**, 3007–12.

GREEN, J., NORCOTT, J. P., LEWIS, D., ARNOLD, C. and BROWN, D. W. G. (1993b). Norwalk-like viruses: detection of genomic diversity by polymerase chain reaction. *J Clin Microbiol* **31**, 3007–12.

GREIG, J. D., TODD, E. C., BARTLESON, C. A. and MICHAELS, B. S. (2007). Outbreaks where food workers have been implicated in the spread of foodborne disease. Part 1. Description of the problem, methods, and agents involved. *J Food Prot* **70**, 1752–61.

GUERRERO, R. A., BALL, J. M., KRATER, S. S., PACHECO, S. E., CLEMENTS, J. D. and ESTES, M. K. (2001). Recombinant norwalk virus-like particles administered intranasally to mice induce systemic and mucosal (fecal and vaginal) immune responses. *J Virol* **75**, 9713–22.

GUNSON, R. N. and CARMAN, W. F. (2005). Comparison of two real-time PCR methods for diagnosis of norovirus infection in outbreak and community settings. *J Clin Microbiol* **43**, 2030–1.

GUO, L., WANG, J., ZHOU, H., SI, H., WANG, M., SONG, J., HAN, B., SHU, Y., REN, L., QU, J. and HUNG, T. (2008). Intranasal administration of a recombinant adenovirus expressing the norovirus capsid protein stimulates specific humoral, mucosal, and cellular immune responses in mice. *Vaccine* **26**, 460–8.

HALE, A., MATTICK, K., LEWIS, D., ESTES, M., JIANG, X., GREEN, J., EGLIN, R. and BROWN, D. (2000). Distinct epidemiological patterns of Norwalk-like virus infection. *J Med Virol* **62**, 99–103.

HARRINGTON, P. R., LINDESMITH, L., YOUNT, B., MOE, C. L. and BARIC, R. S. (2002). Binding of norwalk virus-like particles to ABH histo-blood group antigens is blocked by antisera from infected human volunteers or experimentally vaccinated mice. *J Virol* **76**, 12335–43.

HARRIS, J. (2010). *Hospital Outbreak Norovirus Reporting*: Health Protection Agency.

HARRIS, J., ALLEN, D. J. and ITURRIZA-GÓMARA, M. (2011). Changing epidemiology, changing virology. The challenges for infection control. *J Infect Prevent* **12**, 102–6.

HARRIS, J. P., EDMUNDS, W. J., PEBODY, R., BROWN, D. W. and LOPMAN, B. A. (2008). Deaths from norovirus among the elderly, England and Wales. *Emerg Infect Dis* **14**, 1546–52.

HARRIS, J. P., LOPMAN, B. A. and O'BRIEN, S. J. (2010). Infection control measures for norovirus: a systematic review of outbreaks in semi-enclosed settings. *J Hosp Infect* **74**, 1–9.

HO, M., MONROE, S. S., STINE, S., CUBITT, D., GLASS, R. I., MADORE, H. P., PINSKY, P. F., ASHLEY, C. and CAUL, E. O. (1989). Viral gastroenteritis aboard a cruise ship. *Lancet* **2**, 961–5.

HOEBE, C. J., VENNEMA, H., HUSMAN, A. M. and VAN DUYNHOVEN, Y. T. (2004). Norovirus outbreak among primary schoolchildren who had played in a recreational water fountain. *J Infect Dis* **189**, 699–705.

HUANG, P., FARKAS, T., MARIONNEAU, S., ZHONG, W., RUVOEN-CLOUET, N., MORROW, A. L., ALTAYE, M., PICKERING, L. K., NEWBURG, D. S., LEPENDU, J. and JIANG, X. (2003). Noroviruses bind to human ABO, Lewis, and secretor histo-blood group antigens: identification of 4 distinct strain-specific patterns. *J Infect Dis* **188**, 19–31.

HUANG, P., FARKAS, T., ZHONG, W., TAN, M., THORNTON, S., MORROW, A. L. and JIANG, X. (2005). Norovirus and histo-blood group antigens: demonstration of a wide spectrum of strain specificities and classification of two major binding groups among multiple binding patterns. *J Virol* **79**, 6714–22.

HUTSON, A. M., ATMAR, R. L., GRAHAM, D. Y. and ESTES, M. K. (2002). Norwalk virus infection and disease is associated with ABO histo-blood group type. *J Infect Dis* **185**, 1335–7.

IRITANI, N., SETO, T., HATTORI, H., NATORI, K., TAKEDA, N., KUBO, H., YAMANO, T., AYATA, M., OGURA, H. and SETO, Y. (2007). Humoral immune responses against norovirus infections of children. *J Med Virol* **79**, 1187–93.

JIANG, X., GRAHAM, D. Y., WANG, K. and ESTES, M. K. (1990). Norwalk virus genome cloning and characterization. *Science* **250**, 1580–3.

JIANG, X., WANG, J., GRAHAM, D. Y. and ESTES, M. K. (1992a). Detection of Norwalk virus in stool by polymerase chain reaction. *J Clin Microbiol* **30**, 2529–34.

JIANG, X., WANG, M., GRAHAM, D. Y. and ESTES, M. K. (1992b). Expression, self-assembly, and antgenicity of the Norwalk virus capsid protein. *J Virol* **66**, 6527–32.

JIANG, X., WILTON, N., ZHONG, W. M., FARKAS, T., HUANG, P. W., BARRETT, E., GUERRERO, M., RUIZ-PALACIOS, G., GREEN, K. Y., GREEN, J., HALE, A. D., ESTES, M. K., PICKERING, L. K. and MATSON, D. O. (2000). Diagnosis of human caliciviruses by use of enzyme immunoassays. *J Infect Dis* **181**(Suppl 2), S349–S359.

KAGEYAMA, T., KOJIMA, S., SHINOHARA, M., UCHIDA, K., FUKUSHI, S., HOSHINO, F. B., TAKEDA, N. and KATAYAMA, K. (2003). Broadly Reactive and Highly Sensitive Assay for Norwalk-Like Viruses Based on Real-Time Quantitative Reverse Transcription-PCR. *J Clin Microbiol* **41**, 1548–57.

KAPIKIAN, A. Z., WYATT, R. G., DOLIN, R., THORNHILL, T. S., KALICA, A. R. and CHANOCK, R. M. (1972). Visualization by immune electron microscopy of a 27-nm particle associated with acute infectious nonbacterial gastroenteritis. *J Virol* **10**, 1075–81.

KAPLAN, J. E., FELDMAN, R., CAMPBELL, D. S., LOOKABAUGH, C. and GARY, G. W. (1982a). The frequency of a Norwalk-like pattern of illness in outbreaks of acute gastroenteritis. *Am J Public Health* **72**, 1329–32.

KAPLAN, J. E., GARY, G. W., BARON, R. C., SINGH, N., SCHONBERGER, L. B., FELDMAN, R. and GREENBERG, H. B. (1982b). Epidemiology of Norwalk gastroenteritis and the role of Norwalk virus in outbreaks of acute nonbacterial gastroenteritis. *Ann Intern Med* **96**, 756–61.

KARST, S. M., WOBUS, C. E., LAY, M., DAVIDSON, J. and VIRGIN, H. W. IV (2003). STAT1-dependent innate immunity to a Norwalk-like virus. *Science* **299**, 1575–8.

KATPALLY, U., VOSS, N. R., CAVAZZA, T., TAUBE, S., RUBIN, J. R., YOUNG, V. L., STUCKEY, J., WARD, V. K., VIRGIN, H. W. IV, WOBUS, C. E. and SMITH, T. J. (2010). High-resolution cryo-electron microscopy structures of murine norovirus 1 and rabbit hemorrhagic disease virus reveal marked flexibility in the receptor binding domains. *J Virol* **84**, 5836–41.

KATPALLY, U., WOBUS, C. E., DRYDEN, K., VIRGIN, H. W. IV, and SMITH, T. J. (2007). The structure of antibody neutralized murine norovirus and unexpected differences to virus-like particles. *J Virol* **82**, 2079–88.

KIMURA, H., NAGANO, K., KIMURA, N., SHIMIZU, M., UENO, Y., MORIKANE, K. and OKABE, N. (2011). A norovirus outbreak associated with environmental contamination at a hotel. *Epidemiol Infect*, **139**, 317–25.

KITAJIMA, M., TOHYA, Y., MATSUBARA, K., HARAMOTO, E., UTAGAWA, E. and KATAYAMA, H. (2010). Chlorine inactivation of human norovirus, murine norovirus and poliovirus in drinking water. *Lett Appl Microbiol* **51**, 119–21.

KVITSAND, H. M. and FIKSDAL, L. (2010). Waterborne disease in Norway: emphasizing outbreaks in groundwater systems. *Water Sci Technol* **61**, 563–71.

LAMBDEN, P. R., CAUL, E. O., ASHLEY, C. R. and CLARKE, I. N. (1993). Sequence and genome organization of a human small round structured (Norwalk-like) virus. *Science* **259**, 516–19.

LAMDEN, K., BINGHAM, C., CHEESBROUGH, J. and GALLIMORE, C. (2004). An outbreak of norovirus associated with oysters in Cumbria and Lancashire. *Health Protection Bulletin* July, 5–6.

LE GUYADER, F. S., BON, F., DEMEDICI, D., PARNAUDEAU, S., BERTONE, A., CRUDELI, S., DOYLE, A., ZIDANE, M., SUFFREDINI, E., KOHLI, E., MADDALO, F., MONINI, M., GALLAY, A., POMMEPUY, M., POTHIER, P. and RUGGERI, F. M. (2006). Detection of multiple noroviruses associated with an international gastroenteritis outbreak linked to oyster consumption. *J Clin Microbiol* **44**, 3878–82.

LE GUYADER, F. S., MITTELHOLZER, C., HAUGARREAU, L., HEDLUND, K. O., ALSTERLUND, R., POMMEPUY, M. and SVENSSON, L. (2004). Detection of noroviruses in raspberries associated with a gastroenteritis outbreak. *Int J Food Microbiol* **97**, 179–86.

LE PENDU, J., RUVOEN-CLOUET, N., KINDBERG, E. and SVENSSON, L. (2006). Mendelian resistance to human norovirus infections. *Semin Immunol* **18**, 375–86.

LINDESMITH, L., MOE, C., MARIONNEAU, S., RUVOEN, N., JIANG, X., LINDBLAD, L., STEWART, P., LEPENDU, J. and BARIC, R. (2003). Human susceptibility and resistance to Norwalk virus infection. *Nat Med* **14**, 14.

LINDESMITH, L. C., DONALDSON, E. F. and BARIC, R. S. (2011). Norovirus GII.4 strain antigenic variation. *J Virol* **85**, 231–42.

LINDESMITH, L. C., DONALDSON, E. F., LOBUE, A. D., CANNON, J. L., ZHENG, D. P., VINJE, J. and BARIC, R. S. (2008). Mechanisms of GII.4 Norovirus Persistence in Human Populations. *PLoS Med* **5**, e31.

LOBUE, A. D., LINDESMITH, L., YOUNT, B., HARRINGTON, P. R., THOMPSON, J. M., JOHNSTON, R. E., MOE, C. L. and BARIC, R. S. (2006). Multivalent norovirus vaccines induce strong mucosal and systemic blocking antibodies against multiple strains. *Vaccine* **24**, 5220–34.

LOBUE, A. D., THOMPSON, J. M., LINDESMITH, L., JOHNSTON, R. E. and BARIC, R. S. (2009). Alphavirus-Adjuvanted Norovirus-Like Particle Vaccines: Heterologous, Humoral, and Mucosal Immune Responses Protect against Murine Norovirus Challenge. *J Virol* **83**, 3212–27.

LOPMAN, B., VENNEMA, H., KOHLI, E., POTHIER, P., SANCHEZ, A., NEGREDO, A., BUESA, J., SCHREIER, E., REACHER, M., BROWN, D., GRAY, J., ITURRIZA, M., GALLIMORE, C., BOTTIGER, B., HEDLUND, K. O., TORVEN, M., VON BONSDORFF, C. H., MAUNULA, L., POLJSAK-PRIJATELJ, M., ZIMSEK, J., REUTER, G., SZUCS, G., MELEGH, B., SVENNSON, L., VAN DUIJNHOVEN, Y. and KOOPMANS, M. (2004a). Increase in viral gastroenteritis outbreaks in Europe and epidemic spread of new norovirus variant. *Lancet* **363**, 682–8.

LOPMAN, B. A., REACHER, M., GALLIMORE, C., ADAK, G. K., GRAY, J. J. and BROWN, D. W. (2003). A summer time peak of "winter vomiting disease": Surveillance of Noroviruses in England and Wales, 1995 to 2002. *BMC Public Health* **3**, 1–4.

LOPMAN, B. A., REACHER, M. H., VIPOND, I. B., HILL, D., PERRY, C., HALLADAY, T., BROWN, D. W., EDMONDS, W. J. and SARANGI, J. (2004b). Epidemiology and cost of nosocomial gastroenteritis, Avon, England, 2002–2003. *Emerg Infect Dis* **10**, 1827–34.

LOPMAN, B. A., REACHER, M. H., VIPOND, I. B., SARANGI, J. and BROWN, D. W. (2004c). Clinical manifestation of norovirus gastroenteritis in health care settings. *Clin Infect Dis* **39**, 318–24.

MALEK, M., BARZILAY, E., KRAMER, A., CAMP, B., JAYKUS, L. A., ESCUDERO-ABARCA, B., DERRICK, G., WHITE, P., GERBA, C., HIGGINS, C., VINJE, J., GLASS, R., LYNCH, M. and WIDDOWSON, M. A. (2009). Outbreak of norovirus infection among river rafters associated with packaged delicatessen meat, Grand Canyon, 2005. *Clin Infect Dis* **48**, 31–7.

MARKS, P. J., VIPOND, I. B., CARLISLE, D., DEAKIN, D., FEY, R. E. and CAUL, E. O. (2000). Evidence for airborne transmission of Norwalk-like virus (NLV) in a hotel restaurant. *Epidemiol Infect* **124**, 481–7.

MASON, H. S., BALL, J. M., SHI, J., JIANG, X., ESTES, M. K. and ARNTZEN, C. J. (1996). Expression of Norwalk virus capsid protein in transgenic tobacco and potato and its oral immunogenicity in mice. *Proceedings of the National Academy of Sciences (USA)* **93**, 5335–40.

MATSUI, S. M. and GREENBERG, H. B. (2000). Immunity to Calicivirus Infection. *J Infect Dis* **181**, S331–S335.

MAUNULA, L. (2005). Norovirus outbreaks from drinking water. *Emerg Infect Dis* **11**, 1716–21.

MAUNULA, L., KALSO, S., VON BONSDORFF, C. H. and PONKA, A. (2004). Wading pool water contaminated with both noroviruses and astroviruses as the source of a gastroenteritis outbreak. *Epidemiol Infect* **132**, 737–43.

MAUNULA, L., ROIVAINEN, M., KERANEN, M., MAKELA, S., SODERBERG, K., SUMMA, M., VON BONSDORFF, C. H., LAPPALAINEN, M., KORHONEN, T., KUUSI, M. and NISKANEN, T. (2009). Detection of human norovirus from frozen raspberries in a cluster of gastroenteritis outbreaks. *Euro Surveill* **14**, 19435.

MAYO, M. A. (2002). A summary of taxonomic changes recently approved by ICTV. *Arch Virol* **147**, 1655–63.

MEEROFF, J. C., SCHREIBER, D. S., TRIER, J. S. and BLACKLOW, N. R. (1980). Abnormal gastric motor function in viral gastroenteritis. *Med J Aust* **1**, 211–13.

MORTER, S., BENNET, G., FISH, J., RICHARDS, J., ALLEN, D. J., NAWAZ, S., ITURRIZA-GÓMARA, M., BROLLY, S. and GRAY, J. (2011). Norovirus in the hospital setting: Virus introduction and spread within the hospital environment. *J Hosp infect* **77**, 106–12.

NENONEN, N. P., HANNOUN, C., LARSSON, C. U. and BERGSTROM, T. (2012). Marked genomic diversity of norovirus genogroup I strains in a waterborne outbreak. *Appl Environ Microbiol* **78**, 1846–52.

NENONEN, N. P., HANNOUN, C., OLSSON, M. B. and BERGSTROM, T. (2009). Molecular analysis of an oyster-related norovirus outbreak. *J Clin Virol* **45**, 105–8.

NICOLAY, N., MCDERMOTT, R., KELLY, M., GORBY, M., PRENDERGAST, T., TUITE, G., COUGHLAN, S., MCKEOWN, P. and SAYERS, G. (2011). Potential role of asymptomatic kitchen food handlers during a food-borne outbreak of norovirus infection, Dublin, Ireland, March 2009. *Euro Surveill* **16**, 19931.

NILSSON, M., HEDLUND, K. O., THORHAGEN, M., LARSON, G., JOHANSEN, K., EKSPONG, A. and SVENSSON, L. (2003). Evolution of human calicivirus RNA in vivo: accumulation of mutations in the protruding P2 domain of the capsid leads to structural changes and possibly a new phenotype. *J Virol* **77**, 13117–24.

NODA, M., FUKUDA, S. and NISHIO, O. (2008). Statistical analysis of attack rate in norovirus foodborne outbreaks. *Int J Food Microbiol* **122**, 216–220.

NORDGREN, J., KINDBERG, E., LINDGREN, P. E., MATUSSEK, A. and SVENSSON, L. (2010). Norovirus gastroenteritis outbreak with a secretor-independent susceptibility pattern, Sweden. *Emerg Infect Dis* **16**, 81–7.

NOWAK, P., TOPPING, J. R., FOTHERINGHAM, V., GALLIMORE, C. I., GRAY, J. J., ITURRIZA-GÓMARA, M. and KNIGHT, A. I. (2011). Measurement of the virolysis of human GII.4 norovirus in response to disinfectants and sanitisers. *J Virol Method* **174**, 7–11.

NYGARD, K., VOLD, L., HALVORSEN, E., BRINGELAND, E., ROTTINGEN, J. A. and AAVITSLAND, P. (2004). Waterborne outbreak of gastroenteritis in a religious summer camp in Norway, 2002. *Epidemiol Infect* **132**, 223–9.

OLIVER, S. L., DASTJERDI, A. M., WONG, S., EL-ATTAR, L., GALLIMORE, C., BROWN, D. W., GREEN, J. and BRIDGER, J. C. (2003). Molecular characterization of bovine enteric caliciviruses: a distinct third genogroup of noroviruses (Norwalk-like viruses) unlikely to be of risk to humans. *J Virol* **77**, 2789–98.

PANG, X. L., HONMA, S., NAKATA, S. and VESIKARI, T. (2000). Human caliciviruses in acute gastroenteritis of young children in the community. *J Infect Dis* **181**(Suppl 2), S288–S294.

PARKER, S. P., CUBITT, W. D. and JIANG, X. (1995). Enzyme immunoassay using baculovirus-expressed human calicivirus (Mexico) for the measurement of IgG responses and determining its seroprevalence in London, UK. *J Med Virol* **46**, 194–200.

PARRA, G. I., BOK, K., TAYLOR, R., HAYNES, J. R., SOSNOVTSEV, S. V., RICHARDSON, C. and GREEN, K. Y. (2012). Immunogenicity and specificity of norovirus Consensus GII.4 virus-like particles in monovalent and bivalent vaccine formulations. *Vaccine* **30**, 3580–6.

PARRINO, T. A., SCHREIBER, D. S., TRIER, J. S., KAPIKIAN, A. Z. and BLACKLOW, N. R. (1977). Clinical immunity in acute gastroenteritis caused by Norwalk agent. *New England J Med* **297**, 86–9.

PHILLIPS, G., TAM, C. C., CONTI, S., RODRIGUES, L. C., BROWN, D., ITURRIZA-GÓMARA, M., GRAY, J. and LOPMAN, B. (2010a). Community incidence of norovirus-associated infectious intestinal disease in England: improved estimates using viral load for norovirus diagnosis. *Am J Epidemiol* **171**, 1014–22.

PHILLIPS, G., TAM, C. C., RODRIGUES, L. C. and LOPMAN, B. (2010b). Prevalence and characteristics of asymptomatic norovirus infection in the community in England. *Epidemiol Infect*, **138**, 1454–8.

PODEWILS, L. J., ZANARDI BLEVINS, L., HAGENBUCH, M., ITANI, D., BURNS, A., OTTO, C., BLANTON, L., ADAMS, S., MONROE, S. S., BEACH, M. J. and WIDDOWSON, M. (2007). Outbreak of norovirus illness associated with a swimming pool. *Epidemiol Infect*, **135**, 827–33.

POMMEPUY, M., DUMAS, F., CAPRAIS, M. P., CAMUS, P., LE MENNEC, C., PARNAUDEAU, S., HAUGARREAU, L., SARRETTE, B., VILAGINES, P., POTHIER, P., KHOLI, E. and LE GUYADER, F. (2004). Sewage impact on shellfish microbial contamination. *Water Sci Technol* **50**, 117–24.

PRASAD, V. B. V., ROTHNAGEL, R., JIANG, X. and ESTES, M. K. (1994). Three-dimensional structure of baculovirus-expressed Norwalk virus capsids. *J Virol* **68**, 5117–25.

REPP, K. K. and KEENE, W. E. (2012). A point-source norovirus outbreak caused by exposure to fomites. *J Infect Dis* **205**, 1639–41.

RICHARDS, G. P., WATSON, M. A., FANKHAUSER, R. L. and MONROE, S. S. (2004). Genogroup I and II noroviruses detected in stool samples by real-time reverse transcription-PCR using highly degenerate universal primers. *Appl Environ Microbiol* **70**, 7179–84.

RIERA-MONTES, M., BRUS SJOLANDER, K., ALLESTAM, G., HALLIN, E., HEDLUND, K. O. and LOFDAHL, M. (2011). Waterborne norovirus outbreak in a municipal drinking-water supply in Sweden. *Epidemiol Infect*, **139**, 1928–35.

ROCKX, B., BARIC, R. S., de GRIJS, I., DUIZER, E. and KOOPMANS, M. P. (2005a). Characterization of the homo- and heterotypic immune responses after natural norovirus infection. *J Med Virol* **77**, 439–46.

ROCKX, B. H., VENNEMA, H., HOEBE, C. J., DUIZER, E. and KOOPMANS, M. P. (2005b). Association of histo-blood group antigens and susceptibility to norovirus infections. *J Infect Dis* **191**, 749–54.

ROHAYEM, J., MUNCH, J. and RETHWILM, A. (2005). Evidence of recombination in the norovirus capsid gene. *J Virol* **79**, 4977–90.

ROY, P. and NOAD, R. (2009). Virus-like particles as a vaccine delivery system: myths and facts. *Adv Exp Med Biol* **655**, 145–58.

RUUSKA, T. and VESIKARI, T. (1990). Rotavirus disease in Finnish children: use of numerical scores for clinical severity of diarrhoeal episodes. *Scand J Infect Dis* **22**, 259–67.

SALA, M. R., CARDENOSA, N., ARIAS, C., LLOVET, T., RECASENS, A., DOMINGUEZ, A., BUESA, J. and SALLERAS, L. (2005). An outbreak of food poisoning due to a genogroup I norovirus. *Epidemiol Infect* **133**, 187–91.

SANTI, L., BATCHELOR, L., HUANG, Z., HJELM, B., KILBOURNE, J., ARNTZEN, C. J., CHEN, Q. and MASON, H. S. (2008). An efficient plant viral expression system generating orally immunogenic Norwalk virus-like particles. *Vaccine* **26**, 1846–54.

SARTORIUS, B., ANDERSSON, Y., VELICKO, I., DE JONG, B., LOFDAHL, M., HEDLUND, K. O., ALLESTAM, G., WANGSELL, C., BERGSTEDT, O., HORAL, P., ULLERYD, P. and SODERSTROM, A. (2007). Outbreak of norovirus in Vastra Gotaland associated with recreational activities at two lakes during August 2004. *Scand J Infect Dis* **39**, 323–31.

SAVINI, G., CASACCIA, C., BARILE, N. B., PAOLETTI, M. and PINONI, C. (2009). Norovirus in bivalve molluscs: a study of the efficacy of the depuration system. *Vet Ital* **45**, 535–9.

SAWYER, L. A., MURPHY, J. J., KAPLAN, J. E., PINSKY, P. F., CHACON, D., WALMSLEY, S., SCHONBERGER, L. B., PHILIPS, A., FORWARD, K., GOLDMAN, C., BRUNTON, J., FRALICK, R. A., CARTER, A. O., GARY, W. G., GLASS, R. I. and LOW, D. E. (1988). 25- to 30-nm virus particle associated with a hospital outbreak of acute gastroenteritis with evidence for airborne transmission. *Am J Epidemiol* **127**, 1261–71.

SCARCELLA, C., CARASI, S., CADORIA, F., MACCHI, L., PAVAN, A., SALAMANA, M., ALBORALI, G. L., LOSIO, M. M., BONI, P., LAVAZZA, A. and SEYLER, T. (2009). An outbreak of viral gastroenteritis linked to municipal water supply, Lombardy, Italy, June 2009. *Euro Surveill* **14**, 19274.

SCHMID, D., KUO, H. W., HELL, M., KASPER, S., LEDERER, I., MIKULA, C., SPRINGER, B. and ALLERBERGER, F. (2011). Foodborne gastroenteritis outbreak in an Austrian healthcare facility caused by asymptomatic, norovirus-excreting kitchen staff. *J Hosp Infect* **77**, 237–41.

SCHMID, D., STUGER, H. P., LEDERER, I., PICHLER, A. M., KAINZ- ARNFELSER, G., SCHREIER, E. and ALLERBERGER, F. (2007). A foodborne norovirus outbreak due to manually prepared salad, Austria 2006. *Infection* **35**, 232–9.

SHANKER, S., CHOI, J. M., SANKARAN, B., ATMAR, R. L., ESTES, M. K. and PRASAD, B. V. (2011). Structural analysis of histo-blood group antigen binding specificity in a norovirus GII.4 epidemic variant: implications for epochal evolution. *J Virol* **85**, 8635–45.

SHIRASAKI, N., MATSUSHITA, T., MATSUI, Y., OSHIBA, A. and OHNO, K. (2010). Estimation of norovirus removal performance in a coagulation-rapid sand filtration process by using recombinant norovirus VLPs. *Water Res* **44**, 1307–16.

SHOEMAKER, G. K., VAN DUIJN, E., CRAWFORD, S. E., UETRECHT, C., BACLAYON, M., ROOS, W. H., WUITE, G. J., ESTES, M. K., PRASAD, B. V. and HECK, A. J. (2010). Norwalk virus assembly and stability monitored by mass spectrometry. *Mol Cell Proteomics* **9**, 1742–51.

SIEBENGA, J. J., VENNEMA, H., RENCKENS, B., DE BRUIN, E., VAN DER VEER, B., SIEZEN, R. J. and KOOPMANS, M. (2007). Epochal evolution of GGII.4 norovirus capsid proteins from 1995 to 2006. *J Virol* **81**, 9932–41.

SUKHRIE, F. H., TEUNIS, P., VENNEMA, H., COPRA, C., THIJS BEERSMA, M. F., BOGERMAN, J. and KOOPMANS, M. (2012). Nosocomial transmission of norovirus is mainly caused by symptomatic cases. *Clin Infect Dis* **54**, 931–7.

TACKET, C. O., MASON, H. S., LOSONSKY, G., ESTES, M. K., LEVINE, M. M. and ARNTZEN, C. J. (2000). Human Immune Responses to a Novel Norwalk Virus Vaccine Delivered in Transgenic Potatoes. *J Infect Dis* **182**, 302–5.

TACKET, C. O., SZTEIN, M. B., LOSONSKY, G. A., WASSERMAN, S. S. and ESTES, M. K. (2003). Humoral, mucosal, and cellular immune responses to oral Norwalk virus-like particles in volunteers. *Clin Immunol* **108**, 241–7.

TAM, C. C., O'BRIEN, S. J., TOMPKINS, D. S., BOLTON, F. J., BERRY, L., DODDS, J., CHOUDHURY, D., HALSTEAD, F., ITURRIZA-GÓMARA, M., MATHER, K., RAIT, G., RIDGE, A., RODRIGUES, L. C., WAIN, J., WOOD, B. and GRAY, J. J. (2012). Changes in causes of acute gastroenteritis in the United Kingdom over 15 years: microbiologic findings from 2 prospective, population-based studies of infectious intestinal disease. *Clin Infect Dis* **54**, 1275–86.

TAM, C. C., RODRIGUES, L. C., VIVIANI, L., DODDS, J. P., EVANS, M. R., HUNTER, P. R., GRAY, J. J., LETLEY, L. H., RAIT, G., TOMPKINS, D. S. and O'BRIEN, S. J. (2011). Longitudinal study of infectious intestinal disease in the UK (IID2 study): incidence in the community and presenting to general practice. *Gut* **61**, 69–77.

TAN, M., FANG, P., CHACHIYO, T., XIA, M., HUANG, P., FANG, Z., JIANG, W. and JIANG, X. (2008). Noroviral P particle: structure, function and applications in virus-host interaction. *Virology* **382**, 115–23.

TAN, M., HUANG, P., MELLER, J., ZHONG, W., FARKAS, T. and JIANG, X. (2003). Mutations within the P2 domain of norovirus capsid affect binding to human histo-blood group antigens: evidence for a binding pocket. *J Virol* **77**, 12562–71.

TAN, M. and JIANG, X. (2005). The p domain of norovirus capsid protein forms a subviral particle that binds to histo-blood group antigen receptors. *J Virol* **79**, 14017–30.

TAN, M., MELLER, J. and JIANG, X. (2006). C-terminal arginine cluster is essential for receptor binding of norovirus capsid protein. *J Virol* **80**, 7322–31.

TER WAARBEEK, H. L., DUKERS-MUIJRERS, N. H., VENNEMA, H. and HOEBE, C. J. (2010). Waterborne gastroenteritis outbreak at a scouting camp caused by two norovirus genogroups: GI and GII. *J Clin Virol* **47**, 268–72.

TEUNIS, P. F., MOE, C. L., LIU, P., MILLER, S. E., LINDESMITH, L., BARIC, R. S., LE PENDU, J. and CALDERON, R. L. (2008). Norwalk virus: how infectious is it? *J Med Virol* **80**, 1468–76.

TURCIOS, R. M., WIDDOWSON, M. A., SULKA, A. C., MEAD, P. S. and GLASS, R. I. (2006). Reevaluation of epidemiological criteria for identifying outbreaks of acute gastroenteritis due to norovirus: United States, 1998–2000. *Clin Infect Dis* **42**, 964–9.

VENKATARAM PRASAD, B. V., HARDY, M. E., DOKLAND, T., BELLA, J., ROSSMANN, M. G. and ESTES, M. K. (1999). X-ray crystallographic structure of the norwalk virus capsid. *Science* **286**, 287–90.

VIVANCOS, R., SHROUFI, A., SILLIS, M., AIRD, H., GALLIMORE, C. I., MYERS, L., MAHGOUB, H. and NAIR, P. (2009). Food related norovirus outbreak among people attending two barbeques in Norfolk, UK: Epidemiological, virological and environmental investigation. *International J Infect Dis* **13**, 629–35.

WADL, M., SCHERER, K., NIELSEN, S., DIEDRICH, S., ELLERBROEK, L., FRANK, C., GATZER, R., HOEHNE, M., JOHNE, R., KLEIN, G., KOCH, J., SCHULENBURG, J., THIELBEIN, U., STARK, K. and BERNARD, H. (2010). Food-borne norovirus-outbreak at a military base, Germany, 2009. *BMC Infect Dis* **10**, 30.

WERBER, D., LAUSEVIC, D., MUGOSA, B., VRATNICA, Z., IVANOVIC-NIKOLIC, L., ZIZIC, L., ALEXANDRE-BIRD, A., FIORE, L., RUGGERI, F. M., DI BARTOLO, I., BATTISTONE, A., GASSILLOUD, B., PERELLE, S., NITZAN KALUSKI, D., KIVI, M., ANDRAGHETTI, R. and POLLOCK, K. G. (2009). Massive outbreak of viral gastroenteritis associated with consumption of municipal drinking water in a European capital city. *Epidemiol Infect* **137**, 1713–20.

WESTRELL, T., DUSCH, V., ETHELBERG, S., HARRIS, J., HJERTQVIST, M., JOURDAN-DA SILVA, N., KOLLER, A., LENGLET, A., LISBY, M. and VOLD, L. (2010). Norovirus outbreaks linked to oyster consumption in the United Kingdom, Norway, France, Sweden and Denmark, 2010. *Euro Surveill* **15**, 19524.

WILLCOCKS, M. M., SILCOCK, J. G. and CARTER, M. J. (1993). Detection of Norwalk virus in the UK by polymerase chain reaction. *FEMS Microbiol Lett* **112**, 7–12.

WYATT, R. G., DOLIN, R., BLACKLOW, N. R., DUPONT, H. L., BUSCHO, R. F., THORNHILL, T. S., KAPIKIAN, A. Z. and CHANOCK, R. M. (1974). Comparison of three agents of acute infectious nonbacterial gastroenteritis by cross-challenge in volunteers. *J Infect Dis* **129**, 709–14.

XERRY, J., GALLIMORE, C. I., ITURRIZA GÓMARA, M., ALLEN, D. J. and GRAY, J. J. (2008). Transmission events within outbreaks of gastroenteritis determined through the analysis of nucleotide sequences of the P2 domain of genogroup II noroviruses. *J Clin Microbiol* **46**, 947–53.

XIA, M., FARKAS, T. and JIANG, X. (2007). Norovirus capsid protein expressed in yeast forms virus-like particles and stimulates systemic and mucosal immunity in mice following an oral administration of raw yeast extracts. *J Med Virol* **79**, 74–83.

ZAHORSKY, J. (1929). Hyperemesis hiemis or winter vomiting disease. *Arch Pediatrics* **46**, 391–5.

ZANINI, B., RICCI, C., BANDERA, F., CASELANI, F., MAGNI, A., LARONGA, A. M. and LANZINI, A. (2012). Incidence of post-infectious irritable bowel syndrome and functional intestinal disorders following a water-borne viral gastroenteritis outbreak. *Am J Gastroenterol* **107**, 891–9.

ZHANG, X., BUEHNER, N. A., HUTSON, A. M., ESTES, M. K. and MASON, H. S. (2006). Tomato is a highly effective vehicle for expression and oral immunization with Norwalk virus capsid protein. *Plant Biotechnol J* **4**, 419–32.

ZHENG, D. P., ANDO, T., FANKHAUSER, R. L., BEARD, R. S., GLASS, R. I. and MONROE, S. S. (2006). Norovirus classification and proposed strain nomenclature. *Virology* **346**, 312–23.

17

Advances in understanding of hepatitis A virus as a food- and waterborne pathogen and progress with vaccine development

R. M. Pintó and A. Bosch, University of Barcelona, Spain

DOI: 10.1533/9780857098870.4.349

Abstract: Despite efficient vaccine and improved hygiene, hepatitis A remains the most common viral hepatitis worldwide. Its genome has evolved to render an extremely quiescent replication phenotype, through slow translation of the genomic region encoding the capsid protein to ensure an accurate capsid folding. This capsid is resistant to many biological and environmental constraints during virus transmission, as it is very stable and has a very low antigenic variability. However, the inherent genetic variability of an RNA virus in association with vaccination campaigns may promote the emergence of new variants.

Key words: hepatitis A virus, fine-tuning translation kinetics, quasispecies, vaccine-escape variants.

17.1 Introduction: hepatitis A infection

Hepatitis A is an acute infection of the liver produced by the hepatitis A virus (HAV). Among children below the age of five the infection mostly develops asymptomatically while in older children and in the adulthood the infection usually proceeds with symptoms (Cuthbert, 2001). In this latter case, the clinical course of hepatitis A is indistinguishable from that of other types of acute viral hepatitis. The clinical case definition of hepatitis A is an acute illness with a moderate onset of symptoms including fever, malaise, anorexia, nausea, abdominal discomfort, dark urine and jaundice, and elevated serum bilirubin and aminotransferases levels later on (Cuthbert, 2001). The incubation period of hepatitis A ranges from 15 to 50 days and clinical illness usually does not last longer than 2 months. There is no evidence of chronicity of the infection although 1.5–15% of patients have prolonged or relapsing signs and symptoms for up to 6 months and, occasionally, the infection may proceed to a fulminant hepatitis, mainly among patients with underlying chronic liver diseases.

Immunoglobulin M (IgM) antibodies are usually detected by the onset of clinical symptoms and later on IgA and IgG are also synthesized (Cuthbert, 2001). The anti-HAV IgM response is usually limited to the initial infection and is used as a marker of acute disease for diagnosis. In contrast, the IgG response is delayed compared with the IgM response but is long-lasting and confers resistance to re-infection.

A high and long-lasting viraemia develops (Bower *et al.*, 2000; Costafreda *et al.*, 2006; Costa-Mattioli *et al.*, 2002c; Normann *et al.*, 2004), with the viral titre reaching a peak of up to 10^7 genome copies/ml of sera at just before the beginning to 2 weeks after the onset of symptoms; it then progressively declines during the following 4 weeks. Peaks of virus faecal shedding reach their maximum during the same period as those in sera but with titres up to 10^8 genome copies/g (Costafreda *et al.*, 2006). Disease diagnoses based on the molecular detection and or quantification of genome copies in clinical specimens are still not widely used, despite their ability to provide information on virus genotype and viraemia level, respectively.

17.2 Susceptibility in different sectors of the population

Hepatitis A, although clinically a moderate illness, still remains the most important acute hepatitis worldwide with regard to the number of cases.

Hepatitis A infection is highly endemic in developing regions and much less frequent in developed regions (Cuthbert, 2001; Pintó *et al.*, 2007a). This epidemiological pattern has important implications for the average age of exposure and hence the severity of the clinical disease. Since hepatitis A infection induces a life-long immunity, severe infections among adults are rare in highly endemic regions where most children are infected early in life. In contrast, in low endemic areas the disease occurs mostly in adulthood, mainly as a consequence of travelling to endemic regions, consuming contaminated water or food or engaging in risky sexual practices (Faber *et al.*, 2009; Frank *et al.*, 2007; Petrignani *et al.*, 2010b; Sanchez *et al.*, 2002), and hence the likelihood of developing severe symptomatic illness is high. Men having sex with men (MSM) are a major group at risk for hepatitis A due to the faecal-oral transmission of the virus and the high excretion levels before the onset of symptoms, associated with some common sexual practices in that group (Stene-Johansen *et al.*, 2007; Tortajada *et al.*, 2009).

An epidemiological shift, from intermediate to low prevalence, has been noticed in recent decades in many countries, particularly in Southern Europe, including Spain, Italy and Greece (Dominguez *et al.*, 2008; Germinario *et al.*, 2000; Van Damme and Van Herck, 2005). In contrast, other Mediterranean countries such as Egypt and Turkey have recently been implicated as the source of intercontinental foodborne outbreaks of hepatitis A (Frank *et al.*, 2007; Petrignani *et al.*, 2010a,b). Consequently, although the first well-documented hepatitis epidemic occurred on the Mediterranean island of Minorca 250 years ago, the Mediterranean basin as a whole should no longer be considered an endemic area (Pintó *et al.*, 2007a, 2010; Previsani *et al.*, 2004).

Additionally, some other Eastern European countries (Cianciara, 2000; Tallo *et al.*, 2003) have also described significant declines in the incidence of hepatitis A. Likewise, in several Asian and American countries a shift from highly to moderately endemic has been described (Barzaga, 2000; Tanaka, 2000).

17.3 Highly effective vaccines for hepatitis A prevention

Inactivated HAV vaccines have been available since the early 1990s and provide long-lasting immunity against hepatitis A infection. The immunity is largely related to the induction of high titres of specific antibodies. Thanks to the existence of a single serotype of HAV, these vaccines are of high efficacy (Cuthbert, 2001; Martin and Lemon, 2006).

As a general rule in low and intermediate endemic regions, where paradoxically the severity of the disease is high, vaccination against hepatitis A should be recommended in high-risk groups, including travellers to high endemic areas, the MSM group, drug users and patients receiving blood products. In addition, the inclusion of hepatitis A vaccines in mass vaccination programs in countries receiving high numbers of immigrants from endemic countries is particularly recommended (Vaccination requirements for US immigration: technical instructions for panel physicians, http://www.immihelp.com/immigrant-visa/technical-instructions-panel-physicians-vaccination-requirements.pdf). The effectiveness of paediatric mass vaccination programs in reducing the incidence of hepatitis A has been evidenced in several countries including Argentina, Catalonia in Spain, Israel and United States (Bjorkholm *et al.*, 1995; Dagan *et al.*, 2005; Dominguez *et al.*, 2008; Samandari *et al.*, 2004; Wasley *et al.*, 2005).

17.4 Risk assessment and risk management in water and food

Huge foodborne outbreaks of hepatitis A due to the consumption of foods such as green onions (Wheeler *et al.*, 2005) and dried tomatoes (Petrignani *et al.*, 2010a,b) have been reported. The vegetables associated with these huge outbreaks were imported from highly/intermediate endemic areas. However, the foodstuff more often associated with outbreaks of hepatitis A is shellfish.

The first documented shellfish-borne outbreak of hepatitis A occurred in Sweden in 1955, when 629 cases were associated with raw oyster consumption (Roos, 1956). However, the most significant outbreak of HAV infection occurred in Shanghai, China, in 1988, when almost 300 000 cases were caused by consumption of clams harvested from a sewage-polluted area (Halliday *et al.*, 1991). In fact, this is so far the largest virus-associated outbreak of food poisoning ever reported. Other smaller outbreaks have been reported worldwide (Conaty *et al.*, 2000; Leoni *et al.*, 1998; Mackowiak *et al.*, 1976; Mele *et al.*, 1989; Pintó *et al.*, 2009; Sanchez *et al.*, 2002; Stroffolini *et al.*, 1990).

Several issues such as the fact that faecal excretion precedes the onset of symptoms, together with the difficulty of completely removing and or inactivating viruses through sewage treatment (Blatchley *et al.*, 2007; Bosch, 2007; Pintó *et al.*, 2010), particularly evidenced by HAV detection in treated sewage samples from different wastewater treatment plants based on primary sedimentation, activated sludge and chlorine treatments (Pintó *et al.*, 2007a), make hepatitis foodborne outbreaks hard to prevent if the virus is circulating among the population. However, in the particular case of shellfish, a significant correlation between hepatitis A cases in the harvesting areas and positive HAV isolation in shellfish has been observed, as well as a direct correlation between the level of viral contamination of shellfish and the attack rate in the consumer population (Pintó *et al.*, 2009). This makes it possible to undertake risk assessment analyses in shellfish when evidence exists that a critical limit of viral contamination has been exceeded in the potential sources of contamination discharging into the shellfish growing beds (Bosch *et al.*, 1994; Pintó and Bosch, 2008; Pintó *et al.*, 2010; Polo *et al.*, 2010). Quantitative risk assessment analysis should be performed in the bivalves, since HAV levels in shellfish are predictors of the magnitude of the outbreak and can be used for risk management purposes (Bosch *et al.*, 1994; Pintó and Bosch, 2008; Pintó *et al.*, 2009) such as recommendations to consumers to cook the shellfish, or closure of contaminated shellfish beds. It should not be forgotten that coliform standards may not be effective as indicators of HAV contamination (Pintó *et al.*, 2009).

17.5 Unique properties of hepatitis A virus

Key factors for the viral biological cycle and infection outcome of HAV are a high stability of the capsid to the acid pH of the stomach during the entry phase and to the action of detergents, particularly biliary salts, during the exit phase. This required extremely resistant phenotype of HAV in the body also explains its high persistence in the environment, with reductions of only 2 logs after 2 months dried in different fomites or after treatment with 1 mg/l of free chlorine for 2 h (Abad *et al.*, 1994a,b and1997a,b), and its high potential for transmission by contaminated foods and drinking water (Bosch *et al.*, 1991; Dentinger *et al.*, 2001; Pintó *et al.*, 2009; Reid and Robinson, 1987; Rosenblum *et al.*, 1990; Sanchez *et al.*, 2002). But how such a highly stable capsid conformation is delineated is still unknown, although it has been proposed that a very accurate and genomically intrinsic control of the kinetics of translation could contribute to the very precise capsid folding (Aragonès *et al.*, 2010) (Fig. 17.1). Several critical issues of the HAV genome composition, structure and coding capacity must be pointed out (Pintó *et al.*, 2007b). First of all, the structure of the 5' non-coding region (5'NCR) and its internal ribosome entry site (IRES). The HAV IRES is unique among picornaviruses and constitutes the type III model (Brown *et al.*, 1994; Ehrenfeld and Teterina, 2002), which shows a very low efficiency in directing translation (Whetter *et al.*, 1994) compared to other picornavirus IRESes. Secondly, HAV encodes only a single protease, 3C, while other picornaviruses code for additional proteases which are involved in the cellular protein shutoff induction (Leong *et al.*, 2002). Since picornaviruses utilize a mechanism of translation that is cap-independent and IRES-dependent, the inhibition of the cap-dependent cellular translation allows an almost exclusive use of the translation machinery for the production of viral proteins (Kuechler *et al.*, 2002). This shutoff of host cell protein synthesis is mediated through the cleavage of the cellular translation initiation factor eIF4G by the above-mentioned additional proteases. The lack of these proteases in HAV is associated with the incapacity to induce the cellular shutoff which otherwise is directly related to the requirement for an intact uncleaved eIF4G factor for the formation of the initiation of translation complex (Borman and Kean, 1997; Jackson, 2002).

The foregoing description demonstrates that HAV must compete inefficiently for the cellular translational machinery, and thus it presents a unique translation strategy. This points to the third difference between HAV and other picornavirus members: the codon usage. HAV presents a higher codon usage bias compared to other members of its family, which manifests itself in the adaptation to use abundant and rare codons (Pintó *et al.*, 2007b; Sanchez *et al.*, 2003b). In fact, 14 amino acid codon families contain rare codons, defined in terms of their frequencies, making a total of 22 used rare codons (Pintó *et al.*, 2007b). But what is more surprising is that the HAV codon usage has evolved to be somewhat the opposite to that of human cells, never adopting as abundant codons those abundant for the host cell, and even in some instances using these latter as rare codons. This disparity, unique to HAV, has been interpreted as a subtle strategy to avoid, as much as possible, competition for the cellular tRNAs in the absence of a precise mechanism to induce shutoff of cellular protein synthesis (Sanchez *et al.*, 2003b). As stated before, a consequence of this special codon bias is an increase in the number of rare codons used by HAV. Overall this increment is the result of the addition to the cellular rare codons, also used as rare by the virus, of the most abundant cellular codons that, being unavailable for the virus, are used by it at low frequencies. The clusters of rare codons would induce a transient stop of the translational complex in order to seek a suitable tRNA present at a very low concentration. It has been suggested that one function of these ribosome stallings is to ensure the proper folding of the nascent protein (Adzhubei *et al.*, 1996; Evans *et al.*, 2005; Gavrilin *et al.*, 2000). Such a function has been postulated for HAV, where highly conserved clusters of rare codons strategically located at the carboxi-ends of the structured elements of the capsid coding region have been reported (Sanchez *et al.*, 2003b). In fact, the critical role of HAV codon usage, and particularly of these clusters of rare codons of the capsid coding region, has been shown during the process of HAV adaptation to conditions of artificially induced cellular shutoff (Aragonès *et al.*, 2010). An overall change in codon usage is necessary to regain viral fitness in these conditions, with a clear re-deoptimization with respect to the cellular codon usage; this particularly affects the rare codons located at the above-mentioned strategic positions of the capsid (Aragonès *et al.*, 2010). This mechanism of adaptation to the cellular shutoff proves that the driving selective force of HAV codon usage is translation kinetics, that is, the right combination of codons (common and rare) that allows a regulated ribosome traffic rate ensuring the proper protein folding. This control of the capsid folding will ensure a highly stable phenotype (Fig. 17.1).

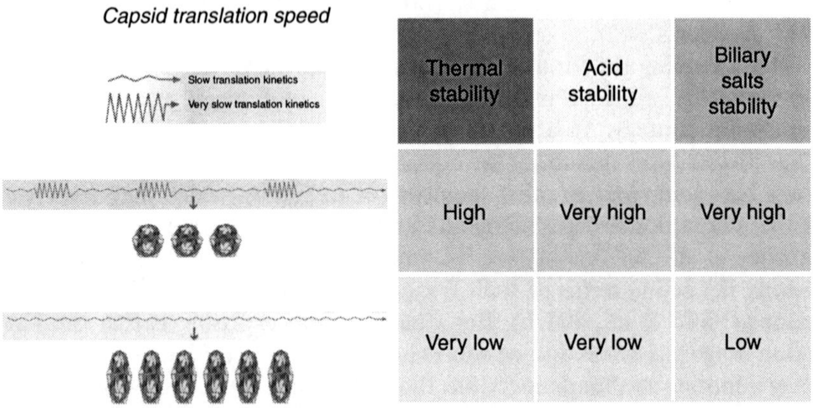

Fig. 17.1 Translation kinetics control in the capsid coding region of hepatitis A virus (HAV) as a measure to ensure the proper capsid folding for a high survival phenotype. Fine-tuning translation kinetics selection, or the right combination of common and rare codons regulates the ribosome traffic rate allowing a sequential folding of the synthesized proteins. The codon usage of hepatitis A virus is highly biased and highly deoptimized with respect to that of the cellular host. Rare codons of the capsid coding region are strategically located at the carboxi borders of the predicted ά-helices and β-sheets where they would likely induce ribosome stallings and consequently a slowdown of the translation speed. Changes in the kinetics of translation are associated with changes of the capsid stability.

A certain contribution of the codon usage to the low variability of the HAV capsid has been proposed (Aragonès *et al.*, 2008). Fifteen per cent of the surface capsid residues are encoded by rare codons. These rare codons are highly conserved among the different HAV strains and their substitution is negatively selected even under specific immune pressure. Many of these capsid residues encoded by rare codons are surface-exposed and located near or at the epitope regions and this negative selection would prevent the emergence of antigenic variants. The clusters of rare codons are maintained in response to the need for a proper capsid folding, which is controlled through the kinetics of translation, as explained above, and a nucleotide substitution giving rise to a new, equally rare, codon with a compatible amino acid is quite unlikely to occur.

17.6 Quasispecies dynamics of evolution and virus fitness

Viral quasispecies are dynamic distributions of non-identical but closely related viral genomes subjected to a continuous process of genetic variation, competition and selection and which act as a unit of selection (Domingo *et al.*, 2001, 2006; Domingo and Vennema, 2008). RNA viruses have the capacity to quickly explore large regions of sequence space thanks to their high mutation rates which are in the range of 10^{-3}–10^{-5} substitutions per nucleotide copied. However, their genome size and diverse selective constraints limit the diversity that is actually expressed (Overbaugh and Bangham, 2001).

HAV as an RNA virus occurs as a swarm of mutants or quasispecies (Sanchez *et al.*, 2003a). The nucleotide diversity is similar to that of other picornaviruses, and allows its differentiation into several genotypes and sub-genotypes. Six genotypes have been defined (Costa-Mattioli *et al.*, 2002b, 2003b; Robertson *et al.*, 1992). Three (I, II and III) are of human origin while the others (IV, V and VI) are of simian origin. Genotypes I, II and III contain sub-genotypes defined by a nucleotide divergence of 7–7.5%. Genetic diversity of HAV is evidenced by the emergence of new sub-genotypes (Perez-Sautu *et al.*, 2011b). Genotypic characterization may be extremely useful in tracing the origin of outbreaks.

However, despite this nucleotide variability, at the amino acid level the diversity is limited and only a few natural antigenic variants have been isolated (Costa-Mattioli *et al.*, 2003a; Gabrieli *et al.*, 2004; Perez-Sautu *et al.*, 2011a; Sanchez *et al.*, 2002), suggesting the occurrence of severe structural and biological constraints of the capsid that would prevent the emergence of new serotypes. Consequently, a single serotype of HAV exists.

The emergence of a new serotype requires extensive substitutions in the capsid that seem quite unlikely to occur in a virus with such severe genomic and structural constraints. However the emergence of new variants may occur if virus populations are forced through bottleneck conditions such as immune selective pressures. The immuno-compromised population, particularly

HIV-positive individuals, have an impaired immunological response to HAV vaccines with lower concentrations of anti-HAV IgG in sera than do healthy individuals after vaccination, and they require additional vaccine doses (Neilsen *et al.*, 1997; Weissman *et al.*, 2006). The normal schedule for HAV vaccination usually requires two doses. In immuno-compromised patients, the IgG level may be too low to completely neutralize the virus between the two doses, particularly if the input virus is high. Conditions of high input virus occur in some risky sexual practices, bearing in mind that the HAV titre in faeces can be of 10^8 particles/g in healthy individuals and as high as 10^{11} particles/g in HIV-positive individuals at two weeks after the onset of symptoms (Costafreda *et al.*, 2006; Costa-Mattioli *et al.*, 2002a), and with even higher peaks during the prodromic phase. Replication of the non-neutralized viruses in the presence of low concentrations of specific IgGs may contribute to select, among the swarm of mutants generated by the quasispecies dynamics of replication, those variants resistant to the effect of the vaccines (Borrego *et al.*, 1993).

Recently, the isolation of several natural antigenic variants of the immunodominant site (Perez-Sautu *et al.*, 2011a) has been described in the MSM community of Barcelona (Perez-Sautu *et al.*, 2011a). Of particular interest is one of these variants with two amino acid substitutions at positions 166 (V to G) and 171 (V to A) of the capsid protein VP1, just in the core of the immunodominant site. Four per cent of the patients affected by this outbreak had been vaccinated and among these 62% were HIV-positive. Additionally, among those who were vaccinated only 12% had followed the complete schedule of vaccination while 88% had received only one dose.

Whether these newly emerged strains will circulate worldwide or disappear is as yet unclear. However, what is imperative is to target the MSM community with more effective information on risky sexual practices and vaccination programs. Additionally efforts should be made, particularly among HIV-positive MSM, to complete the HAV vaccination schedule so as to avoid as far as possible the potential emergence of antigenic variants.

17.7 Conclusion

Hepatitis A is still the most common viral hepatitis worldwide in spite of improved hygiene conditions and an efficient vaccine. Contaminated food and transmission through risky sexual practices are the main cause of outbreaks nowadays. HAV has evolved to render a very controlled translation kinetics which ensures a very accurate capsid folding. Such a capsid is resistant to many biological and environmental constraints during virus transmission and the need to maintain such a phenotype contributes to the low antigenic variability of HAV. However, the inherent genetic variability of an RNA virus in association with vaccination campaigns may promote the emergence of new variants.

17.8 References

ABAD, F. X., PINTÓ, R. M. and BOSCH, A. 1994a. Survival of enteric viruses on environmental fomites. *Applied and Environmental Microbiology*, **60**, 3704–10.

ABAD, F. X., PINTÓ, R. M. and BOSCH, A. 1997a. Disinfection of human enteric viruses on fomites. *Fems Microbiology Letters*, **156**, 107–11.

ABAD, F. X., PINTÓ, R. M., DIEZ, J. M. and BOSCH, A. 1994b. Disinfection of human enteric viruses in water by copper and silver in combination with low levels of chlorine. *Applied and Environmental Microbiology*, **60**, 2377–83.

ABAD, F. X., PINTÓ, R. M., GAJARDO, R. and BOSCH, A. 1997b. Viruses in mussels: public health implications and depuration. *Journal of Food Protection*, **60**, 677–81.

ADZHUBEI, A. A., ADZHUBEIB, I. A., KRASHENINNIKOV, I. A. and NEIDLE, S. 1996. Non-random usage of "degenerate"codons is related to protein three-dimensional structure. *FEBS Letters*, **399**, 78–82.

ARAGONÈS, L., BOSCH, A. and PINTÓ, R. M. 2008. Hepatitis A virus mutant spectra under the selective pressure of monoclonal antibodies: codon usage constraints limit capsid variability. *The Journal of Virology*, **82**, 1688–700.

ARAGONÈS, L., GUIX, S., RIBES, E., BOSCH, A. and PINTÓ, R. M. 2010. Fine-tuning translation kinetics selection as the driving force of codon usage bias in the hepatitis A virus capsid. *PLoS Pathogens*, **6**, e1000797.

BARZAGA, N. G. 2000. Hepatitis A shifting epidemiology in South-East Asia and China. *Vaccine*, **18**, S61–S64.

BJORKHOLM, M., CELSING, F., RUNARSSON, G. and WALDENSTROM, J. 1995. Successful intravenous immunoglobulin therapy for severe and persistent astrovirus gastroenteritis after fludarabine treatment in a patient with Waldenstrom's macroglobulinemia. *International Journal of Hematology*, **62**, 117–20.

BLATCHLEY, E. R., III, GONG, W. L., ALLEMAN, J. E., ROSE, J. B., HUFFMAN, D. E., OTAKI, M. and LISLE, J. T. 2007. Effects of wastewater disinfection on waterborne bacteria and viruses. *Water Environmental Research*, **79**, 81–92.

BORMAN, A. M. and KEAN, K. M. 1997. Intact eukaryotic initiation factor 4G is required for hepatitis A virus internal initiation of translation. *Virology*, **237**, 129–36.

BORREGO, B., NOVELLA, I. S., GIRALT, E., ANDREU, D. and DOMINGO, E. 1993. Distinct repertoire of antigenic variants of foot-and-mouth disease virus in the presence or absence of immune selection. *The Journal of Virology*, **67**, 6071–9.

BOSCH, A. 2007. *Human viruses in water*. Amsterdam, The Netherlands: Elsevier.

BOSCH, A., ABAD, F. X., GAJARDO, R. and PINTÓ, R. M. 1994. Should shellfish be purified before public consumption? *Lancet*, **344**, 1024–5.

BOSCH, A., LUCENA, F., DIEZ, J. M., GAJARDO, R., BLASI, M. and JOFRE, J. 1991. Waterborne viruses associated with hepatitis outbreak. *Journal of the American Water Works Association JAWWA5*, **83**(3), 80–3.

BOWER, W. A., NAINAN, O. V., HAN, X. and MARGOLIS, H. S. 2000. Duration of viremia in hepatitis A virus infection. *Journal of Infectious Diseases*, **182**, 12–17.

BROWN, E. A., ZAJAC, A. J. and LEMON, S. M. 1994. In vitro characterization of an internal ribosomal entry site (IRES) present within the 5' nontranslated region of hepatitis A virus RNA: comparison with the IRES of encephalomyocarditis virus. *The Journal of Virology*, **68**, 1066–74.

CIANCIARA, J. 2000. Hepatitis A shifting epidemiology in Poland and Eastern Europe. *Vaccine*, **18**(Suppl 1), S68–S70.

CONATY, S., BIRD, P., BELL, G., KRAA, E., GROHMANN, G. and MCANULTY, J. M. 2000. Hepatitis A in New South Wales, Australia from consumption of oysters: the first reported outbreak. *Epidemiology and Infection*, **124**, 121–30.

COSTAFREDA, M. I., BOSCH, A. and PINTÓ, R. M. 2006. Development, evaluation, and standardization of a real-time TaqMan reverse transcription-PCR assay for quantification of hepatitis A virus in clinical and shellfish samples. *Applied and Environmental Microbiology*, **72**, 3846–55.

COSTA-MATTIOLI, M., ALLAVENA, C., POIRIER, A. S., BILLAUDEL, S., RAFFI, F. and FERRE, V. 2002a. Prolonged hepatitis A infection in an HIV-1 seropositive patient. *Journal of Medical Virology*, **68**, 7–11.

COSTA-MATTIOLI, M., CRISTINA, J., ROMERO, H., PEREZ-BERCOF, R., CASANE, D., COLINA, R., GARCIA, L., VEGA, I., GLIKMAN, G., ROMANOWSKY, V., CASTELLO, A., NICAND, E., GASSIN, M., BILLAUDEL, S. and FERRE, V. 2002b. Molecular evolution of hepatitis A virus: a new classification based on the complete VP1 protein. *The Journal of Virology*, **76**, 9516–25.

COSTA-MATTIOLI, M., FERRÉ, V., CASANE, D., PEREZ-BERCOFF, R., COSTE-BUREL, M., IMBERT-MARCILLE, B. M., ANDRE, E. C., BRESSOLLETTE-BODIN, C., BILLAUDEL, S. and CRISTINA, J. 2003a. Evidence of recombination in natural populations of hepatitis A. *Virology*, **311**, 51–9.

COSTA-MATTIOLI, M., MONPOEHO, S., NICAND, E., ALEMAN, M. H., BILLAUDEL, S. and FERRE, V. 2002c. Quantification and duration of viraemia during hepatitis A infection as determined by real-time RT-PCR. *Journal of Viral Hepatitis*, **9**, 101–6.

COSTA-MATTIOLI, M., NAPOLI, A. D., FERRE, V., BILLAUDEL, S., PEREZ-BERCOFF, R. and CRISTINA, J. 2003b. Genetic variability of hepatitis A virus. *Journal of General Virology*, **84**, 3191–201.

CUTHBERT, J. A. 2001. Hepatitis A: old and new. *Clinical Microbiology Reviews*, **14**, 38–58.

DAGAN, R., LEVENTHAL, A., ANIS, E., SLATER, P., ASHUR, Y. and SHOUVAL, D. 2005. Incidence of hepatitis A in Israel following universal immunization of toddlers. *JAMA-Journal of the American Medical Association*, **294**, 202–10.

DENTINGER, C. M., BOWER, W. A., NAINAN, O. V., COTTER, S. M., MYERS, G., DUBUSKY, L. M., FOWLER, S., SALEHI, E. D. and BELL, B. P. 2001. An outbreak of hepatitis A associated with green onions. *The Journal of Infectious Diseases*, **183**, 1273–6.

DOMINGO, E., BIEBRICHER, C., EIGEN, M. and HOLLAND, J. 2001. *Quasispecies and RNA virus Evolution: Principles and Consequences*, Austin: Landes Bioscience.

DOMINGO, E., MARTIN, V., PERALES, C., GRANDE-PEREZ, A., GARCIA-ARRIAZA, J. and ARIAS, A. 2006. Viruses as quasispecies: biological implications. *Current Topics in Microbiology and Immunology*, **299**, 51–82.

DOMINGO, E. and VENNEMA, H. 2008. Viral evolution and its relevance for food-borne virus epidemiology. In: KOOPMANS, M., CLIVER, D. O. and BOSCH, A. (eds) *Foodborne viruses: Progress and challenges*. Washington, DC, USA: ASM Press.

DOMINGUEZ, A., OVIEDO, M., CARMONA, G., BATALLA, J., BRUGUERA, M., SALLERAS, L. and PLASENCIA, A. 2008. Impact and effectiveness of a mass hepatitis A vaccination programme of preadolescents seven years after introduction. *Vaccine*, **26**, 1737–41.

EHRENFELD, E. and TETERINA, N. L. 2002. Initiation of translation of picornavirus RNAs: structure and function of the internal ribosome entry site. In: SEMLER, B. L. and WIMMER, E. (eds) *Molecular biology of picornaviruses.* Washington, DC: ASM Press.

EVANS, M. S., CLARKE, T. F. and CLARK, P. L. 2005. Conformations of co-translational folding intermediates. *Protein and Peptide Letters*, **12**, 189–95.

FABER, M. S., STARK, K., BEHNKE, S. C., SCHREIER, E. and FRANK, C. 2009. Epidemiology of hepatitis A virus infections, Germany, 2007–2008. *Emerging Infectious Diseases*, **15**, 1760–8.

FRANK, C., WALTER, J., MUEHLEN, M., JANSEN, A., van TREECK, U., HAURI, A. M., ZOELLNER, I., RAKHA, M., HOEHNE, M., HAMOUDA, O., SCHREIER, E. and STARK, K. 2007. Major outbreak of hepatitis A associated with orange juice among tourists, Egypt, 2004. *Emerging Infectious Diseases*, **13**, 156–8.

GABRIELI, R., SANCHEZ, G., MACALUSO, A., CENKO, F., BINO, S., PALOMBI, L., BUONOMO, E., PINTÓ, R. M., BOSCH, A. and DIVIZIA, M. 2004. Hepatitis in Albanian children: molecular analysis of hepatitis A virus isolates. *Journal of Medical Virology*, **72**, 533–7.

GAVRILIN, G. V., CHERKASOVA, E. A., LIPSKAYA, G. Y., KEW, O. M. and AGOL, V. I. 2000. Evolution of circulating wild poliovirus and vaccine-derived poliovirus in an immunodeficient patient: a unifying model. *The Journal of Virology*, **74**(16), 7381–90.

GERMINARIO, C., LUIGI LOPALCO, P., CHICANNA, M. and da VILLA, G. 2000. From hepatitis B to hepatitis A and B prevention: the Puglia (Italy) experience. *Vaccine*, **18**, 3326.

HALLIDAY, M. L., KANG, L. Y., ZHOU, T. Z., HU, M. D., PAN, Q. C., FU, T. Y., HUANG, Y. S. and HU, S. L. 1991. An epidemic of hepatitis A attributable to the ingestion of raw clams in Shanghai, China. *The Journal of Infectious Diseases*, **164**, 852–9.

JACKSON, T. A. 2002. Proteins involved in the function of picornavirus Internal Ribosome Entry sites. In: SEMLER, B. L. and WIMMER, E. (eds) *Molecular biology of picornaviruses*. Washington, DC: ASM Press.

KUECHLER, E., SEIPELT, J., LIEBIG, H. D. and SOMMERGRUBER, W. 2002. Picornavirus proteinase-mediated shutoff of host cell translation: direct cleavage of a cellular initiation factor. In: SEMLER, B. L. and WIMMER, E. (eds) *Molecular biology of picornaviruses*. Washington, DC: ASM Press.

LEONG, L. E. C., CORNELL, C. T. and SEMLER, B. L. 2002. Processing determinants and functions of cleavage products of picornavirus. In: SEMLER, B. L. and WIMMER, E. (eds) *Molecular biology of picornaviruses*. Washington, DC: ASM Press.

LEONI, E., BEVINI, C., DEGLI ESPOSTI, S. and GRAZIANO, A. 1998. An outbreak of intrafamiliar hepatitis A associated with clam consumption: epidemic transmission to a school community. *European Journal of Epidemiology*, **14**, 187–92.

MACKOWIAK, P. A., CARAWAY, C. T. and PORTNOY, B. L. 1976. Oyster-associated hepatitis: lessons from the Louisiana experience. *American Journal of Epidemiology*, **103**, 181–91.

MARTIN, A. and LEMON, S. M. 2006. Hepatitis A virus: from discovery to vaccines. *Hepatology*, **43**, S164–S172.

MELE, A., RASTELLI, M. G., GILL, O. N., DI BISCEGLIE, D., ROSMINI, F., PARDELLI, G., VALTRIANI, C. and PATRIARCHI, P. 1989. Recurrent epidemic hepatitis A associated with consumption of raw shellfish, probably controlled through public health measures. *American Journal of Epidemiology*, **130**, 540–6.

NEILSEN, G. A., BODSWORTH, N. J. and WATTS, N. 1997. Response to hepatitis A vaccination in human immunodeficiency virus-infected and -uninfected homosexual men. *Journal of Infectious Diseases*, **176**, 1064–7.

NORMANN, A., JUNG, C., VALLBRACHT, A. and FLEHMIG, B. 2004. Time course of hepatitis A viremia and viral load in the blood of human hepatitis A patients. *Journal of Medical Virology*, **72**, 10–16.

OVERBAUGH, J. and BANGHAM, C. R. M. 2001. Selection forces and constraints on retroviral sequence variation. *Science*, **292**, 1106–9.

PEREZ-SAUTU, U., COSTAFREDA, M. I., CAYLA, J., TORTAJADA, C., LITE, J., BOSCH, A. and PINTÓ, R. M. 2011a. Hepatitis A virus vaccine escape variants and potential new serotype emergence. *Emerging and Infectious Diseases*, **17**, 734–7.

PEREZ-SAUTU, U., COSTAFREDA, M. I., LITE, J., SALA, R., BARRABEIG, I., BOSCH, A. and Pintó, R. M. 2011b. Molecular epidemiology of hepatitis A virus infections in Catalonia, Spain, 2005–2009: Circulation of newly emerging strains. *Journal of Clinical Virology*, **52**, 98–102.

PETRIGNANI, M., HARMS, M., VERHOEF, L., VAN HUNEN, R., SWAAN, C., VAN STEENBERGEN, J., BOXMAN, I., PERAN I., SALA, R., OBER, H., VENNEMA, H., KOOPMANS, M. and VAN PELT, W. 2010a. Update: a food-borne outbreak of hepatitis A in the Netherlands related to semi-dried tomatoes in oil, January-February 2010. *Euro Surveillance*, **15**, 19572.

PETRIGNANI, M., VERHOEF, L., VAN HUNEN, R., SWAAN, C., VAN STEENBERGEN, J., BOXMAN, I., OBER, H.J., VENNEMA, H. and KOOPMANS, M. 2010b. A possible foodborne outbreak of hepatitis A in the Netherlands, January-February 2010. *Euro Surveillance*, **15**, 19512.

PINTÓ, R. M., ALEGRE, D., DOMINGUEZ, A., EL SENOUSY, W. M., SANCHEZ, G., VILLENA, C., COSTAFREDA, M. I., ARAGONES, L. and BOSCH, A. 2007a. Hepatitis A virus in urban sewage from two Mediterranean countries. *Epidemiology and Infection*, **135**, 270–3.

PINTÓ, R. M., ARAGONES, L., COSTAFREDA, M. I., RIBES, E. and BOSCH, A. 2007b. Codon usage and replicative strategies of hepatitis A virus. *Virus Res*, **127**, 158–63.

PINTÓ, R. M. and BOSCH, A. 2008. Rethinking virus detection in food. In: KOOPMANS, M., CLIVER, D. O. and BOSCH, A. (eds) *Foodborne viruses: Progress and challenges*. Washington, DC, USA: ASM Press.

PINTÓ, R. M., COSTAFREDA, M. I. and BOSCH, A. 2009. Risk assessment in shellfish-borne outbreaks of hepatitis A. *Applied and Environmental Microbiology*, **75**, 7350–5.

PINTÓ, R. M., COSTAFREDA, M. I., PEREZ RODRIGUEZ, F. J., D'ANDREA, L. and BOSCH, A. 2010. Hepatitis A virus: state of the art. *Food and Environmental Virology*, **2**, 127–35.

POLO, D., VILARINO, M. L., MANSO, C. F. and ROMALDE, J. L. 2010. Imported mollusks and dissemination of human enteric viruses. *Emerging and Infectious Diseases*, **16**, 1036–8.

PREVISANI, N., LAVANCHY, D. and SIEGL, G. 2004. Hepatitis A. In: MUSHAHWAR, I. K. (ed.) *Viral Hepatitis Molecular Biology, Diagnosis, Epidemiology and Control*. Amsterdam: Elsevier.

REID, T. M. S. and ROBINSON, H. G. 1987. Frozen raspberries and hepatitis A. *Epidemiology and Infection*, **98**, 109–12.

ROBERTSON, B. H., JANSEN, R. W., KHANNA, B., TOTSUKA, A., NAINAN, O. V., SIEGL, G., WIDELL, A., MARGOLIS, H. S., ISOMURA, S., ITO, K., ISHIZU, T., MORITSUGU, Y. and LEMON, S. M. 1992. Genetic relatedness of hepatitis A virus strains recovered from different geographical regions. *Journal of General Virology*, **73**, 1365–77.

ROOS, B. 1956. Hepatitis epidemic transmitted by oysters. *Svenska Lakartidningen*, **53**, 989–1003.

ROSENBLUM, L. S., MIRKIN, I. R., ALLEN, D. T., SAFFORD, S. and HADLER, S. C. 1990. A multifocal outbreak of hepatitis A traced to commercially distributed lettuce. *American Journal of Public Health*, **80**, 1075–9.

SAMANDARI, T., BELL, B. P. and ARMSTRONG, G. L. 2004. Quantifying the impact of hepatitis A immunization in the United States, 1995–2001. *Vaccine*, **22**, 4342–50.

SANCHEZ, G., BOSCH, A., GOMEZ-MARIANO, G., DOMINGO, E. and PINTÓ, R. M. 2003a. Evidence for quasispecies distributions in the human hepatitis A virus genome. *Virology*, **315**, 34–42.

SANCHEZ, G., BOSCH, A. and PINTÓ, R. M. 2003b. Genome variability and capsid structural constraints of hepatitis A virus. *Journal of Virology*, **77**, 452–9.

SANCHEZ, G., PINTÓ, R. M., VANACLOCHA, H. and BOSCH, A. 2002. Molecular characterization of hepatitis a virus isolates from a transcontinental shellfish-borne outbreak. *Journal of Clinical Microbiology*, **40**, 4148–55.

STENE-JOHANSEN, K., TJON, G., SCHREIER, E., BREMER, V., BRUISTEN, S., NGUI, S. L., KING, M., PINTÓ, R. M., ARAGONES, L., MAZICK, A., CORBET, S., SUNDQVIST, L., BLYSTAD, H., NORDER, H. and SKAUG, K. 2007. Molecular epidemiological studies show that hepatitis A virus is endemic among active homosexual men in Europe. *Journal of Medical Virology*, **79**, 356–65.

STROFFOLINI, T., BIAGINI, W., LORENZONI, L., PALAZZESI, G. P., DIVIZIA, M. and FRONGILLO, R. 1990. An outbreak of hepatitis A in young adults in central Italy. *European Journal of Epidemiology*, **6**, 156–9.

TALLO, T., NORDER, H., TEFANOVA, V., OTT, K., USTINA, V., PRUKK, T., SOLOMONOVA, O., SCHMIDT, J., ZILMER, K., PRIIMÄGI, L., KRISPIN, T. and MAGNIUS, L. O. 2003. Sequential changes in hepatitis A virus genotype distribution in Estonia during 1994 to 2001. *Journal of Medical Virology*, **70**, 187–93.

TANAKA, J. 2000. Hepatitis A shifting epidemiology in Latin America. *Vaccine*, **18**, S57–S60.

TORTAJADA, C., de OLALLA, P. G., PINTÓ, R. M., BOSCH, A. and CAYLA, J. 2009. Outbreak of hepatitis A among men who have sex with men in Barcelona, Spain, September 2008-March 2009. *Euro Surveillance*, **14**, 19175.

VAN DAMME, P. and VAN HERCK, K. 2005. Effect of hepatitis A vaccination programs. *JAMA-Journal of the American Medical Association*, **294**, 246–8.

WASLEY, A., SAMANDARI, T. and BELL, B. P. 2005. Incidence of hepatitis A in the United States in the era of vaccination. *JAMA-Journal of the American Medical Association*, **294**, 194–201.

WEISSMAN, S., FEUCHT, C. and MOORE, B. A. 2006. Response to hepatitis A vaccine in HIV-positive patients. *Journal of Viral Hepatitis*, **13**, 81–6.

WHEELER, C., VOGT, T. M., ARMSTRONG, G. L., VAUGHAN, G., WELTMAN, A., NAINAN, O. V., DATO, V., XIA, G. L., WALLER, K., AMON, J., LEE, T. M., HIGHBAUGH-BATTLE, A., HEMBREE, C., EVENSON, S., RUTA, M. A., WILLIAMS, I. T., FIORE, A. E. and BELL, B. P. 2005. An outbreak of hepatitis A associated with green onions. *New England Journal of Medicine*, **353**, 890–7.

WHETTER, L. E., DAY, S. P., ELROYSTEIN, O., BROWN, E. A. and LEMON, S. M. 1994. Low efficiency of the 5' nontranslated region of hepatitis A virus RNA in directing cap-independent translation in permissive monkey kidney cells. *Journal of Virology*, **68**, 5253–63.

18

Advances in understanding of rotaviruses as food- and waterborne pathogens and progress with vaccine development

F. M. Ruggeri and L. Fiore, Istituto Superiore di Sanità, Italy

DOI: 10.1533/9780857098870.4.362

Abstract: Rotavirus is a double-stranded RNA virus causing severe acute gastroenteritis in children; except for the elderly, adults are generally asymptomatically infected. The virus undergoes gene reassortment during mixed infection with more human or animal strains. Mechanisms of rotavirus pathogenesis and host restriction may involve complex inter-species genome rearrangement. Molecular detection and genotyping are available for clinical, food, and water samples. Human and animal rotavirus shed into sewage and the environment can contaminate surface waters, soft fruits, vegetables and seafood. Health care professionals and some food workers may be exposed to human and/or animal rotavirus.

Key words: rotavirus, gastroenteritis, molecular detection, zoonotic, waterborne.

18.1 Introduction

The discovery of rotavirus in the early 1970s represented a major breakthrough in the understanding of the etiology of infectious acute gastroenteritis (Bishop *et al.*, 1973). It soon became clear that these agents represented an enteropathogen of primary importance, particularly in childhood (Black, 1993; Kapikian *et al.*, 1976; Ryder *et al.*, 1976). Rotavirus remains the major etiological agent of acute childhood gastroenteritis worldwide, being associated with more deaths attributed to diarrhea than any other viral, bacterial or parasitic organism (Butler, 2011; Glass, 2006; Tate *et al.*, 2012). There has been much research into rotavirus biology and epidemiology over the last four decades, and the high mortality rate (especially but not only in developing countries) and economic burden of disease have driven science toward controlling rotavirus disease through the implementation of vaccine strategies (Glass, 2006; Kapikian *et al.*, 1996; Parashar and Glass, 2006).

Recently, increasing evidence of zoonotic transmission of animal rotaviruses to humans and host adaptation involving reassortment mechanisms during dual infection has been found. This has strengthened the threat of emerging novel rotavirus strains from domestic and wild animals (Martella *et al.*, 2010). The spread of animal rotaviruses in the environment and the consequent contamination of raw food could favor the introduction of rare rotaviruses into human populations, which might eventually reduce the efficacy of current vaccines. The asymptomatic infection of adults and their possible role in virus epidemiology also needs re-thinking. This chapter will summarise the current state of research on rotavirus biology and will discuss the possible role of food as a vehicle for rotavirus transmission.

18.2 Background

Despite its relatively small size, the rotavirus virion presents a complex three-dimensional structure, where the ordered interaction between six distinct proteins and eleven RNA segments finely regulates antigens specificity, virus replication and resistance in different hosts and hard environments.

18.2.1 The virion

Rotavirus constitutes a genus of the *Reoviridae* family and derives its name from its wheel-like shape (*rota* in Latin). The virion is a particle approximately 75 nm in diameter, made of three concentric protein layers protecting a segmented double-stranded RNA (dsRNA) genome, and presents no lipid envelope (Estes and Kapikian, 2007). This complete form of the infectious rotavirus is also known as a TLP (triple-layered particle). Detailed studies using cryo-electron microscopy and image reconstruction or X-ray crystallography have disclosed the relationships between different proteins in the complete rotavirus particle, as well as the location of amino acid residues involved in sialic-acid or antibody binding (Dormitzer *et al.*, 2002; Settembre *et al.*, 2011).

The outermost shell of rotavirus comprises two proteins, VP7 and VP4, both of which contain epitopes recognized by neutralizing antibodies. The glycosylated protein VP7 forms a continuous layer of trimers arranged in a $T = 13$ lattice crossed by 60 spike-like projections of the trypsin-sensitive protein VP4. VP7 is stabilized on the rotavirus virion by the presence of Ca^{2+}, where it represents the major neutralization antigen, and was initially acknowledged as the only protein defining the virus serotype (Estes and Kapikian, 2007; Midthun *et al.*, 1986).

VP4 is the viral attachment protein (Ruggeri and Greenberg, 1991), and is activated by pancreatic trypsin via a cleavage yielding two fragments, VP8* and VP5*, resulting in enhanced infectivity (Estes *et al.*, 1981). During viral assembly, uncleaved VP4 is incorporated into virions as a trimer, but upon trypsin cleavage it undergoes remarkable modification, and the trypsin-activated viral spike exhibits two lectin-like VP8* attachment domains at its tip, supported by three VP5* stems. The third VP8* is lost during this conformational modification process (Settembre *et al.*, 2011).

VP4 spikes protrude through the VP7 layer, which contributes to anchor them to the inner rotavirus shell, which also follows a $T = 13$ symmetry (Mathieu *et al.*, 2001) and is made of the VP6 protein only, which is recognized as the 'group antigen' of rotavirus (Tarlow and McCrae, 1990). Removal of VP7 and VP4 takes place on lowering Ca^{2+} concentration, as it physiologically occurs after virus entry into susceptible gut cells by endocytic process (Baker and Prasad, 2010; Ruiz *et al.*, 1996). The resulting double-layered particle (DLP) is released into the cytosol and is transcriptionally active, subsequently initiating synthesis and extrusion of mRNA transcribed from each of the 11 genome segments (Charpilienne *et al.*, 2002; Lawton *et al.*, 1997). The viral RNA-dependent RNA-polymerase VP1 and the capping enzyme VP3 are also structural constituents of the complete virion, and are present in the 'core' of rotavirus, which is made of 12 decamers of VP2 outlining the inner space where the 11 segments of genomic dsRNA are engulfed. Besides a structural role, VP2 also plays a critical role in activating replication selectively in progeny virions containing packaged plus-strand RNA templates (McDonald and Patton, 2011; Patton *et al.*, 1997). Altogether proteins in the three layers identify 132 channels (type I–III) for entry and release of molecules involved in genome replication and transcription (Estes and Kapikian, 2007; Prasad *et al.*, 1988).

18.2.2 Rotavirus genome

The genome of rotavirus consists of 11 independent segments of double-stranded RNA, with molecular size ranging between approximately 670 and 3300 base pairs (www.iah.bbsrc.ac.uk/dsRNA virus proteins/Rotavirus.htm). Each segment is monocistronic except the shortest, segment 11, which contains two functional open reading frames encoding non-structural proteins NSP5 and NSP6, respectively. Besides the six viral proteins (VPs) mentioned above, the rotavirus genome also codes for six non-structural proteins (NSP1-6), involved in either RNA transcription or progeny virus maturation. When the 12 viral 5'-capped mRNAs are released into the cytosol by activated DLPs, they are translated into viral proteins. In the first stages of infection, electron-dense structures named viroplasms appear in the cytoplasm of infected cells. Viroplasms are the structures where synthesis of dsRNA (genome replication) and the initial steps of assembly of new particles take place (Altenburg *et al.*, 1980). Besides several rotavirus structural proteins (e.g., VP1, VP2, VP3, and VP6) accumulating in viroplasms, non-structural proteins NSP2 and NSP5 are believed to be essential for viroplasm formation, dsRNA synthesis, and virus replication (Campagna *et al.*, 2005; Silvestri *et al.*, 2004).NSP6 also localizes in the viroplasm, where its ssRNA and dsRNA binding properties probably contribute to RNA packaging within assembling virions (Rainsford and McCrae, 2007). NSP3 and NSP1 appear to have specific roles in rotavirus virulence and possibly host range restriction. The former controls the protein response of the cell but has no inhibitory effect on viral protein synthesis (Trujillo-Alonso *et al.*, 2011); NSP1 is an antagonist of the host innate immune response, especially the type I interferon (IFN) response (Barro and Patton, 2005). Finally, NSP4 has different properties. It is functional in virus final morphogenesis, since it interacts with newly assembled DLPs, assisting their budding into the endoplasmic reticulum lumen. At this stage DLPs acquire a transient envelope, which is then replaced with the VP7-VP4 layer (Estes and Kapikian, 2007). A second independent property of NSP4 is an enterotoxin-like effect associated with protein secretion into the lumen and subsequent binding to specific cell receptors (Seo *et al.*, 2008). Virus-free NSP4 can induce diarrhea in infant mice (Ball *et al.*, 1996). Possible further involvement of NSP4 in cell functions impairment is related to Ca^{2+} homeostasis, extracellular Ca^{2+} signaling pathways, and cell membrane permeability (Dong *et al.*, 1997; Newton *et al.*, 1997). At the end of the maturation processes, release of mature viral particles from the cell occurs either by cell lysis (Estes and Kapikian, 2007) or by a non-classical, Golgi apparatus-independent, vesicular transport pathway (Chwetzoff and Trugnan, 2006; Cuadras *et al.*, 2006).

18.2.3 Viral antigens

Of all structural proteins of rotavirus, three are also important antigens: the spike-protein VP4, the major neutralization antigen VP7, and the VP6 protein forming the intermediate layer of the virion, defining the DLPs. The VP6 antigen is important in antibody-mediated diagnosis since it exhibits cross-reactive antigenic sites among all rotavirus belonging to as the same 'group' (Ramig *et al.*, 2005). Rotaviruses are classified into seven groups designated A–G, among which group A contains by far the most important viral strains pathogenic for humans and many other animal species. Humans are also infected with group B and C rotaviruses, known to be implicated in large waterborne outbreaks in Asia and to cause sporadic or epidemic gastroenteritis cases in children and adults, respectively (Bridger *et al.*, 1986; Chen *et al.*, 1985). However, the real burden of rotavirus disease across the world is caused by group A rotavirus, which has consequently been the focus of scientific and public health interest during the past 40 years. There are considerable differences between rotaviruses belonging to distinct groups, not only in VP6 but also in all other proteins and genome segments sequences. For these reasons, there is no cross-reactive immune response between rotavirus groups.

Group A rotaviruses can be divided into serotypes according to antibody reactivity with hyperimmune sera or monoclonal antibodies directed at either VP7 or VP4 (Hoshino and Kapikian, 1996). Early studies have shown that neutralizing antibodies recognize several epitopes in both proteins and exert a protective effect against rotavirus diarrhea in an infant mouse model of disease (Giammarioli *et al.*, 1996; Hoshino *et al.*, 1994; Shaw *et al.*, 1986). The biological mechanisms by which antibodies combat rotavirus infection are unclear, although it seems that anti-VP4 antibodies can interfere with early interaction of viruses with the cell surface (Ruggeri and Greenberg, 1991) whereas anti-VP7 antibodies may block virus decapsidation (Ludert *et al.*, 2002). In addition to and in synergy with neutralizing antibodies to VP4 and VP7, other effectors are likely to contrast rotavirus *in vivo*, such as T-cell response and innate immunity (Desselberger and Huppertz, 2011; Ward, 2009) or non-neutralizing antibodies to other viral proteins, such as VP6 or NSP4 (Burns *et al.*, 1996; Hou *et al.*, 2008). As the two genes encoding VP4 and VP7 segregate independently, and both proteins elicit neutralizing antibody responses believed to be protective for man and animals, a binomial serotype system has been adopted to identify strains, as is used for influenza viruses (Gentsch *et al.*, 1996). Today, classical serotyping is only rarely performed due to the cumbersome nature of immune assays and lack of specific reagents availability. Viral characterization is thus currently carried out with molecular approaches using semi-nested RT-PCR and large panels of genotype-specific oligonucleotide primers (Gouvea *et al.*, 1990a, 1994; Iturriza-Gomara *et al.*, 2004). Genotyping is considered to be a valid proxy for serotyping, is much more user-friendly and is widely adopted throughout the world. An enormous increase in knowledge of the rotavirus sero/genotypes circulating in five continents has thus been made possible (Santos and Hoshino, 2005), allowing instruments to evaluate the adequacy of current vaccines in relation to circulating viral strains and their evolution (Desselberger *et al.*, 2006; Glass *et al.*, 2006; Heaton *et al.*, 2005; O'Ryan, 2009).

18.3 Clinical manifestation

Rotavirus is a typical enteropathogen with a fecal–oral transmission, inducing symptoms that are mostly related to viral replication in the intestinal tract of susceptible hosts. However, its epidemiological cycle and pathogenic mechanisms present peculiar aspects.

18.3.1 Fundamentals of rotavirus epidemiology

Rotavirus is the major cause of acute infantile gastroenteritis worldwide, causing approximately 530 000 deaths every year, mainly in developing countries (Glass, 2006; Parashar *et al.*, 2006b; Tate *et al.*, 2012; Widdowson *et al.*, 2009). Among all known enteric pathogens, rotavirus infection causes severe disease most frequently, particularly in children up to 5 years of age (Bishop, 2009). Virus transmission follows the fecal–oral route, and is thought to occur mainly by direct inter-human passage. However, as with many other enteric pathogens, rotaviruses are shed in the environment by both humans and animals, can contaminate water, food and feed, and airborne transmission is also possible (Estes and Kapikian, 2007; Prince *et al.*, 1986). Rotavirus disease is seasonal, infections being more common in cooler, drier months in most settings, particularly in temperate climates, although seasonal peaks can occur year-round in different countries and vary over time in the same country (Atchison *et al.*, 2010; Bresee *et al.*, 2004; Cook *et al.*, 1990; Levy *et al.*, 2009). A recent study suggests that differences in rotavirus seasonality between areas may also be related to individual country birth rate and virus transmission dynamics, and not simply to environmental conditions favoring virus survival (Pitzer *et al.*, 2011).

The age-group at the highest risk of rotavirus gastroenteritis requiring medical attention is from 6 to 24 months (Ruuska and Vesikari, 1990; Velazquez *et al.*, 1996), although the disease is not necessarily more serious at this age than in older children. Globally, the estimated annual number of rotavirus diarrhea cases is approximately 110 million, 2 million of which require hospitalization. Although with reduced mortality, rotavirus burden is also high in industrialized countries, and as many as 410 000 doctors' visits, 70 000 hospitalizations, and 272 000 emergency department visits due to rotavirus were recorded annually in the US before mass vaccination was introduced in 2006, resulting in a cost to society of at least $1 billion a year (Parashar *et al.*, 2006a; Widdowson *et al.*, 2007). This and the ubiquitousness of rotavirus in nature have driven scientists, industry and public health authorities to implement vaccines to reduce mortality and morbidity globally (Glass *et al.*, 2005), and two live-attenuated vaccines have been successfully introduced in a growing number of countries worldwide since 2006 (Parashar *et al.*, 2006a, 2011; Vesikari *et al.*, 2009; Jit *et al.*, 2010; CDC, 2011).

The mechanisms and duration of protection in rotavirus infection are not completely understood, although clinical protection is likely to involve both gut mucosal and systemic antibody response, as well as cell-mediated immunity. Intestinal secretory IgA antibodies may be one of the major effectors of long-term protection from rotavirus infection (Coulson *et al.*, 1992; Franco *et al.*, 2006). The extent to which different serotypes of rotavirus, and particularly their proxies 'genotypes', reflect antigenic diversity relevant for the reaction to infection by the immune system remains an unanswered question. Although several studies have shown serotype- or strain-specific neutralizing activity of polyclonal or monoclonal antibodies (MAbs) raised against different rotavirus strains (Bell *et al.*, 1988; Dyall-Smith *et al.*, 1986; Urasawa *et al.*, 1984), the existence of antigenic sites cross-reactive between distinct strains and serotypes has also been demonstrated (Mackow *et al.*, 1988; Taniguchi *et al.*, 1988a,b). However, most of this information derives from either *in vitro* studies in cell cultures or from using the mouse model of passive protection that may not realistically reflect *in vivo* protection from natural infection in humans and other mammals. Suggestions of some degree of cross-reactive protection following natural infection or vaccination in humans or animals were first made several years ago (Taniguchi *et al.*, 1991; Ward *et al.*, 1992). As a matter of fact, both major human vaccines presently in use (RotaTeq®, by Merck; Rotarix®, by GlaxoSmithKline) appear to be largely protective against co-circulating rotavirus genotypes in human populations, although they are either pentavalent or monovalent (Bernstein and Ward, 2006; Clark *et al.*, 1996; Jiang *et al.*, 2010). On the other hand, reduced efficacy of vaccination has been noted in less developed countries (Jiang *et al.*, 2010), where uncommon rotavirus genotypes circulate more frequently among humans and may pose problems of unsatisfactory herd immunity (Banerjee *et al.*, 2007; Santos and Hoshino, 2005; Sharma *et al.*, 2008). There are enormous differences between continents with respect to the circulating rotavirus types, probably reflecting different local virus evolution, virus spillover from different animal species and environmental conditions (Santos and Hoshino, 2005). In this respect, it is remarkable that as of today at least 27 different G-types and 35 P-types of rotavirus have been reported from humans and animals worldwide (Matthijnssens *et al.*, 2011a), underpinning the need for surveillance of circulating rotavirus strains so that the emergence of novel strains or reassortants with unusual serotype characteristics can be promptly identified.

18.3.2 Susceptibility to rotavirus in different sectors of the population

Rotavirus in childhood

The main target population of rotavirus, when infection is responsible for the highest number of cases requiring medical care, is children aged from a few months to 4–5 years (Kapikian *et al.*, 1976; Monroe, 2011). Rotavirus infection is also more frequent and severe at younger ages in most animals, and vaccine development has always been considered an important line of defense for several animal species with an economic value (Barrandeguy *et al.*, 1998; Cashman *et al.*, 2010; Dewey *et al.*, 2003; Otto *et al.*, 2006). The decrease in rotavirus disease in humans and animals in older age groups suggests that susceptibility may change with the maturation of the intestinal epithelial cells or of the individual's immune system (Ciarlet *et al.*, 2002).

During the first few weeks of life, in humans and other big mammals rotavirus is only rarely associated with overt symptoms (Kim *et al.*, 2009). This has been correlated with passive transfer of maternal immunity by secretory immunoglobulins present in breast milk helping protect the neonate's gut lining from ingested virus (McLean and Holmes, 1981; Parreno *et al.*, 2010), but also with biochemical changes in the gut microenvironment (McLean and Holmes, 1981; Wold and Adlerberth, 2000), and cell surface receptors (Ciarlet *et al.*, 1998; Hempson *et al.*, 2010; Londrigan *et al.*, 2000). Viral binding to the cell occurs via VP4 and a series of events (Mendez *et al.*, 1999) that may involve glycosylated cell proteins presenting sialic acid or other types of sugars or glycans (Ciarlet and Estes, 1999; Haselhorst *et al.*, 2009; Isa *et al.*, 2006). This first interaction occurs between the VP8* apical part of the spike-protein VP4; afterwards other portions of the same proteins react with other cell receptors, most likely including α2β1-integrins recognizing a DGE integrin-binding motif present in the VP5* domain (Zarate *et al.*, 2000). Other integrin types may act at a later stage, favoring rotavirus entry (Coulson *et al.*, 1997).

Protection against disease can be afforded by both symptomatic and asymptomatic first infection (Bernstein *et al.*, 1991; Bishop *et al.*, 1983) which is also observed with present live attenuate oral vaccines (Butler, 2011; Glass *et al.*, 2006; Wang *et al.*, 2010a).

The efficacy of present vaccines and the suitability of future vaccines are universally evaluated against the strains affecting children as the target population of vaccine prophylaxis, as well as consideration of antigenic cross-reactivity between different rotavirus strains (Ward, 2009). In fact, both RotaTeq® and Rotarix®, as well as other vaccines (Kapikian *et al.*, 2005), exhibit the same VP4 and VP7 genotype as one or more of the rotavirus wild strains affecting children globally (Santos and Hoshino, 2005). Even the monovalent GSK vaccine shares its G1 and/or P[8] types with the majority of strains causing moderate and severe rotavirus diarrhea in developed countries (Iturriza-Gómara *et al.*, 2011; Santos and Hoshino, 2005). In developing countries, virus-presenting genotypes other than G1-4, or G9, and P[4] or P[8] may circulate widely in the child population (Castello *et al.*, 2009; Kabue *et al.*, 2010; Santos and Hoshino, 2005; Than *et al.*, 2011), and this might help explain the lower rate of vaccine efficacy in some areas of the world (Fischer Walker and Black, 2011; Jiang *et al.*, 2010).

Rotavirus in adults and the elderly

There are no data indicating that different serotypes of rotavirus may be associated with a different virulence in children compared to adults. It is however clear that animal strains are normally less pathogenic for humans than they are in their species of origin, which led to a Jennerian approach to rotavirus vaccines for children (Kapikian *et al.*, 1996). Also, genetic reassortment is considered necessary for heterologous rotaviruses to spread into a new animal or human community (Ghosh *et al.*, 2011; Matthijnssens *et al.*, 2011b; McDonald *et al.*, 2009; Park *et al.*, 2011).

However, volunteer studies and field investigations indicate that at least some strains deriving from sick children can infect normal adults (Awachat and Kelkar, 2006; Stelzmueller *et al.*, 2006), despite the presence of pre-existing immunity, usually asymptomatically (Ward *et al.*, 1986, 1990). It is worth noting that in the UK the rotavirus-specific IgM rate among the normal adult population does not show any seasonality (Cox and Medley, 2003), suggesting that adults are not infected solely by children, who are mostly infected during winter months. Since it has long been considered that immunity to rotavirus gastroenteritis was life-long, no real study has been performed to assess to what extent the adult can otherwise represent a healthy carrier of pathogenic rotavirus strains. More cautiously, we have to assume that adults can act as reservoir for rotavirus as well as for other pathogens (de Wit *et al.*, 2001), sometimes in the absence of symptoms, and consider possible rotavirus transmission pathways linking them to disease-susceptible children.

The elderly, as well as adults with immunosuppression, show an increased susceptibility to rotavirus that can require hospitalization and rehydration therapy, and exhibit a seasonal epidemic pattern similar to ill children (Anderson *et al.*, 2012; Faruque *et al.*, 2004). Elderly susceptibility to rotavirus might be related to decreasing immune system activity with ageing, particularly mucosal secretory IgA production (Sakamoto *et al.*, 2009). It has been suggested that mass vaccination of children against rotavirus in the United States might provide indirect protection to older children and adults (Lopman *et al.*, 2011), and might also help reduce the threat of rotavirus-induced severe dehydration in the elderly.

18.3.3 Disease and pathogenesis

The main symptoms associated with rotavirus infection are diarrhea and vomiting, the intensity of which can vary largely from mild disease up to severe dehydration, osmotic shock and death. The clinical severity of cases is conveniently determined by the modified Vesikari scoring system, which assigns a total score from 0 to 20 based on duration of diarrhea (days), number of stools per 24-h period, vomiting duration (days), number of vomiting episodes per 24-h period, maximal fever, health care provider visits, and type of treatment (none, outpatient and hospitalization) (Freedman *et al.*, 2010; Ruuska and Vesikari, 1990).

Rotavirus infects almost exclusively mature enterocytes of the apical portion of villi in the small intestine, and neither the restricted cell population affected nor the epithelial damage is normally proportional to the extent of symptoms, both in humans and in experimentally or naturally infected animals. Common histopathological observations included blunting of villi, inflammatory infiltration, and appearance of apoptotic bodies, in the upper villi but not in the crypts (Ciarlet *et al.*, 2002; Eisengart *et al.*, 2009; Ward *et al.*, 1996). Thus, it appears that other pathophysiological mechanisms intervene to cause the life-threatening watery diarrhea of rotavirus, possibly including: (i) loss of epithelial cell absorptive functions induced by intracellular virus replication; (ii) the action of the rotaviral enterotoxin NSP4 protein; and (iii) the interactions between the enteric nervous system, serotonin and virus-induced chemical mediators, with a possible central role of enterochromaffin cells (Ball *et al.*, 1996; Kordasti *et al.*, 2004; Lundgren *et al.*, 2000). Recently, it has been proposed that serotonin also acts on the second characteristic symptom of rotavirus infection, namely vomiting (Hagbom *et al.*,

2011), which opens up the possibility of therapeutic approaches to reducing water loss. Asymptomatic rotavirus infections also occur frequently in both children and adults, and infection is thought to be milder because of partial immune protection established during the first exposure to the virus (Karsten *et al.*, 2009; Phillips *et al.*, 2010; Velazquez, 2009). However, these studies also confirm and highlight the risk that rotavirus shedding by otherwise healthy children or adults may be transmitted to susceptible non-immune subjects, such as newborns, directly or via environmental or foodstuff contamination circuits (Abad *et al.*, 1994a; Cheong *et al.*, 2009; Gallimore *et al.*, 2005; Le Guyader *et al.*, 1994).

Since group A rotaviruses infect many animal species, including domestic animals and pets, besides transmission of human strains susceptible subjects are also exposed to a large variety of strains of animal origin, through a zoonotic transmission pathway that may include feces-contaminated environmental fomites and foods (Cook *et al.*, 2004; Martella *et al.*, 2010; Midgley *et al.*, 2012; Ziemer *et al.*, 2010). Reassortment is probably crucial in favoring adaptation to humans of animal rotavirus strains transmitted zoonotically via the environment or food (Banyai *et al.*, 2010; Iturriza-Gomara *et al.*, 2001; Matthijnssens *et al.*, 2006b, 2011b). Reassortment occurs not only between animal and human strains, but also between rotaviruses originating from different animal species (Park *et al.*, 2011), which may have a major impact in generating genetic and phenotypic variability of pathogens that may eventually reach man (Matthijnssens *et al.*, 2006b, 2011b).

A growing number of reports support the suggestion that rotavirus leaves the gut mucosa and infects secondary target organs, sometimes with overt pathological manifestations ranging from encephalopathy and meningism to pancreatitis and hepatitis to arthritis (De La Rubia *et al.*, 1996; McMaster *et al.*, 2001; Medici *et al.*, 2011; Venuta, 2012). Rotavirus antigenemia and viremia following symptomatic or asymptomatic intestinal infection are in fact detected more frequently than previously thought (Blutt *et al.*, 2007; Moon *et al.*, 2012). Additional findings on rotavirus systemic infections derive from studies on animals experimentally infected with rotavirus (Blutt and Conner, 2007; Crawford *et al.*, 2006; Graham *et al.*, 2007; Ramig, 2007), and it is in fact from the mouse model that the putative association of NSP3 with the ability to support extra-intestinal rotavirus localization was established (Mossel and Ramig, 2002).

18.4 Rotavirus detection in different samples

There is no univocal approach at the laboratory identification of rotavirus, although most of diagnostic activities concern clinical samples from children presenting at the hospital with acute diarrhea. If these latter can benefit of a variety of validated methods, nonetheless environmental or food analysis for rotavirus presents higher difficulties and require different methodological protocols. Detailed typing of viral strains may also be required for epidemiological investigation of outbreaks, as well as for determining the possible zoonotic origin of strains.

18.4.1 Laboratory diagnosis in clinical and veterinary cases

The typical clinical specimen for rotavirus search during routine diagnostics is feces, where viral load can be as high as 10^{11} particle per gram. In particular cases, vomit or specific tissues can also be used. The first approach to laboratory diagnosis of rotavirus infection was electron microscopy (Bishop *et al.*, 1974); cell-culture methods have never gained the same usefulness for human rotavirus diagnosis and detection as in the case of other enteric viruses. Nonetheless, cell cultures, which have been providing information on different rotavirus serotypes since the early 1980s (Wyatt *et al.*, 1983), offered a method for strain differentiation and posed the basis for development of modern vaccines (Kapikian *et al.*, 1986; Wyatt *et al.*, 1983).

Among the several methods implemented to allow diagnostic and epidemiological studies on rotavirus, several ELISAs and latex-agglutination tests produced at commercial level have been widely used since the 1980s (Dennehy *et al.*, 1988; Lipson and Zelinsky-Papez, 1989; Miotti *et al.*, 1985) for both rapid diagnosis and patient management and for epidemiological investigations. In particular, the introduction of hybridoma cell technology for production of monoclonal antibodies also gave impulse to the exploitation of antibody-based testing for viral detection in clinical samples (Gerna *et al.*, 1989).

With the availability of fast and sensitive RT-PCR methods for virus detection, in the early 1990s specific protocols for rotavirus detection and serotype characterization of rotavirus directly from stool samples were established (Gomara *et al.*, 2000; Nakagomi *et al.*, 1991). Over the last decade, real-time RT-PCR tests in qualitative, quantitative or multiplex formats have been proposed, and some are available commercially (Higgins *et al.*, 2011; Jothikumar *et al.*, 2009; van Maarseveen *et al.*, 2010). These may be conveniently optimized for routine use in hospital or research laboratories possessing adequate laboratory facilities for molecular methods. However, even today many laboratories performing viral diagnosis of gastroenteritis more often use the antigen-detection methods mainly present on the market as ELISA, immunochromatography, dipstick, or latex-agglutination kits (Khamrin *et al.*, 2011). An interesting application of rapid test strips from rotavirus-positive reactions is storage of archival dsRNA for up to several years at room temperature, which enables proper conservation of samples, for later molecular analysis, in peripheral laboratories lacking adequate cold storage systems or even during field studies in remote areas (Shulman *et al.*, 2011).

Given the strict correlations between the gene 6 of all group A rotaviruses of either human or animal origin, virtually all antibody-based assays target cross-reactive group antigenic determinants in the viral VP6, and tests developed for human use are also suitable for veterinary virological diagnostics (Al-Yousif *et al.*, 2001; Nemoto *et al.*, 2010). This, however, is an exception, and in fact antibody-based assays established for detection and characterization of other human rotaviral proteins (e.g., VP4, VP7, NSP4) are normally inefficient in the case of animal strains.

Detection of rotavirus antigen in the stools of a patient may guide subsequent clinical management procedures as well as possible outbreak containment measures. However, in the case of both humans and animals, information on the type of infecting or circulating rotavirus is valuable for both research and epidemiological purposes, as well as for the production or evaluation of vaccines (Chang *et al.*, 2010; Hoshino *et al.*, 2003; Hoshino and Kapikian, 2000; Desselberger *et al.*, 2006).

For rotavirus strain typing, molecular protocols appear significantly more feasible and effective than serological testing by either hyperimmune sera or monoclonal antibodies in cell-culture neutralization assays or ELISA (Gentsch *et al.*, 1992; Gomara *et al.*, 2000; Gouvea *et al.*, 1990a,b; Simmonds *et al.*, 2008). Although several typing approaches have been proposed, a widely used protocol can be downloaded from the website of the Rotavirus Surveillance Network, EuroRotaNet (http://www.eurorota.net/); it uses a multiplex nested PCR system following an RT phase with random primers. The protocol for VP7 (G-genotyping) is optimized for use with a combination of eight different genotype-specific oligonucleotide primers, embracing common human rotavirus G-types G1-4, and G9, and emerging strains G8, G10, and G12 (Banerjee *et al.*, 2007; Iturriza-Gomara *et al.*, 2009, 2011; Steyer *et al.*, 2007). A second VP4-specific multiplex uses six genotyping primers for the two common human P-types P(4) and P(8), and for types P(6), and P(9–11) which may indicate possible animal origin of the strain (Banerjee *et al.*, 2007; Khamrin *et al.*, 2009; Martella *et al.*, 2006). Although VP4 and VP7 gene typing yields substantial information to verify the human or animal origin of capsid antigens, and may thus indicate circulation of strains possibly emerging from the animal world, there is now consensus about the need for a full-genome typing approach to understand the evolution of rotaviruses (Matthijnssens *et al.*, 2008, 2011a). It has been clearly demonstrated that a rotavirus strain may undergo several reassortment events before and after entering human populations (Matthijnssens *et al.*, 2006b, 2011b; Mukherjee *et al.*, 2011). Although the complete mechanisms governing the zoonotic passage of rotavirus are unclear, it is remarkable that apart from a few cases where a genuine animal strain infected a human being, uncommon rotaviruses causing human disease appear to carry most of a typical human genome with the introduction of one or two genes of animal origin (Banyai *et al.*, 2010; Steyer *et al.*, 2007; Wang *et al.*, 2010b). Besides the VP4 and VP7 genes, VP6 and NSP4 genes are also particularly useful in suggesting rotavirus origin, and phylogenetic trees based on the nucleotide sequences of genome segments 6 and 10 show that most human versus animal strains cluster separately (Benati *et al.*, 2010; Iturriza-Gomara *et al.*, 2003; Tatte *et al.*, 2010). In addition to the work of several international rotavirus surveillance networks (de Oliveira *et al.*, 2009; Iturriza-Gomara *et al.*, 2009; Mirzayeva *et al.*, 2009; Nelson *et al.*, 2009; Tate *et al.*, 2012), it is worth mentioning the establishment of the Rotavirus Classification Working Group in 2007 (Matthijnssens *et al.*, 2008, 2011a). The RCWG is an international panel of rotavirus experts seeking consensus and offering expertise on recognition of novel rotavirus types at global level, encouraging full-genome sequencing of rotavirus strains, and making use of the rotavirus web-based genotyping platform RotaC (Maes *et al.*, 2009). Organization of the growing amount of information worldwide derived from detailed genomic characterization of rotaviruses, via genotyping and sequencing, permits the increasingly accurate discrimination of rotavirus strains according to the scheme Gx-P[x]-Ix-Rx-Cx-Mx-Ax-Nx-Tx-Ex-Hx, which represent the genotypes of the VP7-VP4-VP6-VP1-VP2-VP3-NSP1-NSP2-NSP3-NSP4-NSP5/6 encoding gene segments, respectively (Matthijnssens *et al.*, 2008). It has thus become clear that although humans and each animal species tend to maintain their own common viral strains, significant changes of the genotypes, variants, or strains of rotavirus circulating in an animal or human community occur with time, and spillover or importation of novel strains from different species or communities will increasingly favor inter-type reassortment and the continuous emergence of novel rotaviruses, with a genome adjusted to prevail among pre-existing viruses (Ghosh *et al.*, 2011; Matthijnssens *et al.*, 2011b; McDonald *et al.*, 2009; Park *et al.*, 2011).

18.4.2 Rotavirus detection in food and water

While antigen- or genome-based diagnosis of rotavirus infection in humans or animals is relatively easy because of the protracted shedding and high concentration of virus in feces (Richardson *et al.*, 1998), rotavirus detection in environmental samples, such as water, or food matrices is more complex. Virus is normally present extensively diluted in surface waters or irrigation waters, and the viral load in water-contaminated vegetables may also consequently be very low although exceptions may occur (Bidawid *et al.*, 2000; Lodder and de Roda Husman, 2005). Adequate concentration and sensitive molecular approaches are essential for a positive laboratory testing, so antigen-based methods are unfeasible. Recommended molecular detection approaches range from conventional and nested RT-PCR to sensitive real-time methods (Jothikumar *et al.*, 2009; Kittigul *et al.*, 2008; Rutjes *et al.*, 2009; Sdiri-Loulizi *et al.*, 2010). It must however be kept in mind that due to the segmented nature of rotavirus genome, the complex of distinct gene types or sequences identified in a particular sample may not necessarily belong to the same rotavirus strain, since each of the 11 segments might belong to any one of the different viral strains possibly simultaneously present in the sample (Di Bartolo *et al.*,

2011; Ferreira *et al.*, 2009). As mixed infections of a human or animal individual with different strains can occur, this problem should not be ignored for rotavirus detection and typing also in feces, but it is definitely more significant in matrices containing fecal micro-organisms originating from an entire human or animal community.

Furthermore, different environmental conditions may strongly influence the integrity and infectivity of rotavirus in watery matrices, including temperature, pH and salinity, presence of biological or chemical contaminants, influencing viability or provoking aggregation and flocculation (Butot *et al.*, 2007; Hurst and Gerba, 1980; Pancorbo *et al.*, 1987). Overall, rotavirus is considered to be as resistant in fresh and salty waters as are enteroviruses (Hurst and Gerba, 1980). Infectivity assays on rotaviruses naturally contaminating waters are normally impracticable since cell culturing applied to wild human strains is not sufficiently sensitive, and detection of viral genomes may be the only way to identify infectious risks for the population (Gassilloud *et al.*, 2003).

The presence of rotavirus in environmental water has been demonstrated since the early 1990s, when molecular detection methods became available (Abad *et al.*, 1998; Dubois *et al.*, 1997). But even before proper environmental virological protocols were in use, evidence for rotavirus as the agent of a community waterborne gastroenteritis outbreak was clear (Hopkins *et al.*, 1984).

Of course, a major source of environmental rotaviruses is urban sewage, where strains of human interest are discharged, which may eventually pass through waste-water treatment plants with only partial viral load reduction, proceeding into receiving water channels, rivers and eventually the sea (Arraj *et al.*, 2008; Rodriguez-Diaz *et al.*, 2009; Tsai *et al.*, 1994). Despite the limitations, environmental rotavirus surveillance can provide useful epidemiological information both on the main G and P rotavirus types shed by a specific population over time, and on novel viral types not yet addressed by clinicians (Villena *et al.*, 2003).

Rotavirus detection in food is not presently required by national and international legislation. Sensitive reliable methods are nonetheless useful for investigating outbreak viral sources, and further efforts should be made to harmonize protocols in view of the possible reassessment of rotavirus as an important foodborne and zoonotic agent (Cardoen *et al.*, 2009).

18.5 Epidemic outbreaks

In countries characterized by a temperate climate, the child population suffers winter-related major epidemics similar to those of seasonal influenza, whereas seasonality is less marked or absent in tropical areas (Cook *et al.*, 1990; Kapikian *et al.*, 1976). Rotavirus can also generate smaller epidemic outbreaks, which can affect all agegroups, although they are more prevalent among children or elderly people, particularly within schools, hospitals, nursing homes, and care centers.

It has been estimated that this agent was responsible for 5% of all diarrhea outbreaks in residential geriatric institutions in England and Wales in 1992–1994 (Ryan *et al.*, 1997). This incidence of rotavirus among elderly adults was mirrored in Illinois during a survey of outbreaks within retirement communities (Lopman *et al.*, 2011), and several other reports on intestinal disease outbreaks in nursing homes in different areas of the world (Luque Fernandez *et al.*, 2008; Marshall *et al.*, 2003; Trop Skaza *et al.*, 2011; van Duynhoven *et al.*, 2005). Deaths potentially caused by reduced state of health, and the direct costs of outbreaks in long-term care institutions, represent a high price for society (Piednoir *et al.*, 2010). Investigation of an outbreak occurring among long-term residents in an institution for rehabilitation in Japan has revealed that rotavirus-affected subjects had a significantly longer duration of stay in the institute than unaffected residents, suggesting that long-term residence in a closed community, possibly associated with absence of immuno-stimulation, is a risk factor for rotavirus illness (Iijima *et al.*, 2006).

In addition to institutional outbreaks, household outbreaks are also reported, where contact with rotavirus-infected children risks transmission of the virus and development of disease to grandparents (Awachat and Kelkar, 2006).

With waterborne transmission, subjects of all ages can develop severe symptoms (Timenetsky *et al.*, 1996). This is possibly related to a higher virus load (often the case in sewage-contaminated water sources) which is suspected to have caused a number of severe cases in both young and older children in an outbreak in Finland related to drinking water (Rasanen *et al.*, 2010). Two-thirds of the patients involved were infected with rotavirus, which belonged to genotype G1 in all cases except a single G4 virus finding. Several other pathogens were also detected (mostly norovirus), but no difference was observed in severity of disease between patients infected with rotavirus and/or other viruses. This unusual finding might be due to either a synergic effect of multiple pathogens or to an undetected rotavirus co-infection.

Due to the absence of a systematic reporting system for rotavirus outbreaks and sporadic cases other than for children, it is not clear whether rotavirus serotypes less common than the widespread G1P[8], such as G2P[4] or G4P[8], might be more frequently involved in outbreaks or cases in adults. Association of the G2 and G4 rotavirus genotype with gastroenteritis in adult elderly communities or older children has been reported (Feeney *et al.*, 2006; Timenetsky *et al.*, 1996). Nonetheless, differential immunity to viral serotypes remains an issue, and other factors may underlie these observations.

In general, it can be assumed that in countries with high standards of hygiene, the infectious dose of rotaviruses transmitted from person-to-person contact is probably small. It is thus tempting to believe that much larger amounts of virus from contaminated drinking water would explain the unusually severe clinical picture and occurrence of disease in subjects outside the target age noted during some outbreaks (Rasanen *et al.*, 2010). The presence of other enteric pathogens along with rotavirus in the course of common-source outbreaks, particularly waterborne episodes, is frequent and, together with the challenging virus detection practice, makes it complicated to assign the proper etiological role to each virus detected in either patient stools or in the environment. Surface waters and also some foodstuffs can be extensively contaminated with a multiplicity of viruses of human and/or animal origin, capable of starting epidemics (Di Bartolo *et al.*, 2011; Gratacap-Cavallier *et al.*, 2000; Le Guyader *et al.*, 2008; Sdiri-Loulizi *et al.*, 2010). For instance, rotavirus-specific VP7 gene sequences were detected in four drinking water samples taken from homes of 56 children with rotavirus diarrhea in France, but in no case did these sequences match with the virus detected in stools (Gratacap-Cavallier *et al.*, 2000). Three water samples contained swine or bovine VP7 sequences and only one a human serotype 4 virus. Although sequences were too short for more detailed phylogeny, nonetheless it is worth noting that pipelines may be a vehicle of rotaviruses of both human and animal origin, leading to simultaneous infection of several subjects with more rotavirus strains and eventually to gene reassortment. Similarly, rotavirus – as well as other viruses – may be carried, to a remarkable extent, by cockles, mussels and clams, which are only lightly cooked or consumed raw (Le Guyader *et al.*, 2000). Although virus concentration may be lower in vegetables and soft fruit (Butot *et al.*, 2007; Plante *et al.*, 2011), this type of food also carries a potential risk and has been sporadically involved in outbreaks (CDC, 2000; Gallimore *et al.*, 2005; Mayr *et al.*, 2009). A quantitative microbial risk assessment model has been proposed for the health risks correlated with waste-water irrigation, still widely used in some parts of the world (Mara *et al.*, 2007).

Particular conditions for exceptional virus spread and major outbreaks may be generated by natural disasters such as flood and earthquakes (Karmakar *et al.*, 2008; Kukkula *et al.*, 1997; Schwartz *et al.*, 2006).

18.6 Zoonotic transmission

Within the ongoing process of globalization, many infectious diseases are emerging or re-emerging, and novel virus variants are daily considered as new threats for naïve populations. A major component of the current risks for public health is posed by animals and food of animal origin, that calls for a one-health one-medicine approach toward an increasing range of viruses, including rotaviruses.

18.6.1 Emerging rotavirus strains in humans

During the past decade or so it has become very clear that inter-species transmission of rotaviruses, once considered to cause overt disease only in the homologous host species, occurs frequently, and that it often results in symptomatic gastroenteritis (Cook *et al.*, 2004; Martella *et al.*, 2010). However, defining the evolutionary history of a viral strain is complicated. For example, it is still questionable whether the origin of G9 rotavirus, one of the five commonest human rotaviruses today, is human or animal. Rotavirus G9 had been first described as a rare cause of infantile gastroenteritis in the US in 1983–1984 (Clark *et al.*, 1987). It was later found frequently in India but only in asymptomatic infections in neonates (Das *et al.*, 1994) and sporadically in other countries, becoming significantly more widespread among symptomatic children in the US in the course of the following decade (Ramachandran *et al.*, 1998). G9 rotavirus had also been found in pigs in Iowa (Paul *et al.*, 1988) and in later in lambs in Scotland (Fitzgerald *et al.*, 1995). Together with G9, the P[6] gene normally found in swine was also circulating, and at the time of G9 emergence throughout the world G9P[6] strains were frequently observed (Iturriza-Gómara *et al.*, 2000). G9 rotaviruses were reported in the early 1990s among piglets with diarrhea in Brazil, where both G9P[6] and G9P[8] as well other typically human G-types were recovered from animals, suggesting natural human–porcine genetic reassortment (Santos *et al.*, 1999). These data and others make it intriguing to speculate whether pigs or humans or both have been a significant mixing vessel for human G9P[8], which has subsequently become a globally widespread virus type (Iturriza-Gómara *et al.*, 2000; Matthijnssens *et al.*, 2010; Santos *et al.*, 1999).

Newly available fast sequencing tools and powerful programs for phylogenetic analysis are generating more knowledge of rotavirus evolution, making it possible to identify and characterize phylogenetically rare and unconventional rotavirus strains with the aim of promptly identifying strains with the potential for rapid global spread, and increasing herd immunity by extending vaccination (Glass *et al.*, 2006). An appropriate example of rotavirus evolution tracking using the full-genome sequencing approach pertains to G8P[8] and G8P[6] strains identified in children with diarrhea in the Democratic Republic of Congo in 2003 (Matthijnssens *et al.*, 2006b). These strains shared ten almost identical genome segments, suggesting a recent reassortment between a common G8P[6] strain of animal origin and a human strain with a P[8] specificity. However, further analysis showed a very close evolutionary relationship between nine of their genes with rotavirus strains belonging to the DS-1-like (G2P[4]) sub-group, indicating at least three, and possibly four, consecutive reassortment events involving both DS-1-like and Wa-like human rotaviruses and more animal strains of bovine (G8) and swine origin. The virus ultimately resulting from these processes appeared to be well equipped to become a predominant strain within the human population. This is not the only route to humanization followed by G8 originating from bovines, as other emerging G8P[8]strains investigated in Europe appear to exhibit an almost complete Wa-like genome background (Steyer *et al.*, 2007; Tcheremenskaia *et al.*, 2007). Interestingly, one of these strains became a predominant cause of infantile gastroenteritis hospital admissions in Croatia in 2006, although it caused only a minority of cases the year before. It is tempting to believe that these strains (Matthijnssens *et al.*, 2006b; Tcheremenskaia *et al.*, 2007) may be obviously favored by their unusual gene repertoire and may displace other predominant human rotaviruses in at least some situations, with implications for future vaccine strategies (Glass *et al.*, 2006; Jiang *et al.*, 2010).

18.6.2 Animals as rotavirus reservoirs

There is increasingly compelling evidence that some animal species may indeed act as a reservoir of rotavirus strains transmitted zoonotically and emerging in human pathology. Several investigations on a global collection of human G6P[14] and animal G6P[14] or G8P[14] showed close genetic relatedness between strains of different origin, and suggest that ruminants and ungulates may be the reservoir of G6 rotaviruses for humans (Banyai *et al.*, 2009; De Grazia *et al.*, 2011; Matthijnssens *et al.*, 2009). The bovine origin of some G12 and G9 human isolates associated with high mortality in children in Africa has been clarified by sequence-independent amplification and pyrosequencing, and RotaC full-genome analysis showing peculiar evolutionary perspectives for the pool of rotavirus circulating in that part of the world (Jere *et al.*, 2011).

Besides possible G9 strains, the pig is also regarded as a likely reservoir for rotavirus strains with G3, G5, G12 and P[6], P[16], P[19] genotype (Esona *et al.*, 2004; Ghosh *et al.*, 2006; Li *et al.*, 2008; Maneekarn *et al.*, 2006).

G12 rotaviruses have been reported in humans by several laboratories across the world and are thought to be one of the possible emerging strains of zoonotic origin (Castello *et al.*, 2009; Rahman *et al.*, 2007; Sharma *et al.*, 2008; Steyer *et al.*, 2007). These viruses may have originated from swine and become established in humans throughout a series of reassortments (Ghosh *et al.*, 2006; Griffin *et al.*, 2002; Steyer *et al.*, 2007). However, enhanced phylogenetic studies support alternative evolution pathways either leading a human G12 to eventually enter and adapt to the swine population or encompassing a common ancestor for both porcine G12 and Wa-like human rotaviruses (Freeman *et al.*, 2009; Ghosh *et al.*, 2010). Some of the rotavirus zoonotic genotypes have been known in human diseases for a long time, and the division of animal and human strains into different lineages with the same G or P type might indicate that distinct evolution routes started many years before in different hosts (da Silva *et al.*, 2011).

It has also been proposed that rabbits harbor rotaviruses with similar characteristics to strains found in gastroenteritis cases in children, and a human case in Belgium was found to be infected with a G3P[14] rotavirus sharing several genes with a typical lapine rotavirus, including a closely related NSP4 (Matthijnssens *et al.*, 2006a) which might contribute to virulence in both species.

Rotaviruses are not regarded as major enteric pathogens in dogs and cats, although pet animals living at close quarters with humans can, uniquely, easily enter pathogen transmission pathways from or to humans. Some rotavirus genotypes known in these animals are very similar to strains from human disease, particularly G3, and the genomic repertoire of strains can range from a fully canine or feline to a reassorted human–animal constellation (Grant *et al.*, 2011; Khamrin *et al.*, 2006; Matthijnssens *et al.*, 2011b; Tsugawa and Hoshino, 2008).

Furthermore, more complex reassortments involving several different animal species can contribute to the creation of novel zoonotic strains involving different animal reservoirs (Park *et al.*, 2011; Wang *et al.*, 2010b), which suggests that more than one animal species or humans can normally act as a mixing vessel for rotaviruses of diverse origin, as long as different species are in frequent direct contact or share an environment or vehicles of fecal contamination.

Unfortunately, although large collections of fully characterized human rotavirus strains covering several decades are available, repertoires of viral strains of animal origin are limited and most of the available information is recent. This will make it quite difficult to trace back the origin of most strains circulating currently in either man or animal species. However, molecular surveillance systems now in place that involve both humans and animals provide a unique opportunity for the timely fingerprinting of novel rotavirus strains and variants on their emergence.

18.6.3 Farm-to-fork chain

Although person-to-person transmission via human feces plays the major role in rotavirus disease, foodborne transmission of this virus can occur, and deserves further attention and characterization. For this reason, rotavirus was included in recent expert advice on foodborne viruses for Codex Alimentarius, more specifically targeting water used for food preparation and processing, including dried or powdered milk for children, infant formula, or juice (www.who.int/foodsafety/publications/micro/mra13/en/index.html). However, a possible bias is that rotavirus is regarded as a disease of neonates and young children, which limits the range of food types representing a possible risk of transmission. In fact, the circulation of virus is likely to be widespread across the whole age range, resulting in asymptomatic or subclinical infection among adults. Foodborne transmission of pathogens is definitely more significant among adults (Newell *et al.*, 2010),who may pass rotavirus to susceptible children via interpersonal transmission or by contaminating food and environmental fomites at work or at home, in which rotavirus is stable and resistant to disinfection (Abad *et al.*, 1994b).

Fresh produce that is consumed raw or minimally processed, such as fruits, vegetables, and seafood, provide an ideal route for the transmission of certain enteric pathogenic bacteria and viruses, including rotavirus (Newell *et al.*, 2010). Resistance of rotavirus in food as well as in drinking or surface waters or on surfaces is high (Butot *et al.*, 2008; Pancorbo *et al.*, 1987), and the role of fecally contaminated irrigation water or organic fertilizers in pre-harvest contamination of fruits and vegetables with enteric viruses has been illustrated (Cheong *et al.*, 2009; Costantini *et al.*, 2007; Gabrieli *et al.*, 2007). Similarly, polluted growing/harvesting water is a potential source of viral contamination of shellfish (Lees, 2000). Rotavirus can be present in filtrating marine mollusks to a remarkable degree, as shown recently by Wang and colleagues, who reported rotavirus in approximately 9% of field oysters in China, compared with HAV in 5% and norovirus in 1.5% (Wang *et al.*, 2008). As many as 18% of natural bank mussels were positive at rotavirus testing in Italy (Gabrieli *et al.*, 2007), and several other field studies conducted elsewhere confirm natural contamination of seafood with rotavirus (Kittigul *et al.*, 2008; Le Guyader *et al.*, 2000; Rigotto *et al.*, 2010). Although primary contamination is the main cause of high viral load in food, as for other viruses (Baert *et al.*, 2009b; Bidawid *et al.*, 2000), post-harvest contamination of food by asymptomatic adult food handlers with rotavirus cannot be overlooked.

Conventional depuration practice may not be fully effective in removing rotavirus from contaminated oysters or other seafood (Bosch *et al.*, 1995; Loisy *et al.*, 2005). Furthermore, rotavirus remains infectious for periods exceeding the shelf lives of most vegetables and fruits, and can resist freezing, disinfection or washing procedures (Baert *et al.*, 2009a; Lees, 2000; Li *et al.*, 2009).

Since control of environmental pollution can be a key factor in preventing contamination of food chains with rotavirus and other viral pathogens, the recent application of quantitative risk assessment models to water, waste-water and manure may be of great help in food safety (Hamilton *et al.*, 2006; Seidu *et al.*, 2008).

18.6.4 Professional risks

Information on rotavirus disease related to professional activities is scarce, probably due to the absence or paucity of symptoms associated with rotavirus infection normally in adulthood (Karsten *et al.*, 2009; Phillips *et al.*, 2010). The risks of disease resulting from professional exposure to the virus are not as relevant as for other viral infections, particularly because immunity to clinical disease is elevated among people of working age. However, asymptomatic re-infections may occur when workers are in close contact with rotavirus, for example, staff within pediatric wards, nursing homes or hospitals (Iijima *et al.*, 2006; Trop Skaza *et al.*, 2011). Hand carriage of virus by uninfected individuals is considered critical (Gleizes *et al.*, 2006). In a recent review paper involving several European countries, rotavirus was detected in the hands of 76–78% of health workers caring for hospitalized children with community-acquired rotavirus gastroenteritis (Gleizes *et al.*, 2006). Virus was also recovered from 20% of health care workers not having direct contact with pediatric patients, suggesting widespread person-to-person circulation and/or environmental spread of rotavirus within the nosocomial workplace (Cone *et al.*, 1988). Rotavirus can survive for a few days on imperfectly cleaned hands and even longer on dry and nonporous surfaces (e.g., toys, medical tools), particularly if the environmental humidity is low (Abad *et al.*, 1994a;

Ansari *et al.*, 1988; Keswick *et al.*, 1983; Wilde *et al.*, 1992). Measures to reduce virus carry-over by personnel include strict hand hygiene and the use of virus-effective disinfectants; cleaning and disinfection of equipment and surfaces; regular room airing; proper disposal of incontinence pads; limiting temporary contacts with infected patients; removal from work of infected staff; and vaccination (Kribs-Zaleta *et al.*, 2011; Trop Skaza *et al.*, 2011). It should be added that hospital databases severely under-report nosocomial rotavirus cases, at least in Europe, so there is limited awareness of the importance of this disease at country level (Gleizes *et al.*, 2006). If the increasing number of norovirus outbreaks in hospitals and residential homes (Greig and Lee, 2012) are also taken into account, it is important to increase awareness of epidemic transmission of gastroenteritis viruses and consequently promptly to develop and adopt control protocols.

No reports have specifically addressed foodborne transmission of rotavirus to food handlers, although it is quite obvious that close manual contact with superficially contaminated food items could pose a risk of infection or cross-contamination, as in the case of other enteric pathogens (Dreyfuss, 2009). In view of this, rotavirus has been considered among the priority foodborne zoonotic diseases of the near future (Cardoen *et al.*, 2009).

One category of professionals who might play an important role in rotavirus epidemiology is workers in the farming industry, due to possible exposure to animal rotaviruses, with an increased risk of mixed infections leading to gene reassortment and introduction of novel strains into the human population (Cook *et al.*, 2004; Martella *et al.*, 2010). In a study of cattle ranchers and their families conducted in Panama in the 1980s, zoonotic transmission of bovine rotavirus was excluded (Ryder *et al.*, 1986). This contrasts however with much recent indirect data on a major zoonotic rotavirus flow from farm animals to humans, which is likely to involve farm workers globally (Cook *et al.*, 2004; Martella *et al.*, 2010). Arable farmers may also be at risk of infection with rotavirus present in unsterilized manure of human or animal origin (Costantini *et al.*, 2007; Pesaro *et al.*, 1995; Ziemer *et al.*, 2010).

Overall, it can be assumed that although professional hazards are present, they would rarely result in a direct health problem for the worker.

18.7 Future trends

Many aspects of rotavirus biology have become familiar to virologists and epidemiologists over the last four decades, and its role as the primary cause of severe and life-threatening gastroenteritis in childhood worldwide is universally recognized. Consequently, a great deal of research has been carried out into vaccine prophylaxis, and efficacious vaccines are now available for human use.

More recently, the existence of main zoonotic streams of rotavirus from the most important farmed animal species to humans has been recognized.

Consequently there have been some changes in concepts of rotavirus epidemiology and evolution, with particular reference to inter-species gene reassortment.

The molecular mechanisms by which an animal rotavirus can adapt to and spread among humans need to be fully addressed, in a similar way to the current investigation into the zoonotic passage and continuing virulence of highly pathogenic avian influenza viruses. New animal studies on the genetics of non-structural proteins possibly involved in pathogenesis and host restriction are needed.

Globalization has increased the chances of international virus circulation via human carriage and animal and food importation. Consequently, Western countries may face an escalation in the introduction of novel rotavirus strains, with a higher possibility of exotic genotypes emerging that have a modified virulence for humans and/or animals. A global program of coordination and sharing of new epidemiological and molecular information generated by national and international surveillance networks should continue to be implemented, since no single laboratory can successfully address the evolutionary trends of rotaviruses. These data are needed for future vaccine development, in preparation for the possible emergence of non-conventional rotaviruses that may not be fully covered by existing vaccinations.

More work is needed to determine the role of adults, particularly asymptomatic subjects, in rotavirus transmission to susceptible children and as a possible virus reservoir between consecutive annual epidemic peaks of infantile diarrhea. Since adults may be exposed to both environmental and food sources of rotavirus more easily than children, their possible role as a mixing vessel, where multiple co-infecting strains may generate novel rotavirus reassortants of pediatric importance, should be properly addressed in future studies.

Finally, there is a need for more research into the impact of rotavirus on the food industry and food safety. It is necessary to know what the real association of infectious rotavirus and the origin of contamination is with the different products along the food chain, particularly in respect of seawater molluscs, vegetables and soft fruits. Food- and waterborne transmission of rotavirus to food handlers should be investigated, independently of the appearance of typical symptoms in adult workers, and the extent of possible virus shedding in stools of these healthy subjects should be determined to inform risk assessment models for guiding possible control measures.

18.8 References

ABAD F X, PINTÓ R M, and BOSCH A (1994a), Survival of enteric viruses on environmental fomites, *Appl Environ Microbiol*, **60**, 3704–10.

ABAD F X, PINTÓ R M, and BOSCH A (1998), Flow cytometry detection of infectious rotaviruses in environmental and clinical samples, *Appl Environ Microbiol*, **64**, 2392–6.

ABAD F X, PINTÓ R M, DIEZ J M, and BOSCH A (1994b), Disinfection of human enteric viruses in water by copper and silver in combination with low levels of chlorine, *Appl Environ Microbiol*, **60**, 2377–83.

ALTENBURG B C, GRAHAM D Y, and ESTES M K (1980), Ultrastructural study of rotavirus replication in cultured cells, *J Gen Virol*, **46**, 75–85.

AL-YOUSIF Y, ANDERSON J, CHARD-BERGSTROM C, BUSTAMANTE A, MUENZENBERGER M, AUSTIN K, and KAPIL S (2001), Evaluation of a latex agglutination kit (Virogen Rotatest) for detection of bovine rotavirus in fecal samples, *Clin Diagn Lab Immunol*, **8**, 496–8.

ANDERSON E J, KATZ B Z, POLIN J A, REDDY S, WEINROBE M H, and NOSKIN G A (2012), Rotavirus in adults requiring hospitalization, *J Infect*, **64**, 89–95.

ANSARI S A, SATTAR S A, SPRINGTHORPE V S, WELLS G A, and TOSTOWARYK W (1988), Rotavirus survival on human hands and transfer of infectious virus to animate and nonporous inanimate surfaces, *J Clin Microbiol*, **26**, 1513–18.

ARRAJ A, BOHATIER J, AUMERAN C, BAILLY J L, LAVERAN H, and TRAORE O (2008), An epidemiological study of enteric viruses in sewage with molecular characterization by RT-PCR and sequence analysis, *J Water Health*, **6**, 351–358.

ATCHISON C, LOPMAN B, and EDMUNDS W J (2010), Modelling the seasonality of rotavirus disease and the impact of vaccination in England and Wales, *Vaccine*, **28**, 3118–26.

AWACHAT P S and KELKAR S D (2006), Dual infection due to simian G3 – human reassortant and human G9 strains of rotavirus in a child and subsequent spread of serotype G9, leading to diarrhea among grandparents, *J Med Virol*, **78**, 134–8.

BAERT L, DEBEVERE J, and UYTTENDAELE M (2009a), The efficacy of preservation methods to inactivate foodborne viruses, *Int J Food Microbiol*, **131**, 83–94.

BAERT L, UYTTENDAELE M, STALS A, VAN COILLIE E, DIERICK K, DEBEVERE J, and BOTTELDOORN N (2009b), Reported foodborne outbreaks due to noroviruses in Belgium: the link between food and patient investigations in an international context, *Epidemiol Infect*, **137**, 316–25.

BAKER M and PRASAD B V (2010), Rotavirus cell entry, *Curr Top Microbiol Immunol*, **343**, 121–48.

BALL J M, TIAN P, ZENG C Q, MORRIS A P, and ESTES M K (1996), Age-dependent diarrhea induced by a rotaviral nonstructural glycoprotein, *Science*, **272**, 101–4.

BANERJEE I, GLADSTONE B P, LE FEVRE A M, RAMANI S, ITURRIZA-GOMARA M, GRAY J J, BROWN D W, ESTES M K, MULIYIL J P, JAFFAR S, and KANG G (2007), Neonatal infection with G10P[11] rotavirus did not confer protection against subsequent rotavirus infection in a community cohort in Vellore, South India, *J Infect Dis*, **195**, 625–32.

BANYAI K, MARTELLA V, MOLNAR P, MIHALY I, VAN RANST M, and MATTHIJNSSENS J (2009), Genetic heterogeneity in human G6P[14] rotavirus strains detected in Hungary suggests independent zoonotic origin, *J Infect*, **59**, 213–15.

BANYAI K, PAPP H, DANDAR E, MOLNAR P, MIHALY I, VAN RANST M, MARTELLA V, and MATTHIJNSSENS J (2010), Whole genome sequencing and phylogenetic analysis of a zoonotic human G8P[14] rotavirus strain, *Infect Genet Evol*, **10**, 1140–4.

BARRANDEGUY M, PARRENO V, LAGOS MARMOL M, PONT LEZICA F, RIVAS C, VALLE C, and FERNANDEZ F (1998), Prevention of rotavirus diarrhoea in foals by parenteral vaccination of the mares: field trial, *Dev Biol Stand*, **92**, 253–7.

BARRO M and PATTON J T (2005), Rotavirus nonstructural protein 1 subverts innate immune response by inducing degradation of IFN regulatory factor 3, *Proc Natl Acad Sci U S A*, **102**, 4114–19.

BELL L M, CLARK H F, OFFIT P A, SLIGHT P H, ARBETER A M, and PLOTKIN S A (1988), Rotavirus serotype-specific neutralizing activity in human milk, *Am J Dis Child*, **142**, 275–8.

BENATI F J, MARANHAO A G, LIMA R S, DA SILVA R C, and SANTOS N (2010), Multiple-gene characterization of rotavirus strains: evidence of genetic linkage among the VP7-, VP4-, VP6-, and NSP4-encoding genes, *J Med Virol*, **82**, 1797–802.

BERNSTEIN D I, SANDER D S, SMITH V E, SCHIFF G M, and WARD R L (1991), Protection from rotavirus reinfection: 2-year prospective study, *J Infect Dis*, **164**, 277–83.

BERNSTEIN D I, and WARD R L (2006), Rotarix: development of a live attenuated monovalent human rotavirus vaccine, *Pediatr Ann*, **35**, 38–43.

BIDAWID S, FARBER J M, and SATTAR S A (2000), Contamination of foods by food handlers: experiments on hepatitis A virus transfer to food and its interruption, *Appl Environ Microbiol*, **66**, 2759–63.

BISHOP R (2009), Discovery of rotavirus: Implications for child health, *J Gastroenterol Hepatol*, **24**(Suppl 3), S81–S85.

BISHOP R F, BARNES G L, CIPRIANI E, and LUND J S (1983), Clinical immunity after neonatal rotavirus infection. A prospective longitudinal study in young children, *N Engl J Med*, **309**, 72–6.

BISHOP R F, DAVIDSON G P, HOLMES I H, and RUCK B J (1973), Virus particles in epithelial cells of duodenal mucosa from children with acute non-bacterial gastroenteritis, *Lancet*, **2**, 1281–3.

BISHOP R F, DAVIDSON G P, HOLMES I H, and RUCK B J (1974), Detection of a new virus by electron microscopy of faecal extracts from children with acute gastroenteritis, *Lancet*, **1**, 149–51.

BLACK R E (1993), Epidemiology of diarrhoeal disease: implications for control by vaccines, *Vaccine*, **11**, 100–6.

BLUTT S E and CONNER M E (2007), Rotavirus: to the gut and beyond!, *Curr Opin Gastroenterol*, **23**, 39–43.

BLUTT S E, MATSON D O, CRAWFORD S E, STAAT M A, AZIMI P, BENNETT B L, PIEDRA P A, and CONNER M E (2007), Rotavirus antigenemia in children is associated with viremia, *PLoS Med*, **4**, e121.

BOSCH A, PINTÓ R M, and ABAD A (1995), Differential accumulation and depuration of human enteric viruses by mussels, *Water Sci Technol*, **31**, 447–51.

BRESEE J, FANG Z Y, WANG B, NELSON E A, TAM J, SOENARTO Y, WILOPO S A, KILGORE P, KIM J S, KANG J O, LAN W S, GAIK C L, MOE K, CHEN K T, JIRAPHONGSA C, PONGUSWANNA Y, NGUYEN V M, PHAN V T, LE T L, HUMMELMAN E, GENTSCH J R, and GLASS R (2004), First report from the Asian Rotavirus Surveillance Network, *Emerg Infect Dis*, **10**, 988–95.

BRIDGER J C, PEDLEY S, and MCCRAE M A (1986), Group C rotaviruses in humans, *J Clin Microbiol*, **23**, 760–3.

BURNS J W, SIADAT-PAJOUH M, KRISHNANEY A A, and GREENBERG H B (1996), Protective effect of rotavirus VP6-specific IgA monoclonal antibodies that lack neutralizing activity, *Science*, **272**, 104–7.

BUTLER D (2011), Vaccine campaign to target deadly childhood diarrhoea, *Nature*, **477**, 519.

BUTOT S, PUTALLAZ T, and SANCHEZ G (2007), Procedure for rapid concentration and detection of enteric viruses from berries and vegetables, *Appl Environ Microbiol*, **73**, 186–92.

BUTOT S, PUTALLAZ T, and SANCHEZ G (2008), Effects of sanitation, freezing and frozen storage on enteric viruses in berries and herbs, *Int J Food Microbiol*, **126**, 30–5.

CAMPAGNA M, EICHWALD C, VASCOTTO F, and BURRONE O R (2005), RNA interference of rotavirus segment 11 mRNA reveals the essential role of NSP5 in the virus replicative cycle, *J Gen Virol*, **86**, 1481–7.

CARDOEN S, VAN HUFFEL X, BERKVENS D, QUOILIN S, DUCOFFRE G, SAEGERMAN C, SPEYBROECK N, IMBERECHTS H, HERMAN L, DUCATELLE R, and DIERICK K (2009), Evidence-based semiquantitative methodology for prioritization of foodborne zoonoses, *Foodborne Pathog Dis*, **6**, 1083–96.

CASHMAN O, LENNON G, SLEATOR R D, POWER E, FANNING S, and O'SHEA H (2010), Changing profile of the bovine rotavirus G6 population in the south of Ireland from 2002 to 2009, *Vet Microbiol*, **146**, 238–44.

CASTELLO A A, NAKAGOMI T, NAKAGOMI O, JIANG B, KANG J O, GLASS R I, GLIKMANN G, and GENTSCH J R (2009), Characterization of genotype P[9]G12 rotavirus strains from Argentina: high similarity with Japanese and Korean G12 strains, *J Med Virol*, **81**, 371–81.

CDC (2000), Foodborne outbreak of Group A rotavirus gastroenteritis among college students – District of Columbia, March-April 2000, *MMWR Morb Mortal Wkly Rep*, **49**, 1931–3.

Centers for Disease Control and Prevention (2011), Progress in the introduction of rotavirus vaccine – latin america and the Caribbean, 2006–2010, *MMWR Morb Mortal Wkly Rep*, **60**, 1611–14.

CHANG J T, LI X, LIU H J, and YU L (2010), Ovine rotavirus strain LLR-85-based bovine rotavirus candidate vaccines: construction, characterization and immunogenicity evaluation, *Vet Microbiol*, **146**, 35–43.

CHARPILIENNE A, LEPAULT J, REY F, and COHEN J (2002), Identification of rotavirus VP6 residues located at the interface with VP2 that are essential for capsid assembly and transcriptase activity, *J Virol*, **76**, 7822–31.

CHEN C M, HUNG T, BRIDGER J C, and MCCRAE M A (1985), Chinese adult rotavirus is a group B rotavirus, *Lancet*, **2**, 1123–4.

CHEONG S, LEE C, SONG S W, CHOI W C, LEE C H, and KIM S J (2009), Enteric viruses in raw vegetables and groundwater used for irrigation in South Korea, *Appl Environ Microbiol*, **75**, 7745–51.

CHWETZOFF S, and TRUGNAN G (2006), Rotavirus assembly: an alternative model that utilizes an atypical trafficking pathway, *Curr Top Microbiol Immunol*, **309**, 245–61.

CIARLET M, CONNER M E, FINEGOLD M J, and ESTES M K (2002), Group A rotavirus infection and age-dependent diarrheal disease in rats: a new animal model to study the pathophysiology of rotavirus infection, *J Virol*, **76**, 41–57.

CIARLET M, and ESTES M K (1999), Human and most animal rotavirus strains do not require the presence of sialic acid on the cell surface for efficient infectivity, *J Gen Virol*, **80**, 943–8.

CIARLET M, GILGER M A, BARONE C, MCARTHUR M, ESTES M K, and CONNER M E (1998), Rotavirus disease, but not infection and development of intestinal histopathological lesions, is age restricted in rabbits, *Virology*, **251**, 343–60.

CLARK H F, HOSHINO Y, BELL L M, GROFF J, HESS G, BACHMAN P, and OFFIT P A (1987), Rotavirus isolate WI61 representing a presumptive new human serotype, *J Clin Microbiol*, **25**, 1757–62.

CLARK H F, OFFIT P A, ELLIS R W, EIDEN J J, KRAH D, SHAW A R, PICHICHERO M, TREANOR J J, BORIAN F E, BELL L M, and PLOTKIN S A (1996), The development of multivalent bovine rotavirus (strain WC3) reassortant vaccine for infants, *J Infect Dis*, **174**(Suppl 1), S73–80.

CONE R, MOHAN K, THOULESS M, and COREY L (1988), Nosocomial transmission of rotavirus infection, *Pediatr Infect Dis J*, **7**, 103–9.

COOK N, BRIDGER J, KENDALL K, GOMARA M I, EL-ATTAR L, and GRAY J (2004), The zoonotic potential of rotavirus, *J Infect*, **48**, 289–302.

COOK S M, GLASS R I, LEBARON C W, and HO M S (1990), Global seasonality of rotavirus infections, *Bull World Health Organ*, **68**, 171–7.

COSTANTINI V P, AZEVEDO A C, LI X, WILLIAMS M C, MICHEL F C, JR., and SAIF L J (2007), Effects of different animal waste treatment technologies on detection and viability of porcine enteric viruses, *Appl Environ Microbiol*, **73**, 5284–91.

COULSON B S, GRIMWOOD K, HUDSON I L, BARNES G L, and BISHOP R F (1992), Role of coproantibody in clinical protection of children during reinfection with rotavirus, *J Clin Microbiol*, **30**, 1678–84.

COULSON B S, LONDRIGAN S L, and LEE D J (1997), Rotavirus contains integrin ligand sequences and a disintegrin-like domain that are implicated in virus entry into cells, *Proc Natl Acad Sci U S A*, **94**, 5389–94.

COX M J and MEDLEY G F (2003), Serological survey of anti-group A rotavirus IgM in UK adults, *Epidemiol Infect*, **131**, 719–26.

CRAWFORD S E, PATEL D G, CHENG E, BERKOVA Z, HYSER J M, CIARLET M, FINEGOLD M J, CONNER M E, and ESTES M K (2006), Rotavirus viremia and extraintestinal viral infection in the neonatal rat model, *J Virol*, **80**, 4820–32.

CUADRAS M A, BORDIER B B, ZAMBRANO J L, LUDERT J E, and GREENBERG H B (2006), Dissecting rotavirus particle-raft interaction with small interfering RNAs: insights into rotavirus transit through the secretory pathway, *J Virol*, **80**, 3935–46.

DAS B K, GENTSCH J R, CICIRELLO H G, WOODS P A, GUPTA A, RAMACHANDRAN M, KUMAR R, BHAN M K, and GLASS R I (1994), Characterization of rotavirus strains from newborns in New Delhi, India, *J Clin Microbiol*, **32**, 1820–2.

DA SILVA M F, TORT L F, GOMEZ M M, ASSIS R M, VOLOTAO EDE M, DE MENDONCA M C, BELLO G, and LEITE J P (2011), VP7 Gene of human rotavirus A genotype G5: Phylogenetic analysis reveals the existence of three different lineages worldwide, *J Med Virol*, **83**, 357–66.

DE GRAZIA S, MARTELLA V, ROTOLO V, BONURA F, MATTHIJNSSENS J, BANYAI K, CIARLET M, and GIAMMANCO G M (2011), Molecular characterization of genotype G6 human rotavirus strains detected in Italy from 1986 to 2009, *Infect Genet Evol*, **11**, 1449–55.

DE LA RUBIA L, HERRERA M I, CEBRERO M, and DE JONG J C (1996), Acute pancreatitis associated with rotavirus infection, *Pancreas*, **12**, 98–9.

DENNEHY P H, GAUNTLETT D R, and TENTE W E (1988), Comparison of nine commercial immunoassays for the detection of rotavirus in fecal specimens, *J Clin Microbiol*, **26**, 1630–4.

DE OLIVEIRA L H, DANOVARO-HOLLIDAY M C, ANDRUS J K, DE FILLIPIS A M, GENTSCH J, MATUS C R, and WIDDOWSON M A (2009), Sentinel hospital surveillance for rotavirus in Latin American and Caribbean countries, *J Infect Dis*, **200**(Suppl 1), S131–S139.

DESSELBERGER U, and HUPPERTZ H I (2011), Immune responses to rotavirus infection and vaccination and associated correlates of protection, *J Infect Dis*, **203**, 188–95.

DESSELBERGER U, WOLLESWINKEL-VAN DEN BOSCH J, MRUKOWICZ J, RODRIGO C, GIAQUINTO C, and VESIKARI T (2006), Rotavirus types in Europe and their significance for vaccination, *Pediatr Infect Dis J*, **25**, S30–S41.

DEWEY C, CARMAN S, PASMA T, JOSEPHSON G, and MCEWEN B (2003), Relationship between group A porcine rotavirus and management practices in swine herds in Ontario, *Can Vet J*, **44**, 649–53.

DE WIT M A, KOOPMANS M P, KORTBEEK L M, VAN LEEUWEN N J, BARTELDS A I, and VAN DUYNHOVEN Y T (2001), Gastroenteritis in sentinel general practices,The Netherlands, *Emerg Infect Dis*, **7**, 82–91.

DI BARTOLO I, MONINI M, LOSIO M N, PAVONI E, LAVAZZA A, and RUGGERI F M (2011), Molecular characterization of noroviruses and rotaviruses involved in a large outbreak of gastroenteritis in Northern Italy, *Appl Environ Microbiol*, **77**, 5545–8.

DONG Y, ZENG C Q, BALL J M, ESTES M K, and MORRIS A P (1997), The rotavirus enterotoxin NSP4 mobilizes intracellular calcium in human intestinal cells by stimulating phospholipase C-mediated inositol 1,4,5-trisphosphate production, *Proc Natl Acad Sci U S A*, **94**, 3960–5.

DORMITZER P R, SUN Z Y, WAGNER G, and HARRISON S C (2002), The rhesus rotavirus VP4 sialic acid binding domain has a galectin fold with a novel carbohydrate binding site, *EMBO J*, **21**, 885–97.

DREYFUSS M S (2009), Is norovirus a foodborne or pandemic pathogen? An analysis of the transmission of norovirus-associated gastroenteritis and the roles of food and food handlers, *Foodborne Pathog Dis*, **6**, 1219–28.

DUBOIS E, LE GUYADER F, HAUGARREAU L, KOPECKA H, CORMIER M, and POMMEPUY M (1997), Molecular epidemiological survey of rotaviruses in sewage by reverse transcriptase seminested PCR and restriction fragment length polymorphism assay, *Appl Environ Microbiol*, **63**, 1794–800.

DYALL-SMITH M L, LAZDINS I, TREGEAR G W, and HOLMES I H (1986), Location of the major antigenic sites involved in rotavirus serotype-specific neutralization, *Proc Natl Acad Sci U S A*, **83**, 3465–8.

EISENGART L J, CHOU P M, IYER K, COHRAN V, and RAJARAM V (2009), Rotavirus infection in small bowel transplant: a histologic comparison with acute cellular rejection, *Pediatr Dev Pathol*, **12**, 85–8.

ESONA M D, ARMAH G E, GEYER A, and STEELE A D (2004), Detection of an unusual human rotavirus strain with G5P[8] specificity in a Cameroonian child with diarrhea, *J Clin Microbiol*, **42**, 441–4.

ESTES M K, GRAHAM D Y, and MASON B B (1981), Proteolytic enhancement of rotavirus infectivity: molecular mechanisms, *J Virol*, **39**, 879–88.

ESTES M K, and KAPIKIAN A Z 2007. Rotaviruses. In: KNIPE D M, HOWLEY P M, GRIFFIN D E, LAMB R A, MARTIN M A, ROIZMAN B and STRAUS S E (eds) *Fields Virology*. Philadelphia, PA: Kluwer/Lippincott, Williams and Wilkins.

FARUQUE A S, MALEK M A, KHAN A I, HUQ S, SALAM M A, and SACK D A (2004), Diarrhoea in elderly people: aetiology, and clinical characteristics, *Scand J Infect Dis*, **36**, 204–8.

FEENEY S A, MITCHELL S J, MITCHELL F, WYATT D E, FAIRLEY D, MCCAUGHEY C, COYLE P V, and O'NEILL H J (2006), Association of the G4 rotavirus genotype with gastroenteritis in adults, *J Med Virol*, **78**, 1119–23.

FERREIRA F F, GUIMARAES F R, FUMIAN T M, VICTORIA M, VIEIRA C B, LUZ S, SHUBO T, LEITE J P, and MIAGOSTOVICH M P (2009), Environmental dissemination of group A rotavirus: P-type, G-type and subgroup characterization, *Water Sci Technol*, **60**, 633–42.

FISCHER WALKER C L and BLACK R E (2011), Rotavirus vaccine and diarrhea mortality: quantifying regional variation in effect size, *BMC Public Health*, **11**(Suppl 3), S16.

FITZGERALD T A, MUNOZ M, WOOD A R, and SNODGRASS D R (1995), Serological and genomic characterisation of group A rotaviruses from lambs, *Arch Virol*, **140**, 1541–8.

FRANCO M A, ANGEL J, and GREENBERG H B (2006), Immunity and correlates of protection for rotavirus vaccines, *Vaccine*, **24**, 2718–31.

FREEDMAN S B, ELTORKY M, and GORELICK M (2010), Evaluation of a gastroenteritis severity score for use in outpatient settings, *Pediatrics*, **125**, e1278–85.

FREEMAN M M, KERIN T, HULL J, TEEL E, ESONA M, PARASHAR U, GLASS R I, and GENTSCH J R (2009), Phylogenetic analysis of novel G12 rotaviruses in the United States: A molecular search for the origin of a new strain, *J Med Virol*, **81**, 736–46.

GABRIELI R, MACALUSO A, LANNI L, SACCARES S, DI GIAMBERARDINO F, CENCIONI B, PETRINCA A R, and DIVIZIA M (2007), Enteric viruses in molluscan shellfish, *New Microbiol*, **30**, 471–5.

GALLIMORE C I, PIPKIN C, SHRIMPTON H, GREEN A D, PICKFORD Y, MCCARTNEY C, SUTHERLAND G, BROWN D W, and GRAY J J (2005), Detection of multiple enteric virus strains within a foodborne outbreak of gastroenteritis: an indication of the source of contamination, *Epidemiol Infect*, **133**, 41–7.

GASSILLOUD B, SCHWARTZBROD L, and GANTZER C (2003), Presence of viral genomes in mineral water: a sufficient condition to assume infectious risk?, *Appl Environ Microbiol*, **69**, 3965–9.

GENTSCH J R, GLASS R I, WOODS P, GOUVEA V, GORZIGLIA M, FLORES J, DAS B K, and BHAN M K (1992), Identification of group A rotavirus gene 4 types by polymerase chain reaction, *J Clin Microbiol*, **30**, 1365–73.

GENTSCH J R, WOODS P A, RAMACHANDRAN M, DAS B K, LEITE J P, ALFIERI A, KUMAR R, BHAN M K, and GLASS R I (1996), Review of G and P typing results from a global collection of rotavirus strains: implications for vaccine development, *J Infect Dis*, **174**(Suppl 1), S30–36.

GERNA G, SARASINI A, DI MATTEO A, PAREA M, TORSELLINI M, and BATTAGLIA M (1989), Rapid detection of human rotavirus strains in stools by single-sandwich enzyme-linked immunosorbent assay systems using monoclonal antibodies, *J Virol Methods*, **24**, 43–56.

GHOSH S, ADACHI N, GATHERU Z, NYANGAO J, YAMAMOTO D, ISHINO M, URUSHIBARA N, and KOBAYASHI N (2011), Whole-genome analysis reveals the complex evolutionary dynamics of Kenyan G2P[4] human rotavirus strains, *J Gen Virol*, **92**, 2201–8.

GHOSH S, KOBAYASHI N, NAGASHIMA S, CHAWLA-SARKAR M, KRISHNAN T, GANESH B, and NAIK T N (2010), Full genomic analysis and possible origin of a porcine G12 rotavirus strain RU172, *Virus Genes*, **40**, 382–8.

GHOSH S, VARGHESE V, SAMAJDAR S, BHATTACHARYA S K, KOBAYASHI N, and NAIK T N (2006), Molecular characterization of a porcine Group A rotavirus strain with G12 genotype specificity, *Arch Virol*, **151**, 1329–44.

GIAMMARIOLI A M, MACKOW E R, FIORE L, GREENBERG H B, and RUGGERI F M (1996), Production and characterization of murine IgA monoclonal antibodies to the surface antigens of rhesus rotavirus, *Virology*, **225**, 97–110.

GLASS R I (2006), New hope for defeating rotavirus, *Sci Am*, **294**, 46–51.

GLASS R I, BRESEE J S, TURCIOS R, FISCHER T K, PARASHAR U D, and STEELE A D (2005), Rotavirus vaccines: targeting the developing world, *J Infect Dis*, **192**(Suppl 1), S160–S166.

GLASS R I, PARASHAR U D, BRESEE J S, TURCIOS R, FISCHER T K, WIDDOWSON M A, JIANG B, and GENTSCH J R (2006), Rotavirus vaccines: current prospects and future challenges, *Lancet*, **368**, 323–32.

GLEIZES O, DESSELBERGER U, TATOCHENKO V, RODRIGO C, SALMAN N, MEZNER Z, GIAQUINTO C, and GRIMPREL E (2006), Nosocomial rotavirus infection in European countries: a review of the epidemiology, severity and economic burden of hospital-acquired rotavirus disease, *Pediatr Infect Dis J*, **25**, S12–S21.

GOMARA M I, GREEN J, and GRAY J (2000), Methods of rotavirus detection, sero- and genotyping, sequencing, and phylogenetic analysis, *Methods Mol Med*, **34**, 189–216.

GOUVEA V, GLASS R I, WOODS P, TANIGUCHI K, CLARK H F, FORRESTER B, and FANG Z Y (1990a), Polymerase chain reaction amplification and typing of rotavirus nucleic acid from stool specimens, *J Clin Microbiol*, **28**, 276–82.

GOUVEA V, HO M S, GLASS R, WOODS P, FORRESTER B, ROBINSON C, ASHLEY R, RIEPENHOFF-TALTY M, CLARK H F, TANIGUCHI K, *et al.* (1990b), Serotypes and electropherotypes of human rotavirus in the USA: 1987–1989, *J Infect Dis*, **162**, 362–7.

GOUVEA V, SANTOS N, and TIMENETSKY MDO C (1994), Identification of bovine and porcine rotavirus G types by PCR, *J Clin Microbiol*, **32**, 1338–40.

GRAHAM K L, O'DONNELL J A, TAN Y, SANDERS N, CARRINGTON E M, ALLISON J, and COULSON B S (2007), Rotavirus infection of infant and young adult nonobese diabetic mice involves extraintestinal spread and delays diabetes onset, *J Virol*, **81**, 6446–58.

GRANT L, ESONA M, GENTSCH J, WATT J, REID R, WEATHERHOLTZ R, SANTOSHAM M, PARASHAR U, and O'BRIEN K (2011), Detection of G3P[3] and G3P[9] rotavirus strains in American Indian children with evidence of gene reassortment between human and animal rotaviruses, *J Med Virol*, **83**, 1288–99.

GRATACAP-CAVALLIER B, GENOULAZ O, BRENGEL-PESCE K, SOULE H, INNOCENTI-FRANCILLARD P, BOST M, GOFTI L, ZMIROU D, and SEIGNEURIN J M (2000), Detection of human and animal rotavirus sequences in drinking water, *Appl Environ Microbiol*, **66**, 2690–2.

GREIG J D and LEE M B (2012), A review of nosocomial norovirus outbreaks: infection control interventions found effective, *Epidemiol Infect*, **140**, 1151–60.

GRIFFIN D D, NAKAGOMI T, HOSHINO Y, NAKAGOMI O, KIRKWOOD C D, PARASHAR U D, GLASS R I, and GENTSCH J R (2002), Characterization of nontypeable rotavirus strains from the United States: identification of a new rotavirus reassortant (P2A[6],G12) and rare P3[9] strains related to bovine rotaviruses, *Virology*, **294**, 256–69.

HAGBOM M, ISTRATE C, ENGBLOM D, KARLSSON T, RODRIGUEZ-DIAZ J, BUESA J, TAYLOR J A, LOITTO V M, MAGNUSSON K E, AHLMAN H, LUNDGREN O, and SVENSSON L (2011),

Rotavirus stimulates release of serotonin (5-HT) from human enterochromaffin cells and activates brain structures involved in nausea and vomiting, *PLoS Pathog*, **7**, e1002115.

HAMILTON A J, STAGNITTI F, PREMIER R, BOLAND A M, and HALE G (2006), Quantitative microbial risk assessment models for consumption of raw vegetables irrigated with reclaimed water, *Appl Environ Microbiol*, **72**, 3284–90.

HASELHORST T, FLEMING F E, DYASON J C, HARTNELL R D, YU X, HOLLOWAY G, SANTEGOETS K, KIEFEL M J, BLANCHARD H, COULSON B S, and VON ITZSTEIN M (2009), Sialic acid dependence in rotavirus host cell invasion, *Nat Chem Biol*, **5**, 91–3.

HEATON P M, GOVEIA M G, MILLER J M, OFFIT P, and CLARK H F (2005), Development of a pentavalent rotavirus vaccine against prevalent serotypes of rotavirus gastroenteritis, *J Infect Dis*, **192**(Suppl 1), S17–S21.

HEMPSON S J, MATKOWSKYJ K, BANSAL A, TSAO E, HABIB I, BENYA R, MACKOW E R, and SHAW R D (2010), Rotavirus infection of murine small intestine causes colonic secretion via age restricted galanin-1 receptor expression, *Gastroenterology*, **138**, 2410–17.

HIGGINS R R, BENIPRASHAD M, CARDONA M, MASNEY S, LOW D E, and GUBBAY J B (2011), Evaluation and verification of the Seeplex Diarrhea-V ACE assay for simultaneous detection of adenovirus, rotavirus, and norovirus genogroups I and II in clinical stool specimens, *J Clin Microbiol*, **49**, 3154–62.

HOPKINS R S, GASPARD G B, WILLIAMS F P, Jr., KARLIN R J, CUKOR G, and BLACKLOW N R (1984), A community waterborne gastroenteritis outbreak: evidence for rotavirus as the agent, *Am J Public Health*, **74**, 263–5.

HOSHINO Y, JONES R W, ROSS J, and KAPIKIAN A Z (2003), Construction and characterization of rhesus monkey rotavirus (MMU18006)- or bovine rotavirus (UK)-based serotype G5, G8, G9 or G10 single VP7 gene substitution reassortant candidate vaccines, *Vaccine*, **21**, 3003–10.

HOSHINO Y and KAPIKIAN A Z (1996), Classification of rotavirus VP4 and VP7 serotypes, *Arch Virol Suppl*, **12**, 99–111.

HOSHINO Y and KAPIKIAN A Z (2000), Rotavirus serotypes: classification and importance in epidemiology, immunity, and vaccine development, *J Health Popul Nutr*, **18**, 5–14.

HOSHINO Y, NISHIKAWA K, BENFIELD D A, and GORZIGLIA M (1994), Mapping of antigenic sites involved in serotype-cross-reactive neutralization on group A rotavirus outercapsid glycoprotein VP7, *Virology*, **199**, 233–7.

HOU Z, HUANG Y, HUAN Y, PANG W, MENG M, WANG P, YANG M, JIANG L, CAO X, and WU K K (2008), Anti-NSP4 antibody can block rotavirus-induced diarrhea in mice, *J Pediatr Gastroenterol Nutr*, **46**, 376–85.

HURST C J and GERBA C P (1980), Stability of simian rotavirus in fresh and estuarine water, *Appl Environ Microbiol*, **39**, 1–5.

IIJIMA Y, IWAMOTO T, NUKUZUMA S, OHISHI H, HAYASHI K, and KOBAYASHI N (2006), An outbreak of rotavirus infection among adults in an institution for rehabilitation: long-term residence in a closed community as a risk factor for rotavirus illness, *Scand J Infect Dis*, **38**, 490–6.

ISA P, ARIAS C F, and LOPEZ S (2006), Role of sialic acids in rotavirus infection, *Glycoconj J*, **23**, 27–37.

ITURRIZA-GÓMARA M, ANDERTON E, KANG G, GALLIMORE C, PHILLIPS W, DESSELBERGER U, and GRAY J (2003), Evidence for genetic linkage between the gene segments encoding NSP4 and VP6 proteins in common and reassortant human rotavirus strains, *J Clin Microbiol*, **41**, 3566–73.

ITURRIZA-GÓMARA M, CUBITT D, STEELE D, GREEN J, BROWN D, KANG G, DESSELBERGER U, and GRAY J (2000), Characterisation of rotavirus G9 strains isolated in the UK between 1995 and 1998, *J Med Virol*, **61**, 510–17.

ITURRIZA-GÓMARA M, DALLMAN T, BANYAI K, BOTTIGER B, BUESA J, DIEDRICH S, FIORE L, JOHANSEN K, KOOPMANS M, KORSUN N, KOUKOU D, KRONEMAN A, LASZLO B, LAPPALAINEN

M, MAUNULA L, MARQUES A M, MATTHIJNSSENS J, MIDGLEY S, MLADENOVA Z, NAWAZ S, POLJSAK-PRIJATELJ M, POTHIER P, RUGGERI F M, SANCHEZ-FAUQUIER A, STEYER A, SIDARAVICIUTE-IVASKEVICIENE I, SYRIOPOULOU V, TRAN A N, USONIS V, VAN RANST M, DE ROUGEMONT A, and GRAY J (2011), Rotavirus genotypes co-circulating in Europe between 2006 and 2009 as determined by EuroRotaNet, a pan-European collaborative strain surveillance network, *Epidemiol Infect*, **139**, 895–909.

ITURRIZA-GÓMARA M, DALLMAN T, BANYAI K, BOTTIGER B, BUESA J, DIEDRICH S, FIORE L, JOHANSEN K, KORSUN N, KRONEMAN A, LAPPALAINEN M, LASZLO B, MAUNULA L, MATTHINJNSSENS J, MIDGLEY S, MLADENOVA Z, POLJSAK-PRIJATELJ M, POTHIER P, RUGGERI F M, SANCHEZ-FAUQUIER A, SCHREIER E, STEYER A, SIDARAVICIUTE I, TRAN A N, USONIS V, VAN RANST M, DE ROUGEMONT A, and GRAY J (2009), Rotavirus surveillance in Europe, 2005–2008: web-enabled reporting and real-time analysis of genotyping and epidemiological data, *J Infect Dis*, **200**(Suppl 1), S215–S221.

ITURRIZA-GÓMARA M, ISHERWOOD B, DESSELBERGER U, and GRAY J (2001), Reassortment in vivo: driving force for diversity of human rotavirus strains isolated in the United Kingdom between 1995 and 1999, *J Virol*, **75**, 3696–3705.

ITURRIZA-GÓMARA M, KANG G, and GRAY J (2004), Rotavirus genotyping: keeping up with an evolving population of human rotaviruses, *J Clin Virol*, **31**, 259–265.

JERE K C, MLERA L, O'NEILL H G, POTGIETER A C, PAGE N A, SEHERI M L, and VAN DIJK A A (2011), Whole genome analyses of African G2, G8, G9, and G12 rotavirus strains using sequence-independent amplification and 454(R) pyrosequencing, *J Med Virol*, **83**, 2018–42.

JIANG V, JIANG B, TATE J, PARASHAR U D, and PATEL M M (2010), Performance of rotavirus vaccines in developed and developing countries, *Hum Vaccin*, **6**, 532–42.

JIT M, MANGEN M J, MELLIEZ H, YAZDANPANAH Y, BILCKE J, SALO H, EDMUNDS W J, and BEUTELS P (2010), An update to 'The cost-effectiveness of rotavirus vaccination: comparative analyses for five European countries and transferability in Europe', *Vaccine*, **28**, 7457–9.

JOTHIKUMAR N, KANG G, and HILL V R (2009), Broadly reactive TaqMan assay for real-time RT-PCR detection of rotavirus in clinical and environmental samples. *J Virol Methods*, **155**, 126–31.

KABUE J P, PEENZE I, de BEER M, ESONA M D, LUNFUNGULA C, BIAMUNGU M, SIMBA T R, MUYEMBE TAMFUM J J, and STEELE A D (2010), Characterization of human rotavirus recovered from children with acute diarrhea in Kinshasa, Democratic Republic Of Congo, *J Infect Dis*, **202**(Suppl), S193–S197.

KAPIKIAN A Z, FLORES J, HOSHINO Y, GLASS R I, MIDTHUN K, GORZIGLIA M, and CHANOCK R M (1986), Rotavirus: the major etiologic agent of severe infantile diarrhea may be controllable by a "Jennerian" approach to vaccination, *J Infect Dis*, **153**, 815–22.

KAPIKIAN A Z, HOSHINO Y, CHANOCK R M, and PEREZ-SCHAEL I (1996), Jennerian and modified Jennerian approach to vaccination against rotavirus diarrhea using a quadrivalent rhesus rotavirus (RRV) and human-RRV reassortant vaccine, *Arch Virol Suppl*, **12**, 163–75.

KAPIKIAN A Z, KIM H W, WYATT R G, CLINE W L, ARROBIO J O, BRANDT C D, RODRIGUEZ W J, SACK D A, CHANOCK R M, and PARROTT R H (1976), Human reovirus-like agent as the major pathogen associated with "winter" gastroenteritis in hospitalized infants and young children, *N Engl J Med*, **294**, 965–72.

KAPIKIAN A Z, SIMONSEN L, VESIKARI T, HOSHINO Y, MORENS D M, CHANOCK R M, La MONTAGNE J R, and MURPHY B R (2005), A hexavalent human rotavirus-bovine rotavirus (UK) reassortant vaccine designed for use in developing countries and delivered in a schedule with the potential to eliminate the risk of intussusception, *J Infect Dis*, **192**(Suppl 1), S22–S29.

KARMAKAR S, RATHORE A S, KADRI S M, DUTT S, KHARE S, and LAL S (2008), Post-earthquake outbreak of rotavirus gastroenteritis in Kashmir (India): an epidemiological analysis, *Public Health*, **122**, 981–9.

KARSTEN C, BAUMGARTE S, FRIEDRICH A W, von EIFF C, BECKER K, WOSNIOK W, AMMON A, BOCKEMUHL J, KARCH H, and HUPPERTZ H I (2009), Incidence and risk factors for community-acquired acute gastroenteritis in north-west Germany in 2004, *Eur J Clin Microbiol Infect Dis*, **28**, 935–43.

KESWICK B H, PICKERING L K, DUPONT H L, and WOODWARD W E (1983), Survival and detection of rotaviruses on environmental surfaces in day care centers, *Appl Environ Microbiol*, **46**, 813–16.

KHAMRIN P, MANEEKARN N, PEERAKOME S, MALASAO R, THONGPRACHUM A, CHAN-IT W, MIZUGUCHI M, OKITSU S, and USHIJIMA H (2009), Molecular characterization of VP4, VP6, VP7, NSP4, and NSP5/6 genes identifies an unusual G3P[10] human rotavirus strain, *J Med Virol*, **81**, 176–82.

KHAMRIN P, MANEEKARN N, PEERAKOME S, YAGYU F, OKITSU S, and USHIJIMA H (2006), Molecular characterization of a rare G3P[3] human rotavirus reassortant strain reveals evidence for multiple human-animal interspecies transmissions, *J Med Virol*, **78**, 986–94.

KHAMRIN P, TRAN D N, CHAN-IT W, THONGPRACHUM A, OKITSU S, MANEEKARN N, and USHIJIMA H (2011), Comparison of the rapid methods for screening of group a rotavirus in stool samples, *J Trop Pediatr*, **57**, 375–7.

KIM C R, OH J W, YUM M K, LEE J H, and KANG J O (2009), Rotavirus infection in neonates at a university hospital in Korea, *Infect Control Hosp Epidemiol*, **30**, 893–5.

KITTIGUL L, POMBUBPA K, RATTANATHAM T, DIRAPHAT P, UTRARACHKIJ F, PUNGCHITTON S, KHAMRIN P, and USHIJIMA H (2008), Development of a method for concentrating and detecting rotavirus in oysters, *Int J Food Microbiol*, **122**, 204–10.

KORDASTI S, SJOVALL H, LUNDGREN O, and SVENSSON L (2004), Serotonin and vasoactive intestinal peptide antagonists attenuate rotavirus diarrhoea, *Gut*, **53**, 952–7.

KRIBS-ZALETA C M, JUSOT J F, VANHEMS P, and CHARLES S (2011), Modeling nosocomial transmission of rotavirus in pediatric wards, *Bull Math Biol*, **73**, 1413–42.

KUKKULA M, ARSTILA P, KLOSSNER M L, MAUNULA L, BONSDORFF C H, and JAATINEN P (1997), Waterborne outbreak of viral gastroenteritis, *Scand J Infect Dis*, **29**, 415–18.

LAWTON J A, ESTES M K, and PRASAD B V (1997), Three-dimensional visualization of mRNA release from actively transcribing rotavirus particles, *Nat Struct Biol*, **4**, 118–21.

LEES D (2000), Viruses and bivalve shellfish, *Int J Food Microbiol*, **59**, 81–116.

LE GUYADER F, DUBOIS E, MENARD D, and POMMEPUY M (1994), Detection of hepatitis A virus, rotavirus, and enterovirus in naturally contaminated shellfish and sediment by reverse transcription-seminested PCR, *Appl Environ Microbiol*, **60**, 3665–71.

LE GUYADER F, HAUGARREAU L, MIOSSEC L, DUBOIS E, and POMMEPUY M (2000), Three-year study to assess human enteric viruses in shellfish, *Appl Environ Microbiol*, **66**, 3241–8.

LE GUYADER F S, LE SAUX J C, AMBERT-BALAY K, KROL J, SERAIS O, PARNAUDEAU S, GIRAUDON H, DELMAS G, POMMEPUY M, POTHIER P, and ATMAR R L (2008), Aichi virus, norovirus, astrovirus, enterovirus, and rotavirus involved in clinical cases from a French oyster-related gastroenteritis outbreak, *J Clin Microbiol*, **46**, 4011–17.

LEVY K, HUBBARD A E, and EISENBERG J N (2009), Seasonality of rotavirus disease in the tropics: a systematic review and meta-analysis, *Int J Epidemiol*, **38**, 1487–96.

LI D, GU A Z, HE M, SHI H C, and YANG W (2009), UV inactivation and resistance of rotavirus evaluated by integrated cell culture and real-time RT-PCR assay, *Water Res*, **43**, 3261–9.

LI D D, DUAN Z J, ZHANG Q, LIU N, XIE Z P, JIANG B, STEELE D, JIANG X, WANG Z S, and FANG Z Y (2008), Molecular characterization of unusual human G5P[6] rotaviruses identified in China, *J Clin Virol*, **42**, 141–8.

LIPSON S M, and ZELINSKY-PAPEZ K A (1989), Comparison of four latex agglutination (LA) and three enzyme-linked immunosorbent assays (ELISA) for the detection of rotavirus in fecal specimens, *Am J Clin Pathol*, **92**, 637–43.

LODDER W J, and DE RODA HUSMAN A M (2005), Presence of noroviruses and other enteric viruses in sewage and surface waters in The Netherlands, *Appl Environ Microbiol*, **71**, 1453–61.

LOISY F, ATMAR R L, LE SAUX J C, COHEN J, CAPRAIS M P, POMMEPUY M, and LE GUYADER F S (2005), Use of rotavirus virus-like particles as surrogates to evaluate virus persistence in shellfish, *Appl Environ Microbiol*, **71**, 6049–53.

LONDRIGAN S L, HEWISH M J, THOMSON M J, SANDERS G M, MUSTAFA H, and COULSON B S (2000), Growth of rotaviruses in continuous human and monkey cell lines that vary in their expression of integrins, *J Gen Virol*, **81**, 2203–13.

LOPMAN B A, CURNS A T, YEN C, and PARASHAR U D (2011), Infant rotavirus vaccination may provide indirect protection to older children and adults in the United States, *J Infect Dis*, **204**, 980–6.

LUDERT J E, RUIZ M C, HIDALGO C, and LIPRANDI F (2002), Antibodies to rotavirus outer capsid glycoprotein VP7 neutralize infectivity by inhibiting virion decapsidation, *J Virol*, **76**, 6643–51.

LUNDGREN O, PEREGRIN A T, PERSSON K, KORDASTI S, UHNOO I, and SVENSSON L (2000), Role of the enteric nervous system in the fluid and electrolyte secretion of rotavirus diarrhea, *Science*, **287**, 491–5.

LUQUE FERNANDEZ M A, GALMES TRUYOLS A, HERRERA GUIBERT D, ARBONA CERDA G, and SANCHO GAYA F (2008), Cohort study of an outbreak of viral gastroenteritis in a nursing home for elderly, Majorca, Spain, February 2008, *Euro Surveill*, **13**(51), pii=19070.

MACKOW E R, SHAW R D, MATSUI S M, VO P T, BENFIELD D A, and GREENBERG H B (1988), Characterization of homotypic and heterotypic VP7 neutralization sites of rhesus rotavirus, *Virology*, **165**, 511–17.

MAES P, MATTHIJNSSENS J, RAHMAN M, and VAN RANST M (2009), RotaC: a web-based tool for the complete genome classification of group A rotaviruses, *BMC Microbiol*, **9**, 238.

MANEEKARN N, KHAMRIN P, CHAN-IT W, PEERAKOME S, SUKCHAI S, PRINGPRAO K, and USHIJIMA H (2006), Detection of rare G3P[19] porcine rotavirus strains in Chiang Mai, Thailand, provides evidence for origin of the VP4 genes of Mc323 and Mc345 human rotaviruses, *J Clin Microbiol*, **44**, 4113–19.

MARA D D, SLEIGH P A, BLUMENTHAL U J, and CARR R M (2007), Health risks in wastewater irrigation: comparing estimates from quantitative microbial risk analyses and epidemiological studies, *J Water Health*, **5**, 39–50.

MARSHALL J, BOTES J, GORRIE G, BOARDMAN C, GREGORY J, GRIFFITH J, HOGG G, DIMITRIADIS A, CATTON M, and BISHOP R (2003), Rotavirus detection and characterisation in outbreaks of gastroenteritis in aged-care facilities, *J Clin Virol*, **28**, 331–40.

MARTELLA V, BANYAI K, CIARLET M, ITURRIZA-GOMARA M, LORUSSO E, DE GRAZIA S, ARISTA S, DECARO N, ELIA G, CAVALLI A, CORRENTE M, LAVAZZA A, BASELGA R, and BUONAVOGLIA C (2006), Relationships among porcine and human P[6] rotaviruses: evidence that the different human P[6] lineages have originated from multiple interspecies transmission events, *Virology*, **344**, 509–19.

MARTELLA V, BANYAI K, MATTHIJNSSENS J, BUONAVOGLIA C, and CIARLET M (2010), Zoonotic aspects of rotaviruses, *Vet Microbiol*, **140**, 246–55.

MATHIEU M, PETITPAS I, NAVAZA J, LEPAULT J, KOHLI E, POTHIER P, PRASAD B V, COHEN J, and REY F A (2001), Atomic structure of the major capsid protein of rotavirus: implications for the architecture of the virion, *EMBO J*, **20**, 1485–97.

MATTHIJNSSENS J, CIARLET M, MCDONALD S M, ATTOUI H, BANYAI K, BRISTER J R, BUESA J, ESONA M D, ESTES M K, GENTSCH J R, ITURRIZA-GOMARA M, JOHNE R, KIRKWOOD C D, MARTELLA V, MERTENS P P, NAKAGOMI O, PARRENO V, RAHMAN M, RUGGERI F M, SAIF L J, SANTOS N, STEYER A, TANIGUCHI K, PATTON J T, DESSELBERGER U, and VAN RANST M (2011a), Uniformity of rotavirus strain nomenclature proposed by the Rotavirus Classification Working Group (RCWG), *Arch Virol*, **156**, 1397–413.

MATTHIJNSSENS J, CIARLET M, RAHMAN M, ATTOUI H, BANYAI K, ESTES M K, GENTSCH J R, ITURRIZA-GOMARA M, KIRKWOOD C D, MARTELLA V, MERTENS P P, NAKAGOMI O, PATTON J T, RUGGERI F M, SAIF L J, SANTOS N, STEYER A, TANIGUCHI K, DESSELBERGER U, and VAN RANST M (2008), Recommendations for the classification of group A rotaviruses using all 11 genomic RNA segments, *Arch Virol*, **153**, 1621–9.

MATTHIJNSSENS J, DE GRAZIA S, PIESSENS J, HEYLEN E, ZELLER M, GIAMMANCO G M, BANYAI K, BUONAVOGLIA C, CIARLET M, MARTELLA V, and VAN RANST M (2011b), Multiple reassortment and interspecies transmission events contribute to the diversity of feline, canine and feline/canine-like human group A rotavirus strains, *Infect Genet Evol*, **11**, 1396–406.

MATTHIJNSSENS J, HEYLEN E, ZELLER M, RAHMAN M, LEMEY P, and VAN RANST M (2010), Phylodynamic analyses of rotavirus genotypes G9 and G12 underscore their potential for swift global spread, *Mol Biol Evol*, **27**, 2431–6.

MATTHIJNSSENS J, POTGIETER C A, CIARLET M, PARRENO V, MARTELLA V, BANYAI K, GARAICOECHEA L, PALOMBO E A, NOVO L, ZELLER M, ARISTA S, GERNA G, RAHMAN M, and VAN RANST M (2009), Are human P[14] rotavirus strains the result of interspecies transmissions from sheep or other ungulates that belong to the mammalian order Artiodactyla?, *J Virol*, **83**, 2917–29.

MATTHIJNSSENS J, RAHMAN M, MARTELLA V, XUELEI Y, DE VOS S, DE LEENER K, CIARLET M, BUONAVOGLIA C, and VAN RANST M (2006a), Full genomic analysis of human rotavirus strain B4106 and lapine rotavirus strain 30/96 provides evidence for interspecies transmission, *J Virol*, **80**, 3801–10.

MATTHIJNSSENS J, RAHMAN M, YANG X, DELBEKE T, ARIJS I, KABUE J P, MUYEMBE J J, and VAN RANST M (2006b), G8 rotavirus strains isolated in the Democratic Republic of Congo belong to the DS-1-like genogroup, *J Clin Microbiol*, **44**, 1801–09.

MAYR C, STROHE G, and CONTZEN M (2009), Detection of rotavirus in food associated with a gastroenteritis outbreak in a mother and child sanatorium, *Int J Food Microbiol*, **135**, 179–82.

MCDONALD S M, MATTHIJNSSENS J, MCALLEN J K, HINE E, OVERTON L, WANG S, LEMEY P, ZELLER M, VAN RANST M, SPIRO D J, and PATTON J T (2009), Evolutionary dynamics of human rotaviruses: balancing reassortment with preferred genome constellations, *PLoS Pathog*, **5**, e1000634.

MCDONALD S M and PATTON J T (2011), Rotavirus VP2 core shell regions critical for viral polymerase activation, *J Virol*, **85**, 3095–105.

MCLEAN B S and HOLMES I H (1981), Effects of antibodies, trypsin, and trypsin inhibitors on susceptibility of neonates to rotavirus infection, *J Clin Microbiol*, **13**, 22–9.

MCMASTER P, HUNT R, WOJTULEWICZ J, and WILCKEN B (2001), An unusual cause of hepatitis, *J Paediatr Child Health*, **37**, 587–8.

MEDICI M C, ABELLI L A, GUERRA P, DODI I, DETTORI G, and CHEZZI C (2011), Case report: detection of rotavirus RNA in the cerebrospinal fluid of a child with rotavirus gastroenteritis and meningism, *J Med Virol*, **83**, 1637–40.

MENDEZ E, LOPEZ S, CUADRAS M A, ROMERO P, and ARIAS C F (1999), Entry of rotaviruses is a multistep process, *Virology*, **263**, 450–9.

MIDGLEY S E, HJULSAGER C K, LARSEN L E, FALKENHORST G, and BOTTIGER B (2012), Suspected zoonotic transmission of rotavirus group A in Danish adults, *Epidemiol Infect*, **140**, 1013–17.

MIDTHUN K, HOSHINO Y, KAPIKIAN A Z, and CHANOCK R M (1986), Single gene substitution rotavirus reassortants containing the major neutralization protein (VP7) of human rotavirus serotype 4, *J Clin Microbiol*, **24**, 822–6.

MIOTTI P G, EIDEN J, and YOLKEN R H (1985), Comparative efficiency of commercial immunoassays for the diagnosis of rotavirus gastroenteritis during the course of infection, *J Clin Microbiol*, **22**, 693–8.

MIRZAYEVA R, CORTESE M M, MOSINA L, BIELLIK R, LOBANOV A, CHERNYSHOVA L, LASHKARASHVILI M, TURKOV S, ITURRIZA-GOMARA M, GRAY J, PARASHAR U D, STEELE

D, and EMIROGLU N (2009), Rotavirus burden among children in the newly independent states of the former union of soviet socialist republics: literature review and first-year results from the rotavirus surveillance network, *J Infect Dis*, **200**(Suppl 1), S203–S214.

MONROE S S (2011), Control and prevention of viral gastroenteritis, *Emerg Infect Dis*, **17**, 1347–8.

MOON S, WANG Y, DENNEHY P, SIMONSEN K A, ZHANG J, and JIANG B (2012), Antigenemia, RNAemia, and innate immunity in children with acute rotavirus diarrhea, *FEMS Immunol Med Microbiol*, **64**, 382–91.

MOSSEL E C and RAMIG R F (2002), Rotavirus genome segment 7 (NSP3) is a determinant of extraintestinal spread in the neonatal mouse, *J Virol*, **76**, 6502–09.

MUKHERJEE A, GHOSH S, BAGCHI P, DUTTA D, CHATTOPADHYAY S, KOBAYASHI N, and CHAWLA-SARKAR M (2011), Full genomic analyses of human rotavirus G4P[4], G4P[6], G9P[19] and G10P[6] strains from North-eastern India: evidence for interspecies transmission and complex reassortment events, *Clin Microbiol Infect*, **17**, 1343–6.

NAKAGOMI O, OYAMADA H, and NAKAGOMI T (1991), Experience with serotyping rotavirus strains by reverse transcription and two-step polymerase chain reaction with generic and type-specific primers, *Mol Cell Probes*, **5**, 285–9.

NELSON E A, WIDDOWSON M A, KILGORE P E, STEELE D, and PARASHAR U D (2009), A decade of the Asian Rotavirus Surveillance Network: achievements and future directions, *Vaccine*, **27**(Suppl 5), F1–F3.

NEMOTO M, HATA H, HIGUCHI T, IMAGAWA H, YAMANAKA T, NIWA H, BANNAI H, TSUJIMURA K, KONDO T, and MATSUMURA T (2010), Evaluation of rapid antigen detection kits for diagnosis of equine rotavirus infection, *J Vet Med Sci*, **72**, 1247–50.

NEWELL D G, KOOPMANS M, VERHOEF L, DUIZER E, AIDARA-KANE A, SPRONG H, OPSTEEGH M, LANGELAAR M, THREFALL J, SCHEUTZ F, VAN DER GIESSEN J, and KRUSE H (2010), Food-borne diseases – the challenges of 20 years ago still persist while new ones continue to emerge, *Int J Food Microbiol*, **139**(Suppl 1), S3–S15.

NEWTON K, MEYER J C, BELLAMY A R, and TAYLOR J A (1997), Rotavirus nonstructural glycoprotein NSP4 alters plasma membrane permeability in mammalian cells, *J Virol*, **71**, 9458–65.

O'RYAN M (2009), The ever-changing landscape of rotavirus serotypes, *Pediatr Infect Dis J*, **28**, S60–S62.

OTTO P, LIEBLER-TENORIO E M, ELSCHNER M, REETZ J, LOHREN U, and DILLER R (2006), Detection of rotaviruses and intestinal lesions in broiler chicks from flocks with runting and stunting syndrome (RSS), *Avian Dis*, **50**, 411–18.

PANCORBO O C, EVANSHEN B G, CAMPBELL W F, LAMBERT S, CURTIS S K, and WOOLLEY T W (1987), Infectivity and antigenicity reduction rates of human rotavirus strain Wa in fresh waters, *Appl Environ Microbiol*, **53**, 1803–11.

PARASHAR U D, ALEXANDER J P, and GLASS R I (2006a), Prevention of rotavirus gastroenteritis among infants and children. Recommendations of the Advisory Committee on Immunization Practices (ACIP), *MMWR Recomm Rep*, **55**, 1–13.

PARASHAR U D, GIBSON C J, BRESSE J S, and GLASS R I (2006b), Rotavirus and severe childhood diarrhea, *Emerg Infect Dis*, **12**, 304–06.

PARASHAR U D and GLASS R I (2006), Public health. Progress toward rotavirus vaccines, *Science*, **312**, 851–2.

PARK S I, MATTHIJNSSENS J, SAIF L J, KIM H J, PARK J G, ALFAJARO M M, KIM D S, SON K Y, YANG D K, HYUN B H, KANG M I, and CHO K O (2011), Reassortment among bovine, porcine and human rotavirus strains results in G8P[7] and G6P[7] strains isolated from cattle in South Korea, *Vet Microbiol*, **152**, 55–66.

PARRENO V, MARCOPPIDO G, VEGA C, GARAICOECHEA L, RODRIGUEZ D, SAIF L, and FERNANDEZ F (2010), Milk supplemented with immune colostrum: protection against rotavirus diarrhea and modulatory effect on the systemic and mucosal antibody responses in calves experimentally challenged with bovine rotavirus, *Vet Immunol Immunopathol*, **136**, 12–27.

PATTON J T, JONES M T, KALBACH A N, HE Y W, and XIAOBO J (1997), Rotavirus RNA polymerase requires the core shell protein to synthesize the double-stranded RNA genome, *J Virol*, **71**, 9618–26.

PAUL P S, LYOO Y S, ANDREWS J J, and HILL H T (1988), Isolation of two new serotypes of porcine rotavirus from pigs with diarrhea, *Arch Virol*, **100**, 139–43.

PESARO F, SORG I, and METZLER A (1995), In situ inactivation of animal viruses and a coliphage in nonaerated liquid and semiliquid animal wastes, *Appl Environ Microbiol*, **61**, 92–7.

PHILLIPS G, LOPMAN B, RODRIGUES L C, and TAM C C (2010), Asymptomatic rotavirus infections in England: prevalence, characteristics, and risk factors, *Am J Epidemiol*, **171**, 1023–30.

PIEDNOIR E, BORDERAN G C, BORGEY F, THIBON P, LESELLIER P, LESERVOISIER R, VERGER P, and LE COUTOUR X (2010), Direct costs associated with a hospital-acquired outbreak of rotaviral gastroenteritis infection in a long term care institution, *J Hosp Infect*, **75**, 295–8.

PITZER V E, VIBOUD C, LOPMAN B A, PATEL M M, PARASHAR U D, and GRENFELL B T (2011), Influence of birth rates and transmission rates on the global seasonality of rotavirus incidence, *J R Soc Interface*, **8**, 1584–93.

PLANTE D, BELANGER G, LEBLANC D, WARD P, HOUDE A, and TROTTIER Y L (2011), The use of bovine serum albumin to improve the RT-qPCR detection of foodborne viruses rinsed from vegetable surfaces, *Lett Appl Microbiol*, **52**, 239–44.

PRASAD B V, WANG G J, CLERX J P, and CHIU W (1988), Three-dimensional structure of rotavirus, *J Mol Biol*, **199**, 269–75.

PRINCE D S, ASTRY C, VONDERFECHT S, JAKAB G, SHEN F M, and YOLKEN R H (1986), Aerosol transmission of experimental rotavirus infection, *Pediatr Infect Dis*, **5**, 218–22.

RAHMAN M, MATTHIJNSSENS J, YANG X, DELBEKE T, ARIJS I, TANIGUCHI K, ITURRIZA-GOMARA M, IFTEKHARUDDIN N, AZIM T, and VAN RANST M (2007), Evolutionary history and global spread of the emerging g12 human rotaviruses, *J Virol*, **81**, 2382–90.

RAINSFORD E W and MCCRAE M A (2007), Characterization of the NSP6 protein product of rotavirus gene 11, *Virus Res*, **130**, 193–201.

RAMACHANDRAN M, GENTSCH J R, PARASHAR U D, JIN S, WOODS P A, HOLMES J L, KIRKWOOD C D, BISHOP R F, GREENBERG H B, URASAWA S, GERNA G, COULSON B S, TANIGUCHI K, BRESEE J S, and GLASS R I (1998), Detection and characterization of novel rotavirus strains in the United States, *J Clin Microbiol*, **36**, 3223–9.

RAMIG R F (2007), Systemic rotavirus infection, *Expert Rev Anti Infect Ther*, **5**, 591–612.

RAMIG R F, CIARLET M, MERTENS P P C, and DERMODY T S 2005. Genus *Rotavirus*. In: FAUQUET C M, MAYO M A, MANILOFF J, DESSELBERGER U and BALL L A (eds) *Virus taxonomy. Eighth Report of the International Commit- tee on Taxonomy of Viruses*. New York: Elsevier Academic Press.

RASANEN S, LAPPALAINEN S, KAIKKONEN S, HAMALAINEN M, SALMINEN M, and VESIKARI T (2010), Mixed viral infections causing acute gastroenteritis in children in a waterborne outbreak, *Epidemiol Infect*, **138**, 1227–34.

RICHARDSON S, GRIMWOOD K, GORRELL R, PALOMBO E, BARNES G, and BISHOP R (1998), Extended excretion of rotavirus after severe diarrhoea in young children, *Lancet*, **351**, 1844–8.

RIGOTTO C, VICTORIA M, MORESCO V, KOLESNIKOVAS C K, CORREA A A, SOUZA D S, MIAGOSTOVICH M P, SIMOES C M, and BARARDI C R (2010), Assessment of adenovirus, hepatitis A virus and rotavirus presence in environmental samples in Florianopolis, South Brazil, *J Appl Microbiol*, **109**, 1979–87.

RODRIGUEZ-DIAZ J, QUERALES L, CARABALLO L, VIZZI E, LIPRANDI F, TAKIFF H, and BETANCOURT W Q (2009), Detection and characterization of waterborne gastroenteritis viruses in urban sewage and sewage-polluted river waters in Caracas, Venezuela, *Appl Environ Microbiol*, **75**, 387–94.

RUGGERI F M, and GREENBERG H B (1991), Antibodies to the trypsin cleavage peptide VP8 neutralize rotavirus by inhibiting binding of virions to target cells in culture, *J Virol*, **65**, 2211–19.

RUIZ M C, CHARPILIENNE A, LIPRANDI F, GAJARDO R, MICHELANGELI F, and COHEN J (1996), The concentration of Ca^{2+} that solubilizes outer capsid proteins from rotavirus particles is dependent on the strain, *J Virol*, **70**, 4877–83.

RUTJES S A, LODDER W J, VAN LEEUWEN A D, and DE RODA HUSMAN A M (2009), Detection of infectious rotavirus in naturally contaminated source waters for drinking water production, *J Appl Microbiol*, **107**, 97–105.

RUUSKA T, and VESIKARI T (1990), Rotavirus disease in Finnish children: use of numerical scores for clinical severity of diarrhoeal episodes, *Scand J Infect Dis*, **22**, 259–67.

RYAN M J, WALL P G, ADAK G K, EVANS H S, and COWDEN J M (1997), Outbreaks of infectious intestinal disease in residential institutions in England and Wales 1992–1994, *J Infect*, **34**, 49–54.

RYDER R W, SACK D A, KAPIKIAN A Z, MCLAUGHLIN J C, CHAKRABORTY J, MIZANUR RAHMAN A S, MERSON M H, and WELLS J G (1976), Enterotoxigenic Escherichia coli and Reovirus-like agent in rural Bangladesh, *Lancet*, **1**, 659–63.

RYDER R W, YOLKEN R H, REEVES W C, and SACK R B (1986), Enzootic bovine rotavirus is not a source of infection in Panamanian cattle ranchers and their families, *J Infect Dis*, **153**, 1139–44.

SAKAMOTO Y, UEKI S, KASAI T, TAKATO J, SHIMANUKI H, HONDA H, ITO T, and HAGA H (2009), Effect of exercise, aging and functional capacity on acute secretory immunoglobulin A response in elderly people over 75 years of age, *Geriatr Gerontol Int*, **9**, 81–8.

SANTOS N, and HOSHINO Y (2005), Global distribution of rotavirus serotypes/genotypes and its implication for the development and implementation of an effective rotavirus vaccine, *Rev Med Virol*, **15**, 29–56.

SANTOS N, LIMA R C, NOZAWA C M, LINHARES R E, and GOUVEA V (1999), Detection of porcine rotavirus type G9 and of a mixture of types G1 and G5 associated with Wa-like VP4 specificity: evidence for natural human-porcine genetic reassortment, *J Clin Microbiol*, **37**, 2734–6.

SCHWARTZ B S, HARRIS J B, KHAN A I, LAROCQUE R C, SACK D A, MALEK M A, FARUQUE A S, QADRI F, CALDERWOOD S B, LUBY S P, and RYAN E T (2006), Diarrheal epidemics in Dhaka, Bangladesh, during three consecutive floods: 1988, 1998, and 2004, *Am J Trop Med Hyg*, **74**, 1067–73.

SDIRI-LOULIZI K, HASSINE M, AOUNI Z, GHARBI-KHELIFI H, CHOUCHANE S, SAKLY N, NEJI-GUEDICHE M, POTHIER P, AOUNI M, and AMBERT-BALAY K (2010), Detection and molecular characterization of enteric viruses in environmental samples in Monastir, Tunisia between January 2003 and April 2007, *J Appl Microbiol*, **109**, 1093–104.

SEIDU R, HEISTAD A, AMOAH P, DRECHSEL P, JENSSEN P D, and STENSTROM T A (2008), Quantification of the health risk associated with wastewater reuse in Accra, Ghana: a contribution toward local guidelines, *J Water Health*, **6**, 461–71.

SEO N S, ZENG C Q, HYSER J M, UTAMA B, CRAWFORD S E, KIM K J, HOOK M, and ESTES M K (2008), Integrins alpha1beta1 and alpha2beta1 are receptors for the rotavirus enterotoxin, *Proc Natl Acad Sci U S A*, **105**, 8811–18.

SETTEMBRE E C, CHEN J Z, DORMITZER P R, GRIGORIEFF N, and HARRISON S C (2011), Atomic model of an infectious rotavirus particle, *EMBO J*, **30**, 408–16.

SHARMA S, RAY P, GENTSCH J R, GLASS R I, KALRA V, and BHAN M K (2008), Emergence of G12 rotavirus strains in Delhi, India, in 2000 to 2007, *J Clin Microbiol*, **46**, 1343–8.

SHAW R D, VO P T, OFFIT P A, COULSON B S, and GREENBERG H B (1986), Antigenic mapping of the surface proteins of rhesus rotavirus, *Virology*, **155**, 434–51.

SHULMAN L M, SILBERSTEIN I, ALFANDARI J, and MENDELSON E (2011), Genotyping rotavirus RNA from archived rotavirus-positive rapid test strips, *Emerg Infect Dis*, **17**, 44–8.

SILVESTRI L S, TARAPOREWALA Z F, and PATTON J T (2004), Rotavirus replication: plus-sense templates for double-stranded RNA synthesis are made in viroplasms, *J Virol*, **78**, 7763–74.

SIMMONDS M K, ARMAH G, ASMAH R, BANERJEE I, DAMANKA S, ESONA M, GENTSCH J R, GRAY J J, KIRKWOOD C, PAGE N, and ITURRIZA-GÓMARA M (2008), New oligonucleotide primers for P-typing of rotavirus strains: Strategies for typing previously untypeable strains, *J Clin Virol*, **42**, 368–73.

STELZMUELLER I, BIEBL M, GRAZIADEI I, WIESMAYR S, MARGREITER R, and BONATTI H (2006), Regarding diarrhea in liver transplant recipients: etiology and management, *Liver Transpl*, **12**, 163–4.

STEYER A, POLJSAK-PRIJATELJ M, BUFON T L, MARCUN-VARDA N, and MARIN J (2007), Rotavirus genotypes in Slovenia: unexpected detection of G8P[8] and G12P[8] genotypes, *J Med Virol*, **79**, 626–32.

TANIGUCHI K, HOSHINO Y, NISHIKAWA K, GREEN K Y, MALOY W L, MORITA Y, URASAWA S, KAPIKIAN A Z, CHANOCK R M, and GORZIGLIA M (1988a), Cross-reactive and serotype-specific neutralization epitopes on VP7 of human rotavirus: nucleotide sequence analysis of antigenic mutants selected with monoclonal antibodies, *J Virol*, **62**, 1870–4.

TANIGUCHI K, MALOY W L, NISHIKAWA K, GREEN K Y, HOSHINO Y, URASAWA S, KAPIKIAN A Z, CHANOCK R M, and GORZIGLIA M (1988b), Identification of cross-reactive and serotype 2-specific neutralization epitopes on VP3 of human rotavirus, *J Virol*, **62**, 2421–6.

TANIGUCHI K, URASAWA T, KOBAYASHI N, AHMED M U, ADACHI N, CHIBA S, and URASAWA S (1991), Antibody response to serotype-specific and cross-reactive neutralization epitopes on VP4 and VP7 after rotavirus infection or vaccination, *J Clin Microbiol*, **29**, 483–7.

TARLOW O, and MCCRAE M A (1990), Nucleotide sequence of group antigen (VP6) of the UK tissue culture adapted strain of bovine rotavirus, *Nucleic Acids Res*, **18**, 4921.

TATE J E, BURTON A H, BOSCHI-PINTO C, STEELE A D, DUQUE J, and PARASHAR U D (2012), 2008 estimate of worldwide rotavirus-associated mortality in children younger than 5 years before the introduction of universal rotavirus vaccination programmes: a systematic review and meta-analysis, *Lancet Infect Dis*, **12**, 136–41.

TATTE V S, RAWAL K N, and CHITAMBAR S D (2010), Sequence and phylogenetic analysis of the VP6 and NSP4 genes of human rotavirus strains: evidence of discordance in their genetic linkage, *Infect Genet Evol*, **10**, 940–9.

TCHEREMENSKAIA O, MARUCCI G, DE PETRIS S, RUGGERI F M, DOVECAR D, STERNAK S L, MATYASOVA I, DHIMOLEA M K, MLADENOVA Z, and FIORE L (2007), Molecular epidemiology of rotavirus in Central and Southeastern Europe, *J Clin Microbiol*, **45**, 2197–204.

THAN V T, LE V P, LIM I, and KIM W (2011), Complete genomic characterization of cell culture adapted human G12P[6] rotaviruses isolated from South Korea, *Virus Genes*, **42**, 317–22.

TIMENETSKY M C, GOUVEA V, SANTOS N, ALGE M E, KISIELLIUS J J, and CARMONA R C (1996), Outbreak of severe gastroenteritis in adults and children associated with type G2 rotavirus. Study Group on Diarrhea of the Instituto Adolfo Lutz, *J Diarrhoeal Dis Res*, **14**, 71–4.

TROP SKAZA A, BESKOVNIK L, and ZOHAR CRETNIK T (2011), Outbreak of rotavirus gastroenteritis in a nursing home, Slovenia, December 2010, *Euro Surveill*, **16**, 19837.

TRUJILLO-ALONSO V, MARURI-AVIDAL L, ARIAS C F, and LOPEZ S (2011), Rotavirus infection induces the unfolded protein response of the cell and controls it through the non-structural protein NSP3, *J Virol*, **85**, 12594–604.

TSAI Y L, TRAN B, SANGERMANO L R, and PALMER C J (1994), Detection of poliovirus, hepatitis A virus, and rotavirus from sewage and ocean water by triplex reverse transcriptase PCR, *Appl Environ Microbiol*, **60**, 2400–07.

TSUGAWA T and HOSHINO Y (2008), Whole genome sequence and phylogenetic analyses reveal human rotavirus G3P[3] strains Ro1845 and HCR3A are examples of direct virion transmission of canine/feline rotaviruses to humans, *Virology*, **380**, 344–53.

URASAWA S, URASAWA T, TANIGUCHI K, and CHIBA S (1984), Serotype determination of human rotavirus isolates and antibody prevalence in pediatric population in Hokkaido, Japan, *Arch Virol*, **81**, 1–12.

VAN DUYNHOVEN Y T, DE JAGER C M, KORTBEEK L M, VENNEMA H, KOOPMANS M P, VAN LEUSDEN F, VAN DER POEL W H, and VAN DEN BROEK M J (2005), A one-year intensified study of outbreaks of gastroenteritis in The Netherlands, *Epidemiol Infect*, **133**, 9–21.

VAN MAARSEVEEN N M, WESSELS E, de BROUWER C S, VOSSEN A C, and CLAAS E C (2010), Diagnosis of viral gastroenteritis by simultaneous detection of Adenovirus group F, Astrovirus, Rotavirus group A, Norovirus genogroups I and II, and Sapovirus in two internally controlled multiplex real-time PCR assays, *J Clin Virol*, **49**, 205–10.

VELAZQUEZ F R (2009), Protective effects of natural rotavirus infection, *Pediatr Infect Dis J*, **28**, S54–S56.

VELAZQUEZ F R, MATSON D O, CALVA J J, GUERRERO L, MORROW A L, CARTER-CAMPBELL S, GLASS R I, ESTES M K, PICKERING L K, and RUIZ-PALACIOS G M (1996), Rotavirus infections in infants as protection against subsequent infections, *N Engl J Med*, **335**, 1022–8.

VENUTA A (2012), Acute arthritis associated with rotavirus gastroenteritis: an unusual extraintestinal manifestation of a common enteric pathogen, *J Clin Rheumatol*, **18**, 49.

VESIKARI T, ITZLER R, KARVONEN A, KORHONEN T, VAN DAMME P, BEHRE U, BONA G, GOTHEFORS L, HEATON P M, DALLAS M, and GOVEIA M G (2009), RotaTeq, a pentavalent rotavirus vaccine: efficacy and safety among infants in Europe, *Vaccine*, **28**, 345–51.

VILLENA C, EL-SENOUSY W M, ABAD F X, PINTO R M, and BOSCH A (2003), Group A rotavirus in sewage samples from Barcelona and Cairo: emergence of unusual genotypes, *Appl Environ Microbiol*, **69**, 3919–23.

WANG D, WU Q, YAO L, WEI M, KOU X, and ZHANG J (2008), New target tissue for food-borne virus detection in oysters, *Lett Appl Microbiol*, **47**, 405–09.

WANG F T, MAST T C, GLASS R J, LOUGHLIN J, and SEEGER J D (2010a), Effectiveness of the pentavalent rotavirus vaccine in preventing gastroenteritis in the United States, *Pediatrics*, **125**, e208–e213.

WANG Y H, KOBAYASHI N, NAGASHIMA S, ZHOU X, GHOSH S, PENG J S, HU Q, ZHOU D J, and YANG Z Q (2010b), Full genomic analysis of a porcine-bovine reassortant G4P[6] rotavirus strain R479 isolated from an infant in China, *J Med Virol*, **82**, 1094–102.

WARD L A, ROSEN B I, YUAN L, and SAIF L J (1996), Pathogenesis of an attenuated and a virulent strain of group A human rotavirus in neonatal gnotobiotic pigs, *J Gen Virol*, **77**(Pt 7), 1431–41.

WARD R (2009), Mechanisms of protection against rotavirus infection and disease, *Pediatr Infect Dis J*, **28**, S57–59.

WARD R L, BERNSTEIN D I, SHUKLA R, MCNEAL M M, SHERWOOD J R, YOUNG E C, and SCHIFF G M (1990), Protection of adults rechallenged with a human rotavirus, *J Infect Dis*, **161**, 440–5.

WARD R L, BERNSTEIN D I, YOUNG E C, SHERWOOD J R, KNOWLTON D R, and SCHIFF G M (1986), Human rotavirus studies in volunteers: determination of infectious dose and serological response to infection, *J Infect Dis*, **154**, 871–80.

WARD R L, CLEMENS J D, KNOWLTON D R, RAO M R, van LOON F P, HUDA N, AHMED F, SCHIFF G M, and SACK D A (1992), Evidence that protection against rotavirus diarrhea after natural infection is not dependent on serotype-specific neutralizing antibody, *J Infect Dis*, **166**, 1251–7.

WIDDOWSON M A, MELTZER M I, ZHANG X, BRESEE J S, PARASHAR U D, and GLASS R I (2007), Cost-effectiveness and potential impact of rotavirus vaccination in the United States, *Pediatrics*, **119**, 684–97.

WIDDOWSON M A, STEELE D, VOJDANI J, WECKER J, and PARASHAR U (2009), Global rotavirus surveillance: determining the need and measuring the impact of rotavirus vaccines, *J Infect Dis*, **200**(Suppl 1), S1–S8.

WILDE J, VAN R, PICKERING L, EIDEN J, and YOLKEN R (1992), Detection of rotaviruses in the day care environment by reverse transcriptase polymerase chain reaction, *J Infect Dis*, **166**, 507–11.

WOLD A E, and ADLERBERTH I (2000), Breast feeding and the intestinal microflora of the infant – implications for protection against infectious diseases, *Adv Exp Med Biol*, **478**, 77–93.

WYATT R G, JAMES H D, Jr., PITTMAN A L, HOSHINO Y, GREENBERG H B, KALICA A R, FLORES J, and KAPIKIAN A Z (1983), Direct isolation in cell culture of human rotaviruses and their characterization into four serotypes, *J Clin Microbiol*, **18**, 310–17.

ZARATE S, ESPINOSA R, ROMERO P, GUERRERO C A, ARIAS C F, and LOPEZ S (2000), Integrin alpha2beta1 mediates the cell attachment of the rotavirus neuraminidase-resistant variant nar3, *Virology*, **278**, 50–4.

ZIEMER C J, BONNER J M, COLE D, VINJE J, CONSTANTINI V, GOYAL S, GRAMER M, MACKIE R, MENG X J, MYERS G, and SAIF L J (2010), Fate and transport of zoonotic, bacterial, viral, and parasitic pathogens during swine manure treatment, storage, and land application, *J Anim Sci*, **88**, E84–E94.

19

Advances in understanding of hepatitis E virus as a food- and waterborne pathogen

W. H. M. Van Der Poel, and A. Berto, Wageningen University and Research Centre, The Netherlands

DOI: 10.1533/9780857098870.4.401

Abstract: Hepatitis E is an acute hepatitis in humans, first recognized in 1980 and caused by the hepatitis E virus (HEV). Mammalian HEVs are divided into four genotypes (1–4) of which gt1 and gt2 are involved in epidemics and gt3 and gt4 are more involved in sporadic cases. The principal mode of spread is faecal–oral from contaminated water supplies, almost exclusively in developing regions. Accumulating evidence indicates that HEV (gt3 and gt4) transmission may be zoonotic in developed regions with swine and perhaps other animal species serving as reservoirs for the virus. The exact transmission routes are unclear, but HEV RNA has been detected in retail pig products.

Key words: hepatitis E, HEV, virus, foodborne, waterborne, zoonosis.

19.1 Introduction

Hepatitis E virus is a single-strand positive-sense RNA virus, designated as the sole member of the genus Hepevirus in the family Hepeviridae. Mammalian HEV genotypes 1 and 2 have been identified in humans whereas genotypes 3 and 4 have been identified in humans and animals.

19.1.1 Hepatitis E virus aetiology

Hepatitis E (HEV) is a hepatotropic virus and the causative agent of hepatitis E, an acute viral hepatitis in humans; it is a single-strand positive-sense

RNA virus. The infection may vary in severity from inapparent infection to fulminant liver failure and death. Although considered acute, chronic infections have been observed in liver and kidney transplant and chronic liver disease (CLD) patients. The mortality rate is between 1% and 4% (Purcell and Emerson, 2008), higher than hepatitis A virus – HAV, a *Picornavirus* – and in people with CLD and in pregnant women it can reach 25–30%.

Hepatitis E is an important public health concern and a major contributor to enterically transmitted hepatitis worldwide (Fig. 19.1) (Meng, 2003). Based on seroprevalence data, it is estimated that one-third of the world's population has been infected with HEV (Purcell and Emerson, 2008).

In endemic regions, hepatitis E occurs in epidemic forms, while in developed regions HEV occurs sporadically (Fig. 19.1).

Hepatitis E is the second most important cause of acute clinical hepatitis in adults throughout Asia, Africa and the Middle East where the infection is endemic. In these countries, the infection mainly spreads through the contamination of water supplies occasionally leading to large-scale outbreaks or epidemics. Hepatitis E is rare in industrialized countries, where infection is historically mostly related to travelling to endemic areas. However, more recently, significant numbers of autochthonous cases have been documented in many developed countries (Dalton *et al.*, 2008).

19.1.2 Taxonomy: evolutionary history and population dynamics of hepatitis E virus

HEV is divided into four genotypes and the characterization is based on the genomic sequence analysis of human and animal isolates (Schlauder and Mushahwar, 2001; Wei *et al.*, 2003). A genetically distinct group has also been identified in avian samples, sharing 50% homology with mammalian isolates.

Genotypes (gts) 1 and 2 appear to be anthroponotic, whereas gts 3 and 4 are zoonotic (Teo, 2010). All four genotypes belong to a single serotype (Emerson *et al.*, 2004). The recent discovery of novel lineages of HEV in rabbits (Geng *et al.*, 2010; Zhao *et al.*, 2009), rats (Johne *et al.*, 2010), and wild boar (Takahashi *et al.*, 2011) has further expanded mammalian HEV diversity. It has been suggested that the HEV sequences found in rabbits represent a novel genotype (Geng *et al.*, 2010; Zhao *et al.*, 2009). However, additional phylogenetic analysis indicated that rabbit HEV is closest to gt 3 (Bilic *et al.*, 2009; Johne *et al.*, 2010) and may have zoonotic potential. In addition, the discovery of a genetically distinct avian HEV (Haqshenas *et al.*, 2001) indicates a very long evolutionary history for the HEV group of viruses. Unlike swine HEV (asymptomatic in pigs), avian HEV shows hepatomegaly in poultry.

The first animal strain of HEV was detected in swine (swine HEV) in 1997 in the USA (Meng *et al.*, 1997). Since then, swine HEV strains have been isolated from all over the world and from several animal species (e.g., wild boar, mongoose and sika deer). In developed regions the human and swine strains show a parallel distribution and circulation (Banks *et al.*, 2004).

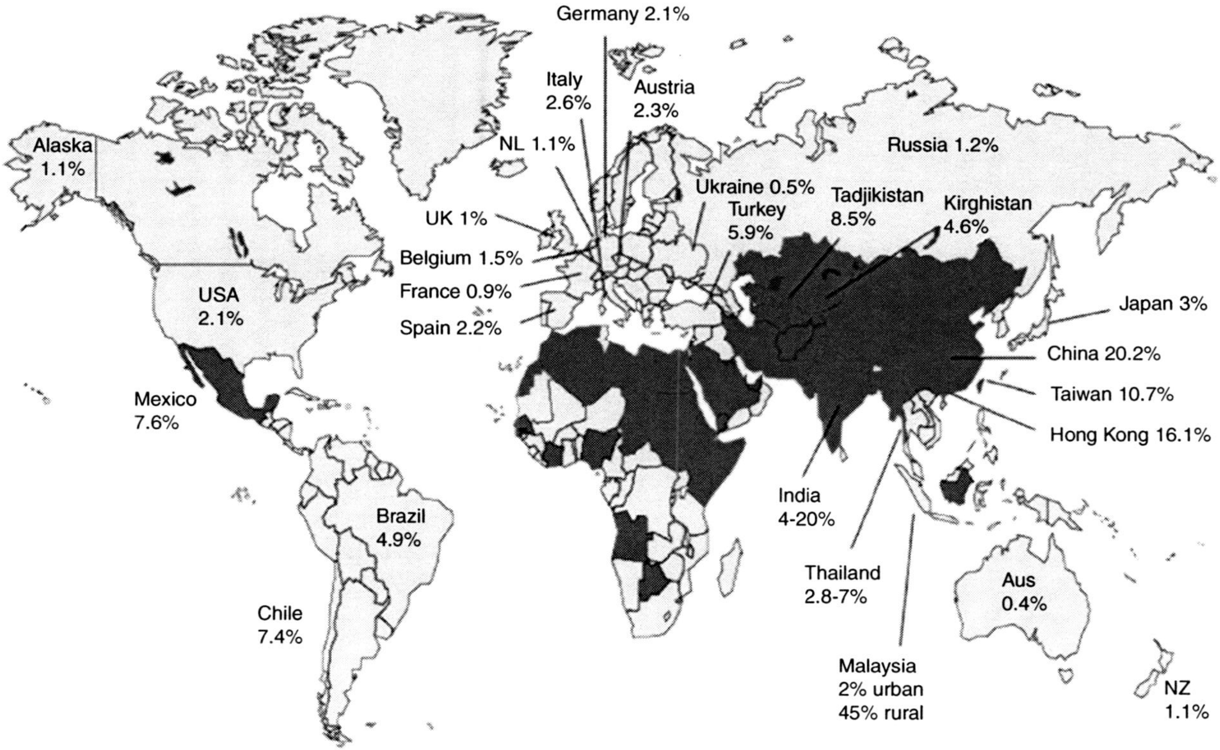

Fig. 19.1 Hepatitis E virus (HEV) endemic areas and global seroprevalence. Dark areas on the map indicate regions of the world that are endemic for hepatitis E and where >25% of acute viral hepatitis is due to HEV infection. Superimposed on the map are seroprevalence rates of HEV from various countries, determined in different studies. Image taken from Chandra *et al.*, 2008.

Purdy and Khudyakov (2010) suggested that HEV can be segregated into two clades. One clade is the enterically transmitted, epidemic form represented by gts 1 and 2, and the other is the zoonotically transmitted, sporadic form exemplified by gts 3 and 4 (Dalton *et al.*, 2008; Reuter *et al.*, 2005; Teo, 2010).

Genotypes 1 and 2 have been identified only in humans; gts 3 and 4 have been identified both in humans and in animals (Goens and Perdue, 2004; Meng *et al.*, 1997; Okamoto, 2007; Panda *et al.*, 2007). Gt 1 HEV has been identified from human cases in Asia and Africa (Lu *et al.*, 2006) while gt 2 was identified first in Mexico and subsequently in Africa. Gt 3 has been identified in humans and animals in several developed countries, such as Japan, Australia, New Zealand and European countries. Gt 4 has been identified in both animals and humans in China, Taiwan, Japan and Vietnam and most recently in the Netherlands (Hakze-van der Honing *et al.*, 2011). HEV strains of gts 1 and 2 have less genomic variability than those of gt 3 and 4 (Okamoto, 2007). This could be due to the differences in the transmission patterns between the genotypes. In addition, the presence of an animal reservoir for gts 3 and 4 could have caused an independent evolution of the virus in specific animal species (Okamoto, 2007).

The clear division of HEV genotypes into two modes of transmission offers an important opportunity for studying molecular evolutionary processes related to the transition from one mode to another. Purdy and Khudyakov (2010) studied the evolutionary history of HEV using several models estimating population dynamics, in terms of time to the most recent common ancestor (TMRCA), and variation in selective pressures acting on different HEV genotypes. They did not analyse HEV gt 2 due to lack of available samples. ORF2 analysis suggests that the mean time of emergence of the ancestor for modern HEV genotypes ranged from 536 to 1344 years ago. For gt 3, the mean time was from 265 to 342 years ago; for gt 4, from 131 to 266 years ago; and for gt 1, from 87 to 199 years ago. Thus, the anthroponotic gt 1 is the most recent compared to the enzootic gts 3 and 4 (Purdy and Khudyakov 2010).

Following Drummond *et al.* (2006), Purdy and Khudyakov (2010) decided to set up a model using ORF2 sequences for gts 1, 3 and 4 to understand the genotype dynamics and to study the demographic history of HEV genotypes. Gt 1 went through an increase in population size between 25 and 35 years ago. Gt 3 population had been stable since 1760, but it grew dramatically over the twentieth century. The effective population size of gt 4 remained constant until 20 years ago, decreasing over 10 years to the original level (Purdy and Khudyakov, 2010).

Purdy and Khudyakov (2010) suggested that HEV has histories dating back tens of thousands to millions of years but early members have been replaced by the modern variants (Purdy and Khudyakov, 2010). An ancient TMRCA is suggested due to contacts between humans and domesticated swine about 11 000 years ago immediately after urbanization started. HEV gt 1 has increased during the last 35 years whereas incidences of gts 3 and 4 seemed lower around 1990 (Purdy and Khudyakov, 2010). This may be due to greater awareness of the HEV health problem around the world and improved diagnostics rather than an actual expansion of the HEV strains (Purdy and Khudyakov, 2010). During the Second World War the increase of HEV cases was probably related to the increasing population size rather than meat consumption (Tanaka *et al.*, 2006). The country-specific HEV evolutionary history observed probably reflects temporal variations in rates of transmission and/or exposure for HEV strains of the same genotype circulating in different geographic regions (Purdy and Khudyakov, 2010).

19.1.3 Genotype classification

Extensive genomic diversity has been observed among HEV isolates, but a single serotype is recognized (Emerson and Purcell, 2003; Okamoto, 2007). Genotype 1 was first identified and subjected to sequencing in 1991 (Tam *et al.*, 1991) from a sample that came from Myanmar (Burma strain) showing more than 88% nucleotide identity with other gt 1 strains isolated in Asia (China, India, Nepal and Pakistan) and Africa (Chad and Morocco) (Okamoto, 2007).

In 1992, a new strain which was completely different from the Burma strain was sequenced from outbreaks in Mexico (1986) and classified as gt 2. Compared to gt 1, which is present in many geographic regions, gt 2 occurs in fewer countries (Lu *et al.*, 2006).

Genotype 3 was identified in 1997 in the USA from an autochthonous infection in a patient with no history of foreign travel; it was sequenced and became the first strain belonging to gt 3 (Kwo *et al.*, 1997). Later on, gt 3 HEV was shown to be distributed in many countries in Asia, Europe, Oceania, North and South America (Banks *et al.*, 2004; Peron *et al.*, 2006; Pina *et al.*, 2000; Takahashi *et al.*, 2003).

Currently, the four genotypes are classified into different subtypes, based on approximately 300–450 nucleotides of sequence in the 5′ end of the ORF2 region which are most conserved among all HEV isolates. The phylogenetic analysis demonstrated that HEV can be divided into a total of 24 subtypes. Gt 1 was divided into five subtypes (1a, 1b, 1c, 1d, 1e); gt 2 into two subtypes (2a, 2b); gt 3 segregates into ten subtypes (3a, 3b, 3c, 3d, 3e, 3f, 3g, 3h, 3i, 3j); and gt 4 into seven subtypes (4a, 4b, 4c, 4d, 4e, 4f and 4g) (Lu *et al.*, 2006) (Fig. 19.2) (Meng *et al.*, 2009).

19.1.4 Morphology and genomic organization

HEV was designated in 2004 as the sole member of the genus *Hepevirus* in the family *Hepeviridae* (Emerson *et al.*, 2004). The HEV genome was first cloned from cDNA libraries prepared from the bile of macaques experimentally inoculated with stool suspensions from human patients (Reyes *et al.*, 1991). A similar PCR was later used to clone the genomes of multiple geographically distinct isolates of HEV (Huang *et al.*, 1992; Panda *et al.*, 2000; Purcell and Emerson, 2001).

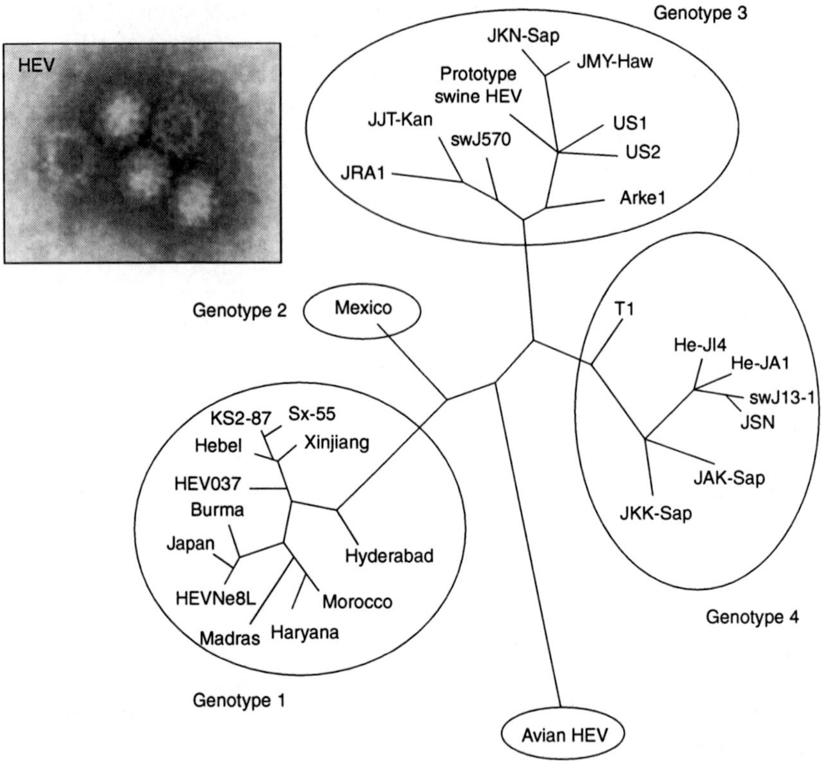

Fig. 19.2 HEV phylogenetic tree and genotypes. Electron micrograph showing HEV particles from the stool of a hepatitis E patient visualized after aggregation with anti-HEV positive serum and negative staining. Phylogenetic tree showing the distribution of human and swine HEV isolates in four distinct genotype numbers 1–4 and the outlier group containing avian HEV. Image taken from Chandra *et al.* (2008).

HEV is a small, non-enveloped, single-stranded, positive-sense RNA virus. The genome size is approximately 7.2 kb (Fauquet *et al.*, 2005; Tam *et al.,* 1991) (Fig. 19.3). The genome of HEV is capped at the 5′ end and polyadenylated at the 3′ end (Fig. 19.3a). It contains short stretches of untranslated regions (UTR) at both ends (Fig. 19.3b, dark grey box). The HEV genome has three open reading frames (ORFs), shown in Fig. 19.3b. ORF1 encodes the non-structural polyprotein (nsp) that contains various functional units: methyltransferase (MeT), papain-like cysteine protease (PCP), RNA helicase (Hel) and RNA-dependent RNA polymerase (RdRp) (Chandra *et al.,* 2008). ORF2 encodes the viral capsid protein, the N-terminal signal sequence and glycosylation loci. ORF3 encodes a small regulatory phosphoprotein. Details of the ORF3 protein are shown in Fig. 19.3. The roles of the ORF3 protein in HEV pathogenesis are promotion of cell survival, modulation of the acute phase response and immunosuppression (Chandra *et al.*, 2008).

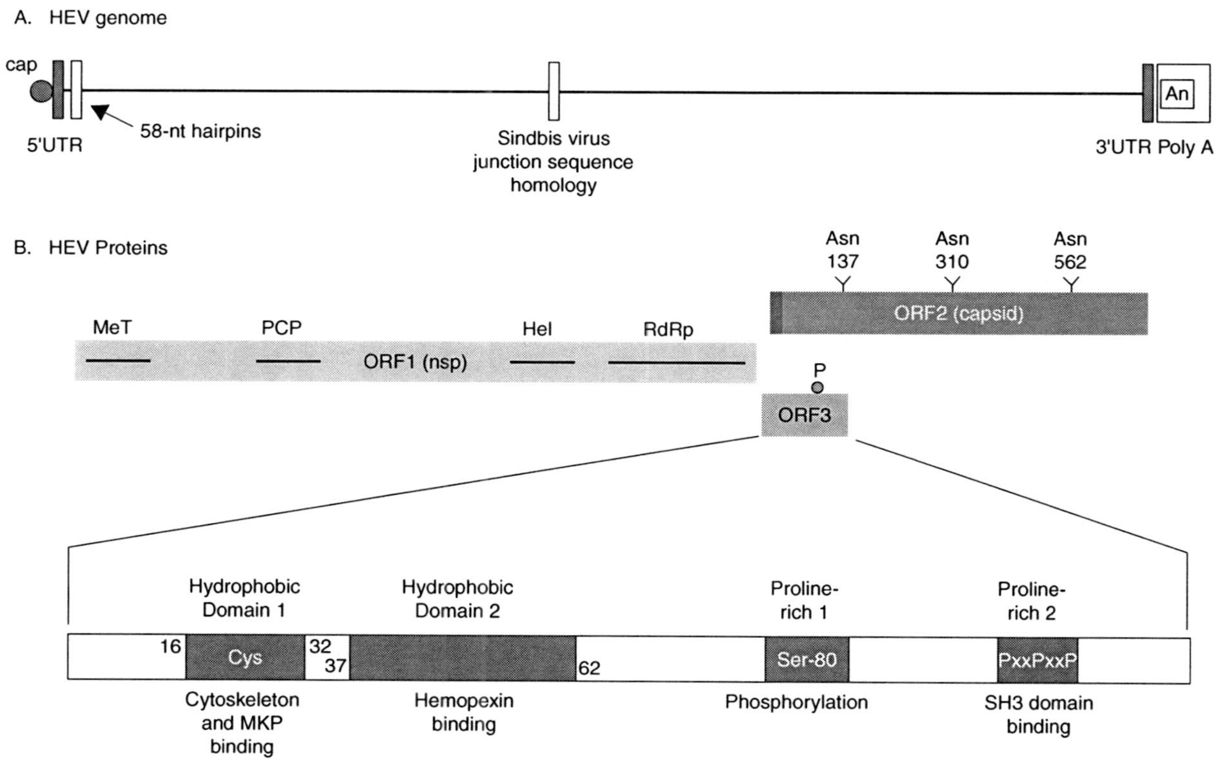

Fig. 19.3 Genome organization and proteins of HEV. (a) The ~7.2 kb positive strand RNA genome of HEV is capped at the 5′ end and polyadenylated at the 3′ end. It contains short stretches of untranslated regions (UTR) at both ends (dark grey box). Other structural features proposed to be important for replication are also indicated. (b) The three open reading frames (ORFs) are shown. ORF1 encodes the non-structural polyprotein (nsp) that contains various functional units – methyltransferase (MeT), papain-like cysteine protease (PCP), RNA helicase (Hel) and RNA-dependent RNA polymerase (RdRp). ORF2 encodes the viral capsid protein; the N-terminal signal sequence (dark grey box) and glycosylation sites are indicated. ORF3 encodes a small regulatory phosphoprotein. Details of the ORF3 proteins are shown, including two N-terminal hydrophobic domains (two dark grey boxes on the left) and two C-terminal proline-rich regions (two dark grey boxes on the right). (Image adapted from Chandra *et al.* (2008).)

19.2 Viral proteins

Open reading frame one (ORF1) is the largest (5079 nt) of the three ORFs and it begins after the 5′ noncoding region (5′-NCR) of 27 to 35 nucleotides (nt). It encodes a 1693 aminoacid polyprotein including viral non-structural proteins such as methyltransferase, a papain-like cysteine protease, a helicase and an RNA-dependent RNA polymerase (RdRp) (Aggarwal and Krawczynski, 2000; Koonin and Senkevich, 1992; Krawczynski *et al.*, 1999, 2000; Magden *et al.*, 2001).

The region between the end of ORF1 and start of ORF3/ORF2 appears to be complex and contains regulatory elements (Tam *et al.*, 1991).

- *Methyltransferase.* The methyltransferase domain has been suggested by computer-assisted assignments to encompass an amino terminal domain between 60 and 240 amino acids. Downstream of the methyltransferase domain there is a Y domain of 200 amino acids but at present no particular function is known.
- *Papain-like cysteine protease.* A Papain-like protease domain follows the Y domain. It is postulated that this viral protease is involved in either co- or post-translational viral polyprotein processing to yield discrete non-structural protein products (Panda *et al.*, 2007).
- *Helicase.* It promotes unwinding of DNA, RNA or DNA duplexes required for genome replication, recombination, repair and transcription (Panda *et al.*, 2007).
- *RNA-dependent RNA polymerase (RdRp).* The RdRp has a crucial role in binding to the 3'UTR (untranslated region) of HEV RNA and directing the synthesis of the complementary strand RNA (Panda *et al.*, 2007).

ORF2 and the major capsid protein. Open reading frame 2 is about 1980 nt in length from nt 5147 to nt 7124, downstream of ORF1. Translation of this region produces the HEV structural polypeptide (pORF2) of 660/599 aminoacids (Okamoto, 2007) and this appears to be highly conserved. The 5' end of ORF2 region presents an average of approximately 350–450 nt most conserved among HEV isolates; recently it has been used for classifying different subgenotypes of HEV (Lu *et al.*, 2006).It is strongly considered that the ORF2 protein transcomplements a replicon that is deficient in capsid protein production and efficiently encapsidates the replicon viral RNA to form stable HEV particles which are infectious for naïve hepatoma cells (Parvez *et al.*, 2011).

- *ORF3 and its product.* Open reading frame 3 (ORF3) partially overlaps with ORF1 by 4 nt and shares most of the remaining nucleotides of ORF2 at the 5' end (Panda *et al.*, 2007). ORF3 encodes for a 123/122 amino acid immunogenic phosphoprotein of 13.5 KDa (pORF3) with yet, a not fully defined function (Tam *et al.*, 1991).

Recent studies using a HEV replicon with a deleted ORF3 in cell culture showed normal RNA replication, suggesting that ORF3 is not required for HEV replication, virion assembly or infection of culture cells (Emerson *et al.*, 2006).

Yamada *et al.* provided evidence that pORF3 is required for virion egress from infected cells (Yamada *et al.*, 2009).

Chandra *et al.* suggested that ORF3 could attenuate inflammatory responses and create an environment for increased viral replication and survival mainly in the liver (Fig. 19.4) (Chandra *et al.*, 2008).

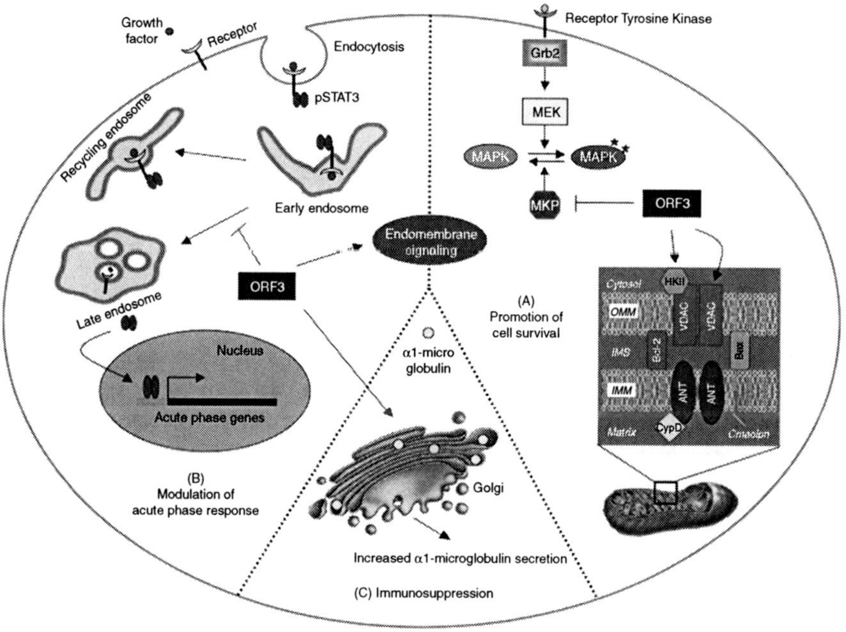

Fig. 19.4 Role of the ORF3 protein in HEV pathogenesis. (a) *Promotion of cell survival.* The ORF3 protein activates MAP kinase by binding and inactivating its cognate phosphatase (MKP). Additionally, it upregulates and promotes homo-oligomerization of the outer mitochondrial membrane porin, VDAC, and increases hexokinase levels, thus reducing mitochondrial depolarization and inhibiting intrinsic cell death. (b) *Modulation of the acute phase response.* The ORF3 protein localizes to early and recycling endosomes, and inhibits the movement of activated growth factor receptors to late endosomes. This prolongs endomembrane growth factor signaling and contributes to cell survival. Through this mechanism, pORF3 also reduces the nuclear transport of pSTAT3, a critical transcription factor for the expression of acute phase response genes. (c) *Immunosuppression.* The ORF3 protein promotes the secretion of α1-microglobulin, an immuno-suppressive protein that could act in the immediate vicinity of the infected cell. Figure taken from Chandra *et al.*, 2008.

19.3 Hepatitis E virus replication, pathogenesis and clinical symptoms

It is believed that the primary site of HEV replication is the liver. Hepatocytes are the most likely cell type but the process of binding to cell surfaces and entry of HEV is still unknown. Common infection symptoms include fever and mild to severe acute hepatitis.

19.3.1 The HEV replication cycle

Viral receptor and entry
The cell surface molecules that bind HEV or its capsid proteins are not known yet. He *et al.* (2008) described a truncated peptide of ORF2 involved in binding and entry of the following cell lines: HepG2, Huh-7, PLC/PRF5 and A549 cells (He *et al.*, 2008).

Model of HEV replication
The process by which HEV RNA enters the target cells is still unknown (Fig. 19.5: 1–2). In the cytoplasm the genomic RNA is translated into non-structural proteins (Fig. 19.5: 3). The genome amplification step involves replication of positive strand genomic RNA into negative strand RNA intermediates (Fig. 19.5a). These are used as template for the synthesis of the genomic positive strands (Fig. 19.5b). This is akin to alphaviruses and a region homologous to alphavirus junction sequences is proposed to serve as the subgenomic promoter. The subgenomic RNA can then be translated into the structural protein(s) (Fig. 19.5: 5). Based on *in vitro* expression and replicon studies, some details have now begun to emerge. The genomic RNA is packaged with the capsid protein to assemble new virions (Fig. 19.5: 6). The mechanism by which the virion is released from the cell has yet to be characterized (Chandra *et al.*, 2008).

It is unclear whether gut cells are infected following ingestion of the virus. It is believed that the primary site of HEV replication is the liver, with hepatocytes being the most likely cell type (Williams *et al.*, 2001). Results support infection and replication in non-hepatic cell types such as A549 lung carcinoma cells and in Caco-2 colon carcinoma cells. Although it is not efficient, viral replication has been demonstrated. In pigs experimentally infected with swine HEV, positive-sense viral RNA was detected in almost all tissues at some point during the infection, but negative-sense RNA intermediates were detected primarily in the small intestine, lymph node, colon and liver (Williams *et al.*, 2001). In a recent report, HEV RNA was detected in peripheral blood mononuclear cells, but due to the lack of an efficient HEV, *in vitro* cell culture verifying the evidence of viral replication in this compartment in patients with HEV infection was not possible (Ippagunta *et al.*, 2010).

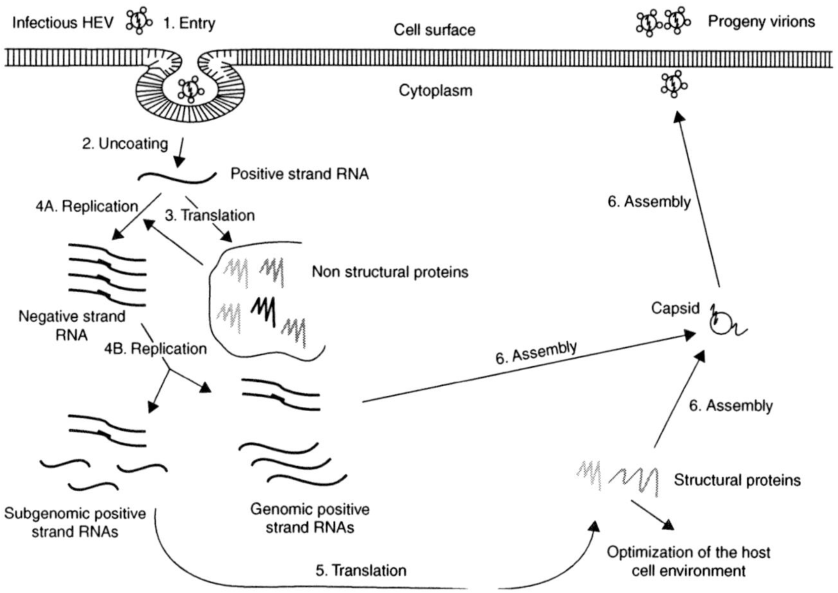

Fig. 19.5 Proposed replication cycle of HEV. The virus enters the target cell (1) and uncoats (2) to release the HEV genomic RNA through uncharacterized processes. The genomic RNA is translated in the cytoplasm into non-structural proteins (3). The replicase replicates the positive strand genomic RNA into negative strand RNA intermediates (a) and then into positive strands (b). This is the genome amplification step. Additionally, the positive strand subgenomic RNA is also synthesized that is translated into structural proteins (5). The capsid protein packages the genomic RNA to assemble new virions (6) that are then released from the cell through an uncharacterized mechanism. (Image adapted from Chandra *et al.* (2008)).

19.3.2 Pathogenesis, clinical signs and symptoms

In humans

Studies on HEV have facilitated the understanding of elements of its replication, host immune response, and liver pathology in HEV-infected patients and primates (Balayan *et al.*, 1983; Krawczynski *et al.*, 1999). It has been estimated that the infectivity titre of HEV for macaques is 10 000-fold higher when inoculated intravenously compared with when it is ingested (Meng, 2003). Clinical signs of hepatitis E are dose-dependent in these animal models and production of disease may require challenge doses 1000 times or more greater than that required for infection (Emerson and Purcell, 2003).

After ingestion, the virus probably replicates in the intestinal tract (the primary site of replication has not been identified yet) and reaches the liver, presumably via the portal vein (Panda *et al.*, 2007). It replicates in the cytoplasm of hepatocytes (Hussaini *et al.*, 1997) and is released into the bile and bloodstream by mechanisms that are still poorly understood, and excreted in the faeces (Emerson and Purcell, 2003). The incubation period is 4–5 weeks based on an oral infection study in human volunteers (Balayan *et al.*, 1983; Chauhan *et al.*, 1993). Viral excretion in faeces begins approximately 1 week prior to the onset of illness and typically persists for 2–4 weeks; in some cases RT-PCR has yielded positive results until 52 days after onset (Arora *et al.*, 1996). The viraemia can be detected in the first 2 weeks after the onset of illness (Chauhan *et al.*, 1993; Clayson *et al.*, 1995; Nanda *et al.*, 1995). Viral excretion and viraemia have also been detected by RT-PCR prior to liver abnormalities, which normally appear with an elevation of aminotransferase levels and reach a peak by the end of the first week of clinical symptoms. Simultaneously the humoral immune responses appear. Anti-HEV IgM or IgG levels are detected by enzyme immunoassay (Aggarwal *et al.*, 2000; Jameel, 1999). Anti-HEV IgM appears during clinical illness and then gradually disappears over 4–5 months). Anti-HEV IgG appears some days later than IgM, persisting for several years (Dawson *et al.*, 1992; Favorov *et al.*, 1992). The persistence of HEV antibody in the sera is still unclear. One study observed that 14 years after acute HEV infection anti-HEV antibodies were still circulating in 47% of patients (Khuroo *et al.*, 1993). Anti-HEV IgM is a useful tool in the diagnosis of acute HEV infection, whereas IgG anti-HEV does not necessarily indicate recent HEV infection (Aggarwal and Krawczynski, 2000).

Hepatitis E symptoms are typical of acute icteric viral hepatitis; the most common recognizable symptom is an initial prodromal phase (preicteric phase) lasting a few days, with a variable combination of flu-like symptoms, fever, mild chills, abdominal pain, anorexia, nausea, aversion to smoking, vomiting, clay-coloured stools, dark or tea-coloured urine, diarrhoea, arthralgia, asthenia and a transient macular skin rash (Aggarwal and Krawczynski, 2000). These symptoms are followed in a few days by lightening of the stool colour and the appearance of jaundice. Itching may also occur. With the onset of jaundice, fever and other prodromal symptoms tend to diminish rapidly and then disappear entirely. Laboratory test abnormalities include bilirubinuria, a variable rise in serum bilirubin (predominantly conjugated), marked elevation in serum alanine aminotransferase (ALT), aspartate aminotransferase, gammaglutamyltransferase activities and a mild rise activity in serum alkaline phosphatase. The magnitude of the increase in transaminase levels does not always correlate closely with the severity of liver injury. The illness is usually self-limiting and typically lasts 1–4 weeks (Aggarwal and Krawczynski, 2000). Recent reports described evidence of chronic HEV infection in transplant patients (Haagsma *et al.*, 2008; Kamar *et al.*, 2008a). A small number of patients have a prolonged clinical illness with marked cholestasis (cholestatic hepatitis), including persistent jaundice and prominent itching. In these cases, laboratories observed a rise in alkaline phosphatase and a persistent bilirubin rise even after transaminase levels returned to normal (Aggarwal and Krawczynski, 2000). The prognosis is good, as jaundice finally resolves spontaneously after 2–6 months. Within the past few years, HEV has been demonstrated to be responsible for chronic hepatitis, which can rapidly evolve to cirrhosis in immuno-compromised patients (Kamar *et al.*, 2008b, c; Dalton *et al.*, 2009). However, little data regarding HEV-related extra-hepatic manifestations has been published, although an association between neurologic manifestations (e.g., Guillain-Barré syndrome, neuralgic amyotrophy, acute transverse myelitis) and acute HEV infection has been suggested (Fong and Illahi, 2009; Kamani *et al.*, 2005; Loly *et al.*, 2009; Mandal and Chopra, 2006; Sood *et al.*, 2000). Previously, the association between neurologic signs and symptoms and HEV infection has been based on detection of anti-HEV immunoglobulin IgM in serum. However, Rianthavorn *et al.* (2010) reported a case of HEV gt 3-induced neurological amyotrophic in which HEV RNA was detected in the serum of patients with neurologic signs and symptoms (Kamar *et al.*, 2011). Recently, Kamar *et al.* (2011) detected HEV RNA in the cerebrospinal fluid (CSF) of a kidney transplant recipient with chronic HEV infection and neurological signs and symptoms (Kamar *et al.*, 2010a). Kamar *et al.* also reported seven chronic HEV gt 3 infections, with development of neurological complications, between January 2004 and April 2009 (Kamar *et al.*, 2010b), and the disappearance of the neurological symptoms correlated with a decreasing HEV titre.

Other infected individuals have a milder clinical course and develop only non-specific symptoms that resemble those of an acute viral febrile illness without jaundice (anicteric hepatitis) (Aggarwal *et al.*, 2000). Histological features of hepatitis E may differ from other forms of acute viral hepatitis. Nearly half of hepatitis E patients have a cholestatic hepatitis, which is characterized by canalicular bile stasis and gland-like transformation of parenchymal cells. In these patients, degenerative changes in hepatocytes are less marked (Aggarwal and Krawczynski, 2000; Gupta and Smetana, 1957). The Kupffer cells appear prominent. Portal tracts are enlarged and contain an inflammatory infiltrate consisting of lymphocytes, a few polymorphonuclear leucocytes and eosinophils. Polymorphonuclear cell volume is particularly increased in the cholestatic type of lesion (Aggarwal and Krawczynski, 2000; Gupta and Smetana, 1957). In cases with severe liver injury, a large proportion of the hepatocytes are affected, leading to sub-massive or massive necrosis with collapse of liver parenchyma (Aggarwal and Krawczynski, 2000). At the beginning, HEV infection is entirely inapparent and asymptomatic. A small percentage of patients have more severe symptoms with fulminant or subacute (or late-onset) hepatic failure. The exact frequencies of asymptomatic infection and of anicteric hepatitis are not known but a large proportion of individuals test positive for anti-HEV IgG (Aggarwal and Krawczynski, 2000). In resource-limited regions hepatitis E is common in young adult and adults (15–40 years of age). Hepatitis E appears to cause more severe disease in pregnant women, particularly during the second and third trimesters (Aggarwal and Krawczynski, 2000). HEV commonly causes intrauterine infection as well as substantial prenatal morbidity and mortality (Khuroo *et al.*, 1995), suggesting that the placenta may be the viral replication site, as with Lassa fever (Hamid *et al.*, 1996; McCormick, 1986). Death is usually due to encephalopathy, haemorrhagic diathesis or renal failure. In a preliminary report (Longer *et al.*, 1993) cynomolgus monkeys infected intravenously with HEV developed acute tubular necrosis with focal haemorrhages, suggesting that HEV may replicate in monkey kidneys. In pregnant monkeys, however, no increased mortality has been observed (Arankalle *et al.*, 1993). In endemic countries such as India, the mortality rate of women with acute gt 1 hepatitis E in the third trimester of pregnancy is usually fairly high (26–64%) (Navaneethan *et al.*, 2008). Why acute HEV infection in pregnant women causes severe liver dysfunction is not known.

Experimentally, HEV transmission has occurred from infected to uninfected in-contact pigs, confirming that the virus is contagious (Kasorndorkbua *et al.*, 2004).

Many reports (Kamar *et al.*, 2008b, 2010a, 2011) have described the rapid progress of chronic HEV infection in immunosuppressed transplant patients in cirrhosis(Haagsma *et al.*, 2008). Established cirrhosis has been shown in two HIV-infected patients, in 2009 in the UK and France (Dalton *et al.*, 2011). HEV and HIV coinfection need to be more extensively studied (Kamar *et al.*, 2011). What it is known to date is that there appears to be no difference in anti-HEV seroprevalence between patients with HIV infection and control group (Dalton *et al.*, 2011).

In pigs

The mechanisms of HEV pathogenesis and replication are poorly understood, due to the absence of a practical animal model and an efficient *in vitro* cell-culture system for HEV. HEV may replicate in tissues and organs other than the liver (Aggarwal, 2011).

Williams *et al.* confirmed the clinical and pathological findings of HEV infection in pigs previously reported by Halbur *et al.* (2001). It is unclear how the virus reaches the liver and extra-hepatic site(s), but HEV is presumably transmitted by the faecal–oral route. The hepatocytes are the only known sites of HEV replication (Williams *et al.*, 2001). It has been hypothesized that liver damage induced by HEV infection may be due to the immune response to the invading virus and may not be a direct cause of viral replication in hepatocytes (Kageyama *et al.*, 2003; Williams *et al.*, 2001). Several studies with naturally infected pigs described HEV RNA detectable in different organs and tissues, even after viraemia was cleared (Kolasa, 2010). For swine HEV-infected pigs, viral RNA was detected in small intestines, colons, lymphnodes, and livers (Kolasa, 2010; Williams *et al.*, 2001). Other extra-hepatic tissues such as kidney, tonsil, and salivary gland had detectable HEV RNA for only 1 or 2 weeks (Williams *et al.*, 2001). It appears that lymphnodes and the intestinal tract are the main extra-hepatic sites of replication, but the significance of identifying these sites is at present unclear.

Experimentally infected pigs do not present any clinical signs. Histological analysis shows signs of mild focal liver necrosis but no fever or other signs (e.g., lack of appetite) are observed.

19.4 Susceptibility and effects in different sectors of the population

In developing countries HEV epidemics occur frequently and are related to unhygienic water conditions. In industrialized regions, HEV sequences from indigenously acquired cases show high homology to swine sequences.

19.4.1 Hepatitis E in non-endemic regions

In developed regions, the transmission of HEV is probably mainly via a zoonotic route. Evidence of this is provided by the many autochthonous (indigenously acquired) cases worldwide where swine isolates show a very high RNA sequence homology to human HEV isolates (Erker *et al.*, 1999). Additional evidence comes in the form of experimental transmission of human isolates to pigs and of swine HEV to primates (Erker *et al.*, 1999). Hepatitis E autochthonous transmission has been recorded in most developed countries and regions including the USA, Europe (including the UK, France, the Netherlands, Austria, Spain, Greece and Italy), and developed countries of the Asia-Pacific region (Japan, Taiwan, Hong Kong, Australia) (Ijaz *et al.*, 2005).

In the UK, the disease appears to be more common among residents of coastal and estuarine areas (Ijaz *et al.*, 2005). Zoonotic transmission has been proposed (Ijaz *et al.*, 2005) and is now widely accepted; in some developed regions transmission appears to be seasonal with peaks in spring and summer (Ijaz *et al.*, 2005).

Patients with unexplained hepatitis are tested by serological tests and these are the cases where the disease is most often recognized. Generally, the symptoms are similar to those in endemic regions. In developed areas, the majority of the cases have been in middle-aged or elderly men, often coexisting with another established disease (Mansuy *et al.*, 2004, Table 19.1).

19.4.2 Hepatitis E in disease-endemic regions

Although the majority of hepatitis E cases in resource-limited countries are sporadic, local epidemic outbreaks occur frequently. They are usually separated by a few years and they can affect several thousand individuals (Naik *et al.*, 1992; Zhuang, 1991). Most outbreaks are due to consumption of drinking water contaminated by human faeces and the longevity varies from a few weeks to over a year (Naik *et al.*, 1992). The outbreaks frequently follow heavy rainfall and floods (Corwin *et al.*, 1999); conflicts leading to concentrations of displaced persons in refugee camps (Corwin *et al.*, 1999); or they are associated with disposal of human excrement into rivers (Corwin *et al.*, 1999). Foodborne transmissions have been described in resource-limited areas, but due to a relatively long incubation period (up to 9 weeks), establishing a correlation between consumption of pork food and occurrence of disease is difficult.

Table 19.1 Hepatitis E virus clinical and epidemiological features. Comparisons of the four different HEVgenotypes

Feature	Genotype 1	Genotype 2	Genotype 3	Genotype 4
Severity of hepatitis	Asymptomatic disease, mild and severe hepatitis, also fulminant hepatitis	Asymptomatic disease, mild and severe hepatitis, also fulminant hepatitis	Compared to Gt1 more asymptomatic cases and less severe hepatitis, also chronic hepatitis	Severity in between Gt1 and Gt3 infections. Chronic hepatitis not reported
Geographical distribution	Asia, Africa	Mid. America, N. and Middle Africa	America, Europe, Australia	China, Japan
Transmission	Faecal–oral, waterborne	Faecal–oral, waterborne	Foodborne transmission, direct zoonotic transmission not demonstrated	Potential foodborne and direct zoonotic transmission not demonstrated
Epidemiological pattern	Large epidemics, small outbreaks and sporadic cases	Epidemics, small outbreaks and sporadic cases	Sporadic cases and small clusters	Sporadic cases and small clusters
Animal reservoir	No indications	No indications	Domestic pigs, potential reservoirs in wildlife	Potential reservoirs in domestic pigs and wildlife
Risk groups	Young and middle-aged men, pregnant women	Young and middle-aged men	Elderly people	Elderly people
Remarks	High fatality rates in pregnant women	Not as many clinical and epidemiological reports compared to other genotypes		

In India HEV is hyper-endemic, the majority of the cases reported are sporadic and 40% of sewage specimens obtained throughout all seasons are HEV-positive (Arankalle *et al.*, 1995; Khuroo, 1980, 2010; Wong *et al.*, 1980, 2004). Interfamilial spread is not common but multiple cases in one family have been reported (Arankalle *et al.*, 1995; Khuroo, 1980, 2010; Wong *et al.*, 1980, 2004). It is suggested that this is due to shared contaminated water rather than person-to-person transmission, as the time interval between cases is too short.

Studies in endemic regions show high seroprevalence rates ranging from 15 to 60% (Arankalle *et al.*, 1995, 2000; Assis *et al.*, 2002; Ataei *et al.*, 2009; Clayson *et al.*, 1997; Tran *et al.*, 2003). Notably, the age-specific seroprevalence profiles for HEV are found to differ from those reported for antibody to HAV, even though in endemic countries the transmission routes for these two viruses are similar (Arankalle *et al.*, 1995). HAV seroprevalence rates are over 95% in children by the age of 10, whereas HEV infection is rarely detected in children (Arankalle *et al.*, 1995).

The peak incidence in sporadic cases of hepatitis E in endemic regions occurs in the 15–35 age group (Arankalle *et al.*, 1995; Khuroo, 1980, 2010; Wong *et al.*, 1980, 2004). Additionally, HEV infections are predominantly reported in men, with a male-to-female ratio ranging from 1/1 to 3/1 (Margolis H). This sex bias is, however, not seen in children presenting with hepatitis E (Balayan, 1997). The reason why men more commonly develop hepatitis E infection is not understood, but males outnumbering females may be due to a greater risk of exposure to HEV infection (Balayan, 1997). Morbidity rates during hepatitis E epidemics have ranged from 1% to 15% (Khuroo *et al.*, 1981). Higher mortality rates and fulminant liver disease have been described among pregnant women during hepatitis E outbreaks (Khuroo *et al.*, 1981). Furthermore, HEV infection during pregnancy is not only associated with severe disease or higher mortality, but also with an increased risk of prenatal mortality and low birth weight. In developing regions neonatal vertical transmission rates have been estimated at 78.9% (Khuroo *et al.*, 2009) but it remains unclear whether high morbidity and mortality rates during pregnancy are also seen in developed regions. The exact cause of this predilection to severe disease in pregnant women needs to be studied in greater depth, including the suspicion that it is due to hormonal or immunological factors (Kar *et al.*, 2008) (Table 19.1).

19.5 Epidemiology of hepatitis E virus

In industrialized countries hepatitis E occurs sporadically, and both travel related and autochtonous infections are reported. The route of transmission is seldom determined and zoonotic spread is assumed. In endemic regions both sporadic and epidemic outbreaks occur through contaminated water and food.

19.5.1 Epidemiology in humans

The epidemiology of HEV differs significantly in industrialized and non-industrialized countries. In resource-limited countries, the infection is endemic and spreads mainly through contamination of water supplies.

Data from sero-surveys have forced re-evaluation of the epidemiology of hepatitis E and has, indirectly, indicated that HEV infections are vocationally acquired in industrialized countries (Khuroo, 2010).

In industrialized countries, hepatitis E occurs sporadically and affects mainly visitors returning from endemic areas. Some of the cases in industrialized countries, however, are non-travel-related and are considered as being autochthonous. Autochthonous cases have been reported in N and S America, many European countries and industrialized countries of the Asia-Pacific area, including Japan, Taiwan, Hong Kong and Australia.

Zoonotic spread of the virus was first suspected when genomic sequences of HEV isolates from two autochthonous cases in the USA were found to be closely related to swine HEV (Kwo *et al.*, 1997).

In 2001 (Lees, 2010) HEV swine strains were identified in the Netherlands, showing close genetic similarity to European human strains. In 2002 field isolates of swine HEV were identified from different geographic areas (Li *et al.*, 2011), demonstrating nucleotide identity between swine (88–100%) and human strains (89–98%). In 2004 in the UK two swine HEV strains were identified with 100% amino acid sequence identity over a partial sequence amplified by PCR, to one autochthonous human case of HEV in the UK (Fig. 19.6) (Banks *et al.*, 2004).

In Spain, 2006, de Deus *et al.* (Martin *et al.*, 2007) identified swine affected by HEV with 85.7–100% nucleotide identity between swine and human strains.

Recently, a hepatitis E outbreak on board a UK cruise ship returning from an 80-night world cruise was investigated. The UK Health Protection Agency (HPA) was informed of four cases of jaundice on board a cruise ship which departed from Southampton on 7 January and returned on 28 March 2008. An epidemiological investigation was launched by HPA to identify any additional cases of hepatitis E and potential risk factors for infection. The investigation was a cohort study including all UK passengers who were on the cruise at any point, and a total of 851 of the 2850 eligible passengers took part. Finally, 33 (4%) individuals were identified with recent acute HEV infection, although only 11 of these were symptomatic cases. A common source in the outbreak was shellfish eaten on board the cruise ship. The causative agent was identified as HEV gt 3 which was closely related to the other gt 3 strains isolated in Europe (Emerson and Purcell, 2003; Hirano *et al.*, 2003; Meng *et al.*, 2002).

The route of transmission has not been determined in most of these cases, although zoonotic spread has been proposed (Dalton *et al.*, 2008). To investigate the possible presence of animal reservoirs, several animal species have been tested for anti-HEV antibodies. HEV antibodies have been detected in different animal species – monkeys, pigs, rodents, chickens, dogs, cats, cattle

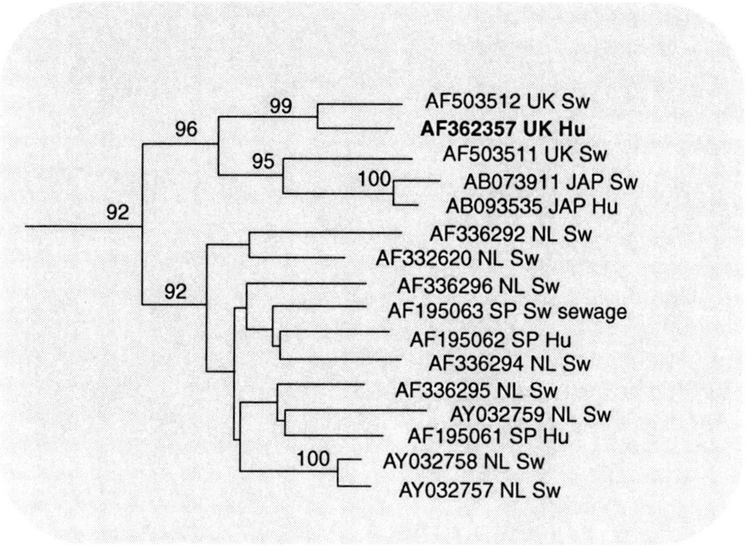

Fig. 19.6 Phylogenetic tree. Human United Kingdom isolate (AY362357) is shown in bold and compared with closely related swine and human hepatitis E virus isolates. Bootstrap values greater than 70% are considered significant and are indicated. Figure taken from Banks *et al.* (2004).

and sheep – both in resource-limited and industrialized countries, suggesting that these animals could be infected by HEV (Emerson and Purcell, 2003; Hirano *et al.*, 2003; Meng *et al.*, 2002).

Waterborne (effectively faecal–oral) and foodborne transmissions, as well as transfusion of infected blood products and vertical (maternal-foetal) transmission (Aggarwal and Naik, 2009), are now established routes of HEV transmission.

Faecal–oral transmission of HEV occurs primarily through contaminated water in endemic regions, where it is responsible for both sporadic and epidemic outbreaks (Balayan *et al.*, 1983) (Fig. 19.7). In epidemic form, the disease may involve tens of thousands of cases and is the cause of considerable morbidity and mortality, posing a major public health problem in endemic regions. In India alone, over 2.2 million cases of hepatitis E are thought to occur annually. Hepatitis E in resource-limited countries has different epidemiological and clinical features and investigation is patchy. Disruption of water supplies in conflict zones has been shown to have caused major outbreaks of hepatitis E amongst displaced persons (Boccia *et al.*, 2006; Guthmann *et al.*, 2006). During the conflict in Darfur, Sudan, over 6 months in 2004, 2621 hepatitis E cases were recorded (incidence 3.3%), with a case-fatality rate of 1.7% (45 deaths, 19 of which were pregnant women). Interestingly in this outbreak, as well as age, a risk factor for infection was drinking chlorinated surface water (odds ratio, 2.49; 95% confidence interval, 1.22–5.08) (Guthmann *et al.*, 2006) (Table 19.2) (Guthmann *et al.*, 2006).

Fig. 19.7 Epidemic region in N. Africa. Drinking water is collected from a canal in which sewage flows, garbage may be dumped and people wash. Photo by TesfaalemTekleghiorghis 2011.

Although on the basis of phylogenetic data it is assumed the disease has been around for many years, hepatitis E was first recognized only during a 1978 epidemic of hepatitis in Kashmir Valley, which involved an estimated 52,000 cases of icteric hepatitis with 1700 deaths (Aggarwal and Naik, 2009).

Based on these data, the possibility of another human hepatitis virus distinct from post-transfusion non-A, non-B hepatitis was postulated. Balayan *et al.* (1983) successfully transmitted the disease to themselves by oral administration of pooled stool extracts of nine patients from a non-A, non-B hepatitis outbreak which had occurred in a Soviet military camp located in Afghanistan. Over the years, hepatitis E has been identified as a major health problem in resource-limited countries with unsafe water supplies and poor sanitation facilities.

Foodborne transmission of HEV was first demonstrated in clusters of Japanese patients that had eaten raw or undercooked pig, wild boar or sika deer meat. The genomic sequences of HEV identified from these patients were identical to those recovered from the frozen leftover meat (Tei *et al.*, 2003).

In addition, Colson *et al.* (2010) reported, based on epidemiological findings, that five cases of autochthonous acute hepatitis E were linked to ingestion of raw figatelli (Colson *et al.*, 2010). Figatelli are traditional sausages from Corsica made from pig liver and commonly eaten uncooked; they can be considered as a possible source of HEV infection in France.

Legrand-Abravanel *et al.* (2011) studied 38 patients in south-western France with HEV gt 3 infection. The patients were compared with matched

Table 19.2 Risk factors for asymptomatic hepatitis E virus infection in a random sample of Mornay population, Darfur, Sudan, September 2004

Exposure	Number of individuals		Risk of asymptomatic HEV infection (%)	Risk ratio (95% CD)
	All (n = 104)	Asymptomatic HEV infection (n = 49)		
Age group (years)				
>45	17	5	29.4	Reference
15–45	51	22	43.1	1.47 (0.48–4.47)
0–14	36	22	61.1	2.08 (0.67–6.43)
Sex				
Female	73	35	47.9	Reference
Male	31	14	45.2	0.94 (0.45–1.99)
Size of the family				
≤6 persons	65	29	44.6	Reference
>6 persons	39	20	51.3	1.15 (0.57–2.30)
Presence of animals in the house				
No	54	23	42.6	Reference
Yes	50	26	52.0	1.22 (0.62–2.41)
Ever collected water from river				
Never	76	35	46.1	Reference
Yes	28	14	50.0	1.09 (0.51–2.31)
Number of water reservoirs in house				
1	19	6	31.6	Reference
2	37	16	43.2	1.37 (0.46–4.07)
>2	48	27	56.3	1.78 (0.63–5.00)
Source of drinking water				
Borehole, unchlorinated	42	17	40.5	Reference
Surface water, chlorinated	55	28	50.9	1.26 (0.61–2.59)
Other	7	4	57.1	1.41 (0.37–5.45)
Use latrines				
At least sometimes	31	37	45.7	Reference
Never	23	12	52.2	1.14 (0.51–2.54)
Wash hands before eating				
At least sometimes	80	35	43.8	Reference
Never	24	14	58.3	1.33 (0.62–2.88)
Wash hands after defecating				
At least sometimes	83	38	45.8	Reference
Never	21	11	52.4	1.14 (0.50–2.61)

Source: Adapted from Guthamann et al. (2006).

control participants in the same region who had no evidence of HEV infection. A questionnaire showed that consumption of game meat, consumption of processed pork and consumption of mussels were all statistically significantly more common among case patients than among control participants.

Eating undercooked pork and pork products is quite common in Europe. Although the study by Legrand-Abravanel *et al.* (2011) did not address the consumption of undercooked meat, other studies have explored its association with hepatitis E. A case-control study by Wichmann *et al.* (2008) in Germany found that consumption of raw or undercooked wild boar meat and offal (liver, kidney, and intestine) was statistically significantly associated with autochthonous HEV infection.

Other direct evidence of zoonotic transmission was recently reported by Kim *et al.* (2011). A sporadic case of acute hepatitis E was confirmed as gt 4 HEV in a 51-year-old Korean female. The case was reported as the first case of presumably zoonotic transmission of HEV identified as gt 4 in a patient with acute hepatitis E after ingestion of raw bile juice from a wild boar living on a mountain in South Korea.

Furthermore, it has been shown that commercial pig livers purchased from local grocery stores as food in Japan, the United States (11%) and Europe (Feagins *et al.*, 2007; Yazaki *et al.*, 2003) are contaminated by HEV and that some of the HEV-contaminated commercial pig livers still contain infectious virus (Feagins *et al.*, 2007).

A study performed in a UK hospital tested 500 blood donors, 336 individuals over the age of 60 years and 126 patients with chronic liver disease for HEV IgG. At the end of the study 40 cases of autochthonous hepatitis E (gt 3) were identified (Dalton *et al.*, 2008). These patients did not have a recent travel history and the major probability was autochthonous hepatitis (Dalton *et al.*, 2008). Autochthonous hepatitis E in developed regions is frequently misdiagnosed as drug-induced liver injury, a common problem that occurs with increased frequency in elderly people. The outcome can be poor in those individuals with underlying chronic liver disease, with mortality approaching 70% (Dalton *et al.*, 2008).

Seroprevalence data from industrialized countries suggests that subclinical or unrecognised infection is common. However, the real incidence of clinical autochthonous hepatitis E in the UK is not known (Lewis *et al.*, 2008), but increased and improved surveillance for hepatitis E has shown it may be more common than hepatitis A. (Dalton *et al.*, 2008; De Silva *et al.*, 2008). Data from France and Japan show similar trends (Mitsui *et al.*, 2006; Peron *et al.*, 2006). The literature contains relatively few reports from the USA regarding autochthonous hepatitis E (Amon *et al.*, 2006). It is known that the HEV human seroprevalence is around 21% in blood donors; it is strongly possible that the majority of the patients with unexplained hepatitis are 'missed' since hepatitis E infection is often not considered a diagnostic possibility in the USA (Amon *et al.*, 2006).

19.5.2 Epidemiology in pigs and other animals

It is now accepted that autochthonous hepatitis E in developed regions has a largely zoonotic source.

Evidence supporting this statement can be found in many reports where HEV sequences derived from pigs are closely related to HEV sequences from humans. Many animals, for example domestic pigs, wild boar, deer, mongoose, trout and bivalves are found to be HEV-positive (Li *et al.*, 2007; Nakamura *et al.*, 2006). In addition, HEV antibodies are detected in domestic and feral animals. The detected strains in the wide range of animals mainly belong to genotypes 3 and 4. Some species host divergent non-zoonotic strains. To date only domestic pigs have been demonstrated to be a true reservoir of HEV (Gt3) (Bouwknegt *et al.*, 2008). Data obtained from animal experiments suggest that genotype 3 (zoonotic) is the most attenuated relative to genotype 1 and 2 (human to human) where they cause more severe pathology (Arankalle *et al.*, 2001; Wang *et al.*, 2002; Yarbough, 1999). Although genotype 3 is considered by some to be the most attenuated for human beings, differences in genotype virulence are still not well understood (Halbur *et al.*, 2001). It is also suggested that gt 4 in India is different from gt 4 subtypes found in China, Japan, and Taiwan. Data show that Indian gt 4 is apparently not able to infect humans and it has been suggested that this is probably due to the substitution of 26 amino acids, 16 in ORF-1, 8 in ORF-2 and 2 in ORF-3 (Chobe *et al.*, 2006).

Autochthonous hepatitis E gt 3 was first observed in the USA by comparing human sequences with pig sequences (Meng *et al.*, 1997). HEV seroprevalence in pig farms is high worldwide and it can be as high as 100% in some pig herds (Balayan *et al.*, 1983). Furthermore, it has been demonstrated in some studies that slaughterhouse workers, farmers, veterinarians and people that work in close contact with pigs may be more exposed and at greater risk of HEV infection, with this category of workers presenting a higher HEV IgG seroprevalence than non-pig workers (Bouwknegt *et al.*, 2008). More than 20% of pigs close to slaughter are excreting HEV in faeces (Rutjes *et al.*, 2007). Watercourses may be HEV-contaminated due to run-off of pig faeces from outdoor pig units. In addition HEV has been detected in slurry lagoons on pig farms, from urban sewage works, and from pig slaughterhouses (Wibawa *et al.*, 2004). The risks of spreading untreated slurry on farmland still need to be quantified but it should be remembered that rhesus monkeys have been infected with HEV recovered from sewage and slurry (Pina *et al.*, 1998).

19.6 Hepatitis E virus stability and inactivation

HEV has proven difficult to propagate *in vivo*, and despite some recent improvements, there is no doubt that the failure to develop a repeatable and efficient *in vitro* propagation system for HEV has hindered attempts to understand the environmental survival and other physical and pathobiological characteristics of HEV. The determination of these qualities would potentially offer much valuable information in understanding the epidemiology and control of HEV infections.

Feagins *et al.* performed an HEV heat inactivation study in a pig animal model in 2008. The objective of the study was to determine if traditional cooking methods are effective in inactivating infectious HEV present in contaminated commercial pig livers. The result obtained was that four of the five pigs inoculated with a pool of two HEV-positive liver homogenates incubated at 56°C (Bouwknegt *et al.*, 2008) for 1 h developed an active HEV infection, shedding virus in the faeces. The pigs inoculated with a pooled homogenate of two HEV-positive livers stir-fried at 191°C (Bouwknegt *et al.*, 2008) for 5 min and the group of pigs inoculated with a pooled homogenate of two HEV-positive livers boiled in water for 5 min showed no evidence of infection as there was no seroconversion, viraemia, or faecal virus shedding in any of the inoculated pigs (Feagins *et al.*, 2007).

What is not clear is how effective the usual processing procedures for uncooked pig products are in inactivating pathogens such as HEV. Moreover, the risk of HEV infection via the consumption of HEV-contaminated pig tissues raises public health concerns, since it is not clear what cooking conditions will be effective in inactivating the virus present in the contaminated pig tissues.

HEV can be found in the liver, blood, intestinal tract and skeletal muscle, all of which are consumed in one form or another and often together, for example in sausages. How safe are these products? The question is difficult to answer because HEV grows poorly in cell culture, and testing HEV viability *in vivo* requires nonstandard laboratory animals.

To date there have not been any other inactivation studies with HEV. Inactivation studies with UV light have been conducted with other viruses such as HAV, calicivirus or other enteric viruses, or with bacteria (Matallana-Surget *et al.*, 2010; Nuanualsuwan *et al.*, 2002). Exposure to solar ultraviolet (UV) radiation is a primary means of virus inactivation in the environment, and germicidal (UVC) light is used to inactivate viruses in hospitals and other critical public and military environments (Giese, 1976; Nicholson *et al.*, 2005). Safety and security constraints have hindered exposing highly virulent viruses to UV and gathering the data needed to assess the risk of environments contaminated with high-consequence viruses (Borio *et al.*, 2002). UV sensitivity has been extrapolated for some viruses from data obtained with UVC (254 nm) radiation by using a model based on the type, size and strandedness of the nucleic acid genomes of the different virus families (Lytle and Sagripanti, 2005; Rzezutka and Cook, 2004). Therefore, there was little information to allow accurate modelling, confident extrapolation, and prediction of the UV sensitivity of viruses deposited on contaminated surfaces, conditions more likely to be relevant to public health or biodefence.

Other inactivation studies with disinfectants such as chlorine have not been performed until now with HEV. Sodium hypochlorite, a derivate of chlorine solution, commonly known as bleach, is frequently used as a disinfectant or a bleaching agent. US government regulations (21 CFR Part 178) allow food-processing equipment and food contact surfaces to be sanitized with solutions containing bleach, provided that the solutions do not exceed 200 parts per million (ppm) available chlorine. A 1-in-5 dilution of household bleach with water is effective against many bacteria and viruses.

19.6.1 Heat inactivation

HEV transmission in industrialized regions is not fully understood. It has been suggested and is now widely accepted that HEV transmission is zoonotic (Meng *et al.*, 2001; Yazaki *et al.*, 2003). Tei *et al.* (2003) reported direct evidence of zoonotic HEV transmission via the consumption of grilled or undercooked commercial pig liver purchased from local grocery stores in Japan (Tei *et al.*, 2003). The majority of the patients in that study had a history of consuming undercooked pig livers prior to the onset of the disease, indicating that consumption of pig livers is a risk factor for hepatitis E (Tei *et al.*, 2003). Eleven percent of livers purchased from local grocery stores in the United States, 6% in the Netherlands (Bouwknegt *et al.*, 2007) and 9.5% in the United Kingdom were found to be contaminated by HEV (Berto *et al.*, 2012). HEV inactivation and environmental resistance is not a well-covered topic and little information is available. As an orally transmitted virus, HEV is most likely resistant to inactivation by the acidic conditions of the stomach. The ability of HEV to survive harsh or extreme environmental conditions can be attributed at least in part to its non-enveloped viral structure (Feagins *et al.*, 2007, 2008).

To study the efficacy of heat inactivation, a homogenate of pig liver known to contain infectious HEV was subjected to heating, simulating some normal cooking conditions, and was applied to 3D cell cultures to determine the effect of the virus inactivation as measured by HEV RNA copy numbers in cell supernatants.

Differences in the Ct values were observed between the supernatant of the cells infected with non- heated liver and supernatant of cells infected with HEV-positive liver heated at 56°C for 1 h. As we can see in Fig. 19.8, the Ct values were lower (ranging between 40 and 29) in the sample infected with the homogenate of non-heated liver compared to the supernatant of cells that received as inoculum the homogenate of liver heated at 56°C for 1 h. The Ct values in the supernatant of cells infected with HEV-positive liver heated at 56°C for 1 h were higher, probably reflecting partial virus inactivation. Full HEV inactivation was observed in the inoculum heated at 100°C, since no HEV RNA was detected by real-time RT-PCR at any point during the experiment. The results are similar to those of Feagins *et al.* (2007, 2008) where the pigs infected with HEV-positive liver heated at 56°C were shedding virus in the faeces, showing that the treatment was not sufficient to inactivate HEV. Furthermore, the similarity of the results of *in vivo* and *in vitro* experiments in this study underlines the potential for the 3D cell-culture system to replace traditional *in vivo* infectivity studies.

In Europe most pork meat is cooked prior to consumption, but there are some exceptions where pork meat is eaten raw, for example, liver sausages

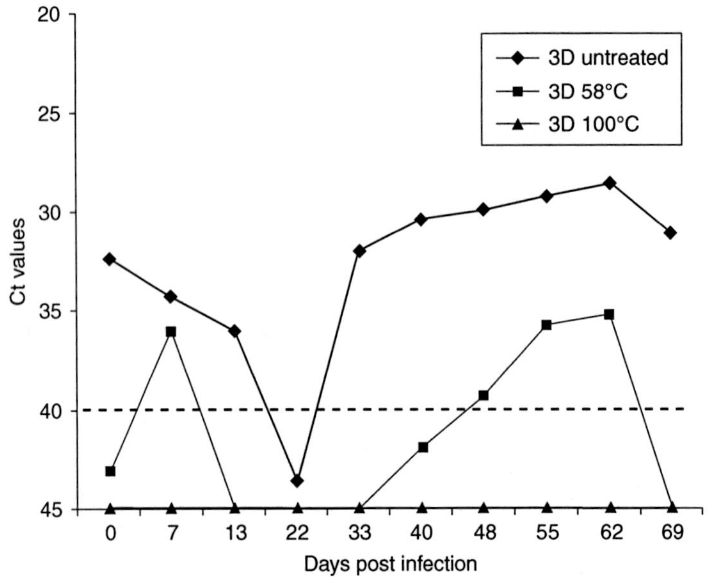

Fig. 19.8 RT-PCR detections (Ct values) of hepatitis E virus samples cultured after different heat treatments. Treatment of HEV-infected liver at 100°C leads to inactivation of the virus.

in France. The United States Department of Agriculture (USDA) and the United States National Pork Board (NPB) recommend a cooking method for fresh pork that will result in a minimum internal cooking temperature of 71°C (http://www.fsis.usda.gov/is_it_done_yet/, accessed on 15 March 2007). A time stipulation is suggested based on the level of heat but many recipes do not specify a minimum cooking temperature. Stir-frying and boiling are the two most widely used and accepted methods for cooking pig livers for consumption. Feagins *et al.* established that stir-frying and boiling of HEV-contaminated pig livers can effectively inactivate the virus by using a swine bioassay to determine the virus infectivity (Feagins *et al.*, 2007, 2008). Using an *in vitro* system, Emerson *et al.* (2005) reported that HEV is approximately 50% inactivated when heated at 56°C for 1 h. Described study in the book performed by the authors (unpublished) that incubation of homogenate of contaminated pig livers at 56°C for 1 h (temperature that produced an internal cooking temperature slightly below the recommended 71°C without burning the tissue) did not fully inactivate the virus, as HEV RNA was detected during the course of the experiment. The results support the *in vitro* results of Emerson *et al.* (2005) and the *in vivo* results of Feagins *et al.* (2007), which show that adequate cooking of HEV-contaminated commercial pig livers will inactivate HEV in the tissue, thereby decreasing the risk of foodborne HEV transmission. Importantly the work confirmed that partial inactivation of HEV (heat at 56°C for 1 h) may allow the virus to initiate an active infection *in vitro* while the treatment of the liver at 100°C appears to be sufficient to inactivate the virus completely.

Other HEV inactivation methods in use include UV light treatment and NaOCl treatment, among others. However most consistent inactivation results in *in vitro* and *in vivo* models are reported using heat inactivation. Thorough cooking of pork meat is an effective means of inactivating HEV, and should therefore be recommended.

19.7 Diagnostic procedures

Enzyme-linked immunosorbent assays (ELISA), conventional reverse transcriptase PCR (RT-PCR) and real-time RT-PCR, cell culture, confocal microscopy and electron microscopy have been used for detection or confirmation of HEV infection. These methods differ significantly in their sensitivity and specificity.

19.7.1 ELISA

HEV recombinant proteins and synthetic peptides, corresponding to immunodominant epitopes of the ORF2 and ORF3 structural proteins of the virus, have been sourced as the capturing antigen (Emerson and Purcell, 2003). Sub-units of ORF2 have been expressed in different systems, such as prokaryotic, insect, animal and plant cells in order to obtain pure antigen for ELISA (Dawson *et al.*, 1992). Recombinant antigens derived from ORF2 generally have a superior sensitivity and specificity. In common with all serological tests, ELISA can only be applied once antibody has developed, in most cases at least 2 weeks after infection. However, serological tests are able to discriminate between IgM and IgG, thus enabling the acute phase to be distinguished from the convalescent phase of infection. HEV antibody prevalence has been reported in several studies in industrialized countries (Dalton *et al.*, 2008; Dawson *et al.*, 1992). Commercially available ELISAs have improved in recent years, but it is suspected that some of the earlier prevalence data reflected subclinical infections and serological cross-reactivity that may have contributed to this high seroprevalence in the non-endemic areas (Emerson and Purcell, 2003; Krawczynski *et al.*, 2000).

19.7.2 Conventional RT-PCR

Conventional RT-PCR assays are currently utilized in direct diagnosis of HEV. The samples collected may be faeces, serum (from animal or human), cultures of infected cells cultivated in 2D and 3D configurations, or post mortem tissue highly positive as bile and liver (Panda *et al.*, 2007; Varma *et al.*, 2011). HEV is an RNA virus and the RNA needs to be extracted before being subjected to the reverse-transcription reaction phase to cDNA. This is a limiting step, because cDNA is easily degradable; if in the samples the viral load is initially too low, a false negative may result. Various sets of sense and antisense synthetic oligonucleotide primers may be used for the detection of the HEV genome, differing according to conservative region targets in the genome against the central or terminal part of ORF1 or C-terminal of ORF2 (Panda *et al.*, 2007). There are reports describing broad-spectrum degenerate primers for identifying positive samples from all genotypes. For example A1/S1 and 3156/7 primers (Li *et al.*, 2008) are used to amplify the ORF2 region. Often the first product of PCR amplification, it is too small to be visualized by electrophoresis. However, if the first product of PCR has been amplified by nested RT-PCR with the internal primers A2S2 (Martelli *et al.*, 2008) and 3158/9 (Li *et al.*, 2008), respectively, the PCR product became clearly visible on the electrophoresis gel through ethidium bromide staining.

19.7.3 Real-time RT-PCR

Real-time RT-PCR is becoming the most popular method for direct detection of HEV in clinical samples. It is a sensitive tool in epidemiological investigations and a fast and reliable technique enabling both detection and confirmation of specificity genotyping. The full viral genome of HEV was cloned in 1991 (Tam *et al.*, 1991). Since then several pairs of primers have been designed to amplify various segments of the genome. The primers are mainly designed to the conserved regions (helicase, polymerase and the terminal fragment of ORF2) of the HEV genome (Panda *et al.*, 2007). The development of real-time RT-PCR, whereby the accumulation of the PCR amplicon can be detected in real time, has enabled the quantification of HEV.

19.7.4 Cell culture and new technology for *in vitro* propagation of the virus

Several cell lines for *in vitro* replication of HEV have been tested in the 2D monolayer culture system (Huang *et al.*, 1992, 1995, 1999). These cell lines were hepatocytes from non-human primates, human embryonic lung diploid cells (2BS), human carcinoma alveolar basal epithelial cells (A549), hepatocarcinoma cells (PLC/PRF/5), hepatocellular human carcinoma (HepG2) and primary hepatocytes from non-human primates. However, the majority of the cell lines did not support replication of HEV or the virus growth was limited, that is, low titre virus. The lack of an efficient and reliable cell-culture system and a practical animal model for HEV have hindered studies on mechanisms of HEV replication, transmission, pathogenesis and environmental survival.

In a recent study, Tanaka *et al.* (2007) tested 21 cell lines including PLC/PRF/5 cells using a faecal suspension with high HEV load as inoculum. A high load of HEV was detected in the culture supernatant of cultivated PLC/PRF/5 cells from day 12 post-inoculation. At AHVLA laboratory, several attempts have been made to reproduce Tanaka's work using field swine

HEV PCR-positive faecal materials as inoculum, but without success. In 2011 Okamoto described for the first time a cell-culture system capable of secreting infectious HEV in high titres into culture media (Okamoto, 2011). The success with the original JE03–1760F strain has been extended to other strains that can support the replication of HEV with an even higher efficiency, and can be passaged through many generations (Okamoto, 2011). Okamoto was able to infect PLC/PRF/5 cells with both sera and faeces of patients and observe high HEV titre in the cell-culture system (Okamoto, 2011). Okamoto has also engineered infectious HEV cDNA clones, and has affirmed that this system, reinforced by reverse genetics, will answer numerous previously unanswered questions surrounding the epidemiology, viral absorption/entry, packaging and delivery of viral particles, providing further understanding of the life cycle of HEV. Okamoto's experiments have not yet been successfully reproduced anywhere.

Hence, an efficient *in vitro* propagation system for HEV is crucial for HEV research in general. There are several reports in the literature demonstrating the potential of the Rotating Wall Vessel (RWV) (Nickerson *et al.*, 2001), a new 3D culture system for the growth of fastidious viruses. The RWV is a cylindrical bioreactor which is rotated on an axis parallel with the ground. A solid body mass rotation of the culture medium is obtained, creating a low fluid-shear environment (Nauman *et al.*, 2007). The RWV culture method has been shown to be applicable to fluid-shear stress-related studies in suspension. The cells are maintained in suspension by the resolution of the centrifugal, gravitational and Coriolis forces, so cells placed in the RWV bioreactor experience minimal mechanical stresses and high mass transport (of nutrients, oxygen, etc.) and are thus able to assemble into tissue-like aggregates. The RWV technology is used to simulate the low-shear environment inherent to microgravity (Klaus, 2001). This 3D culture system has been used to grow fastidious norovirus from faecal materials (Kageyama *et al.*, 2003), and recently it was demonstrated that the system can also be used for *in vitro* cultivation of HEV.

19.8 Hepatitis E virus prevention and control

Several hepatitis E virus vaccine candidates have been produced and may contribute to HEV prevention in future. HEV vaccination in swine may be considered to block epidemic spread in pig populations.

19.8.1 Vaccination

In the absence of a reliable cell-culture system for HEV, vaccine development has been difficult. Two candidate vaccines have successfully completed phase 3 clinical trials in humans. Baculovirus-expressed ORF2 protein from a Pakistani strain of HEV has been licensed by Smith Kline-Beecham (Shrestha *et al.*, 2007). In the Royal Nepalese Army, a vaccination trial to prevent HEV clinical disease showed 95.5% efficacy (95% CI). In another trial in China, the vaccine was prepared with a recombinant protein from the HEV ORF2 viral capsid expressed in *Escherichia coli* (HEV 239) (Huang *et al.*, 2010). Vaccine efficacy after three doses was 100% (95% CI 72.1–100.0). It was considered that these two vaccines could prevent HEV morbidity and mortality in pregnant women, patients with chronic liver disease in endemic areas, organ transplant patients and other immuno-compromised subjects who may contract HEV gt 3 in industrialized countries.

As far as we know, these two vaccines cover gt 1, but nothing is known about prevention of gt 3, and it would be impossible to set up a vaccination plan against HEV gt 3 for the entire world population, mostly because in non-endemic areas HEV is sporadic and incidence is generally still very low. The production of an HEV vaccine for pigs would be more feasible and cheaper, but it is acknowledged that in the absence of any disease in pigs, it might not be justified or practicable to vaccinate pigs. However, when considering the options for control of autochthonous acquired gt 3 and gt 4 HEV in humans, it is important to have some data on the estimated impact and optional timing of HEV vaccination of pigs. This would be useful feasibility data in case of changes in the incidence of human gt 3 infections in developed regions or other events that may require the vaccination of pigs.

Few studies have been done on the dynamics of HEV transmission. Bouwknegt *et al.* (2008) described HEV transmission among pigs from chains of one-to-one transmission. The model describes HEV transmission in pigs and it can be used both with animal contact-exposure experiments and in the field. Each age group or contact-exposure animal is subdivided into three distinct categories: pigs that are susceptible (S), infectious (I) or recovered (R). The system described by this SIR model is assumed to be in an endemic equilibrium. This endemic equilibrium can only exist when the virus is sufficiently transmissible. The transmissibility is expressed by the reproduction number (R_0) and it is the number of infections by an infectious individual during its entire infectious period (in an infinite susceptible population). The endemic equilibrium assumes that $R_0 > 1$. In the SIR model it is assumed that the infected animals reach immunity after infection. The latent period between infection and excretion of infectious virus was observed to be 3 days in intravenously inoculated pigs. R_0 for contact-exposure was estimated to be 8.8 (CI 95%), showing the potential of HEV to cause epidemics in populations of pigs (Bouwknegt *et al.*, 2008).

Casas *et al.* reported a longitudinal survey study on swine HEV infection dynamics conducted in different herds (Casas *et al.*, 2011), but the dynamics of HEV transmission were analysed using SPSS 15.1 software (SPSS Inc., Chicago, IL, USA) and not a mathematical model such as SIR. Mathematical modelling helps to understand HEV transmission dynamics (Satou and Nishiura, 2007) and can also be used to find out what vaccination strategy would be most effective to eradicate an endemic virus such as HEV.

19.8.2 Potential targets for the development of antiviral drugs

Various steps in the HEV life cycle can be potential targets for the development of antiviral drugs. The methyltransferase and guanyltransferase activities in the ORF1 protein are strictly virus-specific and thus good targets for antiviral development (Magden *et al.*, 2001). The RNA helicase of HEV has been biochemically characterized and it is essential for replication of the viral RNA genome (Karpe and Lole, 2010), but it is not clear how distinct a potential drug target it is from human helicases. The HEV RdRp expressed in *E. coli* was shown to bind the 3' end of the viral RNA genome (Aggarwal *et al.*, 2001), but its biochemical activity has so far not been characterized. Since the RdRp is unique to RNA viruses, it would again be a good drug target, and perhaps some viral inhibitors against this target can be explored. Interference with HEV RNA replication has been attempted using ribozymes and small interfering RNAs. Mono- and di- hammerhead ribozymes designed against the 3' end of the HEV genomic RNA were shown to inhibit expression from a reporter construct in HepG2 cells (Sriram *et al.*, 2003). In A549 cells infected with HEV, small interfering RNAs (siRNAs) against the ORF2 region were also shown to offer protection (Huang *et al.*, 2010). While such approaches are feasible *in vitro*, the delivery and targeting of such inhibitors *in vivo* would be the real challenge. At least one study in immuno-compromised transplant patients with chronic HEV infection has also shown the efficacy of Ribavarin monotherapy (Kamar *et al.*, 2010b). Again, the utility of this approach among the vast majority of HEV infections that are acute remains questionable.

19.8.3 HEV control

The VITAL project (integrated monitoring and control of foodborne viruses in the European food chain) (http://www.eurovital.org/) has published guidelines that should be considered by pork consumers and pork products handlers. These are summarized below.

- Recent information indicates that a very large proportion of commercial and domestic pig herds are infected by the hepatitis E virus (HEV), which can cause hepatitis in humans.
- There are no official control policies regarding HEV in pigs.
- At any given time, it is possible that most pigs within a herd have an active infection.
- Thoroughly cooked pig meat and liver should be safe to eat as the virus should be destroyed by cooking; however, consumption of these products raw or undercooked leads to the risk of infection for the consumer.

In slaughterhouses contamination of pig skin with faeces may occur during movement of the animals through the corridor to the slaughter point. If skinning is carried out without hot steaming or burning, contaminating viruses could be transferred to the surface of meat and increase the risk of cross-contamination. Pork meat and fat, equipment and utensils, workers and the environment could also be contaminated especially in the case of inappropriate removal or accidental perforation of the intestine. Special care should be taken to prevent direct contact of pig muscle and fat with faeces or bile.

19.9 References

AGGARWAL, R. (2011). 'Hepatitis E: historical, contemporary and future perspectives'. *Journal of gastroenterology and hepatology* **26**(Suppl 1(April 1979)): 72–82.

AGGARWAL, R., S. KAMILI, J. SPELBRING and K. KRAWCZYNSKI. (2001). 'Experimental studies on subclinical hepatitis E virus infection in cynomolgus macaques'. *J Infect Dis* **184**(11): 1380–5.

AGGARWAL, R., D. KINI, S. SOFAT, S. R. NAIK and K. KRAWCZYNSKI. (2000). 'Duration of viraemia and faecal viral excretion in acute hepatitis E'. *Lancet* **356**(9235): 1081–2.

AGGARWAL, R. and K. KRAWCZYNSKI. (2000). 'Hepatitis E: an overview and recent advances in clinical and laboratory research'. *J Gastroenterol Hepatol* **15**(1): 9–20.

AGGARWAL, R. and S. NAIK (2009). 'Epidemiology of hepatitis E: current status'. *J Gastroenterol Hepatol* **24**(9): 1484–93.

AMON, J. J., J. DROBENIUC, W. A. BOWER, J. C. MAGAÑA, M. A. ESCOBEDO, I. T. WILLIAMS, B. P. BELL and G. L. ARMSTRONG. (2006). 'Locally acquired hepatitis E virus infection, El Paso, Texas'. *J Med Virol* **78**(6): 741–6.

ARANKALLE, V. A., M. S. CHADHA, K. BANERJEE, M. A. SRINIVASAN and L. P. CHOBE. (1993). 'Hepatitis E virus infection in pregnant rhesus monkeys'. *Indian J Med Res* **97**: 4–8.

ARANKALLE, V. A., M. S. CHADHA, S. M. MEHENDALE and S. P. TUNGATKAR. (2000). 'Epidemic hepatitis E: serological evidence for lack of intrafamilial spread'. *Indian J Gastroenterol* **19**(1): 24–8.

ARANKALLE, V. A., M. V. JOSHI, A. M. KULKARNI, S. S. GANDHE, L. P. CHOBE, S. S. RAUTMARE, A. C. MISHRA and V. S. PADBIDRI. (2001). 'Prevalence of anti-hepatitis E virus antibodies in different Indian animal species'. *J Viral Hepat* **8**(3): 223–7.

ARANKALLE, V. A., S. A. TSAREV, M. S. CHADHA, D. W. ALLING, S. U. EMERSON, K. BANERJEE and R. H. PURCELL. (1995). 'Age-specific prevalence of antibodies to hepatitis A and E viruses in Pune, India, 1982 and 1992'. *J Infect Dis* **171**(2): 447–50.

ARORA, N. K., S. K. NANDA, S. GULATI, I. H. ANSARI, M. K. CHAWLA, S. D. GUPTA and S. K. PANDA. (1996). 'Acute viral hepatitis types E, A, and B singly and in combination in acute liver failure in children in north India'. *J Med Virol* **48**(3): 215–21.

ASSIS, S. B., F. J. SOUTO, C. J. FONTES and A. M. GASPAR. (2002). 'Prevalence of hepatitis A and E virus infection in school children of an Amazonian municipality in Mato Grosso State'. *Rev Soc Bras Med Trop* **35**(2): 155–8.

ATAEI, B., Z. NOKHODIAN, A. A. JAVADI, N. KASSAIAN, P. SHOAEI, Z. FARAJZADEGAN and P. ADIBI. (2009). 'Hepatitis E virus in Isfahan Province: a population-based study'. *Int J Infect Dis* **13**(1): 67–71.

BALAYAN, M. S. (1997). 'Epidemiology of hepatitis E virus infection'. *J Viral Hepat* **4**(3): 155–65.

BALAYAN, M. S., A. G. ANDJAPARIDZE, S. S. SAVINSKAYA, E. S. KETILADZE, D. M. BRAGINSKY, A. P. SAVINOV and V. F. POLESCHUK. (1983). 'Evidence for a virus in non-A, non-B hepatitis transmitted via the fecal-oral route'. *Intervirology* **20**(1): 23–31.

BANKS, M., R. BENDALL, S. GRIERSON, G. HEATH, J. MITCHELL and H. DALTON (2004). 'Human and porcine hepatitis E virus strains, United Kingdom'. *Emerg Infect Dis* **10**(5): 953–5.

BANKS, M., G. S. HEATH, S. S. GRIERSON, D. P. KING, A. GRESHAM, R. GIRONES, F. WIDEN and T. J. HARRISON. (2004). 'Evidence for the presence of hepatitis E virus in pigs in the United Kingdom'. *Vet Rec* **154**(8): 223–7.

BERTO, A., F. MARTELLI, S. GRIERSON and M. BANKS. (2012). 'Hepatitis E virus in pork food chain, United Kingdom, 2009–2010'. *Emerg Infect Dis* **18**(8): 1358–60.

BILIC, I., B. JASKULSKA, A. BASIC, C. J. MORROW and M. HESS. (2009). 'Sequence analysis and comparison of avian hepatitis E viruses from Australia and Europe indicate the existence of different genotypes'. *J Gen Virol* **90**(Pt 4): 863–73.

BOCCIA, D., J. P. GUTHMANN, H. KLOVSTAD, N. HAMID, M. TATAY, I. CIGLENECKI, J.-Y. NIZOU, E. NICAND and P. JEAN GUERIN. (2006). 'High mortality associated with an outbreak of hepatitis E among displaced persons in Darfur, Sudan'. *Clin Infect Dis* **42**(12): 1679–84.

BORIO, L., T. INGLESBY, C. J. PETERS, A. L. SCHMALJOHN, J. M. HUGHES, P. B. JAHRLING, T. KSIAZEK, K. M. JOHNSON, A. MEYERHOFF, T. O'TOOLE, M. S. ASCHER, J. BARTLETT, J. G. BREMAN, E. M. EITZEN JR, M. HAMBURG, J. HAUER, D. A. HENDERSON, R. T. JOHNSON, G. KWIK, M. LAYTON, S. LILLIBRIDGE, G. J. NABEL, M. T. OSTERHOLM, T. M. PERL, P. RUSSELL, K. TONAT and WORKING GROUP ON CIVILIAN BIODEFENSE. (2002). 'Hemorrhagic fever viruses as biological weapons: medical and public health management'. *JAMA* **287**(18): 2391–405.

BOUWKNEGT, M., B. ENGEL, M. M. HERREMANS, M. A. WIDDOWSON, H. C. WORM, M. P. KOOPMANS, K. FRANKENA, A. M. DE RODA HUSMAN, M. C. DE JONG and W.H. VAN DER POEL. (2008). 'Bayesian estimation of hepatitis E virus seroprevalence for populations with different exposure levels to swine in The Netherlands'. *Epidemiol Infect* **136**(4): 567–76.

BOUWKNEGT, M., K. FRANKENA, S. A. RUTJES, G. J. WELLENBERG, A. M. DE RODA HUSMAN, W. H. VAN DER POEL and M. C. DE JONG. (2008). 'Estimation of hepatitis E virus transmission among pigs due to contact-exposure'. *Vet Res* **39**(5): 40.

BOUWKNEGT, M., F. LODDER-VERSCHOOR, W. H. VAN DER POEL, S. A. RUTJES and A. M. DE RODA HUSMAN. (2007). 'Hepatitis E virus RNA in commercial porcine livers in The Netherlands'. *J Food Prot* **70**(12): 2889–95.

CASAS, M., R. CORTES, S. PINA, B. PERALTA, A. ALLEPUZ, M. CORTEY, J. CASAL and M. MARTÍN. (2011). 'Longitudinal study of hepatitis E virus infection in Spanish farrow-to-finish swine herds'. *Vet Microbiol* **148**(1): 27–34.

CHANDRA, V., A. KAR-ROY, S. KUMARI, S. MAYOR and S. JAMEEL. (2008). 'The hepatitis E virus ORF3 protein modulates epidermal growth factor receptor trafficking, STAT3 translocation, and the acute-phase response'. *J Virol* **82**(14): 7100–10.

CHANDRA, V., S. TANEJA, M. KALIA and S. JAMEEL. (2008). 'Molecular biology and pathogenesis of hepatitis E virus'. *J Biosci* **33**(4): 451–64.

CHAUHAN, A., S. JAMEEL, J. B. DILAWARI, Y. K. CHAWLA, U. KAUR and N. K. GANGULY. (1993). 'Hepatitis E virus transmission to a volunteer'. *Lancet* **341**(8838): 149–50.

CHOBE, L. P., K. S. LOLE and V. A. ARANKALLE. (2006). 'Full genome sequence and analysis of Indian swine hepatitis E virus isolate of genotype 4'. *Vet Microbiol* **114**(3–4): 240–51.

CLAYSON, E. T., B. L. INNIS, K. S. MYINT, S. NARUPITI, D. W. VAUGHN, S. GIRI, P. RANABHAT and M. P. SHRESTHA. (1995). 'Detection of hepatitis E virus infections among domestic swine in the Kathmandu Valley of Nepal'. *Am J Trop Med Hyg* **53**(3): 228–32.

CLAYSON, E. T., M. P. SHRESTHA, D. W. VAUGHN, R. SNITBHAN, K. B. SHRESTHA, C. F. LONGER and B. L. INNIS. (1997). 'Rates of hepatitis E virus infection and disease among adolescents and adults in Kathmandu, Nepal'. *J Infect Dis* **176**(3): 763–6.

COLSON, P., P. BORENTAIN, B. QUEYRIAUX, M. KABA, V. MOAL, P. GALLIAN, L. HEYRIES, D. RAOULT and R. GEROLAMI. (2010). 'Pig liver sausage as a source of hepatitis E virus transmission to humans'. *J Infect Dis* **202**(6): 825–34.

CORWIN, A. L., N. T. TIEN, K. BOUNLU, J. WINARNO, M. P. PUTRI, K. LARAS, R. P. LARASATI, N. SUKRI, T. ENDY, H. A. SULAIMAN and K. C. HYAMS. (1999). 'The unique riverine ecology of hepatitis E virus transmission in South-East Asia'. *Trans R Soc Trop Med Hyg* **93**(3): 255–60.

DALTON, H. R., R. BENDALL, S. IJAZ and M. BANKS. (2008). 'Hepatitis E: an emerging infection in developed countries'. *Lancet Infect Dis* **8**(11): 698–709.

DALTON, H. R., R. P. BENDALL, F. E. KEANE, R. S. TEDDER and S. IJAZ. (2009). 'Persistent carriage of hepatitis E virus in patients with HIV infection'. *N Engl J Med* **361**(10): 1025–7.

DALTON, H. R., F. E. KEANE, R. BENDALL, J. MATHEW and S. IJAZ. (2011). 'Treatment of chronic hepatitis E in a patient with HIV infection'. *Ann Intern Med* **155**(7): 479–80.

DALTON, H. R., W. STABLEFORTH, P. THURAIRAJAH, S. HAZELDINE, R. REMNARACE, W. USAMA, L. FARRINGTON, N. HAMAD, C. SIEBERHAGEN, V. ELLIS, J. MITCHELL, S. H. HUSSAINI, M. BANKS, S. IJAZ and R. P. BENDALL. (2008). 'Autochthonous hepatitis E in Southwest England: natural history, complications and seasonal variation, and hepatitis E virus IgG seroprevalence in blood donors, the elderly and patients with chronic liver disease'. *Eur J Gastroenterol Hepatol* **20**(8): 784–90.

DAWSON, G. J., K. H. CHAU, C. M. CABAL, P. O. YARBOUGH, G. R. REYES and I. K. MUSHAHWAR. (1992). 'Solid-phase enzyme-linked immunosorbent assay for hepatitis E virus IgG and IgM antibodies utilizing recombinant antigens and synthetic peptides'. *J Virol Methods* **38**(1): 175–86.

DAWSON, J., A. D. SEDGWICK, J. C. EDWARDS and P. LEES. (1992). 'The monoclonal antibody MEL-14 can block lymphocyte migration into a site of chronic inflammation'. *Eur J Immunol* **22**(6): 1647–50.

DE SILVA, A. N., A. K. MUDDU, J. P. IREDALE, N. SHERON, S. I. KHAKOO and E. PELOSI. (2008). 'Unexpectedly high incidence of indigenous acute hepatitis E within South Hampshire: time for routine testing?', *J Med Virol* **80**(2): 283–8.

DRUMMOND, A. J., S. Y. HO, M. J. PHILLIPS and A. RAMBAUT. (2006). 'Relaxed phylogenetics and dating with confidence'. *PLoS Biol* **4**(5): e88.

EMERSON, S. U., V. A. ARANKALLE and R. H. PURCELL. (2005). 'Thermal stability of hepatitis E virus'. *J Infect Dis* **192**(5): 930–3.

EMERSON, S. U., P. CLEMENTE-CASARES, N. MOIDUDDIN, V. A. ARANKALLE, U. TORIAN and R. H. PURCELL. (2006). 'Putative neutralization epitopes and broad cross-genotype neutralization of Hepatitis E virus confirmed by a quantitative cell-culture assay'. *J Gen Virol* **87**(Pt 3): 697–704.

EMERSON, S. U., H. NGUYEN, J. GRAFF, D. A. STEPHANY, A. BROCKINGTON and R. H. PURCELL. (2004). 'In vitro replication of hepatitis E virus (HEV) genomes and of an HEV replicon expressing green fluorescent protein'. *J Virol* **78**(9): 4838–46.

EMERSON, S. U. and R. H. PURCELL (2003). 'Hepatitis E virus'. *Rev Med Virol* **13**(3): 145–54.

ERKER, J. C., S. M. DESAI, G. G. SCHLAUDER, G. J. DAWSON and I. K. MUSHAHWAR. (1999). 'A hepatitis E virus variant from the United States: molecular characterization and transmission in cynomolgus macaques'. *J Gen Virol* **80**(Pt 3): 681–90.

FAUQUET, C. M., M.A. MAYO, J. MANILOFF, U. DESSELBERGER and L. A. BALL (eds). (2005). *Taxonomy*. Elsevier/Academic Press, London.

FAVOROV, M. O., H. A. FIELDS, M. A. PURDY, T. L. YASHINA, A. G. ALEKSANDROV, M. J. ALTER, D. M. YARASHEVA, D. W. BRADLEY and H. S. MARGOLIS. (1992). 'Serologic identification of hepatitis E virus infections in epidemic and endemic settings'. *J Med Virol* **36**(4): 246–50.

FEAGINS, A. R., T. OPRIESSNIG, D. K. GUENETTE, P. G. HALBUR and X. J. MENG. (2007). 'Detection and characterization of infectious hepatitis E virus from commercial pig livers sold in local grocery stores in the USA'. *J Gen Virol* **88**(Pt 3): 912–17.

FEAGINS, A. R., T. OPRIESSNIG, *et al.*, (2008). 'Inactivation of infectious hepatitis E virus present in commercial pig livers sold in local grocery stores in the United States'. *Int J Food Microbiol* **123**(1–2): 32–7.

FONG, F. and M. ILLAHI (2009). 'Neuralgic amyotrophy associated with hepatitis E virus'. *Clin Neurol Neurosurg* **111**(2): 193–5.

GENG, J., L. WANG, H. FU, Q. BU, Y. ZHU and H. ZHUANG. (2010). 'Study on prevalence and genotype of hepatitis E virus isolated from Rex Rabbits in Beijing, China'. *J Viral Hepat* **18**(9): 661–7.

GENG, J. B., H. W. FU, L. WANG, X. J. WANG, J. M. GUAN, Y. B. CHANG, L. J. LI, Y. H. ZHU, H. ZHUANG, Q. H. LIU and X. C. PENG. (2010). 'Hepatitis E virus (HEV) genotype and the prevalence of anti-HEV in 8 species of animals in the suburbs of Beijing'. *Zhonghua Liu Xing Bing Xue Za Zhi* **31**(1): 47–50.

GIESE, A. C. (1976). *Living with the Sun's Ultraviolet Rays*. Plenum Press, New York.

GOENS, S. D. and M. L. PERDUE (2004). 'Hepatitis E viruses in humans and animals'. *Anim Health Res Rev* **5**(2): 145–56.

GUPTA, D. N. and H. F. SMETANA (1957). 'The histopathology of viral hepatitis as seen in the Delhi epidemic (1955–56)'. *Indian J Med Res* **45**(Suppl.): 101–13.

GUTHMANN, J. P., H. KLOVSTAD, D. BOCCIA, N. HAMID, L. PINOGES, J. Y. NIZOU, M. TATAY, F. DIAZ, A. MOREN, R. F. GRAIS, I. CIGLENECKI, E. NICAND and P. J. GUERIN. (2006). 'A large outbreak of hepatitis E among a displaced population in Darfur, Sudan, 2004: the role of water treatment methods'. *Clin Infect Dis* **42**(12): 1685–91.

HAAGSMA, E. B., A. P. VAN DEN BERG, R. J. PORTE, C. A. BENNE, H. VENNEMA, J. H. REIMERINK and M. P. KOOPMANS. (2008). 'Chronic hepatitis E virus infection in liver transplant recipients'. *Liver Transpl* **14**(4): 547–53.

HAKZE-VAN DER HONING, R. W., E. VAN COILLIE, A. F. ANTONIS and W. H. VAN DER POEL. (2011). 'First isolation of hepatitis E virus genotype 4 in Europe through swine surveillance in the Netherlands and Belgium'. *PLoS One* **6**(8): e22673.

HALBUR, P. G., C. KASORNDORKBUA, C. GILBERT, D. GUENETTE, M. B. POTTERS, R. H. PURCELL, S. U. EMERSON, T. E. TOTH and X. J. MENG. (2001). 'Comparative pathogenesis of infection of pigs with hepatitis E viruses recovered from a pig and a human'. *J Clin Microbiol* **39**(3): 918–23.

HAMID, S. S., S. M. JAFRI, H. KHAN, H. SHAH, Z. ABBAS and H. FIELDS. (1996). 'Fulminant hepatic failure in pregnant women: acute fatty liver or acute viral hepatitis?', *J Hepatol* **25**(1): 20–7.

HAQSHENAS, G., H. L. SHIVAPRASAD, P. R. WOOLCOCK, D. H. READ and X. J. MENG. (2001). 'Genetic identification and characterization of a novel virus related to human hepatitis E virus from chickens with hepatitis-splenomegaly syndrome in the United States'. *J Gen Virol* **82**(Pt 10): 2449–62.

HE, S., J. MIAO, Z. ZHENG, T. WU, M. XIE, M. TANG, J. ZHANG, M. H. NG and N. XIA. (2008). 'Putative receptor-binding sites of hepatitis E virus'. *J Gen Virol* **89**(Pt 1): 245–9.

HIRANO, M., X. DING, T. C. LI, N. TAKEDA, H. KAWABATA, N. KOIZUMI, T. KADOSAKA, I. GOTO, T. MASUZAWA, M. NAKAMURA, K. TAIRA, T. KUROKI, T. TANIKAWA, H. WATANABE and K. ABE. (2003). 'Evidence for widespread infection of hepatitis E virus among wild rats in Japan'. *Hepatol Res* **27**(1): 1–5.

HUANG, F., J. ZHOU, Z. YANG, L. CUI, W. ZHANG, C. YUAN, S. YANG, J. ZHU and X. HUA. (2010). 'RNA interference inhibits hepatitis E virus mRNA accumulation and protein synthesis in vitro'. *Vet Microbiol* **142**(3–4): 261–7.

HUANG, R., D. LI, *et al.*,. (1999). 'Cell culture of sporadic hepatitis E virus in China'. *Clin Diagn Lab Immunol* **6**(5): 729–33.

HUANG, R., N. NAKAZONO, K. ISHII, D. LI, O. KAWAMATA, R. KAWAGUCHI and Y. TSUKADA. (1995). 'Hepatitis E virus (87A strain) propagated in A549 cells'. *J Med Virol* **47**(4): 299–302.

HUANG, R. T., D. R. LI, J. WEI, X. R. HUANG, X. T. YUAN and X. TIAN. (1992). 'Isolation and identification of hepatitis E virus in Xinjiang, China'. *J Gen Virol* **73**(Pt 5): 1143–8.

HUSSAINI, S. H., S. J. SKIDMORE, P. RICHARDSON, L. M. SHERRATT, B. T. COOPER and J. G. O'GRADY. (1997). 'Severe hepatitis E infection during pregnancy'. *J Viral Hepat* **4**(1): 51–4.

IJAZ, S., E. ARNOLD, M. BANKS, R. P. BENDALL, M. E. CRAMP, R. CUNNINGHAM, H. R. DALTON, T. J. HARRISON, S. F. HILL, L. MACFARLANE, R. E. MEIGH, S. SHAFI, M. J. SHEPPARD, J. SMITHSON, M. P. WILSON and C. G. TEO. (2005). 'Non-travel-associated hepatitis E in England and Wales: demographic, clinical, and molecular epidemiological characteristics'. *J Infect Dis* **192**(7): 1166–72.

IPPAGUNTA, S. K., S. NAIK, S. JAMEEL, K. N. RAMANA and R. AGGARWAL. (2010). 'Viral RNA but no evidence of replication can be detected in the peripheral blood mononuclear cells of hepatitis E virus-infected patients'. *J Viral Hepat* **18**(9): 668–72.

JAMEEL, S. (1999). 'Molecular biology and pathogenesis of hepatitis E virus'. *Expert Rev Mol Med* **1999**: 1–16.

JOHNE, R., A. Plenge-Bonig, M. HESS, R. G. ULRICH, J. REETZ and A. SCHIELKE. (2010). 'Detection of a novel hepatitis E-like virus in faeces of wild rats using a nested broad-spectrum RT-PCR'. *J Gen Virol* **91**(Pt 3): 750–8.

KAGEYAMA, T., S. KOJIMA, M. SHINOHARA, K. UCHIDA, S. FUKUSHI, F. B. HOSHINO, N. TAKEDA and K. KATAYAMA. (2003). 'Broadly reactive and highly sensitive assay for Norwalk-like viruses based on real-time quantitative reverse transcription-PCR'. *J Clin Microbiol* **41**(4): 1548–57.

KAMANI, P., R. BAIJAL, D. AMARAPURKAR, P. GUPTE, N. PATEL, P. KUMAR and S. AGAL. (2005). 'Guillain-Barre syndrome associated with acute hepatitis E'. *Indian J Gastroenterol* **24**(5): 216.

KAMAR, N., F. ABRAVANEL, J. M. MANSUY, J. M. PERON, J. IZOPET and L. ROSTAING. (2010a). 'Hepatitis E infection in dialysis and after transplantation'. *Nephrol Ther* **6**(2): 83–7.

KAMAR, N., R. P. BENDALL, J. M. PERON, P. CINTAS, L. PRUDHOMME, J. MICHEL MANSUY, L. ROSTAING, F. KEANE, S. IJAZ, J. IZOPET and H. R. DALTON. (2011). 'Hepatitis E virus and neurologic disorders'. *Emerg Infect Dis* **17**(2): 173–9.

KAMAR, N., J. GUITARD, D. RIBES, L. ESPOSITO and L. ROSTAING. (2008a). 'A monocentric observational study of darbepoetin alfa in anemic hepatitis-C-virus transplant patients treated with ribavirin'. *Exp Clin Transplant* **6**(4): 271–5.

KAMAR, N., J. IZOPET, P. CINTAS, C. GARROUSTE, E. Uro-Coste, O. COINTAULT and L. ROSTAING. (2010b). 'Hepatitis E virus-induced neurological symptoms in a kidney-transplant patient with chronic hepatitis'. *Am J Transplant* **10**(5): 1321–4.

KAMAR, N., J. M. MANSUY, O. COINTAULT, J. SELVES, F. ABRAVANEL, M. DANJOUX, P. OTAL, L. ESPOSITO, D. DURAND, J. IZOPET and L. ROSTAING. (2008b). 'Hepatitis E virus-related cirrhosis in kidney- and kidney-pancreas-transplant recipients'. *Am J Transplant* **8**(8): 1744–8.

KAMAR, N., J. SELVES, J. M. MANSUY, L. OUEZZANI, J. M. PÉRON, J. GUITARD, O. COINTAULT, L. ESPOSITO, F. ABRAVANEL, M. DANJOUX, D. DURAND, J. P. VINEL, J. IZOPET and L. ROSTAING. (2008c). 'Hepatitis E virus and chronic hepatitis in organ-transplant recipients'. *N Engl J Med* **358**(8): 811–17.

KAR, P., N. JILANI, S. A. HUSAIN, S. T. PASHA, R. ANAND, A. RAI and B. C. DAS. (2008). 'Does hepatitis E viral load and genotypes influence the final outcome of acute liver failure during pregnancy?', *Am J Gastroenterol* **103**(10): 2495–501.

KARPE, Y. A. and K. S. LOLE (2010). 'RNA 5'-triphosphatase activity of the hepatitis E virus helicase domain'. *J Virol* **84**(18): 9637–41.

KASORNDORKBUA, C., D. K. GUENETTE, F. F. HUANG, P. J. THOMAS, X. J. MENG and P. G. HALBUR. (2004). 'Routes of transmission of swine hepatitis E virus in pigs'. *J Clin Microbiol* **42**(11): 5047–52.

KHUROO, M. S. (1980). 'Chronic liver disease after non-A, non-B hepatitis'. *Lancet* **2**(8199): 860–1.

KHUROO, M. S. (2010). 'Seroepidemiology of a second epidemic of hepatitis E in a population that had recorded first epidemic 30 years before and has been under surveillance since then'. *Hepatol Int* **4**(2): 494–9.

KHUROO, M. S., S. KAMILI, M. Y. DAR, R. MOECKLII and S. JAMEEL. (1993). 'Hepatitis E and long-term antibody status'. *Lancet* **341**(8856): 1355.

KHUROO, M. S., S. KAMILI and S. JAMEEL. (1995). 'Vertical transmission of hepatitis E virus'. *Lancet* **345**(8956): 1025–6.

KHUROO, M. S., S. KAMILI, and M. S. KHUROO. (2009). 'Clinical course and duration of viremia in vertically transmitted hepatitis E virus (HEV) infection in babies born to HEV-infected mothers'. *J Viral Hepat* **16**(7): 519–23.

KHUROO, M. S., M. R. TELI, S. SKIDMORE, M. A. SOFI and M. I. KHUROO. (1981). 'Incidence and severity of viral hepatitis in pregnancy'. *Am J Med* **70**(2): 252–5.

KIM, Y. M., S. H. JEONG, J. Y. KIM, J. C. SONG, J. H. LEE, J. W. KIM, H. YUN and J. S. KIM. (2011). 'The first case of genotype 4 hepatitis E related to wild boar in South Korea'. *J Clin Virol* **50**(3): 253–6.

KLAUS, D. M. (2001). 'Clinostats and bioreactors'. *Gravit Space Biol Bull* **14**(2): 55–64.

KOLASA, K. (2010). 'Food hypersensitivity: diagnosing and managing food allergies and intolerance'. *J Nutr Education Behav* **42**(2): 142.e7.

KOONIN, E. V. and T. G. SENKEVICH. (1992). 'Evolution of thymidine and thymidylate kinases: the possibility of independent capture of TK genes by different groups of viruses'. *Virus Genes* **6**(2): 187–96.

KRAWCZYNSKI, K., R. AGGARWAL, and S. KAMILI (2000). 'Hepatitis E'. *Infect Dis Clin North Am* **14**(3): 669–87.

KRAWCZYNSKI, K., E. E. MAST and M. A. PURDY. (1999). 'Hepatitis E: an overview'. *Minerva Gastroenterol Dietol* **45**(2): 119–30; discussion 130–115.

KWO, P. Y., G. G. SCHLAUDER, H. A. CARPENTER, P. J. MURPHY, J. E. ROSENBLATT, G. J. DAWSON, E. E. MAST, K. KRAWCZYNSKI and V. BALAN. (1997). 'Acute hepatitis E by a new isolate acquired in the United States'. *Mayo Clin Proc* **72**(12): 1133–6.

LEES, D. (2010). 'International standardisation of a method for detection of human pathogenic viruses in molluscan shellfish'. *Food Environ Virol* **2**(3): 146–55.

LEGRAND-ABRAVANEL, F., N. KAMAR, K. SANDRES-SAUNE, S. LHOMME, J. M. MANSUY, F. MUSCARI, F. SALLUSTO, L. ROSTAING and J. IZOPET. (2011). 'Hepatitis E virus infection without reactivation in solid-organ transplant recipients, France'. *Emerg Infect Dis* **17**(1): 30–7.

LEWIS, H. C., S. BOISSON, S. IJAZ, K. HEWITT, S. L. NGUI, E. BOXALL, C. G. TEO and D. MORGAN. (2008). 'Hepatitis E in England and Wales'. *Emerg Infect Dis* **14**(1): 165–7.

LI, D., L. BAERT, E. VAN Coillie, J. RYCKEBOER, F. DEVLIEGHERE and M. UYTTENDAELE. (2011). 'Inactivation of murine norovirus 1, coliphage phiX174, and Bacillus fragilis phage B40–8 on surfaces and fresh-cut iceberg lettuce by hydrogen peroxide and UV light'. *Appl Environ Microbiol* **77**(4): 1399–404.

LI, T. C., T. MIYAMURA and N. TAKEDA. (2007). 'Detection of hepatitis E virus RNA from the bivalve Yamato-Shijimi (Corbicula japonica) in Japan'. *Am J Trop Med Hyg* **76**(1): 170–2.

LI, T.-C., Y. SUZAKI, Y. AMI, H. TSUNEMITSU, T. MIYAMURA and N. TAKEDA. (2008). 'Mice are not susceptible to hepatitis E virus infection'. *J Vet Med SciIJapanese Soc Vet Sci* **70**(12): 1359–62.

LOLY, J. P., E. RIKIR, M. SEIVERT, E. LEGROS, P. DEFRANCE, J. BELAICHE, G. MOONEN and J. DELWAIDE. (2009). 'Guillain-Barre syndrome following hepatitis E'. *World J Gastroenterol* **15**(13): 1645–7.

LONGER, C. F., S. L. DENNY, J. D. CAUDILL, T. A. MIELE, L. V. S. ASHER, K. S. A. MYINT, C.-C. HUANG, W. F. ENGLER, J. W. LEDUC, L. N. BINN and J. R. TICEHURST. (1993). 'Experimental hepatitis E: pathogenesis in cynomolgus macaques (*Macaca fascicularis*)'. *J Infect Dis* **168**(3): 602–9.

LU, L., C. LI and C. H. HAGEDORN. (2006). 'Phylogenetic analysis of global hepatitis E virus sequences: genetic diversity, subtypes and zoonosis'. *Rev Med Virol* **16**(1): 5–36.

LYTLE, C. D. and J. L. SAGRIPANTI. (2005). 'Predicted inactivation of viruses of relevance to biodefense by solar radiation'. *J Virol* **79**(22): 14244–52.

MAGDEN, J., N. TAKEDA, T. LI, P. AUVINEN, T. AHOLA, T. MIYAMURA, A. MERITS and L. KÄÄRIÄINEN. (2001). 'Virus-specific mRNA capping enzyme encoded by hepatitis E virus'. *J Virol* **75**(14): 6249–55.

MANDAL, K. and N. CHOPRA (2006). 'Acute transverse myelitis following hepatitis E virus infection'. *Indian Pediatr* **43**(4): 365–6.

MANSUY, J. M., J. M. PERON, F. ABRAVANEL, H. POIRSON, M. DUBOIS, M. MIEDOUGE, F. VISCHI, L. ALRIC, J. P. VINEL and J. IZOPET. (2004). 'Hepatitis E in the south west of France in individuals who have never visited an endemic area'. *J Med Virol* **74**(3): 419–24.

MARGOLIS H. A. M. AND HADLER, S. C. Viral hepatitis. In EVANS, A.S., KASLOW, R.A. (eds). (1997). *Viral Infections of Humans: Epidemiology and Control*, 4th edn. New York, Plenum Medical Books Co, 363–418.

MARTELLI, F., A. CAPRIOLI, M. ZENGARINI, A. MARATA, C. FIEGNA, I. DI Bartolo, F. M. RUGGERI, M. DELOGU and F. OSTANELLO. (2008). 'Detection of hepatitis E virus (HEV) in a demographic managed wild boar (Sus scrofa scrofa) population in Italy'. *Vet Microbiol* **126**(1–3): 74–81.

MARTIN, M., J. SEGALES, F. F. HUANG, D. K. GUENETTE, E. MATEU, N. DE DEUS and X. J. MENG. (2007). 'Association of hepatitis E virus (HEV) and postweaning multisystemic wasting syndrome (PMWS) with lesions of hepatitis in pigs'. *Vet Microbiol* **122**(1–2): 16–24.

MATALLANA-SURGET, S., T. DOUKI, J. A. MEADOR, R. CAVICCHIOLI and F. JOUX. (2010). 'Influence of growth temperature and starvation state on survival and DNA damage induction in the marine bacterium Sphingopyxis alaskensis exposed to UV radiation'. *J Photochem Photobiol B* **100**(2): 51–6.

MCCORMICK, J. B. (1986). 'Clinical, epidemiologic, and therapeutic aspects of Lassa fever'. *Med Microbiol Immunol* **175**(2–3): 153–5.

MENG, J., X. DAI, J. C. CHANG, E. LOPAREVA, J. PILLOT, H. A. FIELDS and Y. E. KHUDYAKOV. (2001). 'Identification and characterization of the neutralization epitope(s) of the hepatitis E virus'. *Virology* **288**(2): 203–11.

MENG, S., S. ZHAN and J. LI. (2009). 'Nuclease-resistant double-stranded DNA controls or standards for hepatitis B virus nucleic acid amplification assays'. *Virol J* **6**: 226.

MENG, X. J. (2003). 'Swine hepatitis E virus: cross-species infection and risk in xenotransplantation'. *Curr Top Microbiol Immunol* **278**: 185–216.

MENG, X. J., R. H. PURCELL, P. G. HALBUR, J. R. LEHMAN, D. M. WEBB, T. S. TSAREVA, J. S. HAYNES, B. J. THACKER, and S. U. EMERSON. (1997). 'A novel virus in swine is closely related to the human hepatitis E virus'. *Proc Natl Acad Sci U S A* **94**(18): 9860–5.

MENG, X. J., B. WISEMAN, F. ELVINGER, D. K. GUENETTE, T. E. TOTH, R. E. ENGLE, S. U. EMERSON and R. H. PURCELL. (2002). 'Prevalence of antibodies to hepatitis E virus in veterinarians working with swine and in normal blood donors in the United States and other countries'. *J Clin Microbiol* **40**(1): 117–22.

MITSUI, T., Y. TSUKAMOTO, A. HIROSE, S. SUZUKI, C. YAMAZAKI, K. MASUKO, F. TSUDA, K. ENDO, M. TAKAHASHI and H. OKAMOTO. (2006). 'Distinct changing profiles of hepatitis A and E virus infection among patients with acute hepatitis, patients on maintenance hemodialysis and healthy individuals in Japan'. *J Med Virol* **78**(8): 1015–24.

NAIK, S. R., R. AGGARWAL, P. N. SALUNKE and N. N. MEHROTRA. (1992). 'A large waterborne viral hepatitis E epidemic in Kanpur, India'. *Bull World Health Organ* **70**(5): 597–604.

NAKAMURA, M., K. TAKAHASHI, K. TAIRA, M. TAIRA, A. OHNO, H. SAKUGAWA, M. ARAI and S. MISHIRO. (2006). 'Hepatitis E virus infection in wild mongooses of Okinawa, Japan: Demonstration of anti-HEV antibodies and a full-genome nucleotide sequence'. *Hepatol Res* **34**(3): 137–40.

NANDA, S. K., I. H. ANSARI, S. K. ACHARYA, S. JAMEEL and S. K. PANDA. (1995). 'Protracted viremia during acute sporadic hepatitis E virus infection'. *Gastroenterology* **108**(1): 225–30.

NAUMAN, E. A., C. M. OTT, E. SANDER, D. L. TUCKER, D. PIERSON, J. W. WILSON and C. A. NICKERSON. (2007). 'Novel quantitative biosystem for modeling physiological fluid shear stress on cells'. *Appl Environ Microbiol* **73**(3): 699–705.

NAVANEETHAN, U., M. AL MOHAJER and M. T. SHATA. (2008). 'Hepatitis E and pregnancy: understanding the pathogenesis'. *Liver Int* **28**(9): 1190–9.

NICHOLSON, W. L., A. C. SCHUERGER and P. SETLOW. (2005). 'The solar UV environment and bacterial spore UV resistance: considerations for Earth-to-Mars transport by natural processes and human spaceflight'. *Mutat Res* **571**(1–2): 249–64.

NICKERSON, C. A., T. J. GOODWIN, J. TERLONGE, C. MARK OTT, K. L. BUCHANAN, W. C. UICKER, K. EMAMI, C. L. LEBLANC, R. RAMAMURTHY, M. S. CLARKE, C. R. VANDERBURG, T. HAMMOND and D. L. PIERSON. (2001). 'Three-dimensional tissue assemblies: novel models for the study of Salmonella enterica serovar Typhimurium pathogenesis'. *Infect Immun* **69**(11): 7106–20.

NUANUALSUWAN, S., T. MARIAM, S. HIMATHONGKHAM and D. O. CLIVER. (2002). 'Ultraviolet inactivation of feline calicivirus, human enteric viruses and coliphages'. *Photochem Photobiol* **76**(4): 406–10.

OKAMOTO, H. (2007). 'Genetic variability and evolution of hepatitis E virus'. *Virus Res* **127**(2): 216–28.

OKAMOTO, H. (2011). 'Efficient cell culture systems for hepatitis E virus strains in feces and circulating blood'. *Rev Med Virol* **21**(1): 18–31.

PANDA, S. K., I. H. ANSARI, H. DURGAPAL, S. AGRAWAL and S. JAMEEL. (2000). 'The in vitro-synthesized RNA from a cDNA clone of hepatitis E virus is infectious'. *J Virol* **74**(5): 2430–7.

PANDA, S. K., D. THAKRAL and S. REHMAN. (2007). 'Hepatitis E virus'. *Rev Med Virol* **17**(3): 151–80.

PARVEZ, K., R. H. PURCELL and S. U. EMERSON. (2011). 'Hepatitis E virus ORF2 protein over-expressed by baculovirus in hepatoma cells, efficiently encapsidates and transmits the viral RNA to naive cells'. *Virol J* **8**(1): 159.

PERON, J. M., J. M. MANSUY, J. IZOPET and J. P. VINEL. (2006). 'Hepatitis E virus: an emerging disease'. *Sante* **16**(4): 239–43.

PERON, J. M., J. M. MANSUY, H. POIRSON, C. BUREAU, E. DUPUIS, L. ALRIC, J. IZOPET and J. P. VINEL. (2006). 'Hepatitis E is an autochthonous disease in industrialized countries. Analysis of 23 patients in South-West France over a 13-month period and comparison with hepatitis A'. *Gastroenterol Clin Biol* **30**(5): 757–62.

PINA, S., M. BUTI, M. COTRINA, J. PIELLA and R. GIRONES. (2000). 'HEV identified in serum from humans with acute hepatitis and in sewage of animal origin in Spain'. *J Hepatol* **33**(5): 826–33.

PINA, S., J. JOFRE, S. U. EMERSON, R. H. PURCELL and R. GIRONES. (1998). 'Characterization of a strain of infectious hepatitis E virus isolated from sewage in an area where hepatitis E is not endemic'. *Appl Environ Microbiol* **64**(11): 4485–8.

PURCELL, R. H. and S. U. EMERSON (2001). 'Animal models of hepatitis A and E'. *ILAR J* **42**(2): 161–77.

PURCELL, R. H. and S. U. EMERSON (2008). 'Hepatitis E: an emerging awareness of an old disease'. *J Hepatol* **48**(3): 494–503.

PURDY, M. A. and Y. E. KHUDYAKOV (2010). 'Evolutionary history and population dynamics of hepatitis E virus'. *PLoS One* **5**(12): e14376.

REUTER, G., D. FODOR, A. KÁTAI and G. SZUCS. (2005). 'Molecular detection of hepatitis E virus in non-imported hepatitis case – identification of a potential new human hepatitis E virus lineage in Hungary'. *Orv Hetil* **146**(47): 2389–94.

REYES, G. R., P. O. YARBOUGH, A. W. TAM, M. A. PURDY, C. C. HUANG, J. S. KIM, D. W. BRADLEY and K. E. FRY. (1991). 'Hepatitis E virus (HEV): the novel agent responsible for enterically transmitted non-A, non-B hepatitis'. *Gastroenterol Jpn* **26**(Suppl 3): 142–7.

RIANTHAVORN, P., C. THONGMEE, N. LIMPAPHAYOM, P. KOMOLMIT, A. THEAMBOONLERS and Y. POOVORAWAN. (2010). 'The entire genome sequence of hepatitis E virus genotype

3 isolated from a patient with neuralgic amyotrophy'. *Scand J Infect Dis* **42**(5): 395–400.

RUTJES, S. A., W. J. LODDER, M. BOUWKNEGT and A. M. DE RODA HUSMAN. (2007). 'Increased hepatitis E virus prevalence on Dutch pig farms from 33 to 55% by using appropriate internal quality controls for RT-PCR'. *J Virol Methods* **143**(1): 112–16.

RZEZUTKA, A. and N. COOK (2004). 'Survival of human enteric viruses in the environment and food'. *FEMS Microbiol Rev* **28**(4): 441–53.

SATOU, K. and H. NISHIURA (2007). 'Transmission dynamics of hepatitis E among swine: potential impact upon human infection'. *BMC Vet Res* **3**: 9.

SCHLAUDER, G. G. and I. K. MUSHAHWAR (2001). 'Genetic heterogeneity of hepatitis E virus'. *J Med Virol* **65**(2): 282–92.

SHRESTHA, M. P., R. M. SCOTT, D. M. JOSHI, M. P. MAMMEN Jr, G. B. THAPA, N. THAPA, K. S. MYINT, M. FOURNEAU, R. A. KUSCHNER, S. K. SHRESTHA, M. P. DAVID, J. SERIWATANA, D. W. VAUGHN, A. SAFARY, T. P. ENDY and B. L. INNIS. (2007). 'Safety and efficacy of a recombinant hepatitis E vaccine'. *N Engl J Med* **356**(9): 895–903.

SOOD, A., V. MIDHA and N. SOOD (2000). 'Guillain-Barre syndrome with acute hepatitis E'. *Am J Gastroenterol* **95**(12): 3667–8.

SRIRAM, B., D. THAKRAL and S. K. PANDA. (2003). 'Targeted cleavage of hepatitis E virus 3' end RNA mediated by hammerhead ribozymes inhibits viral RNA replication'. *Virology* **312**(2): 350–8.

TAKAHASHI, K., J. H. KANG, S. OHNISHI, K. HINO, H. MIYAKAWA, Y. MIYAKAWA, H. MAEKUBO and S. MISHIRO. (2003). 'Full-length sequences of six hepatitis E virus isolates of genotypes III and IV from patients with sporadic acute or fulminant hepatitis in Japan'. *Intervirology* **46**(5): 308–18.

TAKAHASHI, M., T. NISHIZAWA, H. SATO, Y. SATO, Jirintai, S. NAGASHIMA and H. OKAMOTO. (2011). 'Analysis of the full-length genome of a hepatitis E virus isolate obtained from a wild boar in Japan that is classifiable into a novel genotype'. *J Gen Virol* **92**(Pt 4): 902–8.

TAM, A. W., M. M. SMITH, M. E. GUERRA, C. C. HUANG, D. W. BRADLEY, K. E. FRY and G. R. REYES. (1991). 'Hepatitis E virus (HEV): molecular cloning and sequencing of the full-length viral genome'. *Virology* **185**(1): 120–31.

TANAKA, T., M. TAKAHASHI, E. KUSANO and H. OKAMOTO. (2007). 'Development and evaluation of an efficient cell-culture system for Hepatitis E virus'. *J Gen Virol* **88**(Pt 3): 903–11.

TANAKA, Y., K. TAKAHASHI, E. ORITO, Y. KARINO, J. H. KANG, K. SUZUKI, A. MATSUI, A. HORI, H. MATSUDA, H. SAKUGAWA, Y. ASAHINA, T. KITAMURA, M. MIZOKAMI and S. MISHIRO. (2006). 'Molecular tracing of Japan-indigenous hepatitis E viruses'. *J Gen Virol* **87**(Pt 4): 949–54.

TEI, S., N. KITAJIMA, K. TAKAHASHI and S. MISHIRO. (2003). 'Zoonotic transmission of hepatitis E virus from deer to human beings'. *Lancet* **362**(9381): 371–3.

TEO, C. G. (2010). 'Much meat, much malady: changing perceptions of the epidemiology of hepatitis E'. *Clin Microbiol Infect* **16**(1): 24–32.

TRAN, H. T., H. USHIJIMA, V. X. QUANG, N. PHUONG, T. C. LI, S. HAYASHI, T. XUAN LIEN, T. SATA and K. ABE. (2003). 'Prevalence of hepatitis virus types B through E and genotypic distribution of HBV and HCV in Ho Chi Minh City, Vietnam'. *Hepatol Res* **26**(4): 275–80.

VARMA, S. P., A. KUMAR, N. KAPUR, H. DURGAPAL, S. K. ACHARYA and S. K. PANDA. (2011). 'HEV replication involves alternating negative and positive sense RNA synthesis'. *J Gen Virol.* 2011 Mar; **92**(Pt 3):572–81. doi: 10.1099/vir.0.027714-0. Epub 2010 Dec 1.

WANG, Y. C., H. Y. ZHANG, N. S. XIA, G. PENG, H. Y. LAN, H. ZHUANG, Y. H. ZHU, S. W. LI, K. G. TIAN, W. J. GU, J. X. LIN, X. WU, H. M. LI and T. J. HARRISON. (2002). 'Prevalence, isolation, and partial sequence analysis of hepatitis E virus from domestic animals in China'. *J Med Virol* **67**(4): 516–21.

WEI, H., J. Q. ZHANG, H. Q. LU, J. H. MENG, X. X. LU and W. XIE. (2003). 'Construction and screening of hepatitis E virus-specific phage antibody combinatorial library'. *Xi Bao Yu Fen Zi Mian Yi Xue Za Zhi* **19**(5): 473–5, 485.

WIBAWA, I. D., D. H. MULJONO, Mulyanto, I. G. SURYADARMA, F. TSUDA, M. TAKAHASHI, T. NISHIZAWA and H. OKAMOTO. (2004). 'Prevalence of antibodies to hepatitis E virus among apparently healthy humans and pigs in Bali, Indonesia: identification of a pig infected with a genotype 4 hepatitis E virus'. *J Med Virol* **73**(1): 38–44.

WICHMANN, O., S. SCHIMANSKI, J. KOCH, M. KOHLER, C. ROTHE, A. PLENTZ, W. JILG and K. STARK. (2008). 'Phylogenetic and case-control study on hepatitis E virus infection in Germany'. *J Infect Dis* **198**(12): 1732–41.

WILLIAMS, T. P., C. KASORNDORKBUA, P. G. HALBUR, G. HAQSHENAS, D. K. GUENETTE, T. E. TOTH and X. J. MENG. (2001). 'Evidence of extrahepatic sites of replication of the hepatitis E virus in a swine model'. *J Clin Microbiol* **39**(9): 3040–6.

WONG, D. C., R. H. PURCELL, M. A. SREENIVASAN, S. R. PRASAD and K. M. PAVRI. (1980). 'Epidemic and endemic hepatitis in India: evidence for a non-A, non-B hepatitis virus aetiology'. *Lancet* **2**(8200): 876–9.

WONG, K. H., Y. M. LIU, P. S. NG, B. W. YOUNG and S. S. LEE. (2004). 'Epidemiology of hepatitis A and hepatitis E infection and their determinants in adult Chinese community in Hong Kong'. *J Med Virol* **72**(4): 538–44.

YAMADA, A., A. SAKO, S. NISHIMURA, R. NAKASHIMA, T. OGAMI, K. FUJIYA, N. TSUDA, N. ASAYAMA, T. YADA, K. SHIRAI, N. AKAZAWA, T. SAKURAI, Y. YAGO, N. NAGATA, T. OSHIMA, C. YOKOI, K. SASAJIMA, M. KOBAYAKAWA, J. AKIYAMA, M. IMAMURA, M. YANASE, N. UEMURA and N. MASAKI. (2009). 'A case of HIV coinfected with hepatitis B virus treated by entecavir'. *Nippon Shokakibyo Gakkai Zasshi* **106**(12): 1758–63.

YARBOUGH, P. O. (1999). 'Hepatitis E virus. Advances in HEV biology and HEV vaccine approaches'. *Intervirology* **42**(2–3): 179–84.

YAZAKI, Y., H. MIZUO, M. TAKAHASHI, T. NISHIZAWA, N. SASAKI, Y. GOTANDA and H. OKAMOTO. (2003). 'Sporadic acute or fulminant hepatitis E in Hokkaido, Japan, may be food-borne, as suggested by the presence of hepatitis E virus in pig liver as food'. *J Gen Virol* **84**(Pt 9): 2351–7.

ZHAO, C., Z. MA, T. J. HARRISON, R. FENG, C. ZHANG, Z. QIAO, J. FAN, H. MA, M. LI, A. SONG and Y. WANG. (2009). 'A novel genotype of hepatitis E virus prevalent among farmed rabbits in China'. *J Med Virol* **81**(8): 1371–9.

ZHAO, K., Q. LIU, R. YU, Z. LI, J. LI, H. ZHU, X. WU, F. TAN, J. WANG and X. TANG. (2009). 'Screening of specific diagnostic peptides of swine hepatitis E virus'. *Virol J* **6**: 186.

ZHUANG, H. (1991). 'Advances in the research on non-A, non-B hepatitis in China'. *Zhonghua Liu Xing Bing Xue Za Zhi* **12**(6): 377–9.

20

Epidemiology, control, and prevention of emerging zoonotic viruses

R. Santos and S. Monteiro, Instituto Superior Técnico, Portugal

DOI: 10.1533/9780857098870.4.442

Abstract: Zoonoses are infections in humans transmitted by animal pathogens or animal infections transmitted to humans. Viruses are the main etiological agents of emerging or re-emerging zoonoses. This chapter will discuss the most relevant foodborne and waterborne viral zoonotic infections along with their specific etiological agent, the issue of global and local infections, climate change, clinical manifestations and epidemiology and possible control and prevention measures.

Key words: zoonoses, epidemiology, viral emerging diseases, control and prevention, global and local issues, foodborne, waterborne.

20.1 Introduction

Zoonosis, from the Greek *zoon* (animal) and *nosos* (disease), is defined as infections that are transmitted from animals (wild and domestic) to humans and from humans to animals, through several different routes: (i) direct contact; (ii) via intermediate vectors such as mosquitoes and ticks; or (iii) through food- and waterborne infections (Hubalek, 2003).

The vast majority of recognized human pathogenic viruses have been circulating for a long time in the environment and also among the population. Their evolution is related to human evolution and *vice versa*, resulting in an equilibrium in which both are able to coexist, sometimes without consequences for human health. However, the crossing over of the species barrier by some pathogens can have devastating consequences in terms of infections, public health and mortality. Several factors have highly impacted the spread of viral zoonoses throughout the world. Increasing globalization, increased mobility, changes in demography, environmental determinants such as ecologic and climatologic influences, and intense farming have increased the potential for transmission of pathogenic viruses between animals and humans (Kuiken *et al.*, 2005; Osterhaus, 2001). Human and livestock populations have increased globally, causing closer contact between animals and humans. The human impact of changes in ecology and climate, together with faster transportation between countries and regions (faster than the incubation period of most of the infectious diseases), have accelerated the emergence or re-emergence of zoonotic pathogens. 'Emergence' means either the appearance of a newly evolved or a newly recognized pathogen, or the appearance of a known existing pathogen in a geographic area where it has never been found before. 'Re-emergence' is used to refer to pathogens whose presence is already recognized in a given area but which have increased in incidence (locally or globally, as for pandemics) or for pathogens which have seen their potential to cause severe infections in humans increase due to a shift in their ecology allowing, for example, the crossing of barriers from wildlife to domestic animals (as in the case of avian influenza) (Cutler *et al.*, 2010).

More than half of the microbial diseases affecting humans (61%) are attributed to zoonotic transmission, with wildlife being one of the biggest sources for the spread of infectious diseases (Daszak *et al.*, 2001; Taylor *et al.*, 2001).The majority of the viral infections in humans newly identified over the last decade had a zoonotic background (e.g., SARS-Coronavirus, avian influenza), indicating that the influence of animals in human diseases has increased and is continuing to evolve.

20.2 Emerging viruses: geographical factors

Throughout human history, the geographical distribution of diseases has undergone considerable change. Several factors, including progress in urban sanitation, increased standards of living and improvement in personal hygiene, accompanied by the development of vaccines and antibiotics, have helped to reduce the level of infectious diseases during the last two centuries. Anthropogenic activities have greatly influenced the environment. The following list summarizes the anthropogenic factors contributing to changes in zoonosis patterns:

- national and international conflicts;
- travel;
- population growth, stress, social inequality, urbanization;
- trade, import of exotic animals, transboundary animal transport;
- water control and irrigation projects;
- changes in infrastructures (such as air conditioning, cooling towers, among others);
- agricultural practices;

- conditions affecting pathogens directly or indirectly (antibiotic-resistant pathogens);
- climate change; and
- demographic change.

20.2.1 Human mobility and population growth

Since the beginning of human history, population movement has shaped the patterns and the spread of infectious diseases. During migration, not only do people carry with them their genetic make-up, but they may also carry pathogens in or on their bodies and they may take with them disease vectors or reservoir animals. Religious pilgrimages, military manoeuvres, and trade caravans have facilitated the spread of many diseases, such as the plague or smallpox. Military manoeuvres have always been a source of epidemics since soldiers can be brought into new environments with different infectious zoonosis (Haggett, 1994). Populations in occupied areas sometimes have to abandon their villages and cities, and many severe outbreaks have been associated with poor water quality and low levels of sanitation among refugees.

However, for most of human history, populations have stayed relatively isolated. This pattern has changed in recent centuries. People started exploring by sea and looking to discover new worlds. Columbus not only discovered the New World; he and other explorers also played a significant role in the emergence of infectious diseases. By the time the ships arrived at the New World, diseases such as measles, tuberculosis and influenza were widespread throughout Europe. Explorers from infected urban areas brought diseases to the Americas, where they killed thousands of people who had mostly evolved from a very small gene pool and had no immunity to the infectious agents. So when the explorers arrived several epidemics occurred, the first ones always being the most severe. For instance, the population of central Mexico fell by an estimated one-third in the decade following contact with Europeans and one-third to half of the local population of Santo Domingo died following smallpox infection which then spread to other areas in the Caribbean and the Americas. The explorers also paid a high price for discovering the New World, as they became infected with tropical diseases that were not indigenous in Europe.

Global economics and advances in technology have changed the world greatly in recent centuries. These two factors have also contributed to the emergence or re-emergence of infectious diseases and to the spread of zoonotic diseases that would otherwise have been confined to local areas. Advances in technology have speeded up transportation, increasing the global mobility of people, and allowing faster trade in products and services that can be obtained rapidly from anywhere in the world. Every day, around the world, more than 1.4 billion people cross international borders in aeroplanes. According to the World Tourism Organization, in 2011 a total of 980 million people crossed international borders worldwide, an increase of 4.4% compared to 2010, and

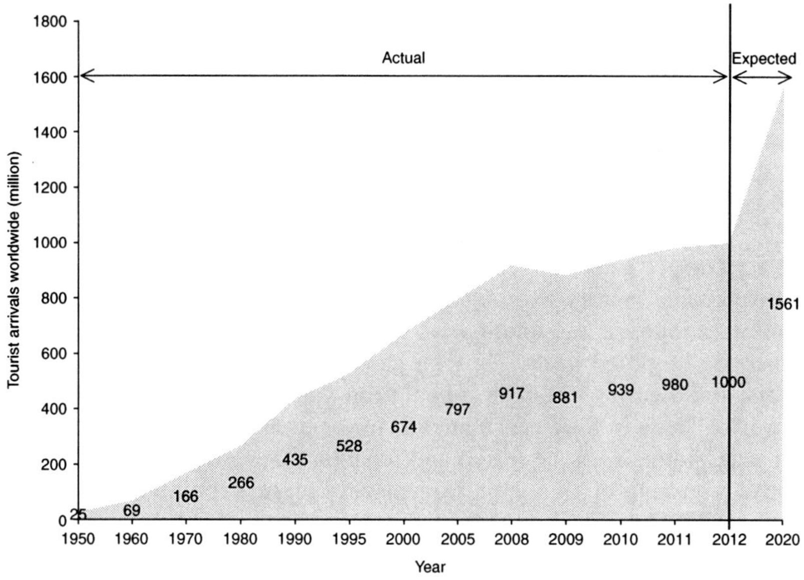

Fig. 20.1 International tourist arrivals (millions). Tourism is the fastest growing industry worldwide over the last century.

it is estimated that by the year 2020 that number will increase to more than 1.5 billion (see Fig. 20.1). Other means of transport carry millions of people, with cruise ships transporting 47 million passengers per year (Wilson, 2003). Today it is possible to travel to virtually any part of the world in a time period that is shorter than the incubation period of most infectious disease agents. Therefore, the traveller acts as a transmitter, a sentinel and a courier of infectious diseases. Wilson (2003) states that 'the traveller can be seen as an interactive biological unit who picks up, processes, carries and drops off microbial genetic material. A traveller can introduce potential pathogens in the absence of signs and symptoms.' Thus, human travel and migration have major consequences for the spread of infectious diseases worldwide.

By the year 2050, it is estimated that world population will increase by 27% from its current 7.0 billion to 8.9 billion. Many different factors – economic, social and political – have caused the human population to start migrating from rural areas to more urban ones. The number of big cities has increased, especially in developing countries where the population has risen steeply. The rapid growth of cities is often related to rapid and unplanned urbanization which can result in the deterioration of air and water quality, inadequate sanitation facilities and services, and the overuse of water resources (Moore *et al.*, 2003). The burden of infectious disease in developing countries with rapid urbanization rates is extremely high, due to both overpopulation and pollution (Crompton and Savioli, 1993). Closer contact between reservoir hosts also results from overpopulation, which then creates a potential risk of the crossing over of species (Haggett, 1994). In developing countries, access to sanitation in urban areas lacking municipal water supply is still restricted, so people often use common water sources that are faecally contaminated (Moore *et al.*, 2003). The rapid and uncontrolled growth of population in urban areas and the extension of agricultural areas to new environments precipitate the risk of the emergence or re-emergence of an infectious disease.

20.2.2 Trade

However, other factors have influenced the emergence and re-emergence of infectious zoonoses, and continue to do so. Globalization is responsible for an increase in global trade. Through global trade, human pathogens, insect vectors, and their intermediate animal hosts can be disseminated throughout the world. Trade in food raw materials in the globalized world is increasing each year. Animals can be reared and fresh produce grown, and both can be shipped thousands of kilometres from place of origin to point of consumption (Tauxe, 1997). Again, advances in technology have enabled mass production, mass processing and a better food distribution network, making it feasible for food grown in one location to be processed somewhere else and consumed in multiple destinations. As a consequence, unwanted micro-organisms can be transported from one place where they are endemic to another where they may adapt and cause major outbreaks of infectious diseases (Wilson, 1995). Vector insects can also be transported along the distribution chain, adapt to a new environment and infect humans through direct contact or by having an amplifying animal host which then can contaminate environmental waters (see Fig. 20.2).

20.3 Clinical manifestations of some emerging types

The majority of zoonotic transmitted viral infections can be categorized into three different groups: (i) diseases with no illness; (ii) severe illness; and (iii) non-specific syndrome. Several emerging or re-emerging zoonotic viral diseases fall into the second category. Despite research and diverse institutional efforts to eradicate viral diseases such as poliomyelitis and smallpox, many infectious diseases, including several caused by emerging or re-emerging zoonotic viruses, are still uncontrolled and far from being eradicated or understood. Some previously known agents have become more infectious due to alterations in disease patterns (e.g.,Oropouche virus or Chikungunya virus). Others, such as Nipah and Hendra viruses or SARS-Coronavirus, were not discovered and described until very recently, and have a high pathogenic potential (Manojkumar and Mrudula, 2006). They are associated with encephalitis (Nipah virus, tick-borne encephalitis), haemorrhagic fevers (Hantavirus), flu-like signs (avian influenza) and respiratory disease (SARS-Coronavirus).

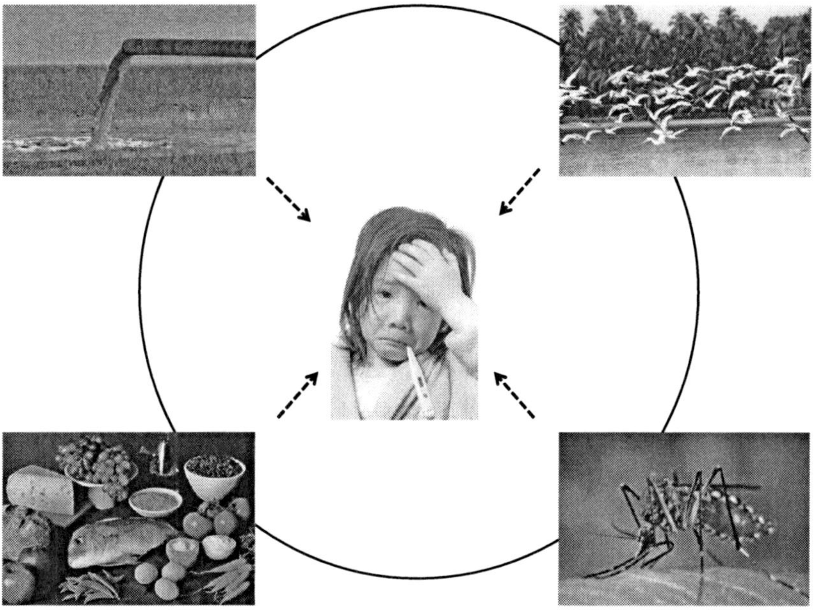

Fig. 20.2 Most common routes of transmission of infectious diseases between animals and humans.

20.3.1 Nipah virus

Nipah virus was discovered recently and belongs to the genus *Henipavirus* in the family *Paramyxoviridae* (Harcourt *et al.*, 2000).It is a single-stranded RNA virus named after the village in peninsular Malaysia in which it was isolated for the first time from a human victim (Chua *et al.*, 2000). Nipah virus was the aetiological agent responsible for a large outbreak in humans in Malaysia during 1998 and 1999 and for five subsequent outbreaks in Bangladesh between 2001 and 2005 (Chua *et al.*, 2000; Hsu *et al.*, 2004). The island flying fox, a fruit bat, (*Pteropushypomelanus*) is the primary reservoir for this virus where it persists in low numbers (Chua *et al.*, 2002), and it has emerged via domestic animal amplifier hosts. Nipah virus has the ability to replicate uncontrollably in pigs, causing respiratory and/or neurological diseases leading ultimately to death (MohdNor *et al.*, 2000). The effects on humans of infection by Nipah virus can range from an asymptomatic infection to fatal encephalitis. The incubation period can vary from four to 45 days. The first stage of the infection is the development of fever, headache, sore throat, myalgia, and vomiting, similar to common 'flu. The infection can then evolve to dizziness, altered consciousness, drowsiness, and some neurological signs indicating acute encephalitis. Moreover, some infected people can also exhibit severe respiratory problems and atypical pneumonia. The most severe cases develop encephalitis and seizures that progress in 24–48 h to a coma situation. Over 20% of the people who survive acute encephalitis suffer subsequent neurological effects, including personality changes and repeated convulsions. Fatality is estimated in between 40% and 75% of cases, with the percentage depending on the outbreak and on the local authorities' surveillance capacity. For example, of a total of 269 human cases of encephalitis infection with Nipah virus reported in 1999 in Malaysia, 40% (108) were fatal (Malaysian Ministry of Health, 2001).

20.3.2 Tick-borne encephalitis (TBE)

TBE virus is a non-fragmented, single-stranded RNA virus that belongs to the genus *Flavivirus* contained in the *Flaviviridae* family (Mandl *et al.*, 1997). TBE is endemic to northern, central and eastern Europe, Russia and the Far East. TBE cases have been reported not only in less developed countries such as Byelorussia, Bulgaria, Lithuania, and Kazakhstan but also in well-developed countries including Austria, Denmark, Germany, and Switzerland, where hygiene and sanitation standards are extremely high (Blaskovic *et al.*, 1967; Korenberg and Kovalevskii, 1999; Lindgren and Gustafson, 2001; Ormaasen *et al.*, 2001). The three main subtypes of TBE virus are the European, Siberian, and Far-Eastern (Ecker *et al.*, 1999). TBE is a vector-borne infectious disease transmitted by the bite of the ticks *Ixodes ricinus* and *Ixodes persulcatus* (eastern Europe). Infections related to the consumption of infected unpasteurized milk and cheese made from such milk have also been reported (Dumpis *et al.*, 1999). Infections as a result of the consumption of contaminated dairy products were first identified in the European part of Russia between 1947 and 1951. In these situations, whole families were being infected. This is contrary to what was observed when the transmission was from tick bites, where only single persons but not the whole family were infected. Research has proven that this form of disease was associated with the consumption of goat's milk containing infectious TBE viruses (Korenberg and Pchelkina, 1975; Popov and Ivanova, 1968). Later reports also demonstrated the transmission of TBE virus through the consumption of non-pasteurized milk from sheep and cows (Gresíková *et al.*, 1975; Leonov *et al.*, 1976).

The clinical manifestations resulting from an infection with TBE virus are well-known and a wide range of symptoms can be observed. The Far-Eastern subtype presents a monophasic progression, whereas the European is biphasic (Dumpis *et al.*, 1999; Gritsun *et al.*, 2003). The incubation period of a TBE virus infection is 7 to 14 days. In the biphasic course, symptoms in the prodromal stage can be mistaken for 'flu, with non-specific moderate fever and myalgia, accompanied by high fever and vomiting (Gritsun *et al.*, 2003). In the European subtype, this is often succeeded by an asymptomatic period lasting 2 to 10 days. The disease can then progress to the nervous system, advancing to the second stage, which is characterized by the development of high temperatures and acute central nervous system (CNS) symptoms.

Acute TBE is characterized by encephalitic symptoms in a high percentage of infected persons, ranging from 45% to 56% (Haglund and Günther, 2003). Several symptoms may be observed, including mild meningitis and severe meningoencephalomyelitis, which progress 5–10 days after the remission of the fever. The acute febrile period correlates with the presence of the viraemia. In the second stage, the virus infects the CNS, where it replicates, resulting in the inflammation, lysis and dysfunction of the cells (Dumpis *et al.*, 1999). The second stage of illness can range from 2 to 20 days. During convalescence, problems with concentration and memory, sleep disturbances, and headache following mental or physical stress have been reported.

Patients with severe infections may show a poliomyelitis-like syndrome and an altered state of awareness which may result in a long-term disability (Dumpis *et al.*, 1999; Gritsun *et al.*, 2003; Kleiter *et al.*, 2007). The Siberian TBE virus subtype is thought to be responsible for severe chronic infections in Siberia and far-eastern Russia (Gritsun *et al.*, 2003). Chronic TBE infections can be divided into two forms. In one form, the infected person develops hyperkinesias (which occurs regularly and can develop during the acute stage), and epileptoid syndrome. The second form of chronic TBE relates to long-term sequelae of one of the types of acute TBE, where the neurological symptoms can appear several years after the infected tick bite.

Latvia, the Urals, and the Western Siberian regions of Russia have the highest incidence of TBE viruses with the rate of attack in these regions reaching up to 199 cases per 100 000 inhabitants per year. Mortality rates due to TBE virus infection are extremely high for the Eastern TBE subtype, ranging from 5% to 20%, and extremely low for the Western subtype where it is between 0.5% and 2%.

20.3.3 Highly pathogenic avian influenza (H5N1)

Avian influenza (AI) is caused by a single-stranded negative-sense RNA virus belonging to the *Influenzavirus* A genus in the *Orthomyxoviridae* family (Swayne and Halvorson, 2008). Even though AI viruses are considered specific to given species and rarely cross the species barrier, since 1959 AI viruses subtypes H5, H7, and H9 have managed to cause sporadic human infections (INFOSAN IFSAN, 2004). H5N1 highly pathogenic AI (HPAI) was first isolated from a domestic goose in Guangdong, China in 1996, followed by an outbreak in the live bird markets in Hong Kong the following year. It was during this outbreak that the first human infection occurred and by the end of 1997, 18 people were hospitalized with six of the cases being fatal. Several other H5N1 HPAI outbreaks in humans have been reported since 2003,the probable cause being close contact with contaminated birds (CDC, 2006; WHO, 2006). Surveillance data have reported the detection of H5N1 HPAI virus in imported frozen duck meat, and both on the surface of and inside contaminated eggs (Beato *et al.*, 2009; Harder *et al.*, 2009;

Tumpey *et al.*, 2002). However, although there is experimental evidence that the ingestion of uncooked poultry blood or meat has transmitted the H5N1 HPAI to carnivorous animals, there is no direct established link between the consumption of contaminated poultry products and human infection (CDC, 2007; WHO, 2005; Writ. Comm. Second WHO Consult. Clin. Aspects Hum. Infect. Avian Influenza A Virus, 2008). Contaminated matrices such as water, and poultry faeces used as fertilizers or fish feed, have been suspected to be the source of transmission of H5N1 HPAI viruses when no direct contact with poultry or poultry products existed (de Jong *et al.*, 2005; Kandun *et al.*, 2006). Human-to-human transmission of AI virus is relatively low but has occurred in Thailand (2004), Vietnam (2004), Azerbaijan (2006), Indonesia (2006), Egypt (2007), and Pakistan (2007), generally after one member of the family has been in direct contact with infected poultry or other infected humans (CDC, 2006; Ungchusak *et al.*, 2005; WHO, 2008).

The first symptoms of infection with H5N1 HPAI viruses can occur 2–4 days after the exposure to infected poultry (Beigel *et al.*, 2005). Longer incubation periods of more than eight days have also been reported (Beigel *et al.*, 2005). It is not known whether and to what extent the virus is shed during this period (Beigel *et al.*, 2005; Chotpitayasunondh *et al.*, 2005; Kandun *et al.*, 2006; Tran *et al.*, 2004). The most common symptoms of an infection with H5N1 HPAI virus in humans include fever, shortness of breath, cough, severe respiratory disease, and pneumonia (Beigel *et al.*, 2005; Chotpitayasunondh *et al.*, 2005; Gilsdorf *et al.*, 2006; Ungchusak *et al.*, 2005). The virus appears to be the sole aetiological agent in the case of pneumonia, with no evidence of infection by any bacteria in most of the cases. In 2–8 days after the infection with H5N1 virus, patients also develop non-respiratory symptoms such as vomiting, diarrhoea, and abdominal pain (Ungchusak *et al.*, 2005; WHO, 2005; Yuen *et al.*, 1998), diarrhoea being in some cases the primary evidence of infection (Apisarnthanarak *et al.*, 2004; de Jong *et al.*, 2005).

H5N1 HPAI virus has been isolated from the cerebrospinal fluid and blood of a patient who developed symptoms of diarrhoea and convulsions, followed by coma (de Jong *et al.*, 2005). It was reported that the sister of this patient had also died of a similar illness just two weeks before, suggesting an existing predisposition to this type of disease (de Jong *et al.*, 2005). As is observed for seasonal influenza virus, infection of the CNS appears to be rare (Morishima *et al.*, 2002). The clinical pathway of an infection with H5N1 HPAI virus usually involves a rapid progression to the lower tract, at which point the patient needs to be mechanically ventilated (Beigel *et al.*, 2005; Chotpitayasunondh *et al.*, 2005; Tam, 2002; Tran *et al.*, 2004). Acute respiratory distress syndrome (ARDS) has also been associated with the progression to respiratory failure. Several other clinical manifestations of multi-organ failure have been diagnosed, with cardiac and renal dysfunction, pulmonary haemorrhage, ventilator-associated pneumonia, and pneumothorax (Beigel *et al.*, 2005; Chotpitayasunondh *et al.*, 2005; Gilsdorf *et al.*, 2006; Ungchusak *et al.*, 2005).The mean time between onset and hospitalization was 4.6 days (median of 4.0 days), ranging from 0 to 22 days (WHO, 2008). An increase in the case fatality rate was directly correlated to time from illness onset to hospitalization: 12% for 0 days, 47 and 55% for days 1 and 2, respectively, and over 70% for 4–6 days (WHO, 2008).

The WHO epidemiological study of the confirmed cases of H5N1 HPAI infection, conducted between November 2003 and May 2008, revealed a marked demographic incidence (WHO, 2008). Of 383 cases, 52% occurred in patients under the age of 20 years with the vast majority of the patients being under 40. The fatality rate was approximately 63%, infection with the H5N1 HPAI virus being more severe than infection with the seasonal virus. The demographic patterns of H5N1 virus are different to those observed for seasonal influenza. A higher fatality rate was found in the group comprising 10–19-year-olds (78%), in contrast to a low rate in those over 50 years. A possible explanation may be an age-related resistance or the paradigm of risk or exposure behaviour (e.g., close contact with infected poultry).

20.3.4 SARS-coronavirus

Severe acute respiratory syndrome (SARS) is a febrile respiratory illness primarily transmitted by respiratory droplets or close personal contact, and is caused by the SARS-Coronavirus (S → ARS-CoV) (Stockman *et al.*, 2006). SARS-CoV is a positive, single-stranded RNA virus belonging to the genus *Coronavirus* within the *Coronaviridae* family. The major clinical features on presentation include persistent fever, chills/rigor, myalgia, dry cough, headache, malaise, dyspnea, sputum production, sore throat, coryza, nausea and vomiting. Dizziness and diarrhoea are less common features. Watery diarrhoea is a prominent extrapulmonary symptom in 40–70% of patients with SARS 1 week into the clinical course of illness (Hui and Chan, 2010).

In February 2003 China reported to the World Health Organization that 305 cases of atypical pneumonia of unknown aetiology had been identified in Guangdong Province since November 2002, and that five people had died. Also in February 2003, a physician from Guangdong Province, ill with atypical pneumonia, visited Hong Kong and stayed overnight in an hotel. The agent that caused his severe acute respiratory syndrome – SARS-CoV – was transmitted to at least ten persons, who subsequently initiated outbreaks in Hong Kong, Singapore, Vietnam, and Canada (Peiris *et al.*, 2003). The incubation period after infection was 4.6 days and the mean time from symptom onset to hospitalization varied between 2 and 8 days, decreasing in the course of the epidemic, while the mean time from onset to death was 23.7 days (Hui and Chang, 2010; Leung *et al.*, 2004). The final outcome of this outbreak was 8000 probable or confirmed cases and 774 deaths, the overall mortality during the outbreak being estimated at 9.6% (Leung *et al.*, 2004).

In the 2002–2003 outbreaks, SARS transmission ceased 4 months after the initiation of global spread in Hong Kong. Subsequently there were three instances of laboratory-acquired infection and one reintroduction from animals in Guangdong Province (Anderson and Tong, 2010), but none of these occurrences had sufficient secondary human-to-human transmission to generate a threat of a recurrent global outbreak.

The earliest cases of SARS in Guangdong involved employees of exotic meat markets in the province. Research has shown that the majority of the infections occurred in people directly handling animals only recently captured from the wild, and that were to be consumed as delicacies (Graham and Bari, 2010; Li *et al.*, 2006; Shi and Hu, 2008). Moreover, infections identified after the primary SARS epidemic was brought under control were associated with restaurants that prepared and served civet meat (Graham and Bari, 2010; Wang *et al.*, 2005). The culling of civets vastly reduced the number of infected animals in Guangdong marketplaces (Zhong, 2004).

However, several studies revealed that palm civets were simply conduits rather than the fundamental reservoirs of SARS-CoV-like viruses in the wild. SARS-CoV-like RNA sequences and anti-SARS nucleocapsid antibodies were found by several studies in an Old World species of horseshoe bats in the genus *Rhinolophus*, especially in *Rhinolophus sinicus* and *Rhinolophus macrotis* (Lau *et al.*, 2005; Li *et al.*, 2005). Further analyses suggest that Bat-SCoVs and SARS-CoV have been evolving independently, presumably in bat hosts, for a long time (Ren *et al.*, 2006). In addition to masked palm civets and bats, other animal species might have been involved in the evolution and emergence of SARS-CoV. At least seven animal species can harbor SARS-CoV in certain circumstances, including raccoon dog, red fox, Chinese ferret, mink, pig, wild boar, and rice field rat (Li *et al.*, 2005). In conclusion, it is likely that another outbreak of SARS-CoV or a similar Coronavirus may occur in the future. Timely measures should be taken that would restrain the virus from causing an epidemic like the one that occurred in 2003.

20.4 Possible control measures

Effective prevention and control measures for most zoonotic viral diseases can be accomplished by means of adequate diagnoses and prophylaxis. The first step in control of any disease is surveillance. The characteristics and transmission patterns of the virus, along with the understanding of vectors and animal reservoirs, and the environment and epidemiology of the disease, have to be fully considered when control and prevention measures are defined. A better understanding of the epidemiology of the diseases associated with wildlife as reservoirs, including the virulence of the agents and their routes of transmission, would contribute to improving eradication measures for such scourges. In addition, better sanitary conditions, including proper treatment and release of human waste, improvement in public water supplies, proper personal hygiene methods and sanitary food preparation, are of extreme importance in control measures. Improved diagnostics and prophylaxis require research deep inside the molecular biology of each virus. A clear knowledge of the migration pattern of birds and the diseases they transmit would help to prevent outbreaks of emerging viruses such as H5N1 HPAI. The emergence and re-emergence of viral zoonoses must unite the efforts of two scientific fields that are currently working separately: public health and veterinary studies. Better, highly sensitive and faster detection techniques, including molecular biology methods such as genomics and proteomics working side-by-side with more conventional methodologies, would enable identification of emerging or re-emerging viruses. Rapid detection would facilitate the timely application of therapeutic/prophylactic/preventive measures. Proper vaccination campaigns can help decrease the incidence and spread of infectious diseases. Travellers should obtain information from the appropriate institutions, such as government medical advisory bodies and local health authorities, about the risks they incur, the prophylactic treatments available and the do's and don'ts while visiting a different country with a different environment and potential hazards. Medical institutions and the media should work together to disclose more knowledge on emerging or re-emerging diseases, the transmission routes, and the prophylactic measures available. Public awareness regarding the handling of wild animals in exotic markets and better food handling and cooking procedures would also help prevent outbreaks from occurring.

20.5 Conclusion

Globalization has enabled rapid development, but it has a darker side, accompanied as it is by the spread of emerging and re-emerging viral zoonoses. Changes in mobility patterns, favoured means of transportation, and land use, frequent and fast trade, and other anthropogenic factors increase the possibility of widespread viral disease. The presence of emerging or re-emerging zoonotic viruses can profoundly affect the environment and human health in a world where there is an ever increasing demand for water and food due to population growth. The implementation of effective prevention and control measures, up-to-date and helpful information for travellers, and better and improved sanitary conditions would greatly help in curbing the spread of emerging and re-emerging viral zoonoses.

20.6 References

ABDEL-GHAFAR A, CHOTPITAYASUNONDH T, GAO Z, HAYDEN F, HIEN N D, JONG M D, NAGHDALIYEV A, PEIRIS J S, SHINDO N, SOEROSO S and UYEKI T M (2008), 'Update on avian influenza A (H5N1) virus infection in humans', *N Engl J Med*, **358**, 261–73.

ANDERSON L and TONG S (2010), 'Update on SARS research and other possibly zoonotic coronaviruses', *Int J Antimicrob Agents*, **36S**, S21–S25.

APISARNTHANARAK A, KITPHATI R, THONGPHUBETH K, PATOOMANUNT P, ANTHANONT P, AUWANIT W, THAWATSUPHA P, CHITTAGANPITCH M, SAENG-AROON S, WAICHAROEN S, APISARNTHANARAK P, STORCH G A, MUNDY L M and FRASER V J (2004),'Atypical avian influenza (H5N1)', *Emerg Infect Dis*, **10**, 1321–4.

BEATO M S, CAPUA I and ALEXANDER D J (2009), 'Avian influenza viruses in poultry products: a review', *Avian Pathol*, **38**, 193–200.

BEIGEL J H, FARRAR J, HAN A M, HAYDEN F G, HYER R, de JONG M D, LOCHINDARAT S, NGUYEN T K, NGUYEN T H, TRAN T H, NICOLL A, TOUCH S and YUEN K Y (2005),'Avian influenza A (H5N1) infection in humans', *N Engl J Med*, **353**, 1374–85.

BLASKOVIC D, PUCEKOVA G, KUBINYI L, STUPALOVA S and ORAVCOCVA V (1967), 'An epidemiological study of tick-borne encephalitis in the Tribec region: 1953–1963', *Bull World Health Org*, **36**, 89–94.

CDC (2006), 'Avian influenza (flu)', In Avian Influenza A Virus Infections of Humans, 1–3. Available from:http://www.cdc.gov/flu/avian/gen-info/avian-flu-humans.htm (Accessed 13 June 2012).

CDC (2007), 'Avian influenza: current H5N1 situation', In Avian Influenza (Bird Flu), 1–4. Available from: http://www.cdc.gov/flu/avian/outbreaks/current.htm (Accessed 13 June 2012).

CHOTPITAYASUNONDH T, UNGCHUSAK K, HANSHAOWORAKUL W, CHUNSUTHIWAT S, SAWANPANYALERT P, KIJPHATI R, LOCHINDARAT S, SRISAN P, SUWAN P, OSOTTHANAKORN Y, ANANTASETAGOON T, KANJANAWASRI S, TANUPATTARACHAI S, WEERAKUL J, CHAIWIRATTANA R, MANEERATTANAPORN M, POOLSAVATHITIKOOL R, CHOKEPHAIBULKIT K, APISARNTHANARAK A and DOWELL S F (2005), 'Human disease from influenza A (H5N1), Thailand, 2004', *Emerg Infect Dis*, **11**, 201–9.

CHUA K B, BELLINI W J, ROTA P A, HARCOURT B H, TAMIN A, LAM S K, KSIAZEK T G, ROLLIN P E, ZAKI S R, SHIEH W, GOLDSMITH C S, GUBLER D J, ROEHRIG J T, EATON B, GOULD A R, OLSON J, FIELD H, DANIELS P, LING A E, PETERS C J, ANDERSON L J and MAHY B W (2000), 'Nipah virus: a recently emergent deadly paramyxovirus', *Science*, **288**, 1432–5.

CHUA K B, KOH C L., HOOI P S, WEE K F, KHONG J H, CHUA B H, CHAN Y P, LIM M E and LAM S K (2002), 'Isolation of Nipah virus from Malaysian Island flying-foxes', *Microbes Infect*, **4**, 145–51.

CROMPTON D W and SAVIOLI L (1993), 'Intestinal parasitic infections and urbanization', *Bull WHO*, **71**, 1–7.

CUTLER S J, FOOKS A R and VAN DER POEL W H (2010), 'Public health threat of new reemerging, and neglected zoonoses in the industrialized world', *Emerg Infect Dis*, **16**, 1–7.

DASZAK P, CUNNINGHAM A A and HYATT A D (2001), 'Anthropogenic environmental change and the emergence of infectious diseases in wildlife', *Acta Trop*, **78**, 103–16.

DE JONG M D, BACH V C, PHAN T Q, VO M H, TRAN T T, NGUYEN B H, BELD M, LE T P, TRUONG H K, NGUYEN V V, TRAN T H, DO Q H and FARRAR J (2005), 'Fatal avian influenza A (H5N1) in a child presenting with diarrhea followed by coma', *N Engl J Med*, **352**, 686–91.

DUMPIS U, CROOK D and OKSI J (1999), 'Tick-borne encephalitis',*Clin Infect Dis*, **28**, 882–90.

ECKER M, ALLISON S L, MEIXNER T and HEINZ F X (1999), 'Sequence analysis and genetic classification of tick-borne encephalitis viruses from Europe and Asia', *J Gen Virol*, **80**, 179–85.

GILSDORF A, BOXALL N, GASIMOV V, AGAYEV I, MAMMADZADE F, URSU P, GASIMOV E, BROWN C, MARDEL S, JANKOVIC D, PIMENTEL G, AYOUB I A, ELASSAL E M, SALVI C, LEGROS D, Pessoa da SILVA C, HAY A, ANDRAGHETTI R, RODIER G and GANTER B (2006), 'Two clusters of human infection with influenza A/H5N1 virus in the Republic of Azerbaijan, February–March', *Eurosurveillance*, **11**, 122–6.

GRAHAM R L and BARI R S (2010), 'Recombination, Reservoirs, and the Modular Spike: Mechanisms of Coronavirus Cross-Species Transmission', *J Virol*, **84**, 3134–46.

GRESÍKOVÁ M, SEKEYOVÁ M, STÚPALOVÁ S and NECAS S (1975), 'Sheep milk-borne epidemic of tick-borne encephalitis in Slovakia', *Intervirology*, **5**, 57–61.

GRITSUN T S, LASHKEVICH V A and GOULD E A (2003), 'Tick-borne encephalitis', *Antiviral Res*, **57**, 129–46.

HAGGETT P (1994), 'Geographical aspects of the emergence of infectious diseases', *Geogr Ann*, **76B**, 91–104.

HAGLUND M and GÜNTHER G (2003), 'Tick-borne encephalitis – pathogenesis, clinical course and long-term follow-up', *Vaccine*, **21**, S11–S18.

HARCOURT B H, TAMIN A, KSIAZEK T G, ROLLIN P E, ANDERSON L J, BELLINI W J and ROTA P A (2000), 'Molecular characterization of Nipah virus, a newly emergent paramyxovirus', *Virology*, **271**, 334–49.

HARDER T C, TEUFFERT J, STARICK E, GETHMANN J, GRUND C, FEREIDOUNI S, DURBAN M, BOGNER K-H, NEUBAUER-JURIC A, REPPER R, HLINAK A, ENGELHARDT A, NÖCKLER A, SMIETANKA K, MINTA Z, KRAMER M, GLOBIG A, METTENLEITER T C, CONRATHS F J and BEER M (2009), 'Highly pathogenic avian influenza virus (H5N1) in frozen duck carcasses, Germany, 2007', *Emerg Infect Dis*, **15**, 1–8.

HSU V P, HOSSAIN M J, PARASHAR U D, ALI M M, KSIAZEK T G, KUZMIN I, NIEZGODA M, RUPPRECHT C E, BRESEE J and BREIMAN R F (2004), 'Nipah virus encephalitis reemergence, Bangladesh', *Emerg Infect Dis*, **10**, 2082–7.

HUBALEK Z (2003), 'Emerging human infectious diseases: anthroponoses, zoonoses, and sapronoses', *Emerg Infect Dis*, **9**, 403–4.

HUI D S and CHAN P K (2010), 'Severe acute respiratory syndrome and coronavirus', *Infect Dis Clin North Am*, **24**, 619–38.

INFOSAN IFSAN (2004), 'Avian influenza', In INFOSAN Inf. Note No. 2/04.

KANDUN I N, WIBISONO H, SEDYANINGSIH E R, YUSHARMEN, HADISOEDARSUNO W, PURBA W, SANTOSO H, SEPTIAWATI C, TRESNANINGSIH E, HERIYANTO B, YUWONO D, HARUN S, SOEROSO S, GIRIPUTRA S, BLAIR P J, JEREMIJENKO A, KOSASIH H, PUTNAM S D, SAMAAN G, SILITONGA M, CHAN K H, POON L L, LIM W, KLIMOV A, LINDSTROM S, GUAN Y, DONIS R, KATZ J, COX N, PEIRIS M and UYEKI T M (2006), 'Three Indonesian clusters of H5N1 virus infection in 2005', *N Engl J Med*, **355**, 2186–94.

KLEITER I, JILG W, BOGDAHN U and STEINBRECHER A (2007), 'Delayed humoral immunity in a patient with severe tick-borne encephalitis after complete active vaccination', *Infection*, **35**, 26–9.

KORENBERG E I and KOVALEVSKII Y V (1999), 'Main features of tick-borne encephalitis eco-epidemiology in Russia', *Zentralbl Bakteriol*, **289**, 525–39.

KORENBERG E I and PCHELKINA A A (1975), 'Multiple viremia in goats following sequential inoculation with tick-borne encephalitis virus', *Med Parazitol Mosk*, **44**, 181–4.

KUIKEN T, LEIGHTON F A, FOUCHIER R A., LEDUC J W, PEIRIS J S,SCHUDEL A, STÖHR K and OSTERHAUS A (2005). 'Public health. Pathogen surveillance in animals', *Science*, **309**, 1680–1.

LAU S K, WOO P C, LI K S, HUANG Y, TSOI H W, WONG B H, WONG S S, LEUNG S Y, CHAN K H and YUEN K Y (2005), 'Severe acute respiratory syndrome coronavirus-like virus in Chinese horseshoe bats', *Proc Natl Acad Sci U. S. A.*, **102**, 14040–5.

LEONOV V A, KOGAN V M, NEKIPELOVA G A, VASENIN A A and SHIKHARBEEV BV (1976), 'Typifying foci of tick-borne encephalitis in the pre-Baikal', *Zh Mikrobiol Epidemiol Immunobiol*, **5**, 56–60.

LEUNG G M, HEDLEY A J, HO L, CHAU P, WONG I O, THACH T Q, GHANI A C, DONNELLY C A, FRASER C, RILEY S, FERGUSON N M, ANDERSON R M, TSANG T, LEUNG P, WONG V, CHAN J C, TSUI E, LO S and LAM T (2004), 'The epidemiology of severe acute respiratory syndrome in the 2003 Hong Kong epidemic: an analysis of all 1755 patients', *Ann Intern Med*, **141**, 662–73.

LI W, SHI Z, YU M, REN W, SMITH C, EPSTEIN J H, WANG H, CRAMERI G, HU Z, ZHANG H, ZHANG J, MCEACHERN J, FIELD H, DASZAK P, EATON B T, ZHANG S and WANG L F (2005), 'Bats are natural reservoirs of SARS-like coronaviruses', *Science*, **310**, 676–9.

LI W, WONG S K, LI F, KUHN J H, HUANG I C, CHOE H and FARZAN M (2006), 'Animal origins of the severe acute respiratory syndrome coronavirus: insight from ACE2-S-protein interactions', *J Virol*, **80**, 4211–19.

LINDGREN E and GUSTAFSON R (2001), 'Tick-borne encephalitis in Sweden and climate change', *Lancet*, **358**, 16–18.

MALAYSIAN MINISTRY OF HEALTH (2001), *Malaysian Annual Reports 2001*.

MANDL C W, ECKER M, HOLZMANN H, KUNZ C and HEINZ F X (1997), 'Infectious cDNA clones of tick-borne encephalitis European subtype prototypic strain Neudorfl and high virulence strain Hypr', *J Gen Virol*, **78**, 1049–57.

MANOJKUMAR R and MRUDULA V (2006), 'Emerging viral diseases of zoonotic importance-review', *Int J Trop Med*, **1**, 162–6.

MOHDNOR M N, GAN C H and ONG B L (2000), 'Nipah virus infection of pigs in peninsular Malaysia', *Rev Sci Tech Off Int Epizoot*, **19**, 160–5.

MOORE D A, MCCRODDAN J, DEKUMYOY P and CHIODINI P L (2003), 'Gnathostomiasis: an emerging imported disease', *Emerg Infect Dis*, **9**, 647–50.

MORISHIMA T, TOGASHI T, YOKOTA S, OKUNO Y, MIYAZAKI C, TASHIRO M, and OKABE N (2002), 'Encephalitis and encephalopathy associated with an influenza epidemic in Japan', *Clin Infect Dis*, **35**, 512–17.

ORMAASEN V, BRANTSAETER A B and MOEN E W (2001), 'Tick-borne encephalitis in Norway', *Tidsskr Nor Laegeforen*, **121**, 807–9.

OSTERHAUS A (2001), 'Catastrophes after crossing species barriers', *Phil Trans Roy Soc Lond B Biol Sci*, **356**, 791–3.

PEIRIS J S, YUEN K Y, OSTERHAUS A D and STOHR K. (2003), 'The severe acute respiratory syndrome', *N Engl J Med*, **349**, 2431–41.

POPOV F and IVANOVA L M (1968), 'Epidemiologic features of the alimentary route of infection of tick-borne encephalitis in the RSFSR territory', *Mikrobiol Epidemiol Immunobiol*, **45**, 36–42.

REN W, LI W, YU M, HAO P, ZHANG Y, ZHOU P, ZHANG S, ZHAO G, ZHONG Y, WANG S, WANG L F and SHI Z, (2006), 'Full-length genome sequences of two SARS-like coronaviruses in horseshoe bats and genetic variation analysis', *J Gen Virol*, **87**, 3355–9.

SHI Z and HU Z (2008), 'A review of studies on animal reservoirs of the SARS coronavirus', *Virus Res*, **133**, 74–87.

STOCKMAN L J, BELLAMY R and GARNER P (2006), 'SARS: systematic review of treatment effects', *PLoS Med*, **3**, e343.

SWAYNE D E and HALVORSON D A (2008), 'Influenza', in SAIF Y M, GLISSON J R, FADLY A M, MCDOUGALD L R and NOLAN L, *In Diseases of Poultry*, Ames, IA, Wiley-Blackwell, 153–84.

TAM J S (2002), 'Influenza A (H5N1) in Hong Kong: an overview', *Vaccine*, **20**, S77–S81.

TAUXE R V (1997), 'Emerging foodborne diseases: an evolving public health challenge', *Emerg Infect Dis*, **3**, 425–41.

TAYLOR L H, LATHAM S M and WOOLHOUSE M E (2001), 'Risk factors for human disease emergence', *Phil Trans Roy Soc Lond B Biol Sci*, **356**, 983–9.

TRAN T H, NGUYEN T L, NGUYEN T D, LUONG T S, PHAM P M, NGUYEN V C, PHAM T S, VO C D, LE T Q, NGO T T, DAO B K, LE P P, NGUYEN T T, HOANG T L, CAO V T, LE T G, NGUYEN D T, LE H N, NGUYEN K T, LE H S, LE V T, CHRISTIANE D, TRAN T T, DE JONG M, SCHULTSZ C, CHENG P, LIM W, HORBY P, THE WORLD HEALTH ORGANIZATION INTERNATIONAL AVIAN INFLUENZA INVESTIGATIVE TEAM and FARRAR J (2004), 'Avian influenza A (H5N1) in 10 patients in Vietnam', *N Engl J Med*, **350**, 1179–88.

TUMPEY T M, SUAREZ D L, PERKINS L E, SENNE D A, LEE J G,LEE Y J, MO I P, SUNG H W and SWAYNE D E (2002), 'Characterization of a highly pathogenic H5N1 avian influenza A virus isolated from duck meat', *J Virol*, **76**, 6344–55.

UNGCHUSAK K, AUEWARAKUL P, Dowell SF, KITPHATI R, AUWANIT W, PUTHAVATHANA P, UIPRASERTKUL M, BOONNAK K, PITTAYAWONGANON C, COX N J, ZAKI S R, THAWATSUPHA P, CHITTAGANPITCH M, KHONTONG R, SIMMERMAN J M and CHUNSUTTHIWAT S (2005), 'Probable person-to-person transmission of avian influenza A (H5N1)', *N Engl J Med*, **352**, 333–40.

WANG M, YAN M, XU H, LIANG W, KAN B, ZHENG B, CHEN H, ZHENG H, XU Y, ZHANG E, WANG H, YE J, LI G, LI M, CUI Z, LIU Y F, GUO R T, LIU X N, ZHAN L H, ZHOU D H, ZHAO A, HAI R, YU D, GUAN Y and XU J (2005), 'SARS-CoV infection in a restaurant from palm civet', *Emerg Infect Dis*, **11**, 1860–5.

WHO (2005), 'Evolution of H5N1 avian influenza viruses in Asia', *Emerg Infect Dis*, **11**, 1515–21.

WHO (2006), 'H5N1 avian influenza: timeline of major events', In *Global Alert and Response*, 10 pp. Available from: http://www.who.int/influenza/human_animal_interface/avian_influenza/H5N1_avian_influenza_update.pdf (Accessed 13 June 2012).

WHO (2008), 'Update: WHO-confirmed human cases of avian influenza A(H5N1) infection, November 2003–May 2008', *Wkly Epidemiol Rec*, **83**, 415–20.

WILSON M E (1995), 'Travel and the emergence of infectious diseases', *Emerg Infect Dis*, **1**, 39–46.

WILSON M E (2003), 'The traveler and emerging infections: sentinel, courier, transmitter', *J Appl Microbiol*, **94**, S1–S11.

YUEN K Y, CHAN P K, PEIRIS M, TSANG D N, QUE T L, SHORTRIDGE K F, CHEUNG P T, TO W K, HO E T, SUNG R and CHENG A F (1998), 'Clinical features and rapid viral diagnosis of human disease associated with avian influenza A H5N1 virus', *Lancet*, **351**, 467–71.

ZHONG, N (2004), 'Management and prevention of SARS in China', *Philos Trans R Soc Lond B Biol Sci*, **359**, 1115–16.

21

Impact of climate change and weather variability on viral pathogens in food and water

C-H. Von Bonsdorff and L. Maunula, University of Helsinki, Finland

DOI: 10.1533/9780857098870.4.458

Abstract: Climate changes, in particular global warming, will have a profound effect on world water resources and food production. Microbial contamination and chemical pollution have a considerable impact on human health and reduce the safety of food production. Regional short-term changes, such as flooding, hurricanes and other extreme weather conditions, are also expected to become more common, and the increasing global population will cause many new demands. Countries with solid economies and strong regional infrastructures will cope better with catastrophe than will developing countries, and increasing global collaboration is needed to meet future challenges.

Key words: climate change, global warming, enteric viruses, hepatitis A virus, hepatitis E virus, foodborne viruses, waterborne viruses.

21.1 Introduction

The availability of safe fresh water and food is a public health concern in every country, and also globally (FAO, 2010; UNFPA, 2007). The current trend of climate change is expected to continue, with global warming and increasing weather disturbances leading to considerable habitat changes. Population growth is expected to create increasing demand for food and water, requiring the adaption of global food and water production. During the last few decades, emerging viruses have been identified as major food and water contaminants responsible for many diseases and outbreaks. The World Health Organization (WHO) has set up an expert group to estimate the global burden of food- and waterborne disease (Hird *et al.*, 2009).

The world population is estimated to increase from its current 7 billion to 8.4–10 billion by the year 2050 (UN, 2009), with the greatest increase in developing countries. At the same time, the area of arable land available in these countries is decreasing at an alarming rate (FAO, 2008). Large areas are becoming arid and suffering from water shortages, limiting crop sizes.

The demand for safe fresh water will also increase constantly. By the year 2025, an estimated 3 billion people will only have access to limited water resources (UNEP, 2002). Approximately 75–85% of the fresh water currently available is used for agricultural purposes (Morison *et al.*, 2008). The balance of water usage between drinking and irrigation will become critical, making high demands on water production and purification systems (de Fraiture *et al.*, 2010; Rosegrant *et al.*, 2009).

The most important source of microbial contamination of both water and foodstuffs is human waste, and sewage in particular. The widespread popularity of the first flush toilet, the 'Marlboro Silent Water Waste Preventer' patented by Joseph Bramah in 1778 (Bryson, 2010), at first seemed the answer to sanitary requirements, but it later turned out to cause major problems, with increasing amounts of human waste being released directly into surface and ground water. Sewage plants were later introduced to sanitize human waste products, but they did little to remove viruses, which are thus released into surface and ground water and transmitted to humans via irrigation and/or drinking water (von Bonsdorff and Maunula, 2008). The environmental load of an infectious virus depends on the efficiency of purification and on the persistence of the contaminating (enteric) virus. The resistance of these viruses to physicochemical stress together with their low infectious dose (for many enteric viruses this is less than 100 viral particles) accentuates the problem.

21.2 Viruses of concern

Viruses of concern as water and food contaminants are presented in earlier chapters of this book. It seems likely that many viruses now prevalent will persist into the future. Geographically, the impact of individual viruses or virus groups will vary. Health authorities in each country will need to take action against those viruses that pose a serious threat to health. The threshold above which action is taken will vary from country to country. Viruses that cause life-threatening diseases may require global action.

21.2.1 Enteric viruses

Viruses that multiply in the cells of the alimentary tract and in particular in the gut cells will always be of importance as environmental contaminants. Here we point out some viruses that may more threat in future than at present. The hepatotrophic viruses HAV and HEV are among those viruses that will need special attention. Vaccine programs will probably be launched, primarily in areas where the viruses are endemic. For HAV, a well-functioning vaccine already exists. A vaccine for HEV is under development (Kamili, 2011). These vaccines, when available, will be used under the auspices of WHO, since it is mostly nations with poor economic resources that are endemic areas. It may also be that industrialized countries which use developing countries as a source of primary food production will be interested in securing safe production sites by participating in actions such as vaccination programs.

Special attention should be given to group B rotaviruses. These viruses are commonly seen among animals, and swine in particular. Human group B rotaviruses have virtually only circulated in Asia, and mainly in the adult population (China, India, Bangladesh and Myanmar) (Yamamoto *et al.*, 2010). It is not known why this virus is distributed over such a limited area. The peculiar limitation in host selection may relate to the receptor specificity of these viruses. If this barrier was overcome by mutation or reassortment events, group B rotaviruses could potentially emerge as the most serious contaminants of food and water globally. A rotavirus that could spread among adults as norovirus does would be a serious threat to public health.

21.2.2 Zoonotic viruses

The possible effects of climate change and global warming have mostly been considered in relation to arboviruses (Woolhouse *et al.*, 2005). The reason for this is obvious: global warming will create temperature zones which will become new habitats for bloodsucking arthropods carrying their respective arboviruses. This will also open up ways for host species jumps, and previously unseen animal diseases. The population in affected areas, be it animal or human, will be naïve to the infections, and therefore susceptible. West-Nile and Dengue viruses provide good examples of such diseases. The spread of these viruses requires penetration into the blood by an insect. The human respiratory and alimentary epithelium is in fact resistant to apical (or alimentary) infection by these viruses (Fuller *et al.*, 1985). The role of these viruses as food or waterborne pathogens is therefore largely unimportant.

A more serious threat lies in the extensive trade in wildlife products. The trade can bring potentially pathogenic viruses into new areas (Karesh *et al.*, 2005). Until now, surveillance of these products has been almost non-existent, and even the capacity to test for 'exotic' viruses, limited. The trade is mainly from Africa to the USA, but other areas can also be seen as potential sources of game (Smith *et al.*, 2012). The increased demand for food in the future may further stimulate this type of commerce.

Special attention must be given to influenza A (Inf A). This virus causes pandemics at varying intervals, since it changes one or both of the two surface glycoproteins HA and N, thus allowing it to overcome current herd immunities. Even though the virus spreads as a person-to-person infection, it usually originates as waterborne. The whole selection of HA and N variants is present in aquatic bird colonies. From here they spread in a species-specific way to different animal hosts (as reviewed by Donatelli *et al.*, 2008). Species specificity relies on the structure of the sialic acid bonding of the host. The recently recognized highly pathogenic Inf A viruses (H5N1, H7N1) that spread directly from birds to humans, but are not adapted for human-to-human spread, pose a risk of serious outbreaks, should relevant modifications of the receptor-binding domain of the HA molecule occur. The risks of water-mediated spread must be recognized.

21.2.3 Seasonality of viruses

Many viruses show a pronounced seasonality that varies according to the climatic conditions. Rota-, noro- and astroviruses appear as winter/spring diseases in global temperate zones (Fig. 21.1), whereas in tropical zones they tend to occur throughout the year, though slightly more during the dry season (Hanlon *et al.*, 1987; Sitbon *et al.*, 1985).

On the other hand, HAV, where endemic, appears to follow the epidemiological pattern of enteroviruses with a peak in late summer/autumn. In countries free of HAV, the outbreaks appear whenever the virus is introduced to the non-immune, susceptible population. In modern societies, the frequency of world travel may introduce HAV at any time. In an 'unstirred', naïve population, the HAV infection is mostly contracted early, conferring lifelong immunity. The seasonality of HAV infections is weak. Shifts in timing may easily occur due to weather conditions (Harley *et al.*, 2011; Harper *et al.*, 2011), population immunity and/or strain variation of the virus in question.

For rotaviruses in the USA, the historical pattern has been for viruses to spread from southwest to northeast during annual outbreak season (Pitzer *et al.*, 2009). This pattern, however, is no longer clear, as a result of the changing birthrate. No such pattern has been found in Europe, where transmission from south to north can be seen, likely reflecting differences in population size and climate conditions, such as temperature and humidity.

21.2.4 Evolution of viruses and climate change

Most gastroenteritis viruses are ssRNA viruses that tend to easily and rapidly evolve. Changing climatic conditions, such as global warming, or indeed cooling, may increase the rate of virus evolution. This has been hypothesized with Inf A viruses. About 1000 years ago, when the Little Ice Age affected migration patterns for the hosts of influenza A viruses (Gatherer, 2010), a subtype diversification may have occurred. Global warming may also cause the migration patterns for hosts of other viruses, such as hepatitis E viruses, to change. Virus evolution combined with invasion of animal viruses has been documented. A well-known example is bluetongue virus, an Orbivirus, the spread of which to Europe has been explained by changes in weather conditions, enabling the spread of midges that serve as carriers for the bluetongue virus (Purse *et al.*, 2008).

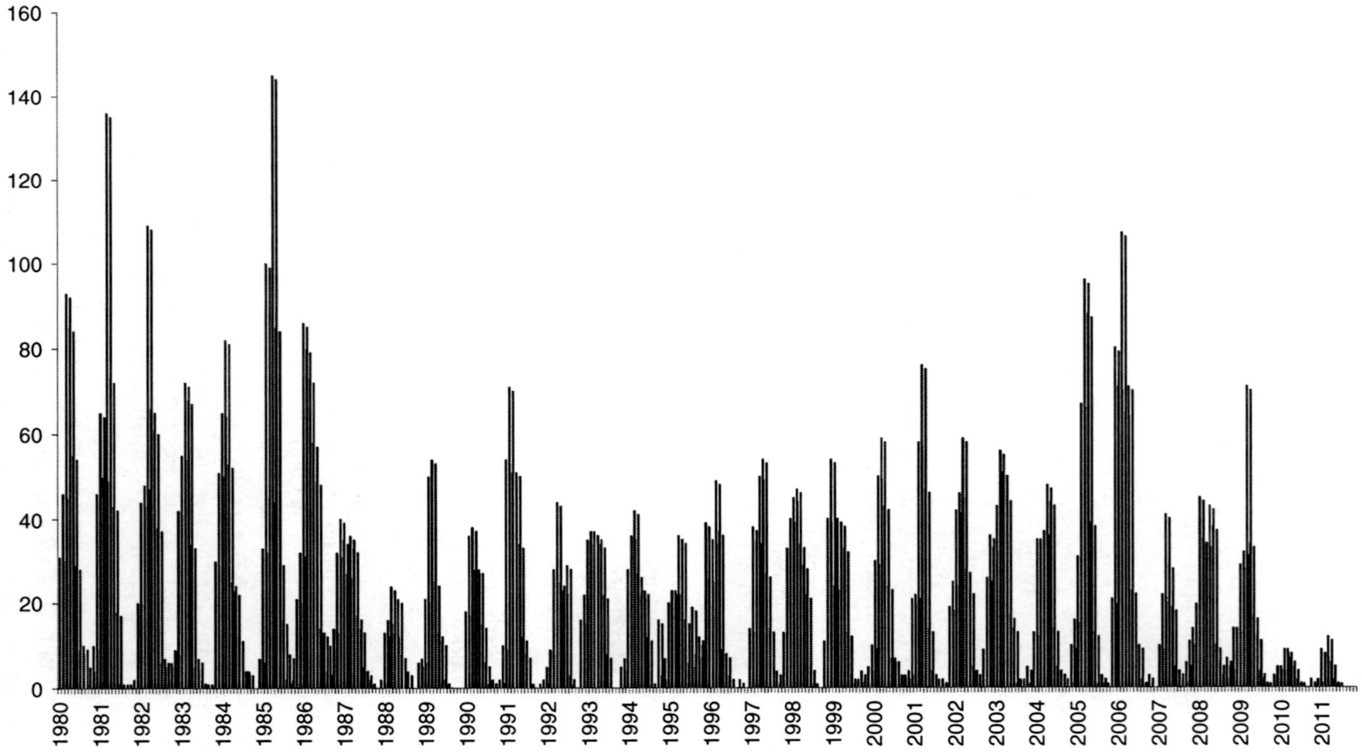

Fig. 21.1 Annual rotavirus diagnoses in Finland from 1995 to 2011 showing the pronounced seasonality typical of this virus (figure is combined from several data sources).

Genetic recombination is one means by which viruses evolve. Viruses within one virus species may change their genetic material between genotypes. It is also possible that viruses of one virus family, that normally infect different species, may exchange genetic material. After this, they may be able to infect new hosts. Recombination may occur in humans or animals following infection after drinking contaminated water or eating shellfish contaminated with multiple viruses. Viruses of different species may infect the same cell in the gut simultaneously. Flooding may increase the likelihood of human and animal sewage mixing, providing more opportunities for multiple infections.

Since the eradication of poliovirus, enteroviruses have not been linked to waterborne outbreaks, although we know of their common presence in sewage-contaminated water. The surveillance of sewage for polioviruses is an example of viral environmental control. However, revertants may appear in the future: a currently non-neurotrophic enterovirus genotype or vaccine strain may mutate and regain its ability to cause neurotrophic symptoms in humans (Yoshida *et al.*, 2000). Viruses that at present cause mild gastroenteritis may change to viruses causing more severe symptoms.

Although the recorded history of noroviruses is relatively short and long-term trends undiscernable, the dynamics of norovirus infection and transmission appear to be changing. Since the sequential appearance of the norovirus GII.4 variants (Siebenga *et al.*, 2007), the number of annual norovirus cases has increased in many countries (Bok *et al.*, 2009). It is unclear why the evolution rate has increased. To our knowledge no study of the relationship of climate change to norovirus cases has been published.

For viruses that possess a spliced genome, such as rotavirus and Inf A viruses, segment exchange leading to mosaic viruses (reassortants), is possible. The G9 rotaviruses that invaded Europe in the 1990s likely originated from animal rotaviruses (Ramachandran *et al.*, 2000) and then when the segment 9 coding for G9 formed a mosaic virus with the rest of human RNA segments, the strain started to spread rapidly from person to person. As pigs and birds are known to be involved in the emergence of novel influenza A viruses, animals may also be involved in the evolution of human rotaviruses.

21.3 Impact of short-term climate changes

The US National Climatic Data Center in Asheville, North Carolina, reports that the frequency of highly expensive weather disasters has doubled since 1980 (Schiermeier, 2011).

21.3.1 Catastrophes and enteric viruses

Under conditions of acute catastrophe, it is often impossible to collect detailed information about the causative agents of gastroenteritis outbreaks, or to make accurate scientific analyses. It may be possible to calculate the number of persons suffering from diarrhoea, but sample collection may be impossible. Thus, in many publications describing outbreaks, the causative agents have remained unknown, or only some among multiple causative agents have been identified. The same reasons may explain why the effect of water, sanitation, and hygiene on the prevention of diarrhoea, has been rigorously studied during catastrophic events according to a systematic review (Cairncross *et al.*, 2010). Only hand washing with soap was consistently found to be effective.

It is important to be especially prepared for investigations in extreme conditions. Methods for rapid identification of potential pathogens in water samples were developed after the experience of Hurricane Katrina in 2005 (LaGier *et al.*, 2007). These hand-held biosensors detect algae and other microbial contaminants electrochemically. Rapidly deployable methodology can be applied to pathogen detection both in patient and in environmental samples.

Climate change is expected to cause more frequent extreme weather conditions in limited geographical regions. If these areas are densely populated, the number of water- or foodborne disease outbreaks is likely to increase, unless adequate preventive steps are taken. Viruses play a major role in waterborne diseases. All enteric viruses including hepatitis A and E viruses are relevant. Extreme weather conditions may increase the likelihood of water supplies becoming contaminated by human sewage.

Not much data is available about survival of viruses in the environment after flooding. Persistence of viruses in soil and in vegetables grown in contaminated soil was examined by Tierney *et al.* (1977) in a study where field plots had been flooded with poliovirus-contaminated sewage wastes. During the Ohio winter, infectious poliovirus 1 could be detected maximally for 96 days, and in summer for 11 days after flooding. In addition, poliovirus was recovered from vegetables 23 days after flooding, the time when lettuce and radish are often harvested.

21.3.2 Flooding

Flooding may occur during or after heavy rain or snowfall, or when the ice cover is melting in northern countries. Water from rivers or lakes rises into neighboring fields, or seawater covers the coastline on islands and the mainland. The damage in rural areas is usually caused to crops, fields, and cattle. In cities buildings may be damaged. Seawater or sewage may contaminate fresh water supplies.

The consequences depend very much on the standard of infrastructure in the country concerned. In some regions flooding occurs regularly and does not cause extensive damage, because timing is predictable and precautions are taken. In developing countries, regular flooding events may cause significant damage to people and crops. In developed countries the consequences of unexpectedly strong flooding may also be devastating, for example, Hurricane Katrina in New Orleans in the USA.

Since we are unable to predict future events, we can only analyze the existing literature to see how flooding has affected the population in terms of enteric virus infection (Table 21.1). In Bangladesh in 1988, an increase in infantile and adult diarrhoea patients was observed in association with nationwide monsoon floods (Ahmed *et al.*, 1991). There was also a different distribution of rotavirus serotypes during the pre-flood period and flood period, with an increase in mixed virus types after flooding. Rotaviruses as well as noro and adenoviruses, together with bacterial pathogens and parasites, caused a gastroenteritis outbreak in Finland in 1994, when a groundwater well was contaminated by river water during the spring flood commonly occurring in northern regions when ice melts rapidly (Kukkula *et al.*, 1997).

In Germany during the Oder river (Odra) flooding in the summer of 1997, water was reported to have the quality of bathing water for the duration of the flood (Fenske *et al.*, 2001). In Dresden, however, the viral load of the river Elbe was high during the flood of 2002, with adeno, astro and enteroviruses (Rohayem *et al.*, 2006). Rohayem *et al.* (2006) reported a clear association between water temperature and the number of different viruses, with as many as 45 viruses when mean water temperature was 7.6°C, compared with 17 viruses when the temperature was 17.8°C. The reason for the level of water contamination in these two regions might have been a difference in population density, with large agricultural areas in the Oder region.

Indonesia offers an example of the consequences of urban flooding, where groundwater became contaminated by viruses in the flooded area (Phanuwan *et al.*, 2006). Both contaminated drinking water and direct contact with floodwaters were found to pose a risk of viral infection in people. Shellfish may become contaminated after flooding, as evidenced in France, where consumption of contaminated oysters caused a gastroenteritis outbreak (Le Guyader *et al.*, 2008). Five different viruses including Aichi-virus were detected in these shellfish, and as many as seven viruses were detected in one patient.

Flood-related hepatitis cases have been reported from South America with HAV, and Africa and Asia with HEV. HAV was detected with viral loads varying from 60 to 5500 copies/l in water samples from the Amazon basin, where the river Negro regularly floods during rainy seasons (De Paula *et al.*, 2007). The prevalence of HAV has remained high in the Amazon region, whereas morbidity has declined in the rest of Brazil (Carrilho *et al.*, 2005). In Somalia, an HEV epidemic peaked as the river rose during rainfall, suggesting waterborne disease (Bile *et al.*, 1994). In the 1990s, a high prevalence of HEV (50%) in humans was found in riverine areas of Borneo (Corwin *et al.*, 1997). The frequency of anti-HEV antibodies in populations in the Mekong river delta region was found to be lower (9%) (Hau *et al.*, 1999). For this reason there is a risk of a major hepatitis E outbreak. The transmission of hepatitis E via water in Far-East Asia was known before the pathogen, HEV, was

Table 21.1 Viral gastroenteritis related to flooding

Continent/Country	Year	Description	Enteric viruses	References
1. Asia/Bangladesh	1988	Nationwide monsoon flooding	Rotaviruses, increase in cases and genotype patterns	Ahmed *et al.*, 1991
2. Africa/Somalia	1988	HEV in Somalia, waterborne?	HEV	Bile *et al.*, 1994
3. Asia/Indonesia	1994	Riverine areas of Borneo	HEV	Corwin *et al.*, 1997
4. Europe/Finland	1994	Waterborne epidemic, groundwater well contaminated by polluted river water during spring flood	Adenovirus, norovirus, group A rotavirus, group C rotavirus	Kukkula *et al.*, 1997
5. Asia/Vietnam	1994	Periodic river flooding of Mekong	Seroprevalence of HAV (9%), HEV (97%)	Hau *et al.*, 1999
6. Europe/Germany	1997	Exceptional flooding of Odra river	Lower no. of infectious enteroviruses than normally	Fenske *et al.*, 2001
7. Europe/Germany	2002	Viral burden in flooded areas of Dresden	Adenovirus 40/41, astrovirus, enterovirus	Rohayem *et al.*, 2006
8. Asia/Indonesia	2005	Comparison between flooding/ non-flooding areas of Ciliwung river	High numbers of HAV, NoV, Adenovirus in floodwaters, contamination of groundwater	Phanuwan *et al.*, 2006
9. South America/ Brazil	2004/2005	HAV load in Amazon basin	HAV 92% of the samples (60–5500 copies/l)	De Paula *et al.*, 2007
10. France	2006	A flooding event close to a shellfish production lagoon	Aichi, NoV, astrovirus, enterovirus, rotavirus	Le Guyader *et al.*, 2008

even found. Why waterborne HEV outbreaks only occur in tropical and subtropical regions is not yet known. It remains to be seen whether the potential expansion of subtropical areas due to climate change will lead to waterborne HEV outbreaks in new regions.

21.3.3 Storms and rainfall

A study in a hospital in Milwaukee, USA, found an association between rainfall and pediatric emergency visits for undefined acute gastrointestinal (GI) illness (Drayna *et al.*, 2010). The authors suggest that a waterborne component may prove an explanation. In several similar studies from Canada, the association between precipitation and gastroenteritis cases has been investigated. However, clear evidence has been difficult to find (Febriani *et al.*, 2010; Teschke *et al.*, 2011; Thomas *et al.*, 2006). In an Australian study, the incidence of norovirus-associated gastroenteritis outbreaks was found to be statistically related to average monthly rainfall as opposed to temperature (Bruggink and Marshall, 2010).

Heavy rainfall may affect the efficiency of wastewater treatment plants, and sewer overflows may increase the number of viruses in the effluent discharged to the sea. Fong *et al.* (2010) show the enteric viral burden in the Michigan River and Katayama *et al.* (2004) that in Tokyo Bay. In neither of these studies, however, is there clear evidence that the increase in the virus burden in the rivers was due to heavy rainfall or sewer overflow.

21.3.4 Hurricanes

Diarrhoea, or viral gastroenteritis, is not usually linked to hurricanes. Virtually the only hurricane after which a multitude of reports about gastroenteritis were published was Hurricane Katrina, which hit New Orleans in 2005. One of its consequences was massive flooding due to the collapse of dam walls. The catastrophe was so extensive and the city destroyed so totally, that the population required evacuation. In one shelter for evacuees, a gastroenteritis outbreak was reported to have occurred, with a peak after five days (Anon, 2005). Norovirus was detected in 50% of stool samples. The outbreak continued for more than a week before declining. Slowing the progression of a norovirus outbreak once it has appeared is very difficult in such catastrophic situations (Yee *et al.*, 2007). The effects of Hurricane Mitch in 1998 were studied in northern Honduras, and fewer diarrheal illnesses were found to have occurred among people living in cement block housing than in poor households (Guill and Shandera, 2001).

21.3.5 Extreme heat

Since human enteric viruses do not multiply outside a host cell, prolonged heat waves do not lead to an increase of viruses in the way that bacteria can increase in unprotected food or water. In high temperatures children, and particularly malnourished children affected by viral diarrhoea, need urgent fluid treatment in order to recover from severe dehydration. The elderly are also prone to many enteric viruses, and extreme heat may worsen the symptoms, as reviewed by Sheffield and Landrigan (2010).

21.3.6 Earthquakes and tsunamis

Although the extent of the relationship between earthquake and tsunami frequency and climate change remains unknown, the presence of enteric viruses in earthquakes is briefly discussed here. Tsunamis are comparable to floods but with more unpredictable catastrophic results. According to a review by Kouadio *et al.* (2012), between 2000 and 2011 a total of 20 natural disasters occurred, and five of them were earthquakes (in El Salvador, Iran, Pakistan, Haiti and Japan). In four a diarrhoea outbreak followed the disaster. Earthquakes usually affect infrastructure, and sewage may contaminate sources of drinking water. Even when preventive efforts are taken regarding water, the infection often continues to spread among evacuees due to difficult living conditions (Fig. 21.2). After two earthquakes that occurred in Japan in 2007 (Nomura *et al.*, 2008) and 2011 (Kouadio *et al.*, 2012), norovirus outbreaks were reported, especially among the elderly. Morbidity of gastroenteritis in evacuees in shelters was highest (6.6%) nine days after the earthquake in 2007. While the number of evacuees in the shelter decreased continuously from over 1000–244 people within a month, the outbreak itself was over in about two weeks. In India a post-earthquake outbreak of rotavirus gastroenteritis in Kashmir has been reported by Karmakar *et al.* (2008). The authors visited villages affected by the earthquake as well as rehabilitation camps and examined cases of acute diarrhoeal disease and drinking water samples. In children the overall attack rate was 20%. In addition to gastroenteritis cases, hepatitis E virus outbreaks following earthquake or tsunami episodes have been reported in endemic regions (Indonesia 2004, Pakistan 2005). Other infections following earthquakes are acute respiratory infections, influenza, measles, meningitis, tetanus and cholera (Kouadio *et al.*, 2012). In Turkey, the seroprevalence of hepatitis A and E was studied in children living in camps after the 1999 earthquakes. HAV prevalence was 44.4% and 68.8% in two camps and HEV prevalence was 4.7% and 17.2%, respectively (Sencan *et al.*, 2004). The authors recommended that the more severe consequences could be avoided by the urgent provision of sufficient sanitary facilities.

21.3.7 Landslides

Geo-morphological disasters, such as landslides and avalanches, may occur near mountains or along rivers, after heavy rainfall, for example. Such disasters may require evacuation and temporary shelter for the affected population. A recent report describes the hygiene and sanitation levels in a camp

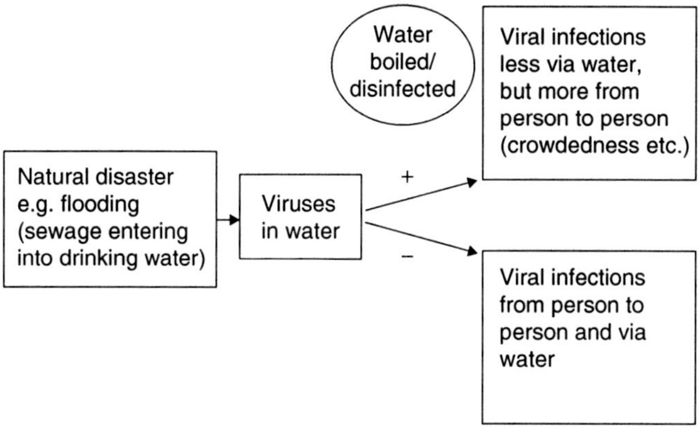

Fig. 21.2 Model of transmission of gastroenteritis viruses during natural disasters.

after a landslide disaster in eastern Uganda (Atuyambe *et al.*, 2011). The main causes of morbidity, 2 weeks after the disaster, were respiratory infection (58.3%) and malaria (47.7%). Diarrhoea was found in 8.8% of the camp residents.

21.3.8 Frost damage to water and sewage lines

Climate change is thought to reduce frost damage, due to an increase in overall temperature. In some regions, however, extreme weather conditions and rapid temperature changes pose a challenge for water and sewage pipelines. Countries that previously expected snow cover throughout the winter may in future have periods of extreme cold without snow, or strong snowstorms combined with periods of mild weather. This could impose a strain on sewage lines, resulting in sewage leakages and, ultimately, an increase in viral gastroenteritis outbreaks.

21.4 Impact of long-term climate changes

Although there are differences of opinion about the projected extent of climate change over the coming century, there is nevertheless uniform agreement that change will happen. In the long term, gradual global warming will have consequences relating to the biosphere and the habitability of land for an ever-increasing population (Fig. 21.3). An increasing population together with an apparently decreasing area able to provide acceptable living conditions is likely to result in higher population densities. At the same time a reduction in arable and pasture land dedicated to food production will lead to intensified agricultural practices, with high environmental stress and contamination as consequences. Under such conditions it is increasingly likely that human and

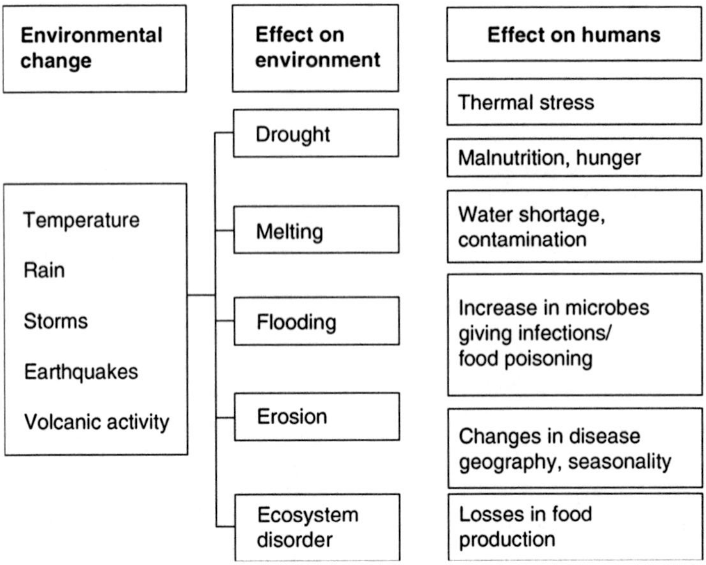

Fig. 21.3 Effects of climate change on human health. (Source: Adapted from McMichael *et al.* (2006), Fig. 21.1.)

animal infections will spread, and as a consequence, a higher environmental microbial contamination load is probable. This will also lead to an increased risk of foodborne disease The role of viruses in such projected scenarios has been largely neglected, or even omitted altogether. Cholera and other enteric bacteria are seen as the main threat, but parasites such as *Cryptosporidium* are also acknowledged. (Bezirtzoglou *et al.*, 2011; Cooney, 2011; Rose *et al.*, 2001). The relative neglect of viruses in this context is largely a consequence of the relative mildness of diseases they cause. Lack of convenient detection methods has also hampered virus monitoring. This situation is, however, rapidly improving, and viral foodborne illness is increasingly acknowledged (Koopmans and Duizer, 2004)

21.4.1 Sewage handling

Globally the monitoring of environmental viral contamination varies greatly. Each country has adopted an 'acceptable disease load' level, depending on both the seriousness of the disease and the prevalence of cases. In developing countries the technical and economic resources for viral monitoring are limited and will at most consist only of registering serious diseases such as hepatitis A and E. Industrialized countries are already paying more attention toward, for example, noroviruses causing milder GI illness. At present the environmental circulation of enteric, especially hepatitis, viruses globally through sewage varies greatly, mirroring the selection of endemic viruses.

Even though the person-to-person spread of these viruses is by far the most common contamination method, sewage will play a key role in their environmental spread.

Human waste, whether from individual households or collected sewage plant effluents, primarily allows contamination of the receiving water bodies (Aw and Gin, 2010; Li *et al.*, 2011; Nordgren *et al.*, 2009). The level will be dependent on the disease load of the population, and on the efficiency of the waste purification methods used. Most sewage plants appear to allow viruses to escape into the effluent (Maunula *et al.*, 2012). Since chlorination will result in the release of unfavourable products, in particular carcinogenic substances, this method has been largely avoided. Currently, the most promising approach appears to be the use of efficient methods to separate solids from sewage, leaving a clear effluent that can be run as thin films over UV tubes, allowing disinfection. With increasing population density, sewage handling will become more centralized, and the relevant plants larger. This will allow monitoring of viral release and an improvement in methods of removing or deactivating viruses from effluent.

In the environmental spread of viruses, sewage is the prime source of contamination. Water in all forms, whether for drinking or food production and preparation, will be an important vehicle for transmitting viruses to the consumer.

21.4.2 Water production

The predicted growth of global population, together with the increase in very densely populated areas, will put a strain on the availability of safe fresh water. Since no significant increase in water resources can be anticipated, more efficient use of water is clearly required.

Global water reserves, as well as efficient use and availability of water, vary greatly all over the world (Gleick and Palaniappan, 2010). Industrialized countries are for the most part located where there are relatively good water reserves. Water management is good, and more money can be invested in maintaining quality. Even then, some areas including parts of Europe and the middle and western USA, are experiencing groundwater overuse (Premanandh, 2011).

As previously stated, agriculture requires the greatest share of fresh water (Parry and Hawkesford, 2010). Even though agricultural areas will continue to be separated from the densely populated areas, a shortage of both drinking and irrigation water can be anticipated. The extent of ground water resource depletion is already being investigated (Konikow and Neuzil, 2007). Pumping contaminated surface water or even sewage water over sandy soils, and then collecting the filtrate from underneath, has been attempted as a method of collecting hygienic artificial groundwater (Horswell *et al.*, 2010). It appears, however, that the filtering capacity for viruses largely depends on the thin layer of soil on top of the sand. This cannot withstand continuous use, and will allow microbes to pass into the end product. Surface water will therefore be used more and more for the production of drinking water. Since this raw water must be considered contaminated by microbes, efficient disinfection and purification procedures, combined with adequate monitoring methods, are needed (Lewis *et al.*, 2011). Sensitive and cost-effective methods for virus detection are particularly in demand.

Agricultural irrigation water presently takes about 80% of available fresh water (Morison *et al.*, 2008). Of available irrigation water, only an estimated 10% is used effectively (Pfister *et al.*, 2011). Estimates suggest the irrigated area will have almost doubled by 2050 (Lobell *et al.*, 2008). In the future, climate changes, both short- and long-term, will increase pressure on the availability of qualitatively acceptable water. With rising temperatures, precipitation is predicted to increase. At the same time, increased evaporation will cause a net loss of water resources. Accurate predictions are required, based on calculations from the recently presented water resources vulnerability index (Jun *et al.*, 2011).

21.4.3 Food production

Global food production has been shown to be sufficient to feed the present population (Parry *et al.*, 2005). This fact does not prevent extensive hunger-related disasters, with large numbers of casualties in areas exposed to extreme weather conditions or catastrophic climate-related events. Such disasters are manageable through international cooperation, with imported water and food supplies reaching the affected population. In the long run, with more pressure on food production, risks of microbial contamination increase. Food production will be affected by the emergence of both animal and plant diseases that can be expected with species translocations (Wilkinson *et al.*, 2011). Even under 'normal' conditions, an ever-increasing number of foodstuffs are transported to consumers in industrialized countries whenever a market appears for them. Under these conditions, the flow of goods is regulated primarily by price. As a consequence, an increased number of foodborne outbreaks has already been seen (Hjertqvist *et al.*, 2006; Sarvikivi *et al.*, 2012). It seems unlikely that such outbreaks will deter commercial interests.

The principal limitations on the global capacity for sustainable food production are availability of arable (together with fertilizers and irrigation water) and grazing land (Strzepek and Boehlert, 2010). Most areas are needed for basic grain production (wheat, rice, and soybeans). The amount of water required for crop production varies depending on the variety grown ('crop per drop') (Marris, 2008). By selecting the optimal variety for each habitat, and by manipulating genes regulating growth characteristics (Salvi and Tuberosa, 2005), crop yields can be improved. Maximizing crop yields will increase the risk that plant diseases will emerge and spread.

These grain commodities are marginal as sources of viral food poisoning. They may play a more critical role in a situation where yields fail; malnutrition and hunger cause people to intensify their use of alternative commodities

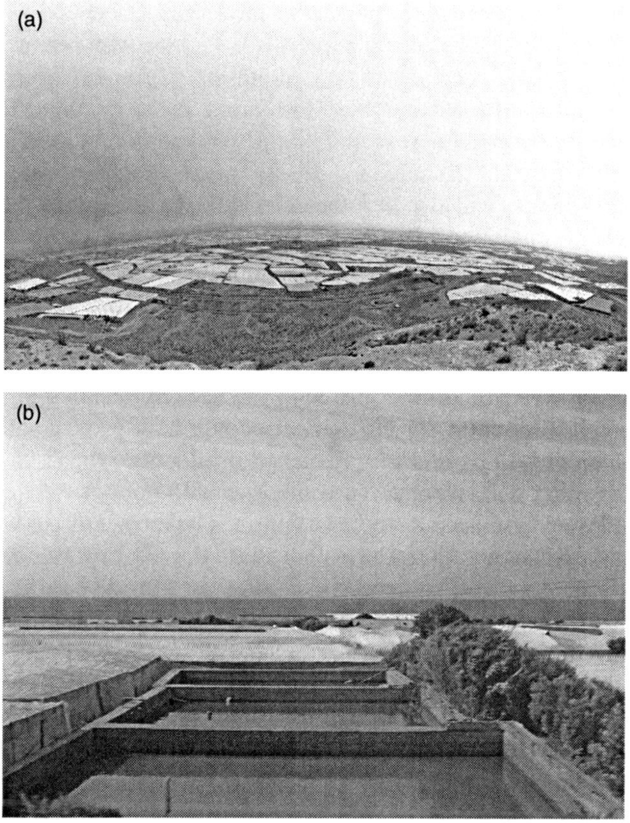

Fig. 21.4 Modern developments in the production of fresh produce. Production units tend to grow larger and products are distributed internationally. Possible contaminations tend to cause major disturbances to the food trade. (a) Covered greenhouses cover hundreds of square miles. (b) Irrigation water distributed from central open water reservoirs. (Source: Pictures Ville K. Teräväinen, with permission.)

that, in turn, are more likely to become contaminated with viruses. Thus fresh produce will become increasingly important. The results of intensifying fresh produce production can already be seen in certain areas (Fig. 21.4). Hundreds of square miles are covered with plastic, under which vegetables are grown in water cultures. The water, supplemented with fertilizer, is distributed from reservoirs which are mostly open. The hands-on production is managed by cheap labour, and in the most troubling cases, by illegal immigrants from poorer countries; altogether, there are numerous risk factors that can lead to (viral) food contamination. These kinds of scenarios can be envisioned when the demand for fresh vegetables increases.

Fruits and berries have emerged as possible sources of viral infections. Even though the mechanisms by which contamination takes place are not fully understood, raspberries in particular have been associated with numerous (noro)virus outbreaks (see Chapter 16). Picking the berries manually seems to be one obvious means of contamination. Here also, both cultivated and wild berries are picked by cheap labourers, often recruited from other countries just for the picking season. Possible viral contamination is then preserved through freezing the berries.

Animal husbandry will also be intensified in the future, due to the demands of a growing population. Increased temperatures will cause geographic changes in the distribution of pastures, primarily in favour of colder climates. The balance between land use for crop growth and pastures will need critical evaluation, and the animal feed production will have to increase. The trend is most likely to be towards larger production units with increasing numbers of animals per farm. At the same time, with the help of genetics, animals with maximal meat production per feed unit used will have to be created. Under the conditions prevailing in such intensified production, the risks of animal diseases spreading will increase. Viral diseases with zoonotic character such as hepatitis E may also become more common. The former diseases will cause a fall in meat production, the latter will increase the risk of foodborne disease in humans.

Seafood is an acknowledged source of viral infections. Varieties consumed uncooked, such as oysters, pose a particular risk. Intensified cultivation will undoubtedly lead to the use of less pure (sewage-contaminated) water in the future. Since bivalve mollusks filter and concentrate all constituents of water quite unselectively, they can be seen as the best 'monitoring devices' for viruses. Since sewage viruses from various origins and species will come together in these animals, they can also be seen as a potential source for recombination and/or reassortment events in the consumers' gut.

An increase in the global demand for food means that the reserves found in the oceans and waters will be increasingly exploited. Many fish and crustacean species have already been harvested to near-extinction. Whaling exemplifies the ruthless decimation of mammals, despite attempts to limit the killing through international agreements. It seems that the natural reserves of commercially attractive species are close to being used up. At the same time, there are increasing numbers of fisheries farming fish and other seafood. Many fish species are successfully farmed already, and production units are growing. It is quite reasonable to predict that this area of food production will grow considerably. Apart from animal production, it is also feasible that plant and algal species will be developed for human consumption and animal fodder.

It is highly probable that these 'monoculture' activities will lead to unwanted microbial invasions, primarily pathogenic for the products, and leading to losses in yields. This has already been recognized in salmon farms both in Norway and along the Chilean coast. At the same time the risk of human pathogens appearing will increase, due to new production chains, and the possible introduction of new pathogens, including viruses.

As previously mentioned, the viral contamination level will reflect the seasonality of respective diseases. As long as the fresh produce remains on the domestic market, there will be few unexpected outbreaks, but the situation will change as soon as the products are introduced to the international market. The epidemiological picture becomes even more disturbed when crops are stored by freezing. In this respect, viruses are in a category of their own in that they survive freezing almost indefinitely. Thus freezing virus-contaminated foodstuffs is an excellent way of introducing viruses to unprotected, naïve populations, with resulting outbreaks.

21.4.4 Surveillance and control of foodborne viruses and viral diseases

There is a general understanding that protecting drinking water and food from microbial contamination is a necessity. Numerous serious outbreaks have led to hygienic improvements throughout the food chain, and lessened the chance of accidents. The introduction of bacterial indicators to identify and monitor risks has been a crucial step in safe food production, resulting in a marked fall in bacterial outbreaks. It has also meant that tests for indicator organisms, mainly *E. coli* or enterobacteria, are strongly established as hygiene markers. The tests are also cost-effective and well-established.

Despite these improvements, a considerable number of outbreaks arise in situations where bacterial indicators do not reveal hygiene failings. Such situations are occasionally found in waterborne outbreaks, which have largely turned out to be viral or parasitic (Nwachcuku and Gerba, 2004). A similar situation is also seen with bivalve mollusks.

By far the most common viruses to contaminate food and water are noroviruses (Maunula, 2007). These may be considered less important due to the relative mildness and brevity of the disease they cause. However they are important in that they reveal weak hygiene points in the food chain. Norovirus outbreaks should therefore be thoroughly investigated. At the same time, there is an urgent need for improvement to the microbial indicator system. It seems likely that supplementing the bacterial indicator(s) with adeno or polyomaviruses may serve this purpose (see Chapter 22). There must, however, be a general agreement on the indicator(s) before one can expect increased safety in the national and international food product trade. Only with a well-functioning indicator system can risk assessments and risk management plans be established. Testing for viruses in foodstuffs remains cumbersome and costly. The problem lies in the difficulty of extracting the viruses from a variety of matrices. The genome-based identification of the viruses is both specific and sensitive. With the increasing threat of viral calamities, the pressure to develop and enforce viral testing can only increase.

A future challenge for safe food production is also posed by the trend to constantly increase the size of production units, be they for fresh produce such as vegetables and berries, or meat production. This trend is driven by cost calculations resulting in cheaper unit prices. The hygiene problem is that large product volumes are lost due to (viral) contamination. It will be important to keep lot sizes limited and labelling adequate. Only by these means can traceability be good and damage limited. The hygienic classification of coastal waters for seafood production and associated regulation of products (depuration) serves as a good model for improved safety (Formiga-Cruz *et al.*, 2002).

For the most part the causes of viral contamination derive from human activity. Good understanding of the importance of hygiene in food production is vital, and correct behaviour springs from adequate training. Hygiene proficiency tests or passports, as enforced in the Nordic countries, contribute to this aim. The passport is obtained on passing an appropriate hygiene skills test (http://www.evira.fi/portal/en/food/hygiene_proficiency/proficiency_test/). Only persons possessing the passport are allowed to work in the food chain, whether producing raw materials or preparing and serving meals.

Apart from the well-known and recognized food-related viral diseases, one can expect the appearance of 'new' diseases. These will only attract attention if they are serious enough or clustered together. They will be identified from patient samples in the first instance, and only thereafter can they be looked for in food. Such viral diseases may originate from recombination or reassortment events of old viruses, or from an animal reservoir that has been encountered when exploiting new areas for food production.

21.5 Conclusion

Among the many effects that global warming will have on human health, food contamination by viruses will have rather modest consequences. A more serious and widespread threat may be caused by viral contamination of water. In the literature concerning climate change and microbial infections, the main concern seems to be bacterial and parasitic infections. In the relevant literature, only arboviruses as representatives of zoonotic viruses appear to be acknowledged, though not as foodborne pathogens.

The change in the epidemiology of foodborne viral infections due to long-term climate change is a web of many interdependent factors (see Table 21.2). On one side are environmental effects increasing the pressure of infection, on the other are both local and global preventive actions taken by health authorities. It is clear that local viral outbreaks due to sudden changes in the environment continue to occur. These should not, however, lead to global disasters. In the longer term, over several decades, there will be sufficient time to take preventive measures.

The expected population increase together with a reduction in the global habitable area will result in higher population densities, a situation favourable to the spread of infections. The increase in GI infections will result in higher concentrations of viruses in sewage, which will in turn favour contamination.

Table 21.2 Factors contributing to the spread of food- and waterborne viral infections in connection with global climate change

Population effects		Effects on food production	
Problems	Solutions	Problems	Solutions
Population increase	Improved hygiene	Decrease in arable land	Effective land use, multiple crops
Ageing	Improved health care	Shortage of irrigation water	Water reuse, sanitation
Crowding	Immunizations	Plant diseases	Better plant variants
Starvation	Viral surveillance of food chains and water	Animal diseases	Improved hygiene, immunization
Waste movement	Sewage sanitation		

The consequences of a rapid, permanent climate change making large areas of the globe non- or less habitable would be more dramatic. Food production would be seriously limited over large areas, with acute famine for a large proportion of the world's population. It is questionable whether large organizations such as the United Nations (UN), its Food and Agriculture Organization (FAO) and WHO would be able to control the situation. These kinds of changes would mean major changes to ecosystems, with extinction and/or relocation of plant and animal species. Under such conditions the epidemiology of infectious diseases is likely to change, and epidemics to appear. This would involve not only humans but also animals and plants. The risks of viral contamination of water and foodstuffs would most likely increase and 'new' diseases appear.

The interaction between viruses and their hosts is an endless struggle in which the viruses have shown great capacity to adapt. Rapid global changes in climate will lead to an imbalance in the host–parasite relation, most likely to the advantage of the latter. In such situations the food- and waterborne route for the spread of viruses will most likely be exploited. Future efficient international environmental surveillance networks and preparedness for disease prevention must therefore include viruses.

21.6 References

AHMED, M. U., URASAWA, S., TANIGUCHI, K., URASAWA, T., KOBAYASHI, N., WAKASUGI, F., ISLAM, A. I. and SAHIKH, H. A. (1991) Analysis of human rotavirus strains prevailing in Bangladesh in relation to nationwide floods brought by the 1988 monsoon. *J Clin Microbiol*, **29**, 2273–9.

ANON. (2005) Norovirus outbreak among evacuees from hurricane Katrina – Houston, Texas, September 2005. *MMWR Morb Mortal Wkly Rep*, **54**, 1016–8.

ATUYAMBE, L. M., EDIAU, M., ORACH, C. G., MUSENERO, M. and BAZEYO, W. (2011) Land slide disaster in eastern Uganda: rapid assessment of water, sanitation and hygiene situation in Bulucheke camp, Bududa district. *Environ Health*, **10**, 38.

AW, T. G. and GIN, K. Y. (2010) Environmental surveillance and molecular characterization of human enteric viruses in tropical urban wastewaters. *J Appl Microbiol*, **109**, 716–30.

BEZIRTZOGLOU, C., DEKAS, K. and CHARVALOS, E. (2011) Climate changes, environment and infection: Facts, scenarios and growing awareness from the public health community within Europe. *Anaerobe*, **17**, 337–40.

BILE, K., ISSE, A., MOHAMUD, O., ALLEBECK, P., NILSSON, L., NORDER, H., MUSHAHWAR, I. K. and MAGNIUS, L. O. (1994) Contrasting roles of rivers and wells as sources of drinking water on attack and fatality rates in a hepatitis E epidemic in Somalia. *Am J Trop Med Hyg*, **51**, 466–74.

BOK, K., ABENTE, E. J., REALPE-QUINTERO, M., MITRA, T., SOSNOVTSEV, S. V., KAPIKIAN, A. Z. and GREEN, K. Y. (2009) Evolutionary dynamics of GII.4 noroviruses over a 34-year period. *J Virol*, **83**, 11890–901.

BRUGGINK, L. D. and MARSHALL, J. A. (2010) The incidence of norovirus-associated gastroenteritis outbreaks in Victoria, Australia (2002–2007) and their relationship with rainfall. *Int J Environ Res Public Health*, **7**, 2822–7.

BRYSON, B. (2010) *At home : a short history of private life*. London: New York, Doubleday.

CAIRNCROSS, S., HUNT, C., BOISSON, S., BOSTOEN, K., CURTIS, V., FUNG, I. C. and SCHMIDT, W. P. (2010) Water, sanitation and hygiene for the prevention of diarrhoea. *Int J Epidemiol*, **39**(Suppl 1), i193–i205.

CARRILHO, F. J., MENDES CLEMENTE, C. and SILVA, L. C. (2005) Epidemiology of hepatitis A and E virus infection in Brazil. *Gastroenterol Hepatol*, **28**, 118–25.

COONEY, C. M. (2011) Climate change and infectious disease: is the future here? *Environ Health Perspect*, **119**, a394–a397.

CORWIN, A., PUTRI, M. P., WINARNO, J., LUBIS, I., SUPARMANTO, S., SUMARDIATI, A., LARAS, K., TAN, R., MASTER, J., WARNER, G., WIGNALL, F. S., GRAHAM, R. and HYAMS, K. C. (1997) Epidemic and sporadic hepatitis E virus transmission in West Kalimantan (Borneo), Indonesia. *Am J Trop Med Hyg*, **57**, 62–5.

DE FRAITURE, C., MOLDEN, D. and WICHELNS, D. (2010) Investing in water for food, ecosystems, and livelihoods: An overview of the comprehensive assessment of water management in agriculture. *Agricultural Water Management*, **97**, 495–501.

DE PAULA, V. S., DINIZ-MENDES, L., VILLAR, L. M., LUZ, S. L., SILVA, L. A., JESUS, M. S., DA SILVA, N. M. and GASPAR, A. M. (2007) Hepatitis A virus in environmental water samples from the Amazon Basin. *Water Res*, **41**, 1169–76.

DONATELLI, F., DE MARCO, M., PASQUA, M. and DELOGU, M. (2008) Why, how and where does interspecies transmission of influenza A viruses occur? In PALOMBO, E. A. and KIRKWOOD, C. D. (eds) *Viruses in the environment*. Kerala, India: Research Signpost.

DRAYNA, P., MCLELLAN, S. L., SIMPSON, P., LI, S. H. and GORELICK, M. H. (2010) Association between rainfall and pediatric emergency department visits for acute gastrointestinal illness. *Environ Health Perspect*, **118**, 1439–43.

FAO (2008) *FAO, Assesment of the World Food Security*. Rome: Food and Agriculture Organization of the United Nations.

FAO (2010) *The State of Food Insecurity in the World*, Rome: Food and Agriculture Organization of the United Nations.

FEBRIANI, Y., LEVALLOIS, P., GINGRAS, S., GOSSELIN, P., MAJOWICZ, S. E. and FLEURY, M. D. (2010) The association between farming activities, precipitation, and the risk of acute gastrointestinal illness in rural municipalities of Quebec, Canada: a cross-sectional study. *BMC Public Health*, **10**, 48.

FENSKE, C., WESTPHAL, H., BACHOR, A., BREITENBACH, E., BUCHHOLZ, W., JULICH, W. D. and HENSEL, P. (2001) The consequences of the Odra flood (summer 1997) for the Odra lagoon and the beaches of Usedom: what can be expected under extreme conditions? *Int J Hyg Environ Health*, **203**, 417–33.

FONG, T. T., PHANIKUMAR, M. S., XAGORARAKI, I. and ROSE, J. B. (2010) Quantitative detection of human adenoviruses in wastewater and combined sewer overflows influencing a Michigan river. *Appl Environ Microbiol*, **76**, 715–23.

FORMIGA-CRUZ, M., TOFINO-QUESADA, G., BOFILL-MAS, S., LEES, D. N., HENSHILWOOD, K., ALLARD, A. K., CONDEN-HANSSON, A. C., HERNROTH, B. E., VANTARAKIS, A., TSIBOUXI, A., PAPAPETROPOULOU, M., FURONES, M. D. and GIRONES, R. (2002) Distribution of human virus contamination in shellfish from different growing areas in Greece, Spain, Sweden, and the United Kingdom. *Appl Environ Microbiol*, **68**, 5990–8.

FULLER, S. D., VON BONSDORFF, C. H. and SIMONS, K. (1985) Cell surface influenza haemagglutinin can mediate infection by other animal viruses. *EMBO J*, **4**, 2475–85.

GATHERER, D. (2010) The Little Ice Age and the emergence of influenza A. *Med Hypotheses*, **75**, 359–62.

GLEICK, P. H. and PALANIAPPAN, M. (2010) Peak water limits to freshwater withdrawal and use. *Proc Natl Acad Sci U S A*, **107**, 11155–62.

GUILL, C. K. and SHANDERA, W. X. (2001) The effects of Hurricane Mitch on a community in northern Honduras. *Prehosp Disaster Med*, **16**, 166–71.

HANLON, P., HANLON, L., MARSH, V., BYASS, P., SHENTON, F., SANDERS, R. C., HASSAN-KING, M. and GREENWOOD, B. M. (1987) Epidemiology of rotavirus in a periurban Gambian community. *Ann Trop Paediatr*, **7**, 238–43.

HARLEY, D., BI, P., HALL, G., SWAMINATHAN, A., TONG, S. and WILLIAMS, C. (2011) Climate change and infectious diseases in Australia: future prospects, adaptation options, and research priorities. *Asia Pac J Public Health*, **23**, 54S–66.

HARPER, S. L., EDGE, V. L., SCHUSTER-WALLACE, C. J., BERKE, O. and MCEWEN, S. A. (2011) Weather, water quality and infectious gastrointestinal illness in two Inuit communities in Nunatsiavut, Canada: potential implications for climate change. *Ecohealth*, **8**, 93–108.

HAU, C. H., HIEN, T. T., TIEN, N. T., KHIEM, H. B., SAC, P. K., NHUNG, V. T., LARASATI, R. P., LARAS, K., PUTRI, M. P., DOSS, R., HYAMS, K. C. and CORWIN, A. L. (1999) Prevalence of enteric hepatitis A and E viruses in the Mekong River delta region of Vietnam. *Am J Trop Med Hyg*, **60**, 277–80.

HIRD, S., STEIN, C., KUCHENMULLER, T. and GREEN, R. (2009) Meeting report: Second annual meeting of the World Health Organization initiative to estimate the global burden of foodborne diseases. *Int J Food Microbiol*, **133**, 210–12.

HJERTQVIST, M., JOHANSSON, A., SVENSSON, N., ABOM, P. E., MAGNUSSON, C., OLSSON, M., HEDLUND, K. O. and ANDERSSON, Y. (2006) Four outbreaks of norovirus gastroenteritis after consuming raspberries, Sweden, June-August 2006. *Euro Surveill*, **11**, E060907 1.

HORSWELL, J., HEWITT, J., PROSSER, J., van SCHAIK, A., CROUCHER, D., MACDONALD, C., BURFORD, P., SUSARLA, P., BICKERS, P. and SPEIR, T. (2010) Mobility and survival of Salmonella Typhimurium and human adenovirus from spiked sewage sludge applied to soil columns. *J Appl Microbiol*, **108**, 104–14.

JUN, K. S., CHUNG, E. S., SUNG, J. Y. and LEE, K. S. (2011) Development of spatial water resources vulnerability index considering climate change impacts. *Sci Total Environ*, **409**, 5228–42.

KAMILI, S. (2011) Toward the development of a hepatitis E vaccine. *Virus Res*, **161**, 93–100.

KARESH, W. B., COOK, R. A., BENNETT, E. L. and NEWCOMB, J. (2005) Wildlife trade and global disease emergence. *Emerg Infect Dis*, **11**, 1000–2.

KARMAKAR, S., RATHORE, A. S., KADRI, S. M., DUTT, S., KHARE, S. and LAL, S. (2008) Post-earthquake outbreak of rotavirus gastroenteritis in Kashmir (India): an epidemiological analysis. *Public Health*, **122**, 981–9.

KATAYAMA, H., OKUMA, K., FURUMAI, H. and OHGAKI, S. (2004) Series of surveys for enteric viruses and indicator organisms in Tokyo Bay after an event of combined sewer overflow. *Water Sci Technol*, **50**, 259–62.

KONIKOW, L. F. and NEUZIL, C. E. (2007) A method to estimate groundwater depletion from confining layers. *Water Resour Res*, **43**, W07417.

KOOPMANS, M. and DUIZER, E. (2004) Foodborne viruses: an emerging problem. *Int J Food Microbiol*, **90**, 23–41.

KOUADIO, I. K., ALJUNID, S., KAMIGAKI, T., HAMMAD, K. and OSHITANI, H. (2012) Infectious diseases following natural disasters: prevention and control measures. *Expert Rev Anti Infect Ther*, **10**, 95–104.

KUKKULA, M., ARSTILA, P., KLOSSNER, M. L., MAUNULA, L., BONSDORFF, C. H. and JAATINEN, P. (1997) Waterborne outbreak of viral gastroenteritis. *Scand J Infect Dis*, **29**, 415–18.

LAGIER, M. J., FELL, J. W. and GOODWIN, K. D. (2007) Electrochemical detection of harmful algae and other microbial contaminants in coastal waters using hand-held biosensors. *Mar Pollut Bull*, **54**, 757–70.

Le GUYADER, F. S., LE SAUX, J. C., AMBERT-BALAY, K., KROL, J., SERAIS, O., PARNAUDEAU, S., GIRAUDON, H., DELMAS, G., POMMEPUY, M., POTHIER, P. and ATMAR, R. L. (2008) Aichi virus, norovirus, astrovirus, enterovirus, and rotavirus involved in clinical cases from a French oyster-related gastroenteritis outbreak. *J Clin Microbiol*, **46**, 4011–17.

LEWIS, S. R., DATTA, S., GUI, M., COKER, E. L., HUGGINS, F. E., DAUNERT, S., BACHAS, L. and BHATTACHARYYA, D. (2011) Reactive nanostructured membranes for water purification. *Proc Natl Acad Sci U S A*, **108**, 8577–82.

LI, D., GU, A. Z., ZENG, S. Y., YANG, W., HE, M. and SHI, H. C. (2011) Monitoring and evaluation of infectious rotaviruses in various wastewater effluents and receiving waters revealed correlation and seasonal pattern of occurrences. *J Appl Microbiol*, **110**, 1129–37.

LOBELL, D. B., BURKE, M. B., TEBALDI, C., MASTRANDREA, M. D., FALCON, W. P. and NAYLOR, R. L. (2008) Prioritizing climate change adaptation needs for food security in 2030. *Science*, **319**, 607–10.

MARRIS, E. (2008) Water: more crop per drop. *Nature*, **452**, 273–7.

MAUNULA, L. (2007) Waterborne norovirus outbreaks. *Future Virol*, **2**, 101–12.

MAUNULA, L., SÖDERBERG, K., VAHTERA, H., VUORILEHTO, V.-P., VON BONSDORFF, C.-H., VALTARI, M., LAAKSO, T. and LAHTI, K. (2012) Presence of human noro- and adenoviruses in river and treated wastewater, a longitudinal study and method comparison. *J Water Health*, **10(1)**, 87–99.

MCMICHAEL, A., WOODRUFF, R. and HALES, S. (2006) Climate change and human health: present and future risks. *Lancet*, **367**, 859–69.

MORISON, J. I., BAKER, N. R., MULLINEAUX, P. M. and DAVIES, W. J. (2008) Improving water use in crop production. *Philos Trans R Soc Lond B Biol Sci*, **363**, 639–58.

NOMURA, K., MURAI, H., NAKAHASHI, T., MASHIBA, S., WATOH, Y., TAKAHASHI, T. and MORIMOTO, S. (2008) Outbreak of norovirus gastroenteritis in elderly evacuees after the 2007 Noto Peninsula earthquake in Japan. *J Am Geriatr Soc*, **56**, 361–3.

NORDGREN, J., MATUSSEK, A., MATTSSON, A., SVENSSON, L. and LINDGREN, P. E. (2009) Prevalence of norovirus and factors influencing virus concentrations during one year in a full-scale wastewater treatment plant. *Water Res*, **43**, 1117–25.

NWACHCUKU, N. and GERBA, C. P. (2004) Emerging waterborne pathogens: can we kill them all? *Curr Opin Biotechnol*, **15**, 175–80.

PARRY, M., ROSENZWEIG, C. and LIVERMORE, M. (2005) Climate change, global food supply and risk of hunger. *Philos Trans R Soc Lond B Biol Sci*, **360**, 2125–38.

PARRY, M. A. and HAWKESFORD, M. J. (2010) Food security: increasing yield and improving resource use efficiency. *Proc Nutr Soc*, **69**, 592–600.

PFISTER, S., BAYER, P., KOEHLER, A. and HELLWEG, S. (2011) Projected water consumption in future global agriculture: scenarios and related impacts. *Sci Total Environ*, **409**, 4206–16.

PHANUWAN, C., TAKIZAWA, S., OGUMA, K., KATAYAMA, H., YUNIKA, A. and OHGAKI, S. (2006) Monitoring of human enteric viruses and coliform bacteria in waters after urban flood in Jakarta, Indonesia. *Water Sci Technol*, **54**, 203–10.

PITZER, V. E., VIBOUD, C., SIMONSEN, L., STEINER, C., PANOZZO, C. A., ALONSO, W. J., MILLER, M. A., GLASS, R. I., GLASSER, J. W., PARASHAR, U. D. and GRENFELL, B. T. (2009) Demographic variability, vaccination, and the spatiotemporal dynamics of rotavirus epidemics. *Science*, **325**, 290–4.

PREMANANDH, J. (2011) Factors affecting food security and contribution of modern technologies in food sustainability. *J Sci Food Agric*, **91**, 2707–14.

PURSE, B. V., BROWN, H. E., HARRUP, L., MERTENS, P. P. and ROGERS, D. J. (2008) Invasion of bluetongue and other orbivirus infections into Europe: the role of biological and climatic processes. *Rev Sci Tech*, **27**, 427–42.

RAMACHANDRAN, M., KIRKWOOD, C. D., UNICOMB, L., CUNLIFFE, N. A., WARD, R. L., BHAN, M. K., CLARK, H. F., GLASS, R. I. and GENTSCH, J. R. (2000) Molecular characterization of serotype G9 rotavirus strains from a global collection. *Virol*, **278**, 436–44.

ROHAYEM, J., DUMKE, R., JAEGER, K., SCHROTER-BOBSIN, U., MOGEL, M., KRUSE, A., JACOBS, E. and RETHWILM, A. (2006) Assessing the risk of transmission of viral diseases in flooded areas: viral load of the River Elbe in Dresden during the flood of August 2002. *Intervirology*, **49**, 370–6.

ROSE, J. B., EPSTEIN, P. R., LIPP, E. K., SHERMAN, B. H., BERNARD, S. M. and PATZ, J. A. (2001) Climate variability and change in the United States: potential impacts on water- and foodborne diseases caused by microbiologic agents. *Environ Health Perspect*, **109**(Suppl 2), 211–21.

ROSEGRANT, M. W., RINGLER, C. and ZHU, T. J. (2009) Water for Agriculture: Maintaining Food Security under Growing Scarcity. *Ann Rev Environ Res*, **34**, 205–22.

SALVI, S. and TUBEROSA, R. (2005) To clone or not to clone plant QTLs: present and future challenges. *Trends Plant Sci*, **10**, 297–304.

SARVIKIVI, E., ROIVAINEN, M., MAUNULA, L., NISKANEN, T., KORHONEN, T., LAPPALAINEN, M. and KUUSI, M. (2012) Multiple norovirus outbreaks linked to imported frozen raspberries. *Epidemiol Infect*, **140**, 260–7.

SCHIERMEIER, Q. (2011) Climate and weather: extreme measures. *Nature*, **477**, 148–9.

SENCAN, I., SAHIN, I., KAYA, D., OKSUZ, S. and YILDIRIM, M. (2004) Assessment of HAV and HEV seroprevalence in children living in post-earthquake camps from Duzce, Turkey. *Eur J Epidemiol*, **19**, 461–5.

SHEFFIELD, P. E. and LANDRIGAN, P. J. (2010) Global climate change and children's health: threats and strategies for prevention. *Environ Health Perspect*, **119**, 291–8.

SIEBENGA, J. J., VENNEMA, H., RENCKENS, B., de BRUIN, E., van der VEER, B., SIEZEN, R. J. and KOOPMANS, M. (2007) Epochal evolution of GGII.4 norovirus capsid proteins from 1995 to 2006. *J Virol*, **81**, 9932–41.

SITBON, M., LECERF, A., GARIN, Y. and IVANOFF, B. (1985) Rotavirus prevalence and relationships with climatological factors in Gabon, Africa. *J Med Virol*, **16**, 177–82.

SMITH, K. M., ANTHONY, S. J., SWITZER, W. M., EPSTEIN, J. H., SEIMON, T., JIA, H., SANCHEZ, M. D., HUYNH, T. T., GALLAND, G. G., SHAPIRO, S. E., SLEEMAN, J. M., MCALOOSE, D., STUCHIN, M., AMATO, G., KOLOKOTRONIS, S. O., LIPKIN, W. I., KARESH, W. B., DASZAK, P. and MARANO, N. (2012) Zoonotic viruses associated with illegally imported wildlife products. *PLoS One*, **7**, e29505.

STRZEPEK, K. and BOEHLERT, B. (2010) Competition for water for the food system. *Philos Trans R Soc Lond B Biol Sci*, **365**, 2927–40.

TESCHKE, K., BELLACK, N., SHEN, H., ATWATER, J., CHU, R., KOEHOORN, M., MACNAB, Y. C., SCHREIER, H. and ISAAC-RENTON, J. L. (2011) Water and sewage systems, socio-demographics, and duration of residence associated with endemic intestinal infectious diseases: a cohort study. *BMC Public Health*, **10**, 767.

THOMAS, K. M., CHARRON, D. F., WALTNER-TOEWS, D., SCHUSTER, C., MAAROUF, A. R. and HOLT, J. D. (2006) A role of high impact weather events in waterborne disease outbreaks in Canada, 1975–2001. *Int J Environ Health Res*, **16**, 167–80.

TIERNEY, J. T., SULLIVAN, R. and LARKIN, E. P. (1977) Persistence of poliovirus 1 in soil and on vegetables grown in soil previously flooded with inoculated sewage sludge or effluent. *Appl Environ Microbiol*, **33**, 109–13.

UN (2009) United Nations. Department of Economic and Social Affairs, Population Division, World population ageing. ESA/P/WP/212.

UNEP (2002) *Synthesis, GEO-3 : Global Environment Outlook 3 : past, present, and future perspectives,* Nairobi, Kenya: United Nations Environment Programme.

UNFPA (2007) *State of world population: Unleashing the potential of Urban Growth.* New York: UNFPA.

VON BONSDORFF, C.-H. and MAUNULA, L. (2008) Waterborne viral infections. In PALOMBO, E. A. K., C. (eds) *Viruses in the Environment*. Kerala, India: Research Signposte.

WILKINSON, K., GRANT, W. P., GREEN, L. E., HUNTER, S., JEGER, M. J., LOWE, P., MEDLEY, G. F., MILLS, P., PHILLIPSON, J., POPPY, G. M. and WAAGE, J. (2011) Infectious diseases of animals and plants: an interdisciplinary approach. *Philos Trans R Soc Lond B Biol Sci*, **366**, 1933–42.

WOOLHOUSE, M. E., HAYDON, D. T. and ANTIA, R. (2005) Emerging pathogens: the epidemiology and evolution of species jumps. *Trends Ecol Evol*, **20**, 238–44.

YAMAMOTO, D., GHOSH, S., GANESH, B., KRISHNAN, T., CHAWLA-SARKAR, M., ALAM, M. M., AUNG, T. S. and KOBAYASHI, N. (2010) Analysis of genetic diversity and molecular evolution of human group B rotaviruses based on whole genome segments. *J Gen Virol*, **91**, 1772–81.

YEE, E. L., PALACIO, H., ATMAR, R. L., SHAH, U., KILBORN, C., FAUL, M., GAVAGAN, T. E., FEIGIN, R. D., VERSALOVIC, J., NEILL, F. H., PANLILIO, A. L., MILLER, M., SPAHR, J. and GLASS, R. I. (2007) Widespread outbreak of norovirus gastroenteritis among evacuees of Hurricane Katrina residing in a large "megashelter" in Houston, Texas: lessons learned for prevention. *Clin Infect Dis*, **44**, 1032–9.

YOSHIDA, H., HORIE, H., MATSUURA, K. and MIYAMURA, T. (2000) Characterisation of vaccine-derived polioviruses isolated from sewage and river water in Japan. *Lancet*, **356**, 1461–3.

22

Virus indicators for food and water

R. Girones and S. Bofill-Mas, University of Barcelona, Spain

DOI: 10.1533/9780857098870.4.483

Abstract: Detailed knowledge of contamination sources is needed for efficient and cost-effective strategies to minimize fecal pollution in water, foods and the environment, and for evaluation and risk assessment. This chapter analyzes the concept of viruses as indices and indicators of fecal pollution and the applicability of specific viral groups as source markers for fecal contamination, usually referred to as microbial source-tracking (MST) tools. Specific human and animal viruses, bacteriophages and plant viruses are described and critically evaluated. Future prospects for technical and scientific developments are considered.

Key words: viral fecal indicator, microbial source-tracking (MST), adenovirus, polyomavirus, F-RNA bacteriophages.

22.1 Introduction

The quality of water used for drinking, irrigation, aquaculture, food processing or recreational purposes has a significant impact on public health worldwide. Fecal pollution is a primary health concern in the environment, in water and in food. The use of index micro-organisms (whose presence points to the possible existence of a similar pathogenic organism) and indicator micro-organisms (whose presence represents a failure affecting the final product) to assess the microbiological quality of waters or food is well established and has been practiced for almost a century (Medema *et al.*, 2003). In this chapter we analyze the concept and applicability of viruses as index and indicators of fecal pollution and the applicability of specific viral groups as markers of the source of fecal contamination, usually referred to as microbial source-tracking (MST) tools.

The most significant viral indicators proposed are human viruses, bacteriophages and plant viruses, and these will be described and critically evaluated according to available data on specificity, abundance, dissemination, geographical and temporal distribution and the feasibility of available techniques for routine analysis. Human DNA viruses – adenoviruses and polyomaviruses – associated with persistent infections are further described as the most promising viral parameters currently quantified by molecular methods.

Recent developments in water and food management, risk assessment and quality monitoring require the exact determination of the source and the extent of fecal pollution. Detailed knowledge of the contamination sources is needed for efficient and cost-effective management strategies to minimize fecal contamination in watersheds and foods, evaluation of the effectiveness of best management practices, and system and risk assessment as part of the water and food safety plans recommended by the World Health Organization (US-EPA, 2005; WHO, 2004). The most significant human and animal viral groups, and bacteriophages proposed as MST tools are described and their role as markers of the source of contamination is critically analyzed. Future perspectives on technical and scientific developments are related to the production of more quantitative information, standardization, multiplex assays and process automation.

22.2 Usage and definition of viral indicators

The pathogens associated with environmental transmission routes, including water and food, encompass hundreds of bacteria, protozoa and viruses. The use of surrogates is essential for investigating water quality, food safety and industrial microbiology.

The concept of fecal indicator was developed at the end of the nineteenth century. Dr Theodor Escherich, a German pediatrician, described in 1885 how within a few weeks of birth the bacterium later named *Escherichia coli* became one of the dominant micro-organisms in the infant colon. Then, in 1892 Dr Franz Schardinger described *E. coli* as a characteristic component of the fecal flora whose presence in water could be taken as an indicator of fecal pollution and therefore of the potential presence of enteric pathogens. Nonpathogenic, easily detectable micro-organisms were used to indicate that contamination had taken place and there was thus a risk of public health (Medema *et al.*, 2003). Although fecal-derived coliforms, thermotolerant coliforms and *E. coli* have several drawbacks and limitations in their role as indicators of fecal pollution, they have proved very useful historically and they are the most commonly used microbial parameters for testing water quality.

22.2.1 Limitations of current microbial standard indicators

Both fecal coliforms/*E. coli* and enterococci are present in relatively high numbers in the mammalian intestinal tract, which is considered their primary habitat. However, some recent findings have questioned the use of the current indicators for regulatory activities, suggesting that there are scenarios in which the presence of indicators does not necessarily correlate with fecal contamination. Some studies have shown that fecal bacterial indicators (i.e., *E. coli* and enterococci) have been isolated in secondary habitats and that they can persist and grow in the environment (Byappanahalli *et al.*, 2003; Fujioka *et al.*, 1998; Pote *et al.*, 2009; Solo-Gabriele *et al.*, 2000). Current water and food microbiological quality criteria focus on the concentration of fecal bacterial indicators and often fail to predict the presence of human pathogenic enteric viruses and protozoa. Various authors have concluded that these indicators could fail to predict the risk of waterborne pathogens, including viruses (Gerba *et al.*, 1979; Lipp *et al.*, 2001). Moreover, the levels of bacterial indicators do not always correlate with the concentration of viruses, especially when these indicators are present at low concentrations (Contreras-Coll *et al.*, 2002; Pina *et al.*, 1998). A summary of the limitations of using classic microbiological indicators as the sole microbiological criterion of the occurrence and concentration of pathogens such as viruses would be that indicator bacteria: (i) are more sensitive to inactivation by treatment processes and by sunlight than viral or protozoan pathogens; (ii) have non-exclusive fecal sources; (iii) are able to multiply in some environments; (iv) are not useful to identify the source of fecal contamination; and (v) do not correlate with the presence of pathogens.

The fact that rapid methods are required and that, moreover, many pathogens cannot be cultivated in the laboratory has led to development of new methodologies for the study of pathogens and new proposed indicators of fecal contamination in water and food. These are based on the implementation of molecular techniques that are rapid and sensitive but may pick up both infectious and non-infectious (dead) types. Hence, the health implications of pathogen detection methods can be ambiguous, such as screening for noroviruses and *Cryptosporidium* oocysts by polymerase chain reaction (PCR) and antibody staining, respectively.

In the last two decades quantitative microbial risk assessment (QMRA) has evolved as a useful tool for guidance in the pursuit of water treatment goals for pathogen removal and assessing the risk from food and other environmental exposures (Haas *et al.*, 1993). In the QMRA paradigm of hazard identification – exposure, dose response and risk characterization – exposure assessment has the greatest variability and uncertainty. The use of suitable surrogates can help reduce the uncertainties associated with exposure assessment (Sinclair *et al.*, 2012). At present there are no routine monitoring techniques for viruses, hence no criteria for viruses have been laid down and the risk of viral disease is not always known when using criteria based on bacteria. The basic premise that the concentration of indicator organisms should be related to the extent of fecal contamination, and by implication to the concentration of pathogens and the incidence of water- and foodborne diseases, cannot be sustained (Pipes, 1982) as an absolute criterion.

To avoid the ambiguity in the term 'microbial indicator', the following three groups are now recognized: (i) process microbial indicators, (ii) fecal indicators, and (iii) index and model organisms. Process indicators comprise a group of organisms that demonstrate the efficacy of a process; fecal indicators are those organisms that indicate the presence of fecal contamination (hence they only imply that pathogens may be present); index and model organisms include a group or species indicative of possible pathogen presence and behavior, respectively. The use of index and indicator organisms to assess the microbiological and sanitary quality of waters or food is well established and has been practiced for almost a century. Viruses are more stable than common bacterial indicators in the environment, and different groups of viruses have been suggested as index or indicators of fecal contamination. Some of the most commonly used viral indicators at present are bacteriophages F + RNA phages (Havelaar *et al.*, 1993) and human DNA viruses producing persistent infections such as adenoviruses (Pina *et al.*, 1998; Puig *et al.*, 1994) and polyomaviruses (Bofill *et al.*, 2000, 2001).

In view of the limitations of current standard fecal indicators, selected viral groups have been proposed as alternative or complementary indicators to improve the control of the microbiological quality of water and to reduce the microbiological risk associated with medium and low levels of fecal contamination in water. Viruses are increasingly becoming recognized as important fecal waterborne pathogens, as improved detection methods establish a strong link between them and reported outbreaks of unknown etiology (i.e., causes or origin of disease). Some of the most important fecal viral pathogens are noroviruses, enteroviruses, adenoviruses, rotaviruses, and hepatitis A and E viruses. Overall, viruses are more resistant to environmental conditions than bacterial indicators, which partly explains the frequent lack of correlation between indicators and the occurrence of enteric viruses (Pina *et al.*, 1998; Wyn-Jones *et al.*, 2011). The high stability of viruses in the environment, their host specificity, and persistence and high prevalence of some viral infections throughout the year strongly support the use of rapid cost-effective sensitive molecular techniques for the identification and quantification of DNA viruses such as adenoviruses and polyomaviruses, which can be used as complementary indicators of fecal and urine (hereinafter 'fecal') contamination and as MST tools (Bofill-Mas *et al.*, 2000; Pina *et al.*, 1998).

22.2.2 Viruses as indices and indicators

Many orally transmitted viruses produce subclinical infections and symptoms in only a small proportion of the population. However, some viruses may give rise to life-threatening conditions, such as acute hepatitis in adults, as well as severe gastroenteritis in small children and the elderly. The development of disease is related to the infective dose of the viral agent, the age, health, immunological and nutritional status of the infected individual (pregnancy, presence of other infections or diseases), and the availability of health care.

Potentially pathogenic viruses in urban wastewater include human adenoviruses (HAdV) and human polyomaviruses (HPyV), which are detected in all geographical areas and throughout the year, and enteroviruses (EV), noroviruses (NoV), rotaviruses (RV), astroviruses, hepatitis A (HAV) and hepatitis E virus (HEV), with variable prevalence in different geographical areas and periods of the year.

The diversity in the epidemiological patterns of infection and excretion between the different viral pathogens clearly shows how difficult it is for one unique viral parameter to fulfill all the criteria of a good indicator. Ideally, indicator levels in environmental waters should bear some relation to the degree or extent of pollution, that is, the indicator should be present in waters whenever the pathogens concerned are present, while absent or at very low levels in clean waters. Indicators should occur in higher numbers than the pathogens and quantification methods should be low-cost and efficient. Indicators must be also more resistant to disinfectants and to the aqueous environment than the pathogens, and should not be able to proliferate to any greater extent than pathogens in the aquatic environment. Because there is no such ideal fecal indicator, it has been concluded that a range of different indicators are probably required to address the range and differing behaviors of microbial pathogens. Accordingly, efficient viral indicators of the level of fecal contamination in water and food could be used as a complement to or substitute for the analysis of a specific viral or protozoan pathogen when this pathogen represents a concern. This may be the case for norovirus which is a highly abundant food- and waterborne pathogen with very variable patterns of excretion in the population. It also is present in highly variable concentrations in urban sewage in various geographical areas and periods of the year (Formiga-Cruz *et al.*, 2002; Hewitt *et al.*, 2011; Kitajima *et al.*, 2010).

The viruses that are transmitted *via* contaminated food or water are typically stable because they lack the lipid envelopes that render other viruses vulnerable to environmental agents. Overall, viruses are more resistant to environmental conditions than bacterial indicators, which partly explains the frequent lack of correlation between indicators and the occurrence of enteric viruses, and the interest in using viruses as alternative or complementary microbial indicators.

Natural populations of viruses are found in virtually all natural aquatic environments, where they reach a concentration up to 10^7–10^8 virus-like particles per ml, which is higher than that of bacteria (Pina *et al.*, 1998a). The human body may also be considered as an environment where many viruses cause latent infections and multiply in the absence of disease, and where viruses may cause persistent infections. Indeed, many human viruses are considered part of the human microbiome and some of them offer protection against infections by other pathogens (Barton *et al.*, 2007). The accumulated data on adenoviral and polyomaviral infections support the concept of some viruses as members of the human microbiome and their use as molecular markers of fecal/urine contamination in water, food and the environment.

Some viruses commonly excreted in feces and/or urine, such as many human adenoviruses and polyomaviruses, infect humans during childhood, thereby establishing persistent infections. In the case of some HAdV that infect the respiratory tract, many viral particles may be excreted in feces for months or even years (Adrian *et al.*, 1988). The human polyomavirus JCPyV and BKPyV persistently infect the kidney and JCPyV is persistently excreted in urine (Shah *et al.*, 1997). Studies on these DNA viruses have changed the paradigm of viral contamination, which was previously considered to be a sporadic event related only to the presence of outbreaks in the population. At present human viruses are expected to be present in practically 100% of urban sewage samples (Albinana-Gimenez *et al.*, 2006; Bofill-Mas *et al.*, 2006).

Monitoring for the presence of human viruses is challenging due to the diversity of viruses and the relatively low concentrations of viral particles in environmental waters or contaminated food. However, this limitation can be overcome by the use of improved molecular methods and selected groups of viruses as indices and indicators of fecal contamination in water and food. However, challenges still remain for the new methods and parameters, and quantitative aspects need to be improved and standardized.

Advances in concentration methods for viruses in water and molecular assays like qPCR provide sensitive, rapid, and quantitative analytical tools with which to study viruses in water and develop new standards for improving the control of the microbiological quality of food and water, to trace the origin of fecal contamination, and to assess the efficiency of virus removal in wastewater treatment plants.

The use of viral groups that are host-specific, highly stable and persistently excreted throughout the year in all geographical areas is expected to overcome many of the limitations associated with current standard fecal indicators.

22.3 Viruses proposed as indicators

Candidate human viruses, groups of bacteriophages and plant viruses have been proposed as indicators, indices or models of human viral contamination.

22.3.1 Adenoviruses and polyomaviruses

The *Adenoviridae* family has a double-stranded DNA genome of approximately 35 000 base pairs (bp) surrounded by a 90–100 nm non-enveloped icosahedral shell (Fig. 22.1) with fiber-like projections from each vertex. Adenovirus infection may be caused by consumption of contaminated water or food, or by inhalation of aerosols from contaminated waters such as those used for recreational purposes. HAdV comprises seven species with 52 types, which are responsible for both enteric illnesses and respiratory and eye infections.

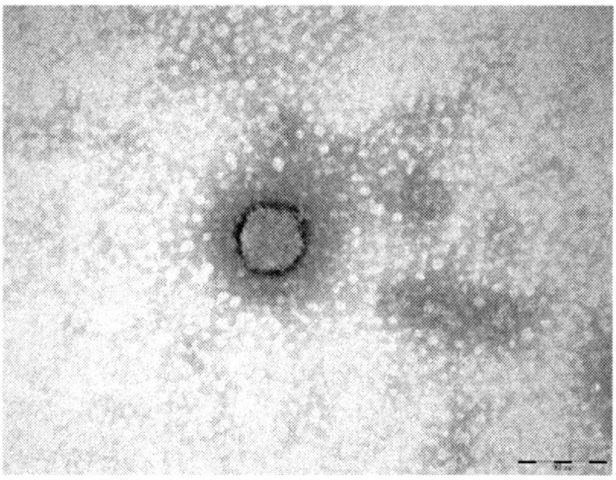

Fig. 22.1 Electron micrograph of human adenovirus type 2. The scale bar represents 100 nm.

The presence of HAdV in the environment has been described for decades, at first based on cell-culture assays then, since the 1990s, based on PCR and more recently on quantitative PCR (qPCR) and Integrated Cell Culture PCR (ICC-PCR). HAdV 1, 2, 5, 7, 12 and 31, responsible for respiratory, ocular and enteric infections, among others, have been detected in contaminated water and shellfish (Bofill-Mas *et al.*, 2006; Formiga-Cruz *et al.*, 2002) (including samples that met current safety standards, based on levels of fecal bacteria) in addition to HAdV 40 and 41, which produce gastroenteritis and are excreted by children with a very high concentration in feces (10^{11} viral particles per gram). HAdV have been included in the US Environmental Protection Agency's contaminant candidate list (US-EPA CCL).

HAdV were proposed as viral fecal indicators in the 1990s (Pina *et al.*, 1998; Puig *et al.*, 1994). Their detection in sewage is widely reported and shows no seasonality (Bofill-Mas *et al.*, 2000; Fong *et al.*, 2010; Kuo *et al.*, 2009; Simmons and Xagoraraki, 2011; and other studies reviewed in Mena and Gerba, 2009). The concentration of HAdV in sewage has been reported to be up to 10^3 infectious units/l (Mena and Gerba, 2009) and of between 5 and 8 genome copy logs/l (Girones *et al.*, 2010). Typical HAdV concentrations in sewage may be exemplified by those obtained in three different wastewater treatment plants from Northern Spain, ranging between two and four logs per ml of raw sewage (Bofill-Mas *et al.*, 2006; Rodriguez-Manzano *et al.*, 2012).

HAdV have also been reported in river and surface waters (Albinana-Gimenez *et al.*, 2009a; Calgua *et al.*, 2008; Hamza *et al.*, 2009; Haramoto *et al.*, 2010 and in studies reviewed in Jiang *et al.*, 2006), in seawater (Bofill-Mas *et al.*, 2010; Souza *et al.*, 2012; Wyn-Jones *et al.*, 2011; Xagoraraki *et al.*, 2007 and in studies reviewed in Jiang *et al.*, 2006), in swimming pools (reviewed in Jiang *et al.*, 2006) and also in groundwater (Guerrero-Latorre *et al.*, 2011; Haramoto *et al.*, 2011; Ogorzaly *et al.*, 2010).

The presence of HAdV in drinking water has been reported by several authors (Albinana-Gimenez *et al.*, 2006, 2009a and articles cited in Jiang *et al.*, 2006) and in shellfish (Bofill-Mas *et al.*, 2001; Formiga-Cruz *et al.*, 2002; Rigotto *et al.*, 2010).

All these data support the applicability of HAdV as a marker of the microbiological quality of water. Moreover, they are detected in nearly of 100% of sewage samples from different geographical areas in high concentrations (Girones *et al.*, 2010); they are highly stable in sewage (Bofill-Mas *et al.* 2006), they have been detected in water samples at different disinfection steps in water treatment plants (Albinana-Gimenez *et al.*, 2006) and they are highly resistant to inactivation by UV radiation, especially adenoviruses 40 and 41 (Gerba *et al.*, 2002; Thurston-Enriquez *et al.*, 2003). They infect exclusively humans and do not replicate in the environment. PCR-based techniques have been applied to the detection of their genome, but to provide any information on their infectivity cell-culture systems are needed. Cell-culture systems have been developed for growing HAdV *in vitro* (Jiang *et al.*, 2009) and cell-culture-based assays such as plaque assays, ICC-PCR and immunofluorescence assays (IFA) are commonly applied (reviewed in Calgua *et al.*, 2011; Jiang 2006) although some serotypes do not grow well in cell culture. DNase treatment of samples previous to nucleic acid extraction and subsequent PCR has proved to be useful to infer viral infectivity (Corrêa *et al.*, 2012). It should be noted that qPCR-based methods designed for HAdV detection may cross-detect, with low efficiency, some adenovirus animal strains (Bofill-Mas *et al.*, 2006). The use of more specific qPCR assays may also entail loss of ability to detect some of the human types.

Polyomaviruses are small and icosahedric (Fig. 22.2) viruses, with a circular double-stranded DNA genome of approximately 5000 bp that infect several species of vertebrates. They include human polyomaviruses (HPyV) which infect humans and can be divided into classical and new HPyV. Classical HPyV were discovered in the 1970s (Gardner *et al*, 1971; Padgett *et al.*, 1971). JCPyV is ubiquitously distributed worldwide and antibodies against it are detected in over 80% of humans (Weber *et al.*, 1997). Kidney and bone marrow are sites of latent infection with JCPyV, which is excreted in the urine of healthy individuals (Kitamura *et al.*, 1990; Koralnik *et al.*, 1999). The pathogenicity of the virus is commonly associated with progressive multifocal leukoencephalopathy (PML) in immuno-compromised states, and it has attracted new attention due to JCPyV reactivation and pathogenesis in some autoimmune disease patients being treated with immunomodulators (Berger *et al.*, 2009; Yousry *et al.*, 2006). BKPyV, the other classical human polyomavirus, causes nephropathy in renal transplant recipients and other immuno-suppressed individuals. It is also excreted in urine and thus is present in wastewater, although its prevalence is lower than that of JCPyV (Bofill-Mas

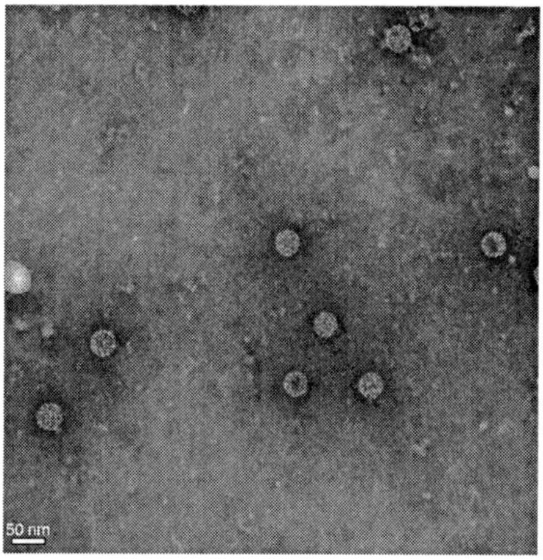

Fig. 22.2 Electron micrograph of a mixed JC and BK polyomavirus suspension. The scale bar represents 50 nm.

et al., 2000). JCPyV was proposed as an indicator of human fecal pollution in 2000 (Bofill-Mas *et al.*, 2000). Recently, new human polyomaviruses have been described. Among them, Merkell cell polyomavirus (MCPyV) has been strongly associated with Merkel Cell carcinomas (Babakir-Mina *et al.*, 2009; Feng *et al.*, 2008; Foulongne *et al.*, 2008). The presence of some of these new polyomaviruses in the environment has been reported (Bofill-Mas *et al.*, 2010; Cantalupo *et al.*, 2011), and MCPyV is the most prevalent in environmental samples (Bofill-Mas *et al.*, 2010).

The presence of JCPyV alone or together with BKPyV under the denomination of HPyV has been established in several types of environmental samples by PCR-based methods. Table 22.1 summarizes all these studies. JCPyV has also been described in shellfish samples (Bofill-Mas *et al.*, 2001; Souza *et al.*, 2012). Concentration of JCPyV in wastewater ranged from 10^4 to 10^7 GC/100 ml of wastewater and up to 10^3 GC/100 ml of river water. JCPyV is more frequently excreted also than other new human polyomaviruses recently described (Bofill-Mas *et al.*, 2010). The main advantages of using JCPyV (alone or together with BKPyV, under the denomination of HPyV) are: (i) it is an exclusively human virus that does not replicate in the environment; (ii) it is excreted by most humans; (iii) it shows no seasonality; and (iv) excreted strains have not been related to pathogenesis. JCPyV is detected in a high percentage of wastewater samples worldwide (Bofill-Mas *et al.*, 2000), is highly stable in sewage samples (Bofill-Mas *et al.*, 2006) and it is resistant to chlorine treatments (Corrêa *et al.*, 2012). Unfortunately, strains commonly present in the environment are difficult to grow in cell culture, so its role as an indicator depends on the presence of its DNA rather than the presence of infectious viruses. However, approaches based on the study of viral capsid integrity using DNase have been performed in order to infer potential infectivity of the

Table 22.1 Summary of published studies reporting the presence and concentration of human polyomaviruses in the environment

Authors	Method of detection	Matrices analyzed	Main results
Bofill-Mas *et al.*, 2000	nPCR for JCPyV, nPCR for BKPyV	Sewage	96% positive for JCPyV (and HAdV), 77.8% positive for BKPyV
Bofill-Mas *et al.*, 2006	qPCR for JCPyV by Pal *et al.*, 2006	Sewage treatment plant raw and effluent sewage and biosolids	Detected in 99% of all samples analyzed in concentrations one log lower than HAdV. Showed high stability in sewage (t99 of 127.3 days)
Brownell *et al.*, 2007	nPCR for HPyVs by Aksamit *et al.*, 1993	Coastal water impacted by storm water	No detected, no correlation with FIB
McQuaig *et al.*, 2006	nPCR for HPyVs by Aksamit *et al.*, 1993	Surface waters, septic tanks, sewage	HPyV not detected in pig and cow waste, not correlated with FIB but significantly correlated with *E. faecium* esp gene and the 16S rRNA of human associated *Bacteroides*
Rafique and Jiang, 2008	nPCR from Bofill-Mas *et al.*, 2000	Primary sewage effluent samples	100% positive for JCPyV
Ahmed *et al.*, 2009	qPCR with McQuaig *et al.*, 2006 primers	Sewage spiked water samples	HPyVs showed 100% host specificity and proved to be useful to detect human fecal contamination
Albinana-Gimenez *et al.*, 2009a	qPCR for JCPyV by Pal *et al.*, 2006	River water	JCPyV detected in 48% of river water samples and in water samples at different treatments steps in a drinking-water treatment plant in absence of FIB
Hamza *et al.*, 2009	qPCR for HPyVs by Biel *et al.*, 2000	River water	Detected (as HAdV) in 97.5% of the samples
Harwood *et al.*, 2009	qPCR for HPyV by McQuaig *et al.*, 2009	Freshwater, animal feces, marine water spiked with sewage	No detection of HPyV in animal feces. More specific than other markers. No correlation with *Enterococcus spp*
Haramoto *et al.*, 2010	qPCR for JCPyV and qPCR for BKPyV	River water	HAdV more prevalent (61.1%) than JCPyV (11.1%) but JCPyV detected in one HAdV-negative sample. BKPyV not detected

(*Continued*)

Table 22.1 Continued

Authors	Method of detection	Matrices analyzed	Main results
McQuaig *et al.*, 2009	qPCR for HPyVs modified from Aksamit el al., 1993	Sewage, environmental samples, animal waste	HPyV detected in dechlorinated tertiary-treated wastewater treatment plant effluents
Abdelzaher *et al.*, 2010	nPCR for HPyV by McQuaig *et al.*, 2006	Subtropical seawater	Detection of HPyV when seawater was impacted by human waste
Ahmed *et al.*, 2010	qPCR for HPyV by McQuaig *et al.*, 2009	Human/animal wastewater, animal feces, coastal river water	HPyV detected in primary and secondary effluent wastewater samples and in coastal river waters. HPyVs showed 0.99 host specificity
Fumian *et al.*, 2010	qPCR for JCPyV by Pal *et al.*, 2006	Raw and treated sewage	JCPyV detected in 96% and 43% of raw and treated wastewater respectively
Jurzik *et al.*, 2010	qPCR for HPyVs by Hamza *et al.*, 2009	Surface waters	68.8% were positive for HPyV (96.3% for HAdV)
Korajkic *et al.*, 2011	nPCR for HPyV by McQuaig *et al.*, 2006 and qPCR for HPyV by McQuaig *et al.*, 2009	Coastal waters	No correlation with FIB. Useful for identification of pollution sources
Gibson *et al.*, 2011	qPCR for HPyV by McQuaig *et al.*, 2009	Surface and drinking water	HPyV were detected in one groundwater, three surface water, and one drinking water sample. No correlation with FIB
Hellerin *et al.*, 2011	qPCR for HPyV by McQuaig *et al.*, 2009	Fresh/marine waters, raw sewage, animal feces	Presence of HPyV in all sewage samples and in one freshwater sample
Kokkinos *et al.*, 2011	nPCR for HPyV by McQuaig *et al.*, 2006	Untreated sewage	HPyV detected in 68.8% of sewage samples, more prevalently than HAdV
Moresco *et al.*, 2012	qPCR for JCPyV by Pal *et al.*, 2006	Coastal waters	3% of the samples analyzed were positive for JCPyV (55% for HAdV). No correlation with FIB

JCPyV detected in environmental samples by PCR (Bofill-Mas *et al.*, 2001; Corrêa *et al.*, 2012). Studies on the presence of this virus in foods have not been reported with the exception of those reporting the presence of JCPyV in shellfish samples mentioned above.

HAdV and JCPyV (or HPyV) may be considered a stable marker of human contamination (Girones *et al.*, 2010). Both HAdV and JCPyV are prevalent in a very high percentage of sewage. Until now, in many studies, human adenoviruses (HAdV) have proved to be more prevalent in the environment than JCPyV (summarized in Table 22.1). Also, in some cases, the numbers of viruses during high excretion periods, such as rotaviruses or noroviruses, may exceed the numbers of human adenoviruses in that specific environment (Miagostovich *et al.*, 2008). Nevertheless, a combination of both viruses might constitute an ideal tool to ascertain human fecal pollution of water and/or food, and the level of HAdV and/or JCPyV is a useful indication of the level of human fecal contamination in these matrices.

22.3.2 Other human viruses

Other human viruses have been proposed as candidates for human fecal indicators, including human enteroviruses (Gantzer *et al.*, 1998; Kopecka *et al.*, 1993), rotaviruses (Miagostovich *et al.*, 2008) and noroviruses (Wolf *et al.*, 2010). However, the presence of these viruses has proved to depend on the rate of infection and shedding within the host populations, and it shows seasonal and geographical patterns (Sellwood *et al.*, 1981).

TT virus (TTV) (Diniz-Mendes *et al.*, 2008; Griffin *et al.*, 2008) and picobirnaviruses (Hamza *et al.*, 2011) have also been proposed as candidates for human markers, but it seems that they are not sufficiently prevalent in the environment for this purpose (Hamza *et al.*, 2011).

22.3.3 Bacteriophages

Some groups of bacteriophages (or phages) are attractive candidates for indicators of fecal pollution since they infect members of the normal bacterial flora of the human gastrointestinal tract. Bacteriophages were first proposed as human fecal indicators in the 1940s (Guelin, 1948). Since then, several studies have focused on the potential use of three groups of phages for this purpose, based on their similarities to enteric viruses (size, morphology, surface charge) and the fact that assay conditions to analyze their presence and viability are easier and cheaper than those used for human viruses.

Three groups of phages have been proposed as potential indicators: somatic coliphages (Kott, 1966), F + RNA phages (male-specific RNA coliphages) (Havelaar and Hogeboom, 1984) and *Bacteroides fragilis* phages (Jofre *et al.*, 1986) (Figs 22.3 and 22.4).

Leclerc *et al.* (2000) reviewed studies on the potential application of these viruses as human fecal indicators and discussed the reasons why, although

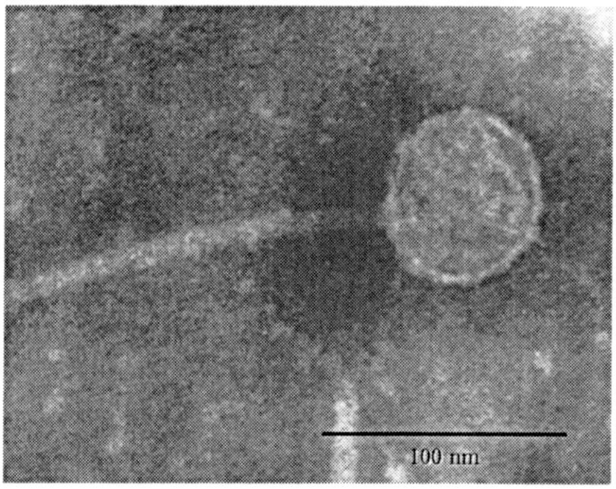

Fig. 22.3 Electron micrograph of phage B40-8 of *Bacteroides fragilis*. (*Source*: Montse Puig, with permission.)

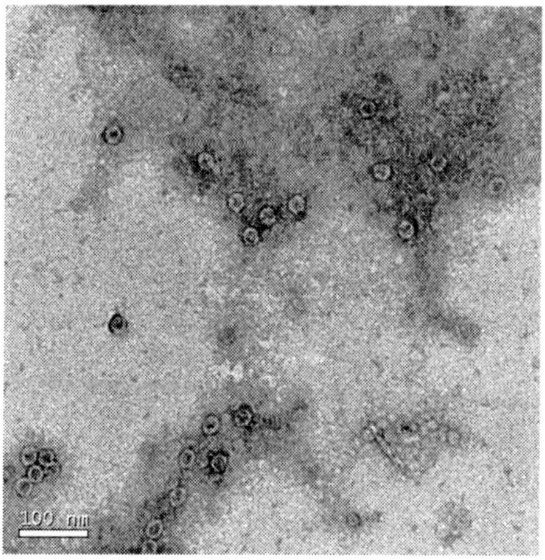

Fig. 22.4 Electron micrograph of bacteriophage MS2.

they have been considered promising candidates, their use requires further evaluation. Their presence does not always correlate with the presence of human enteric viruses (Hot *et al.*, 2003) and there is a lack of association between the detection of phages and the occurrence of disease. Nevertheless, bacteriophages are appealing indicators as model organisms to monitor the removal of human enteric viruses in disinfection treatments, especially those such as somatic coliphages that can be detected and enumerated by applying easy and standardizable assays. Recently, new phages infecting *Bacteroides* have shown new perspectives in the use of phages as indicators of human fecal contamination, once it has been established that different *Bacteroides* host strains are needed for different geographic areas, and once the methods for the detection of these strains have been provided (Payan *et al.*, 2005). For instance, phages of *Bacteroides tethaiataomicron* showed good performance in a study in southern Europe (Payan *et al.*, 2005).

22.3.4 Plant viruses

Food-derived plant viruses are excreted at concentrations up to 10^6 virion/g of human fecal material (Zhang *et al.*, 2006). Pepper mild mottle virus (PMMoV) belongs to the *Tobamovirus* genus, infects pepper plants and is a rod-shaped particle with a positive-sense RNA genome. PMMoV has been reported to be the most abundant RNA virus in human feces (Zhang *et al.*, 2006). A high percentage of pepper-based foods have tested positive for PMMoV (Colson *et al.*, 2010; Zhang *et al.*, 2006) and human fecal samples also contained PMMoV in 66.7 and 95% of the samples analyzed (Hamza *et al.*, 2011; Zhang *et al.*, 2006). PMMoV was first suggested as a human fecal indicator in marine environments in 2009 (Rosario *et al.*, 2009).

In a study by Hamza *et al.* (2011) PMMoV was detected in 100% of river water samples at concentrations ranging from 10^3–10^6 genome copies (GC)/l and 97.3% of the same samples presented HAdV (10^2–10^4 GC/l). In the same study, PMMoV was also detected in 100% of sewage samples (as HAdV, HPyV and somatic coliphages) at levels of 10^8 GC/l (equivalent levels were observed for HPyV and HAdV). In addition, PMMoV showed no seasonality and was more stable in spiked river water than HAdV and HPyV, probably due to its capsid structure. The high stability of this virus suggests it might be used as a conservative indicator and would probably be less suitable for discriminating between recent and old fecal pollution. The presence of PMMoV in seagull, chicken, cow and goose fecal samples has been reported, showing this marker is not 100% specific to human fecal contamination (Hamza *et al.*, 2011; Rosario *et al.*, 2009). Moreover, since plant viruses do not use specific receptors to infect their hosts the possibility of these viruses as parasites of other plants not ingested by humans should also be further evaluated.

22.4 Viruses as microbial source-tracking (MST) tools

Environmental waters including lakes, streams, and coastal marine waters are often susceptible to fecal contamination from a range of point and nonpoint sources, with potential contributions from many individual sources belonging to wildlife, domesticated animals, and humans. MST includes a group of methodologies that aim to identify, and in some cases quantify, the dominant sources of fecal contamination in the environment and, especially, in water resources (Field *et al.*, 2004; Fong and Lipp, 2005). MST plays a very important role in enabling effective management and remediation strategies and has received growing attention in recent years; and the number of methods for source-tracking has increased spectacularly during the last decade. Molecular markers for MST can target sequences in host-associated micro-organisms or sequences derived directly from the host, and can come from prokaryotes, eukaryotes and viruses.

Human- and animal-associated markers have been identified in the order *Bacteroidales* (Bernhard and Field 2000a, 2000b), and other bacteria and are widely used. PCR has been developed and validated for markers of fecal contamination from humans and a diversity of animals (reviewed in Roslev and Bukh, 2011). Although library-independent molecular markers represent some of the most promising methods for MST, a number of limitations have been identified for most of the proposed markers. These limitations include: (i) lack of absolute host specificity among human- and animal-associated microbial markers; (ii) lack of temporal stability of some host-associated microbial markers in different host groups; (iii) horizontal gene transfer of markers associated with toxin and/or virulence genes; (iv) low or unknown abundance of microbial markers in some host individuals and/or populations; and (v) potential carryover of mitochondrial DNA (mtDNA) and existence of nonfecal mtDNA sources (Roslev and Bukh, 2011). The use of highly specific, prevalent and stable viruses producing persistent excretions in their hosts could overcome these limitations.

Environmental samples are characterized as complex matrices and the different variables regarding microbial survival and host specificity have a significant impact on the efficacy of all MST approaches. Furthermore, the choice of MST methods and approaches is largely dependent on the objectives of the study, considering that the ultimate MST goal is the identification of fecal microbial contamination and its sources in the environment, water and food. The detection of some viruses and/or their genomes has been suggested as a potential tool for MST purposes. A summary of different studies describing the detection and concentration of human and animal viruses proposed as MST tools in environmental water samples is presented in Table 22.2.

22.4.1 Human viruses

Among the human viruses discussed in Section 22.4.3, the use of HAdV and human polyomavirus as JCPyV has been widely evaluated and compared to other human markers, with promising results (Harwood *et al.*, 2009). Both markers have proved to be specific to human feces and/or urine. Raw wastewater samples from hospitals in the region of Catalonia, North-eastern Spain, presented HAdV and JCPyV concentrations similar to those reported in urban and rural raw wastewater (Girones *et al.*, 2010), which indicates that these viruses are consistently excreted by the human population, and can be detected even when small numbers of subjects are studied (Bofill-Mas *et al.*, 2012).

Table 22.2 Human and animal viruses proposed as MST tools in freshwater samples

Host	Viruses	Type of samples	Geographic area	Concentration or % of positive samples	References
Human	Adenovirus	River water	USA	2–4 GC/l, 16%	Choi and Jiang, 2005
		River water	France	3 GC logs/l, 100%	Ogorzaly *et al.*, 2010
		River water	Spain	1–4 GC logs/l, 90%	Albinana-Gimenez *et al.*, 2009b
		Seawater	Spain	1–3 GC logs/l	Calgua *et al.*, 2008
		River water	Germany	2–4 GC logs/l, 97.5%	Hamza *et al.*, 2009
	Adenovirus F	River water	Japan	3–5 GC logs/l, 61%	Haramoto *et al.*, 2010
		Surface waters	USA	2–4 GC/l, 36%	Aslan *et al.*, 2011
Human	Enteroviruses	Seawater	Italy	1–3 GC logs/l	Rose *et al.*, 2006
		Surface waters	USA	2–4 GC/l, 20%	Aslan *et al.*, 2011
Human	JC polyomavirus	River water		0–3 GC logs/l, 90%	Albinana-Gimenez *et al.*, 2009a
		River water		2–3 GC logs/l, 11%	Haramoto *et al.*, 2010
		River water		2–3 GC logs/l, 33%	Hundesa *et al.*, 2010
	JC/BK polyomaviruses	River water	Germany	2–4 GC/l, 97.5%	Hamza *et al.*, 2009
Bovine	Polyomaviruses	River water	Spain	2–3 GC/l, 50%	Hundesa *et al.*, 2010
Porcine	Adenoviruses	River water	Spain	0–1 GC/l, 100%	Hundesa *et al.*, 2010
Porcine	Teschoviruses	River water	Spain	Presence	Jiménez-Clavero *et al.*, 2003

22.4.2 Animal viruses

Porcine viruses

Teschoviruses are RNA virus members of the *Picornaviridae* family that specifically infect pigs and are shed in pig feces. They were proposed in 2003 as potential indicators of porcine pollution (Jiménez-Clavero *et al.*, 2003) and quantitative PCR assays for their quantification have recently been described (Cano-Gómez *et al.*, 2011). Teschoviruses have been reported to be highly abundant in swine feces in Spain (Buitrago *et al.*, 2010) and are potential candidates for porcine fecal indicators. Further studies should examine their stability in the environment, their specificity and their presence in different geographical areas. The same approach should be followed for porcine circoviruses, which have also been proposed as porcine fecal contamination indicators on the basis of their high prevalence in porcine fecal samples in Brazil (Viancelli *et al.*, 2012).

Porcine adenoviruses had also been proposed as porcine fecal indicators in 2004 (Maluquer de Motes *et al.*, 2004). They are widely disseminated in the swine population but they do not produce clinically severe diseases. They have been quantified by qPCR in 100% of the wastewater samples from slaughterhouses (mean values of 10^6 GC/l) and also in river water samples near farms (10^3 GC/ml), according to Hundesa *et al.* (2009). They have been consistently detected in swine fecal samples tested from two different areas in Spain in quantities of 10^5 GC/g (Hundesa *et al.*, 2009). Samples such as bovine slaughterhouse wastewater and urban sewage, collected in areas without agricultural activities (Hundesa *et al.*, 2009), as well as hospital wastewater, have tested negative for the presence of porcine adenoviruses (Bofill-Mas *et al.*, submitted for publication). The presence of PAdV has also been reported in manure treatment system waters in Brazil (Viancelli *et al.*, 2012).

Bovine viruses

Bovine adenoviruses (BAdV) were first proposed as bovine fecal markers in 2004 (Maluquer de Motes *et al.*, 2004). Several studies have established their prevalence in bovine waste samples and manure using PCR-based methods (Hundesa *et al.*, 2009; Wong *et al.*, 2009, 2010, 2011b, 2012).

In 2006 bovine polyomaviruses (BPyV) were suggested as potential bovine markers based on their high prevalence in bovine urine samples and in a high percentage of bovine slaughterhouse wastewater (Hundesa *et al.*, 2006). Comparative studies have shown BPyV to be more prevalent than BAdV in environmental samples affected by bovine waste (Hundesa *et al.*, 2009; Wong *et al.*, 2011b). BPyV has been reported to be excreted up to 10^4 GC/l of urine in nearly 30% of the animals studied and concentrations of 10^3 GC/l have been described in wastewater from slaughterhouses where cattle are killed (Hundesa *et al.*, 2009).

22.4.3 Bacteriophages

The four subgroups of F-specific RNA bacteriophages (I–IV) have been proposed as MST tools based on the predominance of groups II and III in human sources, while groups I and IV are more abundant in animal sources (Havelar *et al.*, 1986; Hsu *et al.*, 1995). However the persistence of these four groups in the environment might be variable and this should be considered when using them for MST purposes (Muniesa *et al.*, 2009). Although many F+RNA coliphages may not be absolutely specific to individual host groups they can be useful as part of larger source-tracking toolboxes (Wolf *et al.*, 2010). Molecular assays have been developed for the quantification of the four subgroups of F-specific RNA bacteriophages using RT-qPCR, a multiplex real-time RT-PCR assay (Kirs and Smith, 2007). Other studies using quantitative RT-PCR assays showed some limitations of these assays when compared to the analysis of HAdV in field samples. There was no correlation with HAdV and the assays were not sensitive for F+RNA (38% plaques were not classified as the F-specific phage genogroups), indicating that real-time PCR assays are not applicable to a wide range of aquatic environmental samples worldwide (Haramoto *et al.*, 2009). Another study showed a correlation with HAdV using PCR assays with somatic coliphages (Ogorzaly *et al.*, 2009). This study also reported that the distribution of the genogroups quantified by PCR differs from that obtained by culture-based methods, and the concentration observed of HAdV was higher than for the groups of F+ RNA phages. The results call into question the interest of analyzing F+ RNA phages by molecular methods when human fecal contamination may be quantified by analyzing directly HAdV. Other molecular assays developed combine RT-PCR and reverse line blot hybridization for detection and genotyping F+ RNA coliphages, providing new information on the genetic diversity of F+ phages and proving to be robust and applicable to oysters, clams, mussels or water (Love *et al.*, 2008).

Bacteriophages infecting selected host strains of *Bacteroides* species have also been proposed as tools to track fecal contamination from humans, poultry, pigs and cows (Gómez-Doñate *et al.*, 2011). There are, however, some limitations to using phages infecting specific *Bacteroides* strains, such as the facts that specific bacterial hosts must be isolated in different geographical areas (Ebdon *et al*, 2007; Gomez-Doñate *et al.*, 2011), the number of specific phages are low and the isolated species may not be absolutely specific to individual host groups; however, they can be useful as part of larger source-tracking toolboxes.

22.5 Future trends

Current microbiological quality assessment of environmental waters is widely based on the concept of fecal indicator bacteria. Although this advance has clearly reduced health risks in many countries, the fecal indicator approach may in the future be combined with or replaced by more direct monitoring of genuine pathogenic micro-organisms (Girones *et al.*, 2010; Sadowsky *et al.*, 2007). The microbiological control of water and food quality may evolve, accordingly, to focus more on the tracking of a specific pathogen of interest, or on the use of some viral groups as indicators. For example, human adenoviruses provide both robust information on the level of fecal contamination, and information on a potentially pathogenic micro-organism in the environment, at affordable cost. Furthermore, sampling strategies should be considered carefully to obtain samples that best represent the water or food material in question. This could mean including hydrological and physicochemical sensors, time and flow integrated automated sampling devices, and perhaps automated nucleic extraction and quantification of the selected markers.

Application of new technologies such as high-throughput mass sequencing to analyze urban sewage from diverse geographical areas has produced very interesting information on new groups of viruses (Cantalupo *et al.*, 2011) and further work is needed for the characterization of new emergent viruses and viruses previously unknown. More information on the stability of selected viral genetic markers and distribution of pathogens and indicators in diverse geographical areas and the diverse matrices is needed for the evaluation of the applicability and significance of the new viral indicators and MST tools. However, several assays and cost-effective methods have been developed that may be validated and standardized, and the technology could be ready for routine implementation and automation in the near future.

Several authors have suggested developing integrated systems for the detection of multiple pathogens and indicators using multiplex amplification assays. Multiplex diagnostic tools are already available and multiplex quantitative PCR assays for MST using a diversity of human and animal viruses have been described by Wolf *et al.* (2010). In this study the authors describe a viral toolbox (VTB) consisting of three multiplex reverse transcription (RT)-qPCR assays (VTB-1 to VTB-3) for the detection of human and animal adenoviruses and noroviruses including viruses found in pigs, cattle, sheep, deer, and goats.

Routine quantitative molecular assays for viral indicators using PCR may also be improved and standardized, considering new methods for the absolute quantification of genome copies without the requirement of independent calibration curves (Pinheiro *et al.*, 2012). Other technical improvements expected include advances in microfluidics and nanobiotechnology, as a result of which miniaturized systems for detection of viral indicators could be based on microchips. Several approaches have been described (Gilbride *et al.*, 2006; Ivniski *et al.*, 2003).

Cost-effective methods for the concentration, detection and quantification of viral pathogens and indicators using molecular methods have been developed and optimized during the last few years with acceptable levels of cost, feasibility, sensitivity and repeatability, especially in the case of the DNA viruses.

The proper control of contamination will depend on regulatory authorities choosing and standardizing effective parameters, and then developing appropriate surveillance systems with which to monitor and more effectively reduce established – and perhaps prevent emergent – diseases. Viral indicators may be used not only as an index of viral contamination but also as complementary indicators of fecal/urine contamination in water and food.

22.6 References

ABDELZAHER A M, WRIGHT M E, ORTEGA C, SOLO-GABRIELE H M, MILLER G, ELMIR S, NEWMAN X, SHIH P, BONILLA J A, BONILLA T D, PALMER C J, SCOTT T, LUKASIK J, HARWOOD V J, MCQUAIG S, SINIGALLIANO C, GIDLEY M, PLANO LR, ZHU X, WANG J D and FLEMING L E (2010), 'Presence of pathogens and indicator microbes at a non-point source subtropical recreational marine beach', *Appl Environ Microbiol*, **76**(3), 724–32.

ADRIAN T, SCHÄFER G, COONEY M K, FOX J P and WIGAND R (1988), 'Persistent enteral infections with adenovirus types 1 and 2 in infants: no evidence of reinfection', *Epidemiol Infect*, **101**(3), 503–9.

AHMED W, GOONETILLEKE A, POWELL D, CHAUHAN K and GARDNER T (2009), 'Comparison of molecular markers to detect fresh sewage in environmental waters', *Water Res*, **43**(19), 4908–17.

AHMED W, WAN C, GOONETILLEKE A and GARDNER T (2010), 'Evaluating sewage-associated JCV and BKV polyomaviruses for sourcing human fecal pollution in a coastal river in Southeast Queensland, Australia', *J Environ Qual*, **39**(5), 1743–50.

AKSAMIT A J JR (1993), 'Nonradioactive in situ hybridization in progressive multifocal leukoencephalopathy', *Mayo Clin Proc*, **68**(9), 899–910.

ALBINANA-GIMENEZ N, CLEMENTE-CASARES P, BOFILL-MAS S, HUNDESA A, RIBAS F and GIRONES R (2006), 'Distribution of human polyomaviruses, adenoviruses, and hepatitis E virus in the environment and in a drinking-water treatment plant', *Environ Sci Technol*, **40**(23), 7416–22.

ALBINANA-GIMENEZ N, CLEMENTE-CASARES P, CALGUA B, COURTOIS S, HUGUET J M and GIRONES R (2009a), 'Comparison of methods for the quantification of human adenoviruses, Polyomavirus JC and Norovirus in source and drinking water', *J Virol Methods*, **158**, 104–9.

ALBINANA-GIMENEZ N, MIAGOSTOVICH M P, CALGUA B, HUGUET J M, MATIA L and GIRONES R (2009b), 'Analysis of adenoviruses and polyomaviruses quantified by qPCR as indicators of water quality in source and drinking-water treatment plants', *Water Res*, **43**(7), 2011–19.

ASLAN A, XAGORARAKI I, SIMMONS F J, ROSE J B and DOREVITCH S (2011), 'Occurrence of adenovirus and other enteric viruses in limited-contact freshwater recreational areas and bathing waters', *J Appl Microbiol*, **111**(5), 1250–61.

BARTON E S, WHITE D W, CATHELYN J S, BRETT-MCCLELLAN K A, ENGLE M, DIAMOND M S, MILLER V L and VIRGIN IV H W (2007), 'Herpesvirus latency protects the host from bacterial infection: latency as mutualistic symbiosis', *Nature*, **447**(7142), 326–9.

BERGER J R, HOUFF S A and MAJOR E O (2009), 'Monoclonal antibodies and progressive multifocal leukoencephalopathy', *MAbs*, **1**(6), 583–9.

BERNHARD A E and FIELD K G (2000a), 'A PCR assay to discriminate human and ruminant feces on the basis of host differences in Bacteroides–Prevotella genes encoding 16S rRNA', *Appl Environ Microbiol*, **66**, 4571–4.

BERNHARD A E and FIELD K G (2000b), 'Identification of nonpoint sources of fecal pollution in coastal waters by using host-specific 16S ribosomal DNA genetic markers from fecal anaerobes', *Appl Environ Microbiol*, **66**, 1587–94.

BIEL S S, HELD T K, LANDT O, NIEDRIG M, GELDERBLOM H R, SIEGERT W and NITSCHE A (2000), 'Rapid quantification and differentiation of human polyomavirus DNA in undiluted urine from patients after bone marrow transplantation', *J Clin Microbiol*, **38**(10), 3689–95.

BOFILL-MAS S, ALBIÑANA-GIMENEZ N, CLEMENTE-CASARES P, HUNDESA A, RODRIGUEZ-MANZANO J, ALLARD A, CALVO M and GIRONES R (2006), 'Quantification and stability of human adenoviruses and polyomavirus JCPyV in wastewater matrices', *Appl Environ. Microbiol*, **72**(12), 7894–6.

BOFILL-MAS S, CALGUA B,RODRIGUEZ-MANZANO J, HUNDESA A, CARRATALA A, RUSIÑOL M, GUERRERO L and GIRONES R (2012), 'Cost-effective Applications of human and animal viruses and microbial source-tracking tools in surface waters and groundwater', in *Faecal Indicators and pathogens. Proceedings of the 2011 FIPs Conference*. Royal Society of Chemistry, London.

BOFILL-MAS S, FORMIGA-CRUZ M, CLEMENTE-CASARES P, CALAFELL F and GIRONES R (2001), 'Potential transmisión of human polyomaviruses through the gastrointestinal tract after exposure to virions or viral DNA', *J Virol*, **75**, 10290–9.

BOFILL-MAS S, PINA S and GIRONES R (2000) 'Documenting the epidemiologic patterns of polyomaviruses in human populations by studying their presence in urban sewage', *Appl Environ Microbiol*, **66**(1), 238–45.

BOFILL-MAS S, RODRIGUEZ-MANZANO J, CALGUA B, CARRATALA A and GIRONES R (2010), 'Newly described human polyomaviruses Merkel Cell, KI and WU are present in urban sewage and may represent potential environmental contaminants', *Virol J*, **28**(7), 141.

BROWNELL MJ, HARWOOD VJ, KURZ RC, MCQUAIG SM, LUKASIK J and SCOTT TM (2007), 'Confirmation of putative stormwater impact on water quality at a Florida beach by microbial source tracking methods and structure of indicator organism populations', *Water Res*, **41**(16), 3747–57.

BUITRAGO D, CANO-GÓMEZ C, AGÜERO M, FERNANDEZ-PACHECO P, GÓMEZ-TEJEDOR C and JIMÉNEZ-CLAVERO M A (2010), 'A survey of porcine picornaviruses and adenoviruses in fecal samples in Spain', *J Vet Diagn Invest*, **22**(5), 763–6.

BYAPPANAHALLI M N, SHIVELY D A, NEVERS M B, SADOWSKY M J and WHITMAN R L (2003) 'Growth and survival of Escherichia coli and enterococci populations in the macro-alga Cladophora (Chlorophyta)', *FEMS Microbiol Ecol*, **46**(2), 203–11.

CALGUA B, BARARDI C R, BOFILL-MAS S, RODRIGUEZ-MANZANO J and GIRONES R (2011), 'Detection and quantitation of infectious human adenoviruses and JC polyomaviruses in water by immunofluorescence assay', *J Virol Methods*, **71**(1), 1–7.

CALGUA, B, MENGEWEIN A, GRÜNERT A, BOFILL-MAS S, CLEMENTE-CASARES P, HUNDESA A, WYN-JONES A P, LÓPEZ-PILA J M and GIRONES R (2008) 'Development and application of a one-step low cost procedure to concentrate viruses from seawater samples', *J Virol Methods*, **153**, 79–83.

CANO-GÓMEZ C, PALERO F, BUITRAGO MD, GARCÍA-CASADO MA, FERNÁNDEZ-PINERO J, FERNÁNDEZ-PACHECO P, AGÜERO M, GÓMEZ-TEJEDOR C and JIMÉNEZ-CLAVERO M Á (2011), 'Analyzing the genetic diversity of teschoviruses in Spanish pig populations using complete VP1 sequences', *Infect Genet Evol*, **11**(8), 2144–50.

CANTALUPO P G, CALGUA B, ZHAO G, HUNDESA A, WIER A D, KATZ J P, GRABE M, HENDRIX R W, GIRONES R, WANG D and PIPAS J M (2011), 'Raw sewage harbors diverse viral populations', *mBio* **2**(5): e00180–11. Doi:10.1128/mBio.00180–11.

COLSON P, RICHET H, DESNUES C, BALIQUE F, MOAL V, GROB JJ, BERBIS P, LECOQ H, HARLÉ JR, BERLAND Y and RAOULT D (2010), 'Pepper mild mottle virus, a plant virus associated with specific immune responses, Fever, abdominal pains, and pruritus in humans', *PLoS One*, **5**(4), e10041.

CONTRERAS-COLL N, LUCENA F, MOOIJMAN K, HAVELAAR A, PIERZ V, BOQUE M, GAWLER A, HÖLLER C, LAMBIRI M, MIROLO G, MORENO B, NIEMI M, SOMMER R, VALENTIN B, WIEDENMANN A, YOUNG V and JOFRE J (2002), 'Occurrence and levels of indicator bacteriophages in bathing waters throughout Europe', *Water Res*, **36**(20), 4963–74.

CORRÊA DE ABREU A, CARRATALA A, BARARDI CR, CALVO M, GIRONES R and BOFILL-MAS S (2012), 'Comparative inactivation of murine norovirus, human adenovirus, and human JC polyomavirus by chlorine in seawater', *Appl Environ Microbiol*, **78**(18), 6450–7.

DINIZ-MENDES L, PAULA V S, LUZ SL and NIEL C (2008), 'High prevalence of human Torque teno virus in streams crossing the city of Manaus, Brazilian Amazon', *J Appl Microbiol*, **105**(1), 51–8.

EBDON J, MUNIESA M and TAYLOR H (2007), 'The application of a recently isolated strain of Bacteroides (GB-124) to identify human sources of faecal pollution in a temperate river catchment', *Water Res*, **41**(16), 3683–90.

FENG H, SHUDA M, CHANG Y and MOORE P S (2008), 'Clonal integration of a polyomavirus in human Merkel cell carcinoma', *Science*, **319**(5866), 1096–1100.

FIELD K G (2004), 'Faecal source identification', in Cotruvo J A, Dufour A, Reese G, Bartram J, Carr, D O. Cliver G F Craun R and Fayer V P J (eds), *Waterborne Zoonosis: Identification, Causes and Control*, Gannon, London, IWA Publishing, 349–66.

FONG T T and LIPP E K (2005), 'Enteric viruses of humans and animals in aquatic environments: health risks, detection, and potential water quality assessment tools', *Microbiol Mol Biol Rev*, **69**, 357–71.

FONG T T, PHANIKUMAR M S, XAGORARAKI I and ROSE J B (2010), 'Quantitative detection of human adenoviruses in wastewater and combined sewer overflows influencing a Michigan river', *Appl Environ Microbiol*, **76**(3), 715–23.

FORMIGA-CRUZ M, TOFIÑO-QUESADA G, BOFILL-MAS S, LEES D N, HENSHILWOOD K, ALLARD A K, CONDIN-HANSSON, A C, HERNROTH B E, VANTARAKIS A, TSIBOUXI A, PAPAPETROPOULOU M, FURONES M D and GIRONES R (2002), 'Distribution of human viral contamination in shellfish from different growing areas in Greece, Spain, Sweden and UK', *Appl Environ Microbiol*, **68**(12), 5990–8.

FOULONGNE V, KLUGER N, DEREURE O, BRIEU N, GUILLOT B and SEGONDY M (2008), 'Merkel cell polyomavirus and Merkel cell carcinoma, France', *Emerg Infect Dis*, **14**(9), 1491–3.

FUJIOKA R, SIAN-DENTON C, BORJA M, CASTRO J and MORPHEW K (1998), 'Soil: the environmental source of Escherichia coli and Enterococci in Guam's streams', *J Appl Microbiol*, **85**(Suppl 1), 83S–89S.

FUMIAN T M, GUIMARÃES F R, PEREIRA VAZ B J, DA SILVA M T, MUYLAERT F F, BOFILL-MAS S, GIRONÉS R, LEITE J P and MIAGOSTOVICH M P (2010), 'Molecular detection, quantification and characterization of human polyomavirus JC from waste water in Rio De Janeiro, Brazil', *J Water Health*, **8**(3), 438–45.

GANTZER C, MAUL A, AUDIC J M and SCHWARTZBROD L (1998), 'Detection of infectious enteroviruses, enterovirus genomes, somatic coliphages, and Bacteroides fragilis phages in treated wastewater', *Appl Environ Microbiol*, **64**(11), 4307–12.

GARDNER S D, FIELD A M, COLEMAN D V and HULME B (1971), 'New human papovavirus (B.K.) isolated from urine after renal transplantation', *Lancet*, **1**(7712), 1253–7.

GERBA C P, GOYAL S M, LABELLE R L, CECH I and BODGAN G F (1979), 'Failure of indicator bacteria to reflect the occurrence of enteroviruses in marine waters', *Am J Public Health*, **69**(11), 1116–19.

GERBA C P, GRAMOS D M and NWACHUKU N (2002), 'Comparative inactivation of enteroviruses and adenovirus 2 by UV light', *Appl Environ Microbiol*, **68**(10), 5167–9.

GIBSON K E, OPRYSZKO M C, SCHISSLER J T, GUO Y and SCHWAB K J (2011), 'Evaluation of human enteric viruses in surface water and drinking water resources in southern Ghana', *Am J Trop Med Hyg*, **84**(1), 20–9.

GILBRIDE K A, LEE D Y and BEAUDETTE L A (2006), 'Molecular techniques in wastewater: understanding microbial communities, detecting pathogens, and real-time process control', *J Microbiol Methods*, **66**(1), 1–20.

GIRONES R, FERRÚS M A, ALONSO J L, RODRIGUEZ-MANZANO J, CALGUA B, CORRÊA ADE A, HUNDESA A, CARRATALA A and BOFILL-MAS S (2010), 'Molecular detection of pathogens in water – the pros and cons of molecular techniques', *Water Res*, **44**(15), 4325–39.

GÓMEZ-DOÑATE M, PAYÁN A, CORTÉS I, BLANCH A R, LUCENA F, JOFRE J and MUNIESA M (2011), 'Isolation of bacteriophage host strains of Bacteroides species suitable for tracking sources of animal faecal pollution in water', *Environ Microbiol*, **3**(6):1622–31.

GRIFFIN J S, PLUMMER J D and LONG S C (2008), 'Torque teno virus: an improved indicator for viral pathogens in drinking waters', *Virol J*, **5**, 112.

GUELIN A (1948), 'Etude quantitative des bacteriophages de la mer', *Ann Inst Pasteur*, **104**, 104–112.

GUERRERO-LATORRE L, CARRATALA A, RODRIGUEZ-MANZANO J, CALGUA B, HUNDESA A and GIRONES R (2011), 'Occurrence of water-borne enteric viruses in two settlements based in Eastern Chad: analysis of hepatitis E virus, hepatitis A virus and human adenovirus in water sources', *J Water Health*, **9**(3), 515–24.

HAAS C N, ROSE J B, GERBA C and REGLI S (1993), ' Risk assessment of virus in drinking water', *Risk Anal*, **13**(5), 545–52.

HAMZA I A, JURZIK L, STANG A, SURE K, UBERLA K and WILHELM M (2009), 'Detection of human viruses in rivers of a densly-populated area in Germany using a virus adsorption elution method optimized for PCR analyses', *Water Res*, **43**(10), 2657–68.

HAMZA I A, JURZIK L, UBERLA K and WILHELM M (2011), 'Evaluation of pepper mild mottle virus, human picobirnavirus and Torque teno virus as indicators of fecal contamination in river water', *Water Res*, **45**(3), 1358–68.

HARAMOTO E, KITAJIMA M, KATAYAMA H, ASAMI M, AKIBA M and KUNIKANE S (2009), 'Application of real-time PCR assays to genotyping of F-specific phages in river water and sediments in Japan', *Water Res*, **43**(15), 3759–64.

HARAMOTO E, KITAJIMA M, KATAYAMA H and OHGAKI S (2010), 'Real-time PCR detection of adenoviruses, polyomaviruses, and torque teno viruses in river water in Japan', *Water Res*, **44**(6):1747–52.

HARAMOTO E, YAMADA K and NISHIDA K (2011), 'Prevalence of protozoa, viruses, coliphages and indicator bacteria in groundwater and river water in the Kathmandu Valley, Nepal', *Trans R Soc Trop Med Hyg*, **105**(12):711–16.

HARWOOD V J, BROWNELL M, WANG S, LEPO J, ELLENDER R D, AJIDAHUN A, HELLEIN K N, KENNEDY E, YE X and FLOOD C (2009), 'Validation and field testing of library-independent microbial source tracking methods in the Gulf of Mexico', *Water Res*, **43**(19), 4812–19.

HAVELAAR A H, FURUSE K and HOGEBOOM W M (1986), 'Bacteriophages and indicator bacteria in human and animal faeces', *J Appl Bacteriol*, **60**(3), 255–62.

HAVELAAR A H and HOGEBOOM W M (1984), 'A method for the enumeration of male-specific bacteriophages in sewage', *J Appl Bacteriol*, **56**(3), 439–47.

HAVELAAR A H, VAN OLPHEN M and DROST Y C (1993), 'F-specific RNA bacteriophages are adequate model organisms for enteric viruses in fresh water', *Appl Environ Microbiol*, **59**(9), 2956–62.

HELLEIN K N, BATTIE C, TAUCHMAN E, LUND D, OYARZABAL O A and LEPO J E (2011), 'Culture-based indicators of fecal contamination and molecular microbial indicators rarely correlate with Campylobacter spp. in recreational waters', *Water Health*, **9**(4), 695–707.

HEWITT J, LEONARD M, GREENING G E and LEWIS G D (2011), 'Influence of wastewater treatment process and the population size on human virus profiles in wastewater', *Water Res*, **45**(18), 6267–76.

HOT D, LEGEAY O, JACQUES J, GANTZER C, CAUDRELIER Y, GUYARD K, LANGE M and ANDRÉOLETTI L (2003), 'Detection of somatic phages, infectious enteroviruses and enterovirus genomes as indicators of human enteric viral pollution in surface water', *Water Res*, **37**(19), 4703–10.

HSU F C, SHIEH Y S, VAN DUIN J, BEEKWILDER M J and SOBSEY M D (1995), 'Genotyping male-specific RNA coliphages by hybridization with oligonucleotide probes', *Appl Environ Microbiol*, **61**(11), 3960–6.

HUNDESA A, BOFILL-MAS S, MALUQUER DE MOTES C, RODRIGUEZ-MANZANO J, BACH A, CASAS M and GIRONES R (2010), 'Development of a quantitative PCR assay for the quantitation of bovine polyomavirus as a microbial source-tracking tool', *J Virol Methods*, **163**(2), 385–9.

HUNDESA A, MALUQUER DE MOTES C, ALBINANA-GIMENEZ N, RODRIGUEZ-MANZANO J, BOFILL-MAS S, SUÑEN E and ROSINA GIRONES R (2009), 'Development of a qPCR assay for the quantification of porcine adenoviruses as an MST tool for swine fecal contamination in the environment', *J Virol Methods*, **58**(1–2), 130–5.

HUNDESA A, MALUQUER DE MOTES C, BOFILL-MAS S, ALBINANA-GIMENEZ N and GIRONES R (2006), 'Identification of human and animal adenoviruses and polyomaviruses for determination of sources of fecal contamination in the environment', *Appl Environ Microbiol*, **72**(12), 7886–93.

IVNISKI D, O'NEIL D J, GATTUSO A, SCHLICHT R, CALIDONNA M and FISHER R (2003), 'Nucleic acid approaches for detection and identification of biological warfare and infectious disease agents', *BioTechniques*, **35**, 862–9.

JIANG S C (2006), 'Human adenoviruses in water: occurrence and health implications: a critical review', *Environ Sci Technol*, **40**(23), 7132–40.

JIANG S C, HAN J, HE J W and CHU W (2009), 'Evaluation of four cell lines for assay of infectious adenoviruses in water samples', *J Water Health*, **7**(4), 650–6.

JIMÉNEZ-CLAVERO MA, FERNÁNDEZ C, ORTIZ JA, PRO J, CARBONELL G, TARAZONA JV, ROBLAS N and LEY V (2003), 'Teschoviruses as indicators of porcine fecal contamination of surface water', *Appl Environ Microbiol*, **69**(10), 6311–15.

JURZIK L, HAMZA IA, PUCHERT W, UBERLA K and WILHELM M (2010), 'Chemical and microbiological parameters as possible indicators for human enteric viruses in surface water', *Int J Hyg Environ Health*, **213**(3), 210–16.

KIRS M and SMITH D C (2007), 'Multiplex quantitative real-time reverse transcriptase PCR for F+-specific RNA coliphages: a method for use in microbial source tracking', *Appl Environ Microbiol*, **73**(3), 808–14.

KITAJIMA M, OKA T, HARAMOTO E, TAKEDA N, KATAYAMA K and KATAYAMA H (2010), 'Seasonal distribution and genetic diversity of genogroups I, II, and IV noroviruses in the Tamagawa River, Japan', *Environ Sci Technol*, **44**(18), 7116–22.

KITAMURA T, ASO Y, KUNIYOSHI N, HARA K and YOGO Y (1990), 'High incidence of urinary JC virus excretion in nonimmunosuppressed older patients', *J Infect Dis*, **161**(6), 1128–33.

KOPECKA H, DUBROU S, PREVOT J, MARECHAL J and LÓPEZ-PILA J M (1993), 'Detection of naturally occurring enteroviruses in waters by reverse transcription, polymerase chain reaction, and hybridization', *Appl Environ Microbiol*, **59**(4), 1213–1219.

KORAJKIC A, BROWNELL MJ and HARWOOD V J (2011), 'Investigation of human sewage pollution and pathogen analysis at Florida Gulf coast beaches', *J Appl Microbiol*, **110**(1), 174–83.

KORALNIK I J, BODEN D, MAI V X, LORD C I and LETVIN N L (1999), 'JC virus DNA load in patients with and without progressive multifocal leukoencephalopathy', *Neurology*, **52**(2), 253–60.

KUO D H, SIMMONS F J, BLAIR S, HART E, ROSE J B and XAGORARAKI I (2009), 'Assessment of human adenovirus removal in a full-scale membrane bioreactor treating municipal wastewater', *Water Res*, **44**(5):1520–30.

LECLERC H, EDBERG S, PIERZO V and DELATTRE J M (2000), 'Bacteriophages as indicators of enteric viruses and public health risk in groundwaters', *J Appl Microbiol*, **88**(1), 5–21.

LIPP E K, FARRAH S A and ROSE J B (2001), 'Assessment and impact of microbial fecal pollution and human enteric pathogens in a coastal community', *Mar Pollut Bull*, **42**, 286–329.

LOVE D C, VINJÉ J, KHALIL S M, MURPHY J, LOVELACE G L and SOBSEY M D (2008), 'Evaluation of RT-PCR and reverse line blot hybridization for detection and genotyping F+ RNA coliphages from estuarine waters and molluscan shellfish', *J Appl Microbiol*, **104**(4):1203–12.

MALUQUER DE MOTES C, CLEMENTE-CASARES P, HUNDESA A, MARTÍN M and GIRONES R (2004), 'Detection of bovine and porcine adenovirus in fecal and environmental samples', *Appl Environ Microbiol*, **70**, 1448–54.

MCQUAIG S M, SCOTT T M, HARWOOD V J, FARRAH S R and LUKASIK J O (2006), 'Detection of human-derived fecal pollution in environmental waters by use of a PCR-based human polyomavirus assay', *Appl Environ Microbiol*, **72**(12), 7567–74.

MCQUAIG S M, SCOTT T M, LUKASIK J O, PAUL J H and HARWOOD V J (2009), 'Quantification of human polyomaviruses JC Virus and BK Virus by TaqMan quantitative PCR and comparison to other water quality indicators in water and fecal samples', *Appl Environ Microbiol*, **75**(11), 3379–88.

MEDEMA G, PAYMENT P, DUFOUR A, ROBERTSON W, WAITE M, HUNTER P, KIRBY R and ANDERSSON Y (2003), '*Safe Drinking Water: An Ongoing Challenge*', Geneva, World Health Organization, 1–20 www.who.int/water_sanitation_health/dwq/9241546301_chap1.pdf.

MENA K D and GERBA C P (2009), 'Waterborne adenovirus', *Rev Environ Contam Toxicol*, **198**, 133–67.

MIAGOSTOVICH M P, FERREIRA F F, GUIMARÃES F R, FUMIAN T M, DINIZ-MENDES L, LUZ S L, SILVA L A and LEITE J P (2008), 'Molecular detection and characterization of gastroenteritis viruses occurring naturally in the stream waters of Manaus, central Amazonia, Brazil', *Appl Environ Microbiol*, **74**(2), 375–82.

MORESCO V, VIANCELLI A, NASCIMENTO M A, SOUZA D S, RAMOS A P, GARCIA L A, SIMÕES C M and BARARDI C R (2012), 'Microbiological and physicochemical analysis of the coastal waters of southern Brazil', *Mar Pollut Bull*, **64**(1), 40–8.

MUNIESA M, PAYAN A, MOCE-LLIVINA L, BLANCH A R and JOFRE J (2009), 'Differential persistence of F-specific RNA phage subgroups hinders their use as single tracers for faecal source tracking in surface water', *Water Res*, **43**(6), 1559–64.

OGORZALY L, BERTRAND I, PARIS M, MAUL A and GANTZER C (2010), 'Occurrence, survival, and persistence of human adenoviruses and F-specific RNA phages in raw groundwater', *Appl Environ Microbiol*, **76**(24), 8019–25.

OGORZALY L, TISSIER A, BERTRAND I, MAUL A and GANTZER C (2009), 'Relationship between F-specific RNA phage genogroups, faecal pollution indicators and human adenoviruses in river water', *Water Res*, **43**(5), 1257–64.

PADGETT B L, WALKER D L, ZURHEIN G M, ECKROADE R J and DESSEL B H (1971), 'Cultivation of papova-like virus from human brain with progressive multifocal leucoencephalopathy', *Lancet*, **1**(7712), 1257–60.

PAL A, SIROTA L, MAUDRU T, PEDEN K and LEWIS JR A M (2006), 'Real-time, quantitative PCR assays for the detection of virus-specific DNA in samples with mixed populations of polyomaviruses', *J Virol Methods*, **135**, 32–42.

PAYAN A, EBDON J, TAYLOR H, GANTZER C, OTTOSON J, PAPAGEORGIOU GT, BLANCH AR, LUCENA F, JOFRE J and MUNIESA M (2005), 'Method for isolation of Bacteroides bacteriophage host strains suitable for tracking sources of fecal pollution in water', *Appl Environ Microbiol*, **71**(9), 5659–5962.

PINA S, CREUS A, GONZÁLEZ N, GIRONÉS R, FELIP M and SOMMARUGA R (1998a), 'Abundance, morphology and distribution of planktonic virus-like particles in two high-mountain lakes', *J Plank Res*, **20**, 2413–21.

PINA S, PUIG M, LUCENA F, JOFRE J and GIRONES R (1998), 'Viral pollution in the environment and in shellfish: human adenovirus detection by PCR as an index of human viruses', *Appl Environ Microbiol*, **64**(9), 3376–82.

PINHEIRO L B, COLEMAN V A, HINDSON C M, HERRMANN J, HINDSON B J, BHAT S and EMSLIE K R (2012), 'Evaluation of a droplet digital polymerase chain reaction format for DNA copy number quantification', *Anal Chem*, **84**(2), 1003–11.

PIPES W O (1982) 'Indicators in water quality', in *Bacterial indicators of pollution*, Pipes W O, Boca Raton, CRC Press, 83–96.

POTE J, HALLER L, KOTTELAT R, SASTRE V, ARPAGAUS P and WILDI W (2009), 'Persistence and growth of faecal culturable bacterial indicators in water column and sediments of Vidy Bay, Lake Geneva, Switzerland', *J Environ Sci (China)*, **21**(1), 62–9.

PUIG M, JOFRE J, LUCENA F, ALLARD A, WADELL G and GIRONES R (1994), 'Detection of adenoviruses and enteroviruses in polluted waters by nested PCR amplification', *Appl Environ Microbiol*, **60**(8), 2963–70.

RAFIQUE A and JIANG S C (2008), 'Genetic diversity of human polyomavirus JCPyV in Southern California wastewater', *J Water Health*, **6**(4), 533–8.

RIGOTTO C, VICTORIA M, MORESCO V, KOLESNIKOVAS C K, CORRÊA A A, SOUZA D S, MIAGOSTOVICH M P, SIMÕES C M and BARARDI C R (2010), 'Assessment of adenovirus, hepatitis A virus and rotavirus presence in environmental samples in Florianopolis, South Brazil', *J Appl Microbiol*, **9**(6), 1979–87.

RODRIGUEZ-MANZANO J, ALONSO JL, FERRÚS MA, MORENO Y, AMORÓS I, CALGUA B, HUNDESA A, GUERRERO-LATORRE L, CARRATALA A, RUSIÑOL M and GIRONES R (2012), 'Standard and new faecal indicators and pathogens in sewage treatment plants, microbiological parameters for improving the control of reclaimed water', *Water Sci Technol*, **66**(12):2517–23.

ROSARIO K, SYMONDS EM, SINIGALLIANO C, STEWART J and BREITBART M (2009), 'Pepper mild mottle virus as an indicator of fecal pollution', *Appl Environ Microbiol*, **75**(22), 7261–67.

ROSE M A, DHAR A K, BROOKS H A, ZECCHINI F and GERSBERG R M (2006), 'Quantitation of hepatitis A virus and enterovirus levels in the lagoon canals and Lido beach of Venice, Italy, using real-time RT-PCR', *Water Res*, **40**(12), 2387–96.

ROSLEV P and BUKH A S (2011), 'State of the art molecular markers for fecal pollution source tracking in water', *Appl Microbiol Biotechnol*, **89**(5), 1341–55.

SADOWSKY M J, CALL D R and SANTO DOMINGO J W (2007), 'The future of microbial source tracking studies', in SANTO DOMINGO J W and SADOWSKY M J (EDS), *Microbial source tracking*, Washington, DC, ASM Press, 235–77.

SELLWOOD J, DADSWELL J V and SLADE J S (1981), 'Viruses in sewage as an indicator of their presence in the community', *J Hyg (Lond)*, **86**(2), 217–25.

SHAH K V, DANIEL R W, STRICKLER H D and GOEDERT J J (1997), 'Investigation of human urine for genomic sequences of the primate polyomaviruses simian virus 40, BK virus, and JC virus', *J Infect Dis*, **176**(6), 1618–21.

SIMMONS F J and XAGORARAKI I (2011), 'Release of infectious human enteric viruses by full-scale wastewater utilities', *Water Res*, **45**(12), 3590–8.

SINCLAIR R G, ROSE J B, HASHSHAM S A, GERBA C P and HAAS C N (2012), 'A criteria for selection of surrogates used to study the fate and control of pathogens in the environment', *Appl Environ Microbiol*, **78**(6), 1969–77 (Epub ahead of print).

SOLO-GABRIELE H M, WOLFERT M A, DESMARAIS T R and PALMER C J (2000), 'Sources of *Escherichia coli* in a coastal subtropical environment', *Appl Environ Microbiol*, **66**(1), 230–7.

SOUZA D S, RAMOS A P, NUNES F F, MORESCO V, TANIGUCHI S, LEAL D A, SASAKI S T, BÍCEGO M C, MONTONE R C, DURIGAN M, TEIXEIRA A L, PILOTTO M R, DELFINO N, FRANCO R M, MELO C M, BAINY A C and BARARDI C R (2012), 'Evaluation of tropical water sources and mollusks in southern Brazil using microbiological, biochemical, and chemical parameters', *Ecotoxicol Environ Saf*, **76**(2), 153–61.

THURSTON-ENRIQUEZ J A, HAAS C N, JACANGELO J, RILEY K and GERBA C P (2003), 'Inactivation of feline calicivirus and adenovirus type 40 by UV radiation', *Appl Environ Microbiol*, **69**(1), 577–82.

US-EPA (2005), 'Microbial Source Tracking Guide Document' USA, Environmental Protection Agency, Technical Report No. EPA/600/R-05/064.

VIANCELLI A, GARCIA LA, KUNZ A, STEINMETZ R, ESTEVES P A and BARARDI C R (2012), 'Detection of circoviruses and porcine adenoviruses in water samples collected from swine manure treatment systems', *Res Vet Sci*, **93**(1), 538–43.

WEBER T, KLAPPER P E, CLEATOR G M, BODEMER M, LUKE W, KNOWLES W, CINQUE P, VAN LOON A M, GRANDIEN M, HAMMARIN A L, CIARDI M and BOGDANOVIC G (1997), 'Polymerase chain reaction for detection of JC virus DNA in cerebrospinal fluid: a quality control study. European Union Concerted Action on Viral Meningitis and Encephalitis', *J Virol Methods*, **69**(1–2), 231–7.

WHO (2004) in *Guidelines for drinking-water quality*, 3rd ed., Geneva, WHO, 48–83. Available at: www.who.int/water_sanitation_health/dwq/gdwq3/en/print.html

WOLF S, HEWITT J and GREENING G E (2010), 'Viral multiplex quantitative PCR assays for tracking sources of fecal contamination', *Appl Environ Microbiol*, **76**(5), 1388–1394.

WONG K and XAGORARAKI I (2010), 'Quantitative PCR assays to survey the bovine adenovirus levels in environmental samples', *J Appl Microbiol*, **109**(2), 605–612.

WONG K and XAGORARAKI I (2011), 'Evaluating the prevalence and genetic diversity of adenovirus and polyomavirus in bovine waste for microbial source tracking', *Appl Microbiol Biotechnol*, **90**(4), 1521–1526.

WONG K and XAGORARAKI I (2012), 'A perspective on the prevalence of DNA enteric virus genomes in anaerobic-digested biological wastes', *Environ Monit Assess*, **184**(8), 5009–16.

WONG K, XAGORARAKI I, WALLACE J, BICKERT W, SRINIVASAN S and ROSE J B (2009), 'Removal of viruses and indicators by anaerobic membrane bioreactor treating animal waste', *J Environ Qual*, **38**(4), 1694–1699.

WYN-JONES A P, CARDUCCI A, COOK N, D'AGOSTINO M, DIVIZIA M, FLEISCHER J, GANTZER C, GAWLER A, GIRONES R, HÖLLER C, DE RODA HUSMAN A M, KAY D, KOZYRA I, LÓPEZ-PILA J, MUSCILLO M, JOSÉ NASCIMENTO M S, PAPAGEORGIOU G, RUTJES S, SELLWOOD J, SZEWZYK R and WYER M (2011), 'Surveillance of adenoviruses and noroviruses in European recreational waters', *Water Res*, **45**, 1025–1038.

XAGORARAKI I, KUO DH, WONG K, WONG M and ROSE J B (2007), 'Occurrence of human adenoviruses at two recreational beaches of the great lakes', *Appl Environ Microbiol*, **73**(24), 7874–7881.

YOUSRY T A, MAJOR E O, RYSCHKEWITSCH C, FAHLE G, FISCHER S, HOU J, CURFMAN B, MISZKIEL K, MUELLER-LENKE N, SANCHEZ E, BARKHOF F, RADUE E W, JÄGER H R and CLIFFORD D B (2006), 'Evaluation of patients treated with natalizumab for progressive multifocal leukoencephalopathy', *N Engl J Med*, **354**(9), 924–933.

ZHANG T, BREITBART M, LEE W H, RUN J Q, WEI C L, SOH S W, HIBBERD M L, LIU E T, ROHWER F and RUAN Y (2006), RNA viral community in human feces: prevalence of plant pathogenic viruses, *PLoS Biol*, **4**(1), e3.

Index

CPSIA information can be obtained at www.ICGtesting.com
Printed in the USA
LVOW072226250613

340258LV00007B/85/P

9 780857 094308